Fruit Fly Research and Development in Africa - Towards a Sustainable Management Strategy to Improve Horticulture

Sunday Ekesi • Samira A. Mohamed
Marc De Meyer
Editors

Fruit Fly Research and Development in Africa - Towards a Sustainable Management Strategy to Improve Horticulture

Editors
Sunday Ekesi
Plant Health Theme
International Centre of Insect Physiology &
  Ecology (*icipe*)
Nairobi, Kenya

Samira A. Mohamed
Plant Health Theme
International Centre of Insect Physiology &
  Ecology (*icipe*)
Nairobi, Kenya

Marc De Meyer
Biology Department
Royal Museum for Central Africa
Tervuren, Belgium

ISBN 978-3-319-43224-3   ISBN 978-3-319-43226-7   (eBook)
DOI 10.1007/978-3-319-43226-7

Library of Congress Control Number: 2016959464

© Springer International Publishing Switzerland 2016
Chapter 29 was created within the capacity of an US governmental employment. US copyright protection does not apply.
This work is subject to copyright. All rights are reserved by the Publisher, whether the whole or part of the material is concerned, specifically the rights of translation, reprinting, reuse of illustrations, recitation, broadcasting, reproduction on microfilms or in any other physical way, and transmission or information storage and retrieval, electronic adaptation, computer software, or by similar or dissimilar methodology now known or hereafter developed.
The use of general descriptive names, registered names, trademarks, service marks, etc. in this publication does not imply, even in the absence of a specific statement, that such names are exempt from the relevant protective laws and regulations and therefore free for general use.
The publisher, the authors and the editors are safe to assume that the advice and information in this book are believed to be true and accurate at the date of publication. Neither the publisher nor the authors or the editors give a warranty, express or implied, with respect to the material contained herein or for any errors or omissions that may have been made.

Printed on acid-free paper

This Springer imprint is published by Springer Nature
The registered company is Springer International Publishing AG Switzerland

This book is dedicated to Dr. Slawomir A. Lux (inSilico-IPM, Poland) in recognition of his contribution to fruit fly research and development, both in Africa and beyond. His research activities and scientific discoveries have been instrumental to the development of management strategies for a number of different African fruit fly species. He pioneered the regional approach to fruit fly research and

management when he started the African Fruit Fly Initiative in 1999. This initiative laid the foundation for several of the current fruit fly management activities on the African continent. We have significantly benefitted from the scientific rigour of Dr. Lux which has helped us to develop and continue to promote the science of tephritid fruit flies globally.

Foreword

This book documents up-to-date information on the taxonomy, bioecology and management of tephritid fruit flies in Africa. This information is essential to improve horticulture – a rapidly developing agricultural subsector on the continent that serves as an engine for economic growth, trade expansion and the development of new income-generating opportunities. Horticulture is, however, a highly technical and knowledge-intensive subsector, and success is dependent upon addressing a myriad of factors that influence the development of the industry. These include market systems, postharvest systems and food safety; conservation and development of genetic resources; sustainable production systems and natural resource management; capacity building; and providing an enabling environment, amongst others. The constraints to sustainable production systems include biotic stresses of which arthropod pests and diseases are critical. Tephritid fruit flies are notoriously important pests because of the direct damage they cause to fruits but also due to their global quarantine status. Infestation by fruit flies reduces the revenues and profits of producers and traders because they increase production costs associated with fulfilling the needs of local domestic and export markets.

Over the last two decades the International Centre of Insect Physiology and Ecology (*icipe*) and partners have pioneered fruit fly research and development on the African continent by studying their basic biology and ecology and developing comprehensive pre- and postharvest management packages. They have disseminated and promoted strategies to minimise the use of synthetic pesticides that leads to unwanted residues, thereby facilitating compliance with the standards required for domestic and export markets. There has been widespread adoption of these proven technologies, and the interventions made have helped increase returns (>40%) and market share, particularly for women, thus helping to reduce gender inequality in income and rural opportunities. The capacity of national systems, private sectors, community-based extension service providers and growers, at various levels of competencies, has been expanded tremendously with regard to fruit fly surveillance and management. Regional systems of quarantine have been significantly improved through the provision of taxonomic tools, quarantine equipment (traps, lures and killing strips) and fruit fly distribution maps for rapid detection and identification of native and invasive species. *icipe* views fruit fly management as a package within the context of integrated pest management, and the socio-economic impacts of the interventions made confirm the usefulness of the concept.

The range of activities in relation to fruit fly research and development, and how they have been brought into pragmatic application at the local, national, regional and international level, has never before been documented in one single book; readers of this book will certainly appreciate the importance of managing fruit flies to improve the economy and livelihood of our people. The authors are renowned authorities in the field providing information on recent innovations, applications and opportunities for future research. They also provide guideposts for international cooperation and show how partnerships in various facets of fruit fly research and development help to minimise the impact of fruit flies on horticulture. The 34 chapters contributed by experts from Africa, Europe, Latin America and the USA demonstrate that there is much to be discovered, learnt and applied in the field of fruit fly research and development. Clearly, this is an outstanding contribution to the field that will remain relevant for researchers, academics and students in horticulture for many years. The hard work and dedication of the authors in putting together such a comprehensive reference book deserves commendation.

International Centre of Insect Physiology and Ecology (*icipe*), Nairobi, Kenya

Segenet Kelemu

Preface

Between 2012 and 2014, it was estimated that nearly 805 million people were chronically undernourished (FAO 2014). The vast majority of these hungry people (791 million) live in developing countries, where it is estimated that 13.5 % of the population are undernourished. Africa, and in particular sub-Saharan Africa (SSA), has the highest numbers; it is estimated that 23.8 %, or approximately one in four people, are undernourished. The region as a whole is extremely susceptible to food crises and famines that are easily and frequently triggered by even the lightest of droughts, floods, pest infestations, economic downturns or conflicts (Kidane et al. 2006). These relatively short-term triggers are aggravated by climate change which makes seasonal weather patterns more unpredictable and is causing long-term changes in the suitability of particular areas for food production. Other contributory factors are slow growth in agriculture, rapid population growth, weak foreign exchange earnings and high transaction costs in domestic and international markets. How to guarantee food security is a challenging problem for many African countries and, shockingly, SSA is the only region of the world where hunger is projected to worsen over the next two decades unless some drastic measures are taken to reverse food insecurity.

A sustainable approach to reverse these trends would be to invest in agriculture; this would reduce poverty and hunger directly and would subsequently encourage economic diversification and growth. Indeed, increasing agricultural productivity on the vast expanse of suitable land available could increase income, provide opportunities for otherwise destitute population groups and offer a recognised way to escape the rural poverty trap. In particular, the horticulture subsector provides excellent opportunities to meet both domestic food and nutritional needs and offers opportunities to diversify income streams and create employment for the populace. It is encouraging to see that it is currently viewed as a major economic development strategy for many SSA countries.

Although growth in the horticulture subsector presents many opportunities for improving food security, growing rural economies and improving the livelihoods of the poor, this growth is jeopardised by, amongst other factors, horticultural pests. Of these, tephritid fruit flies (Diptera: Tephritidae) cause the greatest economic losses throughout Africa. Sub-Saharan Africa alone is the native range of about 1000 fruit fly species from 148 genera, of which ~400 species can develop in wild and/or cultivated fruit. The latter belong mainly to the following genera: *Ceratitis*, *Dacus* and *Trirhithrum*. The problem has been further compounded by invasion of alien invasive fruit fly species including *Bactrocera dorsalis, Zeugodacus cucurbitae, Bactrocera latifrons* and *Bactrocera zonata*. These invasions have caused extensive economic and ecological damage; they have negative effects on native populations of fruit flies including displacement, altered successional patterns, mutualistic relationships, community dynamics and resource distributions.

Since the arrival of *B. dorsalis* on the African continent, direct yield losses in mango production across Africa have varied from 40 to 90% depending on locale, cultivar and season, which is significantly higher than earlier losses attributed just to native fruit fly species. In addition to these direct losses, the indirect losses attributed to quarantine restrictions are enormous – estimated to be in excess of US$2 billion annually (Ekesi et al. 2016). In fact, this figure is an underestimation given that export of South African citrus, deciduous and subtropical fruits alone is worth US$ 4.7 billion annually. Postharvest treatments have added considerably to production costs, both with respect to infrastructure and labour. Planning and investment in increasing production have been suspended, and trade with new areas has been inhibited. Disequilibrium in the import-export trade balance has put governments under severe political and industry pressure. The direct and indirect impact of fruit fly pests continue to have wide-reaching socio-economic implications – loss of jobs, income and trade alongside the associated personal emotional stress – for millions of rural and urban populations involved in the horticulture value chain across Africa.

The *icipe*-led African Fruit Fly Programme (AFFP) was established in response to requests from African fruit growers, national authorities and regional commodity and quarantine bodies, to address the fruit fly problem on the continent. The programme (initiated as the African Fruit Fly Initiative, AFFI) started in 1999 with the broad objective to:

1. Assess the impact of fruit fly infestation on key crops in Africa
2. Develop and evaluate affordable fruit fly management methods based on locally produced tools, materials and practices (traps, attractants, biopesticides and field sanitation)
3. Explore, identify and release natural enemies (in particular parasitoids) of exotic fruit fly species in Africa
4. Establish parameters for postharvest treatment for key fruit fly species on export fruits and vegetables
5. Produce and disseminate tools for strengthening fruit fly quarantine in Africa, such as distribution maps and pest identification keys
6. Train personnel in participating African countries to ensure there would be a cadre of next-generation fruit fly experts in place

Inspired by AFFP, other projects also started, including the Belgian Development Cooperation fruit fly projects in Tanzania and Mozambique and the West African

Fruit Fly Initiative (WAFFI), amongst others. Running in parallel to these activities was a major programme on the sterile insect technique (SIT) targeted at *Ceratitis* species in the Hex Valley, South Africa. All these projects were generously supported by the German Federal Ministry for Economic Cooperation and Development, implemented by the Deutsche Gesellschaft für Internationale Zusammenarbeit (GIZ), the International Fund for Agricultural Development (IFAD), the Biovision Foundation, the European Union (EU), Belgian Development Cooperation, Belgian Science Policy (BELSPO), ERAfrica, the International Atomic Energy Agency (IAEA), UK Department for International Development (DFID), the US Agency for International Development (USAID), the US Department of Agriculture, Foreign Agriculture Service (USDA-FAS) in coordination with USDA Animal and Plant Health Inspection Service (USDA-APHIS), the Food and Agriculture Organization (FAO) of the UN, the Dutch Programme for Cooperation with International Institutions (Netherland-SII) and several private sector partners.

This book collates the series of important and diverse achievements made over the last two decades as a result of these research activities and includes advances in fruit fly taxonomy, biology, ecology and management with a focus on native and exotic species present in Africa. It analyses the successes that have been made in the identification of different species using both morphological and molecular tools. The invasion histories of exotic species are documented. Information on behaviour, abundance, dynamics, host plants and damage levels of different species are presented. Management methods based on the use of baiting and male annihilation techniques, biopesticides, parasitoids, ant technology and field sanitation are discussed in line with the demand for socio-economic impact and ecosystem sustainability. The technical knowledge presented in this book is not unique to Africa, and lessons learnt from other successful fruit fly eradication/management programmes in the USA, Latin America and South America are also captured. This book provides state-of-the-art information on fruit fly research and development in Africa and beyond that complements existing knowledge in other systems. In what we hope is a turning point in fruit fly research and development on the continent, this book brings together authors from across the globe, all with vast experience in the myriad facets of fruit flies and the horticulture value chain – from academic research to the practical needs of the growers, consumers, policymakers and investors. For this reason, the 34 chapters cover a wide spectrum of topics grouped into: Part I 'Biology and Ecology' (Chaps. 1, 2, 3, 4, 5, 6, 7, 8, 9, 10 and 11), Part II 'Preharvest and Postharvest Management Measures' (Chaps. 11, 12, 13, 14, 15, 16, 17, 18, 19, 20 and 21), Part III 'Country-Specific Action Programmes and Case Studies' (Chaps. 22, 23, 24, 25, 26, 27 and 28), Part IV 'Experiences from Action Programmes Outside Africa' (Chaps. 29, 30, 31 and 32), Part V 'Socio-economic Impact Assessment' (Chap. 33) and Part VI 'Lessons Learnt and Future Perspectives' (Chap. 34). The book begins with taxonomy, development and application of identification tools in Chaps. 1, 2, 3 and 4 including examples of studies on invasion history and population genetics. Chapters 5, 6 and 7 examine the species composition, abundance, dynamics and impact of native and exotic fruit fly species that have been introduced into Africa and those that have been dispersed from Africa to other continents. The feeding and mating behaviour and chemical ecology of African fruit fly species are described in Chaps. 8 and 9. Fruit fly nutrition, rearing and quality control and the application of ontological modelling for fruit fly control and management knowledge are covered in Chaps. 10 and 11. Various pre- and postharvest management technologies are covered in Chaps. 12, 13, 14, 15, 16, 17, 18, 19, and 20. A pictorial plate of some native and exotic fruit fly species in Africa and their parasitoids is provided in Chap. 21. Examples of seven country-specific operational activities are comprehensively dealt with in Chaps. 22, 23, 24, 25, 26, 27 and 28. Experiences from successful action programmes in Hawaii, Mexico and Suriname alongside the importance of taking a systems approach are discussed in Chaps. 29, 30, 31 and 32. A clear example of how fruit fly IPM interventions can benefit growers and the need for socio-economic impact assessments is captured in Chap. 33. Finally, a synthesis of the lessons learnt and outlooks on the future in terms of research gaps are highlighted in Chap. 34. As the development sector refocuses its attention on poverty alleviation and food security, we hope that this book will provide the reader with information on the impact of fruit flies on horticulture in Africa and beyond. As such, it is expected to lead to the management practices necessary to develop technological and market-oriented solutions that deliver fruit fly-free horticultural production on the continent and also encourage further research and investment in tephritid research and development.

Nairobi, Kenya   Sunday Ekesi
Nairobi, Kenya   Samira A. Mohamed
Tervuren, Belgium   Marc De Meyer

References

Ekesi S, De Meyer M, Mohamed SA, Virgilio M, Borgemeister C (2016) Taxonomy, ecology and management of native and exotic fruit fly species in Africa. Annu Rev Entomol. doi:10.1146/annurev-ento-010715-023603

FAO (2014) The state of food insecurity in the world – strengthening the enabling environment for food security and nutrition. FAO, Rome

Kidane W, Maetz M, Dardel P (2006) Food security and agricultural development in sub-Saharan Africa – building a case for more public support. Policy Assistance Series 2. FAO, Rome

Disclaimer

Each of the articles in this book has been prepared from the original manuscript submitted by the authors. Views expressed do not necessarily reflect those of the editors or the institutions with which they are affiliated. The mention of names of specific companies or products (whether indicated as registered or not) does not imply any intention to infringe property rights, nor should it be construed as an endorsement or recommendation on the part of the editors or the institutions with which they are affiliated. All the authors are responsible for having obtained the necessary permissions to reproduce, translate or use materials from sources already protected by copyright.

Acknowledgements

The publication of this book was made possible by funding from the German Federal Ministry for Economic Cooperation and Development, implemented by the Deutsche Gesellschaft für Internationale Zusammenarbeit (GIZ), to the *icipe*-led Mango IPM project for two phases (March 2007–December 2010; March 2011–December 2014). We appreciate the contribution of the International Fund for Agricultural Development (IFAD) for laying down the foundation of fruit fly research and development at the International Centre of Insect Physiology and Ecology (*icipe*). The Biovision Foundation, the International Atomic Energy Agency (IAEA), the UK Department for International Development (DFID), the US Agency for International Development (USAID) and the US Department of Agriculture, Foreign Agriculture Service (USDA-FAS) in coordination with USDA Animal and Plant Health Inspection Service (UDA-APHIS), the Food and Agriculture Organization (FAO) of the UN, the Dutch Programme for Cooperation with International Institutions (Netherland-SII) and the European Union (EU) are all appreciated for their generous support to *icipe*'s African Fruit Fly Programme (AFFP) over the years. In addition to the above mentioned, our collaborators received additional funding from various sources including the Belgian Development Cooperation (through the Framework Agreement with the Royal Museum for Central Africa), the Belgian Science Policy (BELSPO), the National Science Foundation Flanders (FWO Vlaanderen) and ERAfrica.

We wish to acknowledge the support of the *icipe* senior management and especially the Director General, Dr. Segenet Kelemu, for initiating the first contact with Springer and for all the encouragement she has given us in bringing this book together. We thank Prof. Christian Borgemeister, former Director General of *icipe* and current Director of the Center for Development Research (ZEF C), University of Bonn, for his continuous support to AFFP and the development of horticulture in Africa. We are also grateful to Ms Annah Njui and Ms Carolyne Akal for their handling of the various contacts and correspondence with GIZ and the authors. We recognise the contribution of several graduate students and technical staff in the fruit fly programme at *icipe*. The close collaboration of various national and private sector partners is greatly appreciated. We thank all the authors and their coauthors for the substantial amount of time they have put into each of the chapters. We sincerely acknowledge the scientific and editorial prowess of Dr. Judith K. Pell of J.K. Pell Consulting. Some of the text would not have been readable without her substantial revision and input and also for her constant encouragement and support which has enabled this book to come to fruition.

Our sincere thanks go to Mariska van der Stigchel of Springer for her patience and guidance in putting this book together. We also thank the entire production team at Springer for their valuable contributions to this book.

Contents

Part I Biology and Ecology

1. Taxonomy and Systematics of African
   Fruit Flies (Diptera: Tephritidae) 3
   Marc De Meyer

2. Identification Tools for African Frugivorous Fruit
   Flies (Diptera: Tephritidae) 19
   Massimiliano Virgilio

3. Population Genetics of African Frugivorous
   Fruit Flies (Diptera, Tephritidae): Current Knowledge
   and Future Perspectives 35
   Massimiliano Virgilio and Hélène Delatte

4. Role of Microsatellite Markers in Molecular
   Population Genetics of Fruit Flies with Emphasis
   on the *Bactrocera dorsalis* Invasion of Africa 53
   Fathiya M. Khamis and Anna R. Malacrida

5. Fruit Fly Species Composition, Distribution
   and Host Plants with Emphasis on Mango-Infesting Species 71
   Ivan Rwomushana and Chrysantus M. Tanga

6. Fruit Fly Species Composition, Distribution and Host Plants
   with Emphasis on Vegetable-Infesting Species 107
   Chrysantus M. Tanga and Ivan Rwomushana

7. Exotic Invasive Fruit Flies (Diptera: Tephritidae):
   In and Out of Africa 127
   Marc De Meyer and Sunday Ekesi

8. Feeding and Mating Behaviour of African Fruit Flies 151
   Aruna Manrakhan

9	**Chemical Ecology of African Tephritid Fruit Flies**............................. Ayuka T. Fombong, Donald L. Kachigamba, and Baldwyn Torto	163
10	**Fruit Fly Nutrition, Rearing and Quality Control**.............................. Samira A. Mohamed, Fathiya M. Khamis, and Chrysantus M. Tanga	207
11	**The Ontological Modelling of Fruit Fly Control and Management Knowledge**.. Caroline C. Kiptoo, Aurona Gerber, and Alta Van der Merwe	235

Part II Pre-harvest and Post-harvest Management Measures

12	**Detection and Monitoring of Fruit Flies in Africa**............................... Aruna Manrakhan	253
13	**Baiting and Male Annihilation Techniques for Fruit Fly Suppression in Africa**... Sunday Ekesi	275
14	**Waste Brewer's Yeast as an Alternative Source of Protein for Use as a Bait in the Management of Tephritid Fruit Flies**............ Sunday Ekesi and Chrysantus M. Tanga	293
15	**Development and Application of Mycoinsecticides for the Management of Fruit Flies in Africa**... Jean N.K. Maniania and Sunday Ekesi	307
16	**In and Out of Africa: Parasitoids Used for Biological Control of Fruit Flies**... Samira A. Mohamed, Mohsen M. Ramadan, and Sunday Ekesi	325
17	**From Behavioural Studies to Field Application: Improving Biological Control Strategies by Integrating Laboratory Results into Field Experiments**.. Katharina Merkel, Valentina Migani, Sunday Ekesi, and Thomas S. Hoffmeister	369
18	**The Use of Weaver Ants in the Management of Fruit Flies in Africa**.. Jean-François Vayssières, Joachim Offenberg, Antonio Sinzogan, Appolinaire Adandonon, Rosine Wargui, Florence Anato, Hermance Y. Houngbo, Issa Ouagoussounon, Lamine Diamé, Serge Quilici, Jean-Yves Rey, Georg Goergen, Marc De Meyer, and Paul Van Mele	389
19	**Sterile Insect Technique (SIT) for Fruit Fly Control – The South African Experience**.. Brian N. Barnes	435

20 Cold and Heat Treatment Technologies for Post-harvest Control of Fruit Flies in Africa 465
Tim G. Grout

21 Photographs of Some Native and Exotic Fruit Fly Species in Africa and Their Parasitoids 475
Sunday Ekesi, Samira A. Mohamed, and Marc De Meyer

Part III Country Specific Action Programmes and Case Studies

22 Integrated Management of Fruit Flies – Case Studies from Uganda 497
Brian E. Isabirye, Caroline K. Nankinga, Alex Mayamba, Anne M. Akol, and Ivan Rwomushana

23 Integrated Management of Fruit Flies – Case Studies from Tanzania 517
Maulid Mwatawala

24 Integrated Management of Fruit Flies – Case Studies from Mozambique 531
Domingos R. Cugala, Marc De Meyer, and Laura J. Canhanga

25 Integrated Management of Fruit Flies: Case Studies from Nigeria 553
Vincent Umeh and Daniel Onukwu

26 Release, Establishment and Spread of the Natural Enemy *Fopius arisanus* (Hymenoptera: Braconidae) for Control of the Invasive Oriental Fruit Fly *Bactrocera dorsalis* (Diptera: Tephritidae) in Benin, West Africa 575
Désiré Gnanvossou, Rachid Hanna, Aimé H. Bokonon-Ganta, Sunday Ekesi, and Samira A. Mohamed

27 Integrated Management of Fruit Flies: Case Studies from Ghana 601
Maxwell K. Billah and David D. Wilson

28 Integrated Management of Fruit Flies: Case Studies from the Indian Ocean Islands 629
Preeaduth Sookar and Jean-Philippe Deguine

Part IV Experiences from Actions Programmes Outside Africa

29 Area-Wide Management of Fruit Flies (Diptera: Tephritidae) in Hawaii 673
Roger I. Vargas, Jaime C. Piñero, Luc Leblanc, Nicholas C. Manoukis, and Ronald F.L. Mau

30	**Management of Fruit Flies in Mexico**	695
	Pablo Liedo	
31	**Overview of the Programme to Eradicate** *Bactrocera carambolae* **in South America**	705
	David Midgarden, Alies van Sauers-Muller, Maria Julia Signoretti Godoy, and Jean-François Vayssières	
32	**Systems Approaches for Managing the Phytosanitary Risk of Trading in Commodities that are Hosts of Fruit Flies**	737
	Eric B. Jang	

Part V Socioeconomic Impact Assessment

33	**Economic Impact of Integrated Pest Management Strategies for the Suppression of Mango-Infesting Fruit Fly Species in Africa**..	755
	Beatrice W. Muriithi, Gracious M. Diiro, Hippolyte Affognon, and Sunday Ekesi	

Part VI Lessons Learnt and Future Perspectives

34	**Lessons Learnt and Future Perspectives**	773
	Sunday Ekesi, Samira A. Mohamed, and Marc De Meyer	

Part I
Biology and Ecology

Chapter 1
Taxonomy and Systematics of African Fruit Flies (Diptera: Tephritidae)

Marc De Meyer

Abstract Work on the taxonomy of African fruit flies began with basic descriptions and alpha taxonomy to understand dipteran and/or insect diversity, and they were based upon material collected from isolated locations in Africa. Only later were more detailed taxonomic studies and descriptions initiated largely due to recognition of the economic impact some fruit fly pests had on horticultural crops. Taxonomic research was largely driven by the work of two entomologists at that time: Mario Bezzi (Italy) and Kenneth Munro (South Africa). In the latter part of the twentieth century and the first decades of this century, more comprehensive publications became available dealing both with generic revisions and higher classification. Again the greatest focus was on those groups comprising fruit flies of economic significance, including studies dealing with cryptic species. Considerable progress was also made with descriptions of other tephritid groups, although they received far less attention and it is expected that numerous new species await description. These taxonomic studies form the basis of studies on the phylogenetic relationships amongst the taxa, using both morphological and molecular data. Increasingly, greater attention is being given to putting this information into a larger evolutionary framework, in particular with respect to host plant associations.

Keywords Afrotropical • Species • Phylogeny • Descriptions

1 Introduction

Correctly naming an organism (by which we mean providing a scientific name and making it available as per the requirements set out by international codes) is a key prerequisite for any subsequent biological research or management activity involving that organism. Providing correct names and descriptions is, however, not a straightforward matter. Scientific species names should represent biological entities

M. De Meyer (✉)
Biology Department, Royal Museum for Central Africa,
Leuvensesteenweg 13, B3080 Tervuren, Belgium
e-mail: demeyer@africamuseum.be

with unique characteristics reflecting their behaviour, ecological requirements and interactions with the environment, that allow them to be clearly differentiated from other entities. What exactly constitutes and defines this differentiation is a point of discussion and has led to numerous species concepts (de Queiroz 2007; Wilkins 2009; Hart 2010; Richards 2010; Hausdorf 2011). Nevertheless, for any practical application, there needs to be agreement on (a) the name that unambiguously identifies a particular species so that all can refer to the same species using the same name, and (b) what organisms and populations can be considered to be included under that name. While the first aspect is governed by clear rules, set out by international commissions on nomenclature (in the case of animals, the International Commission on Zoological Nomenclature; ICZN 1999, 2012) and largely followed by the majority of users, the second point leads to more controversy. Opinions and arguments for or against including particular organisms or populations under the same scientific name, or applying different names, is often a point of lengthy discussion. While these may be the biological equivalents of 'councils on the gender of the angels' for the majority of species, it does have far reaching implications when it concerns pest species. An obvious example of this, with relevance to African fruit flies, is the identity of the invasive species *Bactrocera invadens* Drew, Tsuruta and White, and whether this is identical to the oriental fruit fly, *Bactrocera dorsalis* (Hendel) or actually represents a different species (Drew and Romig 2013; Schutze et al. 2014a). Next to naming and describing species, understanding the relationships between species is the next most important consideration in research. Closely related species may share a number of characteristics such as similar host range, demographic characteristics, symbiotic relationships, adjacent or overlapping geographical distribution etc. Systematic and phylogenetic studies, unraveling the interrelationships between species within a larger group (genus, tribe, subfamily, family) provide a stable classification for known species but also a solid framework for further studies. This paper will present past and current developments both in the field of alpha taxonomy and systematic and phylogenetic research on African tephritid fruit flies. In particular, we will highlight the advent and implications of using novel techniques such as genetic markers, and the added advantages of integrated approaches.

2 Historical Account

2.1 Early Taxonomy

The earliest taxonomic descriptions of African fruit flies date back to the late eighteenth and early nineteenth centuries. These descriptions were mainly based upon material collected at trading posts along the African coastline and include species described by the German entomologist, Wiedemann, from specimens collected at the Cape of Good Hope Peninsula in South Africa and the Danish settlement of

Christiansborg in current-day Ghana (Wiedemann 1818, 1819, 1824, 1830; see Pont 1995 for review). Others were collected during organized expeditions. For example, during the German African Society Expedition to the southern Congo Basin, the naturalist, von Homeyer, collected fruit flies which were later described by Karsch, the curator of the Zoological Museum in Berlin (Karsch 1887; reviewed by Evenhuis 1997). In addition, species of Afrotropical origin, but also occurring in the Mediterranean Region, were described at that time (such as the earliest known species, the olive fruit fly, *Bactrocera oleae*, originally described by Rossi from Tuscany, Italy [Rossi 1790]). Throughout much of the nineteenth century, specimens housed in private or institutional natural history collections were the main source material for descriptions of African fruit flies, often as part of larger studies of insect diversity. These include Walker's endeavors to describe the entomological holdings of the British Museum (reviewed by Evenhuis 1997) and the studies of Loew on African Diptera (Loew 1852, 1861, 1862).

Only at the start of the twentieth century did more focused research begin to describe the tephritid fruit flies of Africa. There were two main reasons for this. Firstly, the economic impact of some fruit fly species on horticultural crops was recognized. Secondly, observations of the first invasive species in other continents triggered a search for control measures including potential classical biological control agents in the pest's country of origin. For example, Lounsbury, a government entomologist in the Cape Colony (part of current-day South Africa) and the initiator of biological control measures against invasive pests in Africa, was one of the first entomologists to collect fruit flies and have them identified by specialists such as Coquillett from the United States National Museum (Coquillett 1901). The Hawaii Board of Commissioners of Agriculture and Forestry became interested in the use of natural enemies to limit the economic damage caused by the Mediterranean fruit fly (medfly), *Ceratitis capitata* (Wiedemann), a species of African origin and invasive in Hawaii. In 1912, they engaged the services of an Italian entomologist, Silvestri, to visit Africa in search of natural enemies of *C. capitata* in its natural range. Silvestri spent almost a full year exploring several West African countries (Silvestri 1913). In addition to the natural enemies he reared, he also collected a number of fruit fly species amongst which some were new to science (Bezzi 1912, 1913; Silvestri 1913). This was the first in a series of expeditions in search of natural enemies during the twentieth century (Bianchi 1937; Bianchi and Krauss 1937; Van Zwaluwenburg 1937; Clausen et al. 1965) that resulted in large collections of known and new species with the associated biological data on their host plants. Similar local collections of flies emerging from infested fruits supplemented these data. The most noteworthy of these was the programme established by van Someren at the then Coryndon Museum (now National Museums of Kenya) in Nairobi, Kenya. In addition to these programmes with applied objectives, there were also a number of field expeditions organized by different institutions to document the rich biodiversity found in particular African ecosystems. Two taxonomists specializing in Tephritidae, Bezzi (Italy) and Munro (South Africa), examined much of this material. Together their research spanned a period of almost eighty years (1908–1984) and they described approximately 68% of all known African tephritid species

(Cogan and Munro 1980; Norrbom et al. 1999). In addition to alpha taxonomy Munro also embarked on larger review papers on the transition genera between Tephritinae and Trypetinae (e.g. Munro 1947), and a series of papers with biological notes on host plant range, larval feeding behaviour, infestation rates and phenology (Munro 1925, 1926, 1929).

2.2 Comprehensive Revisions

From the 1980s onwards, more comprehensive publications became available that focused on taxonomic revisions of particular higher taxa at the genus (Freidberg 1985, 1991, 1999; Freidberg and Hancock 1989; Freidberg and Merz 2006) or higher level (Freidberg and Kaplan 1992, 1993; Munro 1984; Hancock 1984, 1985, 1986, 1990, 1999, 2000, 2005, 2012; Hancock et al. 2003). Most recently the focus has shifted to providing thorough revisions of the African representatives of economically important genera, in particular the frugivorous groups within the tribe Dacini that attack the fresh fruits of both commercial and wild host plants (De Meyer 1996, 1998, 2000, 2001; De Meyer and Copeland 2001, 2005, 2009; De Meyer and Freidberg 2005, 2006; White 2006; White and Goodger 2009; White et al. 2003) and closely related genera (De Meyer 2006, 2009; De Meyer and Freidberg 2012). Unrelated frugivorous groups, such as *Munromyia* Bezzi (Adramini), have also been revised recently (Copeland 2009; Copeland et al. 2004). Some of these revisions coincided with local, national or regional surveys (Copeland et al. 2005; De Meyer et al. 2012, 2013; Hancock 2003; Hancock et al. 2003) or biological studies (Copeland et al. 2002, 2006, 2009) that provided new material and data. These studies aimed to incorporate all available specimens to achieve, as much as possible, a global view on intraspecific variability, and resulted in extensive descriptions and redescriptions. In addition, these studies provided a summary of all known information regarding occurrence, distribution and reliability of host plant data.

2.3 Current Status

The current taxonomy of African fruit flies has advanced significantly. As a result of the taxonomic studies described in Sect. 2.2, the majority of genera that contain frugivorous species, which comprises approximately two-fifths of all the fruit fly species recorded from the Afrotropical region, have been revised over the last 20 years. This has provided, and still does, a sound basis for all subsequent research in the fields of ecology and biogeography, but also in applied research on fruit fly control. Most of the information from these revisions is also publicly available through dedicated websites (e.g. http://projects.bebif.be/fruitfly/index.html) and also more general websites (e.g. Global Biodiversity Information Facility [www.gbif.org];

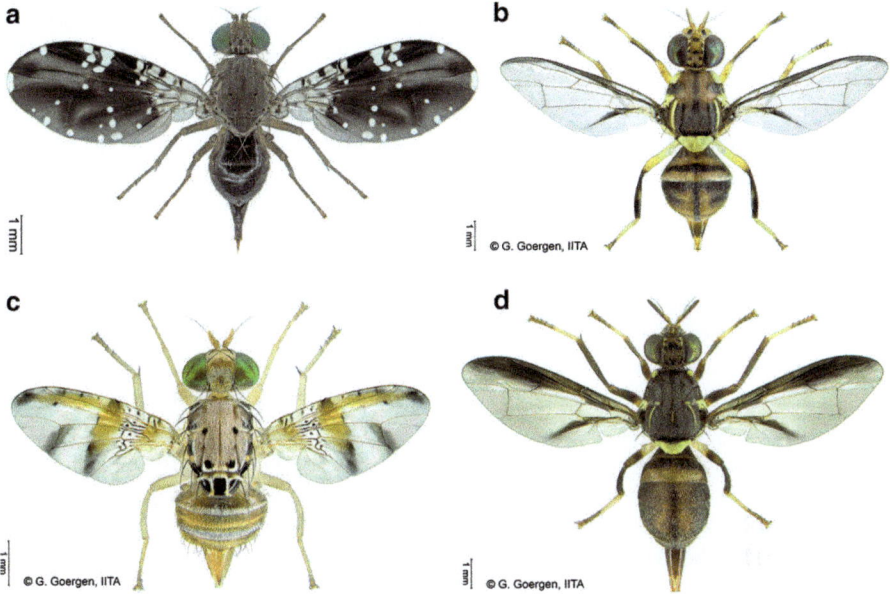

Fig. 1.1 Habitus images of African fruit flies (**a**) *Pseudafreutreta diaphasis* (Bigot); (**b**) *Bactrocera dorsalis* (Hendel); (**c**) *Ceratitis cosyra* (Walker); (**d**) *Dacus bivittatus* (Bigot) (All images copyright and courtesy of Georg Goergen)

Encyclopedia of Life [www.eol.org]). It has also enabled the development of modern multi-entry and electronic identification tools (Virgilio 2016; Virgilio et al. 2014), providing easy-to-use tools with different filters and levels with which to identify specimens, and hyperlinks to associated data.

However, non-frugivorous groups have received far less attention. Most of the tephritids belonging to the subfamily Tephritinae infest flower heads and attack particular plant families: Asteraceae, Acanthaceae, Lamiaceae and Verbenaceae (Freidberg and Han 2012) (Fig. 1.1a). Others, are saprophagous, leaf or stem miners, or gall formers (Norrbom et al. 1999). In Africa, approximately 600 species are known but numerous undescribed species are thought to exist. Besides some larger genera, like *Campiglossa* Rondani and *Trupanea* Schrank, the majority are grouped under about 130 different genera of which half only contain one or two described species. This makes identification, even to genus level, difficult. Some information is available but it is dispersed over numerous articles published over the last 30 years. Other non-frugiverous species are found in the subfamily Trypetinae for which Hancock (1986) provided a key to the Afrotropical genera (excluding the Dacini). There have subsequently been keys published for Afrotropical genera in several tribes: Trypetini (Hancock 2005); Schistopterini (Freidberg 2002); Tephritini (Merz 1999 [*Tephritis* Latreille group]; Merz and Dawah 2005 [*Campiglossa* group]). Other than these there are no comprehensive keys available for the majority of the other genera. However, a genus-level key is currently under development within the framework of the Manual of Afrotropical Diptera (http://afrotropicalmanual.org/), which will include all genera occurring in Africa. This is expected to drastically improve the identification process.

Despite the lack of a comprehensive key to genus level, keys to species level for a number of particular genera have been revised over the past decades. This includes genera within the subtribe Gastrozonina, a group that breed in bamboo but are closely related to the main frugivorous tephritids. They were revised by Hancock (1999) with keys to genus and species level. Within the Trypetinae, species-level keys are available for the genera *Conradtina* Enderlein, *Celidodacus* Hendel and *Coelotrypes* Bezzi, all within the tribe Adramini (Hancock 1986). For the Tephritinae, recent keys have been established for species in the genera: *Cryptophorellia* Freidberg and Hancock (Freidberg and Hancock 1989); *Tanaica* Munro (Merz and Dawah 2005); *Deroparia* Munro, *Euryphalara* Munro, *Xenodorella* Munro (Hancock 2000); *Oedaspis* Loew (Freidberg and Kaplan 1992); *Elgonina* Munro, *Gymnosagena* Munro, *Marriottella* Munro (Freidberg and Merz 2006); *Afreutreta* Bezzi, *Cosmetothrix* Munro and *Tarchonanthea* Freidberg and Kaplan (Freidberg and Kaplan 1993). Also within the Tephritinae the genus *Dicheniotes* Munro (tribe Tephrellini) has been revised and a key for all species provided by Hancock (2012), while the genus *Manicomyia* Hancock (also tribe Tephrellini) was revised by Freidberg and Han (2012). Several other groups within the Tephritinae are also under evaluation at this moment.

3 Cryptic Species and Population Genetics

Although one of the simplest definitions of a species is quoted as "a species is what a competent taxonomist says it is" (Regan (1926) as quoted in Froese (1999)), the truth is usually more complicated. As indicated in the introduction, establishing what exactly represents a species, is a complicated matter and myriad theories exist. For this reason, there are currently more than 20 different species concepts (Mayden 1997). Without going into the details of the philosophical, evolutionary and phylogenetic aspects of the debate (see Kunz 2012 for a recent review with regard to species and the principles of taxonomic classification), taxonomists traditionally work with the concept of morphological species, i.e. a number of individuals or populations that are similar in morphological (and/or anatomical) appearance and that can be separated from another species by discrete but consistent differences in morphology. However, this approach has been shown to have its limitations, especially in the case of cryptic species where closely related species are very similar morphologically, but demonstrate distinctly different biological traits (in particular pre- and postzygotic incompatibility). With species of economic significance this can have important consequences. If a morphologically assigned pest species actually consists of a complex of biologically distinct species that vary in host range, geographic spread, developmental physiology and ecology, then this will have very significant impacts on any proposed management plan. With regard to African fruit fly pests, two groups of economic importance were recently the subject of a coordinated research project to elucidate this issue. The project was initiated by the joint FAO/IAEA Insect Pest Control Section (De Meyer et al. 2015a) and entitled 'resolution of cryptic species complexes of tephritid pests to overcome constraints to SIT (sterile insect technology) applications and international trade'. Firstly, the specific status of *B. invadens* (Fig. 1.1b) was studied in relation to closely related species in the *B. dorsalis* species complex as defined by Drew and Hancock (1994). Secondly, within the genus *Ceratitis* MacLeay the so-called FAR complex (Barr and McPheron 2006; Barr and Wiegmann 2009; Barr et al. 2006), comprising *Ceratitis fasciventris* (Bezzi), *Ceratitis anonae* Graham and *Ceratitis rosa* Karsch (the Natal fruit fly), was studied in detail. In both cases an integrative approach was applied, using different diagnostic approaches to independently investigate whether consistent differences or similarities could be detected. These different approaches included adult and larval morphology, morphometrics, chemical ecology (pheromones and cuticular hydrocarbons), cytology, molecular markers, developmental physiology and pre- and postzygotic compatibility. A review was presented by Schutze et al. (2014a) for *B. invadens*, concluding that this should be considered as a junior synonym of *B. dorsalis*. Several other species of the *dorsalis* complex were also placed in synonymy (Schutze et al. 2014b). In contrast, the study on the FAR complex indicated that these three taxa should be considered as separate species (De Meyer et al. 2015b). For *C. rosa*, there is a strong indication that it actually consists of two different species that are morphologically very similar and can occur in sympatry, but that have different ecological requirements and biological traits that separate them (De Meyer et al. 2015b; Virgilio et al. 2013).

In addition to these two cryptic species complexes, it is possible that there are other cryptic species complexes present in Africa. Recent research on molecular diversity (Barr et al. 2012) and population genetics (Virgilio et al. 2015a) of the mango fruit fly, *Ceratitis cosyra* (Walker) (Fig. 1.1c) indicate that there is high intraspecific variability and a genetic structuring within this species. Morphological differences had been described previously (De Meyer 1998) but do not seem to be directly correlated with the genetic differences observed. Other species, such as the melon fly, *Zeugodacus cucurbitae* (Coquillett), show some differences in biology but are very homogenous with regard to morphology and genetic markers (Virgilio et al. 2010; De Meyer et al. 2015c). The same seems to be the case for other cucurbit feeders such as *Dacus bivittatus* (Bigot) (Fig. 1.1d) and *Dacus punctatifrons* (Karsch). However, the number of population genetic studies is limited and there is a need for more thorough revision with better geographic coverage.

4 Higher Phylogeny

Korneyev (1999a) provided a rather robust ground plan for the phylogenetic relationships between Tephritidae and other families within the superfamily Tephritoidea. Evolutionary developments in larval feeding strategies (from saprophagous life styles in the more primitive tephritoids to specialized phytophagous feeding habits), in conjunction with the key innovation of the tephritid ovipositor and, possibly, oviposition-meditated bacterial transfer (to better exploit food resources; Behar et al. 2008) have played a major role in tephritoid phylogeny and radiation in the Tephritidae (Diaz-Fleischer et al. 1999). Although a phylogenomic study combining morphological and molecular characteristics largely confirmed Tephritoidea as a superfamily (Wiegmann et al. 2011), the interrelationships amongst the families did partially deviate from the phylogeny proposed by Korneyev (1999a).

Within the family Tephritidae, the phylogenetic relationships amongst the major higher taxa are becoming clearer due to several studies investigating morphological (Korneyev 1999b) and genetic (Han and McPheron 1999; Han et al. 2006; Han and Ro 2009; Krosch et al. 2012; Virgilio et al. 2015b) data. Nevertheless, the classification is still not completely stable. The assignment of genera to higher taxa is in a state of flux and several genera remain unplaced in the currently accepted classification (Norrbom et al. 1999). Recently, joint phylogenetic studies of fruit flies and other organisms, such as microbionts, have been conducted which may provide valuable insights regarding the relationships between these higher taxa (Mazzon et al. 2008, 2010; Morrow et al. 2015).

Only a few publications have dealt specifically with phylogenetic relationships within African representatives of particular genera. With respect to frugivorous species the genus *Ceratitis* and the species groups within this genus have been studied (Barr and McPheron 2006; Barr and Wiegmann 2009; Barr et al. 2006; De Meyer 2005; Erbout et al. 2011), as have the interrelationships within the genus *Dacus* Fabricius (Hancock and Drew 2006; Virgilio et al. 2009; White 2006). In both cases, there is to a greater or lesser extent an incongruence between morphological and molecular data. Han and Ro (2005) already demonstrated that it is very difficult to identify phylogenetic signals in groups like these because of the limited number of informative characters. In genera with a relatively speciose fauna and morphological uniformity, such as *Dacus*, the homoplasy of character states is extremely high, resulting in unstable subgeneric divisions (White 2006). Only for genera with a restricted number of species, can a more stable phylogeny be produced (e.g. De Meyer and Freidberg 2005).

5 Taxonomy and Host Plant Relationships

Being phytophagous leads to close relationships between fruit flies and their host plants. The fact that some fruit flies (or Tephritoidea in general) have made the transition to feed upon living plant tissue in their larval stage (cf. the saprophagous lifestyle of related groups) has led to novel resource use and, at least in the case of Tephritidae, to evolutionary diversification and radiation (Diaz-Fleischer et al. 1999). Examples range from extreme specialization to unprecedented polyphagy. As phylogenetic reconstructions reflect an evolutionary process of species diversification throughout time, it is expected that host plant relationships will also be reflected in the phylogeny. With respect to the frugivorous tephritids of Africa, some genera have a stenophagous or oligophagous host range at the genus level; the best known examples are *Perilampsis* Bezzi, *Capparimyia* Bezzi and *Neoceratitis* Hendel. The genus *Perilampsis* comprises 17 species restricted to the Afrotropical region. Host plants are known for eight species and they all come from the Loranthaceae (De Meyer 2009); larval development takes place in the seeds, rather than the pulp of the berries (De Meyer 2009). Strangely enough the same niche (seed feeders in berries of Loranthaceae) is occupied by another ceratitidine genus, *Ceratitella* Malloch, in Asia and Australia (Hardy 1967; Hancock et al. 2000) but the actual phylogenetic affinity between these genera remains to be studied. Representatives of the genus *Capparimyia*, a predominantly Afrotropical genus (with only one representative from the Mediterranean region and the Middle East), feed on flower buds and fruits of Capparidaceae, including commercially grown capers (De Meyer and Freidberg 2005). *Neoceratitis* is also a predominantly Afrotropical genus (with one species from Asia) for which all confirmed host records belong to the plant genus *Lycium* L. (Solanaceae). Only the tomato fruit fly, *Neoceratitis cyanescens* (Bezzi), a species occurring in Madagascar and other islands in the Western Indian Ocean, has been reported from other Solanaceae (De Meyer and Freidberg 2012).

Other genera with larger numbers of species are not necessarily restricted to a single plant genus or family. However, groups within the fruit fly genera can be associated with particular hosts. For example, White et al. (2003) indicated that, within the genus *Trirhithrum* Bezzi, some species with morphological similarities were associated with Rubiaceae while other species with different morphological similarities were associated with Araceae; within this genus other species were known to have a large diversity of host plants. The genus *Dacus* also shows particular associations between species groups and hosts. Munro, in his biological notes on Tephritidae, observed two groups within the genus *Dacus*, i.e. those species infesting Cucurbitaceae and those species infesting Asclepiadaceae (now part of Apocynaceae) (Munro 1925). White (2006) subsequently summarized all host records available, demonstrating the presence of three biological groups, i.e. species feeding on Cucurbitaceae, Apocynaceae and Passifloraceae respectively. However, subgeneric classifications based solely on morphological characters do not seem to reflect the host plant associations of the three biological groups. Virgilio et al. (2009) provided a molecular phylogeny for a subset of species which more clearly reflected the host plant relationship. This study supports the argument for developing an integrated approach in systematic research, especially for groups that are characterized by high homoplasy in character states and high morphological similarity with few informative characters.

Other tephritid genera show a more complex pattern of host plant associations. The Afrotropical genus *Ceratitis* comprises about 100 species, amongst which are some of the most destructive pest species in Africa (White and Elson-Harris 1994). The genus includes representatives of both specialist feeders and generalists. In order to categorize the niche breadth of host plant utilization, we can differentiate monophagy (attacking a single species), stenophagy (attacking several species belonging to the same genus), oligophagy (attacking species of different genera belonging to the same family) and polyphagy (attacking representatives of several, unrelated, families) (White et al. 1999). All these different feeding types can be found in the genus *Ceratitis* (De Meyer 2005). Phylogenetic analyses indicate that stenophagous species form monophyletic clusters based on the host plant genus that they attack, e.g. *Solanum*, *Strychnos* or *Podocarpus* (De Meyer 2005; Erbout et al. 2011). The exact mechanism driving this monophyletic clustering is not clear. However, the host plant genera that support monophyletic clusters of stenophagous fruit fly species are generally not infested by generalist feeders. The same applies to some of the monophagous species like *C. flexuosa* (Walker) (De Meyer et al. 2002). In addition, these hosts are known for having secondary metabolites that could influence the development and fitness of fruit fly larvae (Erbout et al. 2009). It is possible that the phylogenetic pattern that is being observed reflects an evolutionary process whereby new pathways were selected to allow larvae to survive in hostile or toxic environments, followed by a radiation process resulting in a number of closely related species exploiting the same genus of host plant.

Acknowledgements Some of the research presented was funded through successive grants from the Belgian Science Policy (BELSPO Action 1 projects) and with support from the Joint Experimental Molecular Unity (JEMU). I would also like to thank Dr Georg Goergen (International Institute of Tropical Agriculture, Cotonou, Benin) for allowing me to include some of his habitus images of fruit flies.

References

Barr NB, McPheron BA (2006) Molecular phylogenetics of the genus *Ceratitis* (Diptera: Tephritidae). Mol Phylogenet Evol 38:216–230
Barr NB, Wiegmann BM (2009) Phylogenetic relationships of *Ceratitis* fruit flies inferred from nuclear CAD and tango/ARNT gene fragments: testing monophyly of the subgenera *Ceratitis* (*Ceratitis*) and *C*. (*Pterandrus*). Mol Phylogenet Evol 53:412–424
Barr NB, Copeland RS, De Meyer M, Masiga D, Kibogo HG, Billah MK, Osir E, Wharton RA, McPheron BA (2006) Molecular diagnostics of economically important *Ceratitis* fruit fly species (Diptera: Tephritidae) in Africa using PCR and RFLP analyses. Bull Entomol Res 96:505–521
Barr NB, Islam MS, De Meyer M, McPheron BA (2012) Molecular identification of *Ceratitis capitata* (Diptera: Tephritidae) using DNA sequences of the COI barcode region. Ann Entomol Soc Am 105:339–350, doi: http://dx.doi.org/10.1603/AN11100
Behar A, Jurkevitch E, Yuval B (2008) Bringing back the fruit into fruit fly–bacteria interactions. Mol Ecol 17:1375–1386
Bezzi M (1912) Intorno ad alcune *Ceratitis* raccolte nell' Africa occidentale dal Prof. F. Silvestri. Boll Lab Zool Gen Agrar R Scuola Super Agric Portici 7:3–16
Bezzi M (1913) Altre *Ceratitis* africane allevate dal Prof. F. Silvestri. Boll Lab Zool Gen Agrar R Scuola Super Agric Portici 7:19–26
Bianchi FA (1937) Impressions and observations in East Africa. Hawaii Planters Rec 41:111–133
Bianchi FA, Krauss NH (1937) Fruit fly investigations in East Africa. Hawaii Planters Rec 41:299–306

Clausen CP, Clancy DW, Chock QC (1965) Biological control of the oriental fruit fly (*Dacus dorsalis* Hendel) and other fruit flies in Hawaii. United States Dept Agric. Agric Res Serv Tech Bull 1322:1–102

Cogan BH, Munro HK (1980) Family Tephritidae. In: Crosskey RW (ed) (Cogan BH, Freeman P, Pont AC, Smith KGV, Oldroyd H assist eds) Catalogue of the Diptera of the Afrotropical region. British Museum (Natural History), London, p 518–554

Copeland RS (2009) A new species of *Munromyia* Bezzi (Diptera: Tephritidae) reared from *Chionanthus battiscombei* (Oleaceaea) in northern Kenya. J Nat Hist 43:2649–2665

Copeland RS, Wharton RA, Luke Q, De Meyer M (2002) Indigenous hosts of *Ceratitis capitata* (Diptera: Tephritidae) in Kenya. Annls Entomol Soc Amer 95:672–694

Copeland RS, White IM, Okumu M, Machera P, Wharton RA (2004) Insects associated with fruits of the Oleaceae (Asteridae, Mamiales) in Kenya, with special reference to the Tephritidae (Diptera). In: Evenhuis NL, Kaneshiro KY (eds) D. Elmo Hardy memorial volume. Contribution to the systematics and evolution of Diptera. Bishop Mus Bull Entomol 12:135–164

Copeland RS, Okeka W, Freidberg A, Merz B, White IM, De Meyer M, Luke Q (2005) A checklist of the Tephritidae (Diptera) of Kakamega Forest, Kenya. J East Afr Nat Hist 94:247–278

Copeland RS, Wharton RA, Luke Q, De Meyer M, Lux S, Zenz N, Machera P, Okumu M (2006) Geographic distribution, host fruits, and parasitoids of African fruit fly pests *Ceratitis anonae, Ceratitis cosyra, Ceratitis fasciventris,* and *Ceratitis rosa* (Diptera: Tephritidae) in Kenya. Annls Entomol Soc Amer 99:261–278

Copeland RS, Luke Q, Wharton RA (2009) Insects reared from the wild fruits of Kenya. J East Afr Nat Hist 98:11–66

Coquillett DW (1901) New Diptera from southern Africa. Proc U S Natl Mus 24:27–32

De Meyer M (1996) Systematic revision of the subgenus *Ceratitis* (*Pardalaspis*) (Diptera, Tephritidae). Syst Entomol 21:15–26

De Meyer M (1998) Systematic revision of the subgenus *Ceratitis* (*Ceratalaspis*) (Diptera, Tephritidae). Bull Entomol Res 88:257–290

De Meyer M (2000) Systematic revision of the subgenus *Ceratitis* MacLeay s.s. (Diptera, Tephritidae). Zool J Linn Soc 128:439–467

De Meyer M (2001) On the identity of the Natal Fruit Fly, *Ceratitis rosa* (Diptera, Tephritidae). Bull Inst r Sci Nat Belg Entomol 71:55–62

De Meyer M (2005) Phylogenetic relationships within the fruit fly genus *Ceratitis* MacLeay (Diptera: Tephritidae), derived from morphological and host plant evidence. Insect Syst Evol 36:459–480

De Meyer M (2006) A systematic revision of the African fruit fly genus *Carpophthoromyia* (Diptera, Tephritidae). Zootaxa 1235:1–48

De Meyer M (2009) Taxonomic revision of the genus *Perilampsis* (Diptera: Tephritidae). J Nat Hist 43:2425–2463

De Meyer M, Copeland RS (2001) Taxonomic notes on the subgenera *Ceratitis (Hoplolophomyia)* and *Ceratitis (Acropteromma)* (Diptera, Tephritidae). Cimbebasia 17:77–84

De Meyer M, Copeland RS (2005) Description of new *Ceratitis* MacLeay (Diptera, Tephritidae) species from Africa. J Nat Hist 39:1283–1297

De Meyer M, Copeland RS (2009) A new *Ceratitis* from Kenya (Diptera: Tephritidae). J Afrotrop Zool 5:21–26

De Meyer M, Freidberg A (2005) A systematic revision of the genus *Capparimyia* (Diptera: Tephritidae). Zool Scr 34:279–303

De Meyer M, Freidberg A (2006) Revision of the subgenus *Ceratitis* (*Pterandrus*) Bezzi (Diptera: Tephritidae). In: Freidberg A (ed) Biotaxonomy of Tephritoidea. Israel J Entomol 35/36:197–315

De Meyer M, Freidberg A (2012) Taxonomic revision of the fruit fly genus *Neoceratitis* Hendel (Diptera: Tephritidae). Zootaxa 3223:24–39

De Meyer M, Copeland RS, Lux S, Mansell M, Wharton R, White IM, Zenz N (2002) Annotated check list of host plants for Afrotropical fruit flies (Diptera: Tephritidae) of the genus *Ceratitis*. Zool Docum Konk Mus Midden Afr 27: 92 pp

De Meyer M, Quilici S, Franck A, Chadhouliati AC, Issimaila MA, Youssoufa MA, Barbet A, Attié M, White IM (2012) Frugivorous fruit flies (Diptera, Tephritidae, Dacini) of the Comoro Archipelago. Afr Invert 53:69–77

De Meyer M, White IM, Goodger KFM (2013) Notes on the frugivorous fruit fly (Diptera: Tephritidae) fauna of western Africa, with description of a new *Dacus* species. Eur J Taxon 50:1–17

De Meyer M, Clarke AR, Vera MT, Hendrichs J (eds) (2015a) Resolution of Cryptic Species Complexes of Tephritid Pests to Enhance SIT Application and Facilitate International Trade. ZooKeys 540:vi+557 pp

De Meyer M, Delatte H, Ekesi S, Jordaens K, Kalinová B, Manrakhan A, Mwatawala M, Steck G, Van Cann J, Vaníčková L, Břízová R, Virgilio M (2015b) An integrative approach to unravel the *Ceratitis* FAR (Diptera, Tephritidae) cryptic species complex: a review. In: De Meyer M, Clarke AR, Vera MT, Hendrichs J (eds) Resolution of Cryptic Species Complexes of Tephritid Pests to Enhance SIT Application and Facilitate International Trade. ZooKeys 540:405–427

De Meyer M, Delatte H, Mwatawala M, Quilici S, Vayssières JF, Virgilio M (2015c) A review of the current knowledge on *Zeugodacus cucurbitae* (Coquillett) (Diptera, Tephritidae) in Africa, with a list of species included in *Zeugodacus*. In: De Meyer M, Clarke AR, Vera MT, Hendrichs J (eds) Resolution of Cryptic Species Complexes of Tephritid Pests to Enhance SIT Application and Facilitate International Trade. ZooKeys 540:539–557

de Queiroz K (2007) Species concepts and species delimitation. Syst Biol 56:879–886

Diaz-Fleischer F, Papaj DR, Prokopy RJ, Norrbom AL, Aluja M (1999) Evolution of fruit fly oviposition behavior. In: Aluja M, Norrbom AL (eds) Fruit flies (Tephritidae): phylogeny and evolution of behavior. CRC Press, Boca Raton, pp 811–841

Drew RAI, Hancock DL (1994) The *Bactrocera dorsalis* complex of fruit flies (Diptera: Tephritidae: Dacinae) in Asia. Bull Entomol Res (Suppl 2): [viii]+68 p

Drew RAI, Romig MC (2013) Tropical fruit flies of South-East Asia. CABI, Wallingford

Erbout N, De Meyer M, Vangestel C, Lens L (2009) Host plant toxicity affects developmental rates in a polyphagous fruit fly – experimental evidence. Biol J Linn Soc 97:728–737

Erbout N, Virgilio M, Lens L, Barr N, De Meyer M (2011) Discrepancies between subgeneric classification and molecular phylogeny of *Ceratitis* (Diptera: Tephritidae), can the evolution of host use provide some clues? Mol Phylogenet Evol 60:259–264

Evenhuis NL (1997) Litteratura Taxonomica Dipterorum (1758–1930). Backhuys Publishers, Leiden (Two Volumes)

Freidberg A (1985) The genus *Craspedoxantha* Bezzi (Diptera: Tephritidae: Terelliinae). Ann Natal Mus 27:183–206

Freidberg A (1991) A new species of *Ceratitis* (*Ceratitis*) (Diptera: Tephritidae), key to species of subgenera *Ceratitis* and *Pterandrus*, and record of *Pterandrus* fossil. Bishop Mus Occas Pap 31:166–173

Freidberg A (1999) A new species of *Craspedoxantha* Bezzi from Tanzania and a revised phylogeny for the genus (Diptera: Tephritidae). Proc Entomol Soc Wash 101:382–390

Freidberg A (2002) Systematics of Schistopterini (Diptera: Tephritidae: Tephritinae), with descriptions of new genera and species. Syst Entomol 27:1–29

Freidberg A, Han HY (2012) A second species of *Manicomyia* Hancock (Diptera: Tephritidae: Tephrellini). Afr Invert 53:143–156

Freidberg A, Hancock DL (1989) *Cryptophorellia*, a remarkable new genus of Afrotropical Tephritinae (Diptera: Tephritidae). Ann Natal Mus 30:5–32

Freidberg A, Kaplan F (1992) Revision of the Oedaspidini of the Afrotropical region (Diptera: Tephritidae: Tephritinae). Ann Natal Mus 33:51–94

Freidberg A, Kaplan F (1993) A study of *Afreutreta* Bezzi and related genera (Diptera: Tephritidae: Tephritinae). Afr Entomol 1:207–228

Freidberg A, Merz B (2006) A revision of the *Gymnosagena* group of genera (Diptera: Tephritidae: Tephritinae). In: Freidberg A (Ed.) Biotaxonomy of Tephritoidea. Israel J Entomol 35/36:367–422

Froese R (1999) The good, the bad, and the ugly: a critical look at species and their institutions from a user's perspective. Rev Fish Biol Fisheries 9:375–378

Han HY, McPheron BA (1999) Nucleotide sequence data as a tool to test phylogenetic relationships among higher groups of Tephritidae: a case study using mitochondrial ribosomal DNA. In: Aluja M, Norrbom AL (eds) Fruit flies (Tephritidae): phylogeny and evolution of behavior. CRC Press, Boca Raton, pp 115–132

Han HY, Ro KE (2005) Molecular phylogeny of the superfamily Tephritoidea (Insecta: Diptera): new evidence from the mitochondrial 12S, 16S, and COII genes. Mol Phylogen Evol 39:416–430

Han HY, Ro KE (2009) Molecular Phylogeny of the Family Tephritidae (Insecta: Diptera): new insight from combined analysis of the mitochondrial 12S, 16S, and COII genes. Mol Cells 27:55–66

Han HY, Ro KE, McPheron BA (2006) Molecular phylogeny of the subfamily Tephritinae (Diptera: Tephritidae) based on mitochondrial 16S rDNA sequences. Mol Cells 22:78–88

Hancock DL (1984) Ceratitinae (Diptera: Tephritidae) from the Malagasy subregion. J Entomol Soc S Afr 47:277–301

Hancock DL (1985) Trypetinae (Diptera: Tephritidae) from Madagascar. J Entomol Soc S Afr 48:283–301

Hancock DL (1986) Classification of the Trypetinae (Diptera: Tephritidae), with a discussion of the Afrotropical fauna. J Entomol Soc S Afr 49:275–305

Hancock DL (1990) Notes on the Tephrellini-Aciurini (Diptera: Tephritidae), with a checklist of the Zimbabwe species. Trans Zimb Sci Assoc 64:41–48

Hancock DL (1999) Grass-breeding fruit flies and their allies of Africa and Asia (Diptera: Tephritidae: Ceratitidinae). J Nat Hist 33:911–948

Hancock DL (2000) Taxonomic affinities of *Deroparia* Munro, *Euryphalara* Munro and *Xenodorella* Munro (Diptera: Tephritidae: Tephritinae). Afr Entomol 8:261–264

Hancock DL (2003) An annotated checklist of the Tephritidae of Zimbabwe (Diptera: Schizophora), with a key to subgenera of *Dacus* Fabricius. Cimbebasia 19:111–148

Hancock DL (2005) A new combination in *Pseudomyoleja* Han & Freidberg (Diptera: Tephritidae: Trypetinae), with a key to Afrotropical species in the tribe Trypetini. Afr Entomol 13:168–171

Hancock DL (2012) Systematic and distributional notes on some Australasian and African species of *Platensina* Enderlein and *Dicheniotes* Munro (Diptera: Tephritidae: Tephritinae), with description of a new species of *Dicheniotes* from Kenya. Aust Entomol 39:305–320

Hancock DL, Drew RAI (2006) A revised classification of subgenera and species groups in *Dacus* Fabricius (Diptera, Tephritidae). Instrumenta Biodiv 7:167–205

Hancock DL, Hamacek EL, Lloyd AC, Elson-Harris MM (2000) The distribution and host plants of fruit flies (Diptera: Tephritidae)in Australia. DPI publications, Brisbane

Hancock DL, Kirk-Spriggs AH, Marais E (2003) New records of Namibian Tephritidae (Diptera: Schizophora), with notes on the classification of subfamily Tephritinae. Cimbebasia 18:49–70

Hardy DE (1967) Studies of fruitflies associated with mistletoe in Australia and Pakistan with notes and descriptions on genera related to *Perilampsis* Bezzi. Diptera: Tephritidae. Beitr Entomol 17:127–149

Hart MW (2010) The species concept as an emergent property of population biology. Evol 65:613–616

Hausdorf B (2011) Progress toward a general species concept. Evol 65:923–931

ICZN (1999) International code of zoological nomenclature, 4th edn. The International Trust for Zoological Nomenclature, London

ICZN (2012) Amendment of Articles 8, 9, 10, 21 and 78 of the International Code of Zoological Nomenclature to expand and refine methods of publication. Zootaxa 3450:1–7

Karsch F (1887) Dipterologisches von der Delagoabai. Entomol Nachr 13:22–26
Korneyev VA (1999a) Phylogenetic relationships among the families of the superfamily Tephritoidea. In: Aluja M, Norrbom AL (eds) Fruit flies (Tephritidae): phylogeny and evolution of behavior. CRC Press, Boca Raton, pp 3–22
Korneyev VA (1999b) Phylogenetic relationships among higher groups of Tephritidae. In: Aluja M, Norrbom AL (eds) Fruit flies (Tephritidae): phylogeny and evolution of behavior. CRC Press, Boca Raton, pp 73–113
Krosch MN, Schutze MK, Armstrong KF, Graham GC, Yeates DK, Clarke AR (2012) A molecular phylogeny for the Tribe Dacini (Diptera: Tephritidae): systematic and biogeographic implications. Mol Phylogenet Evol 64:513–523
Kunz W (2012) Do species exist?: Principles of taxonomic classification. Wiley, Weinheim
Loew H (1852) Hr. Peters legte Diagnosen und Abbildungen der von ihm in Mossambique neu entdeckten Dipteren vor, welche von Hrn. Professor Loew bearbeitet worden sind. Sitzungsber Akad Wiss Berl 1852:658–661
Loew H (1861) Ueber die afrikanischen Trypetina. Berl Entomol Z 5:253–306
Loew H (1862) Bidrag till kannedomen om Afrikas Diptera. Ofvers. K. Svenska Vetenskapakad Forh 19:3–14
Mayden RL (1997) A hierarchy of species concepts: the denoument in the saga of the species problem. In: Claridge MF, Ha D, Wilson MR (eds) Species: the units of diversity. Chapman and Hall, London, pp 381–423
Mazzon L, Piscedda A, Simonato M, Martinez-Sañudo I, Squartini A, Girolami V (2008) Presence of specific symbiotic bacteria in flies of the subfamily Tephritinae (Diptera Tephritidae) and their microbiome of Australian tephritid fruit fly species phylogenetic relationships: proposal of 'Candidatus Stammerula tephritidis'. Int J Syst Evol Microbiol 58:1277–1287
Mazzon L, Martinez-Sañudo I, Simonato M, Squartini A, Savio C, Girolami V (2010) Phylogenetic relationships between flies of the Tephritinae subfamily (Diptera, Tephritidae) and their symbiotic bacteria. Mol Phylogenet Evol 56:312–326
Merz B (1999) Phylogeny of the Palearctic and Afrotropical genera of the *Tephritis* group (Tephritinae: Tephritini). In: Aluja M, Norrbom AL (eds) Fruit flies (Tephritidae): phylogeny and evolution of behavior. CRC Press, Boca Raton, pp 629–669
Merz B, Dawah HA (2005) Fruit flies (Diptera, Tephritidae) from Saudi Arabia, with descriptions of a new genus and six new species. Rev Suisse Zool 112:983–1028
Morrow JL, Frommer M, Shearman DCA, Riegler M (2015) The microbiome of field-caught and laboratory-adapted Australian tephritid fruit fly species with different host plant use and specialization. Microb Ecol. doi:10.1007/s00248-015-0571-1
Munro HK (1925) Biological notes on South African Trypaneidae (fruitflies). I. Entomol Mem S Afr Dep Agric [1] 3:39–67
Munro HK (1926) Biological notes on the South African Trypaneidae (Trypetidae: fruit-flies). II. Entomol Mem S Afr Dep Agric [1] 5:17–40
Munro HK (1929) Biological notes on the South African Trypetidae. (Fruit-flies. Diptera) III. Entomol Mem S Afr Dep Agric [1] 6: 9–17
Munro HK (1947) African Trypetidae (Diptera). A review of the transition genera between Tephritinae and Trypetinae, with a preliminary study of the male terminalia. Mem Entomol Soc S Afr 1: [viii]+284 p
Munro HK (1984) A taxonomic treatise on the Dacidae (Tephritoidea, Diptera) of Africa. Entomol. Mem S Afr Dep Agric 61: [i]+ii-ix+313 p
Norrbom AL, Carroll LE, Thomopson FC, White IM, Freidberg A (1999) Status of knowledge. In: Thompson FC (ed) Fruit fly expert identification system and systematic information database. Myia 9 (1998): 65–251
Pont AC (1995) The dipterist C.R.W. Wiedemann (1770–1840). His life, work and collections. Steenstrupia 21:125–154
Regan CT (1926) Organic evolutions. Rep 93rd Meet Brit Assoc. Adv Sci 1925:75–86

Richards RA (2010) The species problem: a philosophical analysis. Cambridge University Press, Cambridge

Rossi P (1790) Fauna etrusca. Sistens insecta quae in provinciis Florentina et Pisana praesertim collegit. vol 2. Liburni [= Livorno].

Schutze MK, Mahmood K, Pavasovic A, Bo W, Newman J, Clarke AR, Krosch M, Cameron SL (2014a) One and the same: integrative taxonomic evidence that *Bactrocera invadens* (Diptera: Tephritidae) is the same species as the Oriental fruit fly *Bactrocera dorsalis*. Syst Entomol. doi:10.1111/syen.12114

Schutze MK, Aketarawong N, Amornsak W, Armstrong KF, Augustinos A, Barr N, Bo W, Bourtzis K, Boykin LM, Caceres C, Cameron SL, Chapman TA, Chinvinijkul S, Chomic A, De Meyer M, Drosopoulou ED, Englezou A, Ekesi S, Gariou-Papalexiou A, Hailstones D, Haymer D, Hee AKW, Hendrichs J, Hasanuzzaman M, Jessup A, Khamis FM, Krosch MN, Leblanc L, Mahmood K, Malacrida AR, Mavragani-Tsipidou P, McInnis DO, Mwatawala M, Nishida R, Ono H, Reyes J, Rubinoff DR, San Jose M, Shelly TE, Srikachar S, Tan KH, Thanaphum S, Ul Haq I, Vijaysegaran S, Wee SL, Yesmin F, Zacharopoulou Z, Clarke AR (2014b) Synonymization of key pest species within the *Bactrocera dorsalis* species complex (Diptera: Tephritidae): taxonomic changes based on a review of 20 years of integrative morphological, molecular, cytogenetic, behavioural, and chemoecological data. Syst Entomol. doi:10.1111/syen.12113

Silvestri F (1913) Viaggio in Africa per cercare parassiti di mosche dei frutti. Boll Lab Zool Gen Agrar R Scuola Agric Portici 8:1–164

Van Zwaluwenburg RH (1937) West African notes. Hawaii Planters Rec 41:57–83

Virgilio M (2016) Identification tools for African frugivorous fruit flies (Diptera: Tephritidae). In: Ekesi S, Mohamed SA, De Meyer M (eds) Fruit fly research and development in Africa – towards a sustainable management strategy to improve horticulture. Springer, Cham, pp xx–xx

Virgilio M, De Meyer M, White IM, Backeljau T (2009) African *Dacus* (Diptera: Tephritidae): molecular data and host plant association do not corroborate morphology based classifications. Mol Phylogenet Evol 51:531–539

Virgilio M, Delatte H, Backeljau T, De Meyer M (2010) Macrogeographic population structuring in the cosmopolitan agricultural pest *Bactrocera cucurbitae* (Diptera: Tephritidae). Mol Ecol 19:2713–2724

Virgilio M, Delatte H, Quilici S, Backeljau T, De Meyer M (2013) Cryptic diversity and gene flow among three African agricultural pests: *Ceratitis rosa*, *Ceratitis fasciventris* and *Ceratitis anonae* (Diptera, Tephritidae). Mol Ecol. doi:10.1111/MEC12278

Virgilio M, White IM, De Meyer M (2014) A set of multi-entry identification keys to African frugivorous flies (Diptera, Tephritidae). ZooKeys 428:97–108

Virgilio M, Delatte H, Beda Y, Simiand C, Quilici S, De Meyer M, Mwatawala M (2015a) Possible cryptic speciation in the mango fruit fly, *Ceratitis cosyra* (Diptera, Tephritidae). In: De Meyer M, Clarke AR, Vera MT, Hendrichs J (eds) Resolution of Cryptic Species Complexes of Tephritid Pests to Enhance SIT Application and Facilitate International Trade. ZooKeys 540:525–538

Virgilio M, Jordaens K, Verwimp C, De Meyer M (2015b) Higher phylogeny of frugivorous fruit flies (Diptera, Tephritidae, Dacini) as inferred from mitochondrial and nuclear gene fragments. Mol Phylogenet Evol. doi, http://dx.doi.org/10.1016/j.ympev.2015.01.007

White IM (2006) Taxonomy of the Dacina (Diptera: Tephritidae) of Africa and the Middle East. Afr Entomol Mem 2:156 pp

White IM, Elson-Harris MM (1994) Fruit flies of economic significance: their identification and bionomics. International Institute of Entomology, London

White IM, Goodger KFM (2009) African *Dacus* (Diptera: Tephritidae); new species and data, with particular reference to the Tel Aviv University Collection. Zootaxa 2127:1–49

White IM, Headrick DH, Norrbom AL, Carroll LE (1999) Glossary. In: Aluja M, Norrbom AL (eds) Fruit flies (Tephritidae): phylogeny and evolution of behavior. CRC Press, Boca Raton, pp 881–924

White IM, Copeland RS, Hancock DL (2003) Revision of the afrotropical genus *Trirhithrum* Bezzi (Diptera: Tephritidae). Cimbebasia 18:71–137

Wiedemann CRW (1818) Neue Insekten vom Vorgebirge der Guten Hoffnung. Zool Mag Wiedemann's 1(2):40–48

Wiedemann CRW (1819) Beschreibung neuer Zweiflugler aus Ostindien und Afrika. Zool Mag Wiedemann's 1(3):1–39

Wiedemann CRW (1824) Munus rectoris in Academia Christiana Albertina aditurus analecta entomologica ex Museo Regio Havniensi maxime congesta profert iconibusque illustrat. Kiliae [= Kiel]

Wiedemann CRW (1830) Aussereuropaische zweiflugelige Insekten, vol 2. Schulz, Hamm

Wiegmann BM, Trautwein MD, Winkler IS, Barr NB, Kim JW, Lambkin C, Bertone MA, Cassel BK, Bayless KM, Heimberg AM, Wheeler BM, Peterson KJ, Pape T, Sinclair BJ, Skevington JH, Blagoderov V, Caravas J, Kutty SN, Schmidt-Ott U, Kampmeier GE, Thompson FC, Grimaldi DA, Beckenbach AT, Courtney GW, Freidrich M, Meier R, Yeates DK (2011) Episodic radiations in the fly tree of life. Proc Natl Acad Sci U S A, http://dx.doi.org/10.1073/pnas.1012675108

Wilkins JA (2009) Species: a history of the idea. University of California Press, Berkley/Los Angeles

Marc De Meyer is an entomologist attached to the Royal Museum for Central Africa in Tervuren (Belgium). Previously he worked in Botswana and Kenya for a number of years. His main field of expertise is the taxonomy and systematics of selected groups of Diptera such as Pipunculidae, Syrphidae and Tephritidae, with emphasis on the Afrotropical fauna. His research includes different aspects, from taxonomic revisions to studies on evolutionary trends and speciation events.

Chapter 2
Identification Tools for African Frugivorous Fruit Flies (Diptera: Tephritidae)

Massimiliano Virgilio

Abstract The current classification of African tephritids is the interim result of a continuous process of minor and major changes that, in the last 20 years, has resulted in the description of more than 60 new species from the seven tephritid genera of main economic relevance in Africa (*Bactrocera*, *Capparimyia*, *Ceratitis*, *Dacus*, *Neoceratitis*, *Trirhithrum* and *Zeugodacus*). In this context of dynamic change, rapid and accurate fruit fly identification is critical, particularly with respect to the early detection of pest invasions. Valuable resources for fruit fly identification include: the tephritid reference collections and repositories distributed within and outside the African continent; publicly available online databases; and the single- and multi-entry keys for the morphological identification of African tephritids. Identification through DNA barcoding represents a cost effective tool for the molecular diagnosis of African fruit flies and it has proved particularly useful for the identification of immature stages, of damaged specimens and of incomplete specimens. The molecular diagnosis of tephritids also represents a partial solution to the gradual loss of taxonomical expertise on this and other insect groups. In this chapter the advantages and limitations of the available identification tools and resources are discussed.

Keywords Morphological identification • Natural history collections • Online databases • Identification keys • Molecular diagnosis • DNA barcoding

1 Introduction

Tephritid fruit flies, or 'true' fruit flies (Diptera: Tephritidae), include approximately 500 genera and 4800 valid species, the vast majority (95%) of which are phytophagous (Aluja and Norrbom 1999). Of all tephritid species 25-30% are frugivorous. In Africa there are approximately 400 species of frugivorous tephritids of which more

M. Virgilio (✉)
Department of Biology and Joint Experimental Molecular Unit (JEMU), Royal Museum for Central Africa, Tervuren, Belgium
e-mail: massimiliano.virgilio@africamuseum.be

than 50 are economically important (list provided in Virgilio et al. 2014). The current classification of African tephritids is the interim result of a continuous process of minor and major updating; in just the last 20 years this has included:

- a monograph on the genera *Dacus* and *Bactrocera* from Africa and the Middle East (White 2006) with the genus (Hancock and Drew 2006)
- a revision of the *Ceratitis* subgenera *Acropteromma* and Hoplolophomyia (De Meyer and Copeland 2001), *Ceratalaspis* (De Meyer 1998), *Ceratitis s.s.* (De Meyer 2000), *Pardalaspis* (De Meyer 1996) and *Pterandrus* (De Meyer and Freidberg 2006)
- a revision of the genera *Capparimyia* (De Meyer and Freidberg 2005), *Carpophthoromyia* (De Meyer 2006), *Neoceratitis* (De Meyer and Freidberg 2012), *Perilampsis* (De Meyer 2009) and *Trirhithrum* (White et al. 2003)
- the description of a new pest species: *Bactrocera invadens* Drew et al. (2005)
- a monograph on the genera *Dacus* and *Bactrocera* from Africa and the Middle East (White 2006)
- a revised classification of subgenera and species groups within the genus *Dacus* (Hancock and Drew 2006)
- the description of 17 new *Dacus* species (White and Goodger 2009)
- an analysis of the biodiversity of the western African fauna including the description of a new *Dacus* species (De Meyer et al. 2013)
- the synonimisation of the key pests *Bactrocera invadens* and *Bactrocera. dorsalis* (Hendel) (Schutze et al. 2015)
- a novel generic combination for *Zeugodacus cucurbitae* (Coquillett) (Virgilio et al. 2015)

In this context of dynamic change, rapid and accurate fruit fly identification is critical, particularly with respect to the early detection of pest invasions. For example, in 1995 the incorrect identification of *Bactrocera zonata* (Saunders) as *Bactrocera pallidus* (Perkins and May) in Egypt lead to a three-year delay in the implementation of phytosanitary measures and resulted in serious damage to the agricultural productivity of the whole Alexandria region (Abuel-Ela et al. 1998).

2 Online Databases

The tephritid reference collections and repositories are a valuable resource for fruit fly identification as well as for the training of specialist and non-specialist taxonomists. African researchers can confidently rely on what is a relatively limited number of comprehensive reference collections in the continent which include: the South African National Collection of Insects, Pretoria, South Africa; the collections of the National Museums of Kenya, Nairobi, Kenya; the collection of the Natural History Museum of Zimbabwe, Bulawayo, Zimbabwe; and the collection of the International Institute of Tropical Agriculture (IITA), Cotonou, Benin.

Outside of Africa, one of the largest collections of African frugivorous flies is held at the Royal Museum for Central Africa (RMCA), Tervuren, Belgium; this collection currently includes some 5000 African specimens from approximately 200 species from ten tephritid genera. Detailed information about vouchers available in the RMCA frugivorous tephritid collection can be directly accessed through the 'True fruit flies of the Afrotropical Region' database (http://projects.bebif.be/fruitfly/index.html). This database is part of the Global Biodiversity Information Facility (GBIF, http://www.gbif.org), a platform that aims to provide open access to biodiversity data and is hosted within the Belgian GBIF node (BeBIF, http://www.bebif.be). This database also has information on reference material from African fruit fly species in the genera *Ceratitis, Dacus, Bactrocera, Capparimyia, Trirhithrum, Carpophthoromyia* and *Perilampsis*, that is available from other European, North American and African museums and research institutions. The BeBIF fruit fly database has 150,000 specimens, in excess of 16,000 block records (ie sets of specimens with identical data), material from 60 institutions and private collections, historical collections (eg type collections and collections from USDA expeditions to Africa) and associated data; the associated data include details of approximately 3000 georeferenced localities, 700 host plant records and more than 1500 digital images and maps of sampling locations. Taxon information in BeBIF includes: (a) the current valid taxonomic name of each species and a list of synonyms where applicable; (b) a short taxonomic description of the species based on available taxonomic revisions; (c) a set of images (photographs or drawings) depicting the main morphological characteristics of the species that are taken in a uniform and standardized way in order to facilitate comparison; and (d) a geographical distribution map for each species that is directly linked to specimen information. All relevant data that are linked to individual vouchers or block records are provided and include: place where and the date when the specimen was collected, the name of the collector, the collection where the specimen is deposited and the status of the specimen (type or non-type). Other additional information that can also prove useful for identification are the response to lures and attractants, (which are generally specific at the genus or subgenus level), and the range of host plants attacked by the species.

3 Keys for Morphological Identification

Morphological identification of African fruit flies can be achieved using a range of methods that differ in their technical complexity and reliability. Dichotomous identification keys are generally only accessible by users with existing background knowledge of tephritid morphology and the often-complicated technical terminology used, and who also have access to specialised equipment such as dissection tools and microscopes. Alternative identification tools of more general use include simplified keys for a number of the economically relevant African pests (Ekesi and Billah 2007), identification sheets and online material for the identification of invasive fruit flies in Africa (eg www.africamuseum.be/fruitfly/AfroAsia.htm). These tools, under certain circumstances, can be useful to the large number of untrained users, such as farmers, who are keen to detect pests on their crops. Of course, although these general tools are rapid, they can also be inaccurate when dealing with the less common species.

Classical single-entry (dichotomous) keys are available for most African Dacini. White (2006) produced a dichotomous key with a revised classification for the African and Middle Eastern species of *Bactrocera* (15 species) and *Dacus* (177 species), a database of digital images for 190 species and a database with notes on the identification of pest species. The revised classification of White (2006) was partly based on a cladistic analysis that explored the subgeneric relationships of a subset of representative species; this facilitated a number of advances including the description of 25 new *Dacus and Bactrocera* species, the synonymisation of 26 species and the removal from synonymy of two species. Since the work of White (2006) more new species and changes in synonymy have occurred which are not in the original dichotomous key. For example 17 new species of *Dacus* have been described and two synonymised by White and Goodger (2009). Similarly, Hancock and Drew (2006) produced a revised classification and a dichotomous key that could be useful for the identification of *Dacus* subgenera and species groups.

Dichotomous identification keys are also available for the genus *Ceratitis*; there are four stand-alone subgeneric keys and revisions including: (a) a key to the subgenus *Ceratitis* (*Pardalaspis*) Bezzi (De Meyer 1996) with ten Afrotropical species and information about species distribution and host plants, (b) a key to the subgenus *Ceratitis* (*Ceratalaspis*) Hancock (De Meyer 1998) with 36 species and illustrations of mesonotal and wing patterns, shape of the aculeus tip, distribution and known host plant data, and tentative species groups within the subgenus, (c) a key to eight species of the subgenus *Ceratitis* Macleay *s.s* with illustrations of cephalic bristles, mesonotal and wing patterns and aculeus shape (De Meyer 2000) and (d) a key to 36 species, of the subgenus *Ceratitis* (*Pterandrus*) Bezzi with information about species distribution and host plant data, tentative species groups within the subgenus and illustrations of male and female terminalia, wing and mesonotal patterns and male leg ornamentation (De Meyer and Freidberg 2006).

The dichotomous key to the genus *Trirhithrum* Bezzi (White et al. 2003) allows identification of 40 *Trirhithrum* species and a further seven taxa of uncertain status. The revision published with the key provides host data, largely from a survey in Kenya. The small genus *Capparimyia* Bezzi was revised by De Meyer and Freidberg (2005) who recognized eight species and provided a dichotomous key with illustrations of mesonotal and wing patterns and male and female terminalia. The 17 species of the genus *Carpophthoromyia* Austen can also be identified using the dichotomous key of De Meyer (2006) that also provides illustrations of wing patterns and both male and female terminalia. The dichotomous key to the genus *Perilampsis* Bezzi (De Meyer 2009) includes 17 species and provides illustrations of wing patterns, female terminalia and information about host specificity. The genus *Neoceratitis* Hendel can be identified using the dichotomous key of De Meyer and Freidberg (2012) and includes six species with illustrations and host information.

One of the main limitations of dichotomous single-entry keys is that species identification is not possible when the user is unable to distinguish between one of the dichotomous options provided by in the key. This can occur if the specimen is damaged so that the morphological character is not present or easily recognisable, if the user has inadequate taxonomic expertise, or if there is a lack of clarity in the key. The terminology used in some published keys can represent a serious obstacle for non-specialists who are not well acquainted with insect morphology and taxonomy. In fact, many terms used to describe morphological variation, such as small/ large, dark/ pale, thick/ thin etc. could be considered subjective and unclear to non-specialist users (while specialist taxonomists generally find these definitions straightforward because they have the necessary and essential experience that comes from examining large numbers of specimens). In this respect, multi-entry identification keys might overcome some of the technical difficulties associated with dichotomous keys. As the name suggests, multi-entry identification keys allow identification via multiple paths such that the user has the ability to 'skip' problematic questions and score alternative characters.

Additionally, there is no comprehensive key to all genera of African fruit flies so that non-specialised users might even find it problematic to assign specimens to the genera for which dichotomous keys are available (but see Hancock and White 1997 for a key to distinguish the genus *Trirhithrum* from others in the *Ceratitis* group of genera). This issue is even more relevant for the genus *Ceratitis*, where additional subgenus identification is also necessary before it is possible to use one of the six dichotomous keys available.

The development of a user-friendly set of multi-entry identification keys for African tephritids began in 1999 with a pilot project supported by the U.S. Agency for International Development (USAID, PCE-G-00-98-0048-00) and by the U.S. Department of Agriculture (USDA) / National Institute of Food and Agriculture (CSREES) / Initiative for Future Agricultural and Food Systems (IFAFS) grants to the Texas A&M University (00- 52,103–9651). This resulted in an initial set of two keys for the identification of *Ceratitis* and *Trirhithrum* species, through the CABIKEY platform. Later on, a project co-funded by the Belgian Directorate-General for Development Cooperation (through a framework agreement with the Royal Museum for Central Africa) and the International Atomic Energy Agency (IAEA – Vienna, project 'Development of a Web Based Multi Entry Key for Fruit Infesting Tephritidae', contract number 16,859) allowed development of a set of multi-entry identification keys for African frugivorous flies (Virgilio et al. 2014). These keys included a 'pre-key' for genus designation (built *ex novo* using a set of 23 characters that were deemed to be informative for separation of genera) as well as seven multi-entry keys for species identification within a genus or a group of genera (*Bactrocera+Dacus+Zeugodacus*, *Capparimyia*, *Carpophthoromyia*, *Ceratitis*, *Neoceratitis*, *Perilampsis*, *Trirhithrum*) and including a total of approximately 390 taxa. In this set of keys species lists and morphological characters were revised and optimised to include only species with (a) valid names under the International Code of Zoological Nomenclature and (b) characters including at least two states in congeneric species (Virgilio et al. 2014). The keys were based on eight matrices containing scores for a total of 368 characters and were compiled from data sets that were used within the framework of the taxonomic revisions described above (De Meyer 1996, 1998, 2000, 2006, 2009; White et al. 2003; De Meyer and Freidberg 2005, 2006, 2012; White 2006; White and Goodger 2009). The keys are regularly updated in order to keep pace with changes in the taxonomic status of species and take into account, for example, the recent synonimisation of *B. invadens* and *B. dorsalis* (Schutze et al. 2015), and the novel generic status of *Z. cucurbitae* (Virgilio et al. 2015). To facilitate identification, morphological characters were grouped as sets from the head, thorax, wings, legs and abdomen respectively. Unfolding characters were also included, i.e. those characters that are initially hidden but appear when only a pre-defined subset of species remain to be identified (unfolding keys). Dependencies between characters were also generated; positive dependencies were defined whenever a character was only meaningful in relation to a previously defined character state (eg in the *Ceratitis* key, the morphological character 'number of frontal setae' is positively dependent on the character state -'frontal setae: yes). Conversely, negative dependencies were generated to discard characters that were not meaningful after a previous character state was selected (eg in the *Ceratitis* key, the character 'females: aculeus tip with small notch' is negatively dependent on the character state 'sex: male'). Embedded within the keys are images that illustrate, name and position each character on the insect body. There are also images showing how the same character appears in different species. The initial set of 2300 images and drawings recovered from the databases of the Royal Museum for Central Africa (RMCA) and from the London Natural History Museum (NHM) were rearranged according to species name and body part (head, thorax dorsal, thorax lateral, abdomen, wings, legs), divided into groups and assigned to each combination of character state and species name. This generated a database of approximately 20,000 images that illustrate the phenotypic variability of the same character across species and provides a 'virtual collection' of images that are rapidly accessible. Furthermore, the largest keys (*Bactrocera/Dacus/Zeugodacus, Ceratitis, Trirhithrum*) allow the user to distinguish between different subsets of morphological characters including (1) characters that are the most straightforward to identify; (2) all characters except those that are the most difficult to identify; and (3) all characters, including the 'easy', 'average' and 'difficult' ones. The user has the opportunity to first consider only characters that are straightforward to use, and then follow this up by using characters that are increasingly more difficult to interpret. This process facilitates identification and reduces the risk of misidentification, particularly for species that can be identified using straightforward characters only. The keys also allow identification to be restricted to species of economic importance only. The use of this option should speed up identification of the more commonly trapped / intercepted taxa. However, when using this option, identification should be further verified (eg through an in-depth analysis of the species description – see below) as less common species not included in this option could be erroneously identified as species of economic importance (false positives). The keys also provide (a) species descriptions as provided by the published scientific literature, (b) images from the RMCA and NHM tephritid collections and (c) hyperlinks to the Encyclopedia of Life (EOL), the Belgian Biodiversity Platform (BeBIF) and, when available, to the Barcoding of Life Database (BOLD).

4 Molecular identification through DNA barcoding

DNA barcoding provides a rapid and often effective tool for the molecular diagnosis of species and it has proved to be particularly useful for specimens (or parts of specimens) where distinguishing morphological characteristics are degraded or missing (Hebert et al. 2003; Nagy et al. 2013). DNA barcoding is a distance-based identification method that relies on reference libraries of DNA sequences from unambiguously identified voucher specimens. The most widely used DNA barcode for animal identification is a standardised 648 base-pair region of the mitochondrial cytochrome c oxidase subunit I (COI) while other gene fragments are used for plants (Ribulose-bisphosphate carboxylase [rbcl] and Maturase K [matK]) and fungi (the Internal Transcribed Spacer Region [ITS]). DNA barcoding identification basically relies on (1) calculating the genetic distance between the target DNA sequence of an unidentified specimen (a query) and sequences from the reference library of DNA barcodes and (2) assigning to the query the species name of the most genetically similar reference DNA barcode (ie having the smallest genetic distance from the query) (Hebert et al. 2003; Ratnasingham and Hebert 2007). A number of DNA barcoding bodies and resources are available and include (1) the Consortium for the Barcode of Life (CBOL; http://www.barcodeoflife.org) which promotes DNA barcoding via institutions from over 50 countries and operates out of the Smithsonian Institute's National Museum of Natural History in Washington; (2) the International Barcode of Life (iBOL, http://www.ibol.org) which involves numerous countries in the global barcoding effort and; (3) BOLD (http://www.boldsystems.org) which is an online workbench and the main platform for DNA barcoding identification (reviewed in Taylor and Harris 2012). BOLD is the main barcode repository and provides analytical tools; an interface for submission of sequences to GenBank; species identification tools; and connectivity for external web developers and bioinformaticians (Ratnasingham and Hebert 2007). Each reference DNA barcode in BOLD is linked to specimen information including, *inter alia*, the species name (or its interim), voucher data (catalogue number and institution storage reference), collection records (collector, collection date and location with GPS coordinates) and the name of the person who identified the specimen. The DNA barcoding data associated with animal specimens includes the COI sequence (of at least 500 bp), the polymerase chain reaction (PCR) primers used to generate the amplicon and the sequence forward and reverse trace files (Ratnasingham and Hebert 2007). The DNA barcoding identification tool in BOLD reports the genetic similarity between the query and a list of the best DNA barcode matches in a table of similarity scores (%) and visualizes the distances between the query and its best matches in a neighbor-joining tree reconstruction.

DNA barcoding of fruit flies (eg Meeyen et al. 2014) might indeed represent a feasible and complementary solution to the gradual loss of taxonomical expertise on this and other insect groups (de Carvalho et al. 2007). For immature stages of most fruit fly species and for damaged specimens DNA barcoding is the only available identification tool; for this reason it has potential for routine identification of fruit

Table 2.1 Reference DNA barcodes available in the Barcoding of Life Database (http://www.boldsystems.org, 02/07/2015) across tephritid subfamilies

Subfamilies	Specimens with barcodes	Number of taxonomic entities with barcodes
Dacinae	4530	395
Phytalmiinae	11	6
Tachiniscinae	1	1
Tephritinae	1426	249
Trypetinae	1443	148
Total	7411	799

fly interceptions (Armstrong and Ball 2005; Barr et al. 2012; Boykin et al. 2012). Despite this potential (but see Moritz and Cicero 2004; Cameron et al. 2006; Taylor and Harris 2012; Kvist 2013; Pečnikar and Buzan 2014), DNA barcoding is still not widely used for tephritid identification due to a number of issues associated with the incomplete taxon coverage of the available reference libraries (Virgilio et al. 2010; Kwong et al. 2012; Virgilio et al. 2012; Smit et al. 2013) as well as with difficulties in resolving important species complexes of economic interest (Frey et al. 2013) such as the *Bactrocera dorsalis* (Hendel) (Jiang et al. 2014) or the *Ceratitis* FAR (Virgilio et al. 2012) complexes or even failure to differentiate between closely related species for which there are distinct morphological characters to separate the adults *(eg Ceratitis capitata and Ceratitis caetrata, see Barr et al.* 2012). In 2007 the Consortium for BOLD initiated and supported the Tephritid Barcoding Initiative (TBI), a two year demonstration project to populate the reference database of DNA barcodes for fruit flies and develop protocols for queries in support of pest management, ecology and taxonomy. An analysis of the status of the BOLD libraries (updated 2nd of July 2015) with respect to current taxon coverage for tephritid fruit flies reveals that more than 7000 tephritid vouchers had been barcoded for a total of approximately 800 taxonomic entities including both valid species and interim identifications (the latter representing a relevant 22 % of the tephritid taxa in BOLD). Almost half of the barcoded taxonomic entities (49.4 %) belong to the subfamily Dacinae and include the seven tephritid genera of main economic relevance in Africa (*Bactrocera, Capparimyia, Ceratitis, Dacus, Neoceratitis, Trirhithrum* and *Zeugodacus*) as well as of the two related genera (*Carpophthoromyia* and *Perilampsis*). These genera alone include 94.9 % of all barcoded Dacinae taxonomic entities (corresponding to 98.9 % of all Dacinae specimens in BOLD) (Table 2.1).

There are more than 2200 reference DNA barcodes for the five genera of major economic importance in Africa viz. *Bactrocera, Zeugodacus, Dacus, Ceratitis* and *Trirhithrum* (see Virgilio et al. 2014 for a list of the main African pests). Economically important *Zeugodacus* and *Bactrocera* species viz. *Z. cucurbitae, B. dorsalis, B. latifrons, B. oleae* and *B. zonata*, are all represented by multiple reference DNA barcodes (with more than 1300 DNA barcodes available in total with an average of 226.8, SD = 220.3 DNA barcodes per species). There are more than 170 reference DNA barcodes in the BOLD libraries (average per species = 10.8, SD = 16.6) for

Table 2.2 Reference DNA barcodes available in the Barcoding of Life Database (http://www.boldsystems.org, 02/07/2015) across genera in the subfamily *Dacinae*

Genera within *Dacinae* available in BOLD (02/07/2015)	Specimens with barcodes	Number of taxonomic entities with barcodes	Number of interim species with barcodes	% interim species	Number of economically important species
Bactrocera	2667	197	66	33.5	5
Zeugodacus	284	1	0	0.0	1
Dacus	423	78	3	3.8	16
Ceratitis	937	57	10	17.5	21
Trirhithrum	82	18	2	11.1	8
Capparimyia	28	7	1	14.3	1
Carpophthoromyia	30	8	2	25.0	0
Neoceratitis	2	2	0	0.0	1
Perilampsis	26	7	0	0.0	0
Acanthiophilus	14	2	1	50	
Acroceratitis	2	2	0	0	
Acrotaeniostola	1	1	0	0	
Bistrispinaria	2	1	0	0	
Capitites	1	1	0	0	
Celidodacus	6	4	1	25	
Clinotaenia	4	3	1	33.3	
Cyrtostola	1	1	0	0	
Dectodesis	9	1	0	0	
Euarestella	2	1	0	0	
Gastrozona	4	1	0	0	
Taeniostola	3	1	0	0	
Urelliosoma	2	1	0	0	
Xanthorrachista	0	0			
Total	4530	395	87	22.0	

fourteen of the sixteen *Dacus* species that are pests in Africa. Despite this, for four species (*D. annulatus, D. limbipennis, D. lounsbutyii, D. persicus*) only a single reference barcode is currently available. Similarly, 18 of the 21 African *Ceratitis* pest species are represented in the BOLD libraries and there are multiple reference sequences (average per species = 31.4, SD = 66.8) for all of them except *C. pennicillata*. Of the eight *Trirhithrum* pest species all but *T. albomaculatum, T. basale* and *T. manganum* are represented in the BOLD libraries, all with multiple reference DNA barcodes (average per species = 4.3, SD = 5.2) (Table 2.2).

The completeness of the reference libraries remains a critical issue as, obviously, any query without a conspecific reference DNA barcode in the library cannot be correctly identified (Virgilio et al. 2010; Smit et al. 2013). A distance threshold can be defined such that a query is discarded (ie its identification considered unreliable) whenever the distance between the query and its best DNA barcode match exceeds the threshold value (according to the Best Close Match criterion, see Meier et al. 2006). This reduces the probability that queries that are not represented in the library by a conspecific will be incorrectly identified with the 'closest' (ie most genetically similar) allospecific match. The outcomes of distance threshold based DNA barcoding can be categorised as: (1) true positives (TP), ie queries that are correctly identified with a genetic distance to their best match that is below the threshold; (2) false positives (FP), ie queries that are misidentified despite the distance to their best match remaining below the threshold; (3) true negatives (TN), misidentified queries that are correctly discarded because the distance to their best match is above the threshold and; (4) false negatives (FN), correctly identified queries that are discarded in error as the distance to their best match is above the threshold. Distinguishing amongst these categories allows the user to quantify the level of accuracy (TP+TN/number of queries), precision (TP/ (TP+FP)), overall identification error (FP+FN/number of queries) and the relative identification error (FP/ TP+FP) of the DNA barcoding identification method. Several criteria for setting the distance thresholds have been proposed (eg Meyer and Paulay 2005). Fixed distance thresholds were common in early barcoding studies (eg Hebert et al. 2003) and were initially implemented in BOLD where a 1 % sequence dissimilarity (ie the fraction of base mismatches between two sequences) represented the cut-off value for identification (Ratnasingham and Hebert 2007). Of course, no single interspecific distance threshold fits all taxonomic groups as coalescent depths amongst species vary due to differences in population size, rate of mutation and time since speciation (Collins and Cruickshank 2013). A number of distance thresholds can be generated directly from the data so that cut-off values change according to the particular reference library / taxon group being considered (Meyer and Paulay 2005; Meier et al. 2006; Puillandre et al. 2011; Virgilio et al. 2012). Initially, a 'ten times' rule was proposed (Hebert et al. 2004) to determine a threshold value as calculated from the distribution of intraspecific distances (but see Hickerson et al. 2006 for criticism). Sonet et al. (2013) developed an R package to calculate *ad hoc* distance thresholds producing identifications with an estimated relative error probability that could be fixed by the user (eg 5 %) (Virgilio et al. 2012). BOLD is now implementing a Barcode Index Number (BIN) algorithm that uses a 2.2 % sequence dissimilarity threshold with subsequent refinement using Markov clustering (Ratnasingham and Hebert 2013). Other statistical approaches, aimed at reducing the limits of distance-based identification have been proposed (eg Nielsen and Matz 2006; Tanabe and Toju 2013; Dowton et al. 2014; Porter et al. 2014). However, performing complex statistics on libraries that include millions of reference barcodes still remains computationally challenging. Furthermore, users willing to adopt alternative approaches and criteria for DNA barcoding identification generally need to build their own reference library, and this is not always possible as not all BOLD reference DNA barcodes are publically available.

Producing fruit fly DNA barcodes is a relatively straightforward process when starting from common, recently collected and adequately preserved fruit fly specimens. In these cases, DNA barcodes can generally be obtained using universal DNA primers (Folmer et al. 1994) and standard or slightly modified protocols for DNA

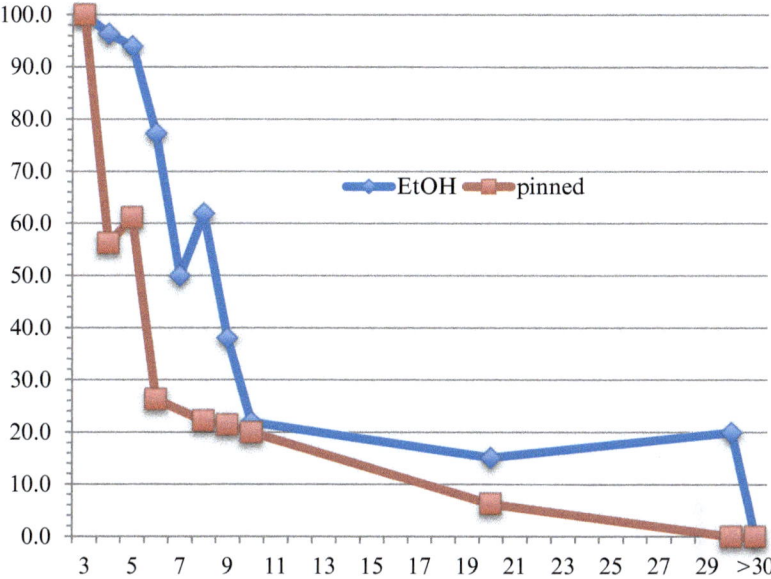

Fig. 2.1 Percentage of DNA barcodes obtained using standard protocols on EtOH – preserved and pinned specimens of different ages (Virgilio and De Meyer, unpublished)

extraction, amplification and sequencing (Barr et al. 2012). However, obtaining DNA barcodes from the less common African fruit flies or for species not commonly found in crop production areas is relatively difficult as many of these species are not regularly trapped / reared in the context of monitoring programmes or sampling campaigns (Virgilio et al. 2011). In this respect, Natural History collections are considered as a valuable source of already referenced vouchers that can be used for DNA barcoding. However, producing DNA barcodes from Natural History collections can also be problematic if the DNA has become degraded during storage (Zimmermann et al. 2008). A screening based on approximately 400 tephritid vouchers from the RMCA collections (Virgilio and De Meyer, unpublished data) confirms that (a) as the age of the specimen increases, standard protocols for DNA extraction, amplification and Sanger sequencing become less and less efficient at producing DNA barcodes and (b) ethanol-preserved specimens tend to be more resistant to DNA degradation than pinned specimens. This screening revealed that, using standard protocols, DNA barcodes could be produced from less than 20 % of voucher specimens when those specimens were more than 10 years old (Fig. 2.1).

A survey of the collections of the RMCA revealed that 51 % of the 16,000 African tephritid vouchers were more than 15 years old (Fig. 2.2) suggesting that standard protocols for Sanger sequencing would be unlikely to produce DNA barcodes (Zimmermann et al. 2008).

An alternative approach for recovery of DNA from pinned museum specimens is the use of internal DNA primers and overlapping amplicons to reconstruct the full

Fig. 2.2 Proportions of vouchers of different age classes in the RMCA collections

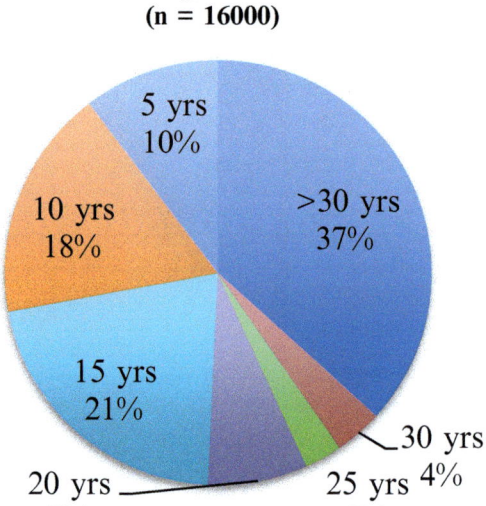

DNA barcode (Mitchell 2015). Van Houdt et al. (2010) and Smit et al. (2013) developed sets of internal primers specifically for tephritid fruit flies in the Natural History Collections. Van Houdt et al. (2010) used two overlapping amplicons successfully for reconstructing the full DNA barcode of specimens that were up to 15 years old and three overlapping amplicons for specimens that were up to 25 years old. However, this approach can be costly and time consuming so it is generally only used for rare collection material.

A more recent and cost-effective approach is the use of high throughput sequencing (next generation sequencing, NGS) that allows millions of DNA fragments from thousands of DNA templates to be sequenced in parallel. The NGS strategy for the mass production of DNA barcodes is promising (Meier et al. 2016) and allows DNA barcode amplicons to be individually tagged (using a set of oligonucleotides with a known sequence) so that multiple individuals can be processed in a single sequencing run and the individual DNA barcodes recovered through bioinformatics (Sucher et al. 2012; Shokralla et al. 2014; Shokralla et al. 2015).

References

Abuel-Ela RG, Hashem AG, Mohamed SMA (1998) *Bactrocera pallidus* (Perkin and May) (Diptera: Tephritidae), a new record in Egypt. J Egyptian German Soc Zool Entomol 27:221–229

Aluja M, Norrbom AL (1999) Fruit flies (Tephritidae) phylogeny and evolution of behavior. CRC Press, Boca Raton, pp 967

Armstrong KF, Ball SL (2005) DNA barcodes for biosecurity: invasive species identification. Phil Trans Roy Soc London Ser B 360:1813–1823

Barr NB, Islam MS, De Meyer M, McPheron BA (2012) Molecular identification of *Ceratitis capitata* (Diptera: Tephritidae) using DNA sequences of the COI barcode region. Anns Entomol Soc Amer 105:339–350

Boykin LM, Armstrong K, Kubatko L, De Barro P (2012) DNA barcoding invasive insects: database roadblocks. Invertebr System 26:506–514

Cameron S, Rubinoff D, Will K (2006) Who will actually use DNA barcoding and what will it cost? System Biol 55:844–847

Collins RA, Cruickshank RH (2013) The seven deadly sins of DNA barcoding. Mol Ecol Res 13:969–975

de Carvalho MR, Bockmann FA, Amorim DS, Brandão CRF, de Vivo M, de Figueiredo JL, Britski HA, de Pinna MCC, Menezes NA, Marques FPL, Papavero N, Cancello EM, Crisci JV, McEachran JD, Schelly RC, Lundberg JG, Gill AC, Britz R, Wheeler QD, Stiassny MLJ, Parenti LR, Page LM, Wheeler WC, Faivovich J, Vari RP, Grande L, Humphries CJ, DeSalle R, Ebach MC, Nelson GJ (2007) Taxonomic impediment or impediment to taxonomy? A commentary on systematics and the cybertaxonomic-automation paradigm. Evol Biol 34:140–143

De Meyer M (1996) Revision of the subgenus *Ceratitis* (*Pardalaspis*) Bezzi, 1918 (Diptera, Tephritidae, Ceratitini). Syst Entomol 21:15–26

De Meyer M (1998) Revision of the subgenus *Ceratitis* (*Ceratalaspis*) Hancock (Diptera: Tephritidae). Bull Entomol Res 88:257–290

De Meyer M (2000) Systematic revision of the subgenus *Ceratitis* Macleay *s.s.* (Diptera, Tephritidae). Zool J Linn Soc 128:439–467

De Meyer M (2006) Systematic revision of the fruit fly genus *Carpophthoromyia* Austen (Diptera, Tephritidae). Zootaxa 1235:1–48

De Meyer M (2009) Taxonomic revision of the fruit fly genus *Perilampsis* Bezzi (Diptera, Tephritidae). J Nat Hist 43:2425–2463

De Meyer M, Copeland R (2001) Taxonomic notes on the subgenera *Ceratitis* (*Hoplolophomyia*) and *Ceratitis* (*Acropteromma*) (Diptera, Tephritidae). Cimbebasia 17:77–84

De Meyer M, Freidberg A (2005) Revision of the fruit fly genus *Capparimyia* (Diptera, Tephritidae). Zool Scripta 34:279–303

De Meyer M, Freidberg A (2006) Revision of the subgenus *Ceratitis* (*Pterandrus*) Bezzi (Diptera: Tephritidae). Israel J Entomol 36:197–315

De Meyer M, Freidberg A (2012) Taxonomic revision of the fruit fly genus *Neoceratitis* Hendel (Diptera: Tephritidae). Zootaxa 3223:24–39

De Meyer M, White IM, Goodger KFM (2013) Notes on the frugivorous fruit fly (Diptera: Tephritidae) fauna of western Africa, with description of a new *Dacus* species. Europ J Taxon 50:1–17

Dowton M, Meiklejohn K, Cameron SL, Wallman J (2014) A preliminary framework for DNA barcoding, incorporating the multispecies coalescent. System Biol 63:639–644

Drew RAI, Tsuruta K, White IM (2005) A new species of pest fruit fly (Diptera : Tephritidae : Dacinae) from Sri Lanka and Africa. Afr Entomol 13:149–154

Ekesi S, Billah MK (2007) A field guide to the management of economically important tephritid fruit flies in Africa. ICIPE Science Press, Nairobi

Folmer O, Black M, Hoeh W, Lutz R, Vrijenhoek R (1994) DNA primers for amplification of mitochondrial cytochrome C oxidase subunit I from diverse metazoan invertebrates. Mol Mar Biol & Biotechnol 3:294–299

Frey J, Guillén L, Frey B, Samietz J, Rull J, Aluja M (2013) Developing diagnostic SNP panels for the identification of true fruit flies (Diptera: Tephritidae) within the limits of COI-based species delimitation. BMC Evol Biol 13:1–19

Hancock DL, Drew RAI (2006) A revised classification of subgenera and species groups in Dacus Fabricius (Diptera: Tephritidae). Instrumenta Biodiversitatis VII:167–205

Hancock DL, White IM (1997) The identity of *Tririthrum nigrum* (Graham) and some new combinations in *Ceratitis* MacLeay (Diptera: Tephritidae). The Entomologist 116:192–197

Hebert PDN, Cywinska A, Ball SL, deWaard JR (2003) Biological identifications through DNA barcodes. Proc Roy Soc B 270:313–321

Hebert PDN, Stoeckle MY, Zemlak TS, Francis CM (2004) Identification of birds through DNA barcodes. PLoS Biol 2:e312

Hickerson MJ, Meyer CP, Moritz C (2006) DNA barcoding will often fail to discover new animal species over broad parameter space. Syst Biol 55:729–739

Jiang F, Jin Q, Liang L, Zhang AB, Li ZH (2014) Existence of species complex largely reduced barcoding success for invasive species of Tephritidae: a case study in *Bactrocera* spp. Mol Ecol Res 14:1114–1128

Kvist S (2013) Barcoding in the dark?: a critical view of the sufficiency of zoological DNA barcoding databases and a plea for broader integration of taxonomic knowledge. Mol Phylog Evol 69:39–45

Kwong S, Srivathsan A, Meier R (2012) An update on DNA barcoding: low species coverage and numerous unidentified sequences. Cladistics 28:639–644

Meeyen K, Nanork Sopaladawan P, Pramual P (2014) Population structure, population history and DNA barcoding of fruit fly *Bactrocera latifrons* (Hendel) (Diptera: Tephritidae). Entomol Sci 17:219–230

Meier R, Shiyang K, Vaidya G, Ng PKL (2006) DNA barcoding and taxonomy in Diptera: a tale of high intraspecific variability and low identification success. Syst Biol 55:715–728

Meier R, Wong W, Srivathsan A, Foo M (2016) $1 DNA barcodes for reconstructing complex phenomes and finding rare species in specimen-rich samples. Cladistics 32:100–110

Meyer CP, Paulay G (2005) DNA barcoding: error rates based on comprehensive sampling. PLoS Biol 3:e422

Mitchell A (2015) Collecting in collections: a PCR strategy and primer set for DNA barcoding of decades-old dried museum specimens. Mol Ecol Res 15:1102–1111

Moritz C, Cicero C (2004) DNA barcoding: promise and pitfalls. PLoS Biol 2:e35

Nagy ZT, Backeljau T, De Meyer M, Jordaens K (2013) DNA barcoding: a practical tool for fundamental and applied biodiversity research. In: ZooKeys p. 410

Nielsen R, Matz M (2006) Statistical approaches for DNA barcoding. Syst Biol 55:162–169

Pečnikar ŽF, Buzan E (2014) 20 years since the introduction of DNA barcoding: from theory to application. J Appl Gen 55:43–52

Porter TM, Gibson JF, Shokralla S, Baird DJ, Golding GB, Hajibabaei M (2014) Rapid and accurate taxonomic classification of insect (class Insecta) cytochrome c oxidase subunit 1 (COI) DNA barcode sequences using a naïve Bayesian classifier. Mol Ecol Res 14:929–942

Puillandre N, Lambert A, Brouillet S, Achaz G (2011) ABGD, Automatic Barcode Gap Discovery for primary species delimitation. Mol Ecol 2:1864–1877

Ratnasingham S, Hebert P (2007) BOLD: The Barcode of Life Data System (http://www.barcodinglife.org). Mol Ecol Notes 7:355–364

Ratnasingham S, Hebert PDN (2013) A DNA-based registry for all animal species: the Barcode Index Number (BIN) system. PLoS One 8:e66213

Schutze MK, Aketarawong N, Amornsak W, Armstrong KF, Augustinos AA, Barr N, Bo W, Bourtzis K, Boykin LM, CÁCeres C, Cameron SL, Chapman TA, Chinvinijkul S, Chomič A, De Meyer M, Drosopoulou E, Englezou A, Ekesi S, Gariou-Papalexiou A, Geib SM, Hailstones D, Hasanuzzaman M, Haymer D, Hee AKW, Hendrichs J, Jessup A, Ji Q, Khamis FM, Krosch MN, Leblanc LUC, Mahmood K, Malacrida AR, Mavragani-Tsipidou P, Mwatawala M, Nishida R, Ono H, Reyes J, Rubinoff D, San Jose M, Shelly TE, Srikachar S, Tan KH, Thanaphum S, Haq I, Vijaysegaran S, Wee SL, Yesmin F, Zacharopoulou A, Clarke AR (2015) Synonymization of key pest species within the Bactrocera dorsalis species complex (Diptera: Tephritidae): taxonomic changes based on a review of 20 years of integrative morphological, molecular, cytogenetic, behavioural and chemoecological data. Syst Entomol 40:456–471

Shokralla S, Gibson JF, Nikbakht H, Janzen DH, Hallwachs W, Hajibabaei M (2014) Next-generation DNA barcoding: using next-generation sequencing to enhance and accelerate DNA barcode capture from single specimens. Mol Ecol Res 14:892–901

Shokralla S, Porter TM, Gibson JF, Dobosz R, Janzen DH, Hallwachs W, Golding GB, Hajibabaei M (2015) Massively parallel multiplex DNA sequencing for specimen identification using an Illumina MiSeq platform. Sci Reps 5:9687

Smit J, Reijnen B, Stokvis F (2013) Half of the European fruit fly species barcoded (Diptera, Tephritidae); a feasibility test for molecular identification. ZooKeys 365:279–305

Sonet G, Jordaens K, Nagy ZT, Breman F, de Meyer M, Backeljau T, Virgilio M (2013) *Adhoc*: an R package to calculate ad hoc distance thresholds for DNA barcoding identification. ZooKeys 365:329–336

Sucher NJ, Hennell JR, Carles MC (2012) DNA fingerprinting, DNA barcoding, and next generation sequencing technology in plants. Methods Mol Biol 862:13–22

Tanabe AS, Toju H (2013) Two new computational methods for universal dna barcoding: a benchmark using barcode sequences of bacteria, archaea, animals, fungi, and land plants. PLoS One 8:e76910

Taylor HR, Harris WE (2012) An emergent science on the brink of irrelevance: a review of the past 8 years of DNA barcoding. Mol Ecol Res 12:377–388

Van Houdt JKJ, Breman FC, Virgilio M, De Meyer M (2010) Recovering full DNA barcodes from natural history collections of Tephritid fruitflies (Tephritidae, Diptera) using mini barcodes. Mol Ecol Res 10:459–465

Virgilio M, Backeljau T, Nevado B, De Meyer M (2010) Comparative performances of DNA barcoding across insect orders. BMC Bioinf 11:206

Virgilio M, Backeljau T, Emeleme R, Juakali JL, De Meyer M (2011) A quantitative comparison of frugivorous tephritids (Diptera: Tephritidae) in tropical forests and rural areas of the Democratic Republic of Congo. Bull Entomol Res 101:591–597

Virgilio M, Jordaens K, Breman FC, Backeljau T, De Meyer M (2012) Identifying insects with incomplete DNA barcode libraries, African fruit flies (Diptera: Tephritidae) as a test case. PLoS One 7:e31581

Virgilio M, White IM, De Meyer M (2014) A set of multi-entry identification keys to African frugivorous flies (Diptera, Tephritidae). ZooKeys 428:97–108

Virgilio M, Jordaens K, Verwimp C, White IM, De Meyer M (2015) Higher phylogeny of frugivorous flies (Diptera, Tephritidae, Dacini): localised partition conflicts and a novel generic classification. Mol Phylog Evol 85:171–179

White IM (2006) Taxonomy of the Dacina (Diptera:Tephritidae) of Africa and the Middle East. Afr Entomol Memoir 2:1–156

White IM, Goodger KFM (2009) African *Dacus* (Diptera: Tephritidae); new species and data, with particular reference to the Tel Aviv University collection. Zootaxa 2127:1–49

White I, Copeland R, Hancock D (2003) Revision of the afrotropical genus *Trirhithrum* (Diptera: Tephritidae). Cimbebasia 18:71–137

Zimmermann J, Hajibabaei M, Blackburn D, Hanken J, Cantin E, Posfai J, Evans T (2008) DNA damage in preserved specimens and tissue samples: a molecular assessment. Front in Zool 5:18

Massimiliano Virgilio is a molecular taxonomist and coordinator of the Joint Experimental Molecular Unit (JEMU, http://jemu.myspecies.info) of the Royal Museum for Central Africa (Tervuren, Belgium). His main interests are in morphological and molecular taxonomy, phylogeny and population genetics of African frugivorous tephritids. He is co-author of a number of publications on DNA barcoding and on the resolution of cryptic species complexes. In collaboration with Marc De Meyer and Ian White, he developed and regularly maintains a set of multi-entry keys for the morphological identification of African fruit flies (http://fruitflykeys.africamuseum.be/en/index.html).

Chapter 3
Population Genetics of African Frugivorous Fruit Flies (Diptera, Tephritidae): Current Knowledge and Future Perspectives

Massimiliano Virgilio and Hélène Delatte

Abstract Population genetics studies provide valuable information about the patterns of connectivity and range expansion of African frugivorous fruit flies. Human-mediated movements related to trade of commodities and transport are generally indicated as one of the primary mechanisms by which tephritid pests expand their contemporary and historical ranges. This results in complex colonisation dynamics, as suggested for the widely distributed pests *Bactrocera dorsalis s.s.* and *Zeugodacus cucurbitae*, and for the cosmopolitan pest of African origin *Ceratitis capitata*. Analysis of the population structure of African fruit flies can also reveal cryptic genetic structures and incipient speciation, as observed for the *Ceratitis* FAR complex (*Ceratitis fasciventris*, *Ceratitis anonae*, *Ceratitis rosa*) and the mango fruit fly, *Ceratitis cosyra*. Here we provide a synthesis of the current knowledge about the population structure of the main frugivorous fruit flies that are pests in Africa.

Keywords Microsatellite markers • Genotypic groups • Range expansion • Cryptic speciation • Inductive/deductive approaches

1 Introduction

Population genetics deals with the ecological and evolutionary processes that affect the population structure of species. Inferences from population genetics studies rely on both inductive and deductive approaches (reviewed in Hamilton 2009). Inductive approaches are typically adopted in descriptive studies when measures of genetic variation (parameters) are collected from representative population samples and

---

M. Virgilio (✉)
Department of Biology and Joint Experimental Molecular Unit (JEMU), Royal Museum for Central Africa, Tervuren, Belgium
e-mail: massimiliano.virgilio@africamuseum.be

H. Delatte
CIRAD, UMR PVBMT, F-97,410, Saint-Pierre, Réunion, France

© Springer International Publishing Switzerland 2016
S. Ekesi et al. (eds.), *Fruit Fly Research and Development in Africa - Towards a Sustainable Management Strategy to Improve Horticulture*,
DOI 10.1007/978-3-319-43226-7_3

used to infer the evolutionary processes that generated the observed population structure. Conversely, the deductive approach uses general population genetics models that describe evolutionary processes (e.g. bottlenecks and genetic drift, mutation, natural selection) to make predictions about spatial and temporal changes in the genetic patterns of the target species.

Allozyme markers were commonly used to describe the population structure of tephritid flies in early studies (e.g. McPheron et al. 1988; Feder et al. 1997; Abreu et al. 2005). Microsatellite markers (or single sequence repeats, SSR) were then widely adopted for the description of native and introduced African tephritids (see below). Microsatellite markers are co-dominant, polymorphic nuclear loci, that are distributed throughout the genome and generally neutral unless linked to loci under selection. They are short repeated sequences of nuclear DNA (one to six base pairs in length) with allelic states that simply correspond to the number of repeats present at each locus that can be scored after electrophoresis of PCR-amplified DNA fragments (Hamilton 2009). These characteristics make microsatellite markers good candidates for comparing different populations and their colonization dynamics (Tautz 1989; Hamilton 2009). The more recent population genomic approaches that rely on high-throughput sequencing techniques (i.e. Next Generation Sequencing, or NGS) now allow the population structure of species to be described in unprecedented detail (Davey and Blaxter 2010; Elshire et al. 2011; Narum et al. 2013); studies using NGS on tephritid fruit flies are becoming more and more common (Shen et al. 2011; Zheng et al. 2012; Nirmala et al. 2013; Geib et al. 2014). During the past three decades a number of studies have been published on the population genetics of African fruit flies. Below we synthesise current knowledge on the population genetics of the main African fruit fly species in the genera *Ceratitis* (i.e. the Mediterranean fruit fly (medfly), *Ceratitis capitata* (Wiedemann), the mango fruit fly, *Ceratitis cosyra* (Walker), and the *Ceratitis* 'FAR' complex); *Bactrocera* (i.e. the oriental fruit fly, *Bactrocera dorsalis* [Hendel]); and *Zeugodacus* (i.e. the melon fruit fly, *Zeugodacus cucurbitae* Coquillett).

2 *Ceratitis capitata*

Ceratitis capitata, is one of the most economically important and widely distributed tephritid pests of African origin (White and Elson-Harris 1994). After considering morphological cladistics, host plant abundances and parasitoid distributions, De Meyer et al. (2002) proposed that Eastern and Southern Africa are the most likely geographic origin of this cosmopolitan pest. Historical records provided important clues to develop hypotheses about the worldwide range expansion of *C. capitata*. For example, *C. capitata* was first reported in Costa Rica (1955) and then Guatemala (1976) before reaching Mexico, possibly due to rapid movement through the so-called 'coffee belt' (Malacrida et al. 1998 and references therein). It has also been reported intermittently in Florida since 1929, in California since 1975, and in Texas since 1966 (Gasparich et al. 1997). *Ceratitis capitata* was introduced into Australia from Europe in around 1897 (Malacrida et al. 1998) where it is currently confined to Western Australia with occasional detections in South Australia and the Northern Territory. Its distribution in Australia has remained unchanged for the last half century and this is likely to be due to the geographical barriers that prevent free movement of this species across Australia and /or to extensive Australian monitoring systems and quarantine restrictions (Dominiak and Daniels 2012).

The first large-scale descriptions of the population structure of *C. capitata* were largely inferred using allozyme markers (e.g. Gasperi et al. 1991). An early reconstruction of the worldwide range expansion of *C. capitata* was attempted when two African populations (from Kenya and La Réunion), two Mediterranean populations (from Procida and Sardinia), and one Central American population (from Guatemala) were genotyped at 27 allozyme loci (Malacrida et al. 1992). Combining these results with historical records, Malacrida et al. (1992) was able to separate *C. capitata* populations in to three groups: ancestral (from sub-Saharan Africa), ancient (Mediterranean) and new (American) and suggested that the colonisation of Central America started from a recent African introduction. Furthermore, they were also able to describe temporal variability in the genotypic patterns of one of the Mediterranean samples, which they attributed to seasonal population fluctuations (see also Gasperi et al. 2002). A more extensive study (Malacrida et al. 1998) used 26 allozyme markers to compare 17 populations from six regions: Africa, Mediterranean, 'extra-Mediterranean islands' (e.g. Gran Canaria and Madeira), Latin America, Pacific and Australia. Levels of genetic variability (as estimated from the number of alleles per locus, percentage of polymorphic loci and mean heterozygosity) suggested that *C. capitata* originated in East Africa (where the highest genetic diversity was observed), and expanded its range to the African–Mediterranean region first (as suggested by a gradual pattern of decreasing genetic variability) and, most recently, to the Latin American–Pacific region. Gene flow estimates, determined from the average frequency of private alleles and the number of migrants, also suggested a route of colonization from South East Africa to northwest Africa and from there to Spain, followed by a West-east Mediterranean range expansion. Additionally, Malacrida et al. (1998) hypothesised that the Latin American and Pacific populations originated from a few, recent and geographically separated colonization events followed by population expansions. In this context, both ancient and recent colonization events involving *C. capitata* were largely attributed to human-mediated transportation and to the history of human trading activities (Malacrida et al. 1998).

Despite the important role that allozyme studies played in the first large scale descriptions of the population structure of *C. capitata*, they could only provide indicative, rather than categorical, information about the chronology of range expansion (Gasparich et al. 1997). It was hoped that new alternative methods and approaches would achieve this and they included: the analysis of intron size polymorphisms (Gomulski et al. 1998); restriction site variation (Haymer et al. 1992; Sheppard et al. 1992; McPheron et al. 1994; Gasparich et al. 1995; Gasparich et al. 1997); Random Amplified Polymorphic DNA (Haymer et al. 1997); and Sanger sequencing (Davies et al. 1999). These approaches did support the African origin of

C. capitata but did not allow any better resolution of its expansion history beyond Africa.

Subsequently microsatellite markers were developed for *C. capitata* (Bonizzoni et al. 2000; Stratikopoulos et al. 2008) and, due to their high levels of polymorphism, provided much better resolution compared to earlier molecular techniques; they were used successfully to investigate the population structure of *C. capitata* (Karsten et al. 2013) and the origin of *C. capitata* infestations in North America (Bonizzoni et al. 2001) and Australia (Bonizzoni et al. 2004). Microsatellites suggested that flies captured in California originated from independent introduction events, including introductions from Central America (Bonizzoni et al. 2001), but also that incomplete eradication might have resulted in endemic Californian populations. The origin of periodic *C. capitata* infestations in California is highly controversial and there remains disagreement as to whether the flies captured over the years represent independent introductions from external sources, or resident populations with sizes fluctuating from non-detectable to detectable levels (Carey 1991; Papadopoulos et al. 2013; Carey et al. 2014; Gutierrez et al. 2014). Conversely, colonization of Australia was more convincingly attributed to secondary colonization from the Mediterranean basin, and the Perth area was indicated as the source for secondary invasion into both Western and South Australia (Bonizzoni et al. 2004). The possible invasion routes of *C. capitata* were reviewed and summarised by Malacrida et al. (2007) who further stressed the importance of human-mediated transportation in the worldwide range expansion of *C. capitata*. Human-mediated movements related to trade of commodities and transport by air, sea and land are generally accepted as the primary mechanism by which *C. capitata*, and other economically important tephritid species, have spread (White and Elson-Harris 1994; see also Karsten et al. 2015 and references therein).

To date, only one study has adopted a deductive approach to investigating the range expansion dynamics of *C. capitata* (Karsten et al. 2015). This approach proved useful, particularly since recent improvements to model-based analyses became available, such as Approximate Bayesian Computation (ABC; Estoup and Guillemaud 2010). ABC modeling allows the complex evolutionary scenarios that are expected in range expansions of cosmopolitan pests to be taken into consideration, and inferences to be made on parameters such as: date of founding of different populations (in numbers of generations); current effective population size (as numbers of diploid individuals); number of founders in the introduced populations; and duration of the initial bottleneck. Results of the Karsten et al. (2015) study suggested that the most likely route of *C. capitata* from Africa closely matched the patterns indicated from historical records, though with much earlier introductions. An initial colonization of Europe, a secondary colonization of Australia from Europe, an introduction from Greece to Central America and, eventually, a back introduction into South Africa from Europe were also implied. This reconstruction did, however, differ from those previously proposed (Malacrida et al. 2007) as it supported secondary colonisation of Central America from admixed European populations (hence, not from Africa) and secondary reintroduction in Africa from Europe.

3 *Ceratitis cosyra*

The mango fruit fly, *C. cosyra*, is possibly the most important pest of mango throughout sub-Saharan Africa (Lux et al. 2003a; Vayssières et al. 2009). Out of the mango season, *C. cosyra* shifts to alternative host plants including wild fruits such as marula, *Sclerocarya birrea* (A. Rich.) Hochst. (Copeland et al. 2006) and soursop, *Annona muricata* L. (Mwatawala et al. 2009). Barr et al. (2006) were the first to suggest that *C. cosyra* was comprised of highly divergent mitochondrial haplotypes; DNA barcodes from two specimens sampled along the coast of southern Kenya (Shimba Hills) were clearly separated from the main haplotype group, thus suggesting cryptic speciation (Barr et al. 2006). In order to further investigate this hypothesis, a set of microsatellite markers was developed (Delatte et al. 2014) and used to describe the population structure of *C. cosyra* across its distribution (Virgilio et al. 2015a). Analysis of 348 specimens from 13 African populations showed that *C. cosyra* was indeed represented by two separate genotypic groups (Fig. 3.1); one included the vast majority of specimens sampled in Burundi and Tanzania as well as a number of outliers from other African countries, while the other included all other specimens sampled. The two genotypic groups were also found, in sympatry, in populations from Kenya, Senegal, Sudan and Tanzania (Virgilio et al. 2015a). Sequential Bayesian assignment of microsatellite genotypes (as described by Coulon et al. 2008) also revealed that, within the second genotypic group, specimens could be further subdivided between a West African cluster (including individuals from Burkina Faso, Ivory Coast, Mali and Nigeria) and an East / South African cluster (including specimens from Ethiopia, Tanzania, Malawi, Mozambique and South Africa) (Virgilio et al. 2015a). This more subtle genetic differentiation was less clear-cut as, for example, specimens from Sudan were genetically closer to the West African samples, and populations from Kenya and Senegal included individuals from both clusters.

4 The 'FAR' Complex

The so call *Ceratitis* 'FAR' complex is a group of African frugivorous flies including the Natal fruit fly, *C. rosa,* and the morphologically similar but less economically important pests, *Ceratitis fasciventris* (Bezzi) and *Ceratitis anonae* Graham. The three species all show clear sexual dimorphism, with the males having distinct leg ornamentation patterns, while in females these are almost indistinguishable (De Meyer 2001). All members of the 'FAR' complex are highly polyphagous with partially overlapping ranges of host plants and geographic distributions (Copeland et al. 2006). Two of these species, *C. rosa* and *C. fasciventris*, have weak reproductive barriers as, when crossed under laboratory conditions, they can produce fertile offspring (Erbout et al. 2008). Phylogenetic analyses of morphological characters (De Meyer 2005) and of mitochondrial and nuclear gene fragments could not fully

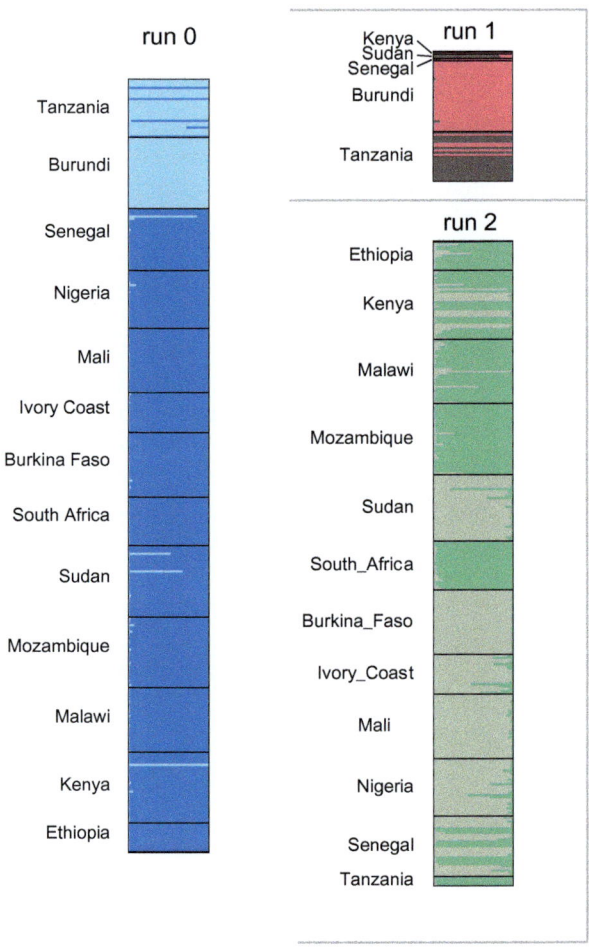

Fig. 3.1 Population structure of *C. cosyra* as inferred from individual Bayesian assignment of multilocus microsatellite genotypes (From Virgilio et al. 2015a)

resolve these three species as distinct monophyletic entities (Virgilio et al. 2008; Barr and Wiegmann 2009). Despite this, genetic differentiation has been reported between samples of *C. fasciventris* from West and East Africa (Virgilio et al. 2008) and between samples of *C. rosa* from Kenya and South Africa (Douglas and Haymer 2001). An earlier study using microsatellites also revealed differences between populations of *C. rosa* from the African mainland and populations of *C. rosa* from the Indian Ocean islands, as well as between populations of *C. fasciventris* from Kenya and populations of *C. fasciventris* from Uganda (Baliraine et al. 2004).

In order to finally resolve the molecular taxonomy and population structure of the 'FAR' complex, a set of 16 microsatellite markers was developed (Delatte et al. 2013) and used to genotype 27 African populations of the three morphospecies (Virgilio et al. 2013). This revealed the presence of five genotypic clusters: two contained *C.*

rosa specimens (R1, R2; allopatric and sympatric populations), two contained *C. fasciventris* specimens (F1, F2; allopatric and parapatric populations) and one contained *C. anonae* specimens (A). Surprisingly, intra- and interspecific genetic diversity was not hierarchically structured; differences in diversity between clusters from the same morphospecies (e.g. between F1 and F2, or between R1 and R2) was greater or comparable with differences between clusters from different morphospecies (e.g. between F1 and A, or between R2 and A). The two *C. fasciventris* genotypic clusters roughly corresponded to West and East African samples, respectively, with the exception of a single population from Tanzania that was more closely related to the West African samples than the East African samples. Relationships amongst the 'FAR' morphospecies and the genotypic clusters were further investigated using an integrative taxonomic approach that included spatial ecology, wing morphometrics, larval morphology, analysis of cuticular hydrocarbons, developmental physiology and pre- and postzygotic mating compatibility. The results of these studies (reviewed in De Meyer et al. 2015a) indicated that the *Ceratitis* 'FAR' complex includes between three and five different taxonomic entities. Males from the two *C. rosa* clusters were morphologically different and were provisionally acknowledged as either 'R1' or 'R2' (De Meyer et al. 2015a) but also, in relation to their different distributional/altitudinal ranges (Mwatawala et al. 2015), as 'lowland' or 'hot' *C. rosa*, and 'highland' or 'cold' *C. rosa*. The integrative approach implemented on the *Ceratitis* FAR complex provided sufficient evidence to consider R1 and R2 as two different biological species, with the type material of *C. rosa* belonging to the R1 type and the R2 type considered as a new species, *Ceratitis quilicii* (De Meyer et al. submitted).

5 *Zeugodacus cucurbitae*

Zeugodacus cucurbitae (Coquillett) stat. rev. (formerly classified as *Bactrocera* (*Zeugodacus*) *cucurbitae* (Coquillett)) was originally described from material collected in Honolulu, Hawaii, USA (Coquillett 1899). Its systematic position was recently revised due to reconstruction of its phylogenetic history. The former subgenus *Zeugodacus* is now considered as a separate genus that is independent from both *Bactrocera* and *Dacus*, and more closely related to the genus *Dacus* than to the genus *Bactrocera* (Krosch et al. 2012; Virgilio et al. 2015b; De Meyer et al. 2015b).

The genus *Zeugodacus* includes approximately 115 species (Norrbom et al. 1999; Drew and Romig 2013) of which the majority are restricted to the Oriental and Australian regions with a few species in the eastern Palearctic regions of China and Japan. The exception is *Z. cucurbitae* which is considered as an invasive pest in Africa and the islands of the Indian Ocean. Jacquard et al. (2013) analysed two mitochondrial gene fragments (COI-ND6 genes, 1297 bp) from 100 specimens of *Z. cucurbitae* sampled from across its distribution (Asia, Hawaii, African mainland and islands of the Indian Ocean). They found remarkably limited intraspecific variability amongst specimens with only 22 haplotypes, 21 polymorphic sites and an average p-distance of 0.003 %. Despite this, a Minimum Spanning Network revealed the occurrence of two clearly distinct haplotype groups corresponding to specimens from (a) Asia and Hawaii, and (b) the African mainland and La Réunion. A finer resolution of the geographic structuring of *Z. cucurbitae* was obtained using microsatellite genotyping of 25 populations sampled from across its entire distribution range (Virgilio et al. 2010). This macrogeographic study of its population genetics revealed the existence of five population groups corresponding to populations from (i) the African continent, (ii) Reunion Island, (iii) Central Asia, (iv) East-Asia and (v) Hawaii. The proportions of inter-regional Bayesian assignments and the high values for genetic diversity in populations from Pakistan, India and Bangladesh suggested that *Z. cucurbitae* originated in Central Asia and expanded its range in one direction to East Asia and Hawaii and in the other direction to Africa and the islands of the Indian Ocean. However, there were a number of outliers with high levels of admixing ($Q>0.70$) amongst populations from different regions which suggested there were more complex patterns of inter-regional gene flow ongoing, possibly as a result of human-mediated transport (Virgilio et al. 2010).

Zeugodacus cucurbitae has also been reported from a series of unrelated host plant families in addition to the main host range represented by *Cucurbitaceae* (see De Meyer et al. 2015b and references therein) and geographic differences in host preferences have also been reported between East and West African populations (Vayssières et al. 2007; Mwatawala et al. 2010; Jacquard et al. 2013). Despite these observations cucurbit hosts are generally preferred and are attacked with higher infestation rates and incidences compared to non-cucurbit hosts. Host records also suggest that feeding preferences differ between populations of *Z. cucurbitae* from the native distribution and populations from the adventive distribution, possibly resulting in locally adapted populations or host races. The fine-scale analysis made on data from 2258 specimens collected from 11 locations in La Réunion elucidated relationships between the genetic structure of *Z. cucurbitae* and environmental factors such as altitude (range 0–400 m, 400–600 m and 600–1200 m), host plant (cultivated and wild cucurbits) and season (subtropical winter and summer) (Jacquard et al. 2013). The presence of three main genetic clusters (with limited inter-cluster genetic structuring) were revealed that could be differentiated from African and Asian populations (although they were of possible African origin) and were distinctly distributed on the eastern and western parts of the island. Abundances of specimens from the three clusters were correlated with the average amount of rainfall while no significant differences were detected in their distribution on wild or cultivated host plants, across altitudinal ranges or across different seasons (Jacquard et al. 2013). Other studies, done in Asia, the South-East Pacific and Hawaii (Clark and Boontop, unpublished data), and in Tanzania (De Meyer et al. 2015b), also showed a lack of consistent genetic differentiation across samples of *Z. cucurbitae* with different feeding preferences.

The results of Jacquard et al. (2013) suggested a common ancestry for the African *Z. cucurbitae* but left a number of questions about the potential colonization pathway open. Two alternative hypotheses for this colonization had been proposed previously by Virgilio et al. (2010) who suggested either a relatively recent invasion of the African continent, roughly corresponding to the first historical records for this species in Africa (*viz*. 1936 in East Africa and 1999 in West Africa), or an older range expansion possibly dating back to the first documented trade contact between Africa and Asia

(100 AD, Gilbert 2004). In order to determine whether either of these hypotheses was correct, Delatte et al. (unpublished data) evaluated a large number of populations (17) from East, West and Central Africa using a larger set of markers than the previous study of Virgilio et al. (2010). This allowed better resolution of the population structure of *Z. cucurbitae* in Africa and, using STRUCTURE analysis as described by Pritchard et al. (2000), showed that the populations from Uganda had diverged from Tanzanian populations and that populations from Burundi and Kenya had traces of admixture with West African samples. The ABC analysis in the DIYABC software (Cornuet et al. 2010, 2014) also suggested that *Z. cucurbitae* had expanded its range in to East and West Africa. Recent studies of the routes of worldwide introductions of alien organisms suggest that many widespread invasions may not have originated from the native range, but from a particularly successful invasive population; these invasive populations could serve as the source of colonists for remote new territories and has subsequently been termed the 'invasive bridgehead effect' (Lombaert et al. 2010). In the case of *Z. cucurbitae*, Central Asia was the most likely native source population, and East Africa the source population that adapted and was the start point of the invasive bridgehead effect for all the colonization events that subsequently occurred in Africa. The parameter estimates from DIYABC suggested that these events occurred soon before the first historical records of *Z. cucurbitae* in the African continent and allow us to exclude alternative hypotheses considering older introductions of *Z. cucurbitae* in to Africa or multiple invasion events (Virgilio et al. 2010).

6 *Bactrocera dorsalis s.s.*

In Africa, *B. dorsalis s.s*, has been reported infesting 72 plant species spread across 28 families (Goergen et al. 2011) and, in mango orchards, causes yield losses of up to 80 % (Ekesi et al. 2006). Due to its major impact on horticultural products, *B. dorsalis s.s.* is one of the most devastating fruit fly pests in Africa (De Meyer et al. 2010). *Bactrocera dorsalis s.s.* is part of the notorious *B. dorsalis* complex that includes almost 100 species (Drew and Hancock 1994; Drew and Romig 2013), is of Asian origin (Clarke et al. 2005) and difficult to identify using morphological or molecular techniques (Khamis et al. 2012; Leblanc et al. in press). Recently, the taxonomy of three important pests within this complex (*Bactrocera papayae* (Drew and Hancock), *Bactrocera philippinensis* (Drew and Hancock) and *Bactrocera invadens* (Drew, Tsuruta and White)) was revised and they were synonymized as *B. dorsalis s.s.* (Schutze et al. 2015). *Bactrocera invadens* was initially described as a novel species native to Asia and introduced into East Africa (Drew et al. 2005). In fact, *B. dorsalis s.s.* was recorded for the first time on the African mainland in 2003 (Lux et al. 2003b) where it had already become a pest species of major concern to fruit growers (see De Meyer et al. 2010 and references therein). The African expansion of *B. dorsalis s.s.* was extremely rapid. After the first record in Kenya, it was subsequently recorded in Tanzania and Nigeria, then it rapidly spread to the west and to the south and it is now distributed throughout sub-Saharan Africa (Table 3.1).

Table 3.1 Range expansion of *B. dorsalis s.s.* in Africa

Country	Year of arrival	Reference
Kenya	2003	Lux et al. (2003a)
Tanzania	2003	Mwatawala et al. (2004)
Nigeria	2003	Umeh et al. (2008)
Uganda	2004	Drew et al. (2005)
Benin	2004	Drew et al. (2005); Vayssières et al. (2005)
Ghana	2005	Drew et al. (2005)
Comoros Archipelago	2005	De Meyer et al. (2012)
Cameroon	2005	Ndzana Abanda et al. (2008)
Guinea	2006	Ekesi et al. (2006)
Senegal	2006	Ekesi et al. (2006)
Sudan	2006	Ekesi et al. (2006)
Togo	2006	Ekesi et al. (2006)
Ivory Coast	2007	Goergen et al. (2011)
Ethiopia	2007	EPPO/CABI (2014)
Mayotte	2007	De Meyer et al. (2012)
Burkina Faso	2007	Goergen et al. (2011)
Mali	2007	Goergen et al. (2011)
Namibia	2007	APHIS (2009)
Mozambique	2008	Correia et al. (2008)
Chad	2008	Goergen et al. (2011)
Angola	2008	Goergen et al. (2011)
Congo	2008	Goergen et al. (2011)
Democratic Republic of Congo	2008	Goergen et al. (2011)
Equatorial Guinea	2008	Goergen et al. (2011)
Gabon	2008	Goergen et al. (2011)
Gambia	2008	EPPO/CABI (2014)
Guinea-Bissau	2008	Goergen et al. (2011)
Liberia	2008	EPPO/CABI (2014)
Mauritania	2008	Goergen et al. (2011)
Niger	2008	Goergen et al. (2011)
Sierra Leone	2008	Goergen et al. (2011)
Central African Republic	2008	Goergen et al. (2011)
Zambia	2009	EPPO/CABI (2014)
Burundi	2009	Liu et al. (2011)
Madagascar	2010	Raoelijaona et al. (2012)
Zimbabwe	2010	Cassidy (2010)
Botswana	2011	EPPO/CABI (2014)
South Africa	2007, 2013	Manrakhan et al. (2015)
Swaziland	2014	EPPO/CABI (2014)

Bactrocera dorsalis s.s. has also reached the islands of the Indian Ocean, beginning with the Comoros archipelago in 2006 and Madagascar in 2010. Other invasive populations of *B. dorsalis s.s.* have been reported in Hawaii, French Polynesia, Japan, Nauru, Guam and the Northern Mariana Islands (Stephens et al. 2007).

After developing a set of 11 polymorphic microsatellite markers, Khamis et al. (2008) published the only study currently available on the African population structure of *B. dorsalis s.s* (Khamis et al. 2009). This study, based on a microsatellite analysis of 13 African populations (from nine countries) and including a population outgroup from Sri Lanka, showed the presence of three main population groups co-occurring across the African distribution of *B. dorsalis s.s.*. One of the three groups included a single population from Nigeria that also shared (limited) co-ancestry with the Asian outgroup. Khamis et al. (2009) hypothesized that the Nigerian population of *B. dorsalis s.s.* could have arisen either from an independent introduction from an unsampled source and/or could represent the outcome of a bottleneck. As a whole these genetic data suggest that the African range expansion of *B. dorsalis s.s.* (resulting from one or more introduction events) was followed by rapid population expansion (Fig. 3.2).

Other studies have investigated the genetic structure of *B. dorsalis s.s.* in Asia (Liu et al. 2007; Shi et al. 2010; Wan et al. 2011), and revealed high levels of genetic diversity between and within samples which supported a South-east Asian origin for *B. dorsalis s.s.* Microsatellite markers also showed relatively high levels of genetic diversity within populations from South-East Asia and high gene flow between

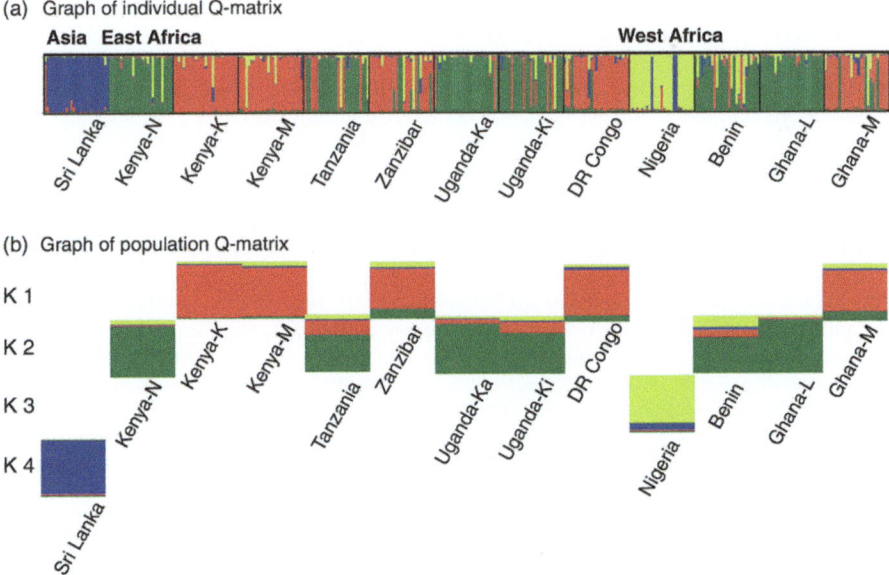

Fig. 3.2 Population structure of *B. dorsalis s.s.* in Africa as inferred from individual Bayesian assignment of multilocus microsatellite genotypes (Modified from Khamis et al. 2009)

population groups but were unable to resolve straightforward geographic patterns (Aketarawong et al. 2007, 2014). Similar results were observed for populations from the Thai/Malay peninsula which were a predominantly panmictic population (Krosch et al. 2013). In adventive Hawaiian populations mitochondrial (Barr et al. 2014) and nuclear markers (Aketarawong et al. 2007) also only detected limited genetic structuring, supporting a recent introduction in to Hawaii followed by genetic differentiation in an environment of isolation.

References

Abreu AG, Prado PI, Norrbom AL, Solferini VN (2005) Genetic and morphological diagnosis and description of two cryptic species of flower head-infesting Tephritidae (Diptera). Ins Syst Evol 36:361–370
Aketarawong N, Bonizzoni M, Thanaphum S, Gomulski LM, Gasperi G, Malacrida AR, Gugliemino CR (2007) Inferences on the population structure and colonization process of the invasive oriental fruit fly, *Bactrocera dorsalis* (Hendel). Mol Ecol 16:3522–3532
Aketarawong N, Isasawin S, Thanaphum S (20 14) Evidence of weak genetic structure and recent gene flow between *Bactrocera dorsalis s.s.* and *B. papayae*, across Southern Thailand and West Malaysia, supporting a single target pest for SIT applications. BMC Genet 15:70
APHIS (2009) Invasive species, Africa, APHIS issues federal import quarantine order for host materials of invasive fruit fly species *Bactrocera invadens*.
Baliraine FN, Bonizzoni M, Guglielmino CR, Osir EO, Lux SA, Mulaa FJ, Gomulski LM, Zheng L, Quilici S, Gasperi G, Malacrida AR (2004) Population genetics of the potentially invasive African fruit fly species, *Ceratitis rosa* and *Ceratitis fasciventris* (Diptera : Tephritidae). Mol Ecol 13:683–695
Barr NB, Wiegmann BM (2009) Phylogenetic relationships of *Ceratitis* fruit flies inferred from nuclear CAD and tango/ARNT gene fragments: Testing monophyly of the subgenera *Ceratitis* (*Ceratitis*) and *C.* (*Pterandrus*). Mol Phyl Evol 53:412–424
Barr NB, Copeland RS, De Meyer M, Masiga D, Kibogo HG, Billah MK, Osir E, Wharton RA, McPheron BA (2006) Molecular diagnostics of economically important *Ceratitis* fruit fly species (Diptera: Tephritidae) in Africa using PCR and RFLP analyses. Bull Entomol Res 96:05–521
Barr NB, Ledezma LA, Leblanc L, San Jose M, Rubinoff D, Geib SM, Fujita B, Bartels DW, Garza D, Kerr P, Hauser M, Gaimari S (2014) Genetic diversity of *Bactrocera dorsalis* (Diptera: Tephritidae) on the Hawaiian Islands: Implications for an introduction pathway into California. J Econ Entomol 107:1946–1958
Bonizzoni M, Malacrida AR, Guglielmino CR, Gomulski LM, Gasperi G, Zheng L (2000) Microsatellite polymorphism in the Mediterranean fruit fly, *Ceratitis capitata*. Ins Mol Biol 9:251–261
Bonizzoni M, Zheng L, Guglielmino CR, Haymer DS, Gasperi G, Gomulski LM, Malacrida AR (2001) Microsatellite analysis of medfly bioinfestations in California. Mol Ecol 10:2515–2524
Bonizzoni M, Guglielmino CR, Smallridge CJ, Gomulski M, Malacrida AR, Gasperi G (2004) On the origins of Medfly invasion and expansion in Australia. Mol Ecol 13:3845–3855
Carey JR (1991) Establishment of the Mediterranean fruit fly in California. Science 253:1369–1373
Carey JR, Plant RE, Papadopoulos NT (2014) Response to commentary by Gutierrez et al. Proc Roy Soc London B Biolog Sci 281:20132989. doi:10.1098/rspb.2013.2989
Cassidy D (2010) Threat and responses to *Bactrocera invadens* in Southern and East African countries exporting to South Africa. USDA-Foreign Agricultural Service, Pretoria, 13 pp

Clarke AR, Armstrong KF, Carmichael AE, Milne JR, Raghu S, Roderick GK, Yeates DK (2005) Invasive phytophagous pests arising through a recent tropical evolutionary radiation: the *Bactrocera dorsalis* complex of fruit flies. Ann Rev Entomol 50:293–319

Copeland R, Wharton R, Luke Q, De Meyer M, Lux S, Zenz N, Machera P, Okumu M (2006) Geographic distribution, host fruit, and parasitoids of African fruit fly pests *Ceratitis anonae*, *Ceratitis cosyra*, *Ceratitis fasciventris*, and *Ceratitis rosa* (Diptera : Tephritidae) in Kenya. Annals Entomol Soc Am 99:261–278

Coquillett D (1899) A new trypetid from Hawaii. Entomol News 10:129–130

Cornuet J-M, Ravigne V, Estoup A (2010) Inference on population history and model checking using DNA sequence and microsatellite data with the software DIYABC (v1.0). BMC Bioinformatics 11:401

Cornuet JM, Pudlo P, Veyssier J, Dehne-Garcia A, Gautier M, Leblois R, Marin JM, Estoup A (2014) DIYABC v2.0: a software package to make approximate Bayesian computation inferences about population history using single nucleotide polymorphism, DNA sequence and microsatellite data. Bioinformatics 30:1187–1189

Correia A, Rego J, Olmi M (2008) A pest of significant economic importance detected for the first time in Mozambique: *Bactrocera invadens* Drew, Tsuruta & White (Diptera: Tephritidae: Dacinae). Boll Zool Agr Bachicoltura 40:9–13

Coulon A, Fitzpatrick JW, Bowman R, Stith BM, Makarewich CA, Stenzler LM, Lovette IJ (2008) Congruent population structure inferred from dispersal behaviour and intensive genetic surveys of the threatened Florida scrub-jay (*Aphelocoma cœrulescens*). Mol Ecol 17:1685–1701

Davey JW, Blaxter ML (2010) RADSeq: next-generation population genetics. Brief Funct Genom 9:416–423

Davies N, Villablanca FX, Roderick GK (1999) Bioinvasions of the medfly *Ceratitis capitata*: source estimation using DNA sequences at multiple intron Loci. Genetics 153:351–360

De Meyer M (2001) On the identity of the Natal fruit fly *Ceratitis rosa* Karsch (Diptera, Tephritidae). Bull Inst Roy Sci Nat Belg Entomol 71:55–62

De Meyer M (2005) Phylogenetic relationships within the fruit fly genus *Ceratitis* MacLeay (Diptera: Tephritidae), derived from morphological and host plant evidence. Ins Syst Evol 36:459–480

De Meyer M, Mwatawala M, Copeland RS, Virgilio M (submitted) Description of new *Ceratitis* species (Diptera: Tephritidae) from Africa, or how morphological and DNA data are complementary in discovering unknown species and matching sexes.

De Meyer M, Copeland RS, Wharton RA, McPheron BA (2002) On the geographic origin of the Medfly *Ceratitis capitata* (Wiedemann) (Diptera: Tephritidae). In: Proceedings of 6th international fruit fly symposium, pp 45–53

De Meyer M, Robertson MP, Mansell MW, Ekesi S, Tsuruta K, Mwaiko W, Vayssières JF, Peterson AT (2010) Ecological niche and potential geographic distribution of the invasive fruit fly *Bactrocera invadens* (Diptera, Tephritidae). Bull Entomol Res 100:35–48

De Meyer M, Quilici S, Franck A, Chadhouliati AC, Issimaila MA, Youssoufa MA, Barbet A, Attié M, White IM (2012) Records of frugivorous fruit flies (Diptera: Tephritidae: Dacini) from the Comoro Archipelago. Afr Invert 53:69–77

De Meyer M, Delatte H, Ekesi S, Jordaens K, Kalinova B, Manrakhan A, Mwatawala M, Steck G, Van Cann J, Vancikova L, Brizova R, Virgilio M (2015a) An integrative approach to unravel the *Ceratitis* FAR (Diptera, Tephritidae) cryptic species complex: a review. ZooKeys 540:405–427

De Meyer M, Delatte H, Mwatawala M, Quilici S, Vayssieres J-F, Virgilio M (2015b) A review of the current knowledge on *Zeugodacus cucurbitae* (Coquillett) (Diptera, Tephritidae) in Africa, with a list of species included in *Zeugodacus*. ZooKeys 540:539–557

Delatte H, Virgilio M, Simiand C, Quilici S, Meyer M (2013) Isolation and characterisation of sixteen microsatellite markers cross-amplifying in a complex of three African agricultural pests (*Ceratitis rosa*, *C. anonae* and *C. fasciventris*, Diptera: Tephritidae). Con Genet Res 5:31–34

Delatte H, Virgilio M, Simiand C, Quilici S, Nzogela YB, Meyer M (2014) Isolation and characterisation of sixteen microsatellite markers amplifying an African agricultural pest, *Ceratitis cosyra* (Walker) (Diptera: Tephritidae). Con Genet Res 6:9–11

Dominiak BC, Daniels D (2012) Review of the past and present distribution of Mediterranean fruit fly (*Ceratitis capitata* Wiedemann) and Queensland fruit fly (*Bactrocera tryoni* Froggatt) in Australia. Austr J Entomol 51:104–115

Douglas L, Haymer DS (2001) Ribosomal ITS1 polymorphisms in *Ceratitis capitata* and *Ceratitis rosa* (Diptera: Tephritidae). Ann Entomol Soc Am 94:726–731

Drew RAI, Hancock DL (1994) The *Bactrocera dorsalis* complex of fruit flies (Diptera: Tephritidae: Dacinae) in Asia. Bull Entomol Res 84:1–68

Drew RAI, Romig MC (2013) Tropical fruit flies of South-East Asia: (Tephritidae: Dacinae). CABI, Wallingford, 664 pp

Drew RAI, Tsuruta K, White IM (2005) A new species of pest fruit fly (Diptera : Tephritidae : Dacinae) from Sri Lanka and Africa. Afr Entomol 13:149–154

Ekesi S, Nderitu PW, Rwomushana I (2006) Field infestation, life history and demographic parameters of the fruit fly *Bactrocera invadens* (Diptera: Tephritidae) in Africa. Bull Entomol Res 96:379–386

Elshire RJ, Glaubitz JC, Sun Q, Poland JA, Kawamoto K, Buckler ES, Mitchell SE (2011) A robust, simple Genotyping-by-Sequencing (GBS) approach for high diversity species. PLoS One 6:e19379

EPPO/CABI (2014) Distribution maps of plant pests - *Bactrocera dorsalis*. doi:https://www.eppo.int

Erbout N, De Meyer M, Lens L (2008) Hybridization between two polyphagous fruit-fly species (Diptera: Tephritidae) causes sex-biased reduction in developmental stability. Biol J Linn Soc 93:579–588

Estoup A, Guillemaud T (2010) Reconstructing routes of invasion using genetic data: why, how and so what? Mol Ecol 19:4113–4130

Feder JL, Roethele JB, Wlazlo B, Berlocher SH (1997) Selective maintenance of allozyme differences among sympatric host races of the apple maggot fly. PNAS 94:11417–11421

Gasparich GE, Sheppard WS, Han H-Y, McPheron BA, Steck GJ (1995) Analysis of mitochondrial DNA and the development of PCR-based diagnostic markers for Mediterranean fruitfly (*Ceratitis capitata*) populations. Ins Mol Biol 4:61–67

Gasparich GE, Silva JG, Han H-Y, McPheron BA, Steck GJ, Sheppard WS (1997) Population genetic structure of Mediterranean fruit fly (Diptera: Tephritidae) and implications for worldwide colonization patterns. Ann Entomol Soc Am 90:790–797

Gasperi G, Guglielmino CR, Malacrida AR, Milani R (1991) Genetic variability and gene flow in geographical populations of *Ceratitis capitata* (Wied.) (medfly). Heredity 67:347–356

Gasperi G, Bonizzoni M, Gomulski LM, Murelli V, Torti C, Malacrida AR, Guglielmino CR (2002) Genetic differentiation, gene flow and the origin of infestations of the medfly, *Ceratitis capitata*. Genetica 116:125–135

Geib SM, Calla B, Hall B, Hou S, Manoukis NC (2014) Characterizing the developmental transcriptome of the oriental fruit fly, *Bactrocera dorsalis* (Diptera: Tephritidae) through comparative genomic analysis with *Drosophila melanogaster* utilizing modENCODE datasets. BMC Genomics 15:942

Gilbert E (2004) Dhows and the colonial economy of Zanzibar, 1860–1970. Ohio University Press, Athens, 176 pp

Goergen G, Vayssières J-F, Gnanvossou D, Tindo M (2011) *Bactrocera invadens* (Diptera: Tephritidae), a new invasive fruit fly pest for the afrotropical region: host plant range and distribution in West and Central Africa. Env Entomol 40:844–854

Gomulski LM, Bourtzis K, Brogna S, Morandi PA, Bonvicini C, Sebastiani F, Torti C, Guglielmino CR, Savakis C, Gasperi G, Malacrida AR (1998) Intron size polymorphism of the Adh1 gene parallels the worldwide colonization history of the Mediterranean fruit fly, *Ceratitis capitata*. Mol Ecol 7:1729–1741

Gutierrez AP, Ponti L, Gilioli G (2014) Comments on the concept of ultra-low, cryptic tropical fruit fly populations. Proc Roy Soc London B Biolog Sci 281:20132825. doi:10.1098/rspb.2013.2825

Hamilton M (2009) Population genetics, Wiley-Blackwell. 422 pp

Haymer DS, Mcinnis DO, Arcangeli L (1992) Genetic variation between strains of the Mediterranean fruit fly, *Ceratitis capitata*, detected by DNA fingerprinting. Genome 35:528–533

Haymer DS, He M, Mcinnis DO (1997) Genetic marker analysis of spatial and temporal relationships among existing populations and new infestations of the mediterranean fruit fly (*Ceratitis capitata*). Heredity 79:302–309

Jacquard C, Virgilio M, David P, Quilici S, Meyer M, Delatte H (2013) Population structure of the melon fly, *Bactrocera cucurbitae*, in Reunion Island. Biol Inv 15:759–773

Karsten M, van Vuuren BJ, Barnaud A, Terblanche JS (2013) Population genetics of *Ceratitis capitata* in South Africa: implications for dispersal and pest management. PLoS One 8:e54281

Karsten M, Jansen van Vuuren B, Addison P, Terblanche JS (2015) Deconstructing intercontinental invasion pathway hypotheses of the Mediterranean fruit fly (*Ceratitis capitata*) using a Bayesian inference approach: are port interceptions and quarantine protocols successfully preventing new invasions? Div Distrib 21:813–825

Khamis F, Karam N, Guglielmino CR, Ekesi S, Masiga D, De Meyer M, Kenya EU, Malacrida AR (2008) Isolation and characterization of microsatellite markers in the newly discovered invasive fruit fly pest in Africa, *Bactrocera invadens*(Diptera: Tephritidae). Mol Ecol Res 8:1509–1511

Khamis FM, Karam N, Ekesi S, De Meyer M, Bonomi A, Gomulski LM, Scolari F, Gabrieli P, Siciliano P, Masiga D, Kenya EU, Gasperi G, Malacrida AR, Guglielmino CR (2009) Uncovering the tracks of a recent and rapid invasion: the case of the fruit fly pest *Bactrocera invadens* (Diptera: Tephritidae) in Africa. Mol Ecol 18:4798–4810

Khamis FM, Masiga DK, Mohamed SA, Salifu D, de Meyer M, Ekesi S (2012) Taxonomic identity of the invasive fruit fly pest, *Bactrocera invadens*: concordance in morphometry and DNA barcoding. PLoS One 7:e44862

Krosch MN, Schutze MK, Armstrong KF, Graham GC, Yeates DK, Clarke AR (2012) A molecular phylogeny for the Tribe Dacini (Diptera: Tephritidae): systematic and biogeographic implications. Mol Phyl Evol 64:513–523

Krosch MN, Schutze MK, Armstrong KF, Boontop Y, Boykin LM, Chapman TA, Englezou A, Cameron SL, Clarke AR (2013) Piecing together an integrative taxonomic puzzle: microsatellite, wing shape and aedeagus length analyses of *Bactrocera dorsalis s.l.* (Diptera: Tephritidae) find no evidence of multiple lineages in a proposed contact zone along the Thai/Malay Peninsula. Syst Entomol 38:2–13

Leblanc L, San Jose M, Barr N, Rubinoff D (in press) A phylogenetic assessment of the polyphyletic nature and intraspecific color polymorphism in the *Bactrocera dorsalis* complex (Diptera, Tephritidae). ZooKeys 540:339–367

Liu J, Shi W, Ye H (2007) Population genetics analysis of the origin of the Oriental fruit fly, *Bactrocera dorsalis* Hendel (Diptera: Tephritidae), in northern Yunnan Province, China. Entomol Sci 10:11–19

Liu L, Liu J, Wang Q, Ndayiragije P, Ntahimpera A, Nkubaye E, Yang Q, Li Z (2011) Identification of *Bactrocera invadens* (Diptera: Tephritidae) from Burundi, based on morphological characteristics and DNA barcode. Afr J Biotechnol 10:13623–13630

Lombaert E, Guillemaud T, Cornuet JM, Malausa T, Facon B, Estoup A (2010) Bridgehead effect in the worldwide invasion of the biocontrol harlequin ladybird. PLoS One 5:e9743

Lux SA, Ekesi S, Dimbi S, Mohamed S, Billah M (2003a) Mango infesting fruit flies in Africa. Perspectives and limitations of biological approaches to their management. In: Neuenschwander P, Borgemeister C, Langewald J (eds) Biological control in integrated pest management systems in Africa. CABI, Wallingford, pp 277–293

Lux SA, Copeland RS, White IM, Manrakhan A, Billah MK (2003b) A new invasive fruit fly species from the *Bactrocera dorsalis* (Hendel) group detected in East Africa. Ins Sci Appl 23:355–360

Malacrida AR, Guglielmino CR, Gasperi G, Baruffi L, Milani R (1992) Spatial and temporal differentiation in colonizing populations of *Ceratitis capitata*. Heredity 69:101–111

Malacrida AR, Marinoni F, Torti C, Gomulski LM, Sebastiani F, Bonvicini C, Gasperi G, Guglielmino CR (1998) Genetic aspects of the worldwide colonization process of *Ceratitis capitata*. J Heredity 89:501–507

Malacrida AR, Gomulski L, Bonizzoni M, Bertin S, Gasperi G, Guglielmino C (2007) Globalization and fruitfly invasion and expansion: the medfly paradigm. Genetica 131:1–9

Manrakhan A, Venter JH, Hattingh V (2015) The progressive invasion of *Bactrocera dorsalis* (Diptera: Tephritidae) in South Africa. Biol Inv 17:1–7

McPheron BA, Smith DC, Berlocher SH (1988) Microgeographic genetic variation in the apple maggot *Rhagoletis pomonella*. Genetics 119:445–451

McPheron BA, Gasparich G, Han H-Y, Steck G, Sheppard W (1994) Mitochondrial DNA restriction map for the Mediterranean fruit fly, *Ceratitis capitata*. Biochem Genetics 32:25–33

Mwatawala MW, White IM, Maerere AP, Senkondo FJ, De Meyer M (2004) A new invasive *Bactrocera* species (Diptera: Tephritidae) in Tanzania. Afr Entomol 12:154–158

Mwatawala MW, De Meyer M, Makundi RH, Maerere AP (2009) Host range and distribution of fruit-infesting pestiferous fruit flies (Diptera, Tephritidae) in selected areas of Central Tanzania. Bull Entomol Res 99:629–641

Mwatawala M, Maerere AP, Makundi R, De Meyer M (2010) Incidence and host range of the melon fruit fly *Bactrocera cucurbitae* (Coquillett) (Diptera: Tephritidae) in Central Tanzania. Int J Pest Man 56:265–273

Mwatawala M, Virgilio M, Joseph J, de Meyer M (2015) Niche partitioning among two *Ceratitis rosa* morphotypes and other *Ceratitis* pest species (Diptera, Tephritidae) along an altitudinal transect in Central Tanzania. ZooKeys 540:429–442

Narum SR, Buerkle CA, Davey JW, Miller MR, Hohenlohe PA (2013) Genotyping-by-sequencing in ecological and conservation genomics. Mol Ecol 22:2841–2847

Ndzana Abanda F-X, Quilici S, Vayssières J-F, Kouodiekong L, Woin N (2008) Inventaire des espèces de mouches des fruits sur goyave dans la région de Yaoundé au Cameroun. Fruits 63:19–26

Nirmala X, Schetelig MF, Yu F, Handler AM (2013) An EST database of the Caribbean fruit fly, *Anastrepha suspensa* (Diptera: Tephritidae). Gene 517:212–217

Norrbom AL, Carroll LE, Thompson FC, White IM, Freidberg A (1999) Systematic database of names. In: Thompson FC (ed) Fruit fly expert identification system and systematic information database. Myia, Leiden, pp 65–251

Papadopoulos NT, Plant RE, Carey JR (2013) From trickle to flood: the large-scale, cryptic invasion of California by tropical fruit flies. Proc Roy Soc B Biolog Sci 280:20131466

Pritchard JK, Stephens M, Donnelly P (2000) Inference of population structure using multilocus genotype data. Genetics 155:945–959

Raoelijaona JCY, Raoelijaona AR, Ratovonomenjanahary TZ, Brunet C, De Meyer M, Vayssières J-F, Quilici S (2012) Situation of *Bactrocera invadens* (Diptera: Tephritidae) in Madagascar (abstract). In: 2nd international TEAM meeting, Kolymbari (Greece), pp 141

Schutze MK, Aketarawong N, Amornsak W, Armstrong KF, Augustinos AA, Barr N, Bo W, Bourtzis K, Boykin LM, CÁCeres C, Cameron SL, Chapman TA, Chinvinijkul S, Chomič A, De Meyer M, Drosopoulou E, Englezou A, Ekesi S, Gariou-Papalexiou A, Geib SM, Hailstones D, Hasanuzzaman M, Haymer D, Hee AKW, Hendrichs J, Jessup A, Ji Q, Khamis FM, Krosch MN, Leblanc LUC, Mahmood K, Malacrida AR, Mavragani-Tsipidou P, Mwatawala M, Nishida R, Ono H, Reyes J, Rubinoff D, San Jose M, Shelly TE, Srikachar S, Tan KH, Thanaphum S, Haq I, Vijaysegaran S, Wee SL, Yesmin F, Zacharopoulou A, Clarke AR (2015) Synonymization of key pest species within the Bactrocera dorsalis species complex (Diptera: Tephritidae): taxonomic changes based on a review of 20 years of integrative morphological, molecular, cytogenetic, behavioural and chemoecological data. Syst Entomol 40:456–471

Shen G-M, Dou W, Niu J-Z, Jiang H-B, Yang W-J, Jia F-X, Hu F, Cong L, Wang J-J (2011) Transcriptome analysis of the oriental fruit fly (*Bactrocera dorsalis*). PLoS One 6:e29127

Sheppard WS, Steck GJ, McPheron BA (1992) Geographic populations of the medfly may be differentiated by mitochondrial DNA variation. Experientia 48:1010–1013

Shi W, Kerdelhue C, Ye H (2010) Population genetic structure of the oriental fruit fly, *Bactrocera dorsalis* (Hendel) (Diptera: Tephritidae) from Yunnan province (China) and nearby sites across the border. Genetica 138:377–385

Stephens AE, Kriticos DJ, Leriche A (2007) The current and future potential geographical distribution of the oriental fruit fly, *Bactrocera dorsalis* (Diptera: Tephritidae). Bull Entomol Res 97:369–378

Stratikopoulos E, Augustinos A, Petalas Y, Vrahatis M, Mintzas A, Mathiopoulos K, Zacharopoulou A (2008) An integrated genetic and cytogenetic map for the Mediterranean fruit fly, *Ceratitis capitata*, based on microsatellite and morphological markers. Genetica 133:147–157

Tautz D (1989) Hypervariability of simple sequences as a general source for polymorphic DNA markers. Nucleic Acids Res 17:6463–6471

Umeh VC, Garcia LE, De Meyer M (2008) Fruit flies of sweet oranges in Nigeria: species diversity, relative abundance and spread in major producing areas. Fruits 63:145–153

Vayssières JF, Goergen G, Lokossou O, Dossa P, Akponon C (2005) A new *Bactrocera* species in Benin among mango fruit fly (Diptera: Tephritidae) species. Fruits 60:371–377

Vayssières JF, Rey JY, Traoré L (2007) Distribution and host plants of *Bactrocera cucurbitae* in West and Central Africa. Fruits 62:391–396

Vayssières JF, Korie S, Ayegnon D (2009) Correlation of fruit fly (Diptera: Tephritidae) infestation of major mango cultivars in Borgou (Benin) with abiotic and biotic factors and assessment of damage. Crop Prot 28:477–488

Virgilio M, Backeljau T, Barr N, De Meyer M (2008) Molecular evaluation of nominal species in the *Ceratitis fasciventris*, *C. anonae*, *C. rosa* complex (Diptera: Tephritidae). Mol Phyl Evol 48:270–280

Virgilio M, Delatte H, Backeljau T, De Meyer M (2010) Macrogeographic population structuring in the cosmopolitan agricultural pest *Bactrocera cucurbitae* (Diptera: Tephritidae). Mol Ecol 19:2713–2724

Virgilio M, Delatte H, Quilici S, Backeljau T, De Meyer M (2013) Cryptic diversity and gene flow among three African agricultural pests: *Ceratitis rosa*, *Ceratitis fasciventris* and *Ceratitis anonae* (Diptera, Tephritidae). Mol Ecol 22:2526–2539

Virgilio M, Delatte H, Nzogela YB, Simiand C, Quilici S, De Meyer M, Mwatawala M (2015a) Population structure and cryptic genetic variation in the mango fruit fly, *Ceratitis cosyra* (Diptera, Tephritidae). ZooKeys 540:525–538

Virgilio M, Jordaens K, Verwimp C, White IM, De Meyer M (2015b) Higher phylogeny of frugivorous flies (Diptera, Tephritidae, Dacini): localised partition conflicts and a novel generic classification. Mol Phyl Evol 85:171–179

Wan X, Nardi F, Zhang B, Liu Y (2011) The oriental fruit fly, *Bactrocera dorsalis*, in China: origin and gradual inland range expansion associated with population growth. PLoS One 6:e25238

White IM, Elson-Harris MM (1994) Fruit flies of economic significance: their identification and bionomics. CAB International, Wallingford. 601 pp.

Zheng W, Peng T, He W, Zhang H (2012) High-throughput sequencing to reveal genes involved in reproduction and development in *Bactrocera dorsalis* (Diptera: Tephritidae). PLoS One 7:e36463

Massimiliano Virgilio is a molecular taxonomist and coordinator of the Joint Experimental Molecular Unit (JEMU; http://jemu.myspecies.info) of the Royal Museum for Central Africa (Tervuren, Belgium). His main interests are in morphological and molecular taxonomy, phylogeny and population genetics of African frugivorous tephritids.

Hélène Delatte is an ecologist and molecular biologist at CIRAD (Centre de Coopération Internationale en Recherche Agronomique pour le Développement) in the UMR PVBMT unit (Saint Pierre, La Réunion). Her main interests are in bioecology, evolution and population genetics of arthropods (from pollinators to agricultural pests).

Chapter 4
Role of Microsatellite Markers in Molecular Population Genetics of Fruit Flies with Emphasis on the *Bactrocera dorsalis* Invasion of Africa

Fathiya M. Khamis and Anna R. Malacrida

Abstract Microsatellites, also referred to as short tandem repeats (STRs) or simple sequence repeats (SSRs) are short sequences of tandem repeats of 1–6 bp in length in clusters of less than 150 bp flanked by sections of non-repetitive unique sequences that are scattered throughout the nuclear genome. These markers are co-dominant and hypervariable, revealing many alleles per locus; they are inherited in Mendelian fashion making them useful for detecting genetic variability within species. Once isolated and characterized, microsatellites can also be used in closely related taxa. Microsatellites can be amplified, even from highly degraded DNA, and are very simple to score. More importantly, these markers are highly polymorphic due to the plethora of variations in the repeat motifs. Several studies have endorsed microsatellite markers as an effective genetic tool to determine the historical distinctiveness of populations, and hence, the designation of species. Being highly polymorphic and selectively neutral, microsatellite markers offer a powerful genetic tool for investigating population structure, colonization processes, temporal and spatial population dynamics and evolutionary trends of insect pests. Furthermore, these markers have been successfully applied to different invasive fruit fly species to infer the evolutionary aspects underlying their invasive processes. Microsatellite markers have offered an analytical tool for the study of fruit fly invasion genetics as exemplified for the Mediterranean fruit fly, *Ceratitis capitata*. Herein, a detailed utility of microsatellite markers in inferring invasion histories of key fruit flies of economic importance is given, with a special focus on invasion into Africa of *Bactrocera dorsalis*.

F.M. Khamis (✉)
International Centre of Insect Physiology & Ecology (*icipe*),
PO Box 30772, Nairobi, Kenya
e-mail: fkhamis@icipe.org

A.R. Malacrida
Dipartimento di Biologia Animale, Universita di Pavia, Pavia 27100, Italy

Keywords Microsatellite markers • Co-dominant • Hypervariable • Population structure • Invasion

1 Introduction to the Use of Molecular Markers

In Africa, and sub-Saharan Africa in particular, the agricultural sector is the backbone of most economies in pursuit of sustainable development. Specifically, the horticultural sub-sector is a precious tool that is contributing to poverty alleviation by increasing food security and income generation for continued economic growth. However, this sub-sector is constrained by a number of biotic and abiotic factors. Amongst the former is infestation by tephritid fruit flies that are well recognised as a group of pests of economic significance (Ekesi and Billah 2007). Sub-Saharan Africa is the native home to about 915 fruit fly species out of which 299 species belong to the genera *Ceratitis* MacLeay, *Trirhithrum* Bezzi and *Dacus* Fabricius (White and Elson-Harris 1992). Due to the globalisation of trade, the emergent tourism industry, porous borders and poor phytosanitary expertise, the likelihood of inadvertent introduction and spread of exotic fruit fly species across the continent is escalating. Therefore, measures to strengthen the phytosanitary/quarantine infrastructure are of paramount importance to avert the establishment of alien invasive pests.

Knowledge of the genetic structure and geographical variability of invasive fruit fly species is a vital pre-requisite to implementing quarantine, control and eradication measures (Roderick and Navajas 2003; Malacrida et al. 2007). In the past, morphological characters were considered sufficient to describe species. However, morphological characters for species delineation have several limitations. The existence of homoplasy amongst characters and cryptic speciation in some insect families, as in the tephritid fruit flies, make species-level descriptions based on adult and larval morphology extremely difficult (Armstrong et al. 1997; De Meyer 1998; McPheron 2000). This has led taxonomists and quarantine officials to seek alternative ways to identify tephritid fruit flies, including the use of molecular markers (Sonvinco et al. 1996; Armstrong et al. 1997; Morrow et al. 2000). Indeed, biochemical (allozymes) and DNA molecular markers have been used to elucidate the variability in population structure of several tephritid species (Baruffi et al. 1995; Malacrida et al. 1996). In addition to this, molecular markers have advantages over biochemical tools and hence have become the current tool of choice. Molecular markers are site specific DNA sequences that are easily detected in the genome. These markers are neutral and can be utilised in a number of ways including, but not limited to: the analysis of genetic variability; to make inferences on population genetic structure; for DNA fingerprinting; for chromosome mapping; and for the identification and description of species. There are a number of molecular markers that have been developed and are in use: RAPDs, RFLPs, PCR-RFLPs, microsatellites, minisatellites and SNPs amongst others (Baruffi et al. 1995; Barr et al. 2006;

Malacrida et al. 1996; Khamis et al. 2008). The best molecular markers have a combination of the following properties: they should occur frequently in the genome; be co-dominant; be highly polymorphic; be vastly reproducible and transferable to many taxa; be cheap to develop and apply; and not be affected by environmental conditions.

In some frugivorous tephritid fruit fly species, diagnostic morphological characters for the identification of adult flies have been made available (Adsavakulchai et al. 1999; De Meyer 2005; Drew et al. 2006, 2008). However, inconsistencies in the limits of fruit fly species identification based on conventional adult morphological features together with overlapping host and geographical ranges, have profound effects on quarantine, management and biological studies of these species (Clarke et al. 2005). These limitations have led to the development and improvement of molecular tools for identification and classification of the fruit flies pest species, and for understanding their population structure (Armstrong et al. 1997; Malacrida et al. 1998; Han and McPheron 1997; Manni et al. 2015). Furthermore, because molecular markers are phenotypically neutral and resistant to environmental cues (unlike morphological characters) they can be used as a single reliable taxonomic tool. Several molecular markers are available to discriminate amongst tephritid fruit fly species. These markers have successfully been used to validate species (Khamis et al. 2012; Schutze et al. 2014), infer phylogenetic relationships (Boykin et al. 2014), verify intra-specific variation between populations (Bonizzoni et al. 2000; Baliraine et al. 2004) and trace the routes of invasion of pest fruit fly species (Aketarawong et al. 2007, 2014; Khamis et al. 2009).

2 Application of Microsatellites: Advantages and Limitations

Several studies have endorsed microsatellite markers as an effective genetic tool to determine the historical distinctiveness of populations, and hence, the designation of species (Hedrick et al. 2001; Wang et al. 2001). Microsatellites are short sequences of tandem repeats of 1–6 bp in length in clusters of less than 150 bp in length flanked by sections of non-repetitive unique sequences that are scattered throughout the nuclear genome and mostly associated with conserved loci containing coding regions (Loxdale and Lushai 1998). Microsatellites have also been referred to as short tandem repeats (STRs) (Edwards et al. 1991) or simple sequence repeats (SSRs) (Jacob et al. 1991). These markers are co-dominant and hypervariable, revealing many alleles per locus; they are inherited in Mendelian fashion making them useful for detecting genetic variability within species. Once these markers have been isolated and characterized, they may also be used in closely related taxa. Microsatellites can be amplified even from highly degraded DNA and are very simple to score (Bruford and Wayne 1993). More importantly, these markers are highly polymorphic due to the plethora of variations in the repeat motifs. In 1989, Weber and May developed a universal method for isolating microsatellites. The polymorphism of these markers was confirmed by Litt and Luty (1989) who detected allelic variants amongst individuals using amplification of the $(TG)_n$ microsatellites in the human actin gene. Although the origin of microsatellite polymorphism is still debated, it is likely to be due to slipped-strand mispairing (Levinson and Gutman 1987), slippage events during DNA replication/repair/recombination (Schlötterer and Tautz 1992) or asymmetrical cross-over between sister chromatids (Innan et al. 1997).

Despite the uncertainty surrounding microsatellite evolution, they have been adopted widely and applied in many fields of study since their initial description by Hamada et al. (1982). Microsatellites markers are an invaluable method for genome mapping in many organisms (Schuler et al. 1996), and are also applicable in various fields ranging from ancient and forensic DNA studies, to population genetics and conservation/management of biological resources (Jarne and Lagoda 1996). Furthermore, microsatellites can be used for species identification, genetic tagging, breeding studies, reproductive biology, taxonomy, phylogenetic studies, disease diagnostics and genetic diversity studies (Abdul-Muneer 2014). Moreover, these markers offer a diagnostic tool that can differentiate between species not easily separated by morphological traits (Kinyanjui et al. 2016). The usefulness of microsatellite markers is evidenced by the rising numbers of mapped genes based on microsatellites (Schuler et al. 1996).

The wide applicability of microsatellite markers is associated with their many advantages. Microsatellite markers are robust and very informative compared with other markers such as RFLPs, RAPDs and AFLPs (He et al. 2003; Lee et al. 2004). Importantly, microsatellite markers are PCR-based and therefore require low quantities of DNA. The primer lengths and high annealing temperatures of microsatellite markers in genotyping guarantee their reproducibility. Once isolated, microsatellite markers can be cross amplified amongst closely related species (Baliraine et al. 2002). Furthermore, these markers can be multiplexed hence reducing the time and cost of analysis. Although microsatellites are a successful tool in genetics, they do have some drawbacks. Firstly, microsatellite markers need to be isolated *de novo* for most species being analysed for the first time; this isolation is expensive, laborious and time consuming (Zane et al. 2002). Secondly, the likelihood of null alleles occuring when using ancient or degraded DNA is very high, leading to difficulties in estimating allelic frequencies and heterozygosity (Kumar et al. 2009). Last but not least, homoplasy in some organisms is a common problem for the application of microsatellite markers in phylogenetics, leading to false identification of species descents (Estoup et al. 2002).

3 Microsatellite Markers: Isolation and Characterization

There are several methods that have been described for the isolation of microsatellite loci; the first was the *de novo* isolation described by Rassmann et al. (1991). This protocol has become the traditional method for microsatellite isolation and it involved identification of microsatellite-containing clones by colony hybridization with their respective specific probes. Following subsequent advances in methodologies and DNA sequencing techniques, isolation of microsatellites can now be achieved either by: (i) Constructing and screening either enriched or non-enriched genomic libraries or by utilizing the products generated by other molecular markers or by the application of next-generation sequencing systems; (ii) Making use of the EST sequences already deposited in the public domain databases or sequencing PCR products generated by consensus/universal primers; and (iii) Testing the amplification potential of other microsatellite markers developed in other related species (i.e. cross-species amplification).

3.1 Isolation of Microsatellites from Enriched Genomic DNA Libraries

This can be achieved through selective hybridization methods where the genomic DNA undergoes fragmentation either by use of restriction enzymes, sonication or, less frequently, nebulisation (Senan et al. 2014). Fragments of DNA in the range of 300–700 bp in length are selected and ligated into a common vector. Ligation can be done directly or after ligation to specific adapters (Zane et al. 2002). The DNA is then denatured and subjected to enrichment by hybridization with either: (i) biotinylated oligos followed by capture of biotinylated hybrids (oligo-bound DNA fragments) in a vectrex-aridin matrix (Kandpal et al. 1994); (ii) oligonucleotides (oligos) bound to nylon membranes (Karagyozov et al. 1993; Chen et al. 1995; Edwards et al. 1996) (iii) 5' biotinylated repeat oligos and subsequent capture by biotinylated hybrids by streptavidin coated magnetic beads (Aketarawong et al. 2006; Khamis et al. 2008); (iv) biotinylated microsatellite-probe-streptavidin coated magnetic bead complex (White and Powell 1997). Screening for positive clones is achieved by means of southern hybridization using the probes mentioned above and after blotting the bacterial colonies onto nylon membranes. Colony transfer is done either by classical replica plating or by picking single colonies and ordering them in new arrayed plates. After successful identification of positive microsatellite-containing clones, specific primers are designed and PCR conditions are optimized to allow the amplification of each locus from different individuals of a population. The fragments are then amplified, cloned and sequenced directly and probed for the presence of microsatellites. The efficiency of this approach entirely depends on the specific binding of streptavidin coated beads to biotin.

Ostrander et al. (1992) and Paetkau (1999) described protocols that allow the selective amplification of microsatellites containing genomic DNA using very specific primers. This is known as the primer extension method and it relies on the construction of a primary genomic DNA library in a phagemid vector to recover the library as single stranded DNA which is then subjected to primer extension using repeat specific non-biotinylated oligos or 5'biotinylated oligos. Ostrander et al. (1992) further demonstrated primer extension steps that selectively generated only double-stranded products from vectors containing the tandem repeats of interest, and then transformed them in to *E. coli* cells. Streptavidin coated magnetic beads were used to selectively pick out the 5'biotinylated hybrids and convert them in to double-stranded DNA via a second round of primer extension for transformation. Pandolfo (1992) described the ligation of a vectorette (i.e. a linker containing a non-complementary region) to restricted yeast artificial chromosome (YAC) DNA. Using microsatellite-specific primers and universal vector primers, the vectorette-ligated DNA could be amplified and the products cloned and sequenced to probe for the desired repeat loci.

There are plenty of enrichment protocols available but the selective hybridization stands out as it allows for enrichment and selection prior to cloning, thereby providing a faster and easier method to handle multiple samples (Senan et al. 2014; Glenn and Schable 2005). This method is very simple, reproducible and cost effective for isolating microsatellites.

3.2 Isolation of Microsatellites from Non-Enriched Genomic DNA Libraries

Golein et al. (2006) demonstrated that genomic DNA could be restricted, ligated in to vectors and transformed to generate a non-enriched genomic DNA library. The clones were then spotted on to gridded nylon filters and screened with radio-labelled microsatellite probes or subjected to enrichment with biotin labelled-probes-streptavidin and sequenced.

3.3 Other Methods for Isolation of Microsatellites

A number of other techniques have been used to isolate microsatellite loci. These include isolation from RAPDs which involves the blotting of RAPD products on to nitrocellulose membranes which are then screened, using digoxygenin-labelled probes, for positive clones that could be detected using autoradiography. Another technique was described by Zane et al. (2002) and is known as FIASCO (Fast isolation by AFLP of sequences containing repeats). In this protocol the AFLP bands were hybridized with biotinylated probes which were then selectively probed using streptavidin-coated beads followed by cloning and sequencing of the enriched DNA fragments, to generate new microsatellite markers. However, due to the labour intensive process required for the *de novo* isolation of microsatellites, and in view of recent advances in DNA sequencing technologies (e.g. next generation sequencing [NGS] and better bioinformatics), these methods represent powerful alternatives to conventional methods for isolation of microsatellite markers. Large amounts of data can be produced via NGS and screening can be done using bioinformatics tools; this avoids the need for construction of microsatellite-enriched DNA libraries and provides a rapid approach for the large-scale generation of microsatellite loci. Reductions in sequencing costs will make the rapid identification of microsatellite markers even easier and cheaper. There are plenty of EST sequences that have been deposited in public domain databases (Rudd 2003). Several tools are available to mine for microsatellite loci and these include TROLL (Castelo et al. 2002), MISA (Thiel et al. 2003), SciRoKo (Kofler et al. 2007), Msatcommander (Faircloth 2008) amongst others. However, generation of these markers is limited to the availability of EST sequences, particularly if EST sequences are not deposited in publicly accessible domains/databases. Closely related individuals tend to have greater DNA conservation in their coding regions hence EST microsatellite markers cannot be used in differentiating such individuals since they show less polymorphism and are therefore less efficient (Gupta et al. 2003). Recently, the high-throughput genomic-sequencing technique has produced millions of base pairs and short fragment reads which can be screened using bioinformatics tools to identify primers that amplify a large number of polymorphic microsatellite loci (Abdelkrim et al. 2009).

4 Prospects for Tracing the Routes of Fruit Fly Invasions out of Africa and the Population Structure of Fruit Flies Using Microsatellite Markers: The Case of *Ceratitis* Species and *Bactrocera oleae* (Rossi)

Being highly polymorphic and selectively neutral, microsatellite markers offer a powerful genetic tool for investigating population structure, colonization processes, temporal and spatial population dynamics and evolutionary trends of insect pests (Wu et al. 2009). Furthermore, these markers have been successfully applied to different invasive fruit fly species to infer the evolutionary aspects underlying their invasive processes (Bonizzoni et al. 2001, 2004; Baliraine et al. 2004; Khamis et al. 2009). Due to their polymorphic nature these markers have also been utilized in the analyses of fruit fly population structure across different geographical areas, and in tracing the origins of adventive populations (Bonizzoni et al. 2000, 2001; Meixner et al. 2002). Moreover, these markers have proven to be useful for cross-species amplification to study the population structure of tephritid species when no previous genetic information was available (Baliraine et al. 2003, 2004; Shearman et al. 2006).

Globally, phytophagous members of the family Tephritidae are amongst the most important pests of fruits and vegetables. With more than 4000 species described, this family is the most diverse and contains 500 genera of which the four most economically important genera are: *Ceratitis*, *Bactrocera*, *Anastrepha* and *Rhagoletis* (White and Elson-Harris 1992). All the genera have native distribution ranges. For example, *Ceratitis* is an Afro-tropical genus, *Bactrocera* is mainly confined to the Oriental and Australasian regions, *Anastrepha* to South and Central America and the West Indies, while *Rhagoletis* has representatives in the Americas, Europe and temperate Asia (White and Elson-Harris 1992). In addition to the polyphagous nature of some species belonging to this family, several are considered highly invasive; aided by globalization of trade and poor quarantine infrastructure in the invaded countries. In recent years a member of these genera have been reported outside their native ranges. The pattern and routes of invasions of these species is of paramount importance for their management hence governments and NRI's have mobilized extensive ecological and evolutionary genetic research on these invasive pest species (Aluja and Norrbom 2000).

Microsatellite markers have offered an analytical tool for the study of fruit fly invasion genetics as exemplified for the Mediterranean fruit fly (medfly), *Ceratitis capitata* (Weidemann) (Diptera: Tephritidae) (Bonizzoni et al. 2000). *Ceratitis capitata* is a tephritid fruit fly of global economic significance (Malacrida et al. 2007). In the past century, this pest has spread from its native Afro-tropical range to several countries including the Mediterranean basin, parts of South and Central America, and Australia (Fletcher 1989). Bonizzoni et al. (2000) isolated 43 microsatellite markers of which they used ten to unravel polymorphisms amongst *C. capitata* samples from six geographical populations (Kenya, La Réunion, Madeira, South Italy, Greece and Peru) from its native and invaded ranges. These markers detected a decrease in the number of polymorphisms from tropical Africa to the Mediterranean basin and to South America. Comparison of the Kenyan population with other populations in the study showed that the Kenyan population had the highest average number of alleles per locus, of which many were at low frequency, and most were private confirming an African origin for *C. capitata*. These results were consistent with the colonization history of *C. capitata* and indicative of a hierarchical migration structure through Spain and subsequently along the Mediterranean basin to the East (Malacrida et al. 1998; Gomulski et al. 1998).

Furthermore, a number of studies have demonstrated the utility of cross amplification of microsatellite markers to closely related species for population genetic studies (Baliraine et al. 2002, 2004; Shearman et al. 2006). For instance, Baliraine et al. (2002) screened 24 microsatallite markers from *C. capitata* for cross species amplification in the Natal fruit fly, *Ceratitis rosa* Karsch, *Ceratitis fasciventris* (Bezzi) and the mango fruit fly, *Ceratitis cosyra* (Walker). The sequence analysis indicated that most *C. capitata*-based microsatellite markers were useful for population genetic studies in the various species tested, a fact that will facilitate the tracing of the geographical origin of adventive pest populations, and assessment of their invasive potential and risk (Baliraine et al. 2002). In a similar study, Baliraine et al. (2004) compared genetic variability data from *C. rosa* and *C. fasciventris* with those derived from *C. capitata* to determine the geographical origin of the *Ceratitis* species. The results from this study confirmed the hypothesis of an East African origin for *Ceratitis* species (De Meyer et al. 2002).

A study by Nardi et al. (2005) used microsatellite markers and mitochondrial sequences to examine the population structure and colonization history of the olive fruit fly, *Bactrocera oleae* (Rossi). Their study revealed that Africa, and not the Mediterranean, is the origin of *B. oleae* infesting cultivated olives, which is supported by the significantly greater genetic diversity of microsatellite loci from samples collected in Africa compared with samples collected from the Mediterranean region. The results also indicated that the recent invasion of *B. oleae* in to the Americas most likely originated from the Mediterranean area.

5 Potential for Use of Microsatellite Markers to Infer Fruit Fly Invasion Histories: The Case of *Bactrocera dorsalis* (Hendel) in Africa

Dacine fruit flies of the genus *Bactrocera* Macquart (Diptera: Tephrtitidae) are also economically important fruit fly pests (White and Elson-Harris 1992). With life-history traits that include high mobility and dispersive powers, high reproductive rates and extreme polyphagy, *Bactrocera* species are well documented invaders and rank high on quarantine lists worldwide (European and Mediterranean Plant Protection Organization 2009). Several *Bactrocera* species have been introduced accidentally in to different parts of the world. This is due to the globalisation of horticultural trade and has had major economic consequences (Clarke et al. 2005; De Meyer et al. 2009). In 2003, a new fruit fly pest was detected in Kenya (Lux et al. 2003) soon after the completion of a programme of monthly fruit collections carried out between 1999 and 2003 (Copeland et al. 2004). The insect was described as *Bactrocera invadens* (Drew et al. 2005) and was rated as "a devastating quarantine pest" by the Inter-African Phytosanitary Council in 2005 (French 2005). Within two years of its detection in the coastal region of Kenya, the species was recorded in several other countries on the African mainland (Mwatawala et al. 2004; Drew et al. 2005; Ekesi and Billah 2007). It is now known to be present in tropical Africa from Senegal to Mozambique, as well as in the Comoro Islands in the Indian Ocean. Through integrative multidisciplinary research efforts this species has now been synonymised with the oriental fruit fly, *Bactrocera dorsalis* (Hendel) (Schutze et al. 2014).

Since this species was a new invader in the African continent, the timing and the pathway of its invasion were unknown. The fact that the first historical records of this pest in Africa were from the East coast may indicate that this area was the port of entry of *B. dorsalis* in Africa, but this hypothesis had not been tested. Moreover the native range was not well defined: it has been suggested that it ranges from Sri Lanka to the Southern Indian sub-continent from where the species may have invaded Africa (Mwatawala et al. 2004; Drew et al. 2005; Khamis et al. 2009). The detrimental effects of this invasive species stimulated several studies to define its ecological niche and invasion potential (Ekesi et al. 2006; De Meyer et al. 2009; Mwatawala et al. 2006). However, due to the 'novelty status' of this fruit fly as a dispersive invader, no data were available on its genetic diversity or on the degree of co-ancestry amongst African populations and the supposed native populations from Southern India and Sri Lanka. Consequently, no inferences based on genetic data

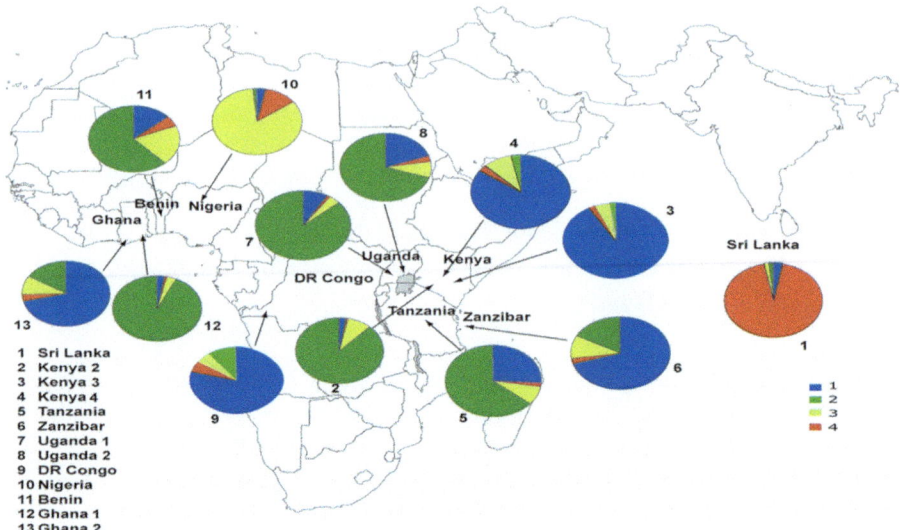

Fig. 4.1 Geographical representation of the clustering outcomes for 13 samples of *Bactrocera dorsalis*. The four colours represent the co-ancestry distribution of 351 individuals in four hypothetical clusters

were possible concerning the invasion route of *B. dorsalis*. To unravel the mystery surrounding this new invader, a set of 11 polymorphic microsatellite markers were isolated, characterised and utilised to evaluate the level of genetic diversity and the extent of common ancestry amongst several African populations collected across the actual invaded area in tropical Africa from East to West (Khamis et al. 2008; 2009).

Using these markers it was possible to successfully infer the dynamic aspects of the invasion of Africa by *B. dorsalis*, confirm its Asian origin, assess the diversity of African populations and its invasion routes in to Africa. The genetic data generated by microsatellite markers left no doubt that Sri Lanka was within the native range of *B. dorsalis* as the sample from there was characterized by all the genetic features expected in a large population from a native area; these included a large number of alleles coupled with a large number of private alleles occurring at high frequency. The Sri Lankan sample was also clearly genetically separate from the African samples and only a small percentage of genomes from Sri Lankan flies could be found in African flies (Fig. 4.1). From throughout the invaded range of East, Central and West Africa, the genetic data also suggested the presence of populations with relatively high levels of genetic diversity associated with limited geographic structure. Furthermore, although the invasion was a relatively recent event in Africa, there was no genetic footprint of bottlenecks although populations appeared large enough to maintain, especially in the West, a relatively large number of alleles with a low frequency. All these genetic features suggest a process of rapid population growth and expansion. The markers also identified genotypes that, when

Table 4.1 Average coefficient of ancestry obtained from a Structure run with K=4 for 351 individuals of *B. dorsalis* from 13 geographical regions

Area	Population	Clusters (K)			
		1	2	3	4
South Asia	Sri Lanka	0.030	0.015	0.014	**0.941**
East Africa	Kenya N	0.025	**0.880**	0.086	0.010
	Kenya K	**0.916**	0.016	0.049	0.019
	Kenya M	**0.847**	0.034	0.090	0.029
	Tanzania	**0.257**	**0.637**	0.085	0.020
	Zanzibar	**0.700**	**0.171**	**0.104**	0.025
	Uganda Ka	0.085	**0.864**	0.038	0.013
	Uganda Ki	**0.198**	**0.700**	0.075	0.027
Central Africa	DR Congo	**0.791**	**0.104**	0.053	0.053
West Africa	Nigeria	0.030	0.014	**0.837**	**0.119**
	Benin	**0.134**	**0.618**	**0.197**	0.051
	Ghana L	0.026	**0.930**	0.034	0.010
	Ghana M	**0.711**	**0.161**	0.092	0.035

Co-ancestry higher than 10% of each population in a cluster is in *bold*
Khamis et al. (2009)

analysed using Structure 2.2 (a program that infers genetic clustering of populations using the Bayesian clustering algorithm; Pritchard et al. 2000) were present throughout Africa, i.e. some genotypes of East African flies could not be distinguished from West African flies (Fig. 4.1). Instead, the populations clustered into four genetic groups, three African clusters and the Sri Lankan cluster. Likewise, the genetic data demonstrated that the invasion and dispersal pattern of *B. dorsalis* in Africa was rapid and apparently chaotic, with the potential for multiple introductions as suggested from hypothetical outbreaks. Also analysis of the genotypes identified by the programs Structure 2.2 and GenClass 2.0 (which assigns or excludes reference populations as possible origins of individuals, on the basis of multi-locus genotypes) (Piry et al. 2004), allowed the main pathway of dispersal of *B. dorsalis* in to Africa to be inferred. Two results were found: (a) a higher or equal rate of co-ancestry of eastern and western flies in two African clusters (Fig. 4.1); (b) the major average assignment probability of eastern flies to the west than vice-versa (Table 4.1) (Khamis et al. 2009). These two results support the fact that the invasion of this pest began in East Africa. Based on their high values of co-ancestry, coastal regions of East Africa, where *B. dorsalis* was first found (i.e. Kenya and / or Tanzania), were consistently identified as the places from where the African invasion probably started. Another result was the major, although low, average assignment probability of East African flies to Sri Lanka compared with West Africa to Sri Lanka. These genetic data were consistent with the supposition that East Africa was the port of entry for *B. dorsalis*. Another very important aspect of the study concerned the Nigerian population of *B. dorsalis* which occupied its own cluster, suggesting that this outbreak in the West arose at the same time as the other two outbreaks in the East.

6 Conclusion

This review, although focused on only a few species of tephritid fruit flies, provides evidence that microsatellite variation can play an important role in the study of fruit fly population dynamics. These markers enable researchers to scrutinize variability within populations and the rate of divergence amongst populations. The information gained provides insights into the population structure of tephritid species present in Africa, allowing inferences to be made concerning their source areas and invasion histories. Moreover, these population data are of paramount importance for implementation of eco-friendly sustainable management strategies against fruit flies. Establishing the origin of pest species has facilitated the identification and introduction of biological control agents (i.e. parasitoids) from the pest's native region, for use in Africa.

Acknowledgements Research activities were funded by grants from GIZ/BMZ, Biovision, the EU, IAEA, DFID and DAAD to the African Fruit Fly Programme (AFFP) of *icipe*.

References

Abdelkrim J, Robertson BC, Stanton J-A L, Gemmell NJ (2009) Fast, cost-effective development of species-specific microsatellite markers by genomic sequencing. BioTechniques 46:185–192

Abdul-Muneer PM (2014) Application of Microsatellite Markers in Conservation Genetics and Fisheries Management: Recent Advances in Population Structure Analysis and Conservation strategies. Hindawi Publishing Corporation, Gen Res Intern Volume 2014, Article ID 691759, 11. http://dx.doi.org/10.1155/2014/691759

Adsavakulchai A, Baimai V, Prachyabrued W, Grote PJ, Lertlum S (1999) Morphometric study for identification of the *Bactrocera dorsalis* complex (Diptera: Tephritidae) using wing image analysis. Biotropica 13:37–48

Aketarawong N, Bonizzoni M, Malacrida AR, Gasperi G, Thanaphum S (2006) Seventeen novel microsatellite markers from an enriched library of the pest species *Bactrocera dorsalis sensu stricto*. Mol Ecol Notes 6:1138–1140

Aketarawong N, Bonizzoni M, Thanaphum S, Gomulski LM, Gasperi G, Malacrida AR, Gugliemino CR (2007) Inferences on the population structure and colonization process of the invasive oriental fruit fly, *Bactrocera dorsalis* (Hendel). Mol Ecol 16:3522–3532

Aketarawong N, Guglielmino CR, Karam N, Falchetto M, Manni M, Scolari F, Gomulski LM, Gasperi G, Malacrida AR (2014) The oriental fruit fly *Bactrocera dorsalis* s.s. in East Asia: disentangling the different forces promoting the invasion and shaping the genetic make-up of populations. Genetica 142:201–213

Aluja M, Norrbom AL (2000) Fruit flies (Tephritidae): phylogeny and evolution of behavior. CRC Press, Boca Raton

Armstrong KF, Cameron CM, Frampton ER (1997) Fruit fly (Diptera: Tephritidae) species identification: a rapid molecular diagnostic technique for quarantine application. Bull Entomol Res 87:111–118

Baliraine FN, Bonizzoni M, Osir EO, Lux SA, Mulaa FJ, Zheng L, Gomulski LM, Gasperi G, Malacrida AR (2002) Comparative analysis of microsatellite loci in four fruit fly species of the genus *Ceratitis* (Diptera: Tephritidae). In: Proceedings of 6th international fruit fly symposium, 6–10 May 2002, Stellenbosch, South Africa

Baliraine FN, Bonizzoni M, Osir EO, Lux SA, Mulaa FJ, Zheng L, Gomulski LM, Gasperi G, Malacrida AR (2003) Comparative analysis of microsatellite loci in four fruit fly species of the genus *Ceratitis* (Diptera: Tephritidae). Bull Entomol Res 93:1–10

Baliraine FN, Bonizzoni M, Guglielmino CR, Osir EO, Lux SA, Mulaa FJ, Gomulski LM, Quilici S, Gasperi G, Malacrida AR (2004) Population genetics of the potentially invasive African fruit fly species, *Ceratitis rosa* and *Ceratitis fasciventris* (Diptera: Tephritidae). Mol Ecol 13:683–695

Barr NB, Copeland RS, De Meyer M, Masiga D, Kibogo HG, Billah MK, Osir E, Wharton RA, McPheron BA (2006) Molecular diagnostics of economically important *Ceratitis* fruit fly species (Diptera: Tephritidae) in Africa using PCR and RFLP analyses. Bull Entomol Res 96:505–521

Baruffi L, Damiani G, Guglielmino CR, Malacrida AR, Bandi C, Gasperi G (1995) Polymorphism within and between populations of *Ceratitis capitata*: comparison between RAPD and multilocus enzyme electrophoresis. Heredity 74:425–437

Bonizzoni M, Malacrida AR, Guglielmino CR, Gomulski LM, Gasperi G, Zheng L (2000) Microsatellite polymorphism in the Mediterranean fruit fly *Ceratitis capitata*. Insect Mol Biol 9:251–261

Bonizzoni M, Zheng L, Guglielmino CR, Haymer DS, Gasperi G, Gomulski LM, Malacrida AR (2001) Microsatellite analysis of medfly bioinfestations in California. Mol Ecol 10:2515–2524

Bonizzoni M, Guglielmino CR, Smallridge CJ, Gomulski LM, Malacrida AR, Gasperi G (2004) On the origins of medfly invasion and expansion in Australia. Mol Ecol 13:3845–3855

Boykin LM, Schutze MK, Krosch MN, Chomic A, Chapman TA, Englezou A, Armstrong KF, Clarke AR, Hailstones D, Cameron SL (2014) Multi-gene phylogenetic analysis of south-east Asian pest members of the *Bactrocera dorsalis* species complex (Diptera: Tephritidae) does not support current taxonomy. J Appl Entomol 138:235–253

Bruford MW, Wayne RK (1993) Microsatellite and their application to population genetic studies. Curr Opin Genet De 3:939–943

Castelo AT, Martins W, Gao GR (2002) TROLL-tandem repeat occurrence locator. Bioinformatics 18:634–636

Chen H, Pulido JC, Duyk GM (1995) MATS: A rapid and efficient method for the development of microsatellite markers from YACs. Genomics 25:1–8

Clarke AR, Armstrong KF, Carmichael AE, Milne JR, Raghu S, Roderick GK, Yeates DK (2005) Invasive phytophagous pests arising through a recent evolutionary radiation: the *Bactrocera dorsalis* complex of fruit flies. Ann Rev Entomol 50:293–319

Copeland RS, White IM, Okumu M, Machera P, Wharton RA (2004) Insects associated with fruits of the Oleaceae (Asteridae, Lamiales) in Kenya, with special reference to the Tephritidae (Diptera). Bishop Mus Bull Entomol 12:135–164

De Meyer M (1998) Revision of the subgenus *Ceratitis* (*Ceratalaspis*) Hancock, 1918 (Diptera: Tephritidae, Ceratitini). Bull Entomol Res 88:257–290

De Meyer M (2005) Phylogenetic relationships within the fruit fly genus *Ceratitis* MacLeay (Diptera: Tephritidae), derived from morphological and host plant evidence. Insect Syst Evol 36:459–480

De Meyer M, Copeland RS, Wharton RA, McPheron BA (2002) On the geographical origin of the medfly, *Ceratitis capitata* (Wiedemann). Abstract presented at the 6th international symposium on fruit flies of economic importance, Stellenbosch, South Africa, 6–10 May 2002.

De Meyer M, Robertson MP, Mansell MW, Ekesi S, Tsuruta K, Mwaiko W, Vayssières JF, Peterson AT (2009) Ecological niche and potential geographic distribution of the invasive fruit fly *Bactrocera invadens* (Diptera, Tephritidae). Bull Entomol Res 100:35–48

Drew RAI, Tsuruta K, White IM (2005) A new species of pest fruit fly (Diptera: Tephritidae: Dacinae) from Sri Lanka and Africa. Afri Entomol 13:149–154

Drew RAI, Dorji C, Romig MC, Loday P (2006) Attractiveness of various combinations of colors and shapes to females and males of *Bactrocera minax* (Diptera: Tephritidae) in a commercial mandarin grove in Bhutan. J Econ Entomol 99:1651–1656

Drew RAI, Raghu S, Halcoop P (2008) Bridging the morphological and biological species concepts: studies on the *Bactrocera dorsalis* (Hendel) complex (Diptera: Tephritidae: Dacinae) in South-east Asia. Biol J Linn Soc 93:217–226

Edwards A, Civitello A, Hammond HA, Caskey CT (1991) DNA typing and genetic mapping with trimeric and tetrameric tandem repeats. Am J Hum Genet 49:746–756

Edwards KJ, Barker JHA, Daly A, Jones C, Karp A (1996) Microsatellite libraries enriched for several microsatellite sequences in plants. Biotechniques 20:758–760

Ekesi S, Billah MK (2007) A field guide to the management of economically important Tephritid fruit flies in Africa. 2nd ed, *icipe* Science press B1–I4

Ekesi S, Nderitu P, Rwomushana I (2006) Field infestation, life history and demographic parameters of the fruit fly *Bactrocera invadens* (Diptera: Tephritidae) in Africa. Bull Entomol Res 96:379–386

Estoup A, Jarne P, Cornuet JM (2002) Homoplasy and mutation model at microsatellite loci and their consequences for population genetics analysis. Mol Ecol 11:1591–1604

European and Mediterranean Plant Protection Organization, EPPO (2009) *Bactrocera invadens* (Diptera: Tephritidae) http://www.eppo.org/QUARANTINE/Alert_List/insects/BCTRIN.htm

Faircloth BC (2008) MSATCOMMANDER: detection of microsatellite repeat arrays and automated, locus-specific primer design. Mol Ecol Resour 8:92–94

Fletcher BS (1989) Movements of tephritid fruit flies. In: Robinson AS, Hooper GH (eds) Fruit flies: their biology, natural enemies and control, vol 3B. Elsevier Science Publications, Amsterdam, pp 209–219

French C (2005) The new invasive *Bactrocera* species. pp. 19–20 In: Insect pest control newsletter, No. 65. International Atomic Energy Agency, Vienna, Austria

Glenn TC, Schable NA (2005) Isolating microsatellite DNA loci. Method Enzymol 395:202–222

Golein B, Talaie A, Zamani Z, Moradi B (2006) Development and characterization of new microsatellite loci from lemon (*Citrus limon*). Int J Agr Biol 8:172–174

Gomulski LM, Bourtzis K, Brogna S, Morandi PA, Bonvicini C, Sebastiani F, Torti C, Guglielmino CR, Savakis C, Gasperi G, Malacrida AR (1998) Intron size polymorphism of the Adh1 gene parallels the worldwide colonization history of the Mediterranean fruit fly, *Ceratitis capitata*. Mol Ecol 7:1729–1742

Gupta PK, Rustgi S, Sharma S, Singh R, Kumar N, Balyan HS (2003) Transferable EST-SSR markers for the study of polymorphism and genetic diversity in bread wheat. Mol Genet Genomics 270:315–323

Hamada HM, Petrino MG, Kakunaga T (1982) A novel repeated element with Z-DNA forming potential is widely found in evolutionarily diverse eukaryotic genomes. PNAS 79:6465–6469

Han HY, McPheron BA (1997) Molecular phylogenetic study of Tephritidae (Insecta: Diptera) using partial sequences of the mitochondrial 16S ribosomal DNA. Mol Phyl Evol 7:17–32

He GH, Meng RH, Newman M, Gao G, Pittman RN, Prakash CS (2003) Microsatellites as DNA markers in cultivated peanut (*Arachis hypogaea* L.). BMC Plant Biol 3:3

Hedrick PW, Parker KM, Lee RN (2001) Using microsatellite and MHC variation to identify species, ESUs, and Mus in endangered Sonoran topminnow. Mol Ecol 10:1399–1412

Innan H, Terauchi R, Miyashita NT (1997) Microsatellite polymorphism in natural populations of the wild plant *Arabidopsis thaliana*. Genetics 146:1441–1452

Jacob HJ, Lindpaintner K, Lincoln SE, Kusumi K, Bunker RK, Mao YP, Ganten D, Dzau VJ, Lander ES (1991) Genetic mapping of a gene causing hypertensive rat. Cell 67:213–224

Jarne P, Lagoda PJL (1996) Microsatellites, from molecules to populations and back. TREE 11:424–429

Kandpal RP, Kandpal G, Weissman SM (1994) Construction of libraries enriched for sequence repeats and jumping clones, and hybridization selection for region-specific markers. PNAS 91:88–92

Karagyozov L, Kalcheva ID, Chapman VM (1993) Construction of random small-insert genomic libraries highly enriched for simple sequence repeats. Nucleic Acids Res 21:3911–3912

Khamis F, Karam N, Guglielmino CR, Ekesi S, Masiga D, De Meyer M, Kenya EU, Malacrida AR (2008) Isolation and characterization of microsatellite markers in the newly discovered invasive fruit fly pest in Africa, *Bactrocera invadens* (Diptera: Tephritidae). Mol Ecol Res 8:1509–1511

Khamis FM, Karam N, Ekesi S, De Meyer M, Bonomi A, Gomulski LM, Scolari F, Gabrieli P, Siciliano P, Masiga D, Kenya EU, Gasperi G, Malacrida AR, Guglielmino CR (2009) Uncovering the tracks of a recent and rapid invasion: the case of the fruit fly pest *Bactrocera invadens* (Diptera: Tephritidae) in Africa. Mol Ecol 18:4798–4810

Khamis FM, Masiga DK, Mohamed SA, Salifu D, De Meyer M, Ekesi S (2012) Taxonomic identity of the invasive fruit fly pest, *Bactrocera invadens*: concordance in morphometry and DNA barcoding. PLoS One 7:e44862

Kinyanjui G, Khamis FM, Mohamed S, Ombura LO, Warigia M, Ekesi S (2016) Identification of aphid (Hemiptera: Aphididae) species of economic importance in Kenya using DNA barcodes. Bull Entomol Res 106:63–72. doi:10.1017/S0007485315000796

Kofler R, Schlotterer C, Lelley T (2007) SciRoKo: a new tool for whole genome microsatellite search and investigation. Biogeosciences 23:1683–1685

Kumar P, Gupta VK, Misra AK, Modi DR, Pandey BK (2009) Potential of molecular markers in plant biotechnology. Plant Omics J 2:141–162

Lee JM, Nahm SH, Kim YM, Kim BD (2004) Characterization and molecular genetic mapping of microsatellite loci in pepper. Theor Appl Genet 108:619–627

Levinson G, Gutman GA (1987) Slipped-strand mispairing: a major mechanism for DNA sequence evolution. Mol Biol Evol 4:203–221

Litt M, Luty JA (1989) A hypervariable microsatellite revealed by *in vitro* amplification of a dinucleotide repeat within the cardiac muscle actin gene. Am J of Hum Gen 44:397–401

Loxdale HD, Lushai G (1998) Molecular markers in entomology. Bull Entomol Res 88:557–600

Lux SA, Copeland RS, White IM, Manrakhan A, Billah MK (2003) A new invasive fruit fly species from the *Bactrocera dorsalis* (Hendel) group detected in East Africa. Insect Sci Appl 23:355–361

Malacrida AR, Guglielmino CR, D'Adamo P, Torti C, Marinoni F, Gasperi G (1996) Allozyme divergence and phylogenetic relationships among species of tephritid flies. Heredity 76:592–602

Malacrida AR, Marinoni F, Torti C, Gomulski LM, Sebastiani F, Bonvicini C, Gasperi G, Guglielmino CR (1998) Genetic aspects of the worldwide colonization process of *Ceratitis capitata*. J Hered 89:501–507

Malacrida AR, Gomulski LM, Bonizzoni M, Bertin S, Gasperi G, Guglielmino CR (2007) Globalization and fruit fly invasion and expansion: the medfly paradigm. Genetica. doi:10.1007/s10709-006-9117-2

Manni M, Lima KM, Guglielmino CR, Lanzavecchia SB, Juri M, Vera T, Cladera J, Scolari F, Gomulski L, Bonizzoni M, Gasperi G, Silva JG, Malacrida AR (2015) Relevant genetic differentiation among Brazilian populations of *Anastrepha fraterculus* (Diptera, Tephritidae). ZooKeys 540:157–173

McPheron BA (2000) Population genetics and cryptic species. In: Tan KH (ed) Area-wide control of fruit flies and other insect pests. Penerbit Universiti Sains Malaysia, Penang, pp 483–490

Meixner MD, McPheron BA, Silva JG, Gasparich GE, Sheppard WS (2002) The Mediterranean fruit fly in California: evidence for multiple introductions and persistent populations based on microsatellite and mitochondrial DNA variability. Mol Ecol 11:891–899

Morrow J, Scott L, Congdon B, Yeates D, Frommer M, Sved J (2000) Close genetic similarity between two sympatric species of tephritid fruit fly reproductively isolated by mating time. Evolution 54:899–910

Mwatawala MW, White IM, Maerere AP, Senkondo FJ, De Meyer M (2004) A new invasive *Bactrocera* species (Diptera: Tephritidae) in Tanzania. Afri Entomol 12:154–156

Mwatawala MW, De Meyer M, Makundi RH, Maerere AP (2006) Biodiversity of fruit flies (Diptera, Tephritidae) at orchards in different agro-ecological zones of the Morogoro region, Tanzania. Fruits 61:321–332

Nardi F, Carapelli A, Dallai R, Roderick GK, Frati F (2005) Population structure and colonization history of the olive fly, *Bactrocera oleae* (Diptera, Tephritidae). Mol Ecol 14:2729–2738

Ostrander EA, Jong PM, Rine J, Duyk G (1992) Construction of small-insert genomic DNA libraries highly enriched for microsatellite repeat sequences. PNAS 89:3419–3423

Paetkau D (1999) Microsatellites obtained using strand extension: an enrichment protocol. Biotechniques 26:690–697

Pandolfo M (1992) A rapid method to isolate (GT)n repeats from yeast artificial chromosomes. Nucleic Acids Res 20:1154

Piry S, Alapetite A, Cornuet JM, Paetkau D, Baudouin L, Estoup A (2004) Geneclass 2: a software for genetic assignment and first s – generation migrant detection. J Hered 95:536–539

Pritchard JK, Stephens M, Donnelly P (2000) Inference of population structure using multilocus genotypic data. Genetics 155:945–959

Rassmann K, Schlotterer C, Tautz D (1991) Isolation of simple sequence loci for use in polymerase chain reaction-based DNA fingerprinting. Electrophoresis 12:113–118

Roderick G, Navajas M (2003) Genotypes in novel environments: genetics and evolution in biological control. Nat Rev Gen 4:889–899

Rudd S (2003) Expressed sequence tags: alternative or complement to whole genome sequences? Trends Plant Sci 8:321–329

Schlötterer C, Tautz D (1992) Slippage synthesis of simple sequence DNA. Nuc Acids Res 20:211–215

Schuler GD, Boguski MS, Stewart EA, Stein LD, Gyapay G, Rice K, White RE, Rodriguez-Tomé P, Aggarwal A, Bajorek E, Bentolila S, Birren BB, Butler A, Castle AB, Chiannilkulchai N, Chu A, Clee C, Cowles S, Day PJ, Dibling T, Drouot N, Dunham I, Duprat S, East C, Edwards C, Fan JB, Fang N, Fizames C, Garrett C, Green L, Hadley D, Harris M, Harrison P, Brady S, Hicks A, Holloway E, Hui L, Hussain S, Louis-Dit-Sully C, Ma J, MacGilvery A, Mader C, Maratukulam A, Matise TC, McKusick KB, Morissette J, Mungall A, Muselet D, Nusbaum HC, Page DC, Peck A, Perkins S, Piercy M, Qin F, Quackenbush J, Ranby S, Reif T, Rozen S, Sanders C, She X, Silva J, Slonim DK, Soderlund C, Sun WL, Tabar P, Thangarajah T, Vega-Czarny N, Vollrath D, Voyticky S, Wilmer T, Wu X, Adams MD, Auffray C, Walter NA, Brandon R, Dehejia A, Goodfellow PN, Houlgatte R, Hudson JR Jr, Ide SE, Iorio KR, Lee WY, Seki N, Nagase T, Ishikawa K, Nomura N, Phillips C, Polymeropoulos MH, Sandusky M, Schmitt K, Berry R, Swanson K, Torres R, Venter JC, Sikela JM, Beckmann JS, Weissenbach J, Myers RM, Cox DR, James MR, Bentley D, Deloukas P, Lander ES, Hudson TJ (1996) A gene map of the human genome. Science 274:540–546

Schutze MK, Aketarawong N, Amornsak W, Armstrong KF, Augustinos AA, Barr N, Bo W, Bourtzis K, Boykin LM, Cáceres C, Cameron SL, Chapman TA, Chinvinijku S, Chomi˘c A, De Meyer M, Drosopoulou E, Englezou A, Ekesi S, Gariou- Papalexiou A, Geib SM, Hailstones D, Hasanuzzaman M, Haymer D, Hee AKW, Hendrichs J, Jessup A, Ji Q, Khamis FM, Krosch MN, Leblanc L, Mahmood K, Malacrida AR, Mavragani-Tsipidou P, Mwatawala M, Nishida R, Ono H, Reyes J, Rubinoff D, San Jose M, Shelly TE, Srikachar S, Tan KH, Thanaphum S, Haq I, Vijaysegaran S, Wee SL, Yesmin F, Zacharopoulou A, Clarke AR (2014) Snonymization of key pest species within the *Bactrocera dorsalis* species complex (Diptera: Tephritidae): Taxonomic changes based on a review of 20 years of integrative morphological, molecular, cytogenetic, behavioural and chemoecological data. System Entomol. doi:10.1111/syen.12113

Senan S, Kizhakayil D, Sasikumar B, Sheeja TE (2014) Methods for development of microsatellite markers: an overview. Not Sci Biol 6:1–13

Shearman DCA, Gilchrist AS, Crisafulli D, Graham G, Lange C, Frommer M (2006) Microsatellite markers for the pest fruit fly, *Bactrocera papayae* (Diptera: Tephritidae) and other *Bactrocera* species. Mol Ecol Notes 6:4–7

Sonvinco A, Manso F, Quesada-Allue LA (1996) Discrimination between the immature stages of *Ceratitis capitata* and *Anastrepha fraterculus* (Diptera: Tephritidae) populations by random amplified polymorphic DNA polymerase chain reaction. J Econ Entomol 89:1208–1212

Thiel T, Michalek W, Varshney RK, Graner A (2003) Exploiting EST databases for the development and characterization of gene-derived SSR-markers in barley (*Hordeum vulgare* L.). Theor Appl Genet 106:411–422

Wang R, Zheng L, Touré YT, Dandekar T, Kafatos F (2001) When genetic distance matters: measuring genetic differentiation at microsatellite loci in whole-genome scans of recent and incipient mosquito species. PNAS 98:10769–10774

Weber JL, May PE (1989) Abundant class of human DNA polymorphism which can be typed using polymerase chain reaction. Am J Hum Gen 44:388–396

White IM, Elson-Harris MM (1992) Fruit flies of economic significance: their identification and bionomics. C.A.B International and Australian Centre for International Agricultural Research, Canberra, p 601

White G, Powell W (1997) Isolation and characterization of microsatellite loci in *Swietenia humilis* (Meliaceae): an endangered tropical hardwood species. Mol Ecol 6:851–860

Wu Y, Li Z, Wu J (2009) Polymorphic microsatellite markers in the melon fruit fly, *Bactrocera cucurbitae* (Coquillett) (Diptera: Tephritidae). Mol Ecol Resour 9:1404–1406

Zane L, Bargelloni L, Patarnello T (2002) Strategies for microsatellite isolation: a review. Mol Ecol 11:1–16

Fathiya M. Khamis Fathiya works in the African fruit fly programme and the Arthropod Pathology Unit at the International Centre of Insect Physiology and Ecology (*icipe*), Nairobi, Kenya and she is co-ordinator of the *icipe* Biovision Foundation (Switzerland). Her main focus is to employ molecular tools for the identification and delineation of species, inference of genetic variability and population genetic structures of pests as a prelude for the development of effective management strategies. She has worked on tephritid fruit flies and other pests of economic importance, both invasive and non-invasive. She has authored or co-authored over 15 peer-reviewed publications in this field.

Anna R. Malacrida Anna is Professor of Insect Biotechnology at the University of Pavia (Italy). She leads an established research group working on the invasion genomics of insects of economic and sanitary importance with the scope of monitoring and controlling their populations. She has experience in the functional analysis of genomic and transcriptomic data to develop tools to interfere with insect reproduction and their capacity to recognize hosts. Her research interests include dipteran insect species of economic importance such as Ceratitis species and Bactrocera species as well as blood-feeding species such as mosquitoes and tsetse flies species. She has co-ordinated the Pavia group within numerous National and International projects funded by the FAO/IAEA, EU, Ministero della Salute and NIH. She is a member of the EU-INFRAVEC project that created a new European Infrastructure for studying the genomics and biology of the tiger mosquito, and to translate research into effective control measures. She has a grant from the Ministero della Salute for the control of Aedes albopictus. She participated in the tsetse fly genome sequencing project together with the University of Yale and has WHO and NIH grants for the control of trypanosomiasis in Africa. She is a member of the EFSA panel for GMO risk assessment and has been a consultant for the IAEA, the Fogarty Foundation and the EU. She is member of the Steering Committee of the International Symposia on fruit flies of economic importance.

Chapter 5
Fruit Fly Species Composition, Distribution and Host Plants with Emphasis on Mango-Infesting Species

Ivan Rwomushana and Chrysantus M. Tanga

Abstract Mango is the most widely cultivated fruit tree in tropical and sub-tropical Africa. However, the sustainability of this lucrative business is threatened by infestations of fruit flies (Diptera: Tephritidae) that annually inflict heavy economic losses on the industry. The nutritional quality of different fruit species can influence the survival and fecundity of adult fruit flies. This host-insect interaction determines the species composition, distribution and abundance of the major frugivorous tephritids. The economic impact of fruit fly pest species includes direct yield losses and the loss of export markets due to quarantine restrictions implemented to prevent the entry and establishment of exotic fruit fly species in importing countries. The economically important tephritid fruit flies attacking mango in Africa can be divided into two major categories based primarily on their origin, i.e., invasive (*Bactrocera dorsalis*, *Bactrocera zonata* and *Zeugodacus cucurbitae*) and indigenous species (*Ceratitis anonae*, *Ceratitis capitata*, *Ceratitis catoirii*, *Ceratitis cosyra*, *Ceratitis ditissima*, *Ceratitis fasciventris*, *Ceratitis quinaria*, *Ceratitis rosa* [recent taxonomic advances have separated *C. rosa* into two species; *C. rosa* and *C. quilicii*], *Ceratitis silvestrii*, *Dacus ciliatus* and some unverified records of *Ceratitis punctata* and *Dacus bivittatus*). These species are known to have a wide host range and distribution across Africa. Their distribution is also influenced by competitive interactions between native and indigenous species. The host plant status and distribution of fruit fly species is an evolving phenomenon largely due to new invasions, misidentification and identification of hitherto unknown species. For this reason this review provides the current situation but should be updated on a regular basis.

Keywords Fruit flies • Mango • Invasive species • Indigenous species • Host plant relationships • Competitive interactions

---

I. Rwomushana (✉) • C.M. Tanga
International Centre of Insect Physiology & Ecology (*icipe*),
PO Box 30772-00100, Nairobi, Kenya
e-mail: irwomushana@icipe.org

© Springer International Publishing Switzerland 2016
S. Ekesi et al. (eds.), *Fruit Fly Research and Development in Africa - Towards a Sustainable Management Strategy to Improve Horticulture*,
DOI 10.1007/978-3-319-43226-7_5

1 Introduction

Mango (*Mangifera indica* L.) is the most widely cultivated fruit tree in the Sahel and one of the most important tree crops in tropical and sub-tropical Africa. West Africa alone produces 1.4 million tonnes of mangoes per year – the 7th largest producer in the world. Although widely grown on the continent, mango is not indigenous to Africa, but native to South-East Asia, from where it was introduced to all other tropical regions. According to the FAO-Intergovernmental Sub-Group on Tropical Fruits, mango is one of the four major high-value commodities and ranks amongst the most internationally traded tropical fruit. FAO estimated mango production in 2013 to be around 42.7 million tons, which accounts for nearly 35% of the world's tropical fruit production (http://www.fao.org/). Mango world imports were forecast to increase by 1.4% annually until 2014, 9% of which would be obtained from Africa (accounting for 2.6 million tons). In Africa over 80% of the produce comes from smallholders who produce for both local and export markets (Jayne et al. 2001). This provides the much-needed cash income to improve the households' food and nutritional security as well as their overall livelihoods. Mango is a highly prized exotic fruit on the European market and one of the most important fruit crops grown in tropical and sub-tropical regions (Nakasone and Paull 1998).

However, several constraints hinder the sector from realizing its full potential, key amongst them being fruit flies (Ekesi et al. 2016). The key insect pests that prevent increased and sustainable production are tephritid fruit flies (Diptera: Tephritidae) (Norrbom et al. 1999). Tephritid fruit flies have been recognized as one of the most economically important groups of insects that pose a serious threat to fruit production in Africa (White and Elson-Harris 1992; Ekesi and Billah 2006; De Meyer et al. 2012). Fruit fly infestation leads to heavy losses in yield and quality of fresh fruits. In Africa, between 30 and 40% of the mangoes produced annually are lost to fruit flies (Ekesi et al. 2006; Goergen et al. 2011). Economically important tephritid fruit flies in Africa are distributed within three genera: *Bactrocera* Macquart, *Ceratitis* MacLeay and *Dacus* Fabricius (White and Elson-Harris 1992). Historically yield losses in mango were due to native fruit flies and estimated to range between 30 and 70% depending on the locality, season and variety (Lux et al. 2003). However, in 2003, a new species, *Bactrocera dorsalis* (Hendel) invaded Africa from the Indian subcontinent (Lux et al. 2003; Mwatawala et al. 2004; Drew et al. 2005). Within only a few years the species had spread across Africa and was detected in more than 30 countries (West, Central, Eastern and Southern Africa) (Drew et al. 2005; Vayssières et al. 2005, 2014; Ekesi et al. 2006; Mwatawala et al. 2004; De Meyer et al. 2007; Correia et al. 2008; Rwomushana et al. 2008; Goergen et al. 2011; Hussain et al. 2015; Isabirye et al. 2015). Mango is considered a primary host of *B. dorsalis* (Drew et al. 2005; Ekesi and Billah 2006) and direct damage has been reported to range between 30 and 80% depending on the cultivar, locality and season (Ekesi et al. 2006; Rwomushana et al. 2008; Vayssières et al. 2009). In addition to *B. dorsalis*, other *Bactrocera* species of Asian origin such as *Zeugodacus cucurbitate* (Coquillett), *Bactrocera zonata* Saunders and *Bactrocera latifrons*

(Hendel) have also been introduced in to mainland Africa, and the islands of the Indian Ocean, thereby aggravating the economic significance of tephritid fruit flies in African horticulture systems (De Meyer et al. 2007; Mwatawala et al. 2004, 2010; Shehata et al. 2008; Elnagar et al. 2010).

In addition to direct losses, indirect losses attributed to quarantine restriction on fruit fly-infested fruits have been enormous and limit export to large lucrative export markets in Europe, the Middle East, Japan and USA, where the insects are quarantine pests. For example, the importation of fruit species that are hosts of *B. dorsalis*, such as mango, from Kenya, Tanzania and Uganda is currently banned in the Seychelles, Mauritius and South Africa. Trade of horticultural produce between Africa and the USA has been severely restricted by a federal order from the USA banning importation of several cultivated fruit species from African countries where *B. dorsalis* has been reported (USDA-APHIS 2008; Ekesi et al. 2016). Interceptions and rejection of African mangoes in the European Union (EU) due to fruit flies have been on the increase since the arrival of *B. dorsalis* (Guichard 2009) with 21 rejections in 2008 increasing to 38 by August 2009. Interceptions have been reported from countries such as Burkina Faso, Côte d'Ivoire, Gambia, Ghana, Guinea, Mali, Senegal, Cameroon, Central Africa Republic, Kenya and Egypt. The direct and indirect damage caused by *B. dorsalis* and other tephritid pests continues to have wide reaching socio-economic implications for millions of rural and urban populations involved in the mango value chain across Africa. This has been further compounded by the introduction of uniform and strict quarantine regulations and maximum residue level (MRL) by the EU which now jeopardizes export of mangoes from Africa estimated at 35,000–40,000 tons annually and worth over US$ 42 million (Lux et al. 2003). This value has gradually been eroded as a result of import bans by several countries due to fruit flies (Ekesi 2010).

2 Relationships Between Host Fruits and Fruit Flies

The relationship between host fruits and fruit flies can strongly influence their species composition and distribution. Usually tephritid fruit flies attack the mature fruit of their host plants that are still on the tree, although in some cases immature fruit are know to be attacked as well. Female fruit flies drill into the fruit using their ovipositor and lay their eggs under the skin. This behaviour causes blemishes on the fruit; the presence of such blemishes means that these fruit do not meet the stringent requirements of the export market. The larvae develop inside the fruit, feed on the tissue and then exit from the fruit completing their developmental cycle in the soil. Fruit fly damage may cause immature ripening and abortion in a wide variety of fruiting species (Stephenson 1981; Sallabanks and Courtney 1992). In many cases mature fruit that are harvested contain developing larvae. Fruits have many important ecological attributes that affect the insects that live, feed, mate, oviposit, grow, rest, and hide on them; these attributes determine whether they are suitable hosts for particular fruit fly species (Fletcher 1987; Robinson and Hooper 1989). Larvae cannot change host plant and therefore depend on both the efficiency of female host-choice and the nutritional quality of the fruit for survival. Fruits have a strong influence on fruit flies at this stage in their life cycle when the quality of nutrition they provide can affect the longevity and fecundity of subsequent adults (Bateman 1972). It is therefore important for any fruit fly species to be able to locate suitable hosts to ensure successful development of their progeny. Most fruit fly species in Africa are highly polyphagous and it is not surprising that mango is one of the fruits most commonly attacked by these pests. Although mango appears to be a preferred host for several fruit fly species on the continent, several other host fruit also act as refugia, often becoming important sources of inoculum at the onset of the mango season.

3   Species Composition of Major Mango-Infesting Fruit Flies in Africa

Globally, at least 5000 tephritid species in 500 genera have been recorded to date (Norrbom et al. 1999). The global species database lists 4710 tephritid fruit fly species (www.globalspecies.org), of which 1400 species are known to develop in fruits. Out of these, about 250 species are pests, inflicting severe damage to fruits of economic importance (White and Elson-Harris 1992; Thompson 1998). The number of recognized tephritid species is constantly evolving as a result of new descriptions, recategorization and genetic analysis. White and Elson-Harris (1992) described 915 fruit fly species in Africa comprising 148 genera, out of which 299 species developed in either wild or cultivated fruit. They belong, mainly, to four genera: *Bactrocera* (562 species), *Ceratitis* (92), *Dacus* (300) and *Trirhithrum* Bezzi (49), although the latter is not economically important. In recent years, the number of species known on the continent has increased largely due to new invasions and identification of hitherto unknown species, although they largely still fall within these four genera.

Most of the fruit fly species in Africa are highly polyphagous with their host ranges overlapping to a varying extent. Mango is one of the most commonly infested fruits that is attacked by a complex of fruit fly species. Several authors have clearly documented that mango is an important host. Economically important tephritid fruit flies attacking mango in Africa can be divided into two categories: invasive species such as the oriental fruit fly, *Bactrocera dorsalis* (Hendel); the melon fruit fly, *Zeugodacus cucurbitae* (Coquillett); and the peach fruit fly, *Bactrocera zonata* (Saunders); and indigenous species such as *Ceratitis anonae* (Graham); the Mediterranean fruit fly (medfly,) *C. capitata* (Wiedemann); the Mascarenes fruit fly, *Ceratitis catoirii* (Guérin-Méneville); the mango fruit fly, *Ceratitis cosyra* (Walker); *Ceratitis ditissima* (Munro); *Ceratitis fasciventris* (Bezzi); the Natal fruit fly, *Ceratitis rosa* (Karsch); *Ceratitis silvestrii* (Bezzi); the five-spotted fruit fly, *Ceratitis quinaria* (Bezzi); the cacao fruit fly, *Ceratitis punctata* (Wiedemann);

Ceratitis flexuosa (Walker); *Dacus bivittatus* (Bigot) and the lesser pumpkin fly, *Dacus* (*Didacus*) *ciliatus* Loew (White and Elson-Harris 1992; Mwatawala et al. 2004, 2009a; Vayssières et al. 2005, 2007; Rwomushana et al. 2008; Isabirye et al. 2016; Goergen et al. 2011; Nboyine et al. 2012; De Meyer et al. 2015). Of these, *B. dorsalis*, wherever it occurs on the continent is ranked as the most important pest of mango followed by *C. cosyra*. Other species of fruit flies are ranked as moderate, and are localized in their distribution with varying degree of infestation on mango depending on the agroecology.

4 Relative Abundance and Seasonal Phenology of Mango-Infesting Fruit Flies in Africa

There have been several studies in Africa examining the relative abundance and seasonal phenology of mango-infesting fruit fly species (Copeland et al. 2006; Mwatawala et al. 2006, 2009b; Virgilio et al. 2011; Ndiaye et al. 2012; N'depo et al. 2013; Rubabura et al. 2015; Vayssières et al. 2014). In general, there is a strong correlation between the availability of fruiting host plants and fruit fly populations. The relative abundance and seasonal phenology of fruit flies is highly dependent on the availability of host plants, prevailing weather conditions and the presence or absence of natural enemies that limit pest population growth (Mohamed et al. 2010). There is a distinct pattern in fruit fly population dynamics with numbers reaching a peak at fruit maturity and ripening stage and declining with fruit harvest. Temperature, relative humidity and rainfall are the major climatic factors influencing fruit fly populations.

In West Africa, fruit flies start appearing in orchards during the dry season (between September and January) reaching a peak in February or March and then a second peak in April or sometimes in June before decreasing in July (Vayssières et al. 2014). Furthermore, as precipitation increases from 50 mm in April to a peak of approximately 240 mm in September, there is a consistent increase in relative humidity (45.5–59%) and a decrease in air temperature (34.9–29.8 °C). These conditions are very conducive to fruit fly population growth. In Ghana, populations of *C. cosyra* predominate during the period between January and April (Badii et al. 2015). The dominance of this fruit fly species at that time coincides with the fruiting of both early- and late-maturing mango varieties. Populations of *C. anonae* begin to build up in the middle of May and reach a peak in June before declining in August. In contrast, *C. fasciventris* and *C. ditissima* appear from late May to early June. It is also noteworthy that mango flowers can be attractive to *Ceratitis* adults, as already recorded for other fly species (Aluja and Mangan 2008) which influences pest abundance during the flowering period.

In West Africa, populations of *B. dorsalis* fluctuate in a similar fashion to native fruit fly species. In the dry season between November and January their populations decline considerably but at the beginning of the rainy season (March-April), populations rapidly build up to reach a peak in April and then a second peak in May or June. In the Guinea Savanna zone of Ghana the period between May and June when *B. dorsalis* populations are peaking coincides with maturation and harvesting of late-maturing mango cultivars such as Keitt and Kent (Badii et al. 2015). Thereafter the populations drop steadily to their lowest levels between October and December. The same trends have been reported in several West African countries (Vayssières et al. 2005, 2011, 2014; Hala et al. 2006; Ndiaye et al 2012; Nboyine et al. 2013; N'depo et al. 2013).

In Eastern Africa, as exemplified in a study in Lake Victoria Crescent, Uganda, *B. dorsalis* was present year-round and all stages of mango fruit development were susceptible to attack (Mayamba et al. 2014). Each year infestations peaked between June and July and again between January and February. Trap catches were larger during the major fruiting season than the minor fruiting season. The highest numbers of *B. dorsalis* were collected when mango was at the physiologically mature or ripening stage (Mayamba et al. 2014).

In Kenya, studies on the seasonal and annual population dynamics of *B. dorsalis* also showed that peak populations coincided with mango fruiting and maturity in the field (Rwomushana 2008). The availability of mango fruits was the most important factor governing population increase in this species. More *B. dorsalis* were captured during the season and throughout the year than any other fruit fly species; abumdance of *B. dorsalis* always significantly exceeded the abundance of the native fruit fly, *C. cosyra*, from all trap collections (Ekesi et al. 2006).

In Tanzania, the general trend of the population dynamics of fruit flies showed that *B. dorsalis* peaks at the end of January and mid February while the lowest abundance was observed between September and October (Mwatawala et al. 2006). This trend was influenced by weather as well as the phenological stage of the fruit (Mwatawala et al. 2006). *Ceratitis rosa* populations peaked between January and March while *C. cosyra* populations had the inverse pattern with a peak in abundance in November, corresponding with the early-mango season, and a second peak between August and September (Mwatawala et al. 2006, 2009b).

In Sudan, which experiences a winter period, fruit fly populations build up gradually from May with two peaks during the humid months of August and November. Thereafter, the populations decline from December until March (Fadlelmula and Ali 2014). The highest populations of *B. dorsalis* occur between July and August, which is associated with late-maturing varieties of mango, and the lowest populations occur in March. The highest population of *C. cosyra* were recorded in August while populations of *C. capitata* increased with the onset of rainfall during autumn, peaking in November. In the Blue Nile State, the seasonal phenology of *B. dorsalis* on mango at Damazine and Rosaries orchards were almost the same. The number of adult males captured was very low during the dry period (March – May) when no rainfall was recorded and temperatures were high (40–43 °C), but increased steadily from the start of the rains in May. Population peaks were observed between June and July and again between December and January, depending on temperature, rainfall and availability of the mango fruits (Fadlelmula and Ali 2014).

In Northern Africa, e.g. Egypt, populations of *C. capitata* occur throughout the year with population peaks reported between September and November and then again between May and June which coincides with ripening of mango, apple and peaches (Hashem et al. 2001). The lowest abundance of *C. capitata* was recorded in winter (between April and July) probably due to low temperatures. In Upper Egypt, Hashem et al. (1986) reported high *C. capitata* populations between August and December that had gradually been building up between January and July. In navel orange orchards, three peaks of *B. zonata* were recorded each year; the highest peak corresponded with the ripening of fruits in November while the other peaks were in April and May. *Bactrocera zonata* populations were completely absent in December and January.

In South Africa, the relative abundance and seasonal phenology of the three main species, *C. capitata*, *C. cosyra* and *C. rosa* were similar, with populations of all three species increasing during late spring (September and October), reaching a peak in the hot summer months (January to March) and declining into the winter (June to August) (De Villiers et al. 2013). Both *C. capitata* and *C. rosa*, population fluctuations varied significantly depending on whether samples were taken in home gardens or commercial orchards (De Villiers et al. 2013). Population peaks occurred earlier in the year in the home gardens (between January and March) compared with commercial orchards where populations peaked between March and May. The peak population levels were also higher for both *C. capitata* and *C. rosa* in home gardens than commercial orchards. Fruit fly populations, in particular *C. rosa*, were sustained in home gardens throughout the year, although during the winter months (June -August) population levels of both species was low.

5 Distribution of Mango-Infesting Fruit Flies in Africa

Regional integration between many African countries allows for trade and free movement of fruits; coupled with the many porous borders between countries, the continent is highly vulnerable to introduction of alien fruit fly species that attack mango. Both invasive and native fruit fly species have been reported to occur all year round, largely due to their ability to infest a wide range of wild host plants and overcome the challenges of geographical barriers (De Meyer et al. 2007; Lux et al. 2003; De Villiers et al. 2013). Here we describe the key fruit fly species that have been reared from mango and their geographic distribution on the continent.

5.1 **Bactrocera dorsalis**

Bactrocera invadens Drew, Tsuruta & White or the 'African Invader Fly', was the name given to the tephritid fruit fly that was introduced to East Africa from Sri Lanka and subsequently invaded the whole of Sub-Saharan Africa. With recent integrative taxonomic studies *B. invadens* was found to exhibit the same biological characteristics as *B. dorsalis* which is a complex of species known to cause extensive damage to fruits globally (Drew 1994). Consequently, *B. invadens* was synonymized with *B. dorsalis* in 2015 (Schutze et al. 2015). The pest arguably ranks first amongst all fruit fly species on the African continent, both native and exotic, and is responsible for causing the most extensive economic losses to horticultural crops. Losses sometimes exceed 80 % resulting in widespread trade restrictions and significant negative economic and social impacts to farming communities. Since its first report in Kenya in 2003 (Lux et al. 2003), *B. dorsalis* has spread rapidly and is now present in more than 30 countries beyond its native range.

In Africa it has been recorded from Angola, Benin, Bostwana, Burkina Faso, Burundi, Cameroon, Central African Republic, Cape Verde, Chad, Comoros Archipelago, Côte d'Ivoire, Mayotte, Republic of the Congo, Democratic Republic of Congo, Ethiopia, Eritrea, Equatorial Guinea, Gabon, Gambia, Ghana, Guinea, Guinea Bissau, Kenya, Liberia, Mali, Mauritania, Mozambique, Namibia, Niger, Nigeria, Senegal, Sierra Leone, South Africa, Sudan, Swaziland,Tanzania, Togo, Uganda, Zimbabwe and Zambia (Drew et al. 2005; Vayssières et al. 2005; Mwatawala et al. 2006; Correia et al. 2008; Rwomushana et al. 2008; Goergen et al. 2011; Manrakhan et al. 2011; Virgilio et al. 2011; De Meyer et al. 2008, 2012; Ibrahim Ali et al. 2013; Aidoo et al. 2014; Fekadu and Zenebe 2015; Hussain et al. 2015; Isabirye et al. 2015; http://www.africamuseum.be/fruitfly/AfroAsia.htm). It was discovered in Sri Lanka soon after it was reported from Africa (Drew et al. 2005). For global distribution and predictions see De Meyer et al. (2010).

5.2 Bactrocera zonata

Bactrocera zonata is native to South and Southeast Asia. In Africa, it occurs in northern Africa (Egypt and Libya). Recently it has been reported from several regions in Sudan, suggesting a southward spread and potential risk of invasion for the Sub-Saharan region (De Meyer et al. 2007; Shehata et al. 2008; Elnagar et al. 2010; El-Samea and Fetoh 2006). It is also become established on the Indian Ocean islands of Mauritius and La Réunion (Quilici et al. 2005).

5.3 Ceratitis anonae

Ceratitis anonae is found in Cameroon, Central African Republic, Côte d'Ivoire, Democratic Republic of Congo (Brazzaville), Gabon, Ghana, Equatorial Guinea, Kenya, São Tomé and Principé, Guinea (Conakry), Mali, Nigeria, Togo, Tanzania and Uganda (White and Elson-Harris 1992; Copeland et al. 2006).

5.4 Ceratitis capitata

Ceratitis capitata is the most widely distributed indigenous fruit fly species. In Africa, it is recorded from Algeria, Angola, Benin, Botswana, Burkina Faso, Burundi, Cameroon, Congo, Côte d'Ivoire, Democratic Republic of Congo, Egypt, Ethiopia, Gabon, Ghana, Guinea, Kenya, Liberia, Libya, Malawi, Morocco, Mozambique, Niger, Nigeria, Senegal, South Africa, Sudan, Tanzania, Togo, Tunisia, Uganda, São Tomé and Principé, Mauritius, Sierra Leone, Seychelles, South Africa, La Réunion, and Zimbabwe (White and Elson-Harris 1992; De Villiers et al. 2013). For global distribution and predictions see De Meyer et al. (2008).

5.5 Ceratitis catoirii

This species has been reported in Mauritius, La Réunion and Seychelles (Duyck et al. 2004).

5.6 Ceratitis cosyra

This species is widespread in Africa and has been reported from Benin, Botswana, Central African Republic, Côte d'Ivoire, Democratic Republic of Congo, Guinea, Ghana, Kenya, Madagascar, Malawi, Mali, Mozambique, Namibia, Nigeria, Sierra Leone, South Africa, Sudan, Tanzania, Togo, Uganda, Zambia, and Zimbabwe (Javaid 1986; White and Elson-Harris 1992; De Meyer 1998; Copeland et al. 2006; De Villiers et al. 2013).

5.7 Ceratitis ditissima

This species is known to be localized mainly in West Africa, particularly Benin, Cameroon, Congo, Côte d'Ivoire, Ghana, Mali, Mozambique Nigeria, Uganda, and Zimbabwe (Vayssières et al. 2007; Foba et al. 2012; Aidoo et al. 2014)

5.8 Ceratitis fasciventris

Ceratitis fasciventris occurs in Côte d'Ivoire, Democratic Republic of Congo, Ethiopia, Ghana, Equatorial Guinea, Kenya, Mali, Nigeria, São Tomé and Principé, Tanzania and Uganda (White and Elson-Harris 1992; Copeland et al. 2006).

5.9 Ceratitis flexuosa

This species occurs in Angola, Cameroon, Congo (D.R), Côte d'Ivoire, Ghana, Guinea, Kenya, Niger, Nigeria, Tanzania, Togo, Uganda (URL: http://ZipcodeZoo.com/index.php/Ceratitis_flexuosa)

5.10 Ceratitis punctata

This species is found in Cameroon, Congo, Democratic Republic of Congo, Côte d'Ivoire, Guinea, Kenya, Rwanda, Senegal, South Africa Tanzania, Uganda, Zambia, and Zimbabwe (De Meyer 2000).

5.11 Ceratitis quinaria

Countries with established infestations of *C. quinaria* include Benin, Botswana, Burkina Faso,, Côte d'Ivoire, Guinea, Ghana, Namibia, Malawi, Mali, Senegal, South Africa, Sudan, Togo, Yemen and Zimbabwe (Hancock et al. 2001; White and Elson-Harris 1992; De Meyer 1998; De Meyer et al. 2002; Vayssières et al. 2005).

5.12 Ceratitis rosa

Ceratitis rosa is not highly invasive showing only limited expansion of its distribution beyond its historical native range, which includes Angola, Ethiopia, Democratic Republic of Congo, Kenya, Malawi, Mali, Mauritius, Mozambique, Nigeria, Islands of Mauritius and La Réunion, Rwanda, Seychelles, Republic of South Africa (KwaZulu Natal), Swaziland, Tanzania, Uganda, Zambia and Zimbabwe (White and Elson-Harris 1992; Copeland et al. 2006; De Villiers et al. 2013). No reliable records from West Africa have been found (De Meyer et al. 2015), although some authors have reported the pest in Côte d'Ivoire (N'depo et al. 2013).

However, recent integrative taxonomy approaches using larval and adult morphology, wing morphometrics, cuticular hydrocarbons, pheromones, microsatellites, developmental physiology, geographical distribution, behavioural and chemoecological data of *Ceratitis rosa* have revealed that this species is made up of two entities: 'R1', 'lowland' or 'hot rosa', and 'R2', 'highland' or 'cold rosa' (De Meyer et al. 2015) with varying distribution patterns. The new data led to the conclusion that these two types should be considered as two different species. Taxonomically, the type material of *C. rosa* belongs to the R1 type (De Meyer et al. 2015), and the R2 type is considered as a new species, which hereinafter is referred to as *Ceratitis quilicii* (De Meyer et al. in press). We should stress here that many publications in the last decades refer only to *C. rosa* and were largely unable to differentiate between the two types as different species although they could have likely been referring to *C. rosa*, *C. quilicii*, or a mixure of the two. Therefore, *C. quilicii* is only used in this chapter where there is a clear distinction between R1 and R2 types of *C. rosa* from published works.

The two species can occur sympatrically in some regions (Malawi, South Africa and Tanzania), but also show a disjunct distribution that appears to be correlated with temperature (Tanga et al. 2015). Only in the Cape and central parts of South Africa is *C. quilicii* alone present, as well as in the adventive populations on the Indian Ocean islands (Virgilio et al. 2013). Therefore, it is likely that the high altitude types were *C. quilicii* and low altitude types probably a mix of the two species. The current distribution of the R2 type or *C. quilicii* includes Botswana, Kenya, La Réunion, Malawi, Mauritius South Africa, Tanzania and Zimbabwe. However, the distribution range of *C. rosa* and *C. quilicii* remains non-exhaustive given that samples from many localities in the above listed countries have not been assigned (De Meyer et al. 2015).

5.13 Ceratitis silvestrii

This species has been reported attacking mango in Nigeria, Senegal, Mali, Burkina Faso and Niger (Vayssières et al. 2005)

5.14 Dacus bivittatus

Dacus bivittatus is known from Angola, Benin, Cameron, Congo, Côte d'Ivoire, Democratic Republic of Congo, Ethiopia, Gabon, Ghana, Guinea, Kenya, Madagascar, Malawi, Mozambique, Nigeria, Senegal, Sierra Leone, South Africa, Tanzania, Togo, Uganda, Zambia, and Zimbabwe (White and Elson-Harris 1992)

5.15 Dacus ciliatus

Dacus ciliatus is widely distributed in Africa occurring in Angola, Benin, Botswana, Burkina Faso, Cameroon, Chad, Democratic Republic of Congo, Côte d'Ivoire, Egypt, Ethiopia, Gabon, Ghana, Guinea, Kenya, Lesotho, Madagascar, Malawi, Mozambique, Namibia, Niger, Nigeria, Senegal, Sierra Leone, Somalia, South Africa, Sudan, Tanzania, Togo, Uganda, Zambia and Zimbabwe (White and Elson-Harris 1992).

5.16 Zeugodacus cucurbitae

Zeugodacus cucurbitae is an invasive pest species in Africa and has been recorded from Benin, Burkina Faso, Burundi, Cameroon, Côte d'Ivoire, Democratic Republic of Congo, Ethiopia, Gambia, Ghana, Kenya, Malawi, Mali, Mozambique Niger, Nigeria, Sierra Leone, Senegal, Sudan, Tanzania, Togo, and Uganda (White and Elson-Harris 1992; Vayssières and Carel 1999; De Meyer et al. 2007, 2015).

6 Relative Abundance of Mango-Infesting Fruit Flies in Africa

Generally, the diversity and species richness of a number of fruit fly species have been shown to increase with altitude while it is the reverse for other species. In addition to climate change, ecological gradients in host plants, parasitoids and predators, as well as physical gradients in temperature, rainfall, and humidity that are encountered along an altitudinal transect can have an impact on the density, diversity and life history of insects including fruit flies; this demands phenotypic flexibility and genotypic adaptability in many species (Bale et al. 2002; Hodkinson 2005; Vayssières et al. 2008). Below we describe the relative density of the major mango infesting-fruit flies in Africa.

6.1 Bactrocera dorsalis

Wherever it is commonly found, *B. dorsalis* is the most abundant pest on mango and in mango orchards generally. In Uganda, 98.9% of trap collections were of *B. dorsalis* (Isabirye et al. 2016) and 97% in Tanzania (Mwatawala et al. 2009b). In Kenya, 15.3 flies/kg and 87.9 flies/kg were recovered from mango fruits in the lowland and the highland respectively (Rwomushana et al. 2008). In Benin, 53.03% of adult fruit flies reared from mango were *B. dorsalis* (Vayssières et al. 2008) and 97.5% of trap catches in Guinea Bissau were of *B. dorsalis* (Ousmane et al. 2014). In West and Central Africa (WCA), the pest infestation index for mango was 13.7 flies/kg (Goergen et al. 2011). Across Africa, *B. dorsalis* has a particuliar affinity for tropical almond and 72 flies/kg have been reported in WCA (Goergen et al. 2011), 264.5 flies/kg in Kenya (Rwomushana et al. 2008) and, in Tanzania, 95.1% of fruit flies recovered from tropical almond were *B. dorsalis* (Mwatawala et al. 2009a).

Currently, *B. dorsalis* is continuing to spread, not only in latitude but also in altitude (Ekesi et al. 2006; Mwatawala et al. 2009a; Geurts et al. 2012). However, the continuous spread and colonization of higher altitudes seems to be limited by climatic conditions, host availability and suitability (Mwatawala et al. 2006; Geurts et al. 2012), and inter-specific competition with cold-tolerant species such as *C. rosa* (Mwatawala et al. 2006). *Bactrocera dorsalis* prefers areas at low altitudes with a warm and humid climate where its preferred cultivated host, mango, is present and where it achieves highest abundances (Rwomushana et al. 2008; De Meyer et al. 2010; Geurts et al. 2012; Vayssières et al. 2014).

6.2 **Bactrocera zonata**

Bactrocera zonata mainly attacks peach, guava and mango (White and Elson-Harris 1992; Allwood et al. 1999; Shehata et al. 2008). It is reported from some of the islands in the Indian Ocean (Mauritius and La Réunion) and is now widespread in northern Africa (Egypt and Libya). There is a potential risk of invasion for Sub-Saharan region (De Meyer et al. 2007). Ni et al. (2012) have predicted that, under current climatic conditions, *B. zonata* would be able to establish itself throughout much of the tropics and subtropics.

In Egypt *B. zonata* reaches significantly higher abundances than any of the other native fruit fly species (Elnagar et al. 2010). It appears to prefer warmer conditions and seems well adapted to hot climates. Since its introduction in Egypt, *B. zonata* has gradually become so widespread that it has surpassed *C. capitata* as the major fruit pest in Egypt. The abundance of *B. zonata* is significantly correlated with temperature and relative humidity and its population growth rate is higher than that of native species. The availability of suitable host plant species plays a role in the abundance of *B. zonata*. (El-Gendy and Nassar Atef 2014). In Mauritius, it mainly feeds on mango, guava, peach and jujube (Sookar et al. 2014). In Egypt, sour orange was the most susceptible host, followed by sweet orange and guava (Amro and Abdel-Galil 2008). At Fayoum governorate (Egypt), *B. zonata* infested 15.5 % of Navel orange, 10 % of grapefruit, 7 % of mandarin, 5.7 % of sour orange, 0.3 % of lemon and 0.6 % of Valencia orange (Saafan et al. 2005). Potato tubers collected from Giza governorate, Egypt, during 2004 were also found to be infested by *B. zonata* (El-Samea and Fetoh 2006).

6.3 **Ceratitis anonae**

Ceratitis anonae is widely distributed throughout western and central Africa and regularly occurs as far east as western Kenya (De Meyer 2001). Its absence from the central highlands of Kenya, an area containing several native and cultivated fruit species that it successfully exploits in western Kenya, suggests that *C. anonae* has become isolated from the common ancestor of all members of the FAR group sometime after the creation of the Gregory Rift. In Kenya, *C. anonae* was only successfully reared from fruit collected in the western highlands at altitudes between 1518 and 1630 m above sea level (a.s.l.), where it was sympatric with *C. fasciventris*

(Copeland et al. 2006). *Ceratitis anonae* was the principal pest of mango in West Africa prior to the arrival of *B. dorsalis* (Badii et al. 2015). In Benin, 0.21% of fruit flies successfully reared from mango were *C. anonae* (Vayssières et al. 2007) and in Uganda *C. anonae* has been successfully reared from mango, albeit only in low numbers (Isabirye et al. 2016). In Uganda, 0.3% of trap collections from mango orchards were of *C. anonae* (Isabirye et al. 2016) and as low as 0.07% in the Democratic Republic of Congo (Virgilio et al. 2011).

6.4 Ceratitis capitata

Ceratitis capitata, is native pest to sub-Saharan Africa. Because of its its ability to tolerate cooler climates better than most other species of tropical fruit flies, and its wide range of hosts, it is often ranked first amongst the economically important fruit fly species in more cooler climates on the continent. *Ceratitis capitata* has a widespread distribution in South Africa (De Villiers et al. 2013) and De Meyer (2001) has described its geographic distribution in Africa extensively, including modelling its potential geographic niches on the continent (De Meyer et al. 2008). In Kenya, the host plant relationships and the geographic distribution of the pest have also been described in detail (Copeland et al. 2002).

6.5 Ceratitis catoirii

Ceratitis catoirii is reported to be an endemic species to Mauritius and La Réunion, found mostly in moist regions at low altitude (Duyck et al. 2006a, b). There are few studies on the distribution of this species although it is believed that its limited host range probably plays a role in determining its distribution and abundance. In La Réunion, *C. catoirii* is very rare and did not seem to have a specific niche, either in terms of climate or in terms of host fruit species. Indeed, there is evidence to suggest that *C. catoirii* is approaching extinction in La Réunion (Duyck et al. 2008). In recent years, there have also been no records of *C. catoirii* from Mauritius either from fruit or from area-wide trapping, suggesting that it might have become extinct there already (Sookar et al. 2008).

6.6 Ceratitis cosyra

Ceratitis cosyra is a native African species mainly found on mango. The economic importance of *C. cosyra* has been growing since the more widespread commercialization of mango and the introduction of exotic mango varieties. Late-maturing varieties of mango reportedly suffer the most due to *C. cosyra* infestation. *Ceratitis cosyra* is widespread in sub-Saharan Africa, occurring in at least 27 countries. It is highly adaptable being recorded from near sea level to 2100 m a.s.l. This makes *C. cosyra* the most widely distributed fruit fly species, particularly on mango (Ekesi et al. 2006). However, studies done in Tanzania by Mwatawala et al. (2006) have shown that *C. cosyra* was the most abundant species at 781 m and 1105 m a.s.l, and has also been reported from mango and marula at Nguruman, Kenya which is 700 m a.s.l (Rwomushana et al. 2008). The abundance of *C. cosyra* is correlated with high temperature, low relative humidity and the presence of mango (Geurts et al. 2012). Despite its wide geographical distribution compared to other *Ceratitis* species, *C. cosyra* has a restricted host range (Copeland et al. 2006). In South Africa and Swaziland, *C. cosyra* distribution generally follows a similar pattern to the distribution of marula, an important wild host (Magagula and Ntonifor 2014; De Villiers et al. 2013).

The abundance of *C. cosyra* is influenced by: the bimodal nature of rainfall in sub-tropical Africa; mixed cultivation of early- and late-maturing mango varieties that ensures mangoes are present in the field for a long time; the fact that mango fruits twice a year in some areas; and the proximity of wild hosts to mango orchards. In eastern Africa, *C. cosyra* is the most abundant fruit fly species on mango after *B. dorsalis*. Vayssières et al (2015) reported that *C. cosyra* was the most abundant species during the dry season in Benin and recovered > 50 pupae per kg of fruit from 15 different mango varieties. Displacement of *C. cosyra* by other mango-infesting species, especially *B. dorsalis* has been reported from Uganda (Isabirye et al. 2015), Tanzania (Mwatawala et al. 2009b) and Kenya (Ekesi et al. 2009). In Mali, *C. cosyra* represented 85.58 % of fruit flies recovered from mango (Vayssières et al. 2007) and 52.25 % in Benin (Vayssières et al. 2005). In Kenya, Copeland et al. (2006) recorded 1723 *C. cosyra* per 1000 fruits. Significant numbers of *C. cosyra* larvae have also been recovered from mango in Tanzania (Geurts et al. 2012).

6.7 Ceratitis fasciventris

In 2006, Copeland et al (2006) demonstrated that *C. fasciventris* was distributed widely throughout the Central Kenyan Highlands, at elevations of up to 2220 m a.s.l., but that it was absent from coastal areas. Populations of *C. fasciventris* on coffee, *Coffea arabica* Linnaeus, have also been reported from the Central Highlands of Kenya at Ruiru (1609 m a.s.l) and Rurima (1228 m a.s.l). *Ceratitis fasciventris* has been reared from fruit collected year-round and is known to be sympatric with *C. anonae*, both species often occuring together in the same sample of wild fruit (Copeland and Wharton 2006). In the Democratic Republic of Congo the largest numbers of fruit flies captured using lures were of *C. fasciventris* in the mid altitudinal areas of South Kivu (Rubabura et al. 2015).

6.8 Ceratitis quinaria

Ceratitis quinaria is widely distributed in West Africa and abundant in mango orchards (Vayssières et al. 2005; 2007; 2009; 2011). Trapping and rearing data indicate that *C. quinaria* is most abundant during the dry season, causing damage only to early-maturing cultivars of mango (Vayssières et al. 2005). There is a positive relationship between high temperature, relative humidity and rainfall with *C. quinaria* populations (Vayssières et al. 2005). In Mali 4.89 % of fruit flies reared from mango were *C. quinaria* (Vayssières et al. 2007) and 5.61 % in Benin (Vayssières et al. 2015).

6.9 Ceratitis rosa

Ceratitis rosa, which is also an indigenous African fruit fly has been reported in coast areas and the Central Highlands of Kenya (Copeland et al. 2006), in the Cape region of South Africa (De Villiers et al 2013) and on the islands of Mauritius and La Réunion (White et al. 2001). This species has also been reported as the dominant fruit fly species in temperate fruit species such as peach, apple and pear (Mwatawala et al. 2009b), which are only grown at high altitudes in Africa. High abundances of *C. rosa* occured during the wet months (February and March) in higher altitude areas in Tanzania (Mwatawala et al. 2009a). For example, at 1305 m there were seven fruit fly species with relatively similar abundances, whereas at 1650 m there were 11 species present but *C. rosa* was the most abundant. Several studies have also confirmed that *C. rosa* is a species that can withstand colder temperatures (Duyck et al. 2004, 2006a, b; Grout and Stoltz 2007; De Meyer et al. 2008, 2010; Duyck and Quilici 2002).

Interestingly, *C. rosa* is considered as potentially invasive as *C. capitata* and feared to be a global threat due to its cold tolerance and its presence at higher altitudes than *C. capitata* in Kenya and La Réunion (Copeland et al. 2006). Some studies have shown greater tolerance of *C. rosa* to lower temperatures than *C. capitata* and *C. catoirii* (Duyck and Quilici 2002). This could explain why *C. rosa* is regularly reared, in small numbers, from fruit of two indigenous plants, two naturalized invasive plants, and an exotic garden ornamental collected in four sites in the Central Highlands of Kenya at altitudes of 1533–1771 m a.s.l. (Copeland et al. 2006). Initially, *C. rosa* was limited mainly to coastal lowland habitats (5–436 m), where it often co-existed with *C. fasciventris* (Copeland et al. 2006). Using genetic algorithms for rule-set prediction (GARP), De Meyer et al. (2008) predicted that much of sub-Saharan Africa and Madagascar were highly suitable for *C. rosa*. In Swaziland, *C. rosa* was the dominant fruit fly species in guava orchards comprising 68.8 % of all fruit flies collected, and regularly co-exists in guava with *C. capitata* and *C. cosyra* (Magagula and Ntonifor 2014). In South Africa *C. rosa* is mostly found in the cooler regions of the country and positively correlated with precipitation (De Villiers et al. 2013). In Côte d'Ivoire, 0.02% of fruit flies recovered from infested mango were *C. rosa* (N'depo et al. 2013).

As described earlier in this chapter (Sect. 5.12), the recently described species *C. quilicii* occurs sympatrically with *C. rosa* in some regions and does not show clear geographic isolation (De Meyer et al. 2015). It is therefore highly likely that some of the records of abundance of *C. rosa* particularly in the highland areas might indeed be of *C. quilicii*. In Tanzania, a gradual shift was observed with *C. rosa* and *C. quilicii* occurring at lower altitudes (with predominance of *C. rosa*) while only *C. quilicii* was observed at the highest elevations (Mwatawala et al. 2015). For instance, *C. quilicii* was more abundant at higher altitudes, reaching a peak at Langali (1268 m asl) while being absent at the lower elevation at Sokoine (550 m asl). However, when examined across an altitudinal transect, *C. rosa* was more abundant (61.2%) than *C. quilicii* (38.8%) (Mwatawala et al. 2015). It can be inferred that the impact of *C. quilicii* might be more pronounced on temperate fruits like peach, avocado and apple and earlier host plants records for *C. rosa* at higher elevations could possibly be *C. quilicii*.

6.10 Ceratitis silvestrii

Ceratitis silvestrii is an important pest of mango in several parts of West Africa, mainly found co-existing with *C. quinaria* (Ouedraogo et al. 2010; Sawadogo et al. 2013). *Ceratitis silvestrii* is most abundant during the dry season causing damage to early-maturing mango cultivars (Vayssières et al. 2005; Vayssières et al. 2009). In Mali, 7.28% of fruit flies reared from mango were *C. silvestrii* (Vayssières et al. 2007) and in Benin 2.77% of fruit flies were *C. silvestrii* (Vayssières et al. 2015).

6.11 Ceratitis punctata

There have only been records of *C. punctata* from mango in Cote d'Ivoire. Hala et al. (2006) reported 0.15% of fruit flies reared from mango were *C. punctata* and N'depo et al. (2013) reported 0.18% of of fruit flies were *C. punctata*.

6.12 Dacus *and* Zeugodacus *species*

On La Réunion (1996–1999), *Z. cucurbitae* (Coquillet) and *D. ciliatus* are reported to mainly infest a range of 16 cucurbit species (Vayssières and Carel 1999). However, there have been recent records of *Z. cucurbitae* also infesting mango (Vayssières et al. 2008; Mwatawala et al. 2010; De Meyer et al. 2015). The altitudinal limits of *Z. cucurbitae* and *D. ciliatus* are 1200 m and 1400 m, respectively during the dry season. These two species overlap on all cucurbit crops up to 600 m during the wet season and up to 1200 m during the dry season. At least one abiotic factor (altitude) and two biotic factors (host availability, interspecific competition) are responsible for the dominance of these species in La Réunion. Studies in Tanzania showed that *Z. cucurbitae* was either absent or less abundant at higher elevations along a transect from approx. 600 m a.s.l to 1650 m a.s.l (Mwatawala et al. 2010). However, the exact relationship between these biotic and abiotic factors and populations of *Z. cucurbitae* and *D. ciliatus* are currently poorly understood and require further investigation. *Dacus bivittatus* has also been reported from mango in Côte d'Ivoire. Approximately 0.42 % and 0.07 % of fruit flies reared from mango in Côte d'Ivoire were *D. bivittatus*, as reported by Hala et al. (2006) and N'depo et al. (2013) respectively.

7 Competitive Displacement Amongst Tephritid Fruit Flies in Mango Agroecosystems

The introduction of species into a new area can alter successional patterns, mutualistic relationships, community dynamics, ecosystem function and resource distribution (Mooney and Cleland 2001). Several studies have shown that, where exotic tephritid species have been introduced into areas already occupied by a native tephritid species, interspecific competition occurs that results in a decrease in numbers and niche shifts of the indigenous species, albeit without leading to complete exclusion (Duyck et al. 2004, 2006a; Ekesi et al. 2009; Mwatawala et al. 2009b). Reitz and Trumble (2002) defined competitive displacement as "the removal of a formerly established species from a habitat through superior use, acquisition or defense of resources by another species". This can occur through many different mechanisms that are often broadly categorized as exploitation or interference. Factors such as superior competitive abilities, resource pre-emption, release from natural enemies and abiotic factors including temperature and anthropogenic disturbances, are amongst the reasons an invasive species could become dominant (Rwomushana et al. 2009). In many cases, larger body size, shorter developmental period and higher realized fecundity, coupled with superior behavioural traits and the absence of coevolved natural enemies, are major factors behind the competitive advantage of alien invasive species over native ones (Reitz and Trumble 2002). Another factor that influences competitive displacement is niche differentiation between tephritid fruit fly species. For example, the large populations of *C. rosa* found in the highlands of La Réunion and Kenya (although reported here and elsewhere as *C. rosa*, recent taxonomic advances suggest this species may be *C quilicii*), where no other species are found, are suggestive of a climate-dependent change in competitive hierarchy. Host fruit preference, although less well studied, might have similar effects.

The most notable examples of competitive displacement outside Africa include displacement of *C. capitata* by the Queensland fruit fly, *Bactrocera tryoni* (Froggatt) in the Sydney area in Australia (Debach 1966) and displacement of the same species by *B. dorsalis* from the coastal areas in Hawaii in 1945 (Duyck et al 2004). According to Vargas et al. (1995), in the latter case, the displacement was, to some extent, mediated by host fruit species in that *C. capitata* persists in the lowlands on coffee, their presumed ancestral host in Africa to which it is better adapted.

A well documented case of this phenomenon in Africa is from the Mascarene Islands where the indigenous species, *C. catoirii*, is reported to have been displaced by *C. capitata* and *C. rosa* in La Réunion occurring in small numbers on the east and south coast of the island, while in Mauritius it seems to have disappeared entirely (Duyck et al. 2004). In a series of fruit fly invasions of La Réunion, Duyck et al. (2006b) further reported that the invasive species *B. zonata*, tended to have a higher rank than the previously established invasive (*C. rosa* and *C. capitata* from mainland Africa) and native (*C. catoirii*) species in the hierarchy. Presumably, the invasion of *B. zonata* in Mauritius in 1987 and La Réunion in 1991 may have further compounded the displacement of the indigenous species. Duyck et al. (2006b) suggested that, because the endemic fruit fly species in La Réunion had no specific climatic niches, they had become very rare species, and could be at risk of extinction due to invasion (Duyck et al. 2008). Large body size and shorter developmental time of the exotic species, *B. zonata*, was associated with superior competitive ability, demonstrating the importance of these traits for its superior competitive response (scramble and interference) compared with all the *Ceratitis* species. (Duyck et al 2006a). Some data also suggest that *C. capitata* and *C. rosa* appear to leave detectable chemical signals that influence the laying behaviour of conspecifics. These two species commonly display the 'dragging ovipositor' behaviour that is classically associated with hostmarking in tephritids and leads to inhibition of oviposition by conspecifics subsequently visiting the same fruit (Nufio and Papaj 2004) in response to the host-marking pheromones that have been deposited (Roitberg and Prokopy 1983; Nufio and Papaj 2001). Interestingly, *B. zonata* is able to detect and avoid signals left by *C. capitata* and *C. rosa*, while the response of *Ceratitis* species to each other's signals is not significant.

In Kenya, Ekesi et al. (2006) speculated that competitive displacement was ongoing because there was a shift in dominance between the native fruit fly *C. cosyra* and the invasive species *B. dorsalis* in mango orchards at Nguruman in the Rift Valley Province of Kenya. The results of their study clearly indicated rapid displacement of *C. cosyra* by *B. invadens* within 4 years of its detection in the country, and was corroborated by Rwomushana et al. (2008, 2009) who showed that *B. dorsalis* constituted up to 98 % of the total fruit flies reared from mango in Kenya. Ekesi et al. (2009) argued that displacement interference could be explained by the aggressive behaviour demonstrated between interacting females of these species at laying sites; this behaviour was highly asymmetrical and this gave *B. dorsalis* a competitive advantage over the resident fruit fly species. Aggressive behaviour has also been been observed by Shelly (1999) who demonstrated that females of *B. dorsalis* defended oviposition sites on mango against conspecific females by lunging at opponents and driving them off through threat displays; occasionally this escalated to head-butting and pushing. As such it is then perhaps not surprising that both sexes of *B. dorsalis* would launch several aggressive behaviours against *Ceratitis* species. In related laboratory experiments, *B. dorsalis* was observed to out compete *C. capitata* and inhibit its development by superior scramble competition (Keiser et al. 1974).

In Tanzania, the Relative Abundance Index (RAI) of *B. dorsalis* to *C. capitata*, *C. cosyra*, *C. rosa* in 19 evaluated hosts was higher (more than 0.5) and in favour of *B. dorsalis*; in some hosts (sweet orange, *Citrus sinensis* (L.) Osbeck; ambarella, *Spondias cytherea* L. and tropical almond, *Terminalia catappa* L.) it reached 1, implying that only *B. dorsalis* was present (Mwatawala et al. 2009a). Certainly, in fruit species such as tropical almond, only *B. dorsalis* emerged. This trend lends credence to the suggestion that the exotic species is slowly displacing other fruit fly species on the same hosts. Trapping data confirms the dominance of *B. dorsalis* (Mwatawala et al. 2004, 2006).

Despite these cases of displacement activity there are several reasons why *Ceratitis* species have not been completely displaced from the mango agroecosystem. *Ceratitis* species have some advantages that allow for some level of coexistence with *B. dorsalis*. *Ceratitis* species have a more specialized host-searching ability and have had close associations with several host plant species over a long period in Africa. Secondly, *Ceratitis* species have been recorded from several hundred plant species in Africa (Lux et al. 2003; Copeland et al. 2006) compared with the host range of *B. dorsalis* that currently stands at just over 40 known cultivated and wild host species, though this is growing (Vayssières et al. 2009). It is likely that *Bactrocera* species can switch to other suitable hosts when there is pressure on the carrying capacity, providing some niche on mango for *Ceratitis* species to survive. High infestations found on wild hosts like *T. catappa*, even when mango is present, attest to this. Generally, most *Bactrocera* species, including *B. dorsalis*, are believed to be lowland residents (Vargas et al. 1983; Wong et al. 1985; Harris et al. 1986; Ekesi et al. 2006), enabling *B. dorsalis* to successfully displace *Ceratitis* species in lowland ecologies. At higher elevations, such as Embu in the Eastern Province of Kenya, *C. cosyra* remains the dominant species, probably because of the poor tolerance of *B. dorsalis* to low temperatures (Ekesi et al. 2006). It is therefore probable that *B. dorsalis* may be restricting populations of *C. cosyra* to the highlands. Indeed, such phenomena have been reported from Hawaii, where *B. dorsalis* largely displaced *C. capitata* from the low-elevation coastal zones and restricted *C. capitata* populations to cooler climates at high altitudes where *B. dorsalis* does not occur (Vargas et al. 1995). Subsequently distribution and abundance of the major mango-infesting fruit flies in Africa will continue to be dependent of the competitive interactions between native and exotic species.

8 Host Plants of Mango-Infesting Fruit Flies in Africa

Despite the economic significance of tephritid fruit flies, the host spectrum throughout their distribution range remains limited or is continuously evolving to include hitherto unknown hosts. Several studies have documented the current fruit fly pests in Africa and their host plants (Liquido et al. 1991; White and Elson-Harris 1992; N'Guetta 1994; Copeland et al. 2002, 2004, 2006; Vayssières and Kalabane 2000; De Meyer et al. 2002; Ekesi et al. 2006; Vayssières et al. 2005; Ndzana Abanda et al. 2008; Rwomushana et al 2008; Vayssières et al. 2010). De Meyer et al. (2002) provided an annotated host check list for all *Ceratitis* species from Africa and Goergen et al. (2011) has provided a detailed listing of host plants for *B. dorsalis* in West and Central Africa. The host plants for each fruit fly species has been documented from published papers on host plants in Africa and insect records that are publicly available from the Royal Museum for Central Africa (http://projects.bebif.be/fruitfly/index.html Table 5.1). The authorities for each species follow the nomenclature of the International Plant Names Index (www.ipni.org) which was cross referenced with the Global Species Database (www.globalspecies.org), the Plant List (www.plantlist.org) and the Herbarium Catalogue www.kew.org/herbcat) (Global Species 2015; Herbarium Catalogue 2015; IPNI 2015; Plant List 2015).We present a summary of the host plant specialization of different fruit fly species, fruit fly species richness and abundance on particular hosts and the compartmentalization of the plant–fruit fly food web.

Table 5.1 Host plants of mango infesting fruit fly species in Africa

Plant family	Host plant species, common name and which species they support
Actinidiaceae	*Actinidia deliciosa* (A. Chev.) C.F. Liang & A.R. Ferguson (kiwifruit)¤
Amaryllidaceae	*Allium cepa* L. (dry onions)ϵ
Amaranthaceae	*Sericostachys scandens* Gilg & Lopr.♣
Anacardiaceae	*Anacardium occidentale* L. (cashew nut)*¤Δ♥♦Ω, *Mangifera indica* L. (mango)*§ß¤‡Δ♦♥♠Ωϵθʊ⊞∞, *Sclerocarya birrea* (A. Rich.) Hochst. (marula)*ΔΩ, *Sorindeia madagascariensis* Thouars ex DC.*, *Spondias dulcis* Parkinson (otaheite apple)*¤, *Spondias mombin* L. (tropical plum)*Δ♦, *Spondias cytherea* Sonner. (hog plum)*, *Harpephyllum caffrum* Bernh. ex C.Krauss¤♣♦, *Spondias purpurea* L. (red mombin)¤, *Spondias tuberosa* Arruda¤, *Spondias* sp (wild plum)Δ
Anisophylleaceae	*Anisophyllea laurina* R.Br. ex SabineΔ♣
Annonaceae	*Annona cherimola* Mill. (cherimoya)*¤Δ♣, *Annona muricata* L. (soursop)*ß¤Δ♣ϵ, *Annona senegalensis* Pers.*ßΔ♣♦Ω, *Annona squamosa* L. (sugar apple)*§¤‡♦ϵ, *Annona reticulata* L. (custard apple)§ß¤‡Δ♦ϵ, *Annona macroprophyllata* Donn. S.M.ß, *Annona montana* Macfad.ß, *Anonidium mannii* (Oliv.) Engl. & Diels ß, *Artabotrys monteiroae* Oliv.ß♣, *Cananga odorata* (Lam.) Hook.f. & Thomson (perfume tree)*¤♣, *Lettowianthus stellatus* Diels♣, *Monanthotaxis parvifolia* (Oliv.) Verdc.♣, *Monanthotaxis fornicata* (Baill.) Verdc.♣, *Monodora grandidieri* Baill.♣, *Monodora* sp., Dunalß, *Rollinia mucosa* (Jacq.) Baill. (wild sweetsop, wild sugar apple)ßΔ, *Rollinia* A.St.-Hil. sp.ß, *Sphaerocoryne gracilis* (Engl. & Diels) Verdc.♣, *Thevetia peruviana* K. Schum. (exile tree, yellow oleander)♣, *Uvaria acuminata* Oliv.♣, *Uvaria catocarpa* Diels♣, *Uvaria lucida* Bojer ex Sweet♣

(continued)

Table 5.1 (continued)

Plant family	Host plant species, common name and which species they support
Apocynaceae	*Acokanthera oppositifolia* (Lam.) Codd☼♣, *Acokanthera schimperi* (A.D.C.) Schweinf. (round-leaved poison bush)☼♣, *Ancylobothrys* sp. Pierre^β, *Carissa carandas* L. (caranda plum, mahakaranda)☼Δ♣, *Carissa edulis* (Forssk.) Vahl (Egyptian carissa)☼, *Carissa grandiflora* (E. Mey.) A. DC (natal plum) ♣, *Carissa macrocarpa* (Eckl.) A.D.C. (natal plum)☼ ♣, *Carpodinus hirsuta* Hua^◊, *Dictyophleba lucida* (K.Schum.) Pierre♣, *Landolphia* P. Beauv.sp*^◊, *Landolphia heudelotii* Stapf^Ω, *Landolphia kirkii*, Dyer ^Δ, *Saba comorensis* (Boj.) Pichon^Δ, *Saba senegalensis* (A.DC.) Pichon*^ΔΩ, *Tabernaemontana longiflora* Rusby^◊, *Tabernaemontana penduliflora* K. Schum^Δ, *Thevetia peruviana* K.Schum. (exile tree, yellow oleander)☼, *Voacanga chalotiana* Pierre ex Stapf. ■, *Voacanga dregei* E.Mey.^◊
Araliaceae	*Irvingia* F. Muell. sp.^β
Arecaceae	*Butia eriospatha* (Mart. ex Drude) Becc.☼, *Cocos plumosa* Hook. f.^β, *Elaeis* Jacq. sp^β, *Phoenix dactylifera* L. (date-palm)^§☼ϵ
Asparagaceae	*Dracaena steudneri* Engl.*
Asteraceae	*Helianthus annuus* L. (sunflower)^ϵ
Boraginaceae	*Cordia sinensis* Lam.*, *Ehretia cymosa* Thonn.☼♣
Brassicaceae	*Brassica oleracea* var. *botrytis* (Caulifolower)^ϵ, *Brassica oleracea* var. *capitata* (Broccoli)^ϵ
Bromeliaceae	*Ananas comosus* (L.) Merr. (pineapple)*, *Ehretia cymosa* Thonn.
Cactaceae	*Cereus peruvianus* (L.) Mill.♣,*Hylocereus undatus* (Haw.) Britton & Rose (dragon fruit)☼♣, *Opuntia ficus-indica* (L.) Mill. (prickly pear)☼♣, *Pereskia aculeata* Mill. (lemon-vine)☼
Caesalpinioideae	*Cordyla pinnata* (Lepr. ex A. Rich.) Milne Redhead (cayor pear tree)*, *Cynometra* L. sp.^β
Canellaceae	*Warburgia salutaris* (Bertol. f.) Chiov. (pepper-bark tree)^Δ♣, *Warburgia ugandensis* Sprague (pepperbark tree, greenheart)^Δ
Capparaceae	*Capparis sepiaria* L. (indian caper)☼, *Crateva tapia* L.☼, *Maerua duchesnei* (De Wild.) F. White*☼,
Caricaceae	*Carica papaya* L. (papaya)*^§☼‡♣ ϵႱ, *Vasconcellea cauliflora* (Jacq.) A. DC.♣
Cecropiaceae	*Myrianthus* P. Beauv. sp.^β, *Myrianthus arboreus* P.Beauv. ■
Celastraceae	*Salacia elegans* Welw. ex Oliv.♣
Chrysobalanaceae	*Chrysobalanus icaco* L. (icaco plum)☼ΔΩ, *Parinari curatellifolia* Planch. ex Benth^Δ
Clusiaceae	*Calophyllum tacamahaca* Willd.☼♣, *Calophyllum* sp. L. (beauty-leaf)☼, *Garcinia livingstonei* T. Anderson (african mangosteen)☼,*Garcinia mannii* Oliv.*, *Garcinia* × *mangostana* L (mangosteen)^β☼♣
Combretaceae	*Terminalia catappa* L. (tropical almond)*^§β☼‡Δ♣

(continued)

Table 5.1 (continued)

Plant family	Host plant species, common name and which species they support
Crassulaceae	*Cotyledon orbiculata* L.ᵝ
Cucurbitaceae	*Benincasa hispida* (Thunb.) Cogn. (Chinese melon)ᵋ, *Cephalendra indica* Naud. (Kundru)ᵋ, *Citrullus colocynthis* (L.) Schrad. (colocynth)*ᵋᵝᵠ, *Citrullus lanatus* (Thunb.) Matsum. & Nakai (watermelon)*ᵋᵝᵠ, *Citrullus vulgaris* Schrad (African melon)ᵋ, *Coccinia grandis* (L.) Voigt (Wild cucurbits)ᵝᵠ, *Coccinia indica* Wight & Arn. (Ivy gourd)ᵋ, *Coccinia dipsaceus* Ehrenb. ex Spach (Wild cucurbits)ᵋ, *Coccinia palmate* M.Roem.ᵝ, *Coccinia quinqueloba* (Thunb.) Cogn.ᵝ, *Coccinia trilobata* (Cogn.) C.Jeffreyᵝᵠ, *Corallocarpus ellipticus* Chiov.ᵝᵠ, *Cucumis aculeatus* Cogniauxᵠ, *Cucumis metuliferus* Naudin (African horned cucumber)ᵠ, *Corallocarpus schimperi* Hook.f.ᵝ, *Cucumeropsis edulis* Cogn. ᵝ, *Cucumis anguria* L. (Wild cucurbit)ᵋᵝ, *Cucumis ficifolius* A.Rich.*, *Cucumis melo* L. (melon)*⌖, *Cucumis sativus* L. (cucumbers, gerkins)*⌖ᵋᵝ, *Cucumis africanus* L.f.ᵝᵠ, *Cucumis dipsaceus* Ehrenb. ex Spach (hedgehog gourd)⌖ᵝᵠ, *Cucumis melo* C. melo var. conomon (Muskmelon)ᵋᵠ, *Cucumis melo* var. momordica (Snap melon)ᵋᵝ, *Cucumis pubescens* Willd. (Wild cucurbit)ᵋ, *Cucumis sativus* L. (cucumber)ᵝ, *Cucumis trigonus* Roxb. (Wild cucurbits)ᵋ, *Cucumis utilissimus* Roxb (Long melon)ᵋ, *Cucumis vulgaris* var *fistulosus* (Squash melon)ᵋ, *Cucurbita maxima* Duchesne (giant pumpkin)*♠ᵋᵝᵠ, *Cucurbita moschata* Duchesne (butternuts)ᵋ, *Cucurbita pepo* L. (ornamental gourd, squash)*ᵋᵝᵠ, *Diplocyclos palmatus* (L.) C.Jeffrey. (Balsam apple)ᵋ, *Kedrostis leloja* (Forsk. ex J.F.Gmel.) C. Jeffreyᵝᵠ, *Kedrostis foetidissima* Cogn.ᵝᵠ, *Lagenaria abyssinica* (Hook.f.) C.Jeffreyᵝ, *Lagenaria amebicana* (Wild cucurbits)ᵋ, *Lagenaria siceraria* (Molina) Standl. (calabash, water bottle)*ᵋᵝᵠ, *Lagenaria sphaerica* E.Mey.ᵝ, *Lagenaria vulgaris* Ser. (Bottle gourd)ᵋ, *Luffa* sp Mill.§, *Luffa aegyptiaca* Mill.ᵝᵠ, *Luffa acutangula* (L.) Roxb. (Ribbed gourd)ᵋᵝᵠ, *Luffa cylindrica* M. Roem. (Sponge gourd)ᵋᵝ, *Momordica balsamina* L.ᵝᵠ, *Momordica charantia* L. (bitter gourd)*§ᵋᵝᵠ, *Momordica calantha* Gilg♠, *Momordica rostrata* Zimm.ᵝᵠ, *Momordica trifoliolata* Hook.fᵝ, *Mukia maderaspatana* (L.) M.Roem.ᵝ, *Telfairia pedata* Hook.ᵝ, *Trichosanthes anguina* Lᵋ, *Trichosanthes cucumeria* (Snake gourd)ᵋᵝ, *Peponium mackenii* Engl.ᵝᵠ, *Peponium vogelii* Engl.ᵝ, *Sycos pachycarpus* (Wild cucurbit)ᵋ, *Sechium edule* (Jacq.) Sw.ᵝ, *Trichosanthes dioica* Roxb. (Pointed gourd)ᵋ, *Trichosanthes cucumerina* Linn. (Wild cucurbit)ᵋᵝᵠ
Dichapetalaceae	*Dichapetalum bangii* (Didr.) Engl.ᵝ
Ebenaceae	*Euclea divinorum* Hiern⌖♠, *Diospyros abyssinica* (Hiern) F. White⌖, *Diospyros kabuyeana* F.White♠ *Diospyros kaki* Thunb. (persimmon)*⌖♠, *Diospyros malabarica* (Desr.) Kostel. (malabar ebony) ⌖, *Diospyros mespiliformis* Hochst. ex A.DC. (ebony diospiros)⌖△, *Diospyros montana* Roxb.*♠, *Diospyros pallens* (Thunb.) F.White⌖, *Diospyros virginiana* L. (persimmon, common)⌖

(continued)

Table 5.1 (continued)

Plant family	Host plant species, common name and which species they support
Ericaceae	*Arbutus unedo* L. (arbutus)⁂, *Vaccinium corymbosum* L. (blueberry)⁂
Euphorbiaceae	*Euphorbia heterophylla* L.ᵾ, *Croton* L. sp.ᵦ, *Drypetes* Vahl sp.ᵦ♣, *Drypetes gerrardii* var. *gerrardii* Hutch.♣♣, *Drypetes natalensis* (Harv.) Hutch.⁂♣, *Drypetes gossweileri* S. Mooreᐃ, *Phyllanthus acidus* (L.) Skeels (star gooseberry)♣, *Ricinus communis* L.ᵾ, *Uapaca kirkiana* (wild loquat) Müll. Arg.ᐃ
Fabaceae	*Cordyla africana* Lour. (wild mango)ᐃ, *Cordyla pinnata* (A. Rich.) Milne-Redh. (cayor pear tree)ᐃ, *Gliricidia maculata* (Humb., Bonpl. & Kunth) Steud.ᐃ, *Inga laurina* (Sw.) Willd. (ice cream bean)♣, *Pericopsis elata* (Harms) Meeuwen■, *Vigna sesquipedalis* (L.) Fruw. (Cowpea)ᵉ, *Vigna sinensis* (L.) Savi (Cowpea)ᵉ, *Vigna unguiculata* (L.) Walp. (Long bean or Cowpea)ᵉ
Flacourtiaceae	*Dovyalis* E. Mey. ex Arn. sp.ᵦ♣, *Dovyalis hebecarpa* (Gardner) Warb. (ceylon gooseberry, ketembilla)⁂♣, *Flacourtia* Comm. ex L'Her. sp.ᵦᐃ, *Rawsonia lucida* Harv. & Sondᵦ♣. *Dovyalis caffra* (Hook. f. & Harv.) Warb. (kei apple)⁂ᐃ♣♣, *Flacourtia indica* (Burm. f.) Merr. (governor's plum),⁂‡♣, *Ludia mauritiana* J.F. Gmel.♣*Rawsonia lucida* Harv. & Sond.♣♣, *Rawsonia usambarensis* Engl. & Gilg.♣,
Flagellariaceae	*Flagellaria guineensis* Schumach⁂
Goodeniaceae	*Scaevola plumieri* (L.) Vahl⁂, *Scaevola sericea* Vahl⁂, *Scaevola taccada* (Gaertn.) Roxb. (beach naupaka)⁂
Guttiferae	*Calophyllum tacamahaca* Willd⁂♣
Hippocrataceae	*Salacia elegans* Oliv.♣
Irvingiaceae	*Irvingia gabonensis* (Aubry-Lecomte ex O'Rorke) Baill. (wild mango)*, *Irvingia smithii* Hook. F.ᵦ
Juglandaceae	*Carya illinoinensis* (Wangenh.) K. Koch (pecan)⁂, *Juglans regia* L. (walnut)⁂
Lauraceae	*Cinnamomum verum* J.Presl (cinnamon)⁂, *Persea americana* Mill. (avocado) *§ᵦ ⁂‡ᐃ♣♣ᵉ
Lecythidaceae	*Careya arborea* Roxb. (tummy wood)§, *Napoleonaea gabonensis* Liben◊■
Leguminosae	*Cajanus cajan* (L.) Millsp. (Pigeon pea)ᵉ, *Cordyla pinnata* (A.Rich.) Milne-Redh.*, *Dolichos lablab* L. (Hyacinth bean)ᵉ, *Faidherbia albida* (Delile) A.Chev.Ω, *Inga laurina* (Sw.) Willd. (Spanish oak)♣, *Pithecellobium dulce* (Roxb.) Benth.⁂, *Angylocalyx braunii* Harms♣, *Phaseolus vulgaris* L. (French bean)ᵉᵠ, *Phaseolus limensis* L. (Lime bean)ᵉ, *Phaseolus radiatus* L. (Green gram)ᵉ
Loganiaceae	*Strychnos decussata* (Pappe) Gilg⁂, *Strychnos henningsii* Gilg⁂♣, *Strychnos mellodora* S. Moore*, *Strychnos potatorum* L.f. ⁂, *Strychnos pungens* Soler⁂, *Strychnos spinosa* Lam.ᐃ♣♣
Lythraceae	*Punica granatum* L. (pomegranate)§⁂‡
Malpighiaceae	*Malpighia glabra* L. (acerola)⁂
Malvaceae	*Abelmoschus esculentus* (L.) Moench (Okra)ᵉᵾᵠ,*Cola natalensis* Oliv.⁂♣, *Durio zibethinus* L. (durian)*⁂, *Grewia asiatica* L. (phalsa)§,
Melastomataceae	*Bellucia* Neck. ex Raf. sp.ᵦ
Meliaceae	*Ekebergia capensis* Sparrm. (dog plum, Cape ash)⁂♣♣, *Sandoricum koetjape* (Burm.f.) Merr. (santol)⁂

(continued)

Table 5.1 (continued)

Plant family	Host plant species, common name and which species they support
Menispermaceae	*Tiliacora funifera* (Miers) Oliv.ᵝ♣
Mimosaceae	*Inga laurina* (Sw.) Willd. (Sackyca)♣, *Pithecellobium dulce* (Roxb.) Benth. (Manila tamarind, guamuchil)♣
Moraceae	*Antiaris toxicaria* Lesch. (antiaris, false iroko, false mvule)*ᵝ♣∞, *Antiaris toxicaria* subsp. *africana* (Engl.) C.C.Berg (upas-tree)ᵝ∞, *Artocarpus* J.R. Forst. & G. Forst. sp.ᵝ, *Artocarpus altilis* (Parkinson ex F.A.Zorn) Fosberg (breadfruit)☼, *Ficus carica* L. (fig)§♣ϵ ☼♣▼, *Ficus ingens* var. *ingens* (Red-leaved fig)Ω *Ficus ottoniifolia* Miq.*, *Ficus sycomorus* L. (sycamore fig)*, *Dorstenia* L. sp.ᵝ∞, *Ficus* L. sp. (fig)ᵝ, *Morus mesozygia* Stapf.ᵝ♣, *Morus nigra* L. (black mulberry)☼
Muntingiaceae	*Muntingia calabura* L. (Jamaica cherry)☼
Musaceae	*Musa acuminata* Colla (cavendish banana)*♣, *Musa × paradisiaca* L. (plantain)*☼, *Musa* sp L. (banana)*, *Musa nana* Lour (banana)♣, *Musa* sp. (Chinese banana)ϵ, *Musa paradisiaca* sp. *sapientum* (Blue field banana)ϵ
Myrtaceae	*Acca sellowiana* (O. Berg) Burret*☼♣, *Eugenia brasiliensis* Lam. (brazil cherry)☼, *Eugenia paniculata* Jacq.☼, *Eugenia uniflora* L. (surinam cherry, pitanga cherry)*ᵝ☼‡Δ♣, *Eugenia* L sp.ᵝ, *Feijoa sellowiana* (O.Berg) O. Berg (Horn of plenty)☼♣, *Psidium araca* Raddi.♣, *Psidium cattleianum* Afzel. ex Sabine (strawberry guava, cherry guava)§ᵝ☼‡♣, *Psidium friedrichsthalianum* (O.Berg) Nied. (wild guava)☼♣, *Psidium guajava* L. (guava)*§ᵝ☼‡Δ♣▼ϵᵾ◊, *Psidium longipes* (O.Berg) McVaugh (strawberry guava)☼♣, *Syzygium aqueum* (Burm.f.) Alston (watery roseapple)‡♣, *Syzygium cumini* (L.) Skeels (black plum)☼♣, *Syzygium jambos* (L.) Alston (rose apple)*§☼‡♣♣, *Syzygium malaccense* (L.) Merr. & L.M.Perry (malay-apple)*☼♣, *Syzygium samarangense* (Blume) Merr. & L.M. Perry (water apple)*§☼♣
Oleaceae	*Olea europaea* subsp. *europaea* L. (olive)☼, *Olea woodiana* Knobl.☼
Opiliaceae	*Opilia amentacea* Roxb.☼♣♣
Oxalidaceae	*Averrhoa carambola* L. (carambola, starfruit)*☼‡Δ♣ϵᵾ, *Averrhoa bilimbi* L. (blimbe)☼♣
Orchidaceae	*Bulbophyllum patens* King ex Hook.f. (Zingerone)ϵ,
Olacaceae	*Strombosia scheffleri* Engl.ᵝ♣♣, *Ximenia americana* L. var *americana* (Hog plum)☼♣ Ω
Passifloraceae	*Passiflora coerulea* Auct. (blue-crown passion flower)☼ ᵾ, *Passiflora edulis* Sims (passionfruit)☼‡ϵ, *Passiflora mollissima* (Kunth) (banana passion)♣, *Passiflora foetida* L.ᵝ, *Passiflora suberosa* L. (corkystem passion flower)☼, *Passiflora subpeltata* Ortega♣, *Passiflora seemannii* Griseb. (passion fruit)ϵ, *Passiflora quadrangularis* L. (giant passion fruit)ϵᵾ
Phyllanthaceae	*Adenia lobata* (Jacq.) Engl.Δ *Antidesma dallachyanum* Baill.☼, *Antidesma venosum* E. Mey. ex Tul.☼, *Flueggea virosa* (Roxb. ex Willd.) Royle☼
Poaceae	*Zea mays* L. (maize, corn)ϵ

(continued)

Table 5.1 (continued)

Plant family	Host plant species, common name and which species they support
Podocarpaceae	*Podocarpus elongatus* (Aiton) L'Hér. ex Pers. (african yellow wood)☼
Polygalaceae	*Carpolobia lutea* G. Don△
Polygonaceae	*Coccoloba uvifera* (L.) L. (seaside grape)☼♦
Proteaceae	*Banksia prionotes* Lindl.☼
Rhamnaceae	*Ziziphus abyssinica* Hochst. ex A.Rich. (indian jujube)§ß♦, *Ziziphus jujuba* Mill. (common jujube)☼‡♦, *Ziziphus joazeiro* Mart.☼, *Ziziphus mauritiana* Lam. (jujube)*§☼♦, *Ziziphus spina-christi* (L.) Willd.▼
Rosaceae	*Cydonia oblonga* Mill.*§☼♦, *Eriobotrya japonica* (Thunb.) Lindl. (loquat) §*☼‡♦♦, *Fragaria chiloensis* (L.) Mill. (Strawberry)ϵ, *Malus communis* Poir. (apple tree)§☼♦, *Malus domestica* Borkh. (apple)*☼♦, *Malus floribunda* Siebold ex Van Houtte☼, *Malus pumila* Mill. (apple)§, *Mespilus germanica* L. (medlar)☼, *Prunus africana* (Hook. f) Kalkmanß♦, *Prunus armeniaca* L. (apricot)§☼▼♦, *Prunus persica* (peach) (L.) Stokes*☼‡△♦▼♦ϵ, *Prunus* sp. L. (stone fruit)☼♦, *Prunus avium* (L.) L. (sweet cherry)☼, *Prunus capuli* Cav. ex Spreng.♦, *Prunus domestica* L. (plum)☼♦, *Prunus salicina* Lindl. (Japanese plum)☼♦, *Pyrus communis* L. (European pear)ϵ♦ϵ, *Pyrus malus* L. (Apple)ϵ*Pyrus pyrifolia* (Burm. f.) Nakai (Oriental pear tree)☼, *Pyrus syriaca* Boiss.☼, *Rubus idaeus* L. (raspberry)☼, *Rubus loganobaccus* L.H. Bailey (loganberry)☼
Rubiaceae	*Sarcocephalus latifolius* (Sm.) E.A. Bruce (pin cushion tree, Guinea peach)*△Ω, *Coffea arabica* L. (arabica coffee)ß☼♦♦, *Coffea canephora* Pierre ex A. Froehner (robusta coffee)ß☼♦, *Leptactina platyphylla* (Hiern) Wernh.ß♦, *Coffea liberica* Hiern (Liberian coffee)☼, *Guettarda speciosa* L.☼, *Vangueria infausta* Burch.☼, *Sarcocephalus esculentus* Sabine△, *Calycosiphonia spathicalyx* (K.Schum.) Robbr.♦, *Tricalysia pallens* Hiern♦
Rutaceae	*Aegle marmelos* (golden apple)§, *Casimiroa edulis* La Llave (white sapote)☼♦, *Citrus aurantiifolia* (Christm.) Swingle (lime)☼, *Citrus aurantium* L. (sour orange)*§ß☼△♦◊, *Citrus grandis* (Linn.) Osbeck (Shaddock/pummel)ϵ, *Citrus hystrix* DC.◊, *Citrus japonica* Thunb. (round kumquat)*, *Citrus limetta* Risso (sweet lemon)☼, *Citrus limon* (L.) Burm. f. (lemon)*☼♦, *Citrus × limon* (L.) Osbeck (mandarin lime)☼, *Citrus maxima* (Burm.) Osbeck (pummelo) ☼, *Citrus medica* L. (citron)☼, *Citrus nobilis* Lour. (tangor)☼♦, *Citrus x paradisi* Macfad. (grapefruit and Orlando)*ß☼♦■, *Citrus reticulata* Blanco (mandarin and Tangelo cv and Ortanique)*§ ß☼‡♦ϵ, *Citrus reticulata x paradisi* (tangelo)☼, *Citrus sinensis* (L.) Osbeck (navel orange and Tangor cv)*ß☼♦▼ϵ■, *Citrus × tangelo* J.W.Ingram & H.E.Moore (tangelo)♦, *Murraya exotica* L. (Chinese box)ß, *Murraya* J.Koenig sp.ß, *Murraya paniculata* (L.) Jack (orange jessamine)☼♦, *Clausena anisata* (Willd.) Hook.f. ex Benth. (horsewood)☼, *Fortunella* sp. Swingle (kumquats)☼, *Fortunella japonica* (Thunb.) Swingle (round kumquat)☼, *Harrisonia abyssinica* Oliv.♦, *Toddalia asiatica* (L.) Lam.♦, *Vepris trichocarpa* (Engl.) Mziray♦, *Vepris undulata* Verdoorn & C. A. Sm.☼
Salicaceae	*Flacourtia indica* (Burm. f.) Merr. (governor's plum)*
Salvadoraceae	*Azima tetracantha* Lam. (beehanger)☼
Santalaceae	*Santalum album* L. (Indian sandalwood)☼

(continued)

Table 5.1 (continued)

Plant family	Host plant species, common name and which species they support
Sapindaceae	*Allophylus ferrugineus* Taub.♣, *Allophylus pervillei* Blume♣, *Blighia sapida* K.D. Koenig (Akee apple)*, *Dimocarpus longan* Lour. (longan tree)☼♣, *Euphoria longan* Lam. (Longan)ᵋ, *Filicium decipiens* (Wight & Arn.) Thwaites (fernleaf)♣, *Filicium decipiens* (Wight & Arn.) Thwaites☼, *Litchi chinensis* Sonn. (lichi)☼Δ♣ᵋ, *Nephelium lappaceum* L. (rambutan)☼ᵝ, *Pancovia laurentii* (De Wild.) Gilg ex De Wildᵝ, *Pancovia turbinata* Radlk♣
Sapotaceae	*Argania spinosa* (L.) Skeels (argan tree)☼, *Chrysophyllum albidum* G. Don (white star-apple)*ᵝ♣♡■, *Chrysophyllum beguei* Aubrév. & Pellegr.■, *Chrysophyllum cainito* L. (common star apple)☼‡♣, *Chrysophyllum carpussum* L.☼♣, *Chrysophyllum imperiale* (Linden ex K.Koch & Fintelm.) Benth. & Hook.f.ᵝ, *Chrysophyllum natalense* Sond.♣, *Chrysophyllum oliviforme* L.☼, *Chrysophyllum pruniforme* Engl■, *Chrysophyllum viridifolium* J.M.Wood & Franks☼, *Englerophytum magalismontanum* (Sond.) T.D.Penn.☼♣, *Englerophytum natalense* (Sond.) T.D.Penn.Δ♣, *Englerophytum oblanceolatum* (S. Moore) T.D. Penn.ᵝ♣, ♣, *Manilkara butugi* Chiov. ☼ᵝ♣, *Manilkara sansibarensis* (Engl.) Dubard☼, *Manilkara zapota* (L.) P. Royen (sapodilla, chicle)*☼‡♣, *Mimusops* L sp.ᵝ, *Mimusops bagshawei* S. Moore☼, *Mimusops caffra* E.Mey. ex A.DC.☼, *Mimusops elengi* L. (Spanish cherry)§☼‡, *Mimusops fruticosa* Bojer☼, *Mimusops obtusifolia* Lam.☼, *Pachystela* sp. Pierre♣, *Pouteria altissima* (A.Chev.) Baehniᵝ♣, *Pouteria caimito* (Ruiz & Pav.) Radlk.☼, *Pouteria sapota* (Jacq.) H.E. Moore & Stearn (mammey sapote)☼, *Pouteria adolfi-friedericii subsp. usambarensis* (J.H.Hemsl.) L.Gaut. ◊, *Pouteria viridis* (Pittier) Cronquist (green sapote)☼, *Richardella campechiana* (Kunth) Pierre☼♣◊, *Sideroxylon inerme* L.☼, *Synsepalum brevipes* (Baker) T.D. Penn.ᵝ♣, *Synsepalum dulcificum* (Schumach. & Thonn.) Daniell (miraculous fruit)☼♣, *Pouteria campechiana* (Kunth) Baehni (canistel)♣, *Synsepalum subverticillatum* (E.A.Bruce) T.D.Penn.♣, *Vitellaria paradoxa* C.F. Gaertn. (shea butter)*ᵝΔ♣▼Ω■
Simaroubaceae	*Brucea antidysenterica* J.F.Mill.☼
Solanaceae	*Capsicum annuum* L. cov. *longum* A. DC. (bell pepper)*☼‡ᶲ, *Capsicum frutescens* L. (chilli)*☼‡ᵋ, *Cyphomandra* sp. Mart. ex Sendtn.☼, *Cyphomandra betacea* (Cav.) Miers (tree tomato)☼, *Lycium* L.sp. (boxthorns)☼, *Lycium barbarum* L. (Matrimonyvine)☼, *Lycium europaeum* L. (european boxthorn)☼, *Physalis peruviana* L. (Cape gooseberry)☼, *Solanum giganteum* Jacq.♣, *Solanum incanum* L. (grey bitter-apple)☼, *Solanum lycopersicum* L. (tomato)*☼‡♣ ᵋʊᶲ, *Solanum* sect. *Lycopersicon* spp.ᶲ, *Solanum macrocarpon* L. (local garden egg)☼, *Solanum mauritianum* Scop. (bugweed, bugtree)ᵝ☼♣♣, *Solanum melongena* L. (eggplant)☼‡ʊ, *Solanum nigrum* L. (black nightshade)☼, *Solanum muricatum* Aiton (melon pear)☼, *Solanum pseudocapsicum* L. (Jerusalem-cherry)☼, *Solanum tuberosum* L. (potato)§, *Solanum seaforthianum* Andrews (Brazilian nightshade)☼
Sterculiaceae	*Cola bruneelii* De Wild.◊, *Cola natalensis* Oliv.ʊ, *Sterculia* L. sp.ᵝ, *Theobroma cacao* L. (cocoa)*ᵝ☼♣♣◊■
Vitaceae	*Vitis vinifera* L. (grapevine)*☼♣, *Vitis trifolia* Linn. (Galls grape vine)ᵋ
Urticaceae	*Myrianthus arboreus* P. Beauv.ᵝ☼♣♣

*B. dorsalis; §B. zonata; ᵝC. anonae; ☼C. capitata; ‡C. catoirii; ΔC. cosyra; ■C. ditissima; ♣C. fasciventris; ∞C. flexuosa; ◊C. punctata; ▼C. quinaria; ♣C. rosa (probably includes C. quilicii in published records); ΩC. silvestrii; ʊD. bivittatus; ᶲD. ciliatus and ᵋZeugodacus cucurbitae

9 Future Perspectives

In most African countries, production of fruits and vegetables is recognized as a major source of income generation for rural communities and has been accorded high priority in national development plans. The domestic demand for fruits and vegetables continues to grow, thereby providing ready market outlets for increased domestic production and exports from Africa thereby generating opportunities for smallholder growers. Trade within Africa in agricultural commodities such as fruits has opened up in recent years due to regional integration that has largely removed many tariff and non tariff barriers that had hitherto restricted regional trade. Despite these inherent advantages, many countries in Africa do harbour a wide diversity of plant species that can support exotic tephritid fruit fly pests, and lack the quarantine and phytosanitary capacity to detect and restrict the entry of invasive insect species into their countries. Therefore, trade has become the principal means for unwitting introduction of invasive pests to new areas and constrained the potential growth of the horticultural industry (Ekesi et al. 2016). Owing to their high reproductive capacity coupled with the lack of competitors and efficient natural enemies, and further compounded with the poor quarantine infrastructure in Africa, invasive pests have spread widely to new locations with far reaching social and economic consequences.

Subsequently, many African countries have taken their own measures to address the fruit fly problem utilizing several Integrated Pest Management (IPM) technologies and innovations that have been tested and proven to be effective in field suppresion (Ekesi et al 2016). However, the resources required for fruit fly management remain enormous and elusive for the small-holder grower. Fruit fly management also requires an area-wide approach, particularly monitoring and surveillance to prevent new invasions (Manrakhan et al. 2011). Countries also need to continously scan the horizon for other emerging invasive fruit fly species already reported elsewhere to prevent their entry and establishment. Key among these is *B. zonata* which has been reported from some of the islands in the Indian Ocean (Mauritius and La Réunion), northern Africa (Egypt and Libya), several countries in the Arabian Peninsula (i.e. Oman, Saudi Arabia, United Arab Emirates and Yemen) and recently from the Gezira region in Sudan, suggesting a southward spread and potential risk of invasion into the sub-Saharan region (De Meyer et al. 2007). The way that *B. zonata* has shown its dominance over *B. dorsalis* in some parts of India is alarming, as such, urgent phytosanitary measures should be enforced to limit further spread. Similarly, *B. latifrons* is another exotic pest only recently established in Tanzania and Kenya (Mwatawala et al. 2007) which has the potential to increase the complex of pests on Solanaceae. *Zeugodacus cucurbitae* recently invaded the African continent and is causing havoc to a wide range of cucurbits (Mwatawala et al. 2010; De Meyer et al. 2015). Therefore, trans-regional invasions by these alien invasive pests would require an integrated and system-wide regional approach for their early detection and management.

The synomymization of *B. invadens*, *Bactrocera philippinensis* Drew and Hancock and *Bactrocera papayae* Drew and Hancock as part of the *B. dorsalis* species complex does suggest that all the 'sub-species' could inhabit similar environments and have the potential to establish if they were ever to invade the continent because they infest similar host plants. Additionally, resolution of the cryptic species within the FAR complex through integrative taxonomy approaches concluded that *C. rosa* belonged to the R1 type and that the R2 type was a new species, *C. quilicii*. This new information does require new research to understand the geographical limits of both species and resolve the host plant status that was, hirtherto, all attributed to *C. rosa*. The cases of *B. dorsalis* and *C. quilicii* will require a different approach to trade, quarantine and field control measures. Strategies and policies to deal with invasive pests need to be put in place to safeguard the entry of such species, but also to restrict the spread to new areas of those already established in Africa. It also calls for a need for regular surveillance and quarantine to restrict introduction. There should be a concerted effort in all African countries in addressing the fruit fly problem particularly with regards to fruit fly surveillance and management.

Significant gaps still exist in human capacity and technological application to the management of fruit flies that need to be addressed. There is also need for capacity building in taxonomy for both the fruit fly pests and their host plant species. Although, the fragmented structure of horticulture across Africa has greatly impeded the application of area-wide IPM, there still exists the potential for implementation in targeted agroecological zones. The use of Sterile Insect Techniques (SIT), for instance, in isolated ecologies is an approach that could be exploited more (Ekesi et al. 2016). The expansion of research activities on postharvest treatments and the need for standardized treatment regimes is crucial. Therefore, long-term protection of African horticulture against new invasive tephritids should be as important as the short-term suppression of invasive pests that are already present in Africa.

Acknowledgements The *icipe*-led African Fruit Fly Programme (AFFP) received funding from GIZ/BMZ, Biovision, EU, and DFID for its activities.

References

Aidoo OF, Kyerematen R, Akotsen-Mensah C, Afreh-Nuamah K (2014) Effect of some climatic factors on insects associated with citrus agro-ecosystems in Ghana. J Biodiversity Environ Sci 5:428–436

Allwood AJ, Chinajariyawong A, Drew RAI, Hamacek EL, Hancock DL, Hengsawad C, Jinapin JC, Jirasurat M, Kong Krong C, Kritsaneepaiboon S, Leong CTS, Vijaysegaran S (1999) Host plant records for fruit flies (Diptera: Tephritidae) in South-East Asia. Raffles Bull Zool Suppl 7:92

Aluja M, Mangan RL (2008) Fruit fly (Diptera: Tephritidae) host status determination: critical conceptual, methodological and regulatory considerations. Ann Rev Entomol 53:473–502

Amro MA, Abdel-Galil FA (2008) Infestation predisposition and relative susceptibility of certain edible fruit crops to the native and invading fruit flies (Diptera: Tephritidae) in the new valley oases, Egypt. Ass Univ Bull Environ Res 11:89–97

Badii KB, Billah MK, Afreh-Nuamah K, Obeng-Ofori D, Nyarko G (2015) Review of the pest status, economic impact and management of fruit-infesting fruit flies (Diptera: Tephritidae) in Africa. Afr J Agric Res 10:1488–1498

Bale JS, Masters GJ, Hodkinson ID, Awmack C, Bezemer TM, Brown VK, Butterfield J, Buse A, Coulson JC, Farrar J, Good JEG, Harrington R, Hartley S, Jones TH, Lindroth RL, Press MC, Symrnioudis I, Watt AD, Whittaker JB (2002) Herbivory in global climate change research: direct effects of rising temperature on insect herbivores. Glob Change Biol 8:1–16

Bateman MA (1972) The ecology of fruit flies. Ann Rev Entomol 17:493–518

Copeland RS, Wharton RA (2006) Year-round production of pest *Ceratitis* species (Diptera: Tephritidae) in fruit of the invasive species *Solanum mauritianum* in Kenya. Ann Entomol Soc Am 99:530–535

Copeland RS, Wharton RA, Luke Q, De Meyer M (2002) Indigenous hosts of *Ceratitis capitata* (Diptera: Tephritidae) in Kenya. Ann Entomol Soc Am 95:672–685

Copeland RS, White IM, Okumu M, Machera P, Wharton RA (2004) Insects associated with fruits of the Oleaceae (Asteridae, Lamiales) in Kenya, with special reference to the Tephritidae (Diptera). Bishop Mus Bull Entomol 12:135–164

Copeland RS, Wharton RA, Luke Q, De Meyer M, Lux S, Zenz N, Machera P, Okumu M (2006) Geographic distribution, host fruit and parasitoids of African fruit fly pests *Ceratitis anonae*, *Ceratitis cosyra*, *Ceratitis fasciventris*, and *Ceratitis rosa* (Diptera: Tephritidae) in Kenya. Ann Entomol Soc Am 99:261–278

Correia ARI, Rego JM, Olmi M (2008) A pest of significant economic importance detected for the first time in Mozambique: *Bactrocera dorsalis* Drew, Tsuruta & White (Diptera: Tephritidae: Dacinae). Boll Zool Agraria Bachicoltura Serie II 40:9–13

De Meyer M (1998) Revision of the subgenus *Ceratitis* (Ceratalaspis) Hancock (Diptera: Tephritidae). Bull Entomol Res 88:257–290

De Meyer M (2000) Phylogeny of the genus *Ceratitis* (Dacinae: Ceratitidini). In: Aluja M, Norrbom AL (eds) Fruit flies (Tephritidae): phylogeny and evolution of behavior. CRC Press, Boca Raton, pp 409–428

De Meyer M (2001) Distribution patterns and host-plant relationships within the genus *Ceratitis* (MacLeay) (Diptera: Tephritidae) in Africa. Cimbebasia 17:219–228

De Meyer M, Copeland RS, Lux SA, Mansell M, Quilici S, Wharton R, White IM, Zenz NJ (2002) Annotated check list of host plants for Afrotropoical fruit flies (Diptera: Tephritidae) of the genus *Ceratitis*. Koninklijk Museum noor Midden-Afrika Tervuren Belge, Zoölogische Documentatie 27:1–91

De Meyer M, Mohamed S, White IM (2007) Invasive fruit fly pests in Africa. Tervuren, Belgium, Royal Museum for Central Africa. http://www.africamuseum.be/fruitfly/AfroAsia.htm. Accessed 15 Sept 2015

De Meyer M, Robertson MP, Peterson AT, Mansell MW (2008) Ecological niches and potential geographical distributions of Mediterranean fruit fly (*Ceratitis capitata*) and Natal fruit fly (*Ceratitis rosa*). J Biogeogr 35:270–281

De Meyer M, Robertson MP, Mansell MW, Ekesi S, Tsuruta K, Mwaiko W, Vayssiéres J-F, Peterson AT (2010) Ecological niche and potential geographic distribution of the invasive fruit fly *Bactrocera dorsalis* (Diptera, Tephritidae). Bull Entomol Res 100:35–48

De Meyer M, Quilici S, Franck A, Chadhouliati AC, Issimaila MA, Youssoufa MA, Abdoul-Karime A-L, Barbet A, Attié M, White IM (2012) Records of frugivorous fruit flies (Diptera: Tephritidae: Dacini) from the Comoro archipelago. Afr Invertebr 53:69–77

De Meyer M, Delatte MH, Mwatawala M, Quilici S, Vayssières JF, Virgilio M (2015) A review of the current knowledge on *Zeugodacus cucurbitae* (Coquillett) (Diptera, Tephritidae) in Africa, with a list of species included in *Zeugodacus*. ZooKeys 540:539–557

De Meyer M, Mwatawala M, Copeland RS, Virgilio M (in press) Description of new *Ceratitis* species (Diptera: Tephritidae) from Africa. Europ J Taxon

De Villiers M, Manrakhan A, Addison P, Hattingh V (2013) The distribution, relative abundance, and seasonal phenology of *Ceratitis capitata*, *Ceratitis rosa*, and *Ceratitis cosyra* (Diptera: Tephritidae) in South Africa. Environ Entomol 42:831–840

Debach P (1966) Competitive displacement and coexistence principles. Ann Rev Entomol 11:183–212
Drew RAI (1994) The *Bactrocera dorsalis* complex of fruit flies (Diptera: Tephritidae: Dacinae) in Asia. Bull Entomol Res Supplement Series Supplement 2, p 68
Drew RAI, Tsuruta K, White IM (2005) A new species of pest fruit fly (Diptera: Tephritidae: Dacinae) from Sri Lanka and Africa. Afr Entomol 13:149–154
Duyck PF, Quilici S (2002) Survival and development of different life stages of three *Ceratitis* spp. (Diptera: Tephritidae) reared at five constant temperatures. Bull Entomol Res 92:461–469
Duyck PF, David P, Quilici S (2004) A review of relationships between interspecific competition and invasions in fruit flies (Diptera: Tephritidae). Ecol Entomol 29:511–520
Duyck PF, David P, Guillemette J, Brunel C, Dupont R, Quilici S (2006a) Importance of competition mechanisms in successive invasions by polyphagous tephritids in la Réunion. Ecology 87:1770–1780
Duyck PF, David P, Quilici S (2006b) Climatic niche partitioning following successive invasions by fruit flies in La Réunion. J Anim Ecol 75:518–526
Duyck PF, David P, Sandrine P, Quilici S (2008) Can host-range allow niche differentiation of invasive polyphagous fruit flies (Diptera: Tephritidae) in La Réunion? Ecol Entomol 33:439–452
Ekesi S (2010) Combating Fruit Flies in Eastern and Southern Africa (COFESA): Elements of a Strategy and Action Plan for a Regional Cooperation Program. Available at:http://www.globalhort.org/network-communities/fruit-flies/
Ekesi S, Billah MK (2006) A field guide to the management of economically important Tephritid fruit flies in Africa. icipe Science Press, Nairobi, p 106. ISBN 92 9064 179 7
Ekesi S, Nderitu PW, Rwomushana I (2006) Field infestation, life history and demographic parameters of *Bactrocera dorsalis* Drew, Tsuruta & White, a new invasive fruit fly species in Africa. Bull Entomol Res 96:379–386
Ekesi S, Billah MK, Nderitu PW, Lux SA, Rwomushana I (2009) Evidence for competitive displacement of the mango fruit fly, *Ceratitis cosyra* by the invasive fruit fly, *Bactrocera dorsalis* (Diptera: Tephritidae) on mango and mechanisms contributing to the displacement. J Econ Entomol 102:981–991
Ekesi S, De Meyer M, Mohamed SA, Virgilio M, Borgemeister C (2016) Taxonomy, ecology, and management of native and exotic fruit fly species in Africa. Ann Rev Entomol 61:219–238
El-Gendy IR, Nassar Atef MK (2014) Delimiting survey and seasonal activity of peach fruit fly, *Bactrocera zonata* and Mediterranean fruit fly, *Ceratitis capitata* (Diptera: Tephritidae) at El-Beheira Governorate, Egypt. Egypt Acad J Biol Sci 7:157–169
Elnagar S, El-Sheikh M, Hashem A, Afia Y (2010) Recent invasion by *Bactrocera zonata* (Saunders) as a new pest competing with Ceratitis capitata (Wiedemann) in attacking fruits in Egypt. Asp Appl Biol 104:97–102
El-Samea SAA, Fetoh BEA (2006) New record of *Bactrocera zonata* (Saundera) (Diptera: Tephritidae) on potatoes in Egypt. Egypt J Agr Res 84:61–63
Fadlelmula AA, Ali EBM (2014) Fruit fly species, their distribution, host range and seasonal abundance in Blue Nile State, Sudan. Persian Gulf Crop Prot 3:17–24
Fekadu M, Zenebe T (2015) Status of *Bactrocera dorsalis* (Diptera: Tephritidae) in mango-producing areas of Arba Minch, Southwestern Ethiopia. J Insect Sci doi. doi:10.1093/jisesa/ieu166
Fletcher BS (1987) The biology of dacine fruit flies. Ann Rev Entomol 32:115–144
Foba CN, Afreh-Nuamah K, Billah MK, Obeng-Ofori D (2012) Species composition of fruit flies (Diptera: Tephritidae) in the citrus museum at the Agricultural Research Centre (ARC), Kade, Ghana. Int J Trop Insect Sci 32:12–23
Geurts K, Mwatawala M, De Meyer M (2012) Indigenous and invasive fruit fly diversity along an altitudinal transect in central Tanzania. J Insect Sci 12:1–18 available online: www.insectscience.org/12.12
Goergen G, Vayssiéres J-F, Gnanvossou D, Tindo M (2011) *Bactrocera dorsalis* (Diptera: Tephritidae), a new invasive fruit fly pest for the Afrotropical region: host plant range and distribution in West and Central Africa. Environ Entomol 40:844–854

Grout TG, Stoltz KC (2007) Developmental rates at constant temperature of three economically important *Ceratitis* spp. (Diptera: Tephritidae) from Southern Africa. Environ Entomol 36:1310–1317

Guichard C (2009) EU interceptions rising in 2009. In: Fighting fruit flies regionally in sub Saharan Africa. Information Letter No. 4 October 2009. COLEACP/CIRAD, p 4

Hala N, Quilici S, Gnago AJ, N'Depo OR, N'Da Adopo A, Kouassi P, Allou K (2006) Status of fruit flies (Diptera: Tephritidae) in Côte d'Ivoire and implications for mango exports. In: Fruit flies of economic importance, from basic to applied knowledge. Proceedings of the 7th international symposium on fruit flies of economic importance, Salvador, Brazil 10–15 September 2006, p 233–239

Hancock DL, Kirk-Spriggs AH, Marais E (2001) An annotated checklist and provisional atlas of Namibian Tephritidae (Diptera: Schizophora). Cimbebasia 17:41–72

Harris EJ, Takara JM, Nishida T (1986) Distribution of the melon fly *Dacus cucurbitae* (Diptera: Tephritidae) and host plants on Kauai, Hawaiian Islands. Environ Entomol 15:488–493

Hashem AG, Saafan MH, Abdelghaffar M (1986) Population fluctuation and rate of infestation of Medfly, *Ceratitis capitata* in Upper Egypt. AlAzher J Agric Res 7:345–353

Hashem AG, Mohamed SMA, El-Wakkad MF (2001) Diversity and abundance of Mediterranean and Peach fruit flies (Diptera: Tephritidae) in different horticultural orchards. Egyptian J App Sci 16:303–314

Hodkinson ID (2005) Terrestrial insects along elevation gradients: species and community responses to altitude. Biol Rev Camb Philos Soc 80:489–513

Hussain MA, Haile A, Ahmad T (2015) Infestation of two tephritid fruit flies, *Bactrocera dorsalis* (syn. *B. invadens*) and *Ceratitis capitata*, in guava fruits from selected regions of Eritrea. Afr Entomol 23:510–513

Ibrahim Ali SA, Mahmoud MEE, Man-Qun W, Mandiana DM (2013) Survey and monitoring of some Tephritidae of fruit trees and their host range in River Nile State, Sudan. Pers Gulf Crop Prot 2:32–39

Isabirye BE, Akol AM, Mayamba A, Nankinga CK, Rwomushana I (2015) Species composition and community structure of fruit flies (Diptera: Tephritidae) across major mango-growing regions in Uganda. Int J Trop Insect Sci 35:69–79

Isabirye BE, Akol AM, Muyinza H, Masembe C, Rwomushana I, Nankinga CK (2016) Fruit fly (Diptera: Tephritidae) host status and relative infestation of selected mango cultivars in three agro ecological zones in Uganda. Int J Fruit Sci 16:23–41

Javaid I (1986) Causes of damage to some wild mango fruit trees in Zambia. Int Pest Control 28:98–99

Jayne TS, Yamano T, Nyoro J, Awuor T (2001). Do farmers really benefit from high food prices? Balancing rural interests in Kenya's maize pricing and marketing policy. Tegemeo Working Paper 2B. Egerton University, Tegemeo Institute of Agricultural Policy and Development. pp 28

Keiser I, Kobayashi RM, Miyashita DH, Harris EJ, Schneider EJ, Chambers DL (1974) Suppression of Mediterranean fruit flies by oriental fruit flies in mixed infestations in guava. J Econ Entomol 67:355–360

Liquido NJ, Shinoda LA, Cunningham RT (1991) Host plants of the Mediterranean fruit fly (Diptera: Tephritidae): an annotated world review. Misc Publ Entomol Soc Am 77:1–52

Lux SA, Copeland RS, White IM, Manrakhan A, Billah MK (2003) A new invasive fruit fly species from the *Bactrocera dorsalis* (Hendel) group detected in east Africa. Insect Sci Appl 23:355–361

Magagula CN, Ntonifor N (2014) Species composition of fruit flies (Diptera: Tephritidae) in feral guavas (*Psidium guajava* Linnaeus) and marula (*Sclerocarya birrea* (A. Richard) Hochstetter) in a subsistence savanna landscape: implications for their control. Afr Entomol 22:320–329

Manrakhan A, Hattingh V, Venter J-H, Holtzhausen M (2011) Eradication of *Bactrocera dorsalis* (Diptera: Tephritidae) in Limpopo province, South Africa. Afr Entomol 19:650–659

Mayamba A, Nankinga CK, Isabirye B, Akol AM (2014) Seasonal population fluctuations of *Bactrocera invadens* (Diptera: Tephritidae) in relation to mango phenology in the Lake Victoria Crescent, Uganda. Fruits 69:473–480

Mohamed SA, Ekesi S, Hanna R (2010) Old and new host-parasitoid associations: parasitism of the invasive fruit fly *Bactrocera dorsalis* (Diptera: Tephritidae) and five African fruit fly species by *Fopius arisanus*, an Asian opiine parasitoid. Biocontrol Sci Technol 20:183–196

Mooney HA, Cleland EE (2001) The evolutionary impact of invasive species. PNAS 98:5446–5451

Mwatawala MW, White IM, Maerere AP, Senkondo FJ, De Meyer M (2004) A new invasive *Bactrocera* species (Diptera; Tephritidae) in Tanzania. Afr Entomol 12:154–156

Mwatawala MW, De Meyer M, Makundi RH, Maerere AP (2006) Seasonality and host utilization of the invasive fruitfly *Bactrocera dorsalis* (Dipt., Tephritidae) in Central Tanzania. J Appl Entomol 130:530–537

Mwatawala M, De Meyer M, White IM, Maerere A, Makundi RH (2007) Detection of the solanum fruit fly, *Bactrocera latifrons* (Hendel) in Tanzania (Dipt., Tephritidae). J Appl Entomol 131:501–503

Mwatawala MW, De Meyer M, Makundi RH, Maerere AP (2009a) Host range and distribution of fruit-infesting pestiferous fruit flies (Diptera, Tephritidae) in selected areas of Central Tanzania. Bull Entomol Res 99:629–641

Mwatawala MW, De Meyer M, Makundi RH, Maerere AP (2009b) An overview of *Bactrocera* (Diptera: Tephritidae) invasions and their speculated dominancy over native fruit fly species in Tanzania. J Entomol 6:18–27

Mwatawala M, Maerere AP, Makund R, De Meyer M (2010) Incidence and host range of the melon fruit fly *Bactrocera cucurbitae* (Coquillett) (Diptera: Tephritidae) in Central Tanzania. Int J Pest Man 56:265–273

Mwatawala M, Virgilio M, Joseph J, De Meyer M (2015) Niche partitioning among two *Ceratitis rosa* morphotypes and other *Ceratitis* pest species (Diptera, Tephritidae) along an altitudinal transect in Central Tanzania. ZooKeys 540:429–442

N'depo OS, Hala N, N'da Adopo A, Coulibaly F, Kouassi KP, Vayssières J-F, De Meyer M (2013) Fruit flies (Diptera: Tephritidae) populations dynamic in mangoes production zone of Côte-d'Ivoire. Agric Sci Res J 3:352–363

N'Guetta K (1994) Inventory of insect fruit pests in northern Cote d'Ivoire. Fruits 49:430–443

Nakasone HY, Paull RE (1998) Tropical fruits. CAB International, Oxon, 443 pp

Nboyine JA, Billah MK, Afreh-Nuamah K (2012) Species range of fruit flies associated with mango in Ghana. J Appl Biosci 52:3696–3703

Nboyine JA, Abudulai M, Nutsugah SK, Badii B, Acheampong A (2013) Population dynamics of fruit fly (Diptera: Tephritidae) species associated with mango in the Guinea savanna agro-ecological zone of Ghana. Int J Agric Sci 3:450–454

Ndiaye O, Vayssieres J-F, Rey JY, Ndiaye S, Diedhiou PM, Ba CT, Diatta P (2012) Seasonality and range of fruit fly (Diptera: Tephritidae) host plants in orchards in Niayes and the Thiès Plateau (Senegal). Fruits 67:311–331

Ndzana Abanda FX, Quilici S, Vayssiéres JF, Kouodiekong L, Woin N (2008) Inventory of fruit fly species on guava in the area of Yaounde, Cameroon. Fruits 63:19–26

Ni WL, Li ZH, Chen HJ, Wan FH, Qu WW, Zhang Z, Kriticos DJ (2012) Including climate change in pest risk assessment: the peach fruit fly, *Bactrocera zonata* (Diptera: Tephritidae). Bull Entomol Res 102:173–183

Norrbom AL, Carroll LE, Thompson FC, White IM, Freidberg A (1999) Systematic database of names. In: Thompson FC (ed) Fruit fly expert identification system and systematic information database. Backhuys Pub. for the North American Dipterists' Society, Leiden, p 524

Nufio CR, Papaj DR (2001) Host marking behavior in phytophagous insects and parasitoids. Entomol Expet Appl 99:273–293

Nufio CR, Papaj DR (2004) Host marking behavior as a quantitative signal of infestation levels in host use by the walnut fly, *Rhagoletis juglandis*. J Ecol Entomol 29:336–344

Ouedraogo SN, Vayssières J-F, Dabiré AR, Rouland-Lefèvre C (2010) Fruitiers locaux hôtes des mouches de fruits (Diptera: Tephritidae) inféodées aux manguiers dans l'ouest du Burkina Faso: identification et taux d'infestation/Wild fruit crops as hosts of mango fruit flies (Diptera:

Tephritidae) in western Burkina Faso: identification and infestation rates (in French). Cereal Sci Bo 4:36–41

Ousmane ZK, Aboubacar K, Da Costa Correia ZA, Kadi Kadi HA, Abdourahamane TD (2014) Agro-ecological management of mango fruit flies in the northern part of Guinea-Bissau. J Appl Biosci 75:6250–6258

Quilici S, Duyck PF, Rousse P, Gourdon F, Simiand C, Franck A (2005) La mouche de la pêche sur mangue, goyave, etc. A La Réunion, évolution des recherches et des méthodes de lutte. Phytoma 584:44–47

Reitz SR, Trumble JT (2002) Competitive displacement among insects and arachnids. Ann Rev Entomol 47:435–465

Robinson AS, Hooper G (1989) Fruit flies: their biology, natural enemies and control, Vol. 3A and 3B. Elsevier, New York, p 81

Roitberg BD, Prokopy RJ (1983) Influence of host fruit deprivation on apple maggot fly response to oviposition deterring pheromone. Physiol Entomol 8:69–72

Rubabura KJA, Munyuli BMT, Bisimwa BE, Kazi KS (2015) Invasive fruit fly, *Ceratitis* species (Diptera: Tephritidae), pests in south Kivu region, eastern of Democratic Republic of Congo. Int J Innov Sci Res 16:403–408

Rwomushana I (2008) Bioecology of the new invasive fruit fly *Bactrocera invadens* (Diptera:Tephritidae) in Kenya and its interaction with indigenous mango-infesting fruit fly species. Ph.D thesis, Kenyatta University. http://ir-library.ku.ac.ke/handle/123456789/155/browse?value=Ivan%2C+Rwomushana&type=author

Rwomushana I, Ekesi S, Gordon I, Ogol CKPO (2008) Host plants and host preference studies for *Bactrocera dorsalis* (Diptera: Tephritidae) in Kenya, a new invasive fruit fly species in Africa. Ann Entomol Soc Am 101:331–340

Rwomushana I, Ekesi S, Gordon I, Ogol CKPO (2009) Mechanisms contributing to the competitive success of the invasive fruit fly *Bactrocera dorsalis* over the indigenous mango fruit fly *Ceratitis cosyra*: the role of temperature and resource pre-emption. Entomol Exp Appl 133:27–37

Saafan MH, Foda SM, Abdel-Hafez TA (2005) Ecological studies on Mediterranean fruit fly, *Ceratitis capitata* (Wied.) and peach fruit fly, *Bactrocera zonata* (Saund.) in apricot orchards. Egypt J Agric Res 83:928–934

Sallabanks R, Courtney SP (1992) Frugivory, seed predation, and insect-vertebrate interactions. Ann Rev Entomol 37:377–400

Sawadogo A, Gnankine O, Badolo A, Ouedraogo A, Ouedraogo S, Dabiré R, Sanon A (2013) First report of the fruits flies, *Ceratitis quinaria* and *Ceratitis silvestri*, on yellow plum *Ximenia americana* in Burkina Faso, West Africa. The Open Ent J 7:9–15

Schutze MK, Mahmood K, Pavasovic A, Bo W, Newman J, Clarke AR, Krosch MN, Cameron SL (2015) One and the same: integrative taxonomic evidence that *Bactrocera dorsalis* (Diptera: Tephritidae) is the same species as the Oriental fruit fly *Bactrocera dorsalis*. Syst Entomol 40:472–486

Shehata NF, Younes MWF, Mahmoud YA (2008) Biological studies on the peach fruit fly, *Bactrocera zonata* (Saunders) in Egypt. J Appl Sci Res 4:1103–1106

Shelly TD (1999) Defense of oviposition sites by female Oriental fruit flies (Diptera: Tephritidae). Fla Entomol 82:339–346

Sookar P, Permalloo S, Gungah B, Alleck M, Seewooruthun SI, Soonnoo AR (2008) An area wide control of fruit flies in Mauritius. In: Fruit flies of economic importance: from basic to applied knowledge. Proceedings of the 7th international symposium on fruit flies of economic 1mportance Salvador, Brazil, 10–15 September 2006, p 261–269

Sookar P, Alleck M, Ahseek N, Permalloo S, Bhagwant S, Chang CL (2014) Artificial rearing of the peach fruit fly *Bactrocera zonata* (Diptera: Tephritidae). Int J Trop Insect Sci 34:99–107

Stephenson AG (1981) Flower and fruit abortion: proximate causes and ultimate functions. Ann Rev Ecol Syst 12:253–279

Tanga CM, Manrakhan A, Daneel JH, Mohamed SA, Fathiya K, Ekesi S (2015) Comparative analysis of development and survival of two Natal fruit fly *Ceratitis rosa* Karsch (Diptera, Tephritidae) populations from Kenya and South Africa. ZooKeys 540:467–487

The Global Species Database (2015) Myers Enterprises II 2009–2015. Available from http://www.globalspecies.org. Accessed 16 Sept 2015

The Herbarium Catalogue (2015) Royal Botanic Gardens, Kew. http://www.kew.org/herbcat. Accessed 16 Sept 2015

The International Plant Names Index (2015) http://www.ipni.org. Accessed 16 Sept 2015

The Plant List (2015). http://www.theplantlist.org. Accessed 16 Sept 2015

Thompson FC (1998) Fruit fly expert identification system and systematic information database. The international journal of the North American Dipterists' Society, vol 9. Backhuys Publishers, Leiden, 524 pp

USDA-APHIS (2008) Federal import quarantine order for host materials of *Bactrocera invadens* (Diptera: Tephritidae), invasive fruit fly species. USDA-APHIS, Riverdale, Maryland, USA

Vargas RI, Nishida T, Beardsley JW (1983) Distribution and abundance of *Dacus dorsalis* (Diptera: Tephritidae) in native and exotic forest areas of Kauai. Environ Entomol 12:1185–1189

Vargas RI, Walsh WA, Nishida T (1995) Colonization of newly planted coffee fields: dominance of Mediterranean fruit fly over Oriental fruit fly (Diptera: Tephritidae). J Econ Entomol 88:620–627

Vayssières JF, Carel Y (1999) Les Dacini (Diptera: Tephritidae) inféodées aux Cucurbitaceae à la Réunion: gamme de plantes hôtes et stades phénologiques préférentiels des fruits au moment de la piqûre pour des espèces cultivées. Annal Soc Entomol France 35:197–202

Vayssières JF, Kalabane S (2000) Inventory and fluctuations of the catches of Diptera Tephritidae associated with mangoes in Coastal Guinea. Fruits 55:259–270

Vayssières JF, Goergen G, Lokossou O, Dossa P, Akponon C (2005) A new *Bactrocera* species in Benin among mango fruit fly (Diptera: Tephritidae) species. Fruits 60:371–377

Vayssières JF, Sanogo F, Noussourou M (2007) Inventory of the fruit fly species (Diptera: Tephritidae) linked to the mango tree in Mali, and tests of integrated control. Fruits 62:329–341

Vayssières JF, Korie S, Coulibaly O, Temple L, Boueyi S (2008) The mango tree in northern Benin: cultivar inventory, yield assessment, infested stages and loss due to fruit flies (Diptera, Tephritidae). Fruits 63:335–348

Vayssières J-F, Korie S, Ayegnon D (2009) Correlation of fruit fly (Diptera Tephritidae) infestation of major mango cultivars in Borgou (Benin) with abiotic and biotic factors and assessment of damage. Crop Prot 28:477–488

Vayssières JF, Adandonon A, Sinzogan A, Korie S (2010) Diversity of fruit fly species (Diptera: Tephritidae) associated with citrus crops (Rutaceae) in southern Benin in 2008–2009. Int J Biol Chem Sci 4:1881–1897

Vayssières JF, Vannière GPS, Barry O, Hanne AM, Niassy A, Ndiaye M, Delhove G, Korie S (2011) Preliminary inventory of fruit fly species (Diptera, Tephritidae) in mango orchards in the Niayes region, Senegal, in 2004. Fruits 66:91–107

Vayssières JF, Sinzogan A, Adandonon A, Rey J-Y, Dieng EO, Camara K, Sangaré M, Ouedraogo S, Hala N, Sidibé A, Keita Y, Gogovor G, Korie S, Coulibaly O, Kikissagbé C, Tossou A, Billah M, Biney K, Nobime O, Diatta P, N'dépo R, Noussourou M, Traoré L, Saizonou S, Tamo M (2014) Annual population dynamics of mango fruit flies (Diptera: Tephritidae) in West Africa: socio-economic aspects, host phenology and implications for management. Fruits 69:207–222

Vayssières JF, De Meyer M, Ouagoussounon I, Sinzogan A, Adandonon A, Korie S, Wargui R, Anato F, Houngbo H, Didier C, De Bon H, Goergen G (2015) Seasonal abundance of mango fruit flies (Diptera: Tephritidae) and ecological implications for their management in mango and cashew Orchards in Benin (Centre & North). J Econ Entomol 108:2213–2230

Virgilio M, Backeljau T, Emeleme R, Juakali J, De Meyer M (2011) A quantitative comparison of frugivorous tephritids (Diptera: Tephritidae) in tropical forests and rural areas of the Democratic Republic of Congo. Bull Entomol Res 101:591–597

Virgilio M, Delatte H, Quilici S, Backeljau T, De Meyer M (2013) Cryptic diversity and gene flow among three African agricultural pests: Ceratitis rosa, Ceratitis fasciventris and Ceratitis anonae (Diptera, Tephritidae). Mol Ecol 22:2526–2539

White IM, Elson-Harris MM (1992) Fruit flies of economic importance: their identification and bionomics. CAB International, Wallingford, p 601

White IM, De Meyer M, Stonehouse J (2001) A review of the native and introduced fruit flies (Diptera, Tephritidae) in the Indian Ocean Islands of Mauritius, Réunion, Rodrigues and Seychelles. In: Proceedings of the Indian Ocean commission regional fruit fly symposium, Mauritius, 5-9th June 2000, p15–21

Wong TMY, McInnis DO, Mochizuki N (1985) Seasonal distribution and abundance of adult male oriental fruit flies (Diptera: Tephritidae) in Kula, Maui, Hawaii. J Econ Entomol 78:1267–1271

I. Rwomushana is a Research Scientist with the African Fruit Fly Program at the International Centre of Insect Physiology and Ecology (*icipe*) with expertise in Agricultural Entomology. He was educated at Makerere University and Kenyatta University. He is involved in the development of sustainable integrated pest management (IPM) options for the management of fruit flies and other arthropod pests that constrain fruit production in sub-Saharan Africa. His research interests involve understanding the bioecology of fruit flies and employing balanced basic and applied research using baiting and male annihilation techniques, biopesticides, natural enemy conservation and classical biological control methods, and soft pesticides for suppression of tree crop pests. He has also coordinated several research projects in Eastern and Central Africa on a number of other arthropod pests

C.M. Tanga is an Agricultural Entomologist/Ecologist working in the Plant Health Theme at the International Centre of Insect Physiology and Ecology (*icipe*), Kenya. He obtained his PhD from the University of Pretoria, South Africa and is currently a postdoctoral research fellow at *icipe*. His area of research interest is sustainable integrated pest management (IPM) that is based on the use of biopesticides, baiting techniques, orchard sanitation and classical biological control for the management of arthropod pests of horticulture. His research activities partly involve the promotion and up-scaling of fruit fly IPM technologies amongst smallholder growers in Africa. He also develops temperature-based phenology models for African fruit flies and other crop pests to improve implementation of IPM strategies for their management

Chapter 6
Fruit Fly Species Composition, Distribution and Host Plants with Emphasis on Vegetable-Infesting Species

Chrysantus M. Tanga and Ivan Rwomushana

Abstract Vegetable crops hold a key position in smallholder agricultural production systems in Africa due to the number of farmers involved, income generation, employment opportunities, and enhancement of food and nutritional security. In many developed countries, demand for vegetables continues to grow due to increased awareness of the nutritional benefits, thereby stimulating increased domestic production and also imports from Africa. However, many pests threaten the productivity of the vegetable sector. Key amongst these are tephritid fruit flies that inflict both direct and indirect losses; alien invasive species are often responsible for severe ecological and economic impacts. The highly invasive and polyphagous melon fly, *Zeugodacus cucurbitae* (Coquillett) (= *Bactrocera cucurbitae*), remains the most damaging vegetable pest and has spread throughout more than 25 countries in sub-Saharan Africa (SSA) since its first detection in East Africa (Kenya and Tanzania). Various other indigenous species belonging to the genus *Dacus* are also known to be notorious pests of vegetables and their distribution overlaps with that of the invasive species. Despite the economic importance of fruit fly species attacking vegetables, little information is available on: their species composition in each country; the damage inflicted on cropsw; their bionomics and population dynamics; and their host plant range. Understanding these parameters is essential for formulating any Integrated Pest Management (IPM) strategy for their control. In this chapter, vegetable-infesting fruit fly species composition, distribution and host range is reviewed.

Keywords Vegetables • Fruit flies • Invasive species

C.M. Tanga (✉) • I. Rwomushana
International Centre of Insect Physiology & Ecology (*icipe*),
PO Box 30772-00100, Nairobi, Kenya
e-mail: ctanga@icipe.org

1 Introduction

Horticulture is unarguably one of the most attractive agricultural sub-sectors owing to the relatively high economic and nutritional value of the products. The sub-sector has experienced tremendous growth globally over the last decade to become the engine for agricultural and economic diversification, especially for smallholders who can gear production to specific local, regional or export markets (USAID 2004, 2005). In tropical and subtropical Africa, production of fruits and vegetables is the mainstay of many economies; it is recognized as a primary source of food security and nutrition, subsistence, employment and income generation in the rural households of over 60% of the population (ILO 2005; WRI 2005). Production and international trade in fresh fruits and vegetables has grew in volume terms by 43% and 37%, respectively between 1996 and 2006 (Legge et al. 2006). At that time the annual volume of fruits and vegetables traded internationally was approximately 73 million tonnes and valued at approximately US$45 billion (Legge et al. 2006). International trade on fresh vegetables during that decade alone increased from $11.5 billion to $18.7 billion, with the fastest growth being in chillies and green peppers (Legge et al. 2006; CBI 2005; Hallam et al. 2004; Stichele et al. 2005).

The European Union (EU) is the world's largest importer of fresh vegetables and one of the highest-priced markets. The largest EU importers of vegetables are Belgium, France, United Kingdom (UK), Netherlands and Germany, together representing more than 80% of the value of EU imports. In 2005, sub-Saharan African (SSA) countries (excluding South Africa) exported 73,788 tonnes of vegetables worth approximately US$ 153.53 million (£ 105 million) to the UK alone, which was of enormous benefit to an estimated 715,390 resource-poor small-scale growers (SSG), mostly women and young adults involved in the production, transport, processing and trade sectors (CBI 2005; Hallam et al. 2004; Stichele et al. 2005). Exports of vegetables to the EU remained relatively stable with steady growth of fresh produce exports from SSA countries, making Europe an increasingly important market (Dolan and Humphrey 2000). The volume of vegetables exported from SSA (South Africa excluded) has been dominated by Kenya, with export figure of about 33,000 tonnes, worth approximately US$ 102.35 million (£70 million) in 2005 (Legge et al. 2006). It has been reported that 75% of produce is destined for supermarkets and 20% for wholesale markets (Jaffee 2003).

In Africa, more than 90% of vegetable exports in the domestic market are destined for South Africa accounting for over 29 million US$ in 2013 (Theyse 2014). Noteworthy is that over 80% of the production is attributed to smallholders who produce for local and regional markets to obtain much-needed cash income for improving household food security and their livelihoods. Thus, vegetable production has been accorded a high priority in the national development plans of most African countries.

Despite the high potential of vegetable production, the intensification of trade has rendered the African continent highly vulnerable to the introduction of alien invasive insect pest species. Amongst the numerous insect pests that attack vegetables, none rank as high as the tephritid fruit flies that cause both direct and indirect damage. Both native and invasive species are highly polyphagous attacking a wide range of host plants such as *Capsicum annuum* L. (bell peppers), *Lagenaria siceraria* (Molina) Standl. (gourd), *Cucumis sativus* L. (cucumber), *Cucurbita maxima* Duchesne (pumpkin) and *Solanum lycopersicum* L. (tomatoes). To date, the melon fly, *Zeugodacus cucurbitae* (Coquillett), which is an invasive species has spread throughout more than 25 African countries and is globally recognized as the most important threat to the vegetable production industry (Allwood and Drew 1997; Barnes 2004; Ekesi and Billah 2007). This fruit fly species causes direct damage through female oviposition and larval feeding that result in rotting of the infested vegetable crop. Indirect losses are as a result of quarantine restrictions imposed by importing countries in response to fruit fly presence in the particular country from which the vegetables were being exported. This restricts free trade with the large lucrative export markets in Europe, the Middle East, Japan, South Africa and USA (Barnes 2004). Earnings from vegetable exports are reported to have gradually been eroded as a result of import bans by several countries due to the invasion by fruit flies (Ekesi 2010). These restrictions threaten the income of smallholder growers and traders that produce and sell fresh vegetables for income generation and improvement of their livelihoods.

2 Species Composition of Vegetable-Infesting Fruit Flies in Africa

The intensification in global trade of fresh produce has rendered the African continent highly vulnerable to the introduction of alien invasive insect species. There are historical ties between the eastern coastal area of Africa and the near East and Indian subcontinent that date back to 100 AD (Gilbert 2004). This accounts for the regular movement and shipment of commodities between these regions through trade. The rapid spread of some invasive fruit fly species across Africa may possibly have been enhanced by trade, porous borders and the fragile nature of the phytosanitary infrastructure coupled with inadequate implementation of quarantine measures. Both invasive and native fruit fly species occur all year round due to their ability to infest a wide range of both cultivated and wild host plants and overcome the challenges of geographical barriers (De Meyer et al. 2007; Lux et al. 2003; De Villiers et al. 2013). Tephritid flies of major economic importance have been well documented and belong mainly to four genera: *Bactrocera* Macquart (562 species), *Ceratitis* MacLeay (92 species), *Dacus* Fabricius (300 species), and *Zeugodacus* Coquillett (192 species). In recent years, the number of species known on the continent has increased largely due to new invasions and the identification of hitherto unknown species, although they largely still fall under the four genera (De Meyer et al. 2014).

The most noteworthy and widespread economically important tephritid flies attacking vegetable crops in Africa can be divided into two categories: invasive (the oriental fruit fly, *Bactrocera dorsalis* (Hendel); the solanum fruit fly, *Bactrocera latifrons* (Hendel); the peach fruit fly, *Bactrocera zonata* [Saunders] and *Z. cucurbitae*; and indigenous: the lesser pumpkin fly, *Dacus ciliatus* Loew; the greater pumpkin fly, *Dacus bivittatus* (Bigot); the jointed pumpkin fly, *Dacus vertebratus* Bezzi; the pumpkin fly, *Dacus frontalis* Becker; the tomato fruit fly, *Dacus punctatifrons* Karsch; *Ceratitis anonae* (Graham); the Mediterranean fruit fly [medfly], *Ceratitis capitata* (Wiedemann); *Ceratitis catoirii* (Guérin-Méneville); *Ceratitis fasciventris* (Bezzi); and the Natal fruit fly, *Ceratitis rosa* (Karsch) (White and Elson-Harris 1992; Mwatawala et al. 2004, 2009a; Vayssiéres et al. 2005, 2007; Rwomushana et al. 2008; Isabirye et al. 2016; Georgen et al. 2011; Nboyine et al. 2013). All the species listed above have large overlaps in geographical range with *Z. cucurbitae* and thus, there is, to a varying extent, interspecific competition for the same larval food sources (De Meyer et al. 2015). Amongst the tephritids attacking vegetables, *Z. cucurbitae* predominates on most vegetable hosts, wherever it occurs, compared with the indigenous *Dacus* and *Ceratitis* species; it is followed by *D. ciliatus*, which predominates on some hosts. The other species of tephritid flies are ranked as moderate, and are localized in their distribution causing varying degrees of infestation on vegetables depending on the season, variety and agroecology. Studies on the interspecific interactions between these tephritid fruit flies on vegetables in Africa are scarce (De Meyer et al. 2015).

3 Distribution of Vegetable-Infesting Fruit Flies in Africa

3.1 Bactrocera dorsalis

In 2003, an invasive species of fruit fly in the genus *Bactrocera* Macquart was detected in Kenya (Lux et al. 2003) and was described as *Bactrocera invadens* Drew, Tsuruta & White (Drew et al. 2005). Recent integrative taxonomic studies of *B. invadens*, the 'African Invader Fly' revealed that it exhibited the same biological characteristics as *B. dorsalis* which is a complex of species known to cause extensive damage globally (Drew 1994). Consequently, *B. invadens* was synonymized with *B. dorsalis* in 2015 (Schutze et al. 2015).

Since the first report, it spread rapidly across the African continent and, in addition to Kenya, it is now known from more than 30 other countries including: Angola, Benin, Burkina Faso, Burundi, Cameroon, Central African Republic, Cape Verde, Comoros Archipelago, Mayotte, Chad, Democratic Republic of Congo, Ethiopia, Eritrea, Equatorial Guinea, Gabon, Gambia, Ghana, Guinea, Guinea Bissau, Côte d'Ivoire, Mali, Mozambique, Namibia, Niger, Nigeria, Senegal, Sierra Leone, Sudan, Tanzania, Togo, Uganda, Liberia, Mali, Mauritania and Zambia (Drew et al. 2005; Vayssières et al. 2005; Mwatawala et al. 2006; Correia et al. 2008; Rwomushana et al. 2008; Goergen et al. 2011; Virgilio et al. 2011; De Meyer et al. 2008, 2012; Ibrahim Ali et al. 2013; Hussain et al. 2015; Isabirye et al. 2015).

The species is thought to have invaded Africa from the Indian subcontinent and it was discovered in Sri Lanka after it was first reported from Africa (Drew et al. 2005), where it has become a significant pest of quarantine and economic importance (Mwatawala et al. 2004; Vayssieres et al. 2005; Ekesi et al. 2006; Ekesi 2006).

3.2 Bactrocera latifrons

This species is native to South and Southeast Asia (e.g., Brunei Darussalam, Pakistan, India, Sri Lanka, Burma, China, Thailand, Laos, Vietnam, Malaysia, Singapore, Indonesia, Laos, Malaysia and Taiwan) but with adventive populations in Hawaii and Japan (Vargas and Nishida 1985; Liquido et al. 1994; White and Elson-Harris 1992, 1994; Carroll et al. 2002; Shimizu et al. 2006). It was detected for the first time in the Morogoro region of eastern central Tanzania in 2006 (Mwatawala et al. 2007), this also being the first record of the pest in Africa. The exact time and point of entry of the pest into Tanzania are unknown and the only additional record of *B. latifrons* was reported in 2007 from southern Kenya (Ekesi, pers. communication; De Meyer et al. 2008) near the border with Tanzania. Apart from this the species has not been reported from any other African country (De Meyer et al. 2007). Despite its narrow geographic distribution in Africa, it remains a pest of quarantine importance and has the potential to permanently establish itself and compete and/or coexist with other native and previously introduced tephritid species (Liquido et al. 1994; Mziray et al. 2010).

3.3 Bactrocera zonata

This is one of the most harmful tephritid species. It is native to India where it was first recorded in Bengal (Kapoor 1993). It is present in South and Southeast Asia: Bangladesh, Bhutan, Iran, Laos, Myanmar, Nepal, Oman, Pakistan, Saudi Arabia, Sri Lanka, Thailand, United Arab Emirates, Vietnam and Yemen. In Africa, *B. zonata* was first recorded in Egypt in 1924 and since then it has spread to Libya, the Indian Ocean islands of Mauritius and La Réunion (Hashem et al. 2001; Quilici et al. 2005). Recently it has been reported from several regions in Sudan, suggesting a southward spread and potential risk of invasion of the sub-Saharan region (El-Samea and Fetoh 2006; De Meyer et al. 2007; Shehata et al. 2008; Elnagar et al. 2010; EPPO 2013). This clearly demonstrates that this pest has the ability to establish outside tropical climates (Hashem et al. 2001), and adapt to local conditions (Iwahashi and Routhier 2001). Pest risk analysis suggests that *B. zonata* would be capable of entering, establishing and spreading in coastal areas of the Mediterranean region, causing significant damage to horticultural production (Delrio and Cocco 2010).

3.4 Ceratitis anonae

Ceratitis anonae has a predominantly equatorial distribution, being widespread in Senegal through to Cameroon, Central African Republic, Côte d'Ivoire, Democratic Republic of Congo (Brazzaville), Gabon, Ghana, Equatorial Guinea, Kenya, Sao Tomé and Principé, Guinea (Conakry), Mali, Nigeria, Togo, Tanzania and Uganda (White and Elson-Harris 1992; Copeland et al. 2006).

3.5 Ceratitis capitata

Ceratitis capitata is one of the world's most widespread and damaging fruit fly pests of horticultural crops. It originated in Africa (White and Elson-Harris 1992) and spread first to the Mediterranean region during the early nineteenth century and from there to the rest of world (Headrick et al. 1996). It occurs in most tropical and temperate regions. Some countries have succeeding in eradicating outbreaks of recently introduced populations (Dowell and Penrose 1995; Clark et al. 1996; Penrose 1996) and even well-established populations (Hendrichs et al. 1983; Fisher et al. 1985). It is considered to be a major quarantine pest throughout its geographic range. In Africa, it has been reported in Algeria, Angola, Benin, Botswana, Burkina Faso, Burundi, Cameroon, Congo, Democratic Republic of Congo, Egypt, Ethiopia, Gabon, Ghana, Guinea, Côte d'Ivoire, Kenya, Liberia, Libya, Malawi, Morocco, Mozambique, Niger, Nigeria, Senegal, South Africa, Sudan, Tanzania, Togo, Tunisia, Uganda, Sao Tome & Principé, Mauritius, Sierra Leone, Seychelles, South Africa, La Réunion and Zimbabwe (White and Elson-Harris 1992; De Villiers et al. 2013). For global distribution and predictions see De Meyer et al. (2008).

3.6 Ceratitis catoirii

Ceratitis catoirii is a less common *Ceratitis* species, reported so far in Mauritius, La Réunion and Seychelles (Duyck et al. 2004).

3.7 Ceratitis fasciventris

Ceratitis fasciventris has a reasonably wide distribution and is found throughout western Africa (Côte d'Ivoire, Democratic Republic of Congo, Ghana, Equatorial Guinea, Mali, Nigeria), with isolated records from Central African Republic, and extensive distribution along the Albertine and Gregory Rifts in eastern Africa (Tanzania, Kenya, Burundi, Rwanda, Uganda), as far north as Ethiopia and South as far as Angola, Namibia and Zambia as well as near the equator (Sao Tomé and Principé) (White and Elson-Harris 1992; Copeland et al. 2006; De Meyer et al. 2015).

3.8 Ceratitis rosa

Ceratitis rosa is largely native to tropical Africa (White and Elson-Harris 1992), with the earliest description being made in 1887 from specimens collected from Mozambique (Anonymous 1963; Weems 1966; Botha et al. 2004). Use of an integrated research approach under the umbrella of the Coordinated Research Project (CRP) on cryptic species revealed that *C. rosa* comprises two entities: 'R1' or 'lowland' and 'R2' or 'highland' rosa (De Meyer et al. 2015). R2 is now regarded as a new species called *C. quilicii* while R1 remains as *C. rosa* (De Meyer, personal communication 13 April 2016). However, throughout this chapter both species will still be referred to as *C. rosa*. *Ceratitis rosa* is a pest of phytosanitary concern that could potentially restrict international trade (Barnes 2000; Barnes et al. 2007; EPPO 2007). The pest is present in the central highlands of Kenya and has spread to the remote Indian Ocean Islands of La Réunion and Mauritius, which seems to be a recent expansion (White et al. 2000, 2001; Copeland and Wharton 2006), and where it has been reported to be strongly competitive against other indigenous and invasive *Ceratitis* species (Hancock 1984; Quilici et al. 2002; Duyck et al. 2006). Thus, *C. rosa* raises major concerns because of the risks of expansion outside its native range (De Meyer et al. 2008; Li et al. 2009; de Villiers et al. 2012) and because of its capacity to attack fruits normally grown in subtropical or temperate climates (Mwatawala et al. 2009b). It is feared that it might become a global threat due to its higher tolerance to low temperatures (Duyck and Quilici 2002; Copeland et al. 2006) compared with *C. capitata*. Elsewhere within the African continent, the species has not demonstrated a high degree of invasiveness, as can be seen by the limited expansion of its distribution beyond its historical native range, which includes Angola, Ethiopia, Democratic Republic of Congo, Kenya, Malawi, Mali, Mozambique, Nigeria, Mauritius and La Réunion, Rwanda, Seychelles, South Africa, Swaziland, Tanzania, Uganda, Zambia and Zimbabwe (White and Elson-Harris 1992; Copeland et al. 2006; De Villiers et al. 2013). The single record of *C. rosa* (one male) from Cameroon in West Africa is probably due to misidentification while the presence of the species in Côte d'Ivoire (N'dépo et al. 2010, 2013) is an erroneous record. No reliable records from western Africa have been found (De Meyer et al. 2015).

3.9 Dacus bivittatus

Dacus bivittatus is a widespread species occurring throughout Africa, especially in Angola, Cameroon, Chad, Egypt, Eritrea, Democratic Republic of Congo, Ghana, Kenya, Malawi, Mozambique, Nigeria, Senegal, Guinea, Seychelles, Mayotte and the Comoros, Sierra Leone, South Africa, Tanzania, Uganda, Botswana, Guinea, Côte d'Ivoire, Togo, Benin, Gabon, Congo, Ethiopia, Sudan, Somalia, Rwanda, Namibia, Zambia, Lesotho, La Réunion, Mauritius, Madagascar and Zimbabwe (White and Elson-Harris 1992; IIE 1995; White et al. 2000; Bordat and Arvanitakis 2004; De Meyer 2009; De Meyer et al. 2015).

3.10 Dacus ciliatus

This species is widespread in Africa but also occurs in the Near and Middle East, and South Asia. On the African continent it appears to thrive in the drier regions, such as the Sahelian, Karoo, Angola, Botswana, Cameroon, Chad, Democratic Republic of Congo, Egypt, Ethiopia, Ghana, Guinea, Kenya, Lesotho, Malawi, Mozambique, Nigeria, Senegal, Sierra Leone, Somalia, South Africa, Sudan, Tanzania, Uganda, Côte d'Ivoire, Burkina Faso, Togo, Benin, Niger, Gabon, Madagascar, Namibia, Zambia and Zimbabwe (White and Elson-Harris 1992). It has also been reported on the Indian Ocean islands of Mauritius and La Réunion (White 2006) and on Madagascar (Mansell 2006).

3.11 Dacus frontalis

Dacus frontalis is also widely distributed in the African continent. Amongst the different tephritid species, it is considered locally as a very serious pest of cucurbits in many countries in Africa and the Middle East (Ekesi and Billah 2007). It was reported in Tunisia for the first time in 2015, where it was found in several locations (Hafsi et al. 2015). It has also been reported in South Africa, Kenya, Zimbabwe, Namibia, Tanzania, Eritrea, Angola, Lesotho, Congo, Botswana, Nigeria, Cape Verde, Benin, Egypt and Sudan. There are also some records of its presence in North Africa in Algeria and Libya. It also occurs outside Africa in the Middle East in Yemen, the United Arab Emirates and Saudi Arabia (White 2000; EPPO 2003; Ekesi and Billah 2007).

3.12 Dacus punctatifrons

Dacus punctatifrons is widespread and has been recorded in several African countries including Angola, Benin, Botswana, Cameroon, Congo, Democratic Republic of Congo, Ethiopia, Gabon, Ghana, Guinea, Côte d'Ivoire, Kenya, Liberia, Malawi, Mozambique, Niger, Nigeria, South Africa, Sudan, Tanzania, Togo, Uganda, Sierra Leone, Zambia, Namibia, Burundi, Rwanda, Swaziland, Madagascar and Zimbabwe, with a (possible) adventive population in Yemen (White and Elson-Harris 1992; Mansell 2006; White and Goodger 2009; De Meyer et al. 2015).

3.13 Dacus vertebratus

This species is a widespread cucurbit feeder, occurring in most Afro-tropical countries including, Angola, Botswana, Cameroon, Chad, Democratic Republic of Congo, Ethiopia, Ghana, Guinea, Kenya, Malawi, Mozambique, Nigeria, Senegal, South Africa, Sudan, Tanzania, Uganda, Côte d'Ivoire, Burkina Faso, Togo, Benin, Niger, Gabon, Madagascar, Namibia, Zambia, Zimbabwe, Gambia, Liberia, Mali, Burkina Faso, Gabon, Eritrea, Madagascar, Rwanda, Swaziland, Mayotte and the Comoros as well as the Arabian Peninsula (Bordat and Arvanitakis 2004; De Meyer et al. 2015).

3.14 Zeugodacus cucurbitae

Zeugodacus cucurbitae is abundant throughout Central and East Asia (including Pakistan, India, Bangladesh, Nepal, China, Indonesia and the Philippines) and Oceania (including New Guinea and the Mariana Islands) (Dhillon et al. 2005; Drew 1989; Drew et al. 1982; Drew and Romig 2013). In some of these regions it has been introduced more than once and subjected to a number of eradication programmes (Suckling et al. 2014). It has been detected a number of times in the Hawaiian Islands and California (Papadopoulos et al. 2013) and its presence there was also the result of accidental human-mediated introduction. The first record of this species from the African mainland was restricted to coastal Tanzania and Kenya, and dates back to 1936 (Bianchi and Krauss 1937). Despite its occurrence in eastern Africa for many decades, *Z. cucurbitae* apparently did not spread rapidly to other parts of the continent until the late 1990s when it was reported from other parts of Africa. Currently, it is present in more than 25 countries namely: Tanzania, Kenya, Mauritius, La Réunion, Gambia, Côte d'Ivoire, Seychelles, Mali, Burkina Faso, Guinea, Nigeria, Cameroon, Senegal, Ghana, Benin, Niger, Democratic Republic of Congo, Togo, Sudan, Sierra Leone, Uganda, Burundi, Ethiopia, Malawi and Mozambique (Vayssières and Carel 1999; White et al. 2001; De Meyer et al.

2012, 2015). The wide dispersal of this insect has increased awareness of its economic significance but not much research has been devoted to this species in comparison to other cucurbit-infesting fruit flies, except in La Réunion (White and Elson-Harris 1994; Vayssières and Carel 1999; Ryckewaert et al. 2010) and Mauritius (Sookar et al. 2012, 2013).

4 Host Plants of Vegetable-Infesting Fruit Flies in Africa

Knowledge of the host range of vegetable-infesting fruit fly species throughout their geographic ranges is generally limited. However, there is continuous adaptation of these flies to new hosts and as such the inclusion of hitherto unknown hosts is likely. Several studies have documented the presence of both exotic and native fruit fly species on vegetables in Africa and their host plants (White and Elson Harris 1992; Copeland et al. 2002; Vayssieres and Kalabane 2000; Copeland et al. 2004, 2006; De Meyer et al. 2002; Ekesi et al. 2006; Vaysiérres et al. 2005; Rwomushana et al. 2008; Ryckewaert et al. 2010). Generally, in vegetable production systems in Africa, cucurbit hosts are preferred over non-cucurbit vegetable host species, with very low infestation rates and incidences in the latter. However, the presence of non-cucurbits in fields/backyards can provide alternative hosts and encourage a population buildup that can then attack other edible vegetables and cause significant losses. Vayssières et al. (2007) indicated that there are geographical differences between vegetable-infesting tephritid flies with some being more oligophagous than others (Jacquard et al. 2013). Also, infestation rates can differ according to region. Although the low preference for non-vegetable hosts has limited impact on actual crop loss, the mere presence of commercial hostplants, such as mango (*Mangifera indica* L.), citrus (*Citrus* spp.) or carambola (*Averrhoa carambola* L.), can have regulatory implications for export of other commodities. Despite this, other polyphagous fruit fly species in Africa, such as *B. dorsalis*, *C. capitata* and *C. rosa*, which attack these commercial non-vegetable hosts, are rarely encountered in vegetable crops (Mwatawala et al. 2009a).

The host plants for each fruit fly species documented below were extracted from published papers on host plants in Africa and insect records that are publicly available from the Royal Museum for Central Africa (http://projects.bebif.be/fruitfly/index.html) (Table 6.1). The author names for each species follow the nomenclature of the International Plant Names Index (www.ipni.org) which was cross referenced with the Global Species Database (www.globalspecies.org) and the Herbarium Catalogue (www.kew.org/herbcat). This represents a summary of host plant specialization of fruit flies, their species richness and abundance on individual hosts and the compartmentalization of the plant–fruit fly food web (Table 6.1).

6 Fruit Fly Species Composition, Distribution and Host Plants with Emphasis...

Table 6.1 Host plant species, common name and infestation status of tephritid fruit flies

Plant family	Host plants, common names and pest status
Cucurbitaceae	*Benincasa hispida* (Thunb.) Cogn. (Chinese melon)[j,k], *Cephalendra indica* Naud. (Kundru)[j], *Citrullus colocynthis* (L.) Schrad. (colocynth)[a,j,h,i,l], *Citrullus lanatus* (Thunb.) Matsum. & Nakai (watermelon)[a,j,h,i,k,n,l], *Citrullus vulgaris* Schrad (African melon)[j], *Coccinia grandis* (L.) Voigt (Wild cucurbits)[j,h,k], *Coccinia indica* Wight & Arn. (Ivy gourd)[j], *Coccinia dipsaceus* Ehrenb. ex Spach (Wild cucurbits)[j], *Coccinia palmate* M.Roem.[h], *Coccinia quinqueloba* (Thunb.) Cogn.[h], *Coccinia trilobata* (Cogn.) C. Jeffrey[h,i], *Corallocarpus ellipticus* Chiov.[h,i], *Cucumis aculeatus* Cogniaux[i], *Cucumis metuliferus* Naudin (African horned cucumber)[i], *Corallocarpus schimperi* Hook. F.[h], *Cucumeropsis edulis* Cogn.[h], *Cucumis anguria* L. (Wild cucurbits)[j,h,l], *Cucumis ficifolius* A.Rich.[a], *Cucumis melo* L. (melon)[a,d,k,l,n,m], *Cucumis sativus* L. (cucumbers, gerkins)[a,d,j,h,i,l,n,m], *Cucumis africanus* L.f.[h,i], *Cucumis dipsaceus* Ehrenb. ex Spach (hedgehog gourd)[d,h,i,k], *Cucumis melo* C. melo var. conomon (Muskmelon)[j,i,k], *Cucumis melo* var. momordica (Snap melon)[j,h], *Cucumis pubescens* Willd. (Wild cucurbit)[j], *Cucumis sativus* L. (cucumber)[h,k], *Cucumis trigonus* Roxb. (Wild cucurbits)[j], *Cucumis utilissimus* Roxb (Long melon)[j], *Cucumis vulgaris* var *fistulosus* (Squash melon)[j], *Cucurbita maxima* Duchesne (giant pumpkin)[a,g,j,h,i,n,k], *Cucurbita moschata* Duchesne (butternuts)[j], *Cucurbita pepo* L. (ornamental gourd, squash)[a,j,h,i,l], *Diplocyclos palmatus* (L.) C.Jeffrey. (Balsam apple)[j,k], *Cucumis melo* L. subsp. melo var. conomon (Thunb.) Makino (pickling melon)[l,k], *Kedrostis leloja* (Forsk. ex J.F.Gmel.) C. Jeffrey[h,i], *Kedrostis foetidissima* Cogn.[h,i], *Lagenaria abyssinica* (Hook.f.) C. Jeffrey[h], *Lagenaria amebicana* (Wild cucurbits)[j], *Lagenaria siceraria* (Molina) Standl. (calabash, water bottle)[a,j,h,i,k], *Lagenaria sphaerica* E.Mey.[h], *Lagenaria vulgaris* Ser. (Bottle gourd)[j], *Luffa* sp Mill.[h], *Luffa aegyptiaca* Mill.[h,i], *Luffa acutangula* (L.) Roxb. (Ribbed gourd)[j,h,i], *Luffa cylindrica* M. Roem. (Sponge gourd)[j,h], *Momordica balsamina* L.[h,i], *Momordica charantia* L. (bitter gourd)[a,b,j,h,i,k], *Momordica calantha* Gilg[f], *Momordica rostrata* Zimm.[h,i], *Momordica trifoliolata* Hook.f[h,k], *Mukia maderaspatana* (L.) M.Roem.[h], *Telfairia pedata* Hook.[h], *Trichosanthes anguina* L[j], *Trichosanthes cucumeria* (Snake gourd)[j,h], *Peponium mackenii* Engl.[h,i], *Peponium vogelii* Engl.[h], *Sycos pachycarpus* (Wild cucurbit)[j], *Sechium edule* (Jacq.) Sw.[h,m], *Trichosanthes dioica* Roxb. (Pointed gourd)[j], *Trichosanthes cucumerina* Linn. var. anguinea (Wild cucurbit or snakegourd)[j,h,i,k], *Momordica cf trifoliata* L.[k], *Cucumeropsis mannii* Naud. (White egusi)[h,n],
Fabaceae	*Inga laurina* (Sw.) Willd. (ice cream bean)[g], *Vigna sesquipedalis* (L.) Fruw. (Cowpea)[j], *Vigna sinensis* (L.) Savi (Cowpea)[j], *Vigna unguiculata* (L.) Walp. (Long bean or Cowpea)[j]

(continued)

Table 6.1 (continued)

Plant family	Host plants, common names and pest status
Leguminosae	*Cajanus cajan* (L.) Millsp. (Pigeon pea)[j], *Cordyla pinnata* (A.Rich.) Milne-Redh.[a], *Dolichos lablab* L. (Hyacinth bean)[j,k], *Faidherbia albida* (Delile) A.Chev.[n], *Inga laurina* (Sw.) Willd. (Spanish oak)[g], *Pithecellobium dulce* (Roxb.) Benth.[d], *Angylocalyx braunii* Harms[g], *Phaseolus vulgaris* L. (French bean)[j], *Phaseolus limensis* L. (Lime bean)[j], *Phaseolus radiatus* L. (Green gram)[j]
Lauraceae	*Cinnamomum verum* J.Presl (cinnamon)[d], *Persea americana* Mill. (avocado) [a,b,c,d,e,f,g,j]
Malvaceae	*Abelmoschus esculentus* (L.) Moench (Okra)[j,h], *Cola natalensis* Oliv.[d,g], *Durio zibethinus* L. (durian)[a,d], *Grewia asiatica* L. (phalsa)[b],
Poaceae	*Zea mays* L. (maize, corn)[j]
Solanaceae	*Capsicum annuum* L. cov. longum A. DC. (bell pepper)[a,d,e,i,k], *Capsicum chinense* Jacq. (Bonnet pepper)[k], *Capsicum frutescens* L. (chilli)[a,d,g,j,k], *Lycianthes biflora* (Lour.) Bitter[k], *Cyphomandra* sp. Mart. ex Sendtn.[d], *Cyphomandra betacea* (Cav.) Miers (tree tomato)[d], *Lycium* L.sp. (boxthorns)[d], *Lycium barbarum* L. (Matrimonyvine)[d], *Lycium europaeum* L. (european boxthorn)[d], *Physalis peruviana* L. (Cape gooseberry)[d,k], *Solanum aculeatissimum* Jacq. (Dutch eggplant)[k], *Solanum giganteum* Jacq.[g], *Solanum incanum* L. (grey bitter-apple)[d,k], *Solanum lycopersicum* L. (tomato)[a,d,e,g,j,h,i,m,k], *Solanum macrocarpon* L. (local garden egg)[d,k], *Solanum mauritianum* Scop. (bugweed, bugtree)[c,d,f,g], *Solanum melongena* L. (eggplant)[d,j,h,k], *Solanum nigrum* L. (black nightshade)[d,k], *Solanum muricatum* Aiton (melon pear)[d], *Solanum pseudocapsicum* L. (Jerusalem-cherry)[d,k], *Solanum tuberosum* L. (potato)[b], *Solanum seaforthianum* Andrews (Brazilian nightshade)[d], *Solanum aethiopicum* L. (Chinese scarlet eggplant)[k], *Solanum anguivi* Lam. (African eggplant)[k], *Solanum donianum* Walp.[k], *Solanum dulcamaroides* Poir.[k], *Solanum erianthum* D. Don (Big eggplant)[k], *Solanum granulosoleprosum* Dunal[k], *Solanum lasiocarpum* Dunal (Indian nightshade)[k], *Solanum linnaeanum* Hepper and P.M. L. Jaeger (Black-spine nightshade)[k], *Solanum lycopersicum* L. var. *cerasiforme* (Alef.) Fosberg (Cherry tomato)[k], *Solanum mammosum* L. (Macawbush)[k], *Solanum nigrescens* M. Martens and Galeotti (Divine nightshade)[k], *Solanum pimpinellifolium* L. (Currant tomato)[k], *Solanum scabrum* Mill. (Garden-huckleberry)[k], *Solanum* sect. *Lycopersicon* spp.[i,k], *Solanum sisymbriifolium* Lam. (Dense-thorn bitter-apple)[k], *Solanum stramoniifolium* Jacq.[k], *Solanum torvum* Sw. (Pea eggplant)[k], *Solanum trilobatum* L. (Purple-fruited pea eggplant)[k], *Solanum viarum* Dunal (Tropical soda-apple)[k], *Solanum violaceum* Ortega[k], *Solanum virginianum* L. (Yellow-fruit nightshade)[k], *Solanum sodomeum* L.[k], *solanum indicum* L. (Indian Nightshade)[k], *Solanum virginianum* L. (yellow-berried nightshade) [k], *Solanum sarmentosum* Nees[k], *Solanum verbascifolium* L. (Mullein nightshade)[k]

[a]*B. dorsalis;* [b]*B. zonata;* [c]*C. anonae;* [d]*C. capitata;* [e]*C. catoirii;* [f]*C. fasciventris;* [g]*C. rosa;* [h]*D. bivitattus* and [i]*D. ciliatus;* [j]*Zeugodacus cucurbitae;* [k]*B. latifrons;* [l]*D. frontalis;* [m]*D. punctatifrons;* [n]*D. vertebratus*

5 Future Perspectives

In Africa production of vegetables is the mainstay of many economies, being one of the primary sources of food security and nutrition, subsistence, employment and income generation for rural households. In many developed countries, domestic demand for vegetables continues to grow, thereby providing ready market outlets for increased domestic production and imports from Africa. Despite the high potential of vegetable production, intensification of trade has rendered the African continent highly vulnerable to the introductions of alien invasive insect species, amongst which none rank as high as tephritid fruit flies causing both direct and indirect damage. Despite research advances in our understanding of the biology of these invasive species on fruits and continued development of management strategies to protect fruit in Africa, the same has not been applied to the primary species that are devastating and jeopardizing vegetable production in the continent. Losses of as much as 80% of tomato and 100% of cucurbit crop harvests have been frequently observed (Ryckewaert et al. 2010). The losses from fruit flies can be caused by a single species or as a result of several species attacking the same vegetable crop at the same time. In addition to the direct losses, fruit fly infestation can result in serious losses in trade and export opportunities due to stricted quarantine regulations imposed by most importing countries.

Beside the tephritid fruit flies that are native to the African continent (e.g. *D. ciliatus*, *D. frontalis*, *D. punctatifrons* and *D. vertebratus*), some are accidentally introduced from other regions, in particular from Asia. So far, four Asian species belonging to the genera *Bactrocera* and *Zeugodacus* have invaded many countries in Africa (*B. dorsalis*, *B. latifrons*, *B. zonata* and *Z. cucurbitae*) with high levels of risk of spreading to new locations. There is, therefore, an urgent need for considerable strengthening of the human and physical quarantine and monitoring infrastructures in Africa, in order to avoid any further unwanted introductions of other highly destructive and polyphagous species such as: *Bactrocera atrisetosa* (Perkins); *Bactrocera caudata* (Fabricius); the cucumber fruit fly, *Bactrocera cucumis* (French); the pumpkin fruit fly, *Bactrocera decipiens* (Drew); also commonly called the pumpkin fruit fly, *Bactrocera depressa* (Shiraki); the three-striped fruit fly, *Bactrocera diversa* (Coquillett); *Bactrocera scutellaris* (Bezzi); the striped fruit fly, *Bactrocera scutellata* (Hendel); *Bactrocera strigifinis* (Walker); *Bactrocera tau* (Walker); *Bactrocera triangularis* (Drew); *Bactrocera trichosanthes* Drew & Romig; *Bactrocera trimaculata* (Hardy & Adachi); and the Queensland fruit fly, *Bactrocera tryoni* (Froggatt) (Vargas et al. 2015).

Although, some commercially available products against fruit flies exist in Africa, the wide-scale unavailability of such technologies for smallholder growers across Africa has forced farmers to resort to synthetic pesticides with all the risks that this entails for human health and the environment. These practices against vegetable fruit flies have proven to be entirely insufficient in limiting the damage to crops below economic threshold. Therefore, there is an urgent need to source alternative environmentally-friendly methods to combat these devastating pests on vegetable crops across Africa. A promising development is the use of female lures based on locally available products such as waste brewer's yeast that can bring quick reprieve to the farmer. There is a great need for increased research on the bioecology of these vegetable crop feeders, new attractants and exploration for efficient biological control agents as well as novel, economic and environmentally-friendly methods aimed at controlling these tephritid flies in Africa to improve trade and enhance the livelihoods of smallholder producers.

Acknowledgements The *icipe*-led African Fruit Fly Programme (AFFP) received funding from GIZ/BMZ, Biovision, EU, and DFID for its activities.

References

Allwood AJ, Drew RAI (1997) Fruit fly management in the Pacific. ACIAR proceedings no 76, 267 pp
Anonymous (1963) Insects not known to occur in the United States: natal fruit fly (*Ceratitis rosa* Karsch). USDA Coop Econ Insect Rep 13:1132–1134
Barnes BN (2000) Monitoring and control of fruit flies in South African fruit orchards. pp. 147–152 In: Price NS, Seewooruthun I (eds) Proceedings of the Indian Ocean Commission, regional fruit fly symposium. Indian Ocean Commission, Flic en Flac, Mauritius, 5–9 June 2000, pp 147–152
Barnes BN (2004) Proceedings of the 6th international symposium on fruit flies of economic importance. Ultra Litho (Pty), Johannesburg, 510 pp
Barnes BN, Rosenberg S, Arnolds L, Johnson J (2007) Production and quality assurance in the SIT Africa Mediterranean fruit fly (Diptera: Tephritidae) rearing facility in South Africa. Fla Entomol 90:41–52
Bianchi FA, Krauss NH (1937) Fruit fly investigations in East Africa. Hawaii Plant Rec 41:299–306
Bordat D, Arvanitakis L (2004) Arthropodes des cultures le´gumie`resd'Afrique de l'Ouest, Centrale, Mayotte et Réunion. CIRAD, Montpellier
Botha J, Hardie D, Barnes B (2004) Natal fruit fly *Ceratitis* (*Pterandrus*) *rosa*. Exotic threat to Western Australia. Department of Agriculture Factsheet No. 6, Government of Western Australia, ISSN 1443–7783
Carroll LE, White IM, Freidberg A, Norrbom AL, Dallwitz MJ, Thompson FC (2002) Pest fruit flies of the world. Version: 8th December 2006. http://delta-intkey.com/ffa/
CBI (2005) EU Market survey: fresh fruit and vegetables, compiled for CBI by ProFound, Netherlands, pp 172
Clark RA, Steck GJ, Weems HV Jr (1996) Detection, quarantine, and eradication of fruit flies in Florida. In: Rosen D, Bennett FD, Capinera JL (eds) Pest management in the sub-tropics: integrated pest management—a Florida perspective. Intercept Ltd., Andover, pp 29–54
Copeland RS, Wharton RA (2006) Year-round production of pest *Ceratitis* species (Diptera: Tephritidae) in fruit of the invasive species *Solanum mauritianum* in Kenya. Anns Entomol Soc America 99:530–535
Copeland RS, Wharton RA, Luke Q, De Meyer M (2002) Indigenous hosts of *Ceratitis capitata* (Diptera: Tephritidae) in Kenya. Anns Entomol Soc Am 95:672–685
Copeland RS, White IM, Okumu M, Machera P, Wharton RA (2004) Insects associated with fruits of the Oleaceae (Asteridae, Lamiales) in Kenya, with special reference to the Tephritidae (Diptera). Mus Bull Entomol 12:135–164

Copeland RS, Wharton RA, Luke Q, De Meyer M, Lux S, Zenz N, Machera P, Okumu M (2006) Geographic distribution, host fruit and parasitoids of African fruit fly pests *Ceratitis anonae*, *Ceratitis cosyra*, *Ceratitis fasciventris*, and *Ceratitis rosa* (Diptera: Tephritidae) in Kenya. Anns Entomol Soc Am 99:261–278

Correia ARI, Rego JM, Olmi M (2008) A pest of significant economic importance detected for the first time in Mozambique: *Bactrocera dorsalis* Drew, Tsuruta & White (Diptera: Tephritidae: Dacinae). Boll Zool Agraria Bachicol, Serie II 40:9–13

De Meyer M (2009) Taxonomic revision of the fruit fly genus *Perilampsis* Bezzi (Diptera, Tephritidae). J Nat Hist 43:2425–2463. http://dx.doi.org/10.1080/00222930903207868

De Meyer M, Copeland RS, Lux SA, Mansell M, Quilici S, Wharton R, White IM, Zenz NJ (2002) Annotated check list of host plants for Afrotropical fruit flies (Diptera: Tephritidae) of the genus Ceratitis. Documentation Zoologique, Musée Royal de l'Afrique Centrale 27: 1–91

De Meyer M, Mohamed S, White IM (2007) Invasive fruit fly pests in Africa. Tervuren, Belgium, Royal Museum for Central Africa. http://www.africamuseum.be/fruitfly/AfroAsia.htm. Accessed 28 May 2016

De Meyer M, Robertson MP, Peterson AT, Mansell MW (2008) Ecological niches and potential geographical distributions of Mediterranean fruit fly (*Ceratitis capitata*) and Natal fruit fly (*Ceratitis rosa*). J Biogeog 35:270–281

De Meyer M, Quilici S, Franck A, Chadhouliati AC, Issimaila MA, Youssoufa MA, Abdoul-Karime A-L, Barbet A, Attié M, White IM (2012) Records of frugivorous fruit flies (Diptera: Tephritidae: Dacini) from the Comoro archipelago. Afri Invertebr 53:69–77

De Meyer M, Mohamed S, White IM (2014) Invasive fruit fly pests in Africa. A diagnostic tool and information reference for the four Asian species of fruit fly (Diptera: Tephritidae) that have become accidentally established as pests in Africa, including the Indian Ocean Islands. http://www.africamuseum.be/fruitfly/AfroAsia.htm. Accessed 28 May 2014

De Meyer M, Delatte MH, Mwatawala M, Quilici S, Vayssières JF, Virgilio M (2015) A review of the current knowledge on *Zeugodacus cucurbitae* (Coquillett) (Diptera, Tephritidae) in Africa, with a list of species included in *Zeugodacus*. ZooKeys 540:539–557

De Villiers M, Hattingh V, Kriticos DJ (2012) Combining field phenological observations with distribution data to model the potential distribution of the fruit fly *Ceratitis rosa* Karsch (Diptera: Tephritidae). Bull Entomol Res 103:60–73

De Villiers M, Manrakhan A, Addison P, Hattingh V (2013) The distribution, relative abundance, and seasonal phenology of *Ceratitis capitata*, *Ceratitis rosa*, and *Ceratitis cosyra* (Diptera: Tephritidae) in South Africa. Environ Entomol 42:831–840

Delrio G, Cocco A (2010) The peach fruit fly, *Bactrocera zonata*: a major threat for mediterranean fruit crops? Acta Horticult 940 doi: 10.17660/ActaHortic.2012.940.80

Dhillon MK, Singh R, Naresh JS, Sharma HC (2005) The melon fruit fly, *Bactrocera cucurbitae*: a review of its biology and management. J Insect Sci 5:40–60

Dolan C, Humphrey J (2000) Governance and trade in fresh vegetables: the impact of the UK supermarkets on the African horticulture industry. J Dev Stud 37:147–176

Dowell RV, Penrose R (1995) Mediterranean fruit fly eradication in California 1994–1995. In: Morse JG, Metcalf RL, Carey J (eds) Mediterranean fruit fly in California: defining critical research. The College of Natural and Agricultural Sciences, University of California, Riverside, pp 161–185

Drew RAI (1989) The tropical fruit flies (Diptera: Tephritidae: Dacinae) of the Australasian and Oceanian regions. Mem Queensland Mus 26:1–521

Drew RAI (1994) The *Bactrocera dorsalis* complex of fruit flies (Diptera: Tephritidae: Dacinae) in Asia. Bull Entomol Res Suppl Ser Suppl 2:68

Drew RAI, Romig MC (2013) Tropical fruit flies of South-East Asia: (Tephritidae: Dacinae). CABI, Wallingford, 653 pp

Drew RAI, Hooper GHS, Bateman MA (1982) Economic fruit flies of the South Pacific region. Queensland Department of Primary Industries, Brisbane, p 139

Drew RAI, Tsuruta K, White IM (2005) A new species of pest fruit fly (Diptera: Tephritidae: Dacinae) from Sri Lanka and Africa. Afri Entomol 13:149–154

Duyck PF, Quilici S (2002) Survival and development of different life stages of three *Ceratitis* spp. (Diptera: Tephritidae) reared at five constant temperatures. Bull Entomol Res 92:461–469

Duyck PF, David P, Quilici S (2004) A review of relationships between interspecific competition and invasions in fruit flies (Diptera: Tephritidae). Ecol Entomol 29:511–520

Duyck PF, David P, Guillemette J, Brunel C, Dupont R, Quilici S (2006) Importance of competition mechanisms in successive invasions by polyphagous tephritids in la Réunion. Ecology 87:1770–1780

Ekesi S (2006) Mass rearing technology for *Bactrocera invadens* and several *Ceratitis* species in Africa. In: Proceedings of the 7th international symposium of fruit flies of economic importance and 6th meeting of the working group on fruit flies of the Western Hemisphere, Salvador, 10–25 Sept 2006

Ekesi S (2010) Combating fruit flies in Eastern and Southern Africa (COFESA): elements of a strategy and action plan for a regional cooperation program. Available at: http://www.globalhort.org/network-communities/fruit-flies/. Accessed 25 May 2016

Ekesi S, Billah MK (2007) A field guide to the management of economically important Tephritid fruit flies in Africa. icipe Science Press, Nairobi, 106 pp. ISBN 92 9064 179 7

Ekesi S, Nderitu PW, Rwomushana I (2006) Field infestation, life history and demographic parameters of *Bactrocera dorsalis* Drew, Tsuruta & White, a new invasive fruit fly species in Africa. Bull Entomol Res 96:379–386

Elnagar S, El-Sheikh M, Hashem A, Afia Y (2010) Recent invasion by Bactrocera zonata (Saunders) as a new pest competing with Ceratitis capitata (Wiedemann) in attacking fruits in Egypt. J Appl Biol 104:97–102

El-Samea SAA, Fetoh BEA (2006) New record of *Bactrocera zonata* (Saundera) (Diptera: Tephritidae) on potatoes in Egypt. Egypt J Agric Res 84:61–63

EPPO (2003) European and Mediterranean Plant Protection Organization Report of EPPO Workshop on *Bactrocera zonata*. UNESCO

EPPO (2007) European and Mediterranean Plant Protection Organization plant quarantine data retrieval system, version 4.6, December 2007. Available online at http://www.eppo.org/DATABASES/pqr/pqr.htm. Accessed 28 May 2016

EPPO (2013) European and Mediterranean Plant Protection Organization. Bulletin OEPP 43:417–424. ISSN 0250–8052. doi:10.1111/epp.12059

Fisher HT, Hill AR, Sproul AN (1985) Eradication of *Ceratitis capitata* (Wiedemann) (Diptera: Tephritidae) in Carnarvon, Western Australia. Aust J Entomol 24:207–208

Gilbert E (2004) Dhows and the colonial economy of Zanzibar, 1860–1970. Ohio University Press, Athens, 167 pp

Goergen G, Vayssiéres J-F, Gnanvossou D, Tindo M (2011) *Bactrocera dorsalis* (Diptera: Tephritidae), a new invasive fruit fly pest for the Afrotropical region: host plant range and distribution in West and Central Africa. Environ Entomol 40:844–854

Hafsi A, Abbes K, Harbi A, Othmen SB, Limem E, Elimem M, Ksantini M, Chermiti B (2015) The pumpkin fly *Dacus frontalis* (Diptera: Tephritidae): a new pest of curcubits in Tunisia. Bull OEPP/EPPO Bull 45:209–213

Hallam D, Liu P, Lavers G, Pilkauskas P, Rapsomanikis G, Claro J (2004) The market for non-traditional agricultural exports FAO Commodities and Trade Technical Paper No. 3

Hancock DL (1984) Ceratitinae (Diptera: Tephritidae) from the Malagasy subregion. J Entomol Soc South Afr 47:277–301

Hashem AG, Mohamed SMA, El-Wakkad MF (2001) Diversity and abundance of Mediterranean and peach fruit fly (Diptera: Tephritidae) in different horticultural orchards. Egypt J Appl Sci 16:303–314

Headrick DH, Goeden RD, Teerink JA (1996) Life history and description of immature stages of *Dioxyna picciola* (Bigot) (Diptera: Tephritidae) on *Coreopsis* spp. (Asteraceae) in southern California. Proc Entomol Soc Wash 98:332–49

Hendrichs J, Ortíz G, Liedo P, Schwarz A, Cavalloro R (1983) Six years of successful medfly program in Mexico and Guatemala. In: Cavalloro R (ed) Fruit flies of economic importance. A. A. Balkema, Rotterdam, pp 353–365

Hussain MA, Haile A, Ahmad T (2015) Infestation of two tephritid fruit flies, *Bactrocera dorsalis* (syn. *B. invadens*) and *Ceratitis capitata*, in guava fruits from selected regions of Eritrea. Afr Entomol 23:510–513

Ibrahim Ali SA, Mahmoud MEE, Man-Qun W, Mandiana DM (2013) Survey and monitoring of some Tephritidae of fruit trees and their host range in River Nile State, Sudan. Pers Gulf Crop Prot 2:32–39

IIE (1995) Distribution maps of pests, Series A No. 323 (1st revision). CAB International, Wallingford

ILO International Labour Organization (2005) Global employment trends model. Available at http://www.ilo.org/public/english/employment/strat/global06.htm. International Labour Office, Geneva

Isabirye BE, Akol AM, Mayamba A, Nankinga CK, Rwomushana I (2015) Species composition and community structure of fruit flies (Diptera: Tephritidae) across major mango-growing regions in Uganda. Int J Trop Insect Sci 35:69–79

Isabirye BE, Akol AM, Muyinza H, Masembe C, Rwomushana I, Nankinga CK (2016) Fruit fly (Diptera: Tephritidae) host status and relative infestation of selected mango cultivars in three agro ecological zones in Uganda. Int J Fruit Sci 16:23–41

Iwahashi O, Routhier W (2001) Aedeagal length and its variation of the peach fruit fly, *Bactrocera zonata*, which recently invaded Egypt. Appl Entomol Zool 36:13–17

Jacquard C, Virgilio M, Quilici S, De Meyer M, Delatte H (2013) Population structuring of an economic important pest: the case of the fruit fly *Bactrocera cucurbitae* on the tropical island of La Réunion. Biol Invasions 15:759–773

Jaffee S (2003) From challenge to opportunity: the transformation of the fresh vegetable trade in the context of emerging food safety and other standards, World Bank PREM trade unit (draft): World Bank, Washington, DC

Kapoor VC (1993) Indian fruit flies (Insecta: Diptera: Tephritidae). Oxford and IBH Publishing Company, New Delhi. 228 pp

Legge A, Orchard J, Graffham A, Greenhalgh P, Kleih U (2006) The production of fresh produce in Africa for export to the United Kingdom: mapping different value chains. Natural Resources Institute, Gillingham, pp 1–87

Li BN, Ma J, Hu XN, Liu HJ, Zhang RJ (2009) Potential geographical distributions of the fruit flies *Ceratitis capitata*, *Ceratitis cosyra*, and *Ceratitis rosa* in China. J Econ Entomol 102:1781–1790

Liquido NJ, Harris EJ, Dekker LA (1994) Ecology of *Bactrocera latifrons* (Diptera: Tephritidae) populations: host plants, natural enemies, distribution and abundance. Anns Entomol Soc Am 87:71–84

Lux SA, Copeland RS, White IM, Manrakhan A, Billah MK (2003) A new invasive fruit fly species from the *Bactrocera dorsalis* (Hendel) group detected in east Africa. Insect Sci Appl 23:355–361

Mansell M (2006). Preliminary report on a visit to Madagascar to evaluate trap catches from a survey of fruit flies (Diptera: Tephritidae) possibly associated with litchis. Unpublished USDA APHIS Report, 3 pp

Mwatawala MW, White IM, Maerere AP, Senkondo FJ, De Meyer M (2004) A new invasive *Bactrocera* species (Diptera; Tephritidae) in Tanzania. Afr Entomol 12:154–156

Mwatawala MW, De Meyer M, Makundi RH, Maerere AP (2006) Seasonality and host utilization of the invasive fruitfly *Bactrocera dorsalis* (Dipt., Tephritidae) in Central Tanzania. J Appl Entomol 130:530–537

Mwatawala MW, De Meyer M, White IM, Maerere A, Makundi RH (2007) Detection of the solanum fruit fly, *Bactrocera latifrons* (Hendel) in Tanzania (Dipt., Tephritidae). J Appl Entomol 131:501–503

Mwatawala MW, De Meyer M, Makundi RH, Maerere AP (2009a) Host range and distribution of fruit-infesting pestiferous fruit flies (Diptera, Tephritidae) in selected areas of Central Tanzania. Bull Entomol Res 99:629–641

Mwatawala MW, De Meyer M, Makundi RH, Maerere AP (2009b) An overview of *Bactrocera* (Diptera: Tephritidae) invasions and their speculated dominancy over native fruit fly species in Tanzania. J Entomol 6:18–27

Mziray HA, Makundi RH, Mwatawala M, Maerere A, De Meyer M (2010) Host use of *Bactrocera latifrons*, a new invasive tephritid species in Tanzania. J Econ Entomol 103:70–76

Nboyine JA, Abudulai M, Nutsugah SK, Badii B, Acheampong A (2013) Population dynamics of fruit fly (Diptera: Tephritidae) species associated with mango in the Guinea savanna agro-ecological zone of Ghana. Int J Agric Sci 3:450–454

N'Dépo OR, Hala N, Gnago A, Allou K, Kouassi KP, Vayssières J-F, De Meyer M (2010) Inventaire des mouches des fruits de trois regions agro-écologiques et des plantes-hôtes associées à l'espèce nouvelle, *Bactrocera* (Bactrocera) *invadens* Drew et al. (Diptera: Tephritidae) en Côte-d'Ivoire. Eur J Sci Res 46:62–72

N'Dépo OR, Hala N, N'da Adopo A, Coulibaly F, Kouassi KP, Vayssières J-F, De Meyer M (2013) Fruit flies (Diptera: Tephritidae) populations dynamic in mangoes production zone of Côte-d'Ivoire. Agric Sci Res J 3:352–363

Papadopoulos N, Plant RE, Carey JR (2013) From trickle to flood: the large-scale, cryptic invasion of California by tropical fruit flies. Proc Roy Soc B 280. doi:10.1098/rspb.2013.1466

Penrose D (1996) California's 1993/1994 Mediterranean fruit fly eradication program. In: McPheron BA, Steck GJ (eds) Fruit fly pests: a world assessment of their biology and management. St. Lucie, Delray Beach, pp 551–554

Quilici S, Franck A, Peppuy A, Correia ED, Mouniama C, Blard F (2002) Comparative studies of courtship behavior of *Ceratitis* spp. (Diptera: Tephritidae) in Réunion Island. Fla Entomol 85:138–142

Quilici S, Duyck PF, Rousse P, Gourdon F, Simiand C, Franck A (2005) La mouche de la pêche sur mangue, goyave, etc. A La Réunion, évolution des recherches et des méthodes de lutte. Phytoma 584:44–47

Rwomushana I, Ekesi S, Gordon I, Ogol CKPO (2008) Host plants and host preference studies for *Bactrocera dorsalis* (Diptera: Tephritidae) in Kenya, a new invasive fruit fly species in Africa. Anns Entomol Soc Am 101:331–340

Ryckewaert P, Deguine J-P, Brévault T, Vayssières JF (2010) Fruit flies (Diptera : Tephritidae) on vegetable crops in Réunion Island (Indian Ocean): state of knowledge, control methods and prospect for management. Fruits 65:113–130. doi:10.1051/fruits/20010006

Schutze MK, Mahmood K, Pavasovic A, Bo W, Newman J, Clarke AR, Krosch MN, Cameron SL (2015) One and the same: integrative taxonomic evidence that *Bactrocera dorsalis* (Diptera: Tephritidae) is the same species as the Oriental fruit fly *Bactrocera dorsalis*. Syst Entomol 40:472–486

Shehata NF, Younes MWF, Mahmoud YA (2008) Biological studies on the peach fruit fly, *Bactrocera zonata* (Saunders). Egypt J Appl Sci Res 4:1103–1106

Shimizu Y, Kohama T, Uesato T, Matsuyama T, Yamagichi M (2006) Invasion of solanum fruit fly *Bactrocera latifrons* (Diptera: Tephritidae) to Yomaguni Island, Okinawa Prefecture, Japan. Appl Entomol Zool 42:269–275

Sookar P, Alleck M, Buldawoo I, Khayrattee FB, Choolun T, Permalloo S, Rambhunjun M (2012) Area-wide management of the melon fly *Bactrocera cucurbitae* (Coquillett) (Diptera: Tephritidae). In: Deguine JP (ed) Actes du séminaire international de clôture Gamour, Saint-Pierre, La Réunion, 21–24 Nov 2011, pp 98–102

Sookar P, Haq I, Jessup A, McInnis D, Franz G, Wornoayporn V, Permalloo S (2013) Mating compatibility among *Bactrocera cucurbitae* (Diptera: Tephritidae) populations from three different origins. J Appl Entomol 137:69–74

Stichele MV, van der Wal S, Oldenziel J (2005) Who reaps the fruit? Critical issues in the fresh fruit and vegetable chain. SOMO – Centre for Research on Multilateral Organizations, Amsterdam, p 176

Suckling DM, Kean JM, Stringer LD, Caceres-Barrios C, Hendrichs J, Reyes-Flores J, Dominiak BC (2014) Eradication of tephritid fruit fly pest populations: outcomes and prospects. Pest Man Sci. doi:10.1002/ps.3905

Theyse M (2014) Trends in South Africa vegetable imports: www.pma.com/~/media/pma-files/fc-south-africa-2014/marianna-theyse.pdf?. Accessed 28 May 2016

USAID (2004) FAS quarterly reference guide to world horticultural trade: 2004 edition. Foreign Agricultural Service. Circular Series FHORT 02–04

USAID (2005) Global horticulture assessment. http://pdf.usaid.gov/pdf_docs/Pnadh769.pdf. Accessed 28 May 2016

Vargas RI, Nishida T (1985) Survey of *Dacus latifrons* (Diptera: Tephritidae). J Econ Entomol 78:1311–1314

Vargas RI, Piñero JC, Leblanc L (2015) An overview of pest species of *Bactrocera* fruit flies (Diptera: Tephritidae) and the integration of biopesticides with other biological approaches for their management with a focus on the Pacific Region. Insects 6:297–318

Vayssières JF, Carel Y (1999) Les Dacini (Diptera: Tephritidae) inféodées aux Cucurbitaceae à la Réunion: gamme de plantes hôtes et stades phénologiques préférentiels des fruits au moment de la piqûre pour des espèces cultivées. Anns Soc Entom France 35:197–202

Vayssières JF, Kalabane S (2000) Inventory and fluctuations of the catches of (Diptera: Tephritidae) associated with mangoes in Coastal Guinea. Fruits 55: 259–270

Vayssières JF, Goergen G, Lokossou O, Dossa P, Akponon C (2005) A new *Bactrocera* species in Benin among mango fruit fly (Diptera: Tephritidae) species. Fruits 60:371–377

Vayssières JF, Sanogo F, Noussourou M (2007) Inventory of the fruit fly species (Diptera: Tephritidae) linked to the mango tree in Mali, and tests of integrated control. Fruits 62:329–341

Virgilio M, Backeljau T, Emeleme R, Juakali J, De Meyer M (2011) A quantitative comparison of frugivorous tephritids (Diptera: Tephritidae) in tropical forests and rural areas of the Democratic Republic of Congo. Bull Entomol Res 101:591–597

Weems HV (1966) Natal fruit fly (*Ceratitis rosa* Karsch) (Diptera: Tephritidae). Florida Department of Agriculture and Consumer Services, Division of Plant Industry, Entomology Circular No. 51, 2 pp

White IM (2000) Morphological features of the tribe Dacini (Dacinae): their significance to behavior and classification. In: Aluja M, Norrbom AL (eds) Fruit flies (Tephritidae): Fruit flies (Tephritidae): phylogeny and evolution of behavior . CRC Press, Boca Raton, pp. 505–546

White IM (2006) Taxonomy of the Dacina (Diptera: Tephritidae) of Africa and the Middle East. Afr Entomol Mem 2:1–156

White IM, Elson-Harris MM (1992) Fruit flies of economic importance: their identification and bionomics. CAB International, Wallingford, p 601

White IM, Elson-Harris MM (1994) Fruit flies of economic significance: their identification and bionomics, 2nd edn. International Institute of Entomology, London, p 601

White IM, Goodger KFM (2009) African *Dacus* (Diptera: Tephritidae); new species and data, with particular reference to the Tel Aviv University Collection. Zootaxa 2127:1–49

White IM, De Meyer M, Stonehouse J (2000) A review of the native and introduced fruit flies (Diptera, Tephritidae) in the Indian Ocean Islands of Mauritius, Réunion, Rodrigues and Seychelles. In: Price NS, Seewooruthun I (eds) Proceedings of the Indian Ocean Commission Regional Fruit Fly Symposium, Mauritius. Indian Ocean Commission, Mauritius, 5–9 June 2000, pp 15–21

White IM, De Meyer M, Stonehouse J (2001) A review of the native and introduced fruit flies (Diptera, Tephritidae) in the Indian Ocean Islands of Mauritius, Réunion, Rodrigues and Seychelles, p 15–21. In: Price NS, Seewooruthun I (eds) Proceedings of the Indian Ocean Commission Regional Fruit Fly Symposium, Mauritius. Indian Ocean Commission, Mauritius, 5–9 June 2000, pp 232

WRI World Resource Institute (2005) The wealth of the poor: managing ecosystems to fight poverty. World Resources Institute, Washington, DC, p 246

C.M. Tanga is an Agricultural Entomologist/Ecologist working with the Plant Health Division at the International Centre of Insect Physiology and Ecology (*icipe*), Kenya. He obtained his PhD from the University of Pretoria, South Africa and is currently a postdoctoral research fellow at *icipe*. His area of research interest is sustainable integrated pest management (IPM) that is based on the use of biopesticides, baiting techniques, orchard sanitation and classical biological control for the management of arthropod pests of horticulture. His research activities involve the promotion and up-scaling of the dissemination of fruit fly integration pest management technologies (IPM) among smallholder growers in Africa. He also develops temperature-based phenology models for African fruit flies and other crop pests to improve implementation of IPM strategies for their management. He also carry out research activities that are focused on filling critical gaps in knowledge aimed to exploit the commercial potential of insects. He seeks to create an effective agribusiness model by establishing profitable community-led agro-enterprises for commercial production of affordable, safe and high quality insect-based protein feeds for poultry, pig and fish industries to improve the livelihood as well as promote youths and women employment through capacity building on the use of insects as a source of protein in feeds for livestock production in Africa.

I. Rwomushana is a Research Scientist with the African Fruit Fly Program at the International Centre of Insect Physiology and Ecology (*icipe*) with expertise in Agricultural Entomology. He was educated at Makerere University and Kenyatta University. He is involved in the development of sustainable Integrated Pest Management options for the management of fruit flies and other arthropod pests that constrain fruit production in sub-Saharan Africa. His research interests involve understanding the bioecology and employing balanced basic and applied research based on the use of baiting and male annihilation techniques, biopesticides, natural enemies' conservation and classical biological control methods and use of soft pesticides for suppression of the tree crop pests. He has also coordinated several research projects in Eastern and Central Africa on several other arthropod pests.

Chapter 7
Exotic Invasive Fruit Flies (Diptera: Tephritidae): In and Out of Africa

Marc De Meyer and Sunday Ekesi

Abstract This paper reviews all the known cases of invasive fruit flies that have originated from Africa and established in other parts of the world, as well as exotic invasive tephritid species that have been introduced accidently into Africa from their natives ranges elsewhere. The former concerns the olive fruit fly, *Bactrocera oleae* (Rossi); the Mediterranean fruit fly (medfly), *Ceratitis capitata* (Wiedemann); and the Ethiopian fruit fly, *Dacus ciliatus* Loew. For the latter category, we include the oriental fruit fly, *Bactrocera dorsalis* (Hendel); the solanum fruit fly, *Bactrocera latifrons* (Hendel); the peach fruit fly, *Bactrocera zonata* (Saunders) and the melon fruit fly, *Zeugodacus cucurbitae* (Coquillett). For each of them we discuss their taxonomic position, their distribution and invasion history, their economic impact and their importance (real and/or potential) in the horticultural industry.

Keywords Introductions • Economic impact • Quarantine

1 Introduction

Tephritid fruit flies (Diptera: Tephritidae) remain one of the most important pest groups of economic significance in horticulture. Malavasi (2014) estimated that losses due to tephritid fruit flies on commercial fruit and vegetable crops amounted to more than two billion dollars annually worldwide. However, sections of the industry believe that the devastating impact on trade, including bans on export goods, results in losses approximating two billion annually on the African continent alone (Ekesi et al. 2016). Their impact is therefore dual: causing direct losses

---

M. De Meyer (✉)
Biology Department, Royal Museum for Central Africa,
Leuvensesteenweg 13, B3080 Tervuren, Belgium
e-mail: demeyer@africamuseum.be

S. Ekesi
Plant Health Theme, International Centre of Insect Physiology & Ecology (*icipe*),
PO Box 30772–00100, Nairobi, Kenya

© Springer International Publishing Switzerland 2016
S. Ekesi et al. (eds.), *Fruit Fly Research and Development in Africa - Towards a Sustainable Management Strategy to Improve Horticulture*,
DOI 10.1007/978-3-319-43226-7_7

through reduction in crop yields, but also indirectly by trade embargoes preventing export to other regions. It is this last activity – international transport – that has also resulted in the introduction of numerous fruit fly species into new territories, either in commercial shipments or by individuals who carry fruit from one place to another. When these exotic species establish, naturalize and spread, they can become invasive pests (Liebhold and Tobin 2008). Numerous articles and books have been written on the topic of invasion biology. With regard to fruit flies an excellent review of the biological, economic and managerial consequences of fruit fly invasions has recently been published by Papadopoulos (2014). It is not our intention to repeat these areas in this article. Rather we want to focus on the African situation specifically.

Africa, and in particular sub-Saharan Africa (SSA), has a very rich indigenous fruit fly fauna including about 400 frugivorous species. Several of these have been recognized as major pest species on a number of commercial crops, specifically subtropical and tropical fruits, as well as vegetables that are biologically fruits such as tomatoes, aubergines and cucurbits. With the increases in trade and movement of humans, some of these fruit fly pests have spread throughout the world. The best known example is the Mediterranean fruit fly (medfly), *Ceratitis capitata* (Wiedemann) which is still considered as one of the most important fruit pests. Information on some of these invasions that have come out of Africa are lost in time, but for others we have actual dates and records, supplemented with modern molecular techniques, that allow us to reconstruct the invasion routes and history.

In contrast, Africa has also been the destination of exotic pests. Over the last decade, the continent has been devastated by the arrival of the oriental fruit fly, formerly known as the invasive fruit fly *Bactrocera invadens* Drew, Tsuruta and White but recently recognized as being a junior synonym of *Bactrocera dorsalis* (Hendel) (Schutze et al. 2014a). This invasive species, which has a competitive advantage over indigenous pest species, has been causing tremendous damage to African fruit production resulting in the closing of trade routes and loss of opportunities that are vital for the survival of the commercial fruit industry on the continent. Fear exists that Africa could be a stepping stone for these invasive species, threatening fruit industries in other regions such as Europe and the USA.

The objective of this article is to review the major pest species that originated in Africa but also those that have been introduced into Africa. For each of them we present the most recent treatment of their taxonomic status, their known distribution and invasion history, and the impact they have on the African horticulture industry.

2 Out of Africa

2.1 Bactrocera oleae

2.1.1 Taxonomic Position

Bactrocera oleae (Rossi) or the olive fruit fly, belongs to the subgenus *Daculus* Speiser. This is an Afrotropical subgenus, comprising nine species (Copeland et al. 2004). Asian species originally placed in *Daculus* or its junior synonym *Afrodacus*, are now all placed in the subgenus *Bactrocera* s.s. (Copeland et al. 2004, Hancock 2015, but see Drew and Romig 2013, who retain *Afrodacus* as a distinct subgenus and who place three species in this subgenus). Four species, including *B. oleae*, are closely associated with Oleaceae (Copeland et al. 2004) and are found on the African continent. The other five species within the subgenus are associated with Verbenaceae (two species from East Africa and South Africa) and Clusiaceae (two species from the Indian Ocean islands and a third species from Madagascar probably also belong to this group, but see Hancock and Drew 2015). Copeland et al. (2004) indicated that, in Kenya, *B. oleae* was exclusively reared from fruit of the wild olive *Olea europaea* ssp. *cuspidata* (Wall. ex G. Don) Cif..

2.1.2 Distribution and Invasion History

Bactrocera oleae is considered to be of African origin. This is supported by the fact that: its hosts, the cultivated olive and its wild relatives, also appear to be of African origin (Besnard et al. 2007a, 2009); its closest relatives are restricted to the Afrotropical region (White 2006); and by the significantly greater genetic diversity in African populations compared with European ones (Nardi et al. 2005). It is currently found throughout southern and eastern Africa. Furthermore, it is found in the Mediterranean region, Pakistan and California. In Pakistan they are considered as relict populations that separated a long time ago (Nardi et al. 2005, 2010) and were earlier described as a morphological variety (*B. oleae* var. *asiatica* [Silvestri 1916]). Occurrence in the Mediterranean region is a result of northward dispersal from Africa. Domestication of the olive tree happened around 4000 BC in the East Mediterranean (Boardman 1976; Lumaret et al. 2004). Augustinos et al. (2005) studied the population genetics of Mediterranean populations of *B. oleae* using microsatellite markers. They discovered three subgroups: Cyprus, Greece + Italy + Turkey, and the Iberian Peninsula. In addition, they observed a gradual decrease in heterozygosity from the eastern to the western Mediterranean. They suggested that this decrease was as a result of historical variation in the dates that olive trees were domesticated and introduced into different parts of the Mediterranean. Cyprus is considered one of the first places where the tree was introduced after domestication in the Levant, followed by intensive trade and cultivation in Hellenistic times and with the Iberian Peninsula being the more recent region where olive trees were introduced. As such, *B. oleae* would have followed the westward expansion of the olive industry (although the large homogenization on the Iberian Peninsula could also be due to other factors like intensive cultivation or later introductions of Arab olive varieties, see Augustinos et al. 2005). However, Nardi et al. (2010) state that the presence of wild olive trees in the Mediterranean pre-dates domestication, with diversification occurring at the start of the Pliocene between African and Asian lineages (*Olea europea cuspidata* vs North African and European lineages of *Olea europea europea*; Besnard et al. [2007b]). Already these wild forms were a suitable host for *B. oleae*. Using full mitochondrial genome data, Nardi et al. (2010) provide evidence for post-glacial arrival of wild olive trees in the Mediterranean region, rather than historic spread as a result of introduction of domesticated olive trees. The former was then followed by a subsequent multi-regional host shift from wild to cultivated olives.

The occurrence of *B. oleae* in California is a recent phenomenon. The first record was from Los Angeles in 1998 (Rice 2000) but it is much more widespread now due to development of the olive industry (primarily for canning but increasingly also for oil) (Yokoyama 2012). The origin of the introduction into California appears to be from the eastern Mediterranean (Zygouridis et al. 2009; Nardi et al. 2010).

2.1.3 Impact

Although *B. oleae* is stenophagous, attacking only the commercial olive and its closest wild relatives, it causes worldwide damage because of the large olive and olive oil industries in regions where the fly occurs, in particular in the Mediterranean Region. *Bactrocera oleae* larvae consume the mesocarp of the olive fruit, thereby reducing the quality of olive oil which becomes more acidic (Kapatos 1989; Torres-Vila and Pérez de Sande 2002). The damaged fruits also drop prematurely, making table olives unfit for consumption (Levinson and Levinson 1984; Manousis and Moore 1987; Burrack and Zalom 2008).

Damage worldwide is estimated at around 800 million US dollars annually in countries around the world where olives are grown (Manousis and Moore 1987; Montiel-Bueno and Jones 2002; Tzanakakis 2003). In parts of the Mediterranean Region, where more than 90 % of olive production is located (Mostakim et al. 2012), it is a major pest and losses can reach 80–100 % of the total harvest in commercial olive groves (Katsoyannos 1992; Broumas et al. 2002). In the USA, olive production is concentrated in the state of California where approximately 15,000 ha are under cultivation (Connell 2005). While the industry faced few pest problems before the arrival of *B. oleae*, the industry is now under great pressure because of *B. oleae* which is now considered as the primary pest of this commodity (Burrack et al. 2008, Daane et al. 2005).

There is no clear evidence that *B. oleae* causes major damage in its region of origin, sub-Saharan Africa. The commercial olive industry is rather limited and, only in countries like South Africa, it is of some economic importance. Although *B. oleae* can cause damage, it is not considered as a serious pest, probably due to the presence of natural enemies that keep it at bay (Hancock 1989; Costa 1998; Mkize et al. 2008).

2.2 Ceratitis capitata

2.2.1 Taxonomic Position

Ceratitis capitata (the Mediterranean fruit fly or medfly) belongs to *Ceratitis*, an Afrotropical genus with approximately 100 different species, and in particular to the subgenus *Ceratitis* s.s.. De Meyer (2000) listed eight species within this subgenus and provided detailed descriptions and a hypothesis for their phylogenetic relationships. The closest relative of *C. capitata* is *Ceratitis caetrata* Munro, a species restricted to Kenya. The two species can be easily differentiated on morphological grounds (especially in males, see De Meyer 2000). However, there is no clear separation between their gene pools and COI barcodes do not allow full differentiation because of the absence of a barcode gap (Barr et al. 2012). The sister clade of the *capitata-caetrata* cluster comprises a number of species only known from the Indian Ocean islands. More basal species are found in southern and eastern Africa. A more detailed phylogenetic analysis for the whole genus (De Meyer 2005) was largely in accord with these findings, albeit with some minor re-arrangements regarding clustering. The molecular phylogeny by Barr and Wiegmann (2009) for a subset of the species within the subgenus confirmed this monophyly, except for *Ceratitis cornuta* (Bezzi).

2.2.2 Distribution and Invasion History

Ceratitis capitata is widespread throughout the Afrotropical region (De Meyer et al. 2008). Given the occurrence of closely related species in the subgenus, its origin seems to be eastern or southern Africa (De Meyer et al. 2004; Malacrida et al. 2007). From Africa, the species spread to other parts of the world; first to the Mediterranean region, and afterwards to the Near and Middle East, the Americas, Australia and Hawaii. For several of these introductions, dates are known with regard to first observations (Gasperi et al. 2002; Papadopoulos 2014) and it is now widely distributed on several continents (Szyniszewska and Tatem 2014). Although recorded from several areas, it also seems to be either eradicated through targeted programmes (for example in Central America, see Enkerlin et al. 1989) or displaced by other fruit flies (for example in eastern Australia by the Queensland fruit fly, *Bactrocera tryoni* [Froggatt], see Dominiak and Daniels 2012). In other areas, there is controversy as to whether the species is present at sub-detectable population levels, or whether the observations on its presence are the result of repeated introductions (Papadopoulos et al. 2013). The invasion pathways have been extensively studied and show a pattern of African origin, with introduction in the Mediterranean region and subsequent introductions to other parts of the world (Gasperi et al. 2002; Bonizzoni et al. 2004). Despite its wide distribution, surveying programmes have been established in several *C. capitata*-free countries to allow early detection, because of its large potential impact on horticulture.

2.2.3 Impact

Ceratitis capitata is still considered one of the most destructive fruit fly pests worldwide (White and Elson-Harris 1994). Its impact is mainly due to its extreme polyphagy (Liquido et al. 2014) and its adaptability to different conditions (Yuval and Hendrichs 2000; Papadopoulos et al. 2001; Terblanche et al. 2010). As a result, intensive surveying projects have been established followed by costly eradication programmes, despite regular outbreaks. For example, each incursion of *C. capitata* into the US (mainly California and Florida) costs between US 300,000 and 200 million to eradicate (APHIS 1992). However, this is considered far less than the cost of that establishment would cause, both in terms of control and loss of export revenue. Ole-MoiYoi and Lux (2004) estimated that more than US 100 million is spent annually on *C. capitata* control and prevention outside of Africa.

In Africa, *C. capitata* also remains a major pest of a wide variety of fruits and crops. Barnes (this volume) reports that the estimated loss in value due to crop loss and control costs in the Western Cape of South Africa alone, reaches US 7.5 million each year. In the Maghreb (Algeria, Libya, Morocco, Tunisia), economic losses due to *C. capitata* are estimated at US67–100 million per annum (IAEA 1995). For most other countries in Africa no figures are available for economic losses, although *C. capitata* is still reported from a wide variety of hosts. Nevertheless, there is an indication that the species has been, or can be, replaced by exotic invasive species. This was observed in La Réunion where *C. capitata* was introduced but has subsequently been largely displaced by other invasive species such as the Natal fruit fly, *Ceratitis rosa* Karsch, and the peach fruit fly, *Bactrocera zonata* (Saunders) (Duyck et al. 2006a). Casual observations in Tanzania also reflect the fact that infestation rates and host range is limited in comparison with the invasive species *B. dorsalis* (Mwatawala et al. 2009).

2.3 Dacus ciliatus

2.3.1 Taxonomic Position

Dacus ciliatus belongs to the subgenus *Didacus* and, in particular, the Ciliatus group within this subgenus. *Didacus* comprises close to 40 species with the Ciliatus group having eight species according to White (2006). Hancock and Drew (2006) provided an alternative classification with the subgenus only comprising one Afrotropical group (Ciliatus group) in addition to the oriental Keiseri group.

However, in both classifications the Ciliatus group was comprised of the same species. Representatives of the Ciliatus group are only recorded from Cucurbitaceae. *Didacus* sensu White, comprises also species reared from Apocynaceae, while *Didacus* sensu Hancock and Drew is only recorded from Cucurbitaceae. A molecular phylogeny based on a subset of African *Dacus* species places *D. ciliatus* with representatives of the subgenus *Lophodacus*, while *Dacus vertebratus* Bezzi (the only other representative of the Ciliatus group included in the analysis) is clustering with species that feed on Apocynaceae (Virgilio et al. 2009). *Dacus ciliatus* has been confused very often with the pumpkin fly, *Dacus frontalis* Becker and *D. vertebratus*. They differ only in coloration of the thoracic katatergite and anatergite, and of the femora (White 2006). It also has a fairly large synonymy and was described independently from specimens originating from the Democratic Republic of Congo, Egypt, Eritrea, India, Mauritius and South Africa.

2.3.2 Distribution and Invasion History

From phylogenetic information it appears that *D. ciliatus* is Afrotropical in origin. However, it is now found throughout the whole of Africa, as well as the Arabian Peninsula, the Near and Middle East and the Indian subcontinent (White and Elson-Harris 1994; Vayssières et al. 2008; Drosopoulou et al. 2011). Although Vayssières et al. (2008) state that the species is native to East Africa, there is no real evidence that this is the area of origin for this species. Other species within the Ciliatus group, and with a more restricted distribution, are mainly from southern Africa (Angola, South Africa, Zimbabwe). Within Africa, it is one of the most widely distributed species, found in all climatic zones. This is unlike other cucurbit-feeding species, like the melon fly, *Zeugodacus cucurbitae* (Coquillett) or even other cucurbit-feeding *Dacus* species, because *D. ciliatus* can also be found in dry areas like the Kalahari semi-desert, Namibia and particular regions on the Arabian peninsula (http://projects.bebif.be/fruitfly/taxoninfo.html?id=272). Overall it is not clear whether *D. ciliatus* actually invaded (by human introduction) into other regions in Africa, or elsewhere, or whether the current occurrence is as a result of natural and ancient spread, although some data of first occurrences may indicate introductions in particular areas outside the African mainland (e.g. first record from India in 1939 [Bhatia 1939], from La Réunion in 1964 [Etienne 1972]).

2.3.3 Impact

White and Elson-Harris (1994) describe *D. ciliatus* as a pest of a wide variety of Cucurbitaceae, in addition to occasional records of non-cucurbit hosts. However, the latter could be the result of casual observations rather than actual rearing of larvae from particular fruit species. Even some of the recorded cucurbit hosts need confirmation because of earlier confusion with *D. frontalis* and *D. vertebratus*.

White (2006) gave a more detailed list of host range as confirmed by rearing *D. ciliatus* from fruit.

Hancock (1989) also described *D. ciliatus* as a serious pest of cucurbits and EPPO lists it as an A1 quarantine species (EPPO 2011). However, there are no actual figures on the impact that *D. ciliatus* has on commercial cucurbit crops. Nevertheless, comparative studies on cucurbit pests have shown that *D. ciliatus* is one of the major competitors of the invasive species, *Z. cucurbitae*, in Africa (Vayssières et al. 2008; Mwatawala et al. 2010). In areas where *Z. cucurbitae* has not (yet) established, *D. ciliatus* can be a serious problem.

3 Into Africa

3.1 Bactrocera dorsalis

3.1.1 Systematic Position

Bactrocera dorsalis is of Asian origin and belongs to the subgenus *Bactrocera* s.s.. It is one species within a complex of more than sixty species (the so-called *B. dorsalis* complex, see Drew and Romig 2013). Until recently the specimens found in Africa where thought to belong to a separate species, called *B. invadens*, that was described based on specimens from Africa and Sri Lanka (the latter being the presumed area of origin) (Drew et al. 2005). However, an integrative taxonomic study, applying a number of different techniques, concluded that a number of entities recognized within the *B. dorsalis* complex actually represent a single species (Schutze et al. 2014a, b). *Bactrocera invadens* is currently, therefore, considered a junior synonym of *Bactrocera dorsalis*. However, most of the relevant literature published before 2015 on this species in Africa will refer to *B. invadens*, rather than *B. dorsalis*.

3.1.2 Distribution and Invasion History

Bactrocera dorsalis is widespread from Central Asia to Southeast Asia. It is also reported from southern China. The species has been introduced to other regions. In some of these regions it has subsequently been eradicated (e.g. Australia where it was known under the junior synonym *Bactrocera papayae* Drew and Hancock [the Asian papaya fruit fly]; see Cantrell et al. 2002). In others, such as Hawaii, *B. dorsalis* has become permanently established.

In Africa, the first specimens were observed on the Kenyan coast in 2003 (Lux et al. 2003). It was then rapidly reported from other parts of the African mainland, first from neighboring countries like Tanzania (Mwatawala et al. 2004) but within 2 years of its first detection it was also reported from as far away as Senegal. The invasion pathway throughout the continent is still only partly resolved (Khamis et al. 2009;

Malavasi et al. 2013) and it is not clear whether the rapid spread in Africa is due to a single introduction or multiple introductions. Over the last decade the species has clearly dispersed further northwards and southwards, with the southward spread throughout Mozambique and into South Africa as the best documented (Cugala et al. 2011; Manrakhan et al. 2015). In South Africa the spread has been analyzed in great detail based on an intensive surveying and monitoring programme. Although initial intrusions into South Africa (first in 2010) were eradicated successfully, the number and geographic spread of detections increased dramatically resulting in an official declaration of presence in 2013 (see Manrakhan et al. 2015 for details). Of the Indian Ocean islands it has become established, thus far, on the Comoros (including Mayotte) and Madagascar (De Meyer et al. 2012). A recent outbreak in Mauritius was eradicated (Sookar et al. 2014). Mauritius did have a much earlier outbreak in the 1990s but this was also eradicated at the time.

3.1.3 Impact

Undoubtedly *B. dorsalis* poses the currently most important threat to fruit production in Africa. Since the first records in Africa, it has had an enormous impact on the indigenous fruit fly population and on horticultural crops. Export embargoes and quarantine restrictions had, and still have, major implications for the fruit and vegetable industry and trade in SSA (French 2005; Ravry 2008; Ekesi 2010; Cugala et al. 2013; Ekesi et al. 2016). In Uganda, the mango (*Mangifera indica* L.) industry alone is estimated to lose more than US$ 116 million annually due to *B. dorsalis*. Following the invasion of *B. dorsalis* in 2003, Kenya lost its entire avocado export market in South Africa, resulting in revenue losses of US$ 1.9 million in 2007 (Otieno 2011) and with the ban still in effect, cumulative losses by 2014 were US$ 15.2 million. Mozambique experienced revenue losses of US$ 2.5 million after South Africa banned imports of fruits from Mozambique due to the presence of *B. dorsalis* (Jose et al. 2013). EPPO established a Pest Risk Analysis (for *B. invadens* at that time) concluding that it was a major risk for the European fruit industry (De Meyer et al. 2009). The fact that *B. dorsalis* is still expanding its distribution range throughout Africa, in particular to South Africa which has the most economically important fruit industry in Africa, is worrying, especially since the export of citrus, deciduous and subtropical fruits from South Africa is worth US$ 1.6 billion in revenues in 2014/2015 (DAFF 2016).

Its impact is mainly due to two factors: its polyphagous nature and its competitive dominance over other fruit fly species. So far it has been reported from 65 host plants in Africa, belonging to 26 families (Malavasi et al. 2013). Also infestation rates are often higher than those observed for indigenous species (Mwatawala et al. 2009). The listing of host plants is based on data before it was acknowledged that *B. invadens* was a junior synonym of *B. dorsalis* and so may well be an underestimation. Non-African hosts for *B. dorsalis* include more than 130 species (Leblanc 2012), so it is likely that it could be more polyphagous in Africa than currently observed. Competitive displacement studies (Ekesi et al. 2009; Rwomushana et al.

2009) demonstrate that *B. dorsalis* has a number of K-selected traits, such as large body size, that give it an advantage over indigenous fruit fly species. Duyck et al. (2004) observed that, where exotic polyphagous tephritid species have been introduced into an area already occupied by indigenous polyphagous tephritids, interspecific competition results in niche shifts and a decrease in the number of pre-established species. From studies of developmental physiology, temperature range and resource acquisition, the mango fruit fly, *Ceratitis cosyra* (Walker), which is the most important indigenous mango-infesting fruit fly species in Africa, is the inferior competitor compared with *B. dorsalis* (Ekesi et al. 2009; Rwomushana et al. 2009; Salum et al. 2013).

3.2 Bactrocera latifrons

3.2.1 Systematic Position

Bactrocera latifrons was originally described as *Chaetodacus latifrons* by Hendel (1915) based on material from Taiwan. It belongs to the subgenus *Bactrocera* s.s. but is not considered to be a part of the *B. dorsalis* species complex, despite the fact that it has similar body colour patterns (Drew and Romig 2013). It can be differentiated from representatives of this complex by the distinctly trilobed apex of the aculeus in females and the uniformly dark orange-brown abdominal tergites in both sexes. Until recently, *Dacus parvulus* Hendel was considered a synonym (although a valid prior name, it was rejected in interest of stability; see White and Liquido [1995]). However, Drew and Romig (2013) considered the latter to be a valid species. *Bactrocera parvula* (Hendel) is also very similar to *B. latifrons* but differs in having dark fuscous patterns around the apices of the femora, and darker colour patterns on the abdominal terga. The most recent description of *B. latifrons* is given in Drew and Romig (2013).

3.2.2 Distribution and Invasion History

Bactrocera latifrons is widespread throughout Central and Southeast Asia, being recorded from China, India, Indonesia, Laos, Malaysia, Singapore, Pakistan, Sri Lanka, Taiwan, Thailand and Vietnam (Norrbom et al. 1999; Drew and Romig 2013). However, given the revised status of *B. parvula*, which was originally described from Taiwan, a revision of all material recognized as *B. latifrons* may be required (Drew and Romig 2013 only mention the Taiwanese type material under *B. parvula*). *Bactrocera latifrons* is an invasive species and was introduced into Hawaii around 1983 (Vargas and Nishida 1985). It has also been reported from Okinawa Prefecture in southern Japan (Shimizu et al. 2007).

The invasion history of *B. latifrons* in Africa is well documented with the first observations dating back to 2006 (Mwatawala et al. 2007). Amongst material collected in traps with protein bait for monitoring *B. dorsalis*, two female specimens were observed in May 2006, in the horticultural unit orchard at Sokoine University of Agriculture, Morogoro, Central Tanzania (Mwatawala et al. 2007). Around the same time, it was also reared from samples of the African eggplant, *Solanum aethiopicum* L. grown in the same orchard. Further observations and sampling in different parts of Tanzania, revealed a larger distribution within the country but mainly in the northeastern part bordering Kenya (Mwatawala et al. 2010). So far, the species has only been reported from Tanzania and Kenya within Africa (Mziray et al. 2010a). However, given the host range (cf below), it has the potential to be much more widespread. Spatiotemporal studies in Central Tanzania showed that *B. latifrons* had a preference for low to medium altitude sites (Mziray et al. 2010a).

3.2.3 Impact

Bactrocera latifrons mainly attacks fruits of plants in the Solanaceae. In Tanzania it has been reared from a number of commercial hosts such as aubergine, *Solanum melongena* L.; tomato, *Solanum lycopersicum* L.; and bell peppers, *Capsicum* sp. (Mziray et al. 2010b) where it outcompeted other generalist species such as *B. dorsalis*. In addition to solanaceous hosts, it is also recorded from a number of cucurbit hosts (Mziray et al. 2010b), although the infestation rates in these was usually low as has also observed elsewhere in the distribution range (Liquido et al. 1994; White and Elson-Harris 1994). Abundance of *B. latifrons* remains low but shows an increase during the rainy season, probably due to higher availability of host plants. Relatively low population densities despite good availability of potential host biomass has also been observed in Hawaii where the species was introduced (Liquido et al. 1994). The life history traits of *B. latifrons* suggest that it is a K-selected species while invaders are generally assumed to be r-strategists (Duyck et al. 2007). This could explain the limited impact and dispersal of this invasive species, so far, in Africa, although K-selected traits have been shown to be a competitive advantage for some fruit fly species (see above under *B. dorsalis*). However, horticultural practices (such as high use of insecticides on solanaceous crops) and the absence of a powerful attractant for monitoring, could also contribute to this observation.

3.3 **Bactrocera zonata**

3.3.1 Systematic Position

Bactrocera zonata belongs to the subgenus *Bactrocera* s.s. but is not directly related to the *B. dorsalis* complex. Nevertheless, it superficially shows some resemblance to representatives of the *B. dorsalis* complex, such as the presence of dark facial spots, lateral postsutural yellow vittae and a 'T' shaped pattern on the abdomen, although the latter is often incomplete in *B. zonata* (Drew and Romig 2013). In

Africa it can be differentiated from *B. dorsalis* by the wing pattern (anal streak and complete costal band absent in *B. zonata* but present in *B. dorsalis*; microtrichia in narrow basal section of cell br absent in *B. zonata* and present, at least anteriorly, in *B. dorsalis*; see White 2006). In its area of origin Drew and Romig (2013) mention *Bactrocera affinis* (Hardy) from India as morphologically very similar.

3.3.2 Distribution and Invasion History

Bactrocera zonata is widespread in oriental regions, being recorded from the Indian subcontinent throughout Southeast Asia, including Bhutan, India, Indonesia (Sumatra), Laos, Sri Lanka, Thailand and Vietnam (White and Elson-Harris 1994; Drew and Romig 2013). Reports of occurrence in Indonesia (Maluku) are considered doubtful by Drew and Romig (2013). Outside its native range, it became established in the Arabian Peninsula in the 1980s where it is now reported from a number of countries including Oman, Saudi Arabia, United Arab Emirates and Yemen (White 2006). On the African mainland, *B. zonata* was first recorded from Egypt in the early 1990s. Initially it was erroneously identified as *Bactrocera pallidus* (Perkins & May) (Abuel-Ela et al. 1998). Consequently, no control measures were taken and the species spread rapidly from Cairo to the Sinai and western Egypt (De Meyer et al. 2007) and became firmly established by 1997 (White 2006). However, there are some suggestions that *B. zonata* was present and possibly established in Egypt before this period (Mohamed and El-Wakkad 2003) but these reports require confirmation. In recent years *B. zonata* spread further to Libya and the Sudan (White 2006; Abdelgader and Salah 2016) but has not been reported from any other African country. On the Indian Ocean islands, *B. zonata* was first found on Mauritius in 1942 (Orian and Moutia 1960). It is not clear whether it became established then but it was found again as an established population in 1987 (White et al. 2001). Since 1991, it has been found occasionally in La Réunion (White et al. 2001) where it is now considered established and one of the dominant fruit fly species (Duyck et al. 2006a, b).

3.3.3 Impact

White and Elson-Harris (1994) list this as a pest on a wide variety of unrelated fruits such as peach, *Prunus persica* (L.) Batsch.; guava, *Psidium guajava* L.; and mango among others. In Egypt, it is considered a serious pest of peach, guava, mango, apricot, *Prunus armeniaca* L., and fig, *Ficus carica* L. (Mosleh et al. 2011) and potentially causing more damage than *C. capitata* (Mohamed 2004) on guava, mango and peach. In La Réunion it is displacing *C. capitata* from several habitats on the island (Duyck et al. 2006a, b). Losses in guava can reach 25–50% (Syed et al. 1970) and it could be more important as a pest species than *B. dorsalis* in Pakistan (Qureshi et al. 1991). The current southward spread into the Sudan represents a future threat to the fruit industry in several parts of Africa. Ecological modeling shows that a large part of the African mainland is climatically suitable (Ni et al. 2012) and that drier areas are even more suitable for *B. zonata* than that they are for *B. dorsalis* (Stephens et al. 2007; De Meyer et al. 2010), thereby creating a separate niche for *B. zonata*. An even greater risk is the further invasion of *B. zonata* from Africa into Europe. EPPO (2013) considers it as an A1 phytosanitary threat for European horticulture. Given the proximity of its occurrence in Africa to the European countries in the Mediterranean Basin, there is a real threat that it could invade from Africa into Europe, especially given it competitive advantage over other established pest species such as *C. capitata*, and the fact that ecological modeling indicates that large parts of the Basin are also climatically suited for the establishment of *B. zonata* (Ni et al. 2012).

3.4 Zeugodacus cucurbitae

3.4.1 Systematic Position

Although *Z. cucurbitae* was originally described from Hawaii (Coquillett 1899) this record was based on material accidently introduced on to the island (Bess et al. 1961). *Zeugodacus cucurbitae* as such is an oriental species, probably originating in Central Asia or the Indian subcontinent and spreading from there to other parts of the world (Virgilio et al. 2010; Wu et al. 2012). It belongs to the genus *Zeugodacus* which comprises about 200 species from the oriental, Australasian and eastern Palaearctic regions. Originally it was placed in the genus *Dacus*, but this genus was split in to two major genera, *Bactrocera* and *Dacus*, by Drew (1989). More recently, the subgenus *Bactrocera* (*Zeugodacus*) and related subgenera (forming the 'Zeugodacus group') were given generic status (Virgilio et al. 2015) (for a more detailed discussion on the generic placement, see De Meyer et al. 2015). *Zeugodacus cucurbitae* is the only species of this genus found in Africa and no other close relatives are found in the region. All other African representatives of the Dacina belong to either *Dacus* or *Bactrocera*. A diagnostic description is given in White (2006) and Drew and Romig (2013). Furthermore, White (2006) provides a key for all African Dacina including *Z. cucurbitae*.

3.4.2 Distribution and Invasion History

Worldwide, *Z. cucurbitae* is widespread throughout Central and East Asia (including Pakistan, India, Bangladesh, Nepal, China, Indonesia and the Philippines) and Oceania (including New Guinea and the Mariana Islands) and has became established in some areas of the Pacific (Dhillon et al. 2005). Populations on the southwestern islands of Japan were eradicated using the Sterile Insect Technique (SIT) (Koyama et al. 2004).

In Africa, the species is currently recorded from several countries in SSA including Benin, Burkina Faso, Burundi, Cameroon, Democratic Republic of Congo, Ethiopia, Gambia, Ghana, Guinea, Ivory Coast, Kenya, Malawi, Mali, Mozambique, Niger, Nigeria, Senegal, Sierra Leone, Sudan, Tanzania, Togo and Uganda (De Meyer et al. 2015). The very first records date back to 1936 from Tanzania and shortly afterwards from Kenya (De Meyer et al. 2015). Collections at the Natural History Museum in London (UK) and the National Museums of Kenya in Nairobi (Kenya) house material from the East African coastal areas (respectively Tanga in Tanzania and Kilifi and Rabai in Kenya). Records from mainland Africa were restricted to these two countries for the next six decades (see specimen data in De Meyer and White 2004).

From the end of the twentieth century onwards, a number of records from Central and western Africa were published (De Meyer et al. 2015). Also in eastern Africa, there were reports of its occurrence in Ethiopia, Sudan, Malawi, Uganda, Burundi and Mozambique (De Meyer et al. 2015). It is unclear whether these new occurrences were as a result of more intensive sampling or due to actual dispersal and/or introduction into new areas within Africa. Virgilio et al. (2010) demonstrated that the African populations (ten studied from Benin, Burkina Faso, Democratic Republic of Congo, Guinea, Ivory Coast, Kenya, Senegal and Tanzania) constituted a genetically well-differentiated group, different from the other groups worldwide, and that there was no evidence to suggest that more than one invasion event originating from outside Africa had taken place. Recent analyses (Delatte et al. unpublished data) demonstrate that both the western and eastern African samples date back to the twentieth century, but that western African ones were more recent. This appears to confirm the idea that the western African records are reflecting intracontinental movement from eastern Africa in recent times. The exact pathways are, however, unknown.

Zeugodacus cucurbitae has also been introduced to islands in the Western Indian Ocean. It was first recorded from Mauritius (1942: see Orian and Moutia 1960), and then from La Réunion (1972: see Vayssières 1999; White et al. 2001). These populations were considered to form a genetically well-defined separate group by Virgilio et al. (2010) and not directly linked to the population cluster from the African continent. Both occurrences appear to be the result of separate invasion events from Asia. However, a more detailed gene flow study of populations from La Réunion showed a higher estimated gene flow between the African mainland and the island than between Asia and the island (Jacquard et al. 2013). This could be the result of two alternative scenarios: either an initial Asian origin of the population at La Réunion followed by secondary contact with population from Africa, or an African origin. *Zeugodacus cucurbitae* was more recently (since 1999) also reported from the island of Mahé in the Seychelles (White et al. 2001), first through interceptions near the airport but currently it is considered established. Virgilio et al. (2010) demonstrated that this population shares the largest proportion of co-ancestry with populations from Africa, but also to some extent from La Réunion and Central Asia, so its origin is also unclear. So far, the presence of *Z. cucurbitae* on the Comoro

Archipelago is questionable (De Meyer et al. 2012) and no records are reported from Madagascar.

3.4.3 Impact

The currently observed dispersal of this species in Africa has increased the awareness of its economic significance. In Africa, little research has been devoted to this species in comparison with other cucurbit-infesting dacines, except for work in La Réunion (White and Elson-Harris 1994; Vayssières 1999; Ryckewaert et al. 2010) and Mauritius (Sookar et al. 2012, 2013). This is currently changing due to the recent observations on its distribution and dominance in particular crops.

White and Elson-Harris (1994) report *Z. cucurbitae* as a very serious pest of cucurbit crops. It has long been considered as a major pest species of commercially grown cucurbits in large parts of Asia (Kapoor 1989; Koyama 1989; Sapkota et al. 2010) and Hawaii (Harris 1989). Worldwide, Dhillon et al. (2005) indicated that losses attributed to *Z. cucurbitae* could be as high as 30–100 % and list 81 plant species, including several non-cucurbits, as possible hosts. However, several of these hosts are considered doubtful, as some of them were probably based on casual observation (c.f. White and Elson-Harris 1994). Nevertheless, observations in Tanzania have shown that *Z. cucurbitae* does attack, and can be reared from, non-cucurbit hosts, predominantly from the Solanaceae, albeit with very low infestation rates or incidence (Mwatawala et al. 2009; Mziray et al. 2010b). In Africa it is also considered the major pest on cultivated cucurbits despite the presence of indigenous cucurbit pests in the genus *Dacus* (Mwatawala et al. 2010). A list of host records from Africa is published in De Meyer et al. (2015). The majority are in the Cucurbitaceae including most of the economically important species such as watermelon (*Citrullus lanatus* (Thunb.) Matsum. & Nakai); *Cucumis* species (in particular cucumber [*Cucumis sativus* L.] and melon [*Cucumis melo* L.]); *Cucurbita* species (in particular pumpkin [*Cucurbita pepo* L.] and *Cucurbita maxima* Duchesne ex Lam.); and *Momordica* species (in particular *Momordica cf trifoliata* Hook. f. and bitter gourd [*M. charantia* L.]). Geographic differences have been observed (e.g. Vayssières et al. 2007) with *Z. cucurbitae* being more oligophagous on La Réunion and having a broader host range than in western Africa, as well as marked differences in infestation rates. However, most of these differences are based on too limited a number of observations and require further extensive sampling.

On La Réunion, Vayssières and Carel (1998), and Vayssières (1999), looked at the interspecific competition between *Z. cucurbitae*, *D. ciliatus*, and *Dacus demmerezi* (Bezzi). They observed that the different species had different altitudinal limits and that overlap was only found at lower elevations. Vayssières et al. (2008) compared in detail the demography of *Z. cucurbitae* and *D. ciliatus* on La Réunion and concluded that these species had a distinctly different life-history pattern; *Z. cucurbitae* was characterized by having a later onset of reproduction, longer oviposition time, longer life span and higher fecundity than *D. ciliatus*.

4 Conclusions and Future Actions

Although Africa has a limited commercial fruit industry (with the exception of South Africa) compared with other continents, it represents an important component of the agricultural industry and there are huge prospects for expansion. For example, in 2008 the horticulture sub-sector generated US$ 1 billion in foreign exchange from exported commodities and over US$ 650 million domestically to Kenya; directly and indirectly over four million people were employed (Ekesi 2010; Irungu 2011). Indeed, horticulture offers one of the most important opportunities for increased food security, improved nutrition and guaranteed income generation and employment opportunities for the rural economy (Ekesi et al. 2016). The numerous pests, amongst which fruit flies feature as one of the more important, poses a real threat to this developing horticultural industry. It is obvious from the examples given in this article that this problem has been aggravated by the introduction of exotic fruit fly species. With the increase in intercontinental trade and travel, the risk of new introductions is high.

Several other *Bactrocera* species, such as the carambola fruit fly, *Bactrocera carambolae* Drew and Hancock; the guava fruit fly, *Bactrocera correcta* (Bezzi); or *B. tryoni*, have spread outside their native ranges while other species, like the Chinese citrus fly, *Bactrocera minax* (Enderlein) or *Zeugodacus tau* (Walker) have the potential to do so. Also *Anastrepha* species, although not found outside the New World, have spread to different parts. Even within Africa, introductions from one region to another will most likely increase in the near future. Furthermore, the displacement of *C. capitata* could re-occur. Already *C. capitata* has been displaced by introductions of *C. rosa* in La Réunion (Duyck et al. 2006a); *C. rosa* has high potential to become established in cooler climates (De Meyer et al. 2008). Climate change may accelerate the spread of some fruit fly species as temperate parts of the world become more suitable for establishment of (sub)tropical species either of African origin (Vera et al. 2002) or of exotic species currently established in Africa (Stephens et al. 2007; Ni et al. 2012).

Clearly there is a need for both sound data on the current occurrence of fruit flies in Africa through monitoring programmes, as well as rapid detection and surveying programmes to quickly identify new intrusions. The former will allow African growers to earmark those areas that are pest free and facilitate international trade. The latter will prevent a repetition of the disastrous introductions into the continent in the past. However, both measures require two essential aspects: funds and regional co-ordination. Monitoring programmes are a costly undertaking but in the end these costs are small in comparison with the economic losses that can be prevented. More importantly, such programmes cannot be conducted solely on a national level if they are to be efficient. Fruit flies know no borders and measures taken by one country can be nullified by the lack of measures in a neighboring country. Trans-border activities are, therefore, to be highly recommended.

Acknowledgements The authors thank the various funding agencies that have facilitated their research activities over the years (see general acknowledgements for this book). Special thanks to Pia Addison and Aruna Manrakhan for providing information on the impact of *B. oleae* in South Africa.

References

Abdelgader H, Salah FEE (2016) Relative abundance of *Bactrocera zonata* in Central Sudan. Abstract book 3rd international symposium of Tephritid workers of Europe, Africa and the Middle East, Stellenbosch, 11–14 April 2016, p 76

Abuel-Ela RG, Hashem AG, Mohamed SMA (1998) *Bactrocera pallidus* (Perkins and May) (Diptera: Tephritidae), a new record in Egypt. J Egypt Germ Soc Zool (Entomol) 27:221–229

APHIS (1992) Risk assessment, Mediterranean fruit fly. Planning and risk analysis systems. Policy and Program Development. APHIS-USDA, Washington, DC

Augustinos AA, Mamuris Z, Stratikopoulos EE, D'Amelio S, Zacharopoulou A, Mathiopoulos KD (2005) Microsatellite analysis of olive fly populations in the Mediterranean indicates a westward expansion of the species. Genetica 125:231–241

Barr NB, Wiegmann BM (2009) Phylogenetic relationships of *Ceratitis* fruit flies inferred from nuclear CAD and *tango*/ARNT gene fragments: testing monophyly of the subgenera *Ceratitis* (*Ceratitis*) and *C.* (*Pterandrus*). Mol Phylogenet Evol 53:412–424

Barr NB, Islam MS, De Meyer M, McPheron BA (2012) Molecular identification of *Ceratitis capitata* (Diptera: Tephritidae) using DNA sequences of the COI barcode region. Ann Entomol Soc Am 105:339–350

Besnard G, Henry P, Wille L, Cooke D, Chapuis E (2007a) The origin of the invasive olives (*Olea europaea* L., Oleaceae). Heredity 99:608–619

Besnard G, Rubio de Casas R, Vargas P (2007b) Plastid and nuclear DNA polymorphism reveals historical processes of isolation and reticulation in the olive tree complex (*Olea europaea*). J Biogeogr 34:736–752

Besnard G, Rubio de Casas R, Christin PA, Vargas P (2009) Phylogenetics of *Olea* (Oleaceae) based on plastid and nuclear ribosomal DNA sequences: tertiary climatic shifts and lineage differentiation times. Ann Bot 104:143–160

Bess H, Van Den Bosch R, Haramoto F (1961) Fruit fly parasites and their activities in Hawaii. Proc Hawaii Entomol Soc 17:367–378

Bhatia HL (1939) New records of two fruits-flies, *Dacus ciliatus* and *Craspedoxantha octopunctata*, from Delhi. Ind J Entomol 1:108

Boardman J (1976) The olive in the Mediterranean: its culture and use. Phil Trans R Soc Lond, B, Biol Sci 275:187–196

Bonizzoni M, Guglielmino CR, Smallridge CJ, Gomulski M, Malacrida AR, Gasperi G (2004) On the origins of medfly invasion and expansion in Australia. Mol Ecol 13:3845–3855

Broumas T, Haniotakis G, Liaropoulos C, Tomazou T, Ragoussis N (2002) The efficacy of an improved form of the mass trapping method for the control of the Olive fruit fly, *Bactrocera oleae* (Gmelin) (Dipt. Tephritidae): pilot-scale feasibility studies. J Appl Entomol 126:217–223

Burrack HJ, Zalom FG (2008) Olive fruit fly (Diptera: Tephritidae) ovipositional preferences and larval performance in several commercially important olive varieties in California. J Econ Entomol 101:750–758

Burrack HJ, Connell JH, Zalom FG (2008) Comparison of olive fruit fly (*Bactrocera oleae* [Rossi]) (Diptera: Tephritidae) captures in several commercial traps in California. Int J Pest Manag 54:227–234

Cantrell B, Chadwick B, Cahill A (2002) Fruit fly fighters – eradication of the Papaya fruit fly. CSIRO, Collingwood

Connell JH (2005) History and scope of the olive industry. In: Sibbit GS, Ferguson L (eds) Olive production manual. University of California Agriculture and Natural Resources, Oakland, pp 1–10

Copeland RS, White IM, Okumu M, Machera P, Wharton RA (2004) Insects associated with fruits of the Oleaceae (Asteridae, Mamiales) in Kenya, with special reference to the Tephritidae (Diptera). In: Evenhuis NL, Kaneshiro KY (eds) D. Elmo hardy memorial volume. Contribution to the systematics and evolution of Diptera. Bishop Mus Bull Entomol 12:135–164

Coquillett DW (1899) A new trypetid from Hawaii. Entomol News 10:129–130

Costa C (1998) Olive production in South Africa: a handbook for Olive growers. Agricultural Research Council [Republic of South Africa], Pretoria

Cugala D, Mansell M, De Meyer M (2011) Situation sur *Bactrocera invadens* au Mozambique. COLEACP Lett Inf 2011(1):3

Cugala D, Ekesi S, Ambasse D, Adamu RS, Mohamed SA (2013) Assessment of ripening stages of Cavendish dwarf bananas as host or non-host to *Bactrocera invadens*. J Appl Entomol. doi:10.1111/jen.12045

Daane KM, Rice RE, Zalom FG, Barnett WW, Johnson MW (2005) Arthropod pests of olive. In: Sibbit GS, Ferguson L (eds) Olive production manual. University of California Agriculture and Natural Resources, Oakland, pp 105–114

DAFF (2016) Department of Agriculture, Forestry and Fisheries: abstract of agricultural statistics – 2016. Directorate statistics and economic analysis. Department of Agriculture, Forestry and Fisheries, Pretoria, 97 pp

De Meyer M (2000) Systematic revision of the subgenus *Ceratitis* MacLeay s.s. (Diptera, Tephritidae). Zool J Linn Soc 128:439–467

De Meyer M (2005) Phylogenetic relationships within the fruit fly genus *Ceratitis* MacLeay (Diptera: Tephritidae), derived from morphological and host plant evidence. Insect Syst Evol 36:459–480

De Meyer M, White IM (2004) True fruit flies (Diptera, Tephritidae) of Afrotropical region. http://projects.bebif.be/fruitfly/index.html

De Meyer M, Copeland RS, Wharton RA, McPheron BA (2004) On the geographical origin of the medfly, *Ceratitis capitata* (Wiedemann). In: Proceedings 6th international symposium fruit flies of economic importance, Stellenbosch, pp 45–53

De Meyer M, Mohamed S, White IM (2007) Invasive fruit fly pests in Africa. http://www.africamuseum.be/fruitfly/AfroAsia.htm

De Meyer M, Robertson M, Peterson T, Mansell M (2008) Climatic modeling for the med fly and Natal fruit fly. J Biogeogr 35:270–281

De Meyer M, Zulma DF, Guichard C, Castrillon JMG, MacLeod A, Plumelle F, Quilici S, Üstün N, Vayssières JF (2009) Pest risk analysis for *Bactrocera invadens*. EPPO, Paris, 125 pp

De Meyer M, Robertson MP, Mansell MW, Ekesi S, Tsuruta K, Mwaiko W, Vayssières JF, Peterson AT (2010) Ecological niche and potential geographic distribution of the invasive fruit fly *Bactrocera invadens* (Diptera, Tephritidae). Bull Entomol Res 100:35–48

De Meyer M, Quilici S, Franck A, Chadhouliati AC, Issimaila MA, Youssoufa MA, Barbet A, Attié M, White IM (2012) Frugivorous fruit flies (Diptera, Tephritidae, Dacini) of the Comoro Archipelago. Afr Invert 53:69–77

De Meyer M, Delatte H, Mwatawala M, Quilici S, Vayssières JF, Virgilio M (2015) A review of the current knowledge on *Zeugodacus cucurbitae* (Coquillett) (Diptera, Tephritidae) in Africa, with a list of species included in *Zeugodacus*. In: De Meyer M, Clarke AR, Vera MT, Hendrichs J (eds) Resolution of Cryptic species complexes of Tephritid pests to enhance SIT application and facilitate international trade. ZooKeys 540:539–557

Dhillon MK, Singh R, Naresh JS, Sharma HC (2005) The melon fruit fly, *Bactrocera cucurbitae*: a review of its biology and management. J Ins Sci 5:40–56

Dominiak BC, Daniels D (2012) Review of the past and present distribution of Mediterranean fruit fly (*Ceratitis capitata* Wiedemann) and Queensland fruit fly (*Bactrocera tryoni* Froggatt) in Australia. Aust J Entomol 51:104–115

Drew RAI (1989) The tropical fruit flies (Diptera: Tephritidae: Dacinae) of the Australasian and Oceanian regions. Mem Queensland Mus 26:1–521

Drew RAI, Romig MC (2013) Tropical fruit flies of South-East Asia: (Tephritidae: Dacinae). CABI, Wallingford

Drew RAI, Tsuruta K, White IM (2005) A new species of pest fruit fly (Diptera: Tephritidae: Dacinae) from Sri Lanka and Africa. Afr Entomol 13:149–154

Drosopoulou E, Nestel D, Nakou I, Kounatidis I, Papadopoulos NT, Bourtzis K, Mavragani-Tsipidou P (2011) Cytogenetic analysis of the Ethiopian fruit fly *Dacus ciliatus* (Diptera: Tephritidae). Genetica 139:723–732. doi:10.1007/s10709-011-9575-z

Duyck PF, David P, Quilici S (2004) A review of relationships between interspecific competition and invasions in fruit flies (Diptera, Tephritidae). Ecol Entomol 29:461–469

Duyck PF, David P, Quilici S (2006a) Climatic niche partitioning following successive invasions by fruit flies in La Réunion. J Anim Ecol 75:518–526

Duyck PF, David P, Junod G, Brunel C, Dupont R, Quilici S (2006b) Importance of competition mechanisms in successive invasions by polyphagous tephritids in La Réunion. Ecology 87:1770–1780

Duyck PF, David P, Quilici S (2007) Can more K-selected species be better invaders? A case study of fruit flies in La Réunion. Div Distrib 13:535–543

Ekesi S (2010) Combating fruit flies in Eastern and Southern Africa (COFESA): elements of a strategy and action plan for a regional cooperation program. http://www.globalhort.org/network-communities/fruit-flies/

Ekesi S, Billah MK, Nderitu PW, Lux SA, Rwomushana I (2009) Evidence for competitive displacement of *Ceratitis cosyra* by the invasive fruit fly *Bactrocera invadens* (Diptera: Tephritidae) on mango and mechanisms contributing to the displacement. J Econ Entomol 102:981–991

Ekesi S, De Meyer M, Mohamed SA, Virgilio M, Borgemeister C (2016) Taxonomy, ecology and management of native and exotic fruit fly species in Africa. Ann Rev Entomol 61:219–238

Enkerlin DL, Garcia R, Lopez M (1989) Mexico, Central and South America. In: Robinson AS, Hooper G (eds) World crop pests: fruit flies, their biology, natural enemies and control, vol 3A. Elsevier Press, Amsterdam, pp 83–90

EPPO (2011) Data sheets on quarantine pests: *Dacus ciliatus*. https://www.eppo.int/QUARANTINE/insects/Dacus_ciliatus/DACUCI_ds.pdf

EPPO (2013) PM 7/114 (1) *Bactrocera zonata*. OEPP/EPPO Bull 43:412–416

Etienne J (1972) Les principales Trypétides nuisibles de l'Ile de La Réunion. Ann Soc Entomol France 8:485–491

French C (2005) The new invasive *Bactrocera* species. Insect pest control Newsletter. International Atomic Energy Agency, Vienna

Gasperi G, Bonizzoni M, Gomulski LM, Murelli V, Torti C, Malacrida AR, Guglielmino AR (2002) Genetic differentiation, gene flow and the origin of infestations of the medfly, *Ceratitis capitata*. Genetica 116:125–135

Hancock DL (1989) Southern Africa. In: Robinson AS, Hooper G (eds) World crop pests: fruit flies, their biology, natural enemies and control, vol 3A. Elsevier Press, Amsterdam, pp 51–58

Hancock DL (2015) A new subgenus for six indo-australian species of *Bactrocera* Macquart (Diptera: Tephritidae: Dacinae) and subgeneric transfer of four other species. Aust Entomol 42:39–44

Hancock DL, Drew RAI (2006) A revised classification of subgenera and species groups in *Dacus* Fabricius (Diptera, Tephritidae). Instr Biodivers 7:167–205

Hancock DL, Drew RAI (2015) A review of the Indo-Australian subgenus *Parazeugodacus* Shiraki of *Bactrocera* Macquart (Diptera: Tephritidae: Dacinae). Aust Entomol 42:91–104

Harris EJ (1989) Hawaiian islands and North Africa. In: Robinson AS, Hooper G (eds) World crop pests: fruit flies, their biology, natural enemies and control, vol 3A. Elsevier Press, Amsterdam, pp 73–80

Hendel F (1915) H. Sauter's Formosa-Ausbeute. Tephritinae. Ann Mus Nat Hung 13:424–467

IAEA (1995) Vienna: International Atomic Energy Agency. http://www-aweb.iaea.org/nafa/ipc/public/ipc-ecnomic-medflymaghreb-TECDOC830.pdf

Irungu J (2011) Contribution of horticulture to food security in Kenya. Acta Hortic 911:27–32

Jacquard C, Virgilio M, Quilici S, De Meyer M, Delatte H (2013) Population structuring of an economic important pest: the case of the fruit fly *Bactrocera cucurbitae* on the tropical island of La Réunion. Biol Invasions 15:759–773

Jose L, Cugala D, Santos L (2013) Assessment of invasive fruit fly infestation and damage in Cabo Delgado Province, northern Mozambique. Afr Crop Sci J 21:21–28

Kapatos ET (1989) Integrated pest management systems of *Dacus oleae*. In: Robinson AS, Hooper G (eds) World crop pests: fruit flies, their biology, natural enemies and control, vol 3A. Elsevier Press, Amsterdam, pp 391–398

Kapoor VC (1989) Indian sub-continent. In: Robinson AS, Hooper G (eds) World crop pests: fruit flies, their biology, natural enemies and control, vol 3A. Elsevier Press, Amsterdam, pp 59–62

Katsoyannos P (1992) Olive pests and their control in the Near East. FAO Plant Prod Prot Paper Ser 115:178 pp

Khamis FM, Karam N, Ekesi S, De Meyer M, Bonom A, Gomulski LM, Scalari F, Gabrieli P, Siriliano P, Masiga D, Kenya EU, Gasperi G, Malacrida AR, Gublielmino CR (2009) Uncovering the tracks of a recent and rapid invasion: the case of the fruit fly pest *Bactrocera invadens* in Africa. Mol Ecol 18:4798–4810

Koyama J (1989) South-east Asia and Japan. In: Robinson AS, Hooper G (eds) World crop pests: fruit flies, their biology, natural enemies and control, vol 3A. Elsevier Press, Amsterdam, pp 63–66

Koyama J, Kakinohana H, Miyatake T (2004) Eradication of the melon fly, *Bactrocera cucurbitae*, in Japan: importance of behavior, ecology, genetics and evolution. Ann Rev Entomol 49:331–349

Leblanc L (2012) Dacine fruit flies of Asia-Pacific. http://www.herbarium.hawaii.edu/fruitfly/index.php

Levinson HZ, Levinson AR (1984) Botanical and chemical aspects of the olive tree with regards to fruit acceptance by *Dacus oleae* (Gmelin) and other frugivorous animals. Z Angew Entomol 98:136–149

Liebhold AM, Tobin PC (2008) Population ecology of insect invasions and their management. Ann Rev Entomol 53:387–408

Liquido NJ, Harris EJ, Dekker LA (1994) Ecology of *Bactrocera latifrons* (Diptera: Tephritidae) populations: host plants, natural enemies, distribution, and abundance. Ann Entomol Soc Am 87:71–84

Liquido NJ, McQuate GT, Suiter KA (2014) MEDHOST: an encyclopedic bibliography of the host plants of the Mediterranean fruit fly, *Ceratitis capitata* (Wiedemann), Version 2.0. United States Department of Agriculture, Center for Plant Health Science and Technology, Raleigh, NC. Available: https://medhostv2.cphst.org/

Lumaret R, Ouazzani N, Michaud H, Vivier G, Deguilloux MF, Di Giusto F (2004) Allozyme variation of oleaster populations (wild olive tree) (*Olea europea* L.) in the Mediterranean Basin. Heredity 92:343–351

Lux SA, Copeland RS, White IM, Manrakhan A, Billah MK (2003) A new invasive fruit fly species form the *Bactrocera dorsalis* (Hendel) group detected in East Africa. Insect Sci Appl 23:355–360

Malacrida A, Gomulski L, Bonizzoni M, Bertin S, Gasperi G, Guglielmino C (2007) Globalization and fruitfly invasion and expansion: the medfly paradigm. Genetica 131:1–9

Malavasi A (2014) Introductory remarks. In: Shelly T, Epsky N, Jang EB, Reyes-Flores J, Vargas R (eds) Trapping and the detection, control, and regulation of Tephritid fruit flies. Springer, Dordrecht, pp ix–x

Malavasi A, Midgarden D, De Meyer M (2013) *Bactrocera* species that pose a threat to Florida: *Bactrocera carambolae* and *B. invadens*. In: Peña JE (ed) Potential invasive pests of agricultural crops. CABI, Wallingford, pp 214–227

Manousis T, Moore N (1987) Control of *Dacus oleae*, a major pest of olives. Insect Sci Appl 8:1–9

Manrakhan A, Venter JH, Hattingh V (2015) The progressive invasion of *Bactrocera dorsalis* (Diptera: Tephritidae) in South Africa. Biol Invasions 17:2803–2809. doi:10.1007/s10530-015-0923-2

Mkize N, Hoelmer KA, Villet MH (2008) A survey of fruit-feeding insects and their parasitoids occurring on wild olives, *Olea europaea* ssp. *cuspidata*, in the Eastern Cape of South Africa. Biocontr Sci Technol 18:991–1004

Mohamed SMA (2004) Competition between Mediterranean fruit fly and peach fruit fly in fruit infestation. J Egypt Germ Soc Zool (Entomol) 43:17–23

Mohamed SMA, El-Wakkad MF (2003) The role of killing bags on the reduction of peach fruit fly in horticultural orchards. J Egypt Germ Soc Zool (Entomol) 42:11–19

Montiel-Bueno A, Jones O (2002) Alternative methods for controlling the olive fly, *Bactrocera oleae*, involving semiochemicals. IOBC/WPRS Bull 25:1–11

Mosleh YY, Moussa SFM, Mohamed LHY (2011) Comparative toxicity of certain pesticides to peach fruit fly, *Bactrocera zonata* Saunders (Diptera: Tephritidae) under laboratory conditions. Plant Prot Sci 47:5–120

Mostakim M, El Abed S, Iraqui M, Benbrahim KF, Houari A, Gounni AS, Ibnsouda SK (2012) Biocontrol potential of a *Bacillus subtilis* strain against *Bactrocera oleae*. Ann Microbiol 62:211–218

Mwatawala MW, White IM, Maerere AP, Senkondo FJ, De Meyer M (2004) A new invasive *Bactrocera* species (Diptera, Tephritidae) in Tanzania. Afr Entomol 12:156–158

Mwatawala M, De Meyer M, White IM, Maerere A, Makundi RH (2007) Detection of the solanum fruit fly, *Bactrocera latifrons* (Hendel) in Tanzania (Dipt., Tephritidae). J Appl Entomol 131:501–503

Mwatawala M, De Meyer M, Makundi R, Maerere A (2009) Host range and distribution of fruit-infesting pestiferous fruit flies (Diptera, Tephritidae) in selected areas of Central Tanzania. Bull Entomol Res 99:629–641

Mwatawala M, Maerere A, Makundi RH, De Meyer M (2010) On the occurrence and distribution of the invasive Solanum fruit fly, *Bactrocera latifrons*, in Tanzania. J Afrotrop Zool 6:83–89

Mziray HA, Makundi RH, Mwatawala M, Maerere A, De Meyer M (2010a) Spatial and temporal abundance the solanum fruit fly, *Bactrocera latifrons* (Hendel) in Morogoro, Tanzania. Crop Prot 29:454–461

Mziray HA, Makundi RH, Mwatawala M, Maerere A, De Meyer M (2010b) Host use of *Bactrocera latifrons*, a new invasive tephritid species in Tanzania. J Econ Entomol 103:70–76

Nardi F, Carapelli A, Dallai R, Roderick GK, Frati F (2005) Population structure and colonization history of the olive fly, *Bactrocera oleae* (Diptera, Tephritidae). Mol Ecol 14:2729–2738

Nardi F, Carapelli A, Boore JL, Roderick GK, Dallai R, Frati F (2010) Domestication of the olive fly through a multiregional host shift to cultivated olives: comparative dating using complete mitochondrial genome. Mol Phylogenet Evol 57:678–686

Ni WL, Li ZH, Chen HJ, Wan FH, Qu WW, Zhang Z, Kriticos DJ (2012) Including climate change in pest risk assessment: the peach fruit fly, *Bactrocera zonata* (Diptera: Tephritidae). Bull Entomol Res 102:173–183

Norrbom AL, Carroll LE, Thompson RC, White IM, Freidberg A (1999) Systematic database of names. In: Thompson FC (ed) Fruit fly expert identification system and systematic information database. Myia 9:65–251

Ole-MoiYoi OK, Lux SA (2004) Fruit flies in sub-Saharan Africa: a long neglected problem devastating local fruit production and a threat to horticulture beyond Africa. In: Proceedings 6th international symposium fruit flies of economic importance, Stellenbosch, pp 5–10

Orian AJE, Moutia LA (1960) Fruit flies (Trypetidae) of economic importance in Mauritius. Rev Agric Sucr Ile Maurice 39:142–150

Otieno W (2011) KEPHIS experience with market access and compliance with official standards. Acta Hortic 911:73–76

Papadopoulos NT (2014) Fruit fly invasion: historical, biological, economic aspects and management. In: Shelly T, Epsky N, Jang EB, Reyes-Flores J, Vargas R (eds) Trapping and the detection, control, and regulation of Tephritid fruit flies. Springer, Dordrecht, pp 219–252

Papadopoulos NT, Katsoyannos BI, Carey JR, Kouloussis NA (2001) Seasonal and annual occurrence of the Mediterranean fruit fly (Diptera: Tephritidae) in northern Greece. Anns Entomol Soc Am 94:41–50

Papadopoulos N, Plant RE, Carey JR (2013) From trickle to flood: the large-scale, cryptic invasion of California by tropical fruit flies. Proc Roy Soc B 280. doi:10.1098/rspb.2013.1466

Qureshi ZA, Hussain T, Siddiqui QH (1991) Relative preference of mango varieties by *Dacus zonatus* and *D. dorsalis*. Pak J Zool 23:85–87

Ravry C (2008) Situation in Botswana, Kenya, Mozambique, Namibia, Zambia and Zimbabwe. Fight Fruit Veg Flies Region West Afr 7:2

Rice RE (2000) Bionomics of the olive fruit ßy *Bactrocera (Dacus) oleae*. University of California Cooperative Extension. UC Plant Prot Q 10:1–5

Rwomushana I, Ekesi S, Ogol CKPO, Gordon I (2009) Mechanisms contributing to the competitive success of the invasive fruit fly *Bactrocera invadens* over the indigenous mango fruit fly, *Ceratitis cosyra*: the role of temperature and resource pre-emption. Entomol Exp et Appl 133:27–37

Ryckewaert P, Deguine JP, Brévault T, Vayssières JF (2010) Fruit flies (Diptera: Tephritidae) on vegetable crops in Réunion Island (Indian Ocean): state of knowledge, control methods and prospect for management. Fruits 65:113–130

Salum JK, Mwatawala MW, Kusolwa PM, De Meyer M (2013) Demographic parameters of the two main fruit fly (Diptera: Tephritidae) species attacking mango in Central Tanzania. J Appl Entomol online doi:10.1111/JEN.12044

Sapkota R, Dahal KC, Thapa RB (2010) Damage assessment and management of cucurbit fruit flies in spring-summer squash. J Entomol Nematol 2:7–12

Schutze MK, Mahmood K, Pavasovic A, Bo W, Newman J, Clarke AR, Krosch M, Cameron SL (2014a) One and the same: integrative taxonomic evidence that *Bactrocera invadens* (Diptera: Tephritidae) is the same species as the Oriental fruit fly *Bactrocera dorsalis*. Syst Entomol 40:472–486. doi:10.1111/syen.12114

Schutze MK, Aketarawong N, Amornsak W, Armstrong KF, Augustinos A, Barr N, Bo W, Bourtzis K, Boykin LM, Caceres C, Cameron SL, Chapman TA, Chinvinijkul S, Chomic A, De Meyer M, Drosopoulou ED, Englezou A, Ekesi S, Gariou-Papalexiou A, Hailstones D, Haymer D, Hee AKW, Hendrichs J, Hasanuzzaman M, Jessup A, Khamis FM, Krosch MN, Leblanc L, Mahmood K, Malacrida AR, Mavragani-Tsipidou P, McInnis DO, Mwatawala M, Nishida R, Ono H, Reyes J, Rubinoff DR, San Jose M, Shelly TE, Srikachar S, Tan KH, Thanaphum S, Ul Haq I, Vijaysegaran S, Wee SL, Yesmin F, Zacharopoulou Z, Clarke AR (2014b) Synonymization of key pest species within the *Bactrocera dorsalis* species complex (Diptera : Tephritidae): taxonomic changes based on a review of 20 years of integrative morphological, molecular, cytogenetic, behavioural, and chemoecological data. Syst Entomol 40:456–471. doi:10.1111/syen.12113

Shimizu Y, Kohama T, Uesato T, Matsuyama T, Yamagishi M (2007) Invasion of solanum fruit fly *Bactrocera latifrons* (Diptera: Tephritidae) to Yonaguni Island, Okinawa Prefecture, Japan. Appl Entomol Zool 42:269–275

Silvestri F (1916) Prima notizia sulla presenza della mosca delle olive e di un parasita di essa in India. Rend Accad Naz Lincei 25:424

Sookar P, Hendrichs J, Alleck M, Buldawoo I, Khayrattee FB, Choolun T, Permalloo S, Rambhunjun M (2012) Area-wide *Bactrocera cucurbitae* (Coquillett) control in selected areas of Mauritius. Actes Semin Int Clôture Gamour 21–24 November 2011, Saint-Pierre, La Réunion, pp 98–102

Sookar P, Haq I, Jessup A, McInnis D, Franz G, Wornoayporn V, Permalloo S (2013) Mating compatibility among *Bactrocera cucurbitae* (Diptera: Tephritidae) populations from three different origins. J Appl Entomol 137:69–74

Sookar P, Bhagwant S, Khayrattee FB, Chooneea Y, Ekesi S (2014) Mating compatibility of wild and sterile melon flies, *Bactrocera cucurbitae* (Diptera: Tephritidae) treated with entomopathogenic fungi. J Appl Entomol 138:409–417

Stephens AEA, Kriticos DJ, Leriche A (2007) The current and future potential geographical distribution of the oriental fruit fly, *Bactrocera dorsalis* (Diptera: Tephritidae). Bull Entomol Res 97:369–378

Syed RA, Ghani MA, Murtaza M (1970) Studies on the tephritids and their natural enemies in West Pakistan. III. *Dacus zonatus* (Saunders). Tech Bull Commonw Inst Biol Control 13:1–16

Szyniszewska AM, Tatem AJ (2014) Global assessment of seasonal potential distribution of Mediterranean fruit fly, *Ceratitis capitata* (Diptera: Tephritidae). PlosOne 9(11), e111582

Terblanche JS, Nyamukondiwa C, Kleynhans E (2010) Thermal variability alters climatic stress resistance and plastic responses in a globally invasive pest, the Mediterranean fruit fly (*Ceratitis capitata*). Entomol Exp et Appl 137:304–315

Torres-Vila LM, Pérez de Sande J (2002) Prospeccion de la resistencia insecticida al dimetoato en la mosca del olivo *Bactrocera oleae* Gmelin (Diptera: Tephritidae) en Extremadura. Bol San Veg Plagas 28:218–286

Tzanakakis ME (2003) Seasonal development and dormancy of insects and mites feeding on olive: a review. Neth J Zool 52:87–224

Vargas RI, Nishida T (1985) Survey for *Dacus latifrons* (Diptera: Tephritidae). J Econ Entomol 78:1311–1314

Vayssières JF (1999) Les relations insectes-plantes chez les Dacini (Diptera Tephritidae) ravageurs des Cucurbitaceae à La Réunion. PhD thesis, University Paris XII, Paris, France

Vayssières JF, Carel Y (1998) Les Dacini (Diptera: Tephritidae) inféodées aux Cucurbitaceae à la Réunion: gamme de plantes hôtes et stades phénologiques préférentiels des fruits au moment de la piqûre pour des espèces cultivées. Anns Soc Entomol France 35:197–202

Vayssières JF, Rey JY, Traoré L (2007) Distribution and host plants of *Bactrocera cucurbitae* in West and Central Africa. Fruits 62:391–396

Vayssières JF, Carel Y, Coubès M, Duyck PF (2008) Development of immature stages and comparative demography of two cucurbit-attacking fruit flies in Réunion island: *Bactrocera cucurbitae* and *Dacus ciliatus*. Environ Entomol 37:307–314

Vera MT, Rodriguez R, Segura DF, Cladera JL, Sutherst RW (2002) Potential geographical distribution of the Mediterranean fruit fly, *Ceratitis capitata* (Diptera: Tephritidae), with emphasis on Argentina and Australia. Pop Ecol 31:1009–1022

Virgilio M, De Meyer M, White IM, Backeljau T (2009) African *Dacus* (Diptera: Tephritidae): molecular data and host plant association do not corroborate morphology based classifications. Mol Phylogenet Evol 51:531–539

Virgilio M, Delatte H, Backeljau T, De Meyer M (2010) Macrogeographic population structuring in the cosmopolitan agricultural pest *Bactrocera cucurbitae* (Diptera: Tephritidae). Mol Ecol 19:2713–2724

Virgilio M, Jordaens K, Verwimp C, White IM, De Meyer M (2015) Higher phylogeny of frugivorous flies (Diptera, Tephritidae, Dacini): localised partition conflicts and a novel generic classification. Mol Phylogen Evol 85:171–179. doi: http://dx.doi.org/10.1016/j.ympev.2015.01.007

White IM (2006) Taxonomy of the Dacina (Diptera:Tephritidae) of Africa and the Middle East. Afr Entomol Mem 2:1–156

White IM, Elson-Harris MM (1994) Fruit flies of economic significance: their identification and bionomics. CAB International, Wallingford

White IM, Liquido NJ (1995) C*haetodacus latifrons* Hendel, 1915 (currently *Bactrocera latifrons*; Insecta, Diptera): proposed precedence of the specific name over that of *Dacus parvulus* Hendel, 1912. Bull Zool Nomncl 52:250–253

White IM, De Meyer M, Stonehouse J (2001) A review of the native and introduced fruit flies (Diptera, Tephritidae) in the Indian Ocean Islands of Mauritius, Réunion, Rodrigues and Seychelles. In: Price NS, Seewooruthun I (eds) Proceedings Indian Ocean Commission Regional Fruit Fly symposium (Mauritius, 5–9 June 2000). Indian Ocean Commission, Mauritius, pp 15–21

Wu Y, McPheron BA, Wu JJ, Li ZH (2012) Genetic relationship of the melon fly, *Bactrocera cucurbitae* (Diptera: Tephritidae) inferred from mitochondrial DNA. Insect Sci 19:195–204

Yokoyama VY (2012) Olive fruit fly (Diptera: Tephritidae) in California: longevity, oviposition, and development in canning olives in the laboratory and greenhouse. J Econ Entomol 105:186–195

Yuval B, Hendrichs J (2000) Behavior of flies in the genus *Ceratitis* (Dacinae: Ceratitidini). In: Aluja M, Norrbom A (eds) Fruit flies (Tephritidae): phylogeny and evolution of behavior. CRC Press, Boca Raton, pp 429–457

Zygouridis NE, Augustinos AA, Zalom FG, Mathiopoulos KD (2009) Analysis of olive fly invasion in California based on microsatellite markers. Heredity 102:402–412

Marc De Meyer is an Entomologist attached to the Royal Museum for Central Africa in Tervuren (Belgium). Previously he worked in Botswana and Kenya for a number of years. His main field of expertise is the taxonomy and systematics of selected groups of Diptera such as Pipunculidae, Syrphidae and Tephritidae, with emphasis on the Afrotropical fauna. His research includes different aspects, from taxonomic revisions to studies on evolutionary trends and speciation events.

Sunday Ekesi is an Entomologist at the International Centre of Insect Physiology and Ecology (*icipe*), Nairobi, Kenya. He heads the Plant Health Theme at *icipe*. Sunday is a professional scientist, research leader and manager with extensive knowledge and experience in sustainable agriculture (microbial control, biological control, habitat management/conservation, managing pesticide use, IPM) and biodiversity in Africa and internationally. Sunday has been leading a continent-wide initiative to control African fruit flies that threaten production and export of fruits and vegetables. The initiative is being done in close collaboration with IITA, University of Bremen, Max Planck Institute for Chemical Ecology together with NARS, private sectors and ARI partners in Africa, Asia, Europe and the USA and focuses on the development of an IPM strategy that encompasses baiting and male annihilation techniques, classical biological control, use of biopesticides, ant technology, field sanitation and postharvest treatment for quarantine fruit flies. The aim is to develop a cost effective and sustainable technology for control of fruit flies on the African continent that is compliant with standards for export markets while also meeting the requirements of domestic urban markets. Sunday has broad perspectives on global agricultural research and development issues, with first-hand experience of the challenges and opportunities in working with smallholder farmers, extension agents, research organizations and the private sector to improve food and nutritional security. He sits on various international advisory and consultancy panels for the FAO, IAEA, WB and regional and national projects on fruit fly, arthropod pests and climate change-related issues. Sunday is a Fellow of the African Academy of Sciences (FAAS).

Chapter 8
Feeding and Mating Behaviour of African Fruit Flies

Aruna Manrakhan

Abstract The majority of African fruit fly species are controlled using behavioural methods (e.g. attract and kill). Sterile Insect Technique (SIT) has also been successfully used for a few African fruit fly pests, mostly outside of the African region. A thorough understanding of the feeding and mating behaviour of fruit fly pests is required when using behavioural control methods and SIT for fruit fly control. The feeding and mating behaviour of key African fruit fly pests are reviewed. Feeding and mating behaviours have been elucidated for only a few species with a wider global distribution. For the remaining *Ceratitis*, *Bactrocera* and *Dacus* species with a limited worldwide distribution, there are still important knowledge gaps in their behavioural ecology. With horticulture expanding in Africa and increasing trade of horticultural produce from the region, it is important that these knowledge gaps are filled so that control methods can be optimised.

Keywords Behaviour • *Ceratitis* • *Bactrocera* • *Dacus* • *Zeugodacus* • Africa

1 Introduction

African fruit fly pests are mainly monitored and controlled using attractive food baits and male lures (Ekesi et al. 2007; Manrakhan 2006). For control, food baits and male lures are combined with a toxin or pathogen in an attract and annihilate approach; adult fruit flies are attracted to the bait/ lure and killed by the toxin or pathogen (Cunningham 1989; Roessler 1989). Such behavioural control methods are effective and limit insecticide use, but their efficacy can be influenced by a number of intrinsic (e.g genetic and physiological factors) and extrinsic (e.g climate, host abundance) factors (Foster and Harris 1997). As such, a thorough understanding of the feeding and mating behaviour of each pest species is essential for the optimisation of control methods.

A. Manrakhan (✉)
Citrus Research International, PO Box 28, Nelspruit 1200, South Africa
e-mail: aruna@cri.co.za

© Springer International Publishing Switzerland 2016
S. Ekesi et al. (eds.), *Fruit Fly Research and Development in Africa - Towards a Sustainable Management Strategy to Improve Horticulture*,
DOI 10.1007/978-3-319-43226-7_8

For some fruit fly pests, the Sterile Insect Technique (SIT) has also been successfully used worldwide as a method for prevention, containment, suppression and eradication (Enkerlin 2005). To date the use of SIT in Africa is still limited. Effectiveness of SIT programmes are dependent on successful mating of sterile males with wild-type females and the subsequent induction of reproductive failure (Perez-Staples et al. 2012). Given that most tropical and subtropical fruit flies have complex mating systems (Burk 1981), a detailed understanding of each mating system is crucial for successful development and optimisation of SIT (Hendrichs et al. 2002).

Here I review the current status of knowledge on the feeding and mating behaviour of key fruit fly pest species in Africa grouped under four genera: *Ceratitis*, *Bactrocera*, *Zeugodacus* and *Dacus*. Future research needs required to fill the gaps in our understanding of feeding and mating behaviour of these African fruit fly pests are also discussed.

2 Feeding Behaviour of African Fruit Flies

Fruit flies only feed during the larval and adult stages (Christenson and Foote 1960). Frugivorous fruit fly larvae feed on the fruit pulp while adult flies have been reported feeding on various sources including extra-floral glandular secretions from plants, plant leachates, fruit juices, bacteria, honeydew and bird faeces in order to acquire sufficient carbohydrate and protein for survival and reproduction (Christenson and Foote 1960; Hendrichs and Hendrichs 1990; Manrakhan and Lux 2009; McQuate et al. 2003; Nishida 1958; Warburg and Yuval 1997). Studies on the mouthparts of adult *Bactrocera* and *Ceratitis* species have revealed that adult flies have a fluid-centered mode of feeding and a labellar-filter-feeding mechanism allowing them to ingest fluids and particles that are less than 0.5 μm in size (Vijaysegaran et al. 1997; Coronado-Gonzalez et al. 2008). All frugivorous female fruit flies are anautogenous, that is they need to feed on protein to realise their reproductive potential (Drew and Yuval 2001). Furthermore, nutrition plays an important role in both the regulation of oocyte maturation and male accessory gland development (Williamson 1989). Foraging behaviour of adult flies occurs at different hierarchical levels: the habitat (fruit-growing area), the patch (host trees or non-host trees) and the specific food item (Hassell and Southwood 1978; Prokopy and Roitberg 1989). The initial identification of a potential food item by a fly is largely achieved through olfaction (Dethier 1976). Olfactory receptors on the antennae and on the palps of fruit flies are sensitive to the gaseous products of amino acid breakdown, mainly ammonia (Rice 1989; Tsiropoulos 1992). The attraction of fruit flies to protein-based odours is exploited in the use of poisoned food baits for fruit fly control (Roessler 1989).

A thorough understanding of adult feeding behaviour has direct implications in optimising control of fruit fly pests using food baits (Hendrichs and Prokopy 1994). In relation to the application of food baits, knowing what attracts flies to food and understanding the factors that influence feeding behaviour can improve the efficacy of control strategies using food baits.

2.1 Ceratitis *Species*

Research on the feeding behaviour of *Ceratitis* species has largely been restricted to a few of the key pest species: the Mediterranean fruit fly (medfly), *Ceratitis capitata* (Wiedemann); *Ceratitis fasciventris* Bezzi; the Natal fruit fly, *Ceratitis rosa* Karsch; and the mango fruit fly, *Ceratitis cosyra* (Walker). Some of these fruit fly pests prefer to reside on host trees compared with non-host trees, and for this reason most of their food foraging activities are likely to occur within host patches (Manrakhan 2009; Hendrichs and Hendrichs 1990).

The type of food ingested influences survival and the reproductive activity of *Ceratitis* species. Survival of some pest species of *Ceratitis* was poor without a sugar source (Hendrichs and Hendrichs 1990; Hendrichs et al. 1991; Manrakhan and Lux 2006). Calling, mating, egg laying and egg fertility was greater in adult *Ceratitis* species feeding on a protein-rich diet than on a protein-poor diet (Kaspi et al. 2000; Kaspi and Yuval 2000; Papadopoulos et al. 1998; Manrakhan and Lux 2006; Taylor and Yuval 1999; Shelly et al. 2002; Hendrichs et al. 1991). The effect of protein on longevity of *C. capitata* varied amongst different studies. Some studies reported better survival of flies maintained on a constant diet of sugar and protein compared with flies maintained on only sugar sources (Cangussu and Zucoloto 1995; Manrakhan and Lux 2006). Other studies showed a reduced mortality in *C. capitata* flies when they were deprived of protein (Carey et al. 1998; Hendrichs et al. 1991; Kaspi and Yuval 2000).

Temporal patterns of feeding have been described for a number of *Ceratitis* species. With respect to time of the day, male *Ceratitis fasciventris* (Bezzi) and male *C. capitata* fed mainly in the morning and late afternoon respectively, after having engaged in reproductive activities (Hendrichs and Hendrichs 1990; Manrakhan and Lux 2006; Warburg and Yuval 1997). With respect to the age of flies, for *C. cosyra*, *C. fasciventris* and *C. capitata*, sugar consumption was greater soon after emergence whilst protein feeding usually occurred either in or after the first week of adult life depending on species and sex (Manrakhan and Lux 2009). Nutritional state had a significant influence on the responses of *C. cosyra*, *C. fasciventris* and *C. capitata* to food odours. Sugar deprivation increased responses of young flies to food odours but there was no preference for particular types of odours in young sugar-deprived flies (Cohen and Voet 2002; Manrakhan and Lux 2008). Protein deprivation in mature flies enhanced their response to protein odours (Barry et al. 2003; Manrakhan and Lux 2008; Prokopy et al. 1992). Not all protein odours are equally attractive to adult *Ceratitis* species. For example, amongst different animal droppings that form part of the natural food complex of *C. capitata*, a preference for droppings from birds and lizards compared with droppings from mammals was shown for this species (Prokopy et al. 1993). Moreover for *C. capitata* and other

Ceratitis species, natural food sources were more attractive than artificial food baits (Manrakhan and Lux 2008; Prokopy et al. 1992). Variation in fly responses to different artificial proteinaceous baits has also been recorded (Katsoyannos et al. 1999; Manrakhan and Kotze 2011). The degree of response to protein baits differed between *Ceratitis* species (Manrakhan and Kotze 2011); this has important implications as baits that are effective for control of one *Ceratitis* species may not be as effective for other *Ceratitis* species. Differences in responses to protein odours amongst *Ceratitis* species could be linked to their protein requirement and therefore their sensitivity to volatiles emanating from protein sources.

2.2 Bactrocera *Species*

The three most important *Bactrocera* pest species in the African region include the oriental fruit fly, *Bactrocera dorsalis* (Hendel) previously known as *B. invadens* Drew, Tsuruta and White (Schutze et al. 2014); the peach fruit fly, *Bactrocera zonata* (Saunders); and the solanum fruit fly, *Bactrocera latifrons* (Hendel); they are all of Asian origin and only recently introduced in to the African region (De Meyer et al. 2014; Ekesi and Muchugu 2006). There is generally a lack of information on the feeding behaviour and food bait preferences of *B. zonata* and *B. latifrons*. In the island of Mauritius, investigations found that *B. zonata* responded equally well to a commercially available protein hydrolysate and processed brewery yeast waste (Gopaul et al. 2000). In contrast to most *Bactrocera* species, more information is available on the feeding behaviour of *B. dorsalis*. Early studies by Bess and Haramoto (1961) in Hawaii suggested that *B. dorsalis* foraged for food and shelter on plants other than one of its preferred host plants, guava. However, subsequent studies in Hawaii have shown that *B. dorsalis* prefers to forage on guava (Vargas et al. 1990). In laboratory studies under controlled conditions, *B. dorsalis* engaged in feeding behaviour during the morning (Arakaki et al. 1984). As with *Ceratitis* species, reproduction of *B. dorsalis* was affected by the type of food ingested. A protein source in the adult diet of *B. dorsalis* was important for reproduction (Shelly et al. 2005). For example, *B. dorsalis* males deprived of protein had fewer matings that those that had access to protein (Shelly et al. 2005). Protein feeding decreased attraction of mature adult *B. dorsalis* females to protein odours whilst immature protein-fed *B. dorsalis* females were equally attracted to fruit and protein odours (Cornelius et al. 2000). In studies conducted recently in Kenya, *B. dorsalis* showed preferences for particular protein baits, and females were more attracted to protein baits than males (Ekesi et al. 2014). In studies conducted in Hawaii, young female *B. dorsalis* had stronger positive responses to protein baits than older females (Barry et al. 2006; Pinero et al. 2011). The results of studies on food bait preferences for *B. dorsalis* have been at times conflicting. For example, in studies done in Tanzania and Kenya, the three-component Biolure (ammonium acetate, putrescine, trimethylamine) was less attractive to *B. dorsalis* than liquid protein baits such as Torula yeast (Ekesi et al. 2014; Mwatawala et al. 2006). However, current trials in South

Africa have found that the three-component Biolure is more attractive to *B. dorsalis* than liquid protein baits such as Torula yeast (Manrakhan et al., unpublished data). A more in depth understanding of the responses of *B. dorsalis* to food attractants is required for optimisation of monitoring and control efforts against this pest.

2.3 Zeugodacus *Species and* Dacus *Species*

In Africa, the melon fly, *Zeugodacus cucurbitae* (Coquillet) previously known as *Bactrocera cucurbitae* (Coquillet) (Virgilio et al. 2015) and a number of *Dacus* fruit fly species are economically important pests of vegetables in the family Cucurbitaceae (Ekesi and Muchugu 2006). Studies on a few cucurbit-infesting fruit fly pests have shown that flies preferred to congregate on non-host plants while host plants were only visited by gravid females for oviposition (Nishida and Bess 1950; Atiama-Nurbel et al. 2012). For this reason control efforts targeted at these species and based on food baits have been done on non-host plants rather than host plants (Prokopy et al. 2003). Other than studies on the distribution of cucurbit-infesting fruit flies within a habitat, there is very little information on their feeding behaviour. *Zeugodacus cucurbitae*; the lesser pumpkin fly, *Dacus ciliatus* (Loew); and *Dacus demmerezi* (Bezzi) respond to protein baits, although the degree of response differed between species and within some species in respect to age and starvation status depending on bait type (Barry et al. 2006; Deguine et al. 2012; Nestel et al. 2004; Pinero et al. 2011). There is a lack of information on the temporal feeding patterns of these species. Moreover, for some *Dacus* species, such as *D. ciliatus*, which do not respond well to commercially available protein baits (Deguine et al. 2012), detailed studies on their feeding requirements and habits are urgently required.

3 Mating Behaviour of African Fruit Flies

Afrotropical fruit flies, in particular those that are polyphagous, have complex mating systems which involve the formation of leks (Sivinski and Burk 1989). Leks are male communal display aggregations, with the purpose of attracting females (Sivinski and Burk 1989). An understanding of the mating behaviour of key African fruit fly pests is important in the development of SIT for control of these pests (Hendrichs et al. 2002). Male lures used for monitoring and control of African fruit fly species are known to influence the mating behaviour of some species (Khoo and Tan 2000; Shelly and Dewire 1994; Tan and Nishida 1998). The use of male lures in combination with insect growth regulators (IGRs) or pathogens will require a better understanding of the effect of these male lures on the mating behaviour of the flies since it is essential that lure-fed males interact and mate with wild females if sterilisation or transmission of pathogens is to be effective.

3.1 Ceratitis *Species*

The mating behaviour of *C. capitata* has been studied extensively over the years (see review by Eberhard [2000]). However, very little is known about the mating behaviour of other *Ceratitis* species. The courtship behaviour of *C. rosa*, *C. cosyra*, *C. fasciventris* and *Ceratitis catoirii* Guérin-Mèneville has been studied in laboratory and field cages (Manrakhan and Lux 2009; Quilici et al. 2002; Myburgh 1962). Unlike *C. capitata* and *C. catoirii* which are day mating species, *C. rosa*, *C. fasciventris* and *C. cosyra* began mating activity at dusk, staying in copula until day break (Manrakhan and Lux 2009; Quilici et al. 2002). In studies conducted under laboratory conditions in Kenya, protein in the adult food diet influenced the calling and mating behaviour of *C. fasciventris* and *C. cosyra* (Manrakhan and Lux 2006). Frequency of calling in *C. fasciventris* males, and mating in *C. fasciventris* and *C. cosyra*, was higher when the flies were fed on a protein-rich diet as compared to a protein-poor diet (Manrakhan and Lux 2006). For *Ceratitis* species other than *C. capitata*, the factors affecting their mating behaviour, remating habits and sperm transfer have yet to be elucidated. This lack of research on *Ceratitis* species other than *C. capitata* may be due to limited interest in developing SIT for control of those pests in Africa. However for other regions in the world which are still free of important polyphagous *Ceratitis* pests such as *C. rosa*, *C. cosyra* and *C. fasciventris*, the development of SIT targeting those species is still relevant and needs to be pursued since SIT can be used either as a preventative control measure against introduction of these pests, or as an eradication tool.

3.2 Bactrocera *Species*

Bactrocera species that are pests in the African continent are dusk-mating species, a characteristic of the majority of dacine fruit flies (Fletcher 1987). There is evidence for male-produced sex pheromones in *B. dorsalis* (Kobayashi et al. 1978). The rectal glands of male fruit flies produce and store sex pheromones (Fletcher 1987) and are present in male *B. dorsalis* (Schultz and Boush 1971). Sexually mature virgin *B. dorsalis* females were highly attracted to mature males and to the excised rectal glands of the males, particularly around dusk (Kobayashi et al. 1978). In *B. dorsalis*, mating begins at dusk when the light intensity falls to 280 Lux and continues until day break (Arakaki et al. 1984). Pairs of *B. dorsalis* remain in copula for ~10 h and copulation was found to be more successful in males that engaged in wing vibration (Arakaki et al. 1984). Under laboratory conditions adults only begin mating on the 11th day after emergence (Arakaki et al. 1984). Under field conditions, leks of 2–12 male *B. dorsalis* were found on host plants with each male defending an individual leaf; when a female arrived on a leaf the male mounts her and copulation proceeds (Shelly and Kaneshiro 1991). Female visits were not influenced by the sizes of leks suggesting a less intense female choice in the mating system of *B. dorsalis* compared with that of *C. capitata* (Shelly 2001). A different facet of female choice was demonstrated by Poramarcom and Boakes (1991), who showed that female *B. dorsalis* preferred dominant males and those that mated twice. Mating success was also greater in *B. dorsalis* males with previous exposure to methyl eugenol and protein (Shelly 2000; Shelly and Dewire 1994; Shelly et al. 2005; Tan and Nishida 1998). This may be because male *B. dorsalis* (reported as *Bactrocera papaya* Drew and Hancock in the publication) are known to convert methyl eugenol to other derivatives, particularly phenyl propanoids, which then act as sex pheromones to attract females during courtship (Hee and Tan 1998; Tan and Nishida 1998); one phenylpropanoid coniferyl alcohol was highly attractive to *B. dorsalis* females (Hee and Tan 1998).

Bactrocera latifrons, another important pest species in Africa, has a similar mating system to *B. dorsalis*. Jackson and Long (1997) described the mating behaviour of this pest species in field cages on coffee trees; leks of 4–9 calling males were observed and mating occurred between 1 h before sunset and 15 min after sunset (Jackson and Long 1997). *Bactrocera latifrons*, like other *Bactrocera* species, sequester male–specific attractants in their rectal glands (Nishida et al. 2009).

In contrast to *B. dorsalis* and *B. latifrons*, information on the mating behaviour of *B. zonata* is largely lacking. *Bactrocera zonata* males have a strong attraction to methyl eugenol (Tan et al. 2010). In the rectal glands of *B. zonata*, methyl eugenol is synthesised to form a blend of phenylpropanoid volatiles that are different to that produced by *B. dorsalis* (Tan et al. 2010).

3.3 Zeugodacus *Species and* Dacus *Species*

Zeugodacus cucurbitae is a dusk-mating species (Suzuki and Koyama 1980), similar to the *Bactrocera* pest species in Africa. Laboratory strains have shorter pre-mating periods and initiate mating at higher light intensities than their wild counterparts (Suzuki and Koyama 1980). Lek formation has been suggested for *Z. cucurbitae* (Kuba et al. 1984; Kuba and Koyama 1985). Raspberry ketone (a derivative of Cuelure which is a known male lure of *Z. cucurbitae*) increased the production of female-attracting sex pheromone, male-aggregation pheromone and was also an allomone (Khoo and Tan 2000).

The mating behaviour of other cucurbit infesting fruit flies has been poorly studied. Rempoulakis et al. (2015) recently described the mating behaviour of *D. ciliatus* but the mating habits of the other *Dacus* species remain undescribed. For *Dacus* species that do not respond to commercially available protein baits, the use of male lures with toxins and pathogens as well as SIT could still be viable control tools. The development of these techniques will, however, require an in depth investigation of the mating behaviour of *Dacus* species.

4 Conclusion

Within the African fruit fly pest complex, feeding and mating behaviours have been elucidated for only a few species with a wider global distribution. For the remaining *Ceratitis*, *Bactrocera* and *Dacus* species with a limited worldwide distribution, there are still important knowledge gaps in their behavioural ecology. With horticulture expanding in Africa and increasing trade of horticultural produce from the region, it is important that these knowledge gaps are filled so that control methods can be optimised.

References

Arakaki N, Kuba H, Soemori H (1984) Mating behaviour of the Oriental fruit fly, *Dacus dorsalis* Hendel (Diptera: Tephritidae). Appl Entomol Zool 19(1):42–51
Atiama-Nurbel T, Deguine J-P, Quilici S (2012) Maize more attractive than Napier grass as non-host plants for *Bactrocera cucurbitae* and *Dacus demmerezi*. Arthropod Plant Interact 6:395–403
Barry JD, Vargas RI, Miller GT, Morse JG (2003) Feeding and foraging of wild and sterile Mediterranean fruit flies (Diptera: Tephritidae) in the presence of spinosad bait. J Econ Entomol 96(5):1405–1411
Barry JD, Miller NW, Pinero JC, Tuttle A, Mau FL, Vargas RI (2006) Effectiveness of protein baits on melon fly and Oriental fruit fly (Diptera: Tephritidae): attraction and feeding. J Econ Entomol 99(4):1161–1167
Bess HA, Haramoto FH (1961) Contributions to the biology and ecology of the Oriental fruit fly, *Dacus dorsalis* Hendel (Diptera: Tephritidae), in Hawaii, vol 44. Hawaii Agricultural Experiment Station, University of Hawaii, Technical Bulletin, Honolulu, pp 5–30
Burk T (1981) Signalling and sex in acalyptrate flies. Fla Entomol 64(1):30–43
Cangussu JA, Zucoloto FS (1995) Self-selection and perception threshold in adult females of *Ceratitis capitata* (Diptera, Tephritidae). J Insect Physiol 41(3):223–227
Carey JR, Liedo P, Muller HG, Wang JL, Vaupel JW (1998) Dual modes of aging in Mediterranean fruit fly females. Science 281:996–998
Christenson LC, Foote RH (1960) Biology of fruit flies. Annu Rev Entomol 5:171–192
Cohen H, Voet H (2002) Effect of physiological state of young *Ceratitis capitata* females, on resource foraging behaviour. Entomol Exp Appl 104:345–351
Cornelius ML, Nergel L, Duan JJ, Messing RH (2000) Responses of female Oriental fruit flies (Diptera:Tephritidae) to protein and fruit odors in field cage and open field tests. Environ Entomol 29(1):14–19
Coronado-Gonzalez PA, Vijaysegaran S, Robinson AS (2008) Functional morphology of the mouthparts of the adult Mediterranean fruit fly, *Ceratitis capitata*. J Insect Sci 8(73):1–11. doi:10.1673/031.008.7301
Cunningham RT (1989) Male Annihilation. In: Robinson AS, Hooper G (eds) Fruit flies: their biology, natural enemies and control, vol 3B. Elsevier, Amsterdam, pp 345–351
De Meyer M, Mohamed S, White IM (2014) Invasive fruit fly pests in Africa. http://www.africa-museum.be/fruitfly/AfroAsia.htm. Accessed 30 Nov 2012
Deguine J-P, Douraguia E, Atiama-Nurbel T, Chiroleu F, Quilici S (2012) Cage study of spinosad-based bait efficacy on *Bactrocera cucurbitae*, *Dacus ciliatus*, and *Dacus demmerezi* (Diptera: Tephritidae) in Reunion Island. J Econ Entomol 105(4):1358–1365
Dethier VG (1976) The hungry fly. Harvard University Press, Cambridge, MA/London

Drew RAI, Yuval B (2001) The evolution of fruit fly feeding behavior. In: Aluja M, Norrbom AL (eds) Fruit flies (Tephritidae): phylogeny and evolution of behaviour. CRC Press, Boca Raton, pp 731–749

Eberhard WG (2000) Sexual behavior and sexual selection in the medfly, Ceratitis capitata. In: Aluja M, Norrbom A (eds) Fruit flies (Tephritidae): phylogeny and evolution of behavior. CRC Press, Boca Raton, pp 459–489

Ekesi S, Muchugu E (2006) Tephritid fruit flies in Africa- fact sheets of some economically important species. In: Ekesi S, Billah MK (eds) A field guide to the management of economically important Tephritid fruit flies in Africa, 2nd edn. icipe Science Press, Nairobi, pp B-1–B-19

Ekesi S, Mohamed SA, Hanna R, Lux SA, Gnanvossou D, Bokonon-Ganta A (2007) Fruit fly suppression- purpose, tools and methodology. In: Ekesi S, Billah MK (eds) A field guide to the management of economically important Tephritid fruit flies in Africa. International Centre of Insect Physiology and Ecology, Nairobi, pp D1–D15

Ekesi S, Mohamed S, Tanga CM (2014) Comparison of food-based attractants for *Bactrocera invadens* (Diptera: Tephritidae) and evaluation of mazoferm-spinosad bait spray for field suppression in mango. J Econ Entomol 107(1):299–309

Enkerlin WR (2005) Impact of fruit fly control programmes using the sterile insect technique. In: Dyck VA, Hendrichs J, Robinson AS (eds) Sterile insect technique. Principles and practice in area-wide integrated pest management. Springer, Dordrecht, pp 651–676

Fletcher BS (1987) The biology of dacine fruit flies. Annu Rev Entomol 32:115–144

Foster SP, Harris MO (1997) Behavioural manipulation methods for insect pest management. Annu Rev Entomol 42:123–146

Gopaul S, Price NS, Soonoo R, Stonehouse J, Stravens R (2000) Local production of protein bait: Mauritius. Technologies of fruit fly monitoring and control in the Indian Ocean region. Indian Ocean Commission Regional Fruit Fly Programme, Reduit

Hassell MP, Southwood TRE (1978) Foraging strategies of insects. Annu Rev Ecol Syst 9:75–78

Hee AK-W, Tan KH (1998) Attraction of female and male *Bactrocera papayae* to conspecific males fed with methyl eugenol and attraction of females to male sex pheromone components. J Chem Ecol 24(4):753–764

Hendrichs J, Hendrichs MA (1990) Mediterranean fruit fly (Diptera: Tephritidae) in nature: location and diel pattern of feeding and other activities on fruiting and nonfruiting hosts and non-hosts. Ann Entomol Soc Am 83(3):632–641

Hendrichs J, Prokopy RJ (1994) Food foraging behaviour of frugivorous fruit flies. In: Calkins CO, Klassen W, Liedo P (eds) Fruit flies and the sterile insect technique. CRC Press, Boca Raton, pp 37–55

Hendrichs J, Katsoyannos BI, Papaj DR, Prokopy RJ (1991) Sexual differences in movement between natural feeding and mating sites and tradeoffs between food consumption, mating success and predator evasion in Mediterranean fruit flies (Diptera: Tephritidae). Oecologia 86:223–231

Hendrichs J, Robinson AS, Cayol JP, Enkerlin W (2002) Medfly areawide sterile insect technique programmes for prevention, suppression or eradication: the importance of mating behaviour studies. Fla Entomol 85(1):1–13

Jackson CG, Long JP (1997) Mating behaviour of *Bactrocera latifrons* (Diptera: Tephritidae) in field cages. Ann Entomol Soc Am 90(6):856–860

Kaspi R, Yuval B (2000) Post-teneral protein feeding improves sexual competitiveness but reduces longevity of mass-reared sterile male Mediterranean fruit flies (Diptera:Tephritidae). Ann Entomol Soc Am 93(4):949–955

Kaspi R, Taylor PW, Yuval B (2000) Diet and size influence sexual advertisement and copulatory success of males in Mediterranean fruit fly leks. Ecol Entomol 25:279–284

Katsoyannos BI, Heath RR, Papadopoulos NK, Epsky ND, Hendrichs J (1999) Field evaluation of Mediterranean fruit fly (Diptera:Tephritidae) female selective attractants for use in monitoring programs. J Econ Entomol 92(3):583–589

Khoo CCH, Tan KH (2000) Attraction of both sexes of melon fly, *Bactrocera cucurbitae* to conspecific males – a comparison after pharmacophagy of cue-lure and a new attractant – zingerone. Entomol Exper Appl 97:317–320

Kobayashi RM, Ohinata K, Chambers DL, Fujimoto MS (1978) Sex pheromones of the Oriental fruit fly and the melon fly: mating behaviour, bioassay method, and attraction of females by live males and by suspected pheromone glands of males. Environ Entomol 7(1):107–112

Kuba H, Koyama J (1985) Mating behaviour of wild melon flies, *Dacus cucurbitae* Coquillett (Diptera: Tephritidae) in a field cage: courtship behaviour. Appl Entomol Zool 20(4):365–372

Kuba H, Koyama J, Prokopy RJ (1984) Mating behaviour of wild melon flies, *Dacus cucurbitae* Coquillett (Diptera: Tephritidae) in a field cage: distribution and behaviour of flies. Appl Entomol Zool 19(3):367–373

Manrakhan A (2006) Fruit fly monitoring – purpose, tools and methodology. In: Ekesi S, Billah MK (eds) A field guide to the management of economically important Tephritid fruit flies in Africa. icipe Science Press, Nairobi, pp c1–c17

Manrakhan A (2009) Diel and lifetime patterns of activities of African fruit flies (Diptera:Tephritidae) (unpublished data)

Manrakhan A, Kotze C (2011) Attraction of *Ceratitis capitata*, *C. rosa* and *C. cosyra* (Diptera: Tephritidae) to proteinaceous baits. J Appl Entomol 135:98–105

Manrakhan A, Lux SA (2006) Contribution of natural food sources to reproductive behaviour, fecundity and longevity of *Ceratitis cosyra*, *Ceratitis fasciventris* and *Ceratitis capitata* (Diptera: Tephritidae). Bull Entomol Res 96:259–268

Manrakhan A, Lux SA (2008) Effect of food deprivation on attractiveness of food sources, containing natural and artificial sugar and protein, to three African fruit flies: *Ceratitis cosyra*, *Ceratitis fasciventris* and *Ceratitis capitata*. Entomol Exper Appl 127:133–143

Manrakhan A, Lux SA (2009) Diel and lifetime patterns of feeding and reproductive activities of three African fruit flies, *Ceratitis cosyra*, *Ceratitis fasciventris* and *Ceratitis capitata* (Diptera: Tephritidae) in semi-field cages of different spatial scales. Afr Entomol 17(1):8–22

McQuate GT, Jones GD, Sylva CD (2003) Assessment of corn pollen as food sources for two Tephritid fruit fly species. Environ Entomol 32(1):141–150

Mwatawala MW, De Meyer M, Makundi RH, Maerere AP (2006) Biodiversity of fruit flies (Diptera: Tephritidae) in orchards in different agro-ecological zones of the Morogoro region, Tanzania. Fruits 61(5):321–332

Myburgh AC (1962) Mating habits of the fruit flies *Ceratitis capitata* (Wied.) and *Pterandus rosa* (Ksh.). South Afr J Agric Sci 5(3):457–464

Nestel D, Nemny-Lavy E, Zilberg L, Weiss M, Akiva R, Gazit Y (2004) The fruit fly PUB: a phagostimulation unit bioassay system to quantitatively measure ingestion of baits by individual flies. J Entomol 128:576–582

Nishida T (1958) Extrafloral glandular secretions, a food source for certain insects. Proc Hawaiian Entomol Soc 16(3):379–386

Nishida T, Bess HA (1950) Applied ecology in melon fly control. J Econ Entomol 43:877–883

Nishida R, Enomoto H, Shelly TE, Ishida T (2009) Sequestration of 3-oxygenated α-ionone derivatives in the male rectal gland of the solanaceous fruit fly, *Bactrocera latifrons*. Entomol Exper Appl 131:85–92

Papadopoulos NK, Katsoyannos BI, Kouloussis NA, Economopoulos AP, Carrey JR (1998) Effect of adult age, food, time of day on sexual calling incidence of wild and mass-reared Ceratitis capitata males. Entomol Exper Appl 89:175–182

Perez-Staples D, Shelly TE, Yuval B (2012) Female mating failure and the failure of 'mating' in sterile insect programs. Entomol Exper Appl 146:66–78

Pinero JC, Mau RFL, Vargas RI (2011) A comparative assessment of the response of three fruit fly species (Diptera: Tephritidae) to a spinosad-based bait: effect of ammonium acetate, female age, and protein hunger. Bull Entomol Res 101(4):373–381

Poramarcom R, Boakes CRB (1991) Behavioural influences on male mating success in the Oriental fruit fly, *Dacus dorsalis* Hendel. Anim Behav 42:453–460

Prokopy RJ, Roitberg BD (1989) Fruit fly foraging behaviour. In: Robinson AS, Hooper G (eds) Fruit flies: their biology, natural enemies and control, vol 3A, World crop pests. Elsevier, Amsterdam, pp 293–304

Prokopy RJ, Papaj DR, Hendrichs J, Wong TTY (1992) Behavioural responses of Ceratitis capitata flies to bait spray droplets and natural food. Entomol Exper Appl 64:247–257

Prokopy RJ, Hsu CL, Vargas RI (1993) Effect of source and condition of animal excrement on attractiveness to adults of Ceratitis capitata (Diptera: Tephritidae). Environ Entomol 22(2):453–458

Prokopy RJ, Miller NW, Pinero JC, Barry JD, Tran LC, Oride L, Vargas RI (2003) Effectiveness of GF-120 fruit fly bait spray applied to border area plants for control of melon flies (Diptera: Tephritidae). J Econ Entomol 96(5):1485–1493

Quilici S, Franck A, Peppuy A, Dos Reis CE, Mouniama C, Blard F (2002) Comparative studies of courtship behaviour of *Ceratitis* spp. (Diptera: Tephritidae) in Reunion Island. Fla Entomol 85:138–142

Rempoulakis P, Nemny-Lavy E, Castro R, Nestel D (2015) Mating behaviour of *Dacus ciliatus* (Loew) [Diptera: Tephritidae]: comparisons between a laboratory and a wild population. J Appl Entomol 140:250–260. doi:10.1111/jen.12252

Rice MJ (1989) The sensory physiology of fruit flies: conspectus and prospectus. In: Robinson AS, Hooper G (eds) Fruit flies, their biology, natural enemies and control, vol 3A. Elsevier, Amsterdam, pp 249–272

Roessler Y (1989) Insecticidal bait and cover sprays. In: Robinson AS, Hooper G (eds) World crop pests, fruit flies: their biology, natural enemies and control, vol 3B. Elsevier, Amsterdam, pp 329–335

Schultz GA, Boush GM (1971) Suspected sex pheromone glands in three economically important species of *Dacus*. J Econ Entomol 64(2):347–349

Schutze MK, Mahmood K, Pavasovic A, Bo W, Newman J, Clarke AR, Krosch MN, Cameron SL (2014) One and the same: integrative taxonomic evidence that *Bactrocera invadens* (Diptera: Tephritidae) is the same species as the Oriental fruit fly *Bactrocera dorsalis*. Syst Entomol 40:472–486. doi:10.1111/syen.12114

Shelly TE (2000) Fecundity of female Oriental fruit flies (Diptera: Tephritidae): effects of methyl-eugenol fed and multiple mates. Ann Entomol Soc Am 93(3):559–564

Shelly TE (2001) Lek size and female visitation in two species of tephritid fruit flies. Anim Behav 62:33–40

Shelly TE, Dewire AM (1994) Chemically mediated mating success in male Oriental fruit flies (Diptera: Tephritidae). Ann Entomol Soc Am 87(3):375–382

Shelly TE, Kaneshiro KY (1991) Lek behaviour of the Oriental fruit fly. Dacus dorsalis, in Hawaii (Diptera: Tephritidae). J Insect Behav 4:235–241

Shelly TE, Kennelly SS, McInnis DO (2002) Effect of adult diet on signalling activity, mate attraction, and mating success in male Mediterranean fruit flies (Diptera: Tephritidae). Fla Entomol 85(1):150–155

Shelly TE, Edu J, Pahio E (2005) Influence of diet and methyl eugenol on the mating success of males of the Oriental fruit fly, *Bactrocera dorsalis* (Diptera: Tephritidae). Fla Entomol 88(3):307–313

Sivinski J, Burk T (1989) Reproductive and mating behaviour. In: Robinson AS, Hooper G (eds) Fruit flies, their biology, natural enemies and control, vol 3A. Elsevier, Amsterdam, pp 343–351

Suzuki Y, Koyama J (1980) Temporal aspects of mating behavior of the melon fly, *Dacus cucurbitae* Coquillett (Diptera: Tephritidae): a comparison between laboratory and wild strains. Appl Entomol Zool 15(3):215–224

Tan KH, Nishida R (1998) Ecological significance of male attractant in the defence and mating strategies of the fruit fly, *Bactrocera papayae*. Entomol Exper Appl 89:155–158

Tan KH, Tokushima I, Ono H, Nishida R (2010) Comparison of phenylpropanoid volatiles in male rectal gland after methyl eugenol consumption, and molecular phylogenetic relationship of four global pest fruit fly species: *Bactrocera invadens*, *B. dorsalis*, *B. correcta* and *B. zonata*. Chemoecology 21(1):25–33

Taylor PW, Yuval B (1999) Postcopulatory sexual selection in Mediterranean fruit flies: advantages for large and protein-fed males. Anim Behav 58:247–254

Tsiropoulos GJ (1992) Feeding and dietary requirements of the tephritid fruit flies. In: Anderson TE, Leppla NC (eds) Westview studies in insect biology: advances in insect rearing for research and pest management. Westview Press, Boulder, pp 93–118

Vargas RI, Stark JD, Nishida T (1990) Population dynamics, habitat preference, and seasonal distribution patterns of Oriental fruit fly and melon fly (Diptera: Tephritidae) in an agricultural area. Environ Entomol 19(6):1820–1828

Vijaysegaran S, Walter GH, Drew RAI (1997) Mouthpart structure, feeding mechanisms, and natural food sources of adult *Bactrocera* (Diptera: Tephritidae). Ann Entomol Soc Am 90(2):184–201

Virgilio M, Jordaens K, Verwimp C, White IM, De Meyer M (2015) Higher phylogeny of frugivorous flies (Diptera, Tephritidae, Dacini): localised partition conflicts and a novel generic classification. Molecular phylogenetics and evolution. doi:http://dx.doi.org/10.1016/j.ympev.2015.01.007

Warburg MS, Yuval B (1997) Circadian patterns of feeding and reproductive activities of Mediterranean fruit flies (Diptera: Tephritidae) on various hosts in Israel. Ann Entomol Soc Am 90(4):487–495

Williamson DL (1989) Oogenesis and spermatogenesis. In: Robinson AS, Hooper G (eds) Fruit flies: their biology, natural enemies and control, vol 3A, World crop pests. Elsevier, Amsterdam, pp 141–151

Aruna Manrakhan is a research entomologist and co-ordinator of the fruit fly research programme at Citrus Research International, South Africa. Soon after completing her MSc degree in Environmental Technology at Imperial College, University of London, in 1996, Aruna began working on fruit flies in Mauritius as part of the Indian Ocean Regional Fruit Fly Programme led by Professor John Mumford from Imperial College. In 2000, she joined *icipe*, Kenya, and completed her PhD on fruit fly feeding behaviour supervised by Dr Slawomir Lux, who contributed to the 'African Fruit Fly Initiative', now known as the 'African Fruit Fly Programme'. In 2006, Aruna began a post-doctoral fellowship with Dr Pia Addison at Stellenbosch University, South Africa where she focussed on the ecology and management of *Ceratitis capitata* and *Ceratitis rosa*. Aruna's current research interests are the behaviour, monitoring and control of fruit fly pests that affect the citrus industry in Southern Africa. Aruna has also been involved in the South African national surveillance and eradication programme targeting *Bactrocera dorsalis*. Her work has been published in several peer-reviewed journals.

Chapter 9
Chemical Ecology of African Tephritid Fruit Flies

Ayuka T. Fombong, Donald L. Kachigamba, and Baldwyn Torto

Abstract African Tephritid fruit flies are distributed in three main genera, *Bactrocera*, *Ceratitis* and *Dacus* constituting both indigenous and invasive species. They use a diverse and complex range of semiochemicals for host location and reproduction. This chapter reviews the identification of these semiochemicals and includes examples of lures developed from some of these chemicals for the management of economically important fruit fly species.

Keywords Tephritid fruit flies • *Bactrocera* • *Ceratitis* • *Dacus* • Semiochemicals • Male lures

1 Introduction

The fruit flies covered in this chapter are those that are of both economic importance in Africa and also have well established chemical communication modes. Before proceeding to discuss fruit fly chemical communication, it is worthwhile defining the chemical classes involved in this mode of communication, which are broadly referred to as semiochemicals (Torto 2004). Semiochemicals are defined as 'chemical signals that convey a message between organisms of the same or different species' and, for the purposes of this chapter, can be divided into three categories: kairomones, allomones and pheromones. Kairomones and allomones mediate 'interspecific' chemical communication, i.e. chemical communication between individuals from different species. Kairomones benefit the receiver and include signals for attraction to feeding or oviposition sites, while allomones benefit the emitter and include some repellents that deter competing or predatory species. Pheromones mediate 'intraspecific' chemical communication, i.e. between

---

A.T. Fombong • B. Torto (✉)
International Centre of Insect Physiology & Ecology,
PO Box 30772-00100, Nairobi, Kenya
e-mail: btorto@icipe.org

D.L. Kachigamba
Department of Agricultural Research Services, Bvumbwe Research Station,
PO Box 5748, Limbe, Malawi

individuals of the same species and include signals that mediate the location of conspecifics (aggregation pheromones), mates (sex pheromones) and oviposition sites (oviposition pheromones). Host-marking pheromones may mediate both intra- and interspecific interactions in some species. These terms for classes of chemical communication are not exhaustive and readers are advised to refer to key reference books on chemical ecology for definitions of additional modes of chemical communication (El-Sayed 2015; Matthews and Matthews 2010).

The biology and chemical ecology of fruit flies have been studied for several decades and have recently been comprehensively reviewed for the six major tephritid fruit fly genera (*Anastrepha, Bactrocera, Ceratitis, Dacus, Rhagoletis* and *Toxotrypana*) (Shelly et al. 2014). The present review considers only the role of semiochemicals mediating behaviour of African tephritid fruit flies, concentrating specifically on inter- and intraspecific chemical communication. Additionally, we provide a summary of the different assays used to study fruit fly behaviour in response to chemical signals, the antennally- and behaviourally-active compounds identified to date, and the candidate lures used in fruit fly control.

2 Interspecific Interactions

Most studies on chemical communication between tephritid fruit flies and other species have been on interactions with various plant species; chemical cues from plants can act as oviposition stimulants or oviposition deterrents to females, and as either attractants or repellents to one or both sexes (Jayanthi et al. 2012). Studies on interspecific chemical interaction between tephritid fruit flies and plants have largely been conducted with the goal of identifying potent plant-based attractants for managing both sexes of fruit fly pests. Thus these studies have mostly focused on genera and species of economically important fruit flies.

In Africa, fruit flies in the genera *Bactrocera, Ceratitis* and *Dacus* are of greatest economic importance since they attack a wide variety of important agricultural fruits including: mango, *Mangifera indica* L.; *Citrus* species; guava, *Psidium guajava* L.; avocado, *Persea americana* Mill.; almond, *Prunus dulcis* (Mill.) Webb; Cucurbitaceae Juss.; tomato, *Solanum lycopersicum* L.; and banana, *Musa* species (Ekesi and Billah 2008). Odours from different parts of these host plants elicit different behavioural responses in fruit flies, acting as attractants, repellents, oviposition stimulants and oviposition deterrents; these behaviours are either olfactory or contact-based, and are characterised by their mode and range of action (Shelly et al. 2014). However, while plant-based attractants and repellents have been studied for a long time, their successful use in integrated fruit fly management has been limited. For instance, despite early recognition that female-biased attractants existed for the oriental fruit fly, *Bactrocera dorsalis* (Hendel) (Jang et al. 1997), research efforts in the subsequent two decades have, thus far, only identified oviposition stimulants (Jayanthi et al. 2012). Furthermore, studies on host plant-fruit fly interactions have been strongly influenced by the economic importance and geographic range of the species in question, favouring research on invasive species over their indigenous counterparts. In this section we summarise research efforts on the identification of plant-based attractants and repellents in species from the genera *Bactrocera*, *Ceratitis* and *Dacus*. Host plants referred to in this chapter are defined as plants on which female fruit flies oviposit, whereas non-host plants are defined as those plants for which there are no records of female oviposition. To the best of our knowledge, and with the exception of the melon fly, *Zeugodacus (=Bactrocera) cucurbitae* (Coquillett), from which potent kairomone-based male and female attractants have been identified (Siderhurst and Jang 2010), similar plant-based attractants for other fruit fly species in these three genera remain to be identified.

2.1 Bactrocera dorsalis *Species Complex*

Amongst the three fruit fly genera known to occur in Africa, members of the genus *Bactrocera*, and in particular the *B. dorsalis* complex, are highly polyphagous and infest over 40 host plant species (Ekesi and Billah 2008; Georgen et al. 2011). Over two decades of work has documented attractants and repellents for both sexes of the *B. dorsalis* species complex. Some of the earliest work on plant-based attractants was done in the 1990s on papaya, *Carica papaya* L. In windtunnel assays, odours from ripe papaya attracted and stimulated oviposition in females of *B. dorsalis* more than odours from unripe papaya. This was attributed to higher levels of esters and monoterpenes being released from ripe fruit compared with unripe fruit (Flath et al. 1990; Jang and Light 1991). However, this preference for ovipositing in ripe papaya decreased with increasing age of flies. In addition to papaya, fruits such as mango, guava, orange (*Citrus sinensis* (L.) Osbeck.), banana, almond and marula (*Sclerocarya birrea* [Rich.] Hochst.), also attracted more male and female *B. dorsalis* when ripe, than when unripe (Cornelius et al. 2000a, b; Kimbokota et al. 2013; Jayanthi et al. 2012; Siderhurst and Jang 2006a; Biasazin et al. 2014) (Tables 9.1 and 9.2). Using coupled gas chromatography/electroantennographic detection (GC/EAD) and coupled gas chromatography/ mass spectrometry (GC/MS) analyses on the antennae of female fruit flies the components present in the attractive fruit odour were identified as a complex mixture of alkanes, esters, green leaf volatiles (GLVs), ketones and terpenes (Biasazin et al. 2014; Light and Jang 1987; Kimbokota 2011; Kimbokota et al. 2013; Jayanthi et al. 2012; Siderhurst and Jang 2006b) (Table 9.3) (Fig. 9.1). In addition to olfactory attractants, 1-octen-3-ol, ethyl tiglate and γ-octalactone from mango have been identified as oviposition stimulants (Jayanthi et al. 2014a) with a particularly strong innate response to γ-octalactone (Jayanthi et al. 2014b) (Fig. 9.1). However, female *B. dorsalis* are repelled by the essential oil from cinnamon, *Cinnamomum osmophloeum* Kaneh, as demonstrated in Petri dish assays (Diongue et al. 2010). The specific components in cinnamon oil that elicit this response are unknown. In a related study, mango cultivars with high levels of phenolics were less infested by *B. dorsalis* than cultivars with low levels of

Table 9.1 Examples of studies that identified host and non-host plants that elicit specific behaviours in African fruit fly species

Fruit fly Scientific name	Common name	Sex studied	Plant source[a]	Plant part	Behavioural role[b]	Extract collection method	Method of analysis[c]	Compound identification[d]	Reference(s)
Bactrocera oleae	Olive fruit fly	Female	Olive	Olive aqueous vegetation extract, olive fruit juice, phenolic extracts,	Oviposition deterrent	Solvent extraction with methanol	GC-MS, TLC, NMR, bioassay	+	Cirio (1971), Fiume and Vita (1977), Girolami et al. (1981), Vita et al. (1977) and Capasso et al. (1994)
				Olive aqueous vegetation extract, olive fruit juice	Oviposition stimulant	Solvent extraction with methanol	GC-MS, TLC, NMR, bioassay	+	Capasso et al. (1994)
				Leaves and fruit	Attractant Repellent Oviposition stimulant and deterrent	Dynamic headspace air sampling of water-based extract	GC-MS, TLC, NMR, bioassay	+	Scarpati et al. (1993) and Scalzo et al. (1994)
				Leaves and fruit (water-based extracts)	Attractant Oviposition stimulant	Solvent extraction with water, dynamic head space sampling, carbon trap 300	GC, GC-MS, colorimetry, bioassay	+	Scarpati et al. (1996)
					Oviposition stimulant	Headspace	GC-MS	+	Panizzi et al. (1960) and Girolami et al. (1974)

Bactrocera dorsalis	Oriental fruit fly	Female	Panax,	Leaf	Attractant	Solvent extraction with dichloromethane and water	–	Jang et al. (1997)
		Female	Papaya	Fruit	Attractant Arrestant Oviposition stimulant	None	Wind tunnel assays	–	Jang and Light (1991)
		Male and Female	Almond		Attractant	Solvent extraction with methanol	GC-MS, bioassays	+	Siderhurst and Jang (2006a)
		Male and Female	Almond	Fruit	Attractant	SPME, dynamic head space collection	GC-EAD, GC-MS, bioassay	+	Siderhurst and Jang (2006b)
		Female	Common and strawberry guava, starfruit, citrus	Fruit	Attractants		bioassays	–	Cornelius et al. (2000a)
		Female	Mango	Fruit	Attractants	SPME, dynamic head space collection	GC-EAD, GC-MS, bioassay	+	Jayanthi et al. (2012)
		Female	Citrus	Fruit, Nu-Lure	Attractants		bioassays	–	Cornelius et al. (2000b)
		Female	Mango (Kent variety), Banana, Citrus Guava	Fruit	Attractant	Dynamic headspace collection	GC-EAD, GC-MS, bioassay	+	Biasazin et al. (2014)
		Female	Mango (Chausa and Alphonso varieties)	Fruit	Oviposition stimulant	Dynamic headspace collection	GC-EAD, GC-MS, bioassay	+	Jayanthi et al. (2014a)
		Male and Female	Mango (Sensation, Apple and Kent varieties), marula, almond	Fruit	Attractant		bioassays	–	Kimbokota (2011)
		Male and Female	Mango (Banganapali, Alphonso, Totapuri, Langra and EC-95862 varieties)	Fruit (peel and pulp)	Repellent	Solvent extraction with methanol	spectrophotometer, bioassays	–	Verghese et al. (2012)
Z. curcubitae	Melon fruit fly	Male and Female	Cucumber	Fruit	Attractant	Dynamic headspace collection, SPME	GC-EAD, GC-MS, bioassay		Siderhurst and Jang (2010)

(continued)

Table 9.1 (continued)

Fruit fly									
Scientific name	Common name	Sex studied	Plant source[a]	Plant part	Behavioural role[b]	Extract collection method	Method of analysis[c]	Compound identification[d]	Reference(s)
B. latifrons	Solanum fruit fly	Male and Female	Various essential oils, formulated aromas and synthetic compounds	–	Attractants	Liquid chromatography	GC-MS, bioassays	+	Flath et al. (1994b)
Ceratitis capitata	Mediterranean fly	Male and Female	Garden Angelica	Seed	Attractants	Fractional distillation	GC, GC-MS, infrared spectroscopy, bioassays	+	Flath et al. (1994a)
		Male	Peruvian peppertree	Leaf	Attractant	solvent extraction with hexane	GC-MS	+	Gikonyo and Lux (2004)
		Male and Female	Citrus (navel variety)	Fruit	–	Dynamic headspace collection	GC-EAD, GC-MS, bioassays	+	Hernandez et al. (1996)
		Female	Coffee	Fruit	Attractant	Solvent extraction with water	bioassays	–	Prokopy et al. (1998)
		Female	Coffee	Fruit	Attractants	Dynamic headspace collection	GC-MS, bioassays	+	Warthen et al. (1997)
		Male and Female	Citrus	Fruit	Insecticidal	Solvent extraction of peel in diethyl ether	GC-MS, UV spectroscopy, bioassays	+	Salvatore et al. (2004)
		Female	Polypod fern (*Elaphoglossum piloselloides*)	–	Oviposition deterrent		Column chromatography, HPLC, infrared spectroscopy, MS, bioassay	+	Socolsky et al. (2008)
		Male	Lychee, fig	Leaves, stem and twigs	Attractant and phagostimulant	Extraction using isooctane, n-hexane, ethylether, ethylacetate dichloromethane		+	Warthen and McInnis (1989)

C. rosa		Male	Peruvian peppertree	Leaf	attractant	Solvent extraction with hexane	GC-MS	+	Gikonyo and Lux (2004)
C. cosyra		Male	Peruvian peppertree	Leaf	Attractant	Solvent extraction with hexane	GC-MS	+	Gikonyo and Lux (2004)
Dacus ciliatus	Ethiopian fruit fly	Male and Female	Galia melon, squash	Fruit	Attractant	Dynamic headspace collection, SPME	GC-EAD, GC-MS, bioassay	+	Alagarmalai et al. (2009)

[a]Common plant names are used except for the polypod fern and represent both host and non-host plant species
[b]Attractants and repellents refer to compounds/extracts whose volatile state elicited these behavioural responses
[c]*GC-MS* coupled gas chromatography mass spectrometry, *GC-EAD* coupled gas chromatography electroantennographic detection, *GC* gas chromatography, *NMR* Nuclear Magnetic Resonance, *TLC* Thin layer chromatography, *GLC* gas liquid chromatography, *HPLC* high pressure liquid chromatography, *UV* ultra violet
[d]Compound identification '+', means the compound has been identified and '−', means that the potential active components have not been identified

Table 9.2 Examples of arena assays used to elucidate specific behaviours in fruit flies

Odour source	Fruit fly species	Bioassay arena	Observed behaviour	References
Squash	*Dacus ciliatus*	Dual choice still air Plexiglass olfactometer (40 × 40 × 30 cm)	Attraction	Alagarmalai et al. (2009)
Olive aqueous vegetation extracts	*Bactrocera oleae*	Plexiglass-tulle cage	Oviposition deterrence	Capasso et al. (1994)
Olive aqueous vegetation extracts	*B. oleae*	Glass test tube (16 cm long × 2.5 cm ID)	Oviposition deterrence	Scalzo et al. (1994)
Olive aqueous vegetation extracts	*B. oleae*	Plexiglass cage (20 × 20 × 60 cm)	Oviposition deterrence	Scarpati et al. (1993)
Olive leaves and fruit, macerated fruit	*B. oleae*	Glass test tube (16 cm long × 2.5 cm ID) with oviposition substrate placed in Plexiglass cage (20 × 20 × 60 cm)	Oviposition cue, Attraction	Scarpati et al. (1993), (1996)
Cucumber, cantaloupe, tomato, kabocha, bittermelon and zuccini squash	*Zeugodacus cucurbitae*	Wood-framed cage covered with 16-mesh black nylon cloth (100 × 100 × 100 cm)	Attraction	Miller et al. (2004)
Cucumber, zucchini, papaya, tomato and ivy gourd	*Z. cucurbitae*	Hemisphere (8 cm diameter) or hemicylinder (4.3 cm in diameter × 15 cm height) placed outdoors		Piñero et al. (2006)
Cucumber	*Z. cucurbitae*	Outdoor multiple-trap rotating olfactometer in a wooden frame cage (3 × 3 × 2.5 m)		Siderhurst and Jang (2010)
Papaya	*Bactrocera dorsalis*	Wind tunnel (261 × 85.5 × 86.5 cm)	Attraction	Jang and Light (1991)
Panax	*B. dorsalis*	Wind tunnel (280 × 90 × 90 cm), Outdoor multiple-trap rotating olfactometer in a wooden frame cage (75 × 75 × 80 cm)	Attraction	Jang et al. (1997)
Common and strawberry guava, starfruit and oranges	*B. dorsalis*	Yellow spheres (7 cm) and McPhail traps placed outdoors	Attraction	Cornelius et al. (2000a, b)
		Multitrap laboratory rotating cage (90 × 90 × 90 cm), Wind tunnel (280 × 90 × 90 cm), Outdoor multiple-trap rotating olfactometer in a wooden frame cage (3 × 3 × 2.5 m), field cage (15 × 6 × 2.5 m)	Attraction	Siderhurst and Jang (2006a, b)

(continued)

Table 9.2 (continued)

Odour source	Fruit fly species	Bioassay arena	Observed behaviour	References
Mountain pepper, cinnamon, eucalyptus and hinoki	B. dorsalis	Conical bug cage (85 × 60 × 60 cm)	Attraction, Repellence	Diongue et al. (2010)
Mango	B. dorsalis	Pulp disc (prepared in 9 cm diameter Petridish)	Oviposition cue	Jayanthi et al. (2014a, b)
Mango	B. dorsalis	Perspex 4-arm olfactometer (12 cm in diameter)	Attraction	Jayanthi et al. (2012)
Mango (Kent variety), guava, banana and orange	B. dorsalis	Glass Y-tube (base arm 14 cm long, side arms 16 m long, internal diameter 3.1 cm) placed in a white box	Attraction	Biasazin et al. (2014)
Orange, guava, papaya, coffee, tomato, cucumber, pumpkin juices and water	B. dorsalis, Z. cucurbitae and Ceratitis capitata	1 Lt polyethylene bottle	Oviposition cue	Vargas and Chang (1991)
Mango (Sensation, Apple and Kent varieties), marula and almond	B. dorsalis	Dual choice olfactometer (100 × 30 × 30 cm)	Attraction	Kimbokota et al. (2013)
Turkey berry and devil's apple eggplants	Bactrocera latifrons	500 ml plastic container with a screen mesh lid which held test odours and insects	Oviposition cue	Peck and McQuate (2004)
Cucumber volatile extract	B. latifrons	Yellow sticky card	Attraction	McQuate et al. (2013)
Artificial orange scent, limonene and Ceratitis lure	Ceratitis rosa	Rearing cage	Attraction	Lebusa et al. (2014)
Garden angelica	C. capitata	Jackson delta trap	Attraction	Flath et al. (1994a)
Coffee	C. capitata	Rectangular cage (30 × 30 × 30 cm) with 3 screen sides	Attraction	Prokopy et al. (1998)

Table 9.3 List of some antennally-active fruit odour components in African fruit fly species

Chemical group	Compound	Plant source	Fruit fly species	References
Carboxylic acid	Acetic acid	Cucumber	Zeugodacus cucurbitae, Bactrocera dorsalis	Siderhurst and Jang (2010)
Carboxylic acid	Formic acid	Peach	Ceratitis capitata	Light et al. (1988)
Carboxylic acid	Acetic acid	Citrus, passion fruit, peach	C. capitata	Light et al. (1988)
Carboxylic acid	Propanoic acid	Citrus	C. capitata	Light et al. (1988)
Carboxylic acid	2-Propenoic acid		C. capitata	Light et al. (1988)
Carboxylic acid	Butanoic acid	Citrus, passion fruit, peach	C. capitata	Light et al. (1988)
Carboxylic acid	Pentanoic acid	Citrus, passion fruit, peach	C. capitata	Light et al. (1988)
Carboxylic acid	Hexanoic acid	Citrus, passion fruit, peach	C. capitata	Light et al. (1988)
Carboxylic acid	(E)-2-Hexenoic acid	Passion fruit, peach	C. capitata	Light et al. (1988)
Carboxylic acid	Heptanoic acid	Citrus, passion fruit	C. capitata	Light et al. (1988)
Carboxylic acid	Octanoic acid	Citrus, passion fruit, peach	C. capitata	Light et al. (1988)
Carboxylic acid	Nonanoic acid	Citrus, passion fruit	C. capitata	Light et al. (1988)
Carboxylic acid	Decanoic acid	Citrus, passion fruit, peach	C. capitata	Light et al. (1988)
Carboxylic acid	Dodecanoic acid	Citrus, passion fruit	C. capitata	Light et al. (1988)
Alcohol	1-Hexanol	Cucumber	Z. cucurbitae, B. dorsalis, C. capitata	Siderhurst and Jang (2010)
Alcohol	1-Octen-3-ol	Cucumber	Z. cucurbitae, B. dorsalis, C. capitata	Siderhurst and Jang (2010)
Alcohol	Benzyl alcohol	Cucumber	Z. cucurbitae	Siderhurst and Jang (2010)

(continued)

Table 9.3 (continued)

Chemical group	Compound	Plant source	Fruit fly species	References
Alcohol	(E)-2-Octen-1-ol	Cucumber	Z. cucurbitae, B. dorsalis	Siderhurst and Jang (2010)
Alcohol	2-Phenylethyl alcohol	Cucumber	Z. cucurbitae	Siderhurst and Jang (2010)
Alcohol	(E,Z)-2,6-Nonadien-1-ol	Cucumber	Z. cucurbitae	Siderhurst and Jang (2010)
Alcohol	(Z)-6-Nonen-1-ol	Cucumber	Z. cucurbitae	Siderhurst and Jang (2010)
Alcohol	1-Nonanol	Cucumber	Z. cucurbitae, B. dorsalis, C. capitata	Siderhurst and Jang (2010)
Alcohol	3-Methyl-1-butanol	Mango (Chausa variety), banana	B. dorsalis	Biasazin et al. (2014) and Jayanthi et al. (2012)
Alcohol	(RS)-1-Octen-3-ol	Mango (Chausa variety)	B. dorsalis	Jayanthi et al. (2012)
Alcohol	Phenylethyl alcohol	Mango (Chausa variety),	B. dorsalis	Jayanthi et al. (2012)
Alcohol	Ethanol	Almond	Z. cucurbitae, B. dorsalis, C. capitata	Siderhurst and Jang (2006a)
Alcohol	3,6-Nonadien-1-ol	Mango	B. dorsalis	Biasazin et al. (2014)
Alcohol	(Z)-3-Octen-1-ol	Marula	B. dorsalis	Kimbokota (2011)
Alcohol	Octan-1-ol	Marula	B. dorsalis	Kimbokota (2011)
Alcohol	(Z)-3-decen-1-ol	Marula	B. dorsalis	Kimbokota (2011)
Alcohol	5-dodecen-1-ol	Marula	B. dorsalis	Kimbokota (2011)
Alcohol	4- pentenol	Almond	B. dorsalis	Kimbokota (2011)
Alcohol	Ethanol	Citrus, papaya, passion fruit, peach	C. capitata	Light et al. (1988)
Alcohol	Propan-l-ol	Citrus, papaya, passion fruit	C. capitata	Light et al. (1988)
Alcohol	2-Propen-1-ol		C. capitata	Light et al. (1988)
Alcohol	Butan-l-ol	Citrus, papaya, passion fruit, peach	C. capitata	Light et al. (1988)
Alcohol	(E)-2-Buten-l-ol		C. capitata	Light et al. (1988)

(continued)

Table 9.3 (continued)

Chemical group	Compound	Plant source	Fruit fly species	References
Alcohol	3-Buten-1-ol		C. capitata	Light et al. (1988)
Alcohol	Pentan-1-ol	Citrus, papaya, passion fruit, peach	C. capitata	Light et al. (1988)
Alcohol	(E)-3-Penten-1-ol		C. capitata	Light et al. (1988)
Alcohol	Hexan-1-ol	Citrus, papaya, passion fruit, peach	C. capitata	Light et al. (1988)
Alcohol	(E)-2-Hexen-1-ol	Citrus, papaya, peach	C. capitata	Light et al. (1988)
Alcohol	(Z)-2-Hexen-1-ol		C. capitata	Light et al. (1988)
Alcohol	(E)-3-Hexen-1-ol	Passion fruit	C. capitata	Light et al. (1988)
Alcohol	(Z)-3-Hexen-1-ol	Citrus, papaya, passion fruit	C. capitata	Light et al. (1988)
Alcohol	Heptan-1-ol	Citrus, papaya, passion fruit	C. capitata	Light et al. (1988)
Alcohol	(±)-Heptan-2-ol	Passion fruit	C. capitata	Light et al. (1988)
Alcohol	Octan-1-ol	Citrus, papaya, passion fruit, peach	C. capitata	Light et al. (1988)
Alcohol	(E)-2-Octen-1-ol		C. capitata	Light et al. (1988)
Alcohol	(E)-3-Octen-1-ol	Passion fruit	C. capitata	Light et al. (1988)
Alcohol	(+)-1-Octen-3-ol	Citrus	C. capitata	Light et al. (1988)
Alcohol	Nonan-1-ol	Citrus	C. capitata	Light et al. (1988)
Alcohol	Decan-1-ol	Citrus	C. capitata	Light et al. (1988)
Alcohol	Undecan-1-ol	Citrus	C. capitata	Light et al. (1988)
Alcohol	Dodecan-1-ol	Citrus	C. capitata	Light et al. (1988)
Alcohol	(±)-Undecan-2-ol	Citrus	C. capitata	Light et al. (1988)
Alcohol	(±)-Nonan-2-ol	Citrus, passion fruit	C. capitata	Light et al. (1988)
Alcohol	Carveol	Citrus (Navel variety)	C. capitata	Hernandez et al. (1996)
Aldehyde	Hexanal	Cucumber	Z. cucurbitae, B. dorsalis, Ceratitis capitata	Siderhurst and Jang (2010)

(continued)

Table 9.3 (continued)

Chemical group	Compound	Plant source	Fruit fly species	References
Aldehyde	(E)-2-Hexenal	Cucumber, almond	Z. cucurbitae, B. dorsalis, C. capitata	Siderhurst and Jang (2006a) and Siderhurst and Jang (2010)
Aldehyde	(E,E)-2,4-Heptadienal	Cucumber	Z. cucurbitae	Siderhurst and Jang (2010)
Aldehyde	(E)-2-Octenal	Cucumber	Z. cucurbitae	Siderhurst and Jang (2010)
Aldehyde	(E)-4-Nonenal	Cucumber	Z. cucurbitae	Siderhurst and Jang (2010)
Aldehyde	(Z)-6-Nonenal	Cucumber	Z. cucurbitae	Siderhurst and Jang (2010)
Aldehyde	(E,Z)-2,6-Nonadienal	Cucumber	Z. cucurbitae	Siderhurst and Jang (2010)
Aldehyde	(E)-2-Nonenal	Cucumber	Z. cucurbitae	Siderhurst and Jang (2010)
Aldehyde	Tetradecanal	Cucumber, marula	Z. cucurbitae, B. dorsalis	Kimbokota (2011) and Siderhurst and Jang (2010)
Aldehyde	Nonanal	Cucumber	Z. cucurbitae, B. dorsalis, C. capitata	Siderhurst and Jang (2010)
Aldehyde	Decanal	Cucumber	Z. cucurbitae, B. dorsalis, C. capitata	Siderhurst and Jang (2010)
Aldehyde	(E,E)-2,4-Nonadienal	Cucumber	Z. cucurbitae	Siderhurst and Jang (2010)
Aldehyde	(E)-2-Decenal	Cucumber	Z. cucurbitae	Siderhurst and Jang (2010)
Aldehyde	Propanal		C. capitata	Light et al. (1988)
Aldehyde	Butanal	Citrus, peach	C. capitata	Light et al. (1988)
Aldehyde	(E)-2-Butenal		C. capitata	Light et al. (1988)
Aldehyde	Pentanal	Citrus	C. capitata	Light et al. (1988)
Aldehyde	Hexanal	Citrus, peach	C. capitata	Light et al. (1988)
Aldehyde	(E)-2-Hexenal	Citrus, peach	C. capitata	Light et al. (1988)
Aldehyde	Heptanal	Citrus, peach	C. capitata	Light et al. (1988)
Aldehyde	Octanal	Citrus	C. capitata	Light et al. (1988)
Aldehyde	Nonanal	Citrus, peach	C. capitata	Light et al. (1988)
Aldehyde	Decanal	Citrus	C. capitata	Light et al. (1988)
Aldehyde	Undecanal	Citrus	C. capitata	Light et al. (1988)
Aldehyde	Dodecanal	Citrus	C. capitata	Light et al. (1988)
Alkane	Heptane	Mango (Alphonso variety)	B. dorsalis	Jayanthi et al. (2012)

(continued)

Table 9.3 (continued)

Chemical group	Compound	Plant source	Fruit fly species	References
Alkane	Octane	Banana, guava, citrus (Tommy variety)	B. dorsalis	Biasazin et al. (2014)
Alkane	Nonane	Guava	B. dorsalis	Biasazin et al. (2014)
Alkane	3-Ethyl-2-methylpentane	Guava	B. dorsalis	Biasazin et al. (2014)
Alkene	4-Undecene	Marula	B. dorsalis	Kimbokota (2011)
Aromatic	1,2-Diethylbenzene	Citrus (Navel variety)	C. capitata	Hernandez et al. (1996)
Aromatic	1,3-Diethylbenzene	Citrus (Navel variety)	C. capitata	Hernandez et al. (1996)
Ester	Ethyl butanoate	Mango (Chausa variety)	B. dorsalis	Jayanthi et al. (2012)
Ester	Ethyl methacrylate	Mango (Chausa variety)	B. dorsalis	Jayanthi et al. (2012) and Biasazin et al. (2014)
Ester	Ethyl crotonate	Mango (Chausa variety), guava	B. dorsalis	Kimbokota (2011), Jayanthi et al. (2012) and Biasazin et al. (2014)
Ester	Ethyl tiglate	Mango (Chausa variety)	B. dorsalis	Jayanthi et al. (2012)
Ester	Ethyl hexanoate	Mango (Chausa variety), almond, citrus (Tommy variety)	B. dorsalis, C. capitata	Jayanthi et al. (2012) and Siderhurst and Jang (2006a)
Ester	Ethyl sorbate	Mango (Chausa variety)	B. dorsalis	Jayanthi et al. (2012) and Biasazin et al. (2014)
Ester	Ethyl octanoate	Mango (Chausa variety)	B. dorsalis	Jayanthi et al. (2012)
Ester	Ethyl acetate	Almond	B. dorsalis, C. capitata	Siderhurst and Jang (2006a)

(continued)

Table 9.3 (continued)

Chemical group	Compound	Plant source	Fruit fly species	References
Ester	Isopentyl acetate	Almond, banana, guava, citrus (Tommy variety)	B. dorsalis, C. capitata	Biasazin et al. (2014), Kimbokota (2011) and Siderhurst and Jang (2006a)
Ester	4-Pentenyl acetate	Almond	B. dorsalis	Siderhurst and Jang (2006a)
Ester	Isopentenyl acetate	Almond	B. dorsalis, C. capitata	Siderhurst and Jang (2006a)
Ester	Hexyl acetate	Almond, citrus (Tommy variety)	B. dorsalis	Biasazin et al. (2014) and Siderhurst and Jang (2006a)
Ester	Linalyl acetate	Almond	B. dorsalis, C. capitata	Siderhurst and Jang (2006a)
Ester	2-Phenylethyl acetate	Almond	B. dorsalis, C. capitata	Siderhurst and Jang (2006a)
Ester	Ethyl nonanoate	Almond	B. dorsalis	Siderhurst and Jang (2006a)
Ester	Nonyl acetate	Almond	B. dorsalis, C. capitata	Siderhurst and Jang (2006a)
Ester	Citronellyl acetate	Almond	B. dorsalis	Kimbokota (2011) and Siderhurst and Jang (2006a)
Ester	Geranyl acetate	Almond	B. dorsalis, C. capitata	Kimbokota (2011) and Siderhurst and Jang (2006a)
Ester	(E)-Ethyl cinnamate	Almond	B. dorsalis	Siderhurst and Jang (2006a)
Ester	Isobutyl acetate	Guava, citrus (Tommy variety)	B. dorsalis	Biasazin et al. (2014)
Ester	Ethyl butyrate	Mango, banana, guava	B. dorsalis	Biasazin et al. (2014) and Kimbokota (2011)
Ester	Butyl acetate	Banana	B. dorsalis	Biasazin et al. (2014)
Ester	Ethyl isovalerate	Mango	B. dorsalis	Biasazin et al. (2014)
Ester	2-Pentyl acetate	Mango, banana, guava	B. dorsalis	Biasazin et al. (2014)
Ester	Propyl butyrate	Mango, banana	B. dorsalis	Biasazin et al. (2014)

(continued)

Table 9.3 (continued)

Chemical group	Compound	Plant source	Fruit fly species	References
Ester	Isobutyl isobutyrate	Banana, guava, citrus (Tommy variety)	B. dorsalis	Biasazin et al. (2014)
Ester	Methyl hexanoate	Guava	B. dorsalis	Biasazin et al. (2014)
Ester	Ethyl 3-methylcrotonate	Mango	B. dorsalis	Biasazin et al. (2014)
Ester	Ethyl tiglate	Mango	B. dorsalis	Biasazin et al. (2014)
Ester	Ethyl-(E)-2-pentanoate	Mango	B. dorsalis	Biasazin et al. (2014)
Ester	Isobutyl butyrate	Banana, guava, citrus (Tommy variety)	B. dorsalis	Biasazin et al. (2014)
Ester	Butyl butyrate	Banana, guava	B. dorsalis	Biasazin et al. (2014)
Ester	Isobutyl isovalerate	Banana, citrus (Tommy variety)	B. dorsalis	Biasazin et al. (2014)
Ester	2-Pentyl butyrate	Banana	B. dorsalis	Biasazin et al. (2014)
Ester	Ethyl-(E)-2-hexenoate	Mango	B. dorsalis	Biasazin et al. (2014)
Ester	Butyl valerate	Guava	B. dorsalis	Biasazin et al. (2014)
Ester	Isoamyl butyrate	Mango, banana, guava, citrus (Tommy variety)	B. dorsalis	Biasazin et al. (2014)
Ester	Isoamyl isovalerate	Banana, guava, citrus (Tommy variety)	B. dorsalis	Biasazin et al. (2014)
Ester	3-Methyl butyl-2-methylbutanoate	Banana	B. dorsalis	Biasazin et al. (2014)
Ester	Heptan-2-yl acetate	Banana	B. dorsalis	Biasazin et al. (2014)
Ester	(Z)-ethyl 4-octenoate	Mango	B. dorsalis	Kimbokota (2011)
Ester	Ethyl octanoate	Mango	B. dorsali, C. capitata	Kimbokota (2011) and Cossé et al. (1995)

(continued)

Table 9.3 (continued)

Chemical group	Compound	Plant source	Fruit fly species	References
Ester	Isopropyl acetate	Marula	*B. dorsalis*	Kimbokota (2011)
Ester	Ethyl propionate	Marula	*B. dorsalis*	Kimbokota (2011)
Ester	Ethyl isobutyrate	Marula	*B. dorsalis*	Kimbokota (2011)
Ester	Ethyl butyrate	Marula	*B. dorsalis*	Kimbokota (2011)
Ester	Ethyl 2-methylbutyrate	Marula	*B. dorsalis*	Kimbokota (2011)
Ester	Ethyl isovalerate	Marula, almond	*B. dorsalis*	Kimbokota (2011)
Ester	Isopropyl valerate	Marula	*B. dorsalis*	Kimbokota (2011)
Ester	Propyl isovalerate	Marula	*B. dorsalis*	Kimbokota (2011)
Ester	Isobutyl acetate	Almond	*B. dorsalis*	Kimbokota (2011)
Ester	2-Butenyl acetate	Almond	*B. dorsalis*	Kimbokota (2011)
Ester	4-Penten-1-yl acetate	Almond	*B. dorsalis*	Kimbokota (2011)
Ester	Prenyl acetate	Almond	*B. dorsalis*	Kimbokota (2011)
Ester	Benzyl acetate	Almond	*B. dorsalis*	Kimbokota (2011)
Ester	Citronellyl acetate	Almond	*B. dorsalis*	Kimbokota (2011)
Ester	Ethyl acetate	Almond, citrus, papaya, passion fruit, peach	*B. dorsalis, C. capitata*	Kimbokota (2011) and Light et al. (1988)
Ester	Propyl acetate	Papaya	*C. capitata*	Light et al. (1988)
Ester	Butyl acetate	Citrus, papaya, passion fruit, peach	*C. capitata*	Light et al. (1988)
Ester	Pentyl acetate	Citrus, papaya, peach	*C. capitata*	Light et al. (1988)
Ester	Hexyl acetate	Papaya, passion fruit, peach	*C. capitata*	Light et al. (1988)
Ester	(*E*)-2-Hexenyl acetate	Peach	*C. capitata*	Light et al. (1988)
Ester	Heptyl acetate		*C. capitata*	Light et al. (1988)
Ester	Octyl acetate	Citrus, passion	*C. capitata*	Light et al. (1988)
Ester	Nonyl acetate	Citrus	*C. capitata*	Light et al. (1988)
Ester	Decyl acetate	Citrus	*C. capitata*	Light et al. (1988)
Ester	(*Z*)-3-hexenyl acetate	Squash, guava	*Dacus ciliatus, B. dorsalis*	Alagarmalai et al. (2009) and Biasazin et al. (2014)
Ester	Hexanyl acetate	Squash	*D. ciliatus*	Alagarmalai et al. (2009)

(continued)

Table 9.3 (continued)

Chemical group	Compound	Plant source	Fruit fly species	References
Ester	Benzyl acetate	Squash, almond	D. ciliatus, B. dorsalis	Alagarmalai et al. (2009) and Kimbokota (2011)
Ester	(Z)-3-octenyl acetate	Squash	D. ciliatus	Alagarmalai et al. (2009)
Ester	Octanyl acetate	Squash	D. ciliatus	Alagarmalai et al. (2009)
Ester	Isopentyl hexanoate	Squash	D. ciliatus	Alagarmalai et al. (2009)
Ester	(Z)-3-decenyl acetate	Squash	D. ciliatus	Alagarmalai et al. (2009)
Ester	(E)-3-decenyl acetate	Squash	D. ciliatus	Alagarmalai et al. (2009)
Ester	Butyl hexanoate	Citrus (Navel variety)	C. capitata	Hernandez et al. (1996)
Ester	Hexyl hexanoate	Citrus (Navel variety))	C. capitata	Hernandez et al. (1996)
Esters	(Z)-3-Dodecenyl acetate	Squash	D. ciliatus	Alagarmalai et al. (2009)
Ketone	3,5-Octadien-2-one	Cucumber	Z. cucurbitae	Siderhurst and Jang (2010)
Ketone	3-Hydroxy-2-butanone	Mango (Chausa variety)	B. dorsalis	Kimbokota (2011) and Jayanthi et al. (2012)
Ketone	2-Hexanone	Banana	B. dorsalis	Biasazin et al. (2014)
Ketone	Dihyrocarvone	Citrus (Navel variety)	C. capitata	Hernandez et al. (1996)
Ketone	Carvone	Citrus (Navel variety)	C. capitata	Hernandez et al. (1996)
Ketone	4-Ethyl acetophenone	Citrus (Navel variety)	C. capitata	Hernandez et al. (1996)
Ketone	Nootkatone	Citrus (Navel variety)	C. capitata	Hernandez et al. (1996)
Ketone	Geranyl acetone	Cucumber	Z. cucurbitae	Siderhurst and Jang (2010)
Lactone	(RS)-γ-Octalactone	Mango (Alphonso variety)	B. dorsalis	Jayanthi et al. (2012)
Lactone	γ-Butyrolactone	Passion fruit	C. capitata	Light et al. (1988)
Lactone	γ-Pentalactone	Peach	C. capitata	Light et al. (1988)
Lactone	γ-Hexalactone	Papaya, passion fruit, peach	C. capitata	Light et al. (1988)

(continued)

Table 9.3 (continued)

Chemical group	Compound	Plant source	Fruit fly species	References
Lactone	γ-Heptalactone	Papaya, passion fruit, peach	C. capitata	Light et al. (1988)
Lactone	γ-Octalactone	Mango, papaya, passion fruit, peach	B. dorsalis, C. capitata	Light et al. (1988) and Jayanthi et al. (2014a)
Lactone	γ-Nonatactone	Papaya, passion fruit, peach	C. capitata	Light et al. (1988)
Lactone	γ-Decalactone	Papaya, passion fruit, peach	C. capitata	Light et al. (1988)
Lactone	γ-Undecalactone	Passion fruit, peach	C. capitata	Light et al. (1988)
Terpene	Myrcene	Mango (Alphonso variety), almond	B. dorsalis	Kimbokota (2011) and Jayanthi et al. (2012)
Terpene	(Z)-β-Ocimene	Mango (Alphonso variety)	B. dorsalis	Jayanthi et al. (2012)
Terpene	(E)-β-Ocimene	Mango (Alphonso variety)	B. dorsalis	Jayanthi et al. (2012)
Terpene	Allo-ocimene	Mango (Alphonso variety)	B. dorsalis	Jayanthi et al. (2012)
Terpene	Myroxide	Mango (Alphonso variety)	B. dorsalis	Jayanthi et al. (2012)
Terpene	(S)-3-Carene	Mango (Chausa variety)	B. dorsalis	Jayanthi et al. (2012) and Biasazin et al. (2014)
Terpene	p-Cymene	Mango (Chausa variety)	B. dorsalis	Jayanthi et al. (2012) and Kimbokota (2011)
Terpene	α-Terpinolene	Mango (Chausa variety)	B. dorsalis	Jayanthi et al. (2012)
Terpene	Menthone	Almond	B. dorsalis	Siderhurst and Jang (2006a)
Terpene	(E)-β-Farnesene	Almond	B. dorsalis, C. capitata	Siderhurst and Jang (2006a)

(continued)

Table 9.3 (continued)

Chemical group	Compound	Plant source	Fruit fly species	References
Terpene	(E,E)-α-Farnesene	Almond, marula	B. dorsalis, C. capitata	Kimbokota (2011) and Siderhurst and Jang (2006a)
Terpene	(Z,E)-α-Farnesene	Almond	B. dorsalis	Siderhurst and Jang (2006a)
Terpene	α-Pinene	Mango	B. dorsali, C. capitata	Biasazin et al. (2014), Kimbokota (2011) and Cosse et al. (1995)
Terpene	β-Pinene	Mango, citrus (Tommy variety)	B. dorsalis	Biasazin et al. (2014) and Kimbokota (2011)
Terpene	β-Myrcene	Mango, citrus (Tommy variety), almond	B. dorsalis	Biasazin et al. (2014) and Kimbokota (2011)
Terpene	(R)-(+)-Limonene	Citrus (Tommy variety), almond	B. dorsalis	Biasazin et al. (2014) and Kimbokota (2011)
Terpene	(Z)-β-Ocimene	Mango, marula	B. dorsalis	Biasazin et al. (2014) and Kimbokota (2011)
Terpene	Tricyclene	Mango	B. dorsalis	Kimbokota (2011)
Terpene	Camphene	Mango	B. dorsalis	Kimbokota (2011)
Terpene	Sabinene	Mango	B. dorsalis	Kimbokota (2011)
Terpene	Terpinolene	Mango, almond	B. dorsalis	Kimbokota (2011)
Terpene	α-Humulene	Mango	B. dorsalis	Kimbokota (2011)
Terpene	δ-3- Carene	Almond	B. dorsalis	Kimbokota (2011)
Terpene	Elemicin	Almond	B. dorsalis	Kimbokota (2011)
Terpene	β-Caryophyllene	Squash, mango	C. capitata, D. ciliatus	Alagarmalai et al. (2009) and Cosse et al. (1995)
Terpene	(E)-β-Farnesene	Squash, almond	D. ciliatus, B. dorsalis, C. capitata	Alagarmalai et al. (2009) and Siderhurst and Jang (2006a)
Terpene	Germacrene D	Squash	D. ciliatus	Alagarmalai et al. (2009)
Terpene	Cadinene	Squash	D. ciliatus	Alagarmalai et al. (2009)
Terpene	Limonene	Citrus (Navel variety)	C. capitata	Hernandez et al. (1996)

(continued)

Table 9.3 (continued)

Chemical group	Compound	Plant source	Fruit fly species	References
Terpene	Limonene oxide	Citrus (Navel variety)	C. capitata	Hernandez et al. (1996)
Terpene alcohol	Linalool	Cucumber, marula	Z. cucurbitae, B. dorsalis, C. capitata	Kimbokota (2011) and Siderhurst and Jang (2010)
Terpene alcohol	Eugenol	Almond	B. dorsalis	Siderhurst and Jang (2006a)
Terpene alcohol	Methyl eugenol	Almond	B. dorsalis	Kimbokota (2011) and Siderhurst and Jang (2006a)
Terpene alcohol	(S)-(+)-Linalool	Citrus (Tommy variety)	B. dorsalis	Biasazin et al. (2014)
Terpene	Caryophyllene oxide	Citrus (Navel variety)	C. capitata	Hernandez et al. (1996)
Terpene	Allo-aromadendrene	Citrus (Navel variety)	C. capitata	Hernandez et al. (1996)
Thiazole	Benzothiazole	Mango (Chausa variety)	B. dorsalis	Jayanthi et al. (2012)

phenolics, a trait that is considered useful in the breeding of resistant mango cultivars (Verghese et al. 2012).

Although visual cues can be attractive to *B. dorsalis* in the absence of odours (Cornelius et al. 1999; Vargas et al. 1991), they also complement olfactory attractants (Jang and Light 1991; Aloykhin et al. 2000). Several authors report that both sexes of *B. dorsalis* prefer yellow-coloured objects over green-coloured objects, the yellow colour being associated with ripening fruit (and the odours they produce) compared with unripe fruit (usually green in these studies) (Seo et al. 1982; Liquido et al. 1989; Liquido and Cunnigham 1990). Several studies using yellow McPhail traps have also demonstrated the importance of the yellow colour in fruit fly attraction (Ekesi and Billah 2008; Mazomenos et al. 2002; Dimou et al. 2003).

2.2 Bactrocera oleae

Unlike members of the *B. dorsalis* complex, *Bactrocera oleae* (Rossi), or the olive fruit fly, is monophagous and a specialist on wild and cultivated olives (Ekesi and Billah 2008). Research on identification of plant-based attractants and repellents for *B. oleae* dates back to the 1960s. One of the first findings on the chemo-ecological interactions between female *B. oleae* and its host plant was the identification of oleuropein as an oviposition stimulant (Panizzi et al. 1960; Girolami et al. 1974).

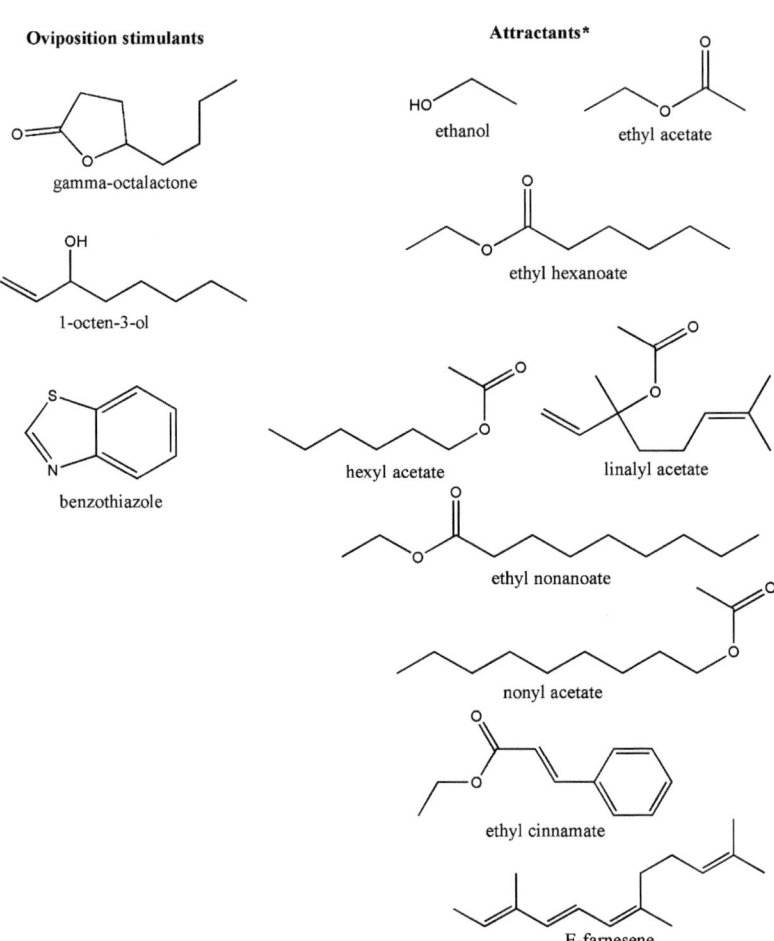

Fig. 9.1 Examples of volatile attractants and oviposition stimulants for the *Bactrocera dorsalis* complex

Catechol, 4-methyl catechol, pyrocatechol and hydroxytyrosol were then identified as oviposition deterrents from methanol extracts of vegetation from olive, *Olea europaea* L. (Capasso et al. 1994; Cirio 1971; Fiume and Vita 1977; Girolami et al. 1981; Vita et al. 1977). Using the same extraction technique, Capasso et al. (1994) identified *o*-quinone as an oviposition stimulant. Scarpati et al. (1993) also identified a number of non-benzenoid and benzenoid compounds as oviposition stimulants. Of these, α-pinene was a more effective oviposition stimulant than *p*-xylene, *o*-xylene, myrcenone and ethylbenzene or n-octane while (*E*)-2-hexenal and hexanal deterred oviposition. In the same study, toluene and ethylbenzene were identified as attractants while (*E*)-2-hexenal and hexanal were identified as repellents to female

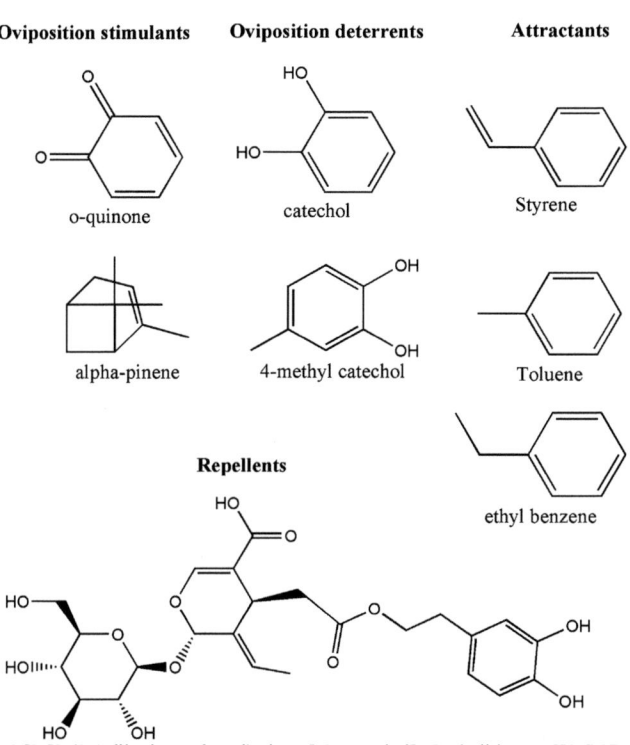

Fig. 9.2 Chemical structures of some attractants, repellents, oviposition stimulants and deterrents in *Bactrocera oleae*

flies (Scarpati et al. 1993) (Fig. 9.2). Styrene from solvent-based extracts of olive fruits and leaves has also been identified as an oviposition stimulant. Ammonia, a by-product of bacterial activity, was also found to be a potent attractant to female *B. oleae* (Scarpati et al. 1996). All these studies were done under controlled laboratory conditions. Their field evaluation will shed more light on their exact role in the chemical communication of this fruit fly species.

In addition to plant odours, there has been increasing interest in elucidating the role of microbial endosymbionts in the chemical ecology of *B. oleae*. Several microbial species have been isolated from *B. oleae* including *Pseudomonas putida* Trevisan and *Candidatus* Erwinia dacicola (Lauzon et al. 2000; Sacchetti et al.

2008). Recently, Liscia et al. (2013) demonstrated that female *B. oleae* were attracted to the volatile thioester, methyl thioacetate, found in odours produced by cultures of *P. putida*; they identified this thioester as a new volatile component within the fruit fly-associated bacterial odour that female *B. oleae* were able to perceive using receptors on their antennae and maxillary palps. The endosymbiont *Ca.* Erwinia dacicolais was more prevalent in wild populations of *B. oleae* than in laboratory-reared populations and it is suspected to affect the reproductive fitness of flies (Estes et al. 2014). These findings suggest that there is potential to develop new attractants based on odours from bacterial endosymbionts that could improve current semiochemical-based management tools for *B. oleae*.

2.3 Zeugodacus cucurbitae

Zeugodacus cucurbitae is considered polyphagous although its preferred host plants are cultivated and wild members of the family Cucurbitaceae. This preference for cucurbits, which has been demonstrated in behavioural assays, explains why *Z. cucurbitae* attacks cucumber, *Cucumis sativis* L., courgette, *Curcurbita pepo* L., bitter gourd, *Momordica charantia* L., kabocha pumpkin, *Curcurbita maxima* Duchesne, cantaloupe, *Cucumis melo* var. *cantalupensis* Naudin, ivy gourd, *Coccinia grandis* (L.) Voigt, although it also attacks tomato (Miller et al. 2004; Piñero et al. 2006). The earliest identification of a kairomone used by *Z. cucurbitae* showed that (*E*)-6-nonenyl acetate was attractive to females (Jacobson et al. 1971) and also stimulated oviposition (Keiser et al. 1973). Three decades later, Siderhurst and Jang (2010) identified several compounds in the headspace odours from ripe fruit of several cucurbit species including the aldehydes (*E,Z*)-2,6-nonadienal, (*E*)-2-nonenal, hexanal and (*E*)-2-hexenal and the alcohols (*Z*)-6-nonen-1-ol, (*E,Z*)-2,6-nonadien-1-ol and 1-hexanol; these compounds were detected by receptors on the antennae of female flies eliciting behavioural responses. Outdoor olfactometer studies demonstrated synergy between some of these headspace compounds when formulated in blends. Both a six-component blend formulated from (*E,Z*)-2,6-nonadienal, (*E*)-2-nonenal, (*Z*)-6-nonenal, nonanal, (*Z*)-6-nonen-1-ol and 1-nonanol, and a nine-component blend comprising of (*E,Z*)-2,6-nonadienal, (*E*)-2-nonenal, (*E*)-2-octenal, (*Z*)-6-nonenal,(*Z*)-6-nonen-1-ol, hexanal, 1-hexanol, acetic acid and 1-octen-3-ol (Fig. 9.3) captured substantial numbers of both male and female flies (Table 9.3) (Siderhurst and Jang 2010). These findings represent the first and most successful attempt to develop a female-biased fruit fly lure that also attracts males (Siderhurst and Jang 2010).

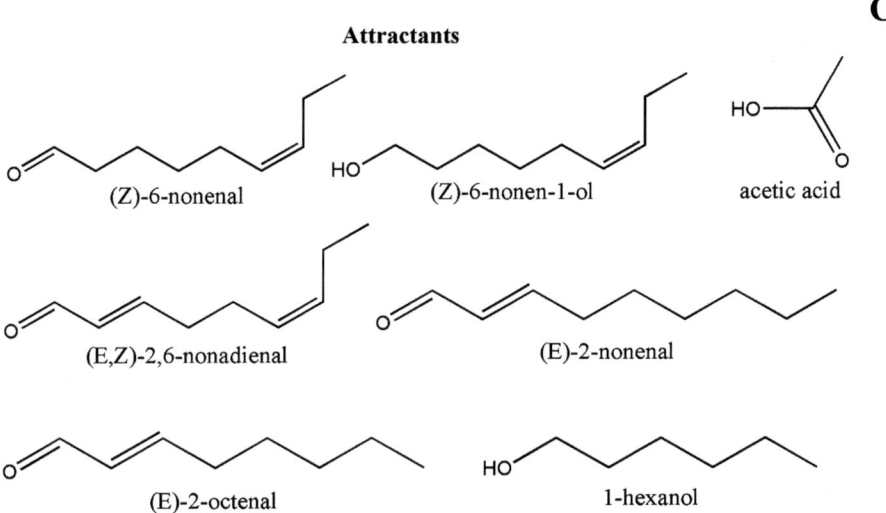

Fig. 9.3 Some examples of components from a cucurbit-based blend attractive to *Zeugodacus cucurbitae*

2.4 Bactrocera latifrons

In Africa, the solanum fruit fly, *Bactrocera latifrons* (Hendel), is an invasive oligophagous species principally infesting solanaceous crops (Ekesi and Billah 2008). Although little is known about the chemical ecology of *B. latifrons* in relation to its host plants, fresh juice from the pepper, *Capsicum annuum* L., does stimulate oviposition (Vargas et al. 1990).

2.5 Ceratitis capitata

The earliest efforts to elucidate the role of host plant odours as attractants for the Mediterranean fruit fly (medfly), *Ceratitis capitata* (Wiedemann), were made by Steiner et al. (1957) who first documented the attractiveness of oil extracts from the seeds of angelica, *Angelica archangelica* L. to male and female *C. capitata*. Following this initial discovery, Beroza and Green (1963) explored the attractiveness to both sexes of extracts from a variety of hosts (i.e. plants that females can oviposit on) and non-host plants (i.e. plants that females do not oviposit on but which may provide exudates on which both sexes can feed) including lesser galangal, *Alpinia officinarum* Hance; *Anaphalis margaritacea* (L.) Benth. & Hook; other *Angelica* species; barberry, *Berberis vulgaris* L.; wintergreen, *Chimaphila umbellata* (L.) Barton; Chinese hemlock-parsley, *Conioselinum chinense* (L.) Britton, Sterne & Poggenb.; cane orchid, *Dendrobium anosmum* var *superbum*; common horsetail, *Equisetum arvense* L.; *Fabiana imbricate* Ruiz & Pav.; *Festuca* species; cow parsnip, *Heracleum lanatum* Michx.; eastern black walnut, *Juglans nigra* L.; tomato; alder buckthorn, *Rhamnus frangula* L.; Provence rose, *Rosa centifolia* L.; and common lime, *Tiliax europaea* L.. Oils from other host and non-host plants have also been evaluated by various authors as cited in Warthen and McInnis (1989) and most were attractive to male *C. capitata*: ylang-ylang, *Cananga oderata* (Lam.) Hook.f. & Thomson; lemon, *Citrus limon* (L.) Burm.f.; grapefruit, *Citrus paradisi* Macfad.; sweet orange, *C. sinensis*; copaiba (an extract from tree bark); gingergrass, *Cymbopogon martinii* (Roxb.) Wats.; hop, *Humulus lupulus* L.; dwarf pine, *Pinus pumilio* Haenke; Scots pine, *Pinus sylvestris* L.; and tailed pepper, *Piper cubeba* L.f.. Oil extracts from lychee, *Litchi chinensis* Sonn.; weeping fig, *Ficus benjamina* L.; Cuban laurel, *Ficus retusa* L.; and Indian banyan, *Ficus benghalensis* L. were all short range attractants and feeding stimulants for male *C. capitata* (Warthen and McInnis 1989). Subsequently, α-copaene from *A. archangelica* seed oils was identified as a potent male attractant. Solvent extracts from a wide range of non-host plants were also attractive to *C. capitata* (sexes not indicated) (Keiser et al. 1975). In field experiments Flath et al. (1994a) found that β-copaene and β-ylangene extracted from *A. archangelica* seeds were less potent male attractants than copaene.

Components of odours from the fruits of host plants can be attractive to both sexes of *C. capitata* under both laboratory and semi-field conditions; although inconsistent, the responses of females towards these odours were stronger than the responses of males. In an electrophysiological study using antennae from both sexes of *C. capitata* and components of the odours from the fruits of several host plant species revealed that the attractive component was heptanal; this was confirmed in field assays (Guerin et al. 1983). Furthermore, when host searching behaviour of females was studied on non-fruiting host plants and non-host plants, it was observed that females spent longer time on the non-fruiting host plants than on non hosts but were able to locate ripe host plant fruits placed on either host or non-host plants (Prokopy et al. 1986). Further studies found that 70 volatile components that are common in the odours of many host plants (mainly aldehydes, alcohols, acids, acetates and lactones), elicited responses in receptors on the antennae of both male and female *C. capitata*; they also responded to components of the odour of nectarines, *Prunus persica* (L.) Batsch (Light et al. 1988, 1992). Receptors on the antennae of female *C. capitata* responded strongly to both odours from 'calling' males (positive control) and from the headspace of mangoes (Cossé et al. 1995). In outdoor field-cage assays Katsoyannos et al. (1997) reported that both male and female *C. capitata* adults responded to odours from sweet oranges and also antennally-active components from another orange variety, the navel orange, that had been identified previously by Hernandez et al. (1996). Surprisingly, at this time there have been no studies on odours from coffee, *Coffea arabica* L., the ancestral host plant of *C. capitata*. However, stimulated by a heavy infestation of coffee berries by *B. dorsalis* in newly planted coffee farms (Vargas et al. 1995; 1997), studies began to identify putative fruit fly attractants associated with coffee; in wind tunnel assays six components of coffee headspace odours were attractive to female *C. capitata* including 3-methyl-1-butanal, decanal, 3-methyl-1-butanol, (*Z*)-2-pentenol, (*E*)-2-hexenol and 2-heptanone (Prokopy et al. 1998). However, in semi-field trials using outdoor

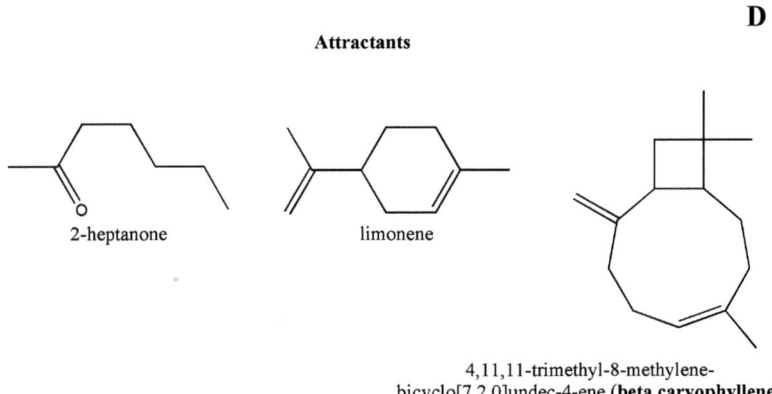

Fig. 9.4 Plant-based compounds reported to attract *Ceratitis capitata* under field conditions either singly or in combination with other attractants

cages, only 2-heptanone showed any potential as either a male or a female attractant (with more consistent results obtained with males) (Prokopy et al. 1998) (Fig. 9.4). Despite these efforts, to date, there have been no field-active plant-based attractants identified for female *C. capitata*.

In mass-rearing experiments Vargas and Chang (1991) demonstrated that, for *C. capitata*, water and the juice extracted from coffee beans were superior oviposition stimulants compared with juice from guava, papaya or orange. Oviposition deterrents have also been reported for *C. capitata* from methanolic extracts of the fern *Elaphoglossum piloselloides* (Presl.) Moore and identified as a mixture of elaphoside-A [p-vinylphenyl (β-D-glucopyranosyl)-(1 → 3)-β-D-allopyranoside] and its racemic derivative p-(1-metoxyethyl) anisole (Socolsky et al. 2003, 2008).

Although *C. capitata* attacks a variety of fruit species, in lemon only overripe or partially decayed fruits are infested suggesting that the peel of the fruit offers some degree of resistance to attack (Quayle 1914, 1929 as cited in Salvatore et al. 2004). Da Silva Branco et al. (2000) have also reported that oxygenated monoterpenes were responsible for the observed resistance to infestation. In a more systematic and organized screening of the effects of extracts from lemon peel on female oviposition, egg hatchability, and larval and adult survival Salvatore et al. (2004) demonstrated that a mixture of citral (itself a mixture of geranial and neral), 5,7-dimethoxycoumarin and linalool was insecticidal to both adults and larvae. This now requires further field evaluation.

2.6 Ceratitis rosa

The Natal fruit fly, *Ceratitis rosa* Karsch, has not been studied extensively, possibly because it is not such a serious pest of fruit compared with other fruit fly species. However, in laboratory assays *C. rosa* (sex not specified) is attracted to limonene and isoamyl acetate, which are volatile compounds from oranges and bananas, respectively (Lebusa et al. 2013) (Fig. 9.4).

2.7 Dacus *Species*

Of the six *Dacus* species found in Africa, host plant attraction has only been studied for the lesser pumpkin fly, *Dacus ciliatus* (Loew), an oligophagous pest of cucurbits (Alagarmalai et al. 2009). Using dual choice air static behavioural assay arenas, electrophysiological recordings of antennae and mass spectrometric analyses, saturated and unsaturated esters and a blend of these were identified as the odour components from Galia melon *Cucumis melo* var. *reticulatus* L. Naud. that were attractive to both male and female *D. ciliatus* (Table 9.3) (Fig. 9.5). When the terpene (*E*)-β-farnesene is added to this blend its activity as an attractant is masked and it becomes a deterrent (Alagarmalai et al. 2009). Field studies are now needed to confirm these laboratory findings.

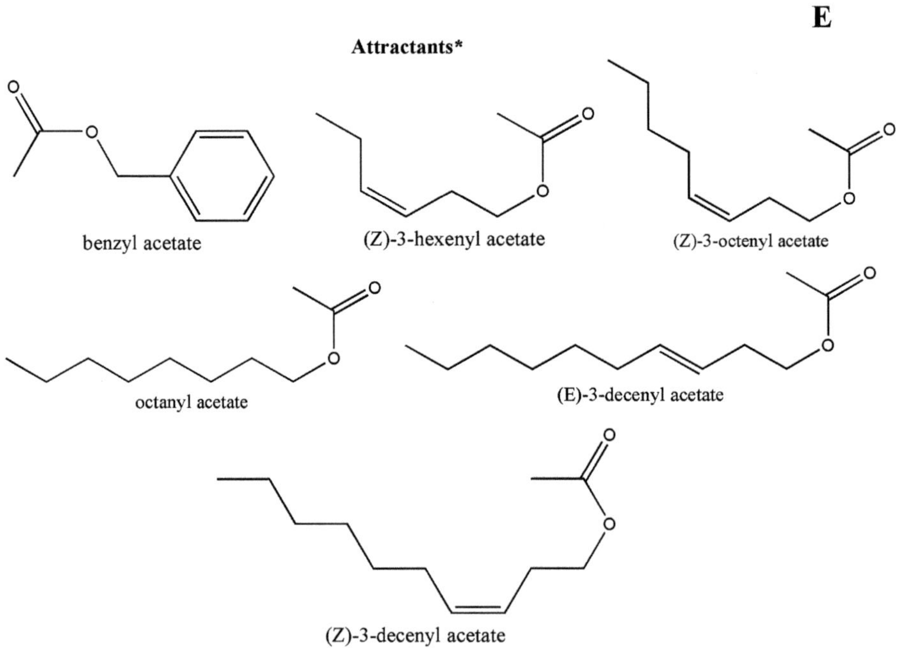

Fig. 9.5 Components of a chemical blend reported attractive to *Dacus ciliatus* in laboratory assays

3 Intraspecific Interactions

Pheromones play a crucial role in both the inter- and intraspecific interactions between fruit flies (Wyatt 2010) and are quite diverse in their behavioural function and chemical structure. There are two types of pheromone that influence the behaviour of conspecifics: sex pheromones and host-marking pheromones, the latter of which are associated with oviposition. In terms of chemical structure, they are mainly alcohols, aldehydes, ketones, esters and aromatics. Sex pheromones are released as volatile airborne chemicals and perceived via olfaction (Regnier and Law 1968), while host-marking pheromones are relatively non-volatile chemicals deposited on the surface of fruit to deter competitors after successful oviposition (Kachigamba et al. 2012). Unlike the highly volatile sex pheromones, host-marking pheromones are more persistent on the surfaces on which they have been deposited (Averill and Prokopy 1987) and, because they are usually polar in nature, they are soluble in water and methanol (Boller 1981; Averill and Prokopy 1982, 1987; Boller and Hurter 1985; Hurter et al. 1987; Aluja et al. 2003).

In African tephritids of economic importance, sex pheromones are predominantly produced by males to attract females (Sivinski and Burk 1989; Tumlinson 1989; Heath et al. 1993; Mavraganis et al. 2010). Females are also known to produce sex pheromones in a few species such as *Z. cucurbitae* (Baker and Bacon 1985), *B. dorsalis* (Baker and Bacon 1985) and *B. oleae* (Gariboldi et al. 1983). Unlike sex pheromones, host-marking pheromones are solely produced by females (Averill and Prokopy 1989). The discovery of the presence of both pheromone types dates back to the late 1950s (Féron 1959). In fruit flies, sex pheromones are secreted and stored in the rectal gland (referred to as the rectal ampulla) (Fletcher 1969; Tan et al. 2002, 2011, 2013), while host-marking pheromones are produced in the posterior half of the midgut and often released along with the faecal matter of females in host-marking species (Averill and Prokopy 1989). Amongst the fruit fly species found in Africa, sex pheromones have been documented for *B. dorsalis*, *B. oleae*, *Z. cucurbitae* and *C. capitata*, while host-marking behaviour, which points to the existence of a host-marking pheromone, has only been recorded in *C. capitata* and its sibling species the mango fruit fly, *Ceratitis cosyra* (Walker), *Ceratitis fasciventris* (Bezzi) and *C. rosa* (Kachigamba et al. 2012). In addition to these, lek site pheromones have also been reported for *C. capitata* (Feron, 1959, 1962; Jang et al. 1989; Siciliano et al. 2014; Shelly et al. 2014). Sex pheromones are likely to exist in every fruit fly species given their ecology. However, not all fruit fly species produce host-marking pheromones (Silva et al. 2012; Kachigamba et al. 2012). In the sub-sections that follow, a summary of the major pheromone types is provided.

3.1 Bactrocera dorsalis *Species Complex*

Males of the *B. dorsalis* complex produce two principal sex pheromone components that have been identified as the phenylpropanoids 2-allyl-4,5-dimethoxyphenol and (*E*)-coniferyl alcohol. These metabolites, present in the rectal gland of males, are synthesized from methyl eugenol ingested by males attracted to specific plants that are rich in this compound. Like most insects producing sex pheromones, the pheromone is released a few hours before the time of mating (Tan and Nishida 1996, 1998; Hee and Tan 1998; Khoo et al. 2000; Nishida et al. 2000; Tan et al. 2013). Females of the *B. dorsalis* complex, including one member of the complex, *Bactrocera invadens* Drew, Tsuruta and White, do not engage in host-marking behaviour after oviposition (Kachigamba 2012).

3.2 **Bactrocera oleae**

Females of *B. oleae* produce 1,7-dioxaspiro [5.5] undecane, α-pinene, n-nonanal and ethyl-dodecanoate as major components of their sex pheromone, with 1,7-dioxaspiro [5.5] undecane being the most attractive to males (Mazomenos and Haniotakis 1985). The identification of 1,7-dioxaspiro [5.5] undecane ('olean') and α-pinene as sex pheromone components was confirmed in recent studies (Levi-Zada et al. 2012; Gerofotis et al. 2013) with the latter compound demonstrated to enhance mating performance and success in both sexes of *B. oleae*. Also, females produce 1,5,7-trioxaspiro [5.5] undecane, as an additional pheromone component. Likewise, males also produce a female attractant contained in their rectal glands with the hydrocarbon (Z)-9-tricosene identified as one of its components (Carpita et al. 2012).

3.3 **Zeugodacus cucurbitae**

Although the rectal glands of male *B. curcubitae* contain a complex mixture of compounds, the most dominant and behaviourally-active of these is the diol, 1,3-nonanediol, which is a component of the sex pheromone. This compound is thought to have a dual function: as a pheromone to attract females and as a chemical defence against predation by the common house gecko, *Hemidactylus frenatus* Schlegel, which is reported to avoid fruit flies as a food source (Tan 2000). The quantity of 1,3-nonanediol in the rectal gland increases with age of the fruit fly (Nishida et al. 1993). A similar defensive phenomenon has been observed in the papaya fruit fly, *Bactrocera papaya* (Drew and Hancock), and the carambola fruit fly, *Bactrocera carambolae* Drew and Hancock, both sibling species of *Z. cucurbitae*. In *B. papaya* and *B. carambolae* the pheromone component, 6-oxo-1-nonanol, also serves as an antifeedant against the predatory spotted house gecko *Gekko monarchus* (Schlegal) (Wee and Tan 2001, 2005). Like females of the *B. dorsalis* species complex, female *Z. cucurbitae* have not been observed depositing a host-marking pheromone after oviposition (Prokopy and Koyama 1982).

3.4 Bactrocera zonata

Nishida et al. (1988) determined that the female-produced sex pheromone of *B. zonata* was composed of the phenylpropanoids 2-allyl-4,5-dimethoxyphenol and (*E*)-coniferyl alcohol. As mentioned previously, these two components are also sex pheromones for *B. dorsalis*.

3.5 Ceratitis capitata

In *C. capitata* sex pheromones are produced by males. The earliest study of this male-produced sex pheromone was done by Jacobson et al. (1973) who isolated, identified and synthesized two components from the headspace odours of males that were attractive to females: methyl (*E*)-6-nonenoate and (*E*)-6-nonen-1-ol. The complexity of the headspace odours of males was revealed in subsequent studies carried out by Baker et al. (1985) who identified nine additional components: (3,4-dihydro-2H-pyrrole, ethyl (*E*)-3-octenoate, (*E*)-3,7-dimethyl-2,6-octadienyl acetate, (*E,E*)-3,7,11-trimethyl-1,3,6,10-dodecatetraene, (*E*)-2-hexenoic acid, dihydro-3-methyl-furan-2(3H)-one, 3,5-dimethyl-2-ethylpyrazine, 2,6-dimethyl-2,7-octadien-6-ol and ethyl acetate. Behavioural assays confirmed the role of the 3,4-dihydro-2H-pyrrole (pyrroline) as a sex pheromone. Further complexity of the headspace odours was reported by Jang et al. (1989, 1994) using electrophysiology and mass spectrometry. In total, 56 components have been identified in the headspace odours of male *C. capitata*. When tested singly the five major compounds (ethyl acetate, 3,4-dihydro-2H-pyrrole, geranyl acetate, ethyl-(E)-3-octenoate and (*E,E*)-α-farnesene) elicited responses in behavioural asaays that were close to the responses to natural male odours. When combined as a blend these five components elicited responses that were indistinguishable from responses to male odours. In a similar study, Heath et al. (1991) identified geranyl acetate, ethyl-(*E*)-3-octenoate and (*E,E*)-α-farnesene as the most abundant components of the *C. capitata* sex pheromone and demonstrated their attractiveness as individual compounds and as a blend in field assays. In addition to the pheromone components, host plant volatiles such as the GLVs (*E*)-2-hexen-1-ol and (*E*)-2-hexenal increase the attractiveness of male odours to females (Dickens et al. 1990). Furthermore, Shelly et al. (2007) observed that female *C. capitata* were more attracted to males exposed to ginger root oil (GRO), with the exposed males having a higher mating frequency than those unexposed to the oil.

Males of *C. capitata* also produce a lekking pheromone which has 2-methyl-6-vinyl pyrazine as one of its components to attract females to lek sites (Chuman et al. 1987; Shelly 2001). Furthermore, female *C. capitata* deposit a host-marking pheromone on fruits after oviposition to deter other gravid females from laying on the same fruit (Prokopy et al. 1978; Arredondo and Diaz-Fletcher 2006).

3.6 Other Ceratitis Species

Until recently, little was known about sex communication in sibling species of the *Ceratitis* genus including *C. cosyra*, *C. fasciventris* and *C. rosa*, commonly known as the *Ceratitis* FAR complex. This complex has now been resolved based on larval morphology, developmental biology under different temperature regimes, and airborne volatiles and cuticular hydrocarbons (CHCs) associated with adults of these species (Steck and Ekesi 2015; Tanga et al. 2015; Břízová et al. 2015; Vaníčková et al. 2015). Břízová et al. (2015) demonstrated that adult male flies of the three species formed leks (Aluja and Norrbom 2001) in response to specific male odours detected also by females, suggesting a pheromonal role of male odours. Using GC/MS and GC/EAD analyses, twelve antennally-active components in male odours were detected by female antennae. Five of these components, namely methyl (*E*)-hex-2-enoate, 6-methyl-5-hepten-2-one, linalool, (*E*)-non-2-enal and methyl (2*E*, 6*E*) farnesoate were common to all three species. On the other hand, whereas (*E*,*E*)-α-farnesene was common to the male odours from both *C. anonae* and *C. rosa*, geranyl acetone was only identified from the male odours of *C. rosa*. The remaining five components (methyl (*E*)-hex-3-enoate, methyl (*E*)-hex-2-enoate, ethyl (*E*)-hex-3-enoate, ethyl (*E*)-hex-2-enoate and methyl (*Z*)-oct-3-enoate) were specific to *C. fasciventris*.

A similar study examined the cuticular hydrocarbon profiles of two *C. rosa* morphotypes derived from lowland (coastal) and highland regions of Kenya and reported their discriminatory potential. Using two-dimensional GC/MS, hydrocarbons with carbon backbones that ranged from C_{14} to C_{37} were identified (Vaníčková et al. 2015). These cuticular hydrocarbons, comprising n-alkanes, monomethyl alkanes, dimethylalkanes and unsaturated hydrocarbons varied qualitatively between the sexes in each morphotype and quantitatively between the morphotypes. Amongst these compounds, C_{29}, C_{31}, C_{33} and C_{35} hydrocarbons were thought to contribute to the differential sensitivity and tolerance of the different morphotypes to different temperature regimes. These authors also reported sexual dimorphism in the cuticular hydrocarbons produced, as potential short-range pheromones involved in the mating of *C. rosa*.

Aside from their apparent display of olfactory and tactile-based communication via volatile compounds and cuticular hydrocarbons, these species also display host-marking behaviours, suggestive of an ability to produce host-marking pheromones (Kachigamba et al. 2012).

4 Male Lures

Worldwide, male lures, also known as male attractants, have been used extensively for fruit fly detection, monitoring and mass trapping. They can be natural or synthetic compounds and are known to attract only male fruit flies. In Africa, fruit fly detection, monitoring and mass-trapping using synthetic pheromones, or parapheromones, has been reported in countries such as Mozambique (Correia, et al. 2008),

Ghana (Appiah, et al. 2009) and Sudan (Ali et al. 2014). A comprehensive list of key male lures used in detection, monitoring or mass-trapping programmes has been compiled by Ekesi and Billah (2008).

4.1 Bactrocera dorsalis *Species Complex*

Nishida et al. (1997) found that males of *B. dorsalis* were strongly attracted to a number of phenylpropanoid compounds from the perfume flower tree, *Fagraea berteroana* Gray ex Benth: (*E*)-3,4-dimethoxycinnamyl alcohol, its acetate, and (*E*)-3,4-dimethoxycinnamaldehyde. Since then many researchers have reported additional male lures. Males of *B. invadens* are attracted to methyl eugenol in the field (Sidahmed et al. 2014) and laboratory assays have identified additional candidate compounds to attract males, such as 4-methyl-3-penten-2-one and 4-hydroxy-4-methyl-2-pentanone (Kimbokota and Torto 2013). In a recent study the odorant receptor for methyl eugenol was identified (Zheng et al. 2012). The authors discovered a cDNA encoding a *Drosophila melanogaster* Meigen odorant receptor co-receptor (Orco) ortholog in *B. dorsalis*. Using qRT-PCR analysis they established that the Orco was abundantly expressed in the antennae of adult male *B. dorsalis*. Interestingly, a number of volatile compounds derived from the fruit of mango, marula and Indian almond, *Terminalia catappa* L., can also be used to attract both males and females of *B. dorsalis* (Kimbokota et al. 2013). Blends of volatile compounds derived from banana, guava and orange fruits also have potential as attractants (Tables 9.1 and 9.3) (Biasazin et al. 2014). Other examples of compounds attractive to male *B. dorsalis* are synthetic fluorinated analogues of methyl eugenol including 1,2-dimethoxy-4-fluoro-5-(2-propenyl) benzene (Khrimian et al. 2009), 1-fluoro-4,5-dimethoxy-2-(3,3-difluoro-2-propenyl) benzene and 1-fluoro-4,5-dimethoxy-2-(3-fluoro-2-propenyl) benzene (Jang et al. 2011).

4.2 Zeugodacus cucurbitae

It is known that male *Z. cucurbitae* are attracted to 4-(4-hydroxyphenyl)-2-butanone (cuelure), which is found in flowers of the orchid *D. anosum* var *superbum* (Nishida et al. 1993). However, attraction of this fruit fly species to cuelure is age dependent (Wong et al. 1991).

4.3 Bactrocera latifrons

Males of this fruit fly species are strongly attracted to α-ionone and α-ionol (Flath et al. 1994b). In field assays, the attractiveness of α-ionone and α-ionol was enhanced when combined with either cade oil or eugenol; however, the key components of cade oil responsible for this increased attractiveness were not determined (McQuate and Peck 2001; McQuate et al. 2004, 2013). Ishida et al. (2008) and Enomoto et al. (2010) demonstrated the improved attractiveness and phagostimulatory activity of α-ionone, α-ionolin combination with 3-oxygenated α-ionone-based synthetic derivatives, and isophorone in combination with isophorol, with the latter two recovered as secondary metabolites from the rectal gland of males. Further, they observed that mixing isophorone or isophorol with α-ionol enhanced the attractiveness of the lure to males.

4.4 Bactrocera zonata

Of the *Bactrocera* species, semiochemical-based host plant interactions in the polyphagous peach fruit fly *Bacterocera zonata* (Saunders) are the least studied (El-Sayed 2015). Methyl eugenol, putrescine, ammonium acetate and protein baits are the known and most potent commercially available male lures for this species (EPPO 2015).

4.5 Ceratitis capitata

Male *C. capitata* are attracted to tert-butyl 4-(and 5)-chloro-(*E*)-2-methylcyclohexanecarboxylate (Doolittle et al. 1990; Khrimian et al. 2003), which is a component of Trimedlure (a synthetic lure comprising eight esters of methylcyclohexanecarboxylic acid), a commercial lure used in the monitoring and mass trapping of this fruit fly species. Male *C. capitata* are differentially attracted to the different trans isomers found in ceralure (CRL), a commercial attractant composed of a mixture of four isomers of methylcyclohexanecarboxylic acid esters (Warthen et al. 1994). These authors showed that one isomer, ethyl (*Z*)-5-iodo-trans-2-methylcyclobexane- 1-carboxylate, or CRL-BI, was more attractive than the other isomers including (1,l-dimethylethyl 4- (and 5-) chloro- (*E*)- 2-methylcyclohexane-l-carboxylate), and (1,1-dimethylethyl (*Z*)-4-chloro-(*E*)-2-methylcyclohexane-l-carboxylate) when they were compared directly on an equal weight basis. In tests with two stereo-selectively-synthesized enantiomers of CRL-B1, which is a potent lure for male *C. capitata*, the (−) CRL-B1 enantiomer attracted significantly more males than the (+) CRL-B1 antipode (Jang et al. 2001). More recently, it has been shown that male *C. capitata* are attracted to α-copaene (Shelly 2013).

5 Summary and Future Directions

This review demonstrates that there are a diverse range of semiochemicals that mediate host location and reproductive biology in African tephritid fruit flies. This has been possible as a result of advances in the chemical analytical techniques available. Some species of fruit flies have been extensively investigated whereas others have received minimal attention, partly because of their economic importance. Nonetheless, irrespective of the fruit fly species, the identified semiochemicals elicited responses in target sexes when presented either singly or in multi-component blends. This review also revealed that a few of the semiochemicals are of microbial origin and that others play a role in fruit fly defence against natural enemies. Notably, only a few of the semiochemicals identified have been evaluated in the field and there are only a limited number of lures targeted at males that have proved to be effective for monitoring and control of fruit flies. The use of semiochemicals for the control of female fruit flies has only been successful for *Z. cucurbitae*. Therefore, much still remains to be done for control of females of other fruit fly species. The elucidation of the molecular mechanisms underlying the olfactory perception of methyl eugenol in *B. dorsalis* is exciting. A similar approach could be employed for some of the other promising semiochemicals identified for other fruit fly species and this would help us move towards development of more effective environmentally-friendly control tools for these species.

Acknowledgements The research activities were partly funded by grants from GIZ/BMZ and USDA.

References

Alagarmalai J, Nestel D, Dragushich D, Nemny-Lavy E, Anshelevich L, Zada A, Soroker V (2009) Identification of host attractants for the Ethiopian fruit fly, *Dacus ciliatus* Loew. J Chem Ecol 35:542–551

Ali SAI, Mohamed SA, Mahmoud MEE, Sabiel SAI, Ali S, Ali A (2014) Monitoring of Tephritidae of fruit trees and their level of infestation in South Kordofan State, Sudan. Int J Agric Innov Res 2:1473–2319

Aluja M, Norrbom AL (2001) Fruit flies (Tephritida) phylogeny and evolution of behavior. CRC Press LLC, Boca Raton

Aluja M, Diaz-Fleischer F, Edmunds AJF, Hagmann L (2003) Isolation, structural determination, synthesis, biological activity and application as control agent of the host-marking pheromone (and derivatives thereof) of the fruit flies of the type *Anastrepha* (Diptera: Tephritidae). United States Patent No. US 6 555 120 BI. 30 p

Alyokhin AV, Messing RH, Duan JJ (2000) Visual and olfactory stimuli and fruit maturity affect trap captures of oriental fruit flies (Diptera: Tephritidae). J Econ Entomol 93:644–649

Appiah EF, Afreh-Nuamah K, Obeng-Ofori D (2009) Abundance and distribution of the Mediterranean fruit fly *Ceratitis capitata* (Diptera: Tephritidae), in Late Valencia citrus orchards in Ghana. Int J Trop Insect Sci 29:11–16

Arredondo J, Diaz-Fletcher F (2006) Oviposition deterrents for the Mediterranean fruit fly, *Ceratitis capitata* (Diptera: Tephritidae) from fly faeces extracts. Bull Entomol Res 96:35–42

Averill AL, Prokopy RJ (1982) Oviposition-deterring fruit-marking pheromone in *Rhagoletis zephyria*. J Ga Entomol Soc 17:315–319

Averill AL, Prokopy RJ (1987) Residual activity of oviposition deterring pheromone in *Rhagoletis pomonella* (Diptera, Tephritidae) and female response to infested fruit. J Chem Ecol 13:167–177

Averill AL, Prokopy RJ (1989) Host-marking Pheromones. In: Robinson AS, Hooper G (eds) World crop pests: fruit flies – their biology, natural enemies and control. Elsevier, New York, pp 207–219

Baker R, Bacon AJ (1985) The identification of spiroacetals in the volatile secretions of two species of fruit fly (*Dacus dorsalis, Dacus cucurbitae*). Experientia 41:1484–1485

Baker R, Herbert RH, Grant GG (1985) Isolation and identification of the sex pheromone of the Mediterranean fruit fly, *Ceratitis capitata* (Wied). J Chem Soc Chem Commun 12:824–825

Beroza M, Green N (1963) Materials tested as insect attractants. Handbook no. 239. USDA Agriculture, Washington, DC

Biasazin TD, Karlsson MF, Hillbur Y, Seyoum E, Dekker T (2014) Identification of host blends that attract the African invasive fruit fly, *Bactrocera invadens*. J Chem Ecol 40:966–976

Boller EF (1981) Oviposition deterring pheromone of the European cherry fruit fly: status of research and potential applications. In: Mitchell ER (ed) Management of insect pests with semiochemicals. Plenum, New York, pp 457–462

Boller EF, Hurter J (1985) Oviposition deterring pheromone in *Rhagoletis cerasi*: behavioural laboratory test to measure pheromone activity. Entomol Exp et Appl 39:163–169

Břízová R, Vaníčková L, Fat'arová M, Ekesi S, Hoskovec M, Kalinová B (2015) Analyses of volatiles produced by the African fruit fly species complex (Diptera, Tephritidae). ZooKeys 540:385–404. doi:10.3897/zookeys.540.9630. http://zookeys.pensoft.net

Capasso R, Evidente A, Tremblay E, Sala A, Santoro C, Cristinzo G, Scognamiglio F (1994) Direct and mediated effects on *Bactrocera oleae* (Gmelin) (Diptera: Tephritidae) of natural polyphenols and some related synthetic compounds: structure-activity relationships. J Chem Ecol 20:1189–1199

Carpita A, Canale A, Raffaelli A, Saba A, Benelli G, Raspi A (2012) (Z)-9-tricosene identified in rectal gland extracts of *Bactrocera oleae* males: first evidence of a male-produced female attractant in olive fruit fly. Naturwissenschaften 99:77–81

Chuman T, Landolt PJ, Heath RR, Tumlinson JH (1987) Isolation, identification, and synthesis of male-produced sex pheromone of papaya fruit fly, *Toxotrypana curvicauda*, Gerstaecker (Diptera: Tephritidae). J Chem Ecol 13:1979–1992

Cirio V (1971) Reperti sul meccanismo stimolo-risposta nell'ovideposizione del *Dacus oleae* Gmelin. Redia 52:577–600

Cornelius ML, Duan JJ, Messing RH (1999) Capture of oriental fruit flies (Diptera: Tephritidae) by protein-baited traps and fruit-mimicking visual traps in a guava orchard. Environ Entomol 28:1140–1144

Cornelius ML, Duan JJ, Messing RH (2000a) Volatile fruit odours as attractants for the oriental fruit fly (Diptera: Tephritidae). J Econ Entomol 93:93–100

Cornelius ML, Nergel L, Duan JJ, Messing RH (2000b) Responses of female oriental fruit flies (Diptera: Tephritidae) to protein and host fruit odors in field cage and open field tests. Environ Entomol 29:14–19

Correia AR, Rego JM, Olmi M (2008) A pest of significant economic importance detected for the first time in Mozambique: *Bactrocera invadens* Drew, Tsuruta and White (Diptera: Tephritidae: Dacinae). Boll Zool Agr Bachicoltura Serie 40:9–13

Cossé AA, Todd JL, Millar JG, Martinez LA, Baker TC (1995) Electroantennographic and coupled gas chromatography-electroantennographic responses of the Mediterranean fruit fly *Ceratitis capitata*, to male-produced volatiles and mango odor. J Chem Ecol 21:1823–1836

Da Silva BE, Vendramin JD, Denardi F (2000) Resistência ás moscas-das-frutas em fruteiras. In: Malavasi A, Zucchi RA (eds) Moscas-das-frutas de Importância Econômica no Brazil. Holos Editora, Riberão Preto, pp 161–167

Dickens JC, Jang EB, Light DM, Alford AR (1990) Enhancement of insect pheromone responses by green leaf volatiles. Naturwissenschaften 77:29–31

Dimou IJ, Koutsikopoulos C, Economopoulos A, Lykakis J (2003) The distribution of olive fruit fly captures with MacPhail traps within an olive orchard. Phys Chem Chem Phys 31:124–131

Diongue A, Yen TB, Lai P-Y (2010) Bioassay studies on the effect of essential oils on the female oriental fruit fly, *Bactrocera dorsalis* (Hendel) (Diptera: Tephritidae). Pest Manag Hortic Ecosyst 16:91–102

Doolittle RE, Cunningham RT, McGovern TP, Sonnet PE (1990) Trimedlure enantiomers: differences in attraction for Mediterranean fruit fly, *Ceratitis capitata* (Wied.) (Diptera: Tephritidae). J Chem Ecol 17:475–484

Ekesi S, Billah MK (2008) A field guide to the management of economically important tephritid fruit flies in Africa, 2nd edn. The International Centre of Insect Physiology and Ecology (icipe) Science Press, Nairobi

El-Sayed AM (2015) The pherobase: database of pheromones and semiochemicals. http://www.pherobase.com. Accessed 17 Dec 2015

Enomoto H, Ishida T, Hamagami A, Nishida R (2010) 3-oxygenated α-ionone derivatives as potent male attractants for the solanaceous fruit fly, *Bactrocera latifrons* (Diptera: Tephritidae), and sequestered metabolites in the rectal gland. Appl Entomol Zool 45:551–556

EPPO (2015) PQR – EPPO database on quarantine pests (available online). http://www.eppo.int. Accessed 27 July 2015

Estes AM, Segura DF, Jessup A, Wornoayporn V, Pierson EA (2014) Effect of the symbiont *Candidatus* Erwinia *dacicola* on mating success of the olive fly *Bactrocera oleae* (Diptera: Tephritidae). Int J Trop Insect Sci 40:123–131

Féron MM (1959) Attraction chimique du mâle de *Ceratitis capitata* Wied. (Dipt. Trypetidae) pour la femelle. C R Acad Sci Paris Ser D 248:2403–2404

Féron M (1962) L'instinct de reproduction chez la mouche Mediterrneene des Fruits *Ceratitis capitata* Wied (Dipt. Trypetidae). Comportment sexual component de ponte. Rev Pathol Veg Entomol Agric Fr 41:1–129

Fiume F, Vita G (1977) L'impiego delle acque di vegetazioce del frutto di olivo per il controllo del *Dacus oleae* Gmel. Boll Lab Entomo Agric 'F Silvestri' Portici 34:25–36

Flath RA, Light DM, Jang EB, Mon TR, John JO (1990) Headspace examination of volatile emissions from ripening papaya (*Carica papaya* L. solo variety). J Agric Food Chem 38:1060–1063

Flath RA, Cunningham RT, Mon TR, John JO (1994a) Additional male Mediterranean fruit fly (*Ceratitis capitata* Wied) attractants from angelica seed oil (*Angelica archangelica* L.). J Chem Ecol 20:1969–1984

Flath RA, Cunningham RT, Mon TR, John JO (1994b) Male lures for Mediterranean fruit fly (*Ceratitis capitata* Wied): structural analogs of α-copaene. J Chem Ecol 20:2595–2609

Fletcher BS (1969) The structure and function of the sex pheromone glands of the male Queensland fruit fly, *Dacus tryoni*. J Insect Physiol 15:1309–1322

Gariboldi P, Verotta L, Fanelli R (1983) Studies on the sex pheromone of *Dacus oleae*. Analysis of the substances contained in the rectal glands. Experientia 39:502–505

Gerofotis CD, Loannou CS, Papadopoulos NT (2013) Aromatized to find mates: α-pinene aroma boosts the mating success of adult olive fruit flies. PLoS One 8:e81336. doi:10.1371/journal.pone.0081336

Gikonyo NK, Lux SA (2004) Evidence of male attractants for fruit flies in the leaf extract of *Schinus molle*. East Cent Afr J Pharm Sci 7:27–33

Girolami V, Pellizzari G, Ragazzi E, Veronese G (1974) Prospects of increased egg production in the rearing of *Dacus oleae* (Gmel.) by the use of chemical stimuli. In: Proceedings of the symposium: sterility principle of insect control, IAEA, Wien, 1975, pp 209–217

Girolami V, Vianello A, Strapazzon A, Ragazzi E, Veronese G (1981) Ovipositional deterrents in *Dacus oleae*. Entomol Exp et Appl 29:177–188

Goergen G, Vayssières J-F, Gnanvossou D, Tindo M (2011) Bactrocera invadens (Diptera: Tephritidae), a new invasive fruit fly pest for the Afrotropical Region: host plant range and distribution in West and Central Africa. Environ Entomol 40:844–854

Guerin PM, Remund U, Boller EF, Katsoyannos B, Delrio G (1983) Fruit fly electroantennogram and behavior responses to some generally occurring fruit volatiles. In: Cavalloro R (ed) Proceedings CEC and IOBC International Symposium Fruit Flies of Economic Importance, Athens, Greece. A. A. Balkema, Rotterdam, pp 248–251

Heath RR, Landolt PJ, Tumlinson JH, Chambers DL, Murphy RE, Doolittle RE, Dueben BD, Sivinski J, Calkins CO (1991) Analysis, synthesis, formulation, and field testing of three major components of male Mediterranean fruit fly pheromone. J Chem Ecol 17:1925–1940

Heath RR, Manukian A, Epsky ND, Sivinski J, Calkins CO, Landolt PJ (1993) A bioassay system for collecting volatiles while simultaneously attracting tephritid fruit flies. J Chem Ecol 19:2395–2410

Hee AK, Tan K (1998) Attraction of female and male *Bactrocera papayae* to conspecific males fed with methyl eugenol and attraction of females to male sex pheromone components. J Chem Ecol 24:753–764

Hernandez MM, Sanz I, Adelantado M, Ballach S, Primo E (1996) Electroantennogram activity from antennae of *Ceratitis capitata* (Wied) to fresh orange airborne volatiles. J Chem Ecol 22:1607–1619

Hurter J, Boller EF, Stadler E, Blattman H, Buser R, Boshard NU, Damm L, Kozlowski MW, Schoni R, Raschdorf F, Dahinden R, Schlumpf E, Fritz H, Richter WJ, Schreiber J (1987) Oviposition-deterring pheromone in *Rhagoletis cerasi* L.: purification and determination of the chemical constitution. Experientia 43:157–164

Ishida T, Enomoto H, Nishida R (2008) New attractants for males of the solanaceous fruit fly *Bactrocera latifrons*. J Chem Ecol 34:1532–1535

Jacobson M, Keiser I, Chambers DL, Miyashita DH, Harding C (1971) Synthetic nonenyl acetates as attractants for female melon flies. J Med Chem 14:236–239

Jacobson M, Ohinata K, Chambers DL, Jones WA, Fujimoto MS (1973) Insect sex attractants. 13. Isolation, identification, and synthesis of sex pheromones of the male Mediterranean fruit fly. J Med Chem 16:248–251

Jang EB, Light DM (1991) Behavioral responses of female Oriental fruit flies to the odor of papayas at three ripeness stages in a laboratory flight tunnel (Diptera: Tephritidae). J Insect Behav 4:751–762

Jang EB, Light DM, Flath RA, Nagata JT, Men TR (1989) Electroantennogram responses of the Mediterranean fruit fly, *Ceratitis capitata* to identify volatile constituentsfrom calling males. Entomol Exp et Appl 50:7–19

Jang EB, Light DM, Binder RG, Flath RA, Carvalho LA (1994) Attraction of female Mediterranean fruit flies to the five major components of male-produced pheromone in a laboratory flight tunnel. J Chem Ecol 20:9–20

Jang EB, Carvalho LA, Stark JD (1997) Atttraction of female oriental fruit fly, *Bactrocera dorsalis*, to volatile semiochemicals from leaves and extracts of a non-host plant, panax (*Polyscias guilfoylei*) in laboratory and olfactometer assays. J Chem Ecol 23:1389–1401

Jang EB, Raw AS, Carvalho LA (2001) Field attraction of Mediterranean fruit fly, *Ceratitis capitata* (Wiedemann) to synthetic stereo-selective enantiomers of the Ceralure B1 isomer. J Chem Ecol 27:235–242

Jang EB, Khrimian A, Siderhurst MS (2011) Di- and tri-fluorinated analogs of methyl eugenol: attraction to and metabolism in the oriental fruit fly, *Bactrocera dorsalis* (Hendel). J Chem Ecol 37:553–564

Jayanthi KPD, Woodcock CM, Caulfield J, Birkett MA, Bruce TJA (2012) Isolation and identification of host cues from mango, *Mangifera indica*, that attract gravid female oriental fruit fly, *Bactrocera dorsalis*. J Chem Ecol 38:361–369

Jayanthi KPD, Kempraj V, Aurade RM, Ravindra KV, Bakthavatsalam N, Verghese A, Bruce TJA (2014a) Oviposition site-selection by *Bactrocera dorsalis* is mediated through an innaterecognition template tuned to γ-octalactone. PLoS One 9:e85764. doi:10.1371/journal.pone.0085764

Jayanthi KPD, Kempraj V, Aurade RM, Venkataramanappa RK, Nandagopal B, Verghese A, Bruce TJA (2014b) Specific volatile compounds from mango elicit oviposition in gravid *Bactrocera dorsalis* females. J Chem Ecol 40:259–266

Kachigamba DL (2012) Host-marking behaviour and pheromones in major fruit fly species (Diptera: Tephritidae) infesting mango (*Mangifera indica*) in Kenya. PhD thesis, Jomo Kenyatta University of Agriculture and Technology, Nairobi, Kenya. 123 pp

Kachigamba DL, Ekesi S, Ndung'u MW, Gitonga LM, Teal PEA, Torto B (2012) Evidence for potential of managing some African fruit fly species (Diptera: Tephritidae) using the mango fruit fly host-marking pheromone. J Econ Entomol 105:2068–2075

Katsoyannos BI, Kouloussis NA, Papadopoulos NT (1997) Response of *Ceratitis capitata* to citrus chemicals under semi-natural conditions. Entomol Exp et Appl 82:181–188

Keiser I, Kobayashi RM, Miyashita DH, Jacobson M, Harris EJ, Chambers DL (1973) Trans-6-nonen-1-ol acetate: an ovipositional attractant and stimulant of the melon fly. J Econ Entomol 66:1355–1356

Keiser I, Harris EJ, Miyashita DH, Jacobsen M, Perdue R (1975) Attraction of ethyl ether extracts of 232 botanicals to oriental fruit flies, melon flies and Mediterranean fruit flies. Lloydia 38:141–152

Khoo CC, Yuen K, Tan K (2000) Attraction of female *Bactrocera papayae* to sex pheromone components with two different release devices. J Chem Ecol 26:2487–2496

Khrimian A, Margaryan A, Schmidt WF (2003) An improved synthesis of ethyl cis-5-iodo-trans-2-methylcyclohexanecarboxylate, a potent attractant for the Mediterranean fruit fly. Tetrahedron 59:5475–5480

Khrimian A, Siderhurst MS, Mcquate GT, Liquido NJ, Nagata J, Carvalho L, Guzman F, Jang EB (2009) Ring-fluorinated analog of methyl eugenol: attractiveness to and metabolism in the Oriental fruit fly, *Bactrocera dorsalis* (Hendel). J Chem Ecol 35:209–218

Kimbokota F (2011) Semiochemicals mediating oviposition and mating behavior of mango infesting fruit fly *Bactrocera invadens*. PhD thesis, University of Dar es Salaam, Tanzania

Kimbokota F, Torto B (2013) Candidate attractants for *Bactrocera invadens* (Diptera: Tephritidae) male flies from *Gynandropsis gynandra* (Capparidaceae). J Nat Sci Res 3:35–44

Kimbokota F, Njagi PGN, Torto B, Ekesi S, Hassanali A (2013) Responses of *Bactrocera invadens* (Diptera: Tephritidae) to volatile emissions of fruits from three hosts. J Biol Agric Healthc 3:53–60

Lauzon CR, Sjogren RE, Prokopy RJ (2000) Enzymatic capabilities of bacteria associated with apple maggot flies: a postulated role in attraction. J Chem Ecol 26:953–967

Lebusa M, Laing M, Miller R (2013) Preference of the natal fruit fly, *Ceratitis rosa* to different fruit scents in a choice trial. In: International conference on entomology. Orlando-Florida. Available at http://www.omicsgroup.com/conferences/entomology-2013/scientific-programme.php?day=1&sid=258&date=2013-09-04

Lebusa M, Laing M, Miller R (2014) Comparison between the attractiveness of an orange scent (Limonene) and a pheromone (Ceratitislure) to the natal fruit fly (*Ceratitis rosa*) in a food bait. In: 9th International Symposium on Fruit Flies of Economic Importance (ISFFEI), p 178

Levi-Zada A, Nestel D, Fefer D, Nemni-Lavy E, Deloya-Kahane I, David M (2012) Analyzing diurnal and age-related pheromone emission of the olive fruit fly, *Bactrocera oleae* by sequential SPME-GCMS analysis. J Chem Ecol 38:1036–1041

Light DM, Jang EB (1987) Electroantennogram responses of the oriental fruit fly, *Dacus dorsalis*, to a spectrum of alcohol and aldehyde plant volatiles. Entomol Exp et Appl 45:55–64

Light DM, Jang EB, Dickens JC (1988) Electroantennogram responses of the Mediterranean fruit fly *Ceratitis capitata*, to a spectrum of plant volatiles. J Chem Ecol 14:159–180

Light DM, Jang EB, Flath RA (1992) Electroantennogram responses of the Mediterranean fruit fly, *Ceratitis capitata*, to volatile constituents of nectarines. Entomol Exp et Appl 63:13–26

Liquido NJ, Cunningham RT (1990) Colorimetry of papaya fruits as an index of infestation rates of oriental fruit fly and melon fly (Diptera: Tephritidae). J Econ Entomol 83:476–484

Liquido NJ, Cunningham RT, Couey HM (1989) Infestation rates of papaya by fruit flies (Diptera: Tephritidae) in relation to the degree of fruit ripeness. J Econ Entomol 82:213–219

Liscia A, Angioni P, Sacchetti P, Poddighe S, Granchietti A, Setzu MD, Belcari A (2013) Characterization of olfactory sensilla of the olive fly: behavioral and electrophysiological responses to volatile organic compounds from the host plant and bacterial filtrate. J Insect Physiol 59:705–716

Matthews RW, Matthews JR (2010) Insect behavior, 2nd edn. Springer, Dordrecht

Mavraganis VG, Papadopoulos NT, Kouloussis NA (2010) Extract of olive fruit fly males (Diptera: Tephritidae) attracts virgin females. Entomol Hellenica 19:14–20

Mazomenos BE, Haniotakis GE (1985) Male olive fruit fly attraction to synthetic sex pheromone components in laboratory and field tests. J Chem Ecol 11:397–405

Mazomenos BE, Pantazi-Mazomenou A, Stefanou D (2002) Attract and kill of the olive fruit fly *Bactrocera oleae* in Greece as a part of an integrated control system. IOBC WPRS Bull. 25:137–146

McQuate GT, Peck SL (2001) Enhancement of attraction of alpha-ionol to male *Bactrocera latifrons* (Diptera: Tephritidae) by addition of a synergist, cadeoil. J Econ Entomol 94:39–46

McQuate GT, Keum Y-S, Sylva CD, Li QX, Jang EB (2004) Active ingredients in cade oil which synergize the attractiveness of alpha-ionol to male *Bactrocera latifrons* (Diptera: Tephritidae). J Econ Entomol 97:862–870

McQuate GT, Jang EB, Siderhurst M (2013) Detection/monitoring of *Bactrocera latifrons* (Diptera: Tephritidae) assessing the potential of new lures. Proc Hawaiian Entomol Soc 45:69–81

Miller NW, Vargas RI, Prokopy RJ, Mackey BE (2004) State dependent attractiveness of protein bait and host fruit odor to *Zeugodacus cucurbitae* (Diptera: Tephritidae) females. Ann Entomol Soc Am 97:1063–1068

Nishida R, Tan KH, Serit M, Lajis NH, Sukari AM, Takahashi S, Fukami H (1988) Accumulation of phenylpropanoids in the rectal glands of males of the oriental fruit fly, *Dacus dorsalis*. Experientia 44:534–536

Nishida R, Iwahashi O, Tan K (1993) Accumulation of *Dendrobium superbum* (Orchidaceae) fragrance in the rectal glands by males of the melon fly, *Dacus cucurbitae*. J Chem Ecol 19:713–722

Nishida R, Shelly TE, Kaneshiro KY (1997) Acquisition of female-attracting fragrance by males of Oriental fruit fly from a Hawaiian lei flower, *Fagraea berteriana*. J Chem Ecol 23:2275–2285

Nishida R, Shelly TE, Whittier TS, Kaneshiro KY (2000) α-Copaene, a potential rendezvous cue forthe mediterranean fruit fly, *Ceratitis capitata*? J Chem Ecol 26:87–100

Panizzi L, Scarpati ML, Oriente G (1960) Costituzione della oleuropeina, glucoside amaro e ad azione ipotensiva dell'olivo. Nota I1. Gazz Chim Ital 90:1449–1485

Peck SL, McQuate GT (2004) Ecological aspects of *Bactrocera latifrons* (Diptera: Tephritidae) on Maui, Hawaii: movement and host preference. Environ Entomol 23:1722–1731

Piñero JC, Jácome I, Vargas RI, Prokopy RJ (2006) Response of female melon fly, *Zeugodacus cucurbitae*, to host-associated visual and olfactory stimuli. Entomol Exp et Appl 121:261–269

Prokopy RJ, Koyama J (1982) Oviposition site partitioning in *Dacus cucurbitae*. Entomol Exp et Appl 31:428–432

Prokopy RJ, Ziegler JR, Wong TTY (1978) Deterrence of repeated oviposition by fruit-marking pheromone in *Ceratitis capitata* (Diptera: Tephritidae). J Chem Ecol 4:55–63

Prokopy RJ, Papaj DR, Wong TTY (1986) Fruit-foraging behavior of Mediterranean fruit fly females on host and no-host plants. Fla Entomol 69:651–657

Prokopy RJ, Hu X, Jang EB, Vargas RI, Warthen JD (1998) Attraction of mature *Ceratitis capitata* females to 2-heptanone, a component of coffee fruit odor. J Chem Ecol 24:1293–1304

Quayle HJ (1914) Citrus fruit insects in Mediterranean countries. U.S. Dept Agric Bull 134:1–36
Quayle HJ (1929) The Mediterranean and other fruit flies. Univ Calif Exp Stn Circ 315:1–19
Regnier FE, Law JH (1968) Insect pheromones. J Lipid Res 9:541–551
Sacchetti P, Granchietti A, Landini S, Viti C, Giovanetti L, Belcari A (2008) Relationships between the olive fly and bacteria. J Appl Entomol 132:682–689
Salvatore A, Borkosky S, Willink E, Bardon A (2004) Toxic effects of lemon peel constituents on *Ceratitis capitate*. J Chem Ecol 30:323–333
Scalzo R, Scarpati ML, Verzegnassi B, Vita G (1994) *Olea europaea* chemicals repellent to *Dacus oleae* females. J Chem Ecol 20:1813–1823
Scarpati ML, Lo Scalzo R, Vita G (1993) *Olea europaea* L. volatiles attractive and repellent to the olive fruit fly (*Dacus oleae*, Gmelin). J Chem Ecol 19:881–891
Scarpati ML, Lo Scalzo R, Vita G, Gambacorta A (1996) Chemiotropic behavior of female olive fly (*Bactrocera oleae* Gmel.) on *Olea europaea* L. J Chem Ecol 22:1027–1036
Seo ST, Farias GJ, Harris EJ (1982) Oriental fruit fly: ripening of fruit and its effect on index of infestation of Hawaiian papayas. J Econ Entomol 75:173–178
Shelly TE (2001) Lek size and female visitation in two species of tephritid fruit flies. Anim Behav 62:33–40
Shelly TE (2013) Detection of male Mediterranean fruit flies (Diptera: Tephritidae): performance of Trimedlure relative to Capilure and enriched ginger root oil. Proc Hawaiian Entomol Soc 45:1–7
Shelly TE, Edu J, Pahio E, Nishimoto J (2007) Scented males and choosy females: does male odor influence female mate choice in the Mediterranean fruit fly? J Chem Ecol 33:2308–2324
Shelly TE, Epsky N, Jang E, Reyes-Flores J, Vargas R (2014) Trapping and the detection, control, and regulation of tephritid fruit flies: Lures, area-wide programs and trade implications. Springer, Berlin. 633 pp
Siciliano P, He XL, Woodcock C, Pickett JA, Field LM, Birkett MA, Kalinova B, Gomulski LM, Scolari F, Gasperi G, Malacrida AR, Zhou JJ (2014) Identification of pheromone components and their binding affinity to the odorant binding protein CcapOBP83a-2 of the Mediterranean fruit fly, *Ceratitis capitata*. Insect Biochem Mol Biol 48:51–62
Sidahmed OAA, Taha AK, Hassan GA, Abdalla IF (2014) Evaluation of pheromone dispenser units in methyl eugenol trap against *Bactrocera invadens* Drew, Tsuruta and White (Diptera: Tephritidae) in Sudan. Sky J Agric Res 3:148–151
Siderhurst MS, Jang EB (2006a) Attraction of female oriental fruit fly, *Bactrocera dorsalis*, to *Terminalia catappa* fruit extracts in wind tunnel and olfactometer tests. Formos Entomol 26:45–55
Siderhurst MS, Jang EB (2006b) Female-biased attraction of oriental fruit fly, *Bactrocera dorsalis* (Hendel), to a blend of host fruit volatiles from *Terminalia catappa* L. J Chem Ecol 32:2513–2524
Siderhurst MS, Jang EB (2010) Cucumber volatile blend attractive to female melon fly, *Zeugodacus cucurbitae* (Coquillett). J Chem Ecol 36:699–708
Silva MA, Bezerra-Silva GCD, Mastrangelo T (2012) The host marking pheromone application on the management of fruit flies. Braz Arch Biol Technol 55:835–842
Sivinski J, Burk T (1989) Reproductive and mating behavior. In: Robinson AS, Hooper G (eds) World crop pests, volume 3A, fruit flies, their biology, natural enemies and control. Elsevier, New York
Socolsky C, Salvatore A, Asakawa Y, Bardón A (2003) Bioactive new bitter-tasting p-hydroxystyrene glycoside and other constituents from the fern *Elaphoglossum spathulatum*. Arkivoc 10:347–355
Socolsky C, Fascio ML, D'Accorso NB, Salvatore A, Willink E, Asakawa Y, Bardon A (2008) Effects of *p*-vinylphenyl glycosides and other related compounds on the oviposition behavior of *Ceratitis capitata*. J Chem Ecol 34:539–548
Steck GJ, Ekesi S (2015) Description of third instar larvae of *Ceratitis fasciventris, C. anonae, C. rosa* (FAR complex) and *C. capita*ta (Diptera, Tephritidae). ZooKeys 540:443–466. doi:10.3897/zookeys.540.10061. http://zookeys.pensoft.net

Steiner LF, Miyashita DH, Christenson LD (1957) Angelica oils as Mediterranean fruit fly lures. J Econ Entomol 50:505

Tan KH (2000) Sex pheromone components in defense of melon fly, *Zeugodacus cucurbitae* against Asian house gecko, *Hemidactylus frenatus*. J Chem Ecol 26:697–704

Tan KH, Nishida R (1996) Sex pheromone and mating competition after methyl eugenol consumption in the *Bactrocera dorsalis* complex. In: McPheron BA, Steck GJ (eds) Fruit fly pests—a world assessment of their biology and management. St. Lucid Press, Florida, pp 147–153

Tan KH, Nishida R (1998) Ecological significance of male attractant in the defence and mating strategies of the fruit fly, *Bactrocera papayae*. Entomol Exp et Appl 89:155–158

Tan KH, Nishida R, Toong YC (2002) Floral synomone of a wild orchid, *Bulbophyllum cheiri*, lures *Bactrocera* fruit flies to perform pollination. J Chem Ecol 28:1161–1172

Tan KH, Tokushima I, Ono H, Nishida R (2011) Comparison of phenylpropanoid volatiles in male rectal pheromone gland after methyl eugenol consumption, and molecular phylogenetic relationship of four global pest fruit fly species: *Bactrocera invadens, B. dorsalis, B. correcta* and *B. zonata*. Chemoecol 21:25–33

Tan KH, Wee SL, Ono H, Nishida R (2013) Comparison of methyl eugenol metabolites, mitochondrial COI, and rDNA sequences of *Bactrocera philippinensis* (Diptera: Tephritidae) with those of three other major pest species within the *dorsalis* complex. Appl Entomol Zool 48:275–282

Tanga CM, Manrakhan A, Daneel JH, Mohamed SA, Fathiya K, Ekesi S (2015) Comparative analysis of development and survival of two Natal fruit fly *Ceratitis rosa* Karsch (Diptera, Tephritidae) populations from Kenya and South Africa. ZooKeys 540:467–487. doi:10.3897/zookeys.540.9906. http://zookeys.pensoft.net

Torto B (2004) Chemical signals as attractants, repellents and aggregation stimulants. In: Encyclopedia of Life Support Systems (EOLSS), developed under the auspices of the UNESCO. Eolss Publishers, Oxford, (http://www.eolss.net)

Tumlinson JH (1989) Insect chemical communication systems. Pure Appl Chem 61:559–562

Vaníčková L, Břízová R, Pompeiano A, Ekesi S, De Meyer M (2015) Cuticular hydrocarbons corroborate the distinction between lowland and highland Natal fruit fly (Tephritidae, *Ceratitis rosa*) populations. ZooKeys 540:507–524. doi:10.3897/zookeys.540.9619. http://zookeys.pensoft.net

Vargas RI, Chang HB (1991) Evaluation of oviposition stimulants for mass production of the melon fly, oriental fruit fly and Mediterranean fruit fly (Diptera: Tephritidae). J Econ Entomol 84:1695–1698

Vargas RI, Mitchell S, Fujita B, Albrecht C (1990) Rearing techniques for *Dacus latifrons* (Hendel) (Diptera: Tephritidae). Proc Hawaiian Entomol Soc 30:71–78

Vargas RI, Stark JD, Prokopy RJ (1991) Response of oriental fruit fly (Diptera: Tephritidae) and associated parasitoids (Hymenoptera: Braconidae) to different-color spheres. J Econ Entomol 84:1503–1507

Vargas RI, Walsh WA, Nishida T (1995) Colonization of newly planted coffee fields: dominance of Mediterranean fruit fly over oriental fruit fly. J Econ Entomol 88:620–627

Vargas RI, Prokopy RJ, Duan JJ, Albrecht C, Li QX (1997) Captures of wild Mediterranean and oriental fruit flies in Jackson traps and McPhail traps baited with coffee juice. J Econ Entomol 90:165–169

Verghese A, Soumya CB, Shivashankar S, Manivannan S, Krishnamurthy SV (2012) Phenolics as chemical barriers to female fruit fly, *Bactrocera dorsalis* (Hendel) in mango. Curr Sci 103:563–566

Vita G, Cirio U, Fedeli E, Iacini G (1977) L'uso di sostanze naturali presenti nell'oliva come prospettiva di lotta contro il *Dacus oleae* Gmelin. Boll Lab Entomol Agric "F Silvestri" Portici 55

Warthen JD, McInnis DO (1989) Isolation and identification of male medfly attractive components in *Litchi chinensis* stems and *Ficus* spp. stem exudates. J Chem Ecol 15:1931–1946

Warthen JD, Cunningham RT, Demilo AB, Spencer S (1994) *Trans*-ceralure isomers: Differences in attraction for Mediterranean fruit fly, *Ceratitis capitata* (Wied.) (Diptera: Tephritidae). J Chem Ecol 20:569–578

Warthen JD, Lee C-J, Jang EB, Lance DR, McInnis DO (1997) Volatile potential attractants from ripe coffee fruit for female Mediterranean fruit fly. J Chem Ecol 23:1891–1900

Wee S, Tan K (2001) Allomonal and hepatotoxic effects following methyl eugenol consumption in *Bactrocera papayae* male against *Gekko monarchus*. J Chem Ecol 27:953–964

Wee S, Tan K (2005) Male endogenous pheromonal component of *Bactrocera carambolae* (Diptera: Tephritidae) deterred Gecko predation. Chemoecology 15:199–203

Wong TTY, McInnis DO, Ramadan MM, Nishimoto JI (1991) Age-related response of male melon flies *Dacus cucurbitae* (Diptera: Tephritidae) to cue-lure. J Chem Ecol 17:2481–2487

Wyatt TD (2010) Pheromones and signature mixtures: defining species-wide signals and variable cues for individuality in both invertebrates and vertebrates. J Comp Physiol 196:685–700

Zheng W, Zhu C, Peng T, Zhang H (2012) Odorant receptor co-receptor Orco is upregulated by methyl eugenol in male *Bactrocera dorsalis* (Diptera: Tephritidae). J Insect Physiol 58:1122–1127

Ayuka T. Fombong is an agricultural entomologist and chemical ecologist with the International Centre of Insect Physiology and Ecology (*icipe*), Nairobi, Kenya. His current research interests focus on understanding the chemistry mediating various insect-insect and insect-plant interactions, and harnessing these findings to develop improved semiochemical-based pest monitoring and management tools. He has authored and co-authored a number of publications in international peer-reviewed journals.

Donald L. Kachigamba graduated from Jomo Kenyatta University of Agriculture and Technology (Kenya) with a PhD in Agricultural Entomology in 2012. He is currently working with the Department of Agricultural Research Services (DARS) in Malawi as a Pest Management Research Scientist. He is Leader of the Pest Management Commodity Team at DARS and also a member of the National Fruit Fly Management Committee in Malawi. His current research focus is on integrated management of field insect pests in Malawi and has, so far, a number of local and international publications.

Baldwyn Torto is an organic chemist and Principal Scientist at the International Centre of Insect Physiology & Ecology (*icipe*), Nairobi, Kenya. His research and teaching experiences span over three decades in various positions in universities and international and national research institutions in Africa, the U.S.A. and the U.K with a focus on both basic and applied chemical ecology. He has published widely including peer-reviewed scientific articles, patents and several book chapters, particularly on the chemical ecology of grasshoppers and locusts, honeybees and the chemistry of anti-arthropod botanicals. He is a Fellow of the African Academy of Sciences, and also affiliated to other professional bodies: International Society of Chemical Ecology, Fellow of the Entomological Society of America and the American Chemical Society. He serves on the editorial board of the International Journal of Tropical Insect Science, Pest Management Science and the Journal of Chemical Ecology.

Chapter 10
Fruit Fly Nutrition, Rearing and Quality Control

Samira A. Mohamed, Fathiya M. Khamis, and Chrysantus M. Tanga

Abstract Tephritid fruit flies are recognized worldwide as an important threat to the horticultural industry. Most of the species belonging to this group are highly polyphagous attacking several important fruits and vegetables. They cause direct damage through larval feeding and indirect losses are associated with quarantine restrictions. The increasing awareness of the damage caused by these fruit flies to the horticultural industry has created a demand for the development of control measures based on integrated pest management (IPM) strategies and the sterile insect technique (SIT). However, success of the majority of these control methods largely depends on the ability to establish cost effective rearing methods of the fruit flies as a pre-requisite to understanding their biology, response to attractants and susceptibility to various biological control agents. In the past decades, considerable advances have been made with regard to formulations of diet for rearing fruit flies and nutritional analyses for both adults and larvae. In general, insects require a diet containing a source of energy, a protein source, vitamins and certain mineral salts. Deficiency in some of these nutrients can influence the quality control parameters of the flies such as body size, survival, pupal weight, adult emergence, longevity, flight ability, fecundity, fertility and mating ability. In this chapter, the role played by nutrition in relation to different quality control parameters is discussed.

Keywords Tephritid fruit flies • IPM • SIT • Mass rearing • Quality control parameters • Nutrition

---

S.A. Mohamed (✉) • F.M. Khamis • C.M. Tanga
Plant Health Theme, International Centre of Insect Physiology and Ecology (*icipe*),
PO Box 30772-00100, Nairobi, Kenya
e-mail: sfaris@icipe.org

© Springer International Publishing Switzerland 2016
S. Ekesi et al. (eds.), *Fruit Fly Research and Development in Africa - Towards a Sustainable Management Strategy to Improve Horticulture*,
DOI 10.1007/978-3-319-43226-7_10

1 Introduction

Adults of the frugivorous tephritid fruit fly species need to feed on a diet rich in amino acids, carbohydrates, vitamins and minerals as well as water for growth, development, survival and reproduction. Being anautogenous, females require protein for egg maturation, while males require protein for production of pheromone and accessory gland secretions as well as for renewal of sperm supplies (Drew and Yuval 2000 and reference therein; Yuval et al. 2007). In nature these species obtain their dietary requirements by feeding on bird droppings, honeydew, plant exudates, extra-floral nectaries, pollen, fruit juice, ripe fruits and microorganisms on both host and non-host trees (Christenson and Foote 1960; Steiner and Mitchell 1966; Bateman 1972). Several species of economic importance belong to this group and necessitating the need to understand their biology, behaviour, host range and other attributes, in order to develop sound management strategies for their suppression, which in turn necessitates maintaining laboratory cultures of these insects on artificial diets. Large-scale mass rearing of insects on artificial diet is also a fundamental requirement for producing good quality flies on a large scale for the sterile insect technique (SIT) and for mass production of parasitoids, which are both essential components of area-wide management of these fruit flies.

Since the beginning of the last century considerable advances have been made with regard to the formulation of diet for rearing fruit flies and nutritional analyses for both adults and larvae (e.g. Tanaka et al. 1965; Tsitsipis 1977; Hooper 1978, 1989; Walker et al. 1997; Vargas et al. 1993; Chang et al. 2001, 2004, 2007; Chang 2009a).

The first meridic adult diet was developed by Hagen and Finney (1950) for the Mediterranean fruit fly (medfly), *Ceratitis capitata* (Wiedemann), the oriental fruit fly, *Bactrocera dorsalis* (Hendel) and the melon fruit fly, *Zeugodacus cucurbitae* (Coquillett) (all Diptera: Tephritidae). This was followed later by the development of a meridic diet for other African species including the olive fruit fly, *Bactrocera oleae* (Gmelin) and the Natal fruit fly, *Ceratitis rosa* Karsch (both Diptera: Tephritidae) (Tsiropoulos 1992 and reference therein). The first chemically defined diet was developed by Hagen (1953) for *B. dorsalis*, *Z. cucurbitae* and *C. capitata*. Tsiropoulos (1981) argued that both the nitrogen in the diet and also the ratio of nitrogen to carbohydrate were important parameters for optimization of a chemically-defined diet for *B. oleae*. He established that a ratio of 1.6:40 of N:C was the best for *B. oleae* reproduction, while a higher nitrogen content in the diet reduced egg production and shortened the life span of the flies.

Like that of the adult diet, development of the larval diet has also been an active area of research in terms of the diet components as well as their effects on fitness parameters of the reared flies. Steiner and Mitchell (1966) provide a detailed account on early studies highlighting the history of the development of the larval media, their modifications to suit the rearing of different fruit flies species, as well as the laboratory techniques for their efficient use. The basic essential ingredients in larval diets are: yeast-based products, sugar, antimicrobial agents, agents for adjusting pH, water and bulking agents. Conventionally, bulking agents used in larval diet are wheat bran, carrot powder, wheat mill feed, wheat shorts, grain corncob, cane and beet bagasse (Vargas et al. 1983), soybean protein and tissue paper (Kakinohana and Yamagishi 1991). However, there are several limitations associated with the use of these bulking agents. These include, but are not limited to, the need for large storage space, waste management, variability in quality, microbial and pesticidal contamination (Hooper 1987) as well as issues related to availability and cost. This necessitated the search for an alternative to bulking agents in larval diets. The first liquid larval diet without a biological bulking agent was developed by Schroeder et al. (1971) for small-scale rearing of *Z. cucurbitae*.

In further efforts to replace the bulking agents with an inert, reusable sponge and to help overcome the limitations associated with solid-based diets, liquid diets that can be used for large-scale rearing have been developed for *C. capitata* (Chang et al. 2007; Chang 2009b), *B. dorsalis* (Chang et al. 2006; Chang and Vargas 2007; Chang 2009b; Khan et al. 2011) and *Z. cucurbitae* (Chang et al. 2004). The liquid diet for *B. dorsalis* developed by Chang et al. (2006) yielded high quality flies, as the adults reared on this diet were identical to those reared on the conventional mill feed diet in terms of fitness parameters as well as overall performance, paving the way for rearing of this species on liquid diet on a large scale. Following this breakthrough, the liquid diet technology developed has been transferred to several countries across the world for assessment for mass rearing of various fruit fly species (Chang 2009a). Amongst the 14 countries that participated in the liquid diet evaluation, three were from Africa. These were, Kenya (for the African populations of *B. dorsalis* and the mango fruit fly, *Ceratitis cosyra* [Walker]), Mauritius (for the peach fruit fly, *Bactrocera zonata* (Saunders) and *Z. cucurbitae*), and South Africa (for *C. capitata*). Both the Stellenbosch mass-rearing facility and the International Centre of Insect Physiology and Ecology (*icipe*) have adopted the technology for their respective fruit flies on which the liquid diet was evaluated.

In Africa, like other parts of the world, diet and its related nutritional components for laboratory reared fruit flies have improved in several phases. For example, at *icipe* colonies of six fruit flies species (*B. dorsalis, C. capitata, C. cosyra, Ceratitis anonae* Graham, *C. rosa* and *Ceratitis fasciventris* [Bezzi]) are currently in culture. They were initiated on their respective host fruits for 3–7 generations depending on the species. Thereafter, the colonies were maintained on solid carrot-based diet, which is a modification of the diet developed by Hooper (1978) (for more details see Ekesi and Mohamed 2011). The first three species were later adapted to rearing on liquid diet the ingredients of which are similar in composition to that described by Chang et al. (2006) for *B. dorsalis*; efforts are underway to adapt the remaining species on the same media.

The Stellenbosch fruit fly rearing facilities in South Africa are the largest in the continent; *C. capitata* is produced on a large scale of 13 million flies/week for the SIT program on grapes in the Hex river valley (Barnes et al. 2007). The colony has been maintained on solid-based larval diet using wheat bran as the bulking agent. However, the inconsistency of the quality of this bulking agent, which is also often contaminated with pesticides, resulted in the colony crashing (Chang 2009a) which prompted the management to evaluate use of the liquid diet developed by Chang et al. (2004) with the aim of future implementation (Chang 2009a).

2 Effect of Fruit Fly Diet on the Fitness Parameters of Different Developmental Stages

2.1 Adult Diet

2.1.1 Effect on Fecundity and Egg Hatchability

Adult diet quality has profound effects on the fecundity of fruit flies. For example, Hagen (1953) demonstrated that amino acids, carbohydrate, vitamins and certain minerals are essential for ovary development in *B. dorsalis*, *Z. cucurbitae* and *C. capitata*. Tsiropoulos (1977) demonstrated that vitamins were crucial for enhancing fertility of *B. oleae* and their absence in the diet led to oviposition of malformed eggs. In a separate study the same author reported that *B. oleae* fed on a vitamin-deficient diet had reduced fecundity and fertility (Tsiropoulos 1980). Interestingly, an excess of either biotin, pyridoxine, inositol or vitamin C also resulted in reduced fecundity in *B. oleae* (Tsiropoulos 1982). Also working with *B. oleae*, Zografou et al. (1998) reported that amino acid analogues affected fecundity and fertility of *B. oleae*. Similarly, addition of 0.25 % Vanderzant's vitamin mixture and 0.05 % cholesterol to the sucrose and yeast hydrolysate diet of the female *B. oleae* diet are thought to increase fecundity and inhibit the production of mottled eggs (George and Ruhm 1977).

Ferro and Zucoloto (1990) and Cangussu and Zucoloto (1997) reported that, although *C. capitata* females produced eggs when fed only sucrose, egg production was significantly enhanced when protein was also consumed. Also Harwood et al. (2013) demonstrated that the egg laying abilities of laboratory-reared *C. capitata* and *Z. cucurbitae* was delayed or suppressed by limiting access to dietary protein. The authors also demonstrated that access to protein at eclosion led to higher reproductive ability in both species.

In a study by Davies et al. (2005) where they varied the content of both yeast and sucrose in the adult diet of *C. capitata*, females laid significantly more eggs when maintained on the highest yeast diet (7.7 %) than when maintained on diets containing lower levels of yeast. Similarly, Manrakhan and Lux (2006) evaluated the effects of three natural food sources, varying in protein and sugar content on, amongst other traits, the reproductive behaviour and fecundity of three African fruit flies: *C. fasciventris*, *C. capitata* and *C. cosyra*. The authors found that females of the first two species had a higher frequency of oviposition when fed on a protein-rich diet, than those fed on a protein-poor diet. Net reproductive rate for these two species also varied with the diet type. In a recent study by Chang (2009b), who tested the effect of various yeasts on production and hatchability of eggs of *C. capitata*, *B. dorsalis* and *Z. curcurbitae*, found that egg production was influenced by yeast type, but that egg hatchability was not different from that of the control (conventional mill feed diet).

Working with *Z. cucurbitae*, Kaur and Srivastava (1994) demonstrated that the absence of essential amino acids negatively affected flies' sexual maturity and fecundity. Similar findings were reported for *C. capitata* by Chang et al. (2001) who found that flies fed on diets containing ten essential amino acids, eight non-essential amino acids or a combination of cholesterol, inositol and choline, produced significantly fewer eggs than flies fed on the control diet. The most prolific age for egg production by adults was 10-d-old when the greatest number of mature eggs were recorded in the ovaries. Also fecundity of *C. capitata* was significantly reduced by omission of 10 essential amino acids or all eight non-essential amino acids from the adult diet (Chang et al. 2004). Removal of arginine, histidine, isoleucine, leucine, lysine, threonine, tryptophan, methionine, or valine also significantly decreased *C. capitata* fecundity. In contrast, increasing the sugar content in the diet had no effect on egg production or hatchability (Chang et al. 2001).

2.1.2 Effect on Male Mating Success

Quality and quantity of food consumed by male fruit flies has a profound effect on male mating success (e.g. Blay and Yuval 1997; Yuval et al. 1998; Field and Yuval 1999; Kaspi et al. 2000; Shelly et al. 2005; Orankanok et al. 2013; Quilici et al. 2013; Haq et al. 2014). The effect of adult food on sexual signaling, an important indicator of mating success, has been extensively documented for both wild and mass-reared male *C. capitata* (Papadopoulos et al. 1998; Kaspi and Yuval 2000; Shelly et al. 2002; Shelly and Kennelly 2002; Shelly and McInnis 2003). Also Manrakhan and Lux (2006) reported that males of *C. fasciventris* and *C. capitata* fed on a protein-rich diet had a higher frequency of calling and mating than those fed on a protein-poor diet; however, diet quality did not influence the mating behaviour of *C. cosyra*. Diamantidis et al. (2008) reported that yeast hydrolysate significantly increased sexual signaling in four populations of *C. capitata*. In a laboratory study using males of a related fruit fly species, *C. rosa* Quilici et al. (2013) demonstrated that the addition of proteins to the adult diet increased mating competitiveness of males for both wild and laboratory reared flies compared to their counterparts fed on sugar only. The authors reported that males fed with a 'full' diet (sugar and hydrolysed yeast) accounted for 85 % of all matings compared with 15 % for those fed with a sugar-only diet. Similarly, *C. capitata* males fed on a high-protein diet achieved a greater number of copulations compared with males fed on a no-protein diet (Joachim-Bravo et al. 2009). More recently, Teal et al. (2013) found that an adult diet enriched with protein hydrolysate and an application of methoprene to adult males or pupae significantly advanced the age at which males of *Z. cucurbitae* become sexually mature and improved the overall reproductive success of the males of this species.

In *B. dorsalis*, immature males deprived of protein (1–12 days old) had very few matings (<5 % total matings), compared with immature males provided with protein

(Shelly et al. 2005). The authors further illustrated that males provided with protein as immature adults, but deprived of protein as mature adults (>12 days old), were competitively inferior to protein-fed males. Orankanok et al. (2013) reported that sterile *B. dorsalis* males fed on sugar–yeast hydrolysate combinations for 2 days post eclosion achieved significantly more matings than males fed only water.

Adult male diet is also an important determining factor in the length of the female sexual refractory period, an aspect that has significant implication in fruit flies management using SIT. For example, Blay and Yuval (1997) found that *C. capitata* females mated with protein-fed males are more likely to refrain from re-mating compared to females mated with protein-deprived males. Also working with *C. capitata*, Gavriel et al. (2009) reported that females mated with males fed on a protein-deficient diet had higher re-mating receptivity than females mated with protein-fed males. Also in a study on the effect of post-teneral nutrition on reproductive success of male *C. capitata* Yuval et al. (2002) found that females whose first mate was protein-deprived, remated sooner than females whose first mate was protein-fed. Similarly, Haq et al. (2014) working on *Z. cucurbitae* found that females mated with protein-deprived males showed higher re-mating receptivity than females first mated with protein-fed males. Contrary to the findings by the previous authors, Shelly and Kennelly (2002) reported that the inclusion of protein in the male diet of *C. capitata* had no apparent effect on female remating tendency.

Other aspects associated with male mating success, such as copula duration, sperm transfer, male participation in leks and male calling, were also found to be affected by male diet in *C. capitata* and other fruit fly species (Yuval et al. 1998; Field and Yuval 1999; Taylor et al. 2000). For instance, protein-fed *C. capitata* males were more likely to emit pheromone in the lek and consequently copulate more than protein-deprived males (Kaspi et al. 2000). Also Kaspi and Yuval (2000), working in field cages, found that protein-fed sterile males of *C. capitata* were significantly more likely to join leks and emit calling pheromone than sterile males fed only sugar.

2.1.3 Effect on Adult Longevity

Results on the effects of adult diet on adult survival are quite variable for different fruit fly species, and even for same species as reported by different authors. For example, in a study to assess the effect of adult diet on longevity of sterile males of *C. capitata*, Barry et al. (2007) found that males fed on diet containing hydrolyzed yeast and sucrose lived longer than those fed on diets containing either sucrose or only water. Also Davies et al. (2005) demonstrated that longevity of *C. capitata* varied with the concentration of the yeast in the adult diet. Similarly, Faria et al. (2008) found that incorporation of protein had a positive impact on laboratory survival of *C. capitata* males. In a study on longevity of *Z. cucurbitae* males fed on hydrolysed yeast and methoprene treatments, it was demonstrated that adult diet quality had a significant effect on survivorship, whereby males fed on the sugar-protein diet throughout showed highest survival compared with those fed on sugar only (Haq and Hendrichs 2013). In a different study, access to protein increased life expectancies of both *C. capitata* and *B. cucurbitae* (Harwood et al. 2013). The authors also emphasized the positive effect of access to protein immediately after eclosion on the longevity of both species. Also sugar concentration in the diet has been reported by Chang et al. (2001) to have a significant effect on *C. capitata* survival.

Kaspi and Yuval (2000) showed that post-teneral protein feeding by mass-reared sterile male *C. capitata*, improved sexual competitiveness but led to a shorter life span. In contrast, Shelly and Kennelly (2002) and Shelly and McInnis (2003) reported no significant difference in longevity of protein-fed and protein-deprived male *C. capitata*. In a separate study, Shelly et al. (2005) demonstrated similar effects for *B. dorsalis* males whereby, males fed on sugar only or sugar and protein and then only sugar had a comparable survival rate. Also Davies et al. (2005) reported no difference in longevity of *C. capitata* males fed on a sugar–protein diet compared with those fed on sugar only. Furthermore, protein intake had a differential effect on adult longevity in different populations of the same species. For instance, Placido-Silva et al. (2006) who studied the effects of different protein concentrations on longevity of two populations of *C. capitata* found that protein intake increased adult longevity of one population but not that of the other. For *B. oleae* survival of both sexes was significantly reduced by overdosing on biotin and pyridoxine with females being more affected (Tsiropoulos 1982).

2.2 Larval Diet

The larval stage is a very important stage of fruit fly growth and development as it dictates many fitness parameters of the adult flies. Therefore, high quality larval diet is crucial for production of healthy adult fruit flies. Many traits of fruit flies, such as adult reproductive success, larval mortality, larval development time, pupal recovery, pupal weight, adult size, egg hatchability and flight ability of the eclosing adults, are functions of various factors that act individually or in tandem. Although fruit flies are anautogenous, nutritional reserves obtained during the larval stage are one of the four factors known to influence various fitness parameters of the adult flies (e.g. Chang 2004; Chang et al. 2000; Nash and Chapman 2014 and reference therein).

2.2.1 Effect on Development of Immature Stages

Chang (2004) assessed the effect of inclusion or removal of amino acids from the larval diet on development of larvae and adults of *C. capitata*. Larval feeding on a diet deficient in ten exogenous essential amino acids (arginine, isoleucine, leucine, lysine, histidine, methionine, phenylalanine, threonine, tryptophan and valine) resulted in larval death. In the same study, larvae reared on diets that lacked all eight of the non-essential exogenous amino acids (alanine, aspartic acid, cystine, glutamic acid, glycine, proline, serine and tyrosine), or either glycine or serine survived. However, they had significantly delayed larval development. Similarly, Nestel et al. (2004) working with *C. capitata*, noted that removal of non-essential amino acids from larval meridic diets delayed larval and pupal development and also resulted in reduced pupal recovery. In a very recent work evaluating the effect of dietary components on the larval life history of *C. capitata*, Nash and Chapman (2014) found that reducing the protein content of larval diet by 40 % resulted in a significant increase in overall egg to adult mortality by up to 66 % in comparison with the standard baseline diet. They also demonstrated that addition of a novel protein source, casein (i.e. milk protein), to the larval diet increased larval mortality by up to 63 % and also lengthened the larval developmental time by 1.93 days in comparison with those larvae reared on the standard diet (mill feed). Significantly higher proportions of larvae reared on protein-rich diets survived to pupation than those reared on low-protein diets or diets amended with casein. Also the proportion of pupae surviving to adult eclosion was significantly lower when larvae were fed casein compared with larvae fed starch. Variation in carbohydrates had no significant effect on larval survival while it did have a significant effect on pupal survival. Larval and pupal development time was significantly longer when larvae were reared on casein. Carbohydrate diets also had no significant effect on the mean duration of larval development. In contrast, pupal developmental time varied amongst the different carbohydrate diets.

Chang (2009b) evaluated the effect on development of *B. dorsalis* of incorporating different yeasts and wheat germ oil in the larval diet. The three hydrolyzed brewer's yeasts evaluated were FNILS65, FNI200 and FNI210, one glutamine-enriched yeast powder (GSH), one vitamin-enriched yeast powder (RDA500), Korean yeast powder, whole cell yeasts, and various combinations of these treatments. In this study the author established that the type of yeast used had a significant effect on pupal recovery of *B. dorsalis*, which was significantly higher when larvae were fed FNI200 and FNIL65 compared with FNI210. In a similar study, Chang et al. (2007) evaluated several yeast-based products as ingredients in the liquid larval diet of *C. capitata*. They assessed the effect on larval duration, pupal recovery, pupal weight and other traits such as adult emergence, mating success, percentage of fliers, egg production of the subsequent generation and egg hatch rates. Larvae reared in a liquid diet with LBI2240:FNILS65 ratios of either 1:1 or 3:1 performed similarly to those reared on conventional mill feed-based control diets in term of pupal recovery. Recently, Ekesi et al. (2014) demonstrated that African populations of *B. dorsalis* had a higher pupal recovery and pupal weight when the larvae were reared on diet containing an imported Lallemand yeast than when the larvae were reared on a diet containing the local waste brewers' yeast. For *Bactrocera latifrons* (Hendel) it was found that incorporation of varying amounts of bell pepper in the larval diet, as a source of ascorbic acid, significantly increasing pupal recovery by up to 21 % (Chang and Kurashima 1999). Despite this the same authors reported that an addition of ascorbic acid phosphate (>15 mg/g of diet) to the larval diet had a negative effect on *B. latifrons* development and resulted in reduced pupal recovery and reduced pupal weight.

Other nutritional components of the larval diet that are essential for fruit fly development are vitamins. For example, Chang et al. (2001) found that the addition of vitamins improved larval development, pupal recovery and pupal weight in *C. capitata*. In a separate study Chang and Li (2004) demonstrated the positive effect of the addition of niacin and other B vitamins to diet on larval development of *C. capitata*.

Pupal weight is an important fitness parameter that determines adult size, which has a significant bearing on fly fertility, fecundity, longevity, flight ability and male mating success. In turn, pupal weight is affected by several factors; the most important is the quality of the larval diet. For instance, Kaspi et al. (2000) reported that protein-fed *C. capitata* males were heavier and emerged with more protein and lipid than protein-deprived males.

2.2.2 Effect on Adult Fitness Traits

Female fecundity and egg hatchability are important quality control parameters that are influenced by the quality and quantity of the larval diet. For example, Kaspi et al. (2002) investigated the effect of larval diets containing varying amounts of protein and sugar on the size, developmental time, nutritional status and reproductive maturation of *C. capitata*. The authors found that protein- and sugar-fed flies were larger in size, developed faster, and had more nutritional reserves than the protein-deprived flies. Furthermore, the protein-fed males became sexually active earlier than the protein-deprived males, while protein-fed females were more fertile than protein-deprived females. Chang (2009b), who tested the effect of various yeasts on production and hatchability of eggs of *C. capitata*, *B. dorsalis* and *Z. curcurbitae*, found that egg production was significantly influenced by the yeast type. Also Ekesi et al. (2014) found that adult emergence, fecundity and egg hatchability of African populations of *B. dorsalis*, were higher for flies from the larvae reared on diets containing an imported Lallemand yeast compared with those from diet containing the local waste brewers' yeast. However, flight ability was not affected by the yeast type. In a recent study, Nash and Chapman (2014) found that a 40 % reduction in the quantity of protein resulted in a significant increase (26.5 %) in the overall egg to adult mortality of *C. capitata* compared with the standard baseline diet. In contrast, Chang and Kurashima (1999) demonstrated a considerable increase in adult emergence of *B. latifrons* after incorporation of bell pepper in the larval diet.

Another important fitness parameter is flight ability, which is of vital significance in flies to be used in SIT programmes (Collins and Taylor 2010). One of the factors that determines the flight ability of fruit flies is the quality of the larval diet. For instance, in a study comparing different yeasts as ingredients of larval diet of *B. dorsalis*, Chang (2009b), found that flies from larvae reared on diet containing FNI200 and FNIL65 had significantly better flight ability and mating success than those reared on Korean yeast. Also the flight ability of *B. dorsalis* increased with the amount of wheat germ oil in the larval diet (Chang and Vargas 2007). The authors postulated that fatty acids and vitamin E in wheat germ oil are responsible for enhancing the flight ability. Similar effects of enhancement of flight ability as a result of inclusion of wheat germ oil in the larval diet has also been reported for the Mexican fruit fly, *Anastrepha ludens* (Loew) (Pascacio-Villafa et al. 2015). Likewise, *C. capitata* reared without fatty acids had compromised flight ability (Cho et al. 2013). Cho et al. (2013) evaluated two larval diets: a conventional mill feed diet and a fatty acid-deficient liquid diet, on the flight ability of *C. capitata*. The authors found that only 20.7 % of flies from larvae reared on the fatty acid-deficient diet displayed full flight ability. However, 97 % of those from larvae reared on mill feed diet displayed full flight ability. The authors speculated that the nutritional deficiency might have induced over-expression of the flightless-I protein (or fli-I gene) resulting in reduced numbers of flies with normal flight ability. In general, Chang and Coudron (2007) demonstrated that wheat germ oil influenced the stage-specific quality of several fruit fly proteins such as stress proteins, detoxification proteins and glutathione-related proteins. Additional reports by Chang et al. (2010) further substantiated the findings that one mechanism of wheat germ oil actions in insect nutrition is the modulation of gene expression. In addition to fatty acids, vitamins were also reported to enhance *C. capitata* adult emergence and flight ability (Chang et al. 2001; Chang and Li 2004).

Larval diet quality also had significant effects on immature development of *B. oleae*. Hanife (2008) reported that 77 % of *B. oleae* larvae reared on an agar-based diet completed development and achieved significantly higher pupal weight compared with those reared on the control cellulose diet.

3 Endosymbionts and Fruit Fly Nutrition

Many insect taxa harbour microbial organisms in their bodies that affect their biology, physiology, nutrition and reproduction. Endosymbiotic gut-associated microbial communities have long been believed to have mutualistic associations with different species of fruit flies (e.g. Petri 1910; Hagen et al. 1963; Buchner 1965; Hagen 1966; Drew et al. 1983; Manousis and Ellar 1988; Tsiropoulos 1989; Behar et al. 2009). These mutualistic associations could enhance the host immune system, or be nutritional, especially for insects that rely on an inadequate food source such as fruit flies. The first symbiotic association in Diptera was described for *B. oleae* (Petri 1904, 1905, 1906, 1907, cited in Tsiropoulos 1992). Later, Hagen et al. (1963) and Hagen (1966) proved that endosymbiotic bacteria have a significant role in fruit fly development. The authors found that *B. oleae* fed on a complete diet produced normal progeny, but addition of antibiotic to the diet resulted in larval mortality. Also using antibiotic in *B. oleae* diet and radiolabeled food Tsiropoulos (1989) was able to establish that four amino acids (alanine, hydroxyproline, proline and tyrosine), which are crucial for *B. oleae* development, were produced by the biosynthetic activities of their gut microflora. Also working with *B. oleae*, Ben Yosef et al.

(2010) tested the hypothesis that symbiotic bacteria contributed to the adult fly's fitness in a diet-dependent manner. When females were fed a diet containing non-essential amino acids as the sole source of amino nitrogen, egg production was significantly enhanced in the presence of bacteria. However, the presence of bacteria did not affect fecundity of adults fed the sucrose-poor diet, or the protein-rich diet. In the light of their results, the authors concluded that bacteria were able to compensate for the skewed amino acid composition of the diet. In a recent study on *B. oleae* Ben Yosef et al. (2014) proved that the predominant gut bacterium, *Candidatus Erwinia dacicola*, was an essential element in the nutritional ecology of this fruit fly species. They demonstrated that the presence of the bacteria significantly enhanced *B. oleae* egg production by contributing essential amino acids and metabolizing urea into an available nitrogen source. The authors also established that bacteria were beneficial to females relying on bird droppings as a food source, but not to those feeding on honeydew. They also highlighted the fact that the evolution of this symbiosis has allowed adult flies to utilize nutritionally unbalanced food in nature.

Another African fruit fly species associated with symbiotic bacteria is *C. capitata* (e.g. Marchini et al. 2002; Yuval et al. 2010 and reference therein; Hamden et al. 2013; Augustinos et al. 2015; Gavriel et al. 2011). Behar et al. (2008) studied the gut bacterial communities in *C. capitata* and their impact on the flies' longevity. The authors found that inoculations with different species in the Enterobacteriaceae increased *C. capitata* longevity. In contrast, longevity was reduced by inoculations with high levels of the pathogenic bacteria, *Pseudomonas aeruginosa* (Schröter) Migula. Based on their results, the authors suggested that the community of Enterobacteriaceae within the gut of *C. capitata* may, in addition to their positive impact on nitrogen and carbon metabolism, development and copulatory success, also have an indirect effect to host fitness by preventing the establishment or proliferation of pathogenic bacteria. Ben-Yosef et al. (2008) compared the mortality rates between antibiotic-treated *C. capitata* and non-treated *C. capitata* when they were either maintained on sugar, or on full diet. They reported that eliminating the gut bacterial population prolonged longevity, but only for the flies fed on sugar, indicating that the effect of bacteria on lifespan was diet dependent.

Ben Ami et al. (2010) demonstrated that the addition of the bacterium *Klebsiella oxytoca* (Flügge) Lautrop (the dominant species of gut bacterium of several tephritid species [Behar et al. 2008, and reference therein]) to the diets of male *C. capitata* being mass-reared for SIT, significantly improved their copulatory performance. They also found that the addition of the bacteria to the post-irradiation diet not only enhanced the level of beneficial bacterial communities within the flies gut, but also resulted in decreased levels of the potentially pathogenic genus *Pseudomonas*.

Gavriel et al. (2011) also tested the effect of diets enriched with *K. oxytoca*, on sexual performance of sterile *C. capitata* males. They established that enriching the sterile male diet with this bacterium considerably improved mating competitiveness in the laboratory as well as in field cages. They also reported that sterile males fed on bacteria-enriched diet had longer life spans and were able to inhibit female remating receptivity more efficiently. Enhancement of sexual performance in male

C. capitata (the Vienna-8 strain) at emergence, as well as increases in male size when mass-reared on a larval diet enriched with *Klebsiella pneumonia* (Schröter) Trevisan, *Enterobacter* spp., or *Citrobacter freundii* (Braak) Werkman and Gillen have also been reported by Hamden et al. (2013).

In a very recent study Augustinos et al. (2015) reported that incorporation of *Enterobacter* sp. as a supplement to the larval diet of *C. capitata* resulted in reduced rearing duration as well as an improvement of both pupal and adult productivity without affecting other fitness traits such as pupal weight, sex ratio, male mating competitiveness, flight ability and longevity under starvation. Yuval et al. (2010) provides a review of recent studies on the effects of endosymbiotic bacteria on fitness of *C. capitata*.

Similar mutualistic associations of bacteria have been reported for non-African-tephritids. For example, as early as 1983 Drew et al. (1983) found that diets of bacteria, sugar and water resulted in increased fecundity of the Queensland fruit fly, *Bactrocera tryoni* (Froggatt), compared with those fed on the conventional diet of autolyzed waste brewer's yeast, sugar and water. The author concluded that the type and abundance of bacteria on leaves and fruit surfaces have an important role in the life history of tropical fruit flies. In a recent study to determine the gut bacterial community in *Bactrocera tau* (Walker) and their effect on fecundity of this species, Khan et al. (2014) identified eight genera and nine species of bacteria in the family Enterobacteriaceae. They further established that *B. tau* females fed on a protein diet supplemented with either *Proteus rettgeri* (Rustigian and Stuart) or *K. oxytoca* had a considerably higher mean number of ovarioles per ovary compared with those fed on a protein diet only.

Ben-Yosef et al. (2015) presented a different aspect of the mutualistic association between bacteria and fruit flies. The authors demonstrated that the bacterium *C. Erwinia dacicola* enabled *B. oleae* larvae to overcome host-plant defences when developing in unripe olive fruits. They suggested that the bacterium counteracted the effect of oleuropein found in unripe olives that confer immunity against fruit fly infestation.

4 Quality Control Parameters and Recording

The concept of quality control is relevant to all kinds of production programmes, regardless of the facilities used. In general, it is defined as 'The sum of all attributes deemed necessary or desirable to achieve a stated objective or expected function' (Boller and Chambers 1977).

Quality control provides a means of optimizing insect mass rearing by identifying and gradually correcting deficient production processes, thereby preserving the genetic variability of the strain (Leppla and Ashley 1989). Therefore, quality control procedures involve development, colonization, maintenance and various other processes that affect the production and use of insects for pest management purposes. Leppla and Ashley (1989) categorized quality control of mass-reared insects into three main interrelated elements: (1) Production quality control which manages the consistency, reliability and timeliness of the production output; (2) Process quality control assures the performance of the production process so that unacceptable deviations do not occur in product quality; and (3) Product quality control regulates the conformity of the product to acceptable standards of quality and predicts the effectiveness of the product in performing its intended function. Aspects of mass-rearing environments that directly affect the quality of fruit flies include artificial larval and adult diets as well as rearing conditions (light intensity, photoperiod, insect density, temperature, humidity etc.), which influence quality control parameters such as body size, survival, pupal recovery, pupal weight, percent adult emergence, percent survival, longevity, flight ability, fecundity, fertility, percent egg hatch and mating ability of the reared insects (Calkins 1989; Vargas 1989; Walker et al. 1997; FAO/IAEA/USDA 2003). Apart from *C. capitata* and *B. dorsalis*, which have been reared for decades in different parts of the world (Vargas 1989) with great levels of success, the mass rearing procedures for the majority of economically important African fruit fly species, such as *C. cosyra*, *C. fasciventris* and *C. rosa*, are little known and documented.

In this regard, biological parameters are evaluated to identify possible deficiencies and to predict insect quality. Production can immediately be improved through testing of key and sensitive parameters and feedback mechanisms. Tests should be practical, uncomplicated, efficient and reproducible. A minimum number of parameters using the smallest sample size are recommended. In fruit fly mass rearing, important parameters include pupal recovery, pupal weight, percent adult emergence, percent survival, flight ability, percent fecundity and percent egg hatch (Calkins 1989; Walker et al. 1997; FAO/IAEA/USDA 2003). Walker et al. (1997) noted that there are variations between individual fruit fly species in any country and within the same species from different countries, but that there is a range that indicates that a colony is healthy. At the *icipe* facility, quality assurance is based on the following parameters: (1) pupal recovery from the number of eggs seeded should be >60%; (2) pupal weight, using measurements of 100 pupae of the same age, should be consistent; (3) adult emergence should be >70%; (4) percent fliers should be >80% and (5) percent egg hatch, using records from 100 eggs, should be >70%. The production and quality assurance in the Stellenbosch *C. capitata* SIT rearing facility in South Africa largely follow the guidelines developed by FAO/IAEA/USDA (2003) which was updated in 2014 (FAO/IAEA/USDA 2014).

4.1 Quality Control of Fruit Flies Reared on Solid Larval Diets

On solid diets, Khan (2013) demonstrated that the parental egg hatch of the fruit fly, *B. tryoni* was significantly lower when reared on carrot diet (75%) and lucerne diet (70%) compared to Hooper's bran diet (82%) and Vargas' bran diet (80%). The F_1

egg hatch follows the same trend observed above on the different diets, with flies reared on Hooper's bran diet achieving the highest egg hatchability (83 %) followed by the Vargas' bran diet (82 %). The developmental duration of the larval stages was was not significantly different on the four different diets (Khan 2013). The larval duration on these diets ranged between 10 and 11 days. The mean pupal weight was significantly greater for pupae recovered from Hooper's bran diet (10 mg) and the lucerne diet (10 mg) than from Vargas' bran diet (8 mg). The Hooper's bran diet (1260 ± 97.1) showed mean pupal yields of approximately double those found from lucerne (658 ± 16.2) and carrot (674 ± 150.8) diets. Despite the high variability, there were no significant differences between the different diets. However, there were substantial differences amongst the diets in pupal recovery. The percentage pupal recovery was not significantly different on Hooper's bran diet and Vargas' bran diet. Although, the pupal recovery was lowest on lucerne and carrot diets, no statistically significant difference were reported. According to the studies conducted by Khan (2013), a greater proportion of flies emerging from pupae reared on carrot diet were males compared with those from pupae reared on any of the bran diets (Hooper's bran diet and Vargas' bran diet), which all had sex ratios closer to 50:50. The mean sex ratio of adults emerging from the lucerne diet was intermediate but it was not significantly different from the bran diets. This observation is supported by Khan (2013) who indicated that it is not uncommon to find that the proportion of each sex surviving to pupation differs on different diets. Adult emergence was reasonably good when reared on the four solid diets ranging from >90 % for the lucerne and carrot diets to <70 % for Vargas' bran diet. Adult emergence was poorest (67 %) on Vargas' bran diet, being significantly lower than any of the other diets. The percentage fliers, which is the number of pupae that either failed to emerge or failed to fly following eclosion, were significantly lower for pupae reared on Hooper's bran diet and Vargas' bran diet than on the lucerne and carrot diets, which had high adult emergence. Rate of fliers, was highest on the lucerne diet (61 %) compared with Hooper's bran diet, Vargas' bran diet and carrot diet. Higher mean egg production per female per day was observed for adult flies from larvae that had been reared on Hooper's bran diet and Vargas' bran diet than for adult flies from larvae that had been reared on lucerne diet or carrot diet. Egg latency (the period between adult emergence and the first egg being laid by a cohort of flies) was significantly shorter for flies reared on the lucerne diet than for those reared on the carrot diet, Hooper's bran diet or Vargas' bran diet. Overall, when compared with other recent studies (Collins et al. 2008; Collins and Taylor 2010) and the recommended standards from the FAO/IAEA/USDA (2003), percentage fliers and rate of fliers in the studies carried out by Khan (2013) on solid diet were low. According to Collins and Taylor (2010) and Collins et al. (2008), this may be due, in part, to the lower light intensity used which may have reduced flight ability.

Ekesi and Mohamed (2011) explored several bulking compounds including wheat, carrot, boiled cassava, sugarcane bagasse for mass-rearing *B. dorsalis*, *C. fasciventris* and *C. rosa*, and carrot supplemented with mango powder for mass-rearing *C. cosyra*. They found that, for *B. dorsalis*, pupal recovery was generally greatest when they were reared on wheat (65 %) followed by sugarcane bagasse

(55%) and carrot-based (54%) diets, with the smallest recoveries in the cassava-based diet (32%). Pupal weight was not significantly different amongst carrot, wheat and sugarcane bagasse-reared flies and ranged between 13.6 and 14.1 mg. However, pupae reared on the cassava-based diet were significantly lighter (12.5 mg) compared with the other three media. Adult emergence was significantly higher on carrot and wheat-based diets (80–86%) compared with the sugarcane bagasse-based diet (70–77%). Fecundity over a 10 days period was also significantly higher on the carrot, wheat and sugarcane bagasse diets (342–366 eggs per female) than on the cassava-based diet (233 eggs per female). For *C. fasciventris* and *C. rosa*, all quality control parameters from carrot and wheat diets outperformed those from sugarcane bagasse and cassava diets. For *C. cosyra*, diets of carrot, carrot supplemented with mango and sugarcane bagasse were used. The study showed that supplementing carrot with mango powder significantly increased pupal weight, egg production and egg hatchability compared with the other diets. The authors recorded a pupal weight of 10.5–11.7 mg when reared on carrot supplemented with mango powder, 9.2 mg on carrot alone and 7.2 mg on sugarcane bagasse. In fruit fly mass rearing, high pupal weight is a desirable characteristic in the production process, as it is a good indicator of the ultimate body size of the eclosing adult. Churchill-Stanland et al. (1986) found that the size of adult *C. capitata* was important for mating success and noted that 8 and 9 mg insects achieved greatest mating success than those that weighed 6 and 4 mg. Fly size is also a determinant of insect fertility and fecundity. Overall, they concluded that quality control parameters from carrot, carrot supplemented with mango and wheat diets were superior for mass rearing of *B. dorsalis*, *C. fasciventris*, *C. rosa* and *C. cosyra* compared with diets based on sugarcane bagasse or boiled cassava.

4.2 Quality Control of Fruit Flies Reared on Liquid Larval Diets

Studies were done by Khan (2013) on *B. tryoni* in three liquid diets: Fay's liquid starter diet (Fay 1989), Chang's 2006 liquid diet (Chang et al. 2006) and Chang's 2007 liquid diet (Chang et al. 2007). Egg hatch was significantly higher for adults reared on Chang's 2006 liquid diet than the other diets and it was similar for adults reared on Fay's starter diet or Chang's 2007 liquid diet. Larval developmental time was significantly (12 days) longer for larvae reared on Chang's 2007 liquid diet than for larvae reared on Chang's 2006 liquid diet or Fay's starter diet. Mean pupal weight and mean pupal yields were similar for all three liquid diets. However, the percentage pupal recovery was significantly higher on Chang's 2006 diet than either Chang's 2007 liquid diet or and Fay's starter diet. From the flies emerging from pupae reared on Fay's starter diet and Chang's 2007 liquid diet, a greater proportion of females were observed in comparison with the number eclosing from pupae reared on Chang's 2006 liquid diet. Also adult emergence was very poor: 50% for

Fay's starter diet, being significantly lower than Chang's 2007 liquid diet (72 %) and Chang's 2006 liquid diet (83 %). Percentage fliers was significantly lower for individuals reared on Fay's starter diet compared with Chang's 2007 liquid diet and Chang's 2006 liquid diet. The lowest adult emergence, flight ability and eggs/female/day were recorded for individuals reared on Fay's starter diet. The percentage egg hatch of the F1 generation was highest for flies reared on Chang's 2006 liquid diet (87 %), followed by Chang's 2007 liquid diet (84 %). These quality control parameters are inferior to those reported by Collins et al. (2008), Collins and Taylor (2010) and the recommended standards from the FAO/IAEA/USDA (2003). The absence of adequate overhead lighting was found to be the major cause of the low values recorded (Collins et al. 2008; Collins and Taylor 2010).

Ekesi and Mohamed (2011) used liquid diet with ingredients that were similar in composition to those described by Chang et al. (2006) for larval rearing of four fruit fly species: *B. dorsalis*, *C. fasciventris*, *C. rosa* and *C. cosyra*. The authors found that *B. dorsalis* reared on this liquid diet had a 60 % percent pupal recovery rate, 14 mg mean pupal weight, 92 % percent adult emergence rate, a mean of 214 eggs laid over a 10-days period (fecundity), a 72 % F_1 egg hatch rate (egg fertility) and a flight ability or flier rate of 82 %. For *C. fasciventris* the mean values recorded were 28 % pupal recovery, 6 mg pupal weight, 65.8 eggs (fecundity, 10 days), 70 % adult emergence, 65 % egg fertility and flight ability (78 %), while for *C. rosa* and *C. cosyra* the mean values recorded were inferior for all these parameters in the liquid-based diet (Ekesi and Mohamed 2011). This was unexpected, given that the nutritional content of the liquid diet was quite high and had previously been found suitable for the development of other *Ceratitis* species such as *C. capitata* (Chang et al. 2007). Since fruit fly adaptation to artificial diets varies with species (Souza et al. 1988; Tsitsipis 1983; Kamikado et al. 1987), it is likely that the *Ceratitis* species in the study conducted by Ekesi and Mohamed (2011) would require a prolonged period of adaptation to the liquid diet to achieve the recommended quality control parameters. This is in accordance with other studies demonstrating that insect adaptation to artificial diet varies with the species and that at least ten generations were required for *C. capitata* to adapt to an artificial diet (Souza et al. 1988). About three or four generations were required for *B. oleae* to adapt (Tsitsipis 1983) and *Z. cucurbitae* required 14 generations to reach a permanent plateau (Kamikado et al. 1987).

Khan et al. (2011) also compared three different liquid diets with the aim of identifying easily available and low-cost protein sources as ingredients for mass production of larval *B. dorsalis*. There were no significant differences in quality control parameters such as total number of pupae produced (3350), larval duration (7 days) male: female ratio (51:49) and percentage egg hatch (88.5 %) of *B. dorsalis* reared on the control liquid diet and the modified diets. Furthermore, the pupal density (0.6), pupal weight (12.5 mg), percentage adult emergence (98 %) and percent fliers (79 %) were almost the same amongst the control diet and the modified diets. The authors concluded that liquid diet containing baking yeast, soy bran and soy protein (2:1:1) used as sources of protein was highly promising for mass rearing of *B. dorsalis* under laboratory conditions. According to the authors, the advantage of this liquid diet is that obviates the need to use a starter diet as described in Fay and Wornoayporn (2002) and a bulking agent, thus saving on labour costs and storage space. Liquid based diets with recyclable substrate systems are cost effective ways to produce high quality mass-produced tephritid fruit flies.

4.2.1 Effects of Yeast in Liquid Diets

Ekesi and Mohamed (2011) compared two yeast types in liquid diet and found that pupal recovery (62 %), pupal weight (13 mg), adult emergence (88 %) fecundity (no. eggs/female/10 days = 368 eggs) and egg hatch (70 %) of *B. dorsalis* reared on diet containing Lallemand yeast was significantly higher than on the diet with the local waste brewers' yeast, though the type of yeast did not affect the number of fliers (81 and 80 % for Lallemand and waste brewers' yeast, respectively). According to the authors, the inferior quality control parameters recorded for the waste brewers' yeast could be attributed to the high protein content of the local waste brewers' yeast that might have been detrimental to the development of *B. dorsalis*. The authors recommended that the quantity of local waste brewers' yeast in liquid diet should be reduced and the impact of this reduction be assessed on *B. dorsalis* production.

4.3 *Quality Control of Fruit Flies Reared on Host Fruit Larval Diets*

Via (1986) reported a positive correlation between preference and offspring performance of fruit flies on different host plants in nature suggesting that these insects have the ability to choose the host plant on which their offspring develop best and fastest. For example, Krainacker et al. (1987) studied the quality control parameters for *C. capitata* reared on 24 different hosts and reported larval development times between 6.9 days for larvae reared on tomato and 11.7 days for those reared on grape (*Vitis vinifera* L.). Carey (1984) and Rivnay (1950) obtained similar results for *C. capitata* reared on many of these same hosts. In contrast, when *C. capitata* was reared on apple (*Malus domestica* Borkh.), which is an unfavourable host, almost all the eggs hatched within a period of 2 days but the larval developmental duration took approximately 18 days and pupal duration took 10 days (Papadopoulos et al. 2002). The developmental time parameters reported by Papadopoulos et al. (2002) were found to be longer than those reported by Krainacker et al. (1987) and Carey (1984) on the fruits of several other hosts, such as blackberry (*Rubus rubrisetus* [Rydb]), cherry (*Prunus avium* [L.]), plum (*Prunus americana* Marsh.), mango (Mangifera indica [L.]), blueberry (Vaccinium corymbosum [L.]) and raspberry (Rubus idaeus [L.]). The high larval mortality as well as the long developmental period on the hosts described above indicates that these hosts were not favourable hosts for C. capitata, probably because of their nutritional elements, the texture of the flesh, and secondary compounds, that are all factors that determine the suitability of a host for development (Krainacker et al. 1987; Zucoloto 1993; Kaspi et al. 2002).

According to Krainacker et al. (1987) C. capitata larval survivorship ranged from 1 % for those reared on apricot (Prunus armeniaca [L.]) and papaya (Carica papaya L.) to 68 % for those reared on blackberry. These values are generally below those reported by Carey (1984) and Papadopoulos et al. (2002) with immature survival rates of >80 % for both the eggs and pupae while the larval survival rate was relative low (>40 %). Pupal survivorship reported by Krainacker et al. (1987) was 59 to 96 %, which is in agreement with the figures reported by Carey (1984). Krainacker et al. (1987) also found that the average pupal size of C. capitata when reared on various host fruits ranged from 1.71 mm (e.g. on tomato) to 1.97 mm (e.g. on lychee [Litchi chinensis Sonn.]). Pupae with the largest mean diameters were recovered from flies reared on fruit from from rutaceous hosts.

Arita and Kaneshiro (1988) compared the mating success of male C. capitata raised on two different host fruits. When competing for mates in the same arena they found that, although significantly smaller in size, males emerging from coffee (Coffea arabica [L.]) copulated more frequently than males emerging from cherry. This observation was further supported by Whittier et al. (1994) and Kaspi et al. (2000) who reported that male copulatory success in C. capitata was not due to their body size, as protein-fed males were more likely to start calling earlier, and, consequently more likely to copulate than protein-deprived males.

Gross fecundity of females reared on the fruits of 24 different hosts by Krainacker et al. (1987), ranged between 490 and 690 eggs/female while net fecundity from half of the hosts was between 350 and 450 eggs/female. The gross fecundity rates are similar to those reported by Shoukry and Hafez (1979) at equivalent temperatures, but are less than those reported by Carey (1984) at 25 °C. This difference may have been due to temperature differences between studies (Rivnay 1950).

Krainacker et al. (1987) reported an average life span of male and female C. capitata of 60 and 50 days, respectively. This is contrary to the figures reported by Papadopoulos et al. (2002), who found that the greatest longevity of male and female C. capitata was 142 and 91 days, respectively. This variation in life expectancy between the two sexes has been attributed to reproductive cost, which is higher in females than in males, hormonal differences, and other behavioural and physiological differences between the two sexes (Vargas and Carey 1989; Carey et al. 1995). This is also in accordance with the report by Bozzini and de Murtas (1975), Rossler (1975) and Shoukry and Hafez (1979).

Laboratory studies of B. zonata on fruits of six host species revealed that guava (Psidium guajava L.) was the most preferred host with pupal recoveries of 434 pupae/fruit, followed by ber (Ziziphus mauritiana Lamk; 177 pupae/fruit), banana (Musa sp.; 120 pupae/fruit), apple (13 pupae/fruit), chikoo (Manikara zapota [L.] Royen; eight pupae/fruit) and citrus (Citrus sp.; five pupae/fruit) (Rauf et al. 2013). Additional studies conducted by Sarwar et al. (2013) on mango, peach (Prunus persica [L.] Baksch) and apple fruits revealed that mango was the most preferred host followed by peach and apple, based on the mean number of pupae (173.2, 150 and ten, respectively) formed. The pupal weight varied significantly across the different host fruits with pupae recovered from mango weighing 6.40 mg, from peach 6.3 mg and from apple 6.1 mg. The percentage emergence of B. zonata was 84.5%, 81.1% and 75.1% on mango, peach and apple, respectively. The results support the observation that oviposition depends upon the decision to select an appropriate host that can support the development of their offspring (Fontellas-Brandalha and Zucoloto 2004; Joachim-Bravo et al. 2001).

Mir et al. (2014) assessed the quality control parameters of Z. cucurbitae when reared on cucumber as a natural host source. The developmental duration of the eggs was 12–24 h with a mean of 16.8 ± 6.2 h. These results are in agreement with those reported by Waseem et al. (2012) and Khan et al. (1993) who reported an incubation period of 24.4 to 38 h on cucumber. Shivarkar and Dumbre (1985) reported an incubation period of 1.2 days on watermelon (Citrullus lanatus [Thunb.] Matsum and Nakai), while Koul and Bhagat (1994) recorded an incubation period of 1.0–5.1 days when Z. cucurbitae was reared on bottle ground (Lagernaria sicer-aria [Molina] Standl.). According to Mir et al. (2014) the mean larval developmental period was 4.5 ± 1.1 days. This is slightly lower compared with studies conducted by several other authors who reported larval developmental periods of 5–22 days (Renjhen 1949), 5–11 days (Singh and Teotia 1970), 3–8 days (Doharey 1983) and 15 days (Shivarkar and Dumbre 1985). The pupal duration reported by Mir et al. (2014) varied between 8 and 9 days with a mean of 8.4 ± 0.51 days. This is consistent with earlier reports on the pupal period of Z. cucurbitae by Narayanan and Batra (1960), Agarwal et al. (1987), Dhillon et al. (2005), Shivayya et al. (2007), Waseem et al. (2012) and Langar et al. (2013) who all reported pupal periods of 8–9 days on cucumber. According to Mir et al. (2014), the copulation period of Z. cucurbitae varied between 2 and 4 h, which is in accordance with the reports by Vishva (2005), Shivayya et al. (2007) and Waseem et al. (2012), who found that a mating period of more than 30 min was required for sperm transfer to occur, and that the amount of sperm transferred increased progressively for up to 4 h. The pre-oviposition period was 12.4 ± 2.4 days and varied from 10 to 15 days, whereas the oviposition period was 18.2 ± 5.61 days and ranged from 12 to 28 days. This is consistent with the findings of Hollingsworth et al. (1997), Khan et al. (1993), Koul and Bhagat (1994) and Langar et al. (2013). Fecundity of sexually mature adult female Z. cucurbitae range between 58 and 92 eggs with a mean of 75.8 ± 12.5 while egg hatchability was found to be $86.1 \pm 0.5\%$ (Mir et al. 2014). These findings are in agreement with those of Atwal (1986) and Langar et al. (2013) who recorded 58–95 and 50–91 eggs per female during her entire life span, respectively.

Additional studies conducted by Sarwar et al. (2013) with Z. cucurbitae on four different vegetable crops: bitter gourd (Momordica charantia L.), aubergine (Solanum melongena L.), muskmelon (Cucumis melo L) and pumpkin (Cucurbita pepo L.) indicated that pupal recovery was generally greatest from bitter gourd (134.1) followed by aubergine (8.3) and least from muskmelon (3.8) and pumpkin (3.8). Pupal weight was also significantly different, ranging between 4.6 mg for individuals reared on pumpkin to 4.19 mg for individuals reared on bitter gourd.

Adult emergence was significantly higher from bitter gourd (82.6%) compared with from aubergine (73.7%), muskmelon (67.2%) and pumpkin (56.4%). A mean total of 110.8 adults emerged from pupae formed when Z. cucurbitae was reared on bitter gourd which was significantly more than when reared on aubergine, muskmelon and pumpkin (6.1, 2.4 and 2.2, respectively).

Khan et al. (2011) also conducted a similar study aimed at understanding the relative suitability of fruits from different host plants for performance of Z. cucurbitae in terms of pupal number and subsequent adult emergence. More pupae developed when larvae were reared on bitter gourd (202) than cucumber (Cucumis sativus [L.]; 193), sponge gourd (Luffa cylindsica L.; 179), aubergine (Solanum melongena [L.]; 124), sweet gourd (Cucurbita maxima [D.]; 118), bottle gourd (Lagenaria siceraria [Mol.] Standl; 115), pointed gourd (Trichosanthes dioica [Roxb.]; 92) and ash gourd (Benincasa hispida [Thunb.]; 64). The fewest pupae developed when larvae were reared on tomato (Solanum lycopersicum [L.]; 49). The percentage emergence of Z. cucurbitae across the different host species was significantly different and ranged from 9.2% on sponge gourd to 85.1% on pointed gourd, 85.4% on ash gourd and 87.0% on bottle gourd. Few studies have been done on the suitability of different host plants for rearing of Zeugodacus species. These findings shed light on the necessity for further research to collect extensive data on demographic patterns and niche differentiation of these polyphagous, invasive fruit flies.

References

Agarwal ML, Sharma DD, Rahman O (1987) Melon fruit fly and its control. Indian Hortic 32:10–11

Arita LH, Kaneshiro KY (1988) Body size and differential mating success between males of two populations of the Mediterranean fruit fly. Pac Sci 42:173–177

Atwal SN (1986) Agriculture pests of India and South East Asia. Kalyani Publication, New Delhi, p 529

Augustinos AA, Kyritsis GA, Papadopoulos NT, Abd-Alla AMM, Cáceres C, Bourtzis K (2015) Exploitation of the Medfly Gut Microbiota for the enhancement of sterile insect technique: use of Enterobacter sp. in Larval diet-based Probiotic applications. PLoS One 10:e0136459. doi:10.1371/journal.pone.0136459

Barnes BN, Rosenberg S, Arnolds L, Johnson J (2007) Production and quality assurance in the SIT Africa Mediterranean fruit fly (Diptera: Tephritidae) rearing facility in South Africa. Fla Entomol 90:41–52

Barry JD, Opp SB, Dragolovich J, Morse JG (2007) Effect of adult diet on Longevity of sterile Mediterranean fruit flies (Diptera: Tephritidae). Fla Entomol 90:650–655

Bateman MA (1972) The ecology of fruit flies. Annu Rev Entomol 17:493–518

Behar A, Yuval B, Jurkevitch E (2008) Gut bacterial communities in the Mediterranean fruit fly (Ceratitis capitata) and their impact on host longevity. J Insect Physiol 54:1377–1383

Behar A, Ben-Yosef M, Lauzon CR, Yuval B, Jurkevitch E (2009) Structure and function of the bacterial community associated with the Mediterranean fruit fly. In: Bourtzis K, Miller TA (eds) Insect symbiosis, vol 3. CRC Press, Boca Raton, pp 251–272

Ben Ami E, Yuval B, Jurkevitch E (2010) Manipulation of the microbiota of mass-reared Mediterranean fruit flies Ceratitis capitata (Diptera: Tephritidae) improves sterile male sexual performance. ISME J 4:28–37

Ben-Yosef M, Behar A, Jurkevitch E, Yuval B (2008) Bacteria-diet interactions affect longevity in the medfly, Ceratitis capitata. J Appl Entomol 132:690–694

Ben-Yosef M, Aharon Y, Jurkevitch E, Yuval B (2010) Give us the tools and we will do the job: symbiotic bacteria affect olive fly fitness in a diet-dependent fashion. Proc Roy Soc B Biol Sci 277:1545–1552

Ben-Yosef M, Pasternak Z, Jurkevitch E, Yuval B (2014) Symbiotic bacteria enable olive flies (Bactrocera oleae) to exploit intractable sources of nitrogen. J Evol Biol 27:2695–2705

Ben-Yosef M, Pasternak Z, Jurkevitch E, Yuval B (2015) Symbiotic bacteria enable olive fly larvae to overcome host defences. Roy Soc Open Sci 2:150170

Blay S, Yuval B (1997) Nutritional correlates of reproductive success of male Mediterranean fruit flies (Diptera: Tephritidae). Anim Behav 54:59–66

Boller EF, Chambers DL (1977) Concepts and approaches. In: Boller EF, Chambers DL (ed) Quality control: an idea book for fruit fly workers. International Organization for Biological Control of Noxious Animals and Plants Bulletin SROP-WPRS, pp 4–13

Bozzini A, de Murtas ID (1975) Insect pest control and insect breeding. Redia 56:473–487

Buchner P (1965) Endosymbiosis of animals with plant microorganisms. Interscience Publishers, New York, p 909

Calkins CO (1989) Quality control. In: Robinson AS, Hooper G (eds) Fruit flies: their biology, natural enemies and control, vol 3B. Elsevier, Amsterdam, pp 153–165

Cangussu JA, Zucoloto FS (1997) Effect of protein sources on fecundity, food acceptance, and sexual choice by Ceratitis capitata (Diptera, Tephritidae). Rev Brasil Biol 57:611–618

Carey JR (1984) Host-specific demographic studies of the Mediterranean fruit fly Ceratitis capitata. Ecol Entomol 9:261–270

Carey JR, Liedo P, Orozco D, Tatar M, Vaupel JW (1995) A male-female longevity paradox in medfly cohorts. JAE 64:107–116

Chang CL (2004) Effect of amino acids on larvae and adults of Ceratitis capitata (Wiedemann) (Diptera:Tephritidae) (2004). Anns Entomol Soc Am 97:529–535

Chang CL (2009a) Fruit fly liquid larval diet technology transfer and update. J Appl Entomol 133:164–173

Chang CL (2009b) Evaluation of yeasts and yeast products in larval and adult diets for the oriental fruit fly, Bactrocera dorsalis, and adult diets for the medfly, Ceratitis capitata, and the melon fly, Bactrocera cucurbitae. J Insect Sci 9:1–9

Chang CL, Coudron TA (2007) Improving fruit fly nutrition and performance through proteomics. International Organization for Biological Control of Noxious Animals and Plants (IOBC). In: Proceedings of 11th meeting of the working group Arthropod Mass Rearing and Quality Control (AMRQC). Bulletin IOBC Global No. 3, pp 24–27

Chang CL, Kurashima R (1999) Effect of ascorbic acid-rich bell pepper on development of Bactrocera latifrons (Diptera: Tephritidae). J Econ Entomol 92:1108–1112

Chang CL, Li QX (2004) Dosage effects between dietary Niacin and other B vitamins on larval development of Ceratitis capitata (Diptera: Tephritidae). Anns Entomol Soc Am 97:536–540

Chang CL, Vargas RI (2007) Wheat germ oil and its effects on a liquid larval rearing diet for Oriental fruit flies (Diptera: Tephritidae). J Econ Entomol 100:322–326

Chang CL, Kurashima R, Albrecht CP (2000) Effect of limiting concentrations of growth factors in mass rearing diets for Ceratitis capitata larvae (Diptera:Tephritidae). Anns Entomol Soc Am 93:898–903

Chang CL, Kurashima R, Albrecht CP (2001) Larval development of Ceratitis capitata (Diptera: Tephritidae) on a meridic diet. Anns Entomol Soc Am 94:433–437

Chang CL, Caceres C, Jang E (2004) A novel liquid larval diet and its rearing system for melon fly, Bactrocera cucurbitae (Diptera: Tephritidae). Anns Entomol Soc Am 97:524–528

Chang CL, Vargas RI, Caceres C, Jang E, Cho IK (2006) Development and assessment of a liquid larval diet for Bactrocera dorsalis (Diptera: Tephritidae). Anns Entomol Soc Am 99:1191–1198

Chang CL, Caceres C, Ekesi S (2007) Life history parameters of Ceratitis capitata (Diptera: Tephritidae) reared on liquid diets. Anns Entomol Soc Am 100:900–906

Chang CL, Coudron TA, Goodman C, Stanley D, An SH, Song QS (2010) Wheat germ oil in larval diet influences gene expression in adult oriental fruit fly. J Insect Physiol 56:356–365

Cho IK, Chang CL, Li QX (2013) Diet-induced over-expression of flightless-I protein and its relation to flightlessness in Mediterranean fruit fly, Ceratitis capitata. PLoS One 8:e81099

Christenson LD, Foote RH (1960) Biology of fruit flies. Annu Rev Entomol 5:171–191

Churchill-Stanland C, Standland R, Wong TT, Tanaka N, McInnis DO, Dowell RV (1986) Size as a factor in the mating propensity of Mediterranean fruit flies, Ceratitis capitata (Diptera: Tephritidae), in the laboratory. J Econ Entomol 79:614–619

Collins SR, Taylor PW (2010) Flight ability procedures for mass-reared Queensland fruit flies, Bactrocera tryoni: an assessment of some variations. Entomol Exp et Appl 136:308–311

Collins SR, Weldon CW, Banos C, Taylor PW (2008) Effects of irradiation dose rate on quality and sterility of Queensland fruit flies, Bactrocera tryoni (Froggatt). J Appl Entomol 132:398–405

Davies S, Kattel R, Bhatia B, Petherwick A, Chapman T (2005) The effect of diet, sex and mating status on longevity in Mediterranean fruit flies (Ceratitis capitata), Diptera: Tephritidae. Exp Gerontol 40:784–792

de Souza HML, Matioli SR, de Souza WN (1988) The adaptation process of Ceratitis capitata to the laboratory analysis of life history traits. Entomol Exp et Appl 49:195–201

Dhillon MK, Singh R, Naresh JS, Sharma HC (2005) The melon fruit fly, Bactrocera cucurbitae: a review of its biology and management. J Insect Sci 5:40–56

Diamantidis AD, Papadopoulos NT, Carey JR (2008) Medfly populations differ in diel and age patterns of sexual. Entomol Exp et Appl 128:389–397

Doharey KL (1983) Bionomics of fruit flies (Bactrocera spp.) on some fruits. Indian J Entomol 45:406–413

Drew RAI, Yuval B (2000) The evolution of fruit fly feeding behavior. In: Aluja M, Norrbom A (eds) Fruit flies. CRC, Boca Raton, pp 731–749

Drew RAI, Courtice AC, Teakle DS (1983) Bacteria as a natural source of food for adult fruit flies (Diptera: Tephritidae). Oecologia 60:279–284

Ekesi S, Mohamed SA (2011) Mass rearing and quality control parameters for Tephritid fruit flies of economic importance in Africa. In: Akyar, I (ed) Wide spectra of quality control. ISBN: 978- 953-307-683-6, InTech. Available from: http://www.intechopen.com/books/wide-spectra-of-qualitycontrol/

Ekesi S, Mohamed SA, Chang CL (2014) A liquid larval diet for rearing Bactrocera invadens and Ceratitis fasciventris (Diptera: Tephritidae). Int J Trop Insect Sci 34:S90–S98

FAO/IAEA/USDA (2003) Food and Agriculture Organization of the United Nations/International Atomic Energy Agency/United States Department of Agriculture manual for product quality control and shipping procedures for sterile mass-reared tephritid fruit flies.Version 5.0. IAEA, Vienna. http://www.iaea.org/programmes/nafa/d4/index.html

FAO/IAEA/USDA (2014) Product quality control for sterile mass-reared and released Tephritid fruit flies, Version 6.0. International Atomic Energy Agency, Vienna. 164 pp

Faria MJ, Pereira R, Dellinger T, Teal PEA (2008) Influence of methoprene and protein on survival, maturation and sexual performance of male *Ceratitis capitata* (Diptera: Tephritidae). J Appl Entomol 132:812–819

Fay HAC (1989) Multi-host species of fruit fly. In: Robinson AS, Hooper G (eds) Fruit flies: their biology, natural enemies and control, vol 3B. Elsevier, Amsterdam, pp 129–140

Fay HAC, Wornoayporn V (2002) Inert reusable substrates as potential replacement for wheat bran in larval diets for Mediterranean fruit fly, *Ceratitis capitata* (Wied.) (Diptera: Tephritidae). J Appl Entomol 126:92–96

Ferro MIT, Zucoloto FS (1990) Effect of the quantity of dietary amino acids on egg production and layings by *Ceratitis capitata* (Diptera, Tephritidae). Braz J Med Biol Res 23:525–532

Field SA, Yuval B (1999) Nutritional status affects copula duration in the Mediterranean fruit fly *Ceratitis capitata* (Insecta: Tephritidae). Ethol Ecol Evol 11:61–70

Fontellas-Brandalha TML, Zucoloto FS (2004) Selection of oviposition sites by wild *Anastrepha obliqua* (Macquart) (Diptera: Tephritidae) based on the nutritional composition. Neotrop Entomol 33:557–562

Gavriel S, Gazit Y, Yuval B (2009) Remating by female Mediterranean fruit flies (*Ceratitis capitata*, Diptera: Tephritidae): temporal patterns and modulation by male condition. J Insect Physiol 55:637–642

Gavriel S, Jurkevitch E, Gazit Y, Yuval B (2011) Bacterially enriched diet improves sexual performance of sterile male Mediterranean fruit flies. J Appl Entomol 135:564–573

George JA, Ruhm EM (1977) Modifications of adult olive fruit fly diet to control the production of mottled eggs and to increase fecundity. J Econ Entomol 70:1–4

Hagen KS (1953) Influence of adult nutrition upon the reproduction of three fruit fly species. In: Third Special report on control of Oriental fruit fly (*Dacus dorsalis*) in Hawaiian Islands. Senate of the State of California, pp 72–76

Hagen KS (1966) Dependence of the olive fly, *Dacus oleae*, larvae on symbiosis with *Pseudomonas savastanoi* for the utilization of olive. Nature 209:423–424

Hagen KS, Finney GL (1950) A food supplement for effectively increasing the fecundity of certain tephritid species. J Econ Entomol 43(5):735

Hagen KS, Santas L, Tsecouras A (1963) A technique of culturing the olive fruit fly, *Dacus oleae* Gmel. on synthetic media under xenic conditions. In: IAEA editorial staff (eds) Proceedings of a symposium on radiation and radioisotopes applied to insects of agricultural importance, International Atomic Energy Agency, Athens/Vienna, 22–26 April 1963, pp 333–356

Hamden H, Guerfali MM, Fadhl S, Saidi M, Chevrier C (2013) Fitness improvement of mass-reared sterile males of *Ceratitis capitata* (Vienna 8 strain) (Diptera: Tephritidae) after gut enrichment with probiotics. J Econ Entomol 106:641–647

Hanife G (2008) Modified agar-based diet for small scale laboratory rearing of olive fruit fly, *Bactrocera oleae*. Fla Entomol 91:651–656

Haq I, Hendrichs J (2013) Pre-release feeding on hydrolysed yeast and methoprene treatment enhances male *Bactrocera cucurbitae* Coquillett (Diptera: Tephritidae) longevity. J Appl Entomol 137:99–102

Haq I, Vreysen MJB, Teal PEA, Hendrichs J (2014) Methoprene application and diet protein supplementation to male melon fly, *Bactrocera cucurbitae*, modifies female remating behavior. Insect Sci 21:637–646

Harwood JF, Chen K, Müller HG, Wang JL, Vargas RI, Carey JR (2013) Effects of diet and host access on fecundity and lifespan in two fruit fly species with different life history patterns. Physiol Entomol 38:81–88

Hollingsworth R, Vagalo M, Tsatsia F (1997) Biology of melon fly, with special reference to Solomon Islands. In: Allwood AJ, Drew RAI (eds) Management of fruit flies in the Pacific. ACIAR proceedings no 76. 267 pp

Hooper GHS (1978) Effects of larval rearing temperature on the development of the Mediterannean fruit fly, *Ceratitis capitata*. Entomol Exp et Appl 23:222–226

Hooper GHS (1987) Effects of pupation environment on the quality of pupae and adults of the Mediterranean fruit fly. Entomol Exp et Appl 44:155–159

Hooper GHS (1989) The effect of ionizing radiation on reproduction. In: Robinson AS, Hooper G (eds) World crop pests, vol 3A, Fruit flies, their biology, natural enemies and control. Elsevier Science, Amsterdam, pp 153–164

Joachim-Bravo IS, Fernandes OA, De-Bortoli SA, Zucoloto FS (2001) Oviposition behavior of *Ceratitis capitata* Wiedemann (Diptera: Tephritidae): association between oviposition preference and larval performance in individual females. Neotrop Entomol 30:559–564

Joachim-Bravo IS, Anjos CS, Costa AM (2009) The role of protein in the sexual behavior ofmales of Ceratitis capitata (Diptera: Tephritidae): mating success, copula duration and number of copulation. Zoologia 26:407–412

Kakinohana H, Yamagishi M (1991) The mass production of the melon fly techniques and problems. In: Kawasaki K, Iwahashi O, Kaneshiro KY (eds) Proceedings, the international symposium on the biology and control of fruit flies. FFTC, Ginowan/Okinawa

Kamikado T, Chisaki N, Kamiwada H, Tanaka A (1987) Mass rearing of the melon fly, *Dacus cucurbitae* Coquillett, by the sterile insect release method. I. Changes in the amount of eggs laid and longevity of mass-reared insects. In: Proceedings No 33 of the association of plant protection of Kyushu, Kagoshima Agriculture, Experiment Station, Naze, Kagoshima

Kaspi R, Yuval B (2000) Post-teneral protein feeding improves sexual competitiveness but reduces longevity of mass reared sterile male Mediterranean fruit flies. Anns Entomol Soc Am 93:949–955

Kaspi R, Taylor PW, Yuval B (2000) Diet and size influence sexual advertisement and copulatory success of males in Mediterranean fruit fly leks. Ecol Entomol 25:279–284

Kaspi R, Mossinson S, Drezner T, Kamensky B, Yuval B (2002) Effects of larval diet on development rates and reproductive maturation of male and female Mediterranean fruit flies. Physiol Entomol 27:29–38

Kaur S, Srivastava BG (1994) Effect of amino acids on various parameters of reproductive potential of *Dacus cucurbitae* (Coquillett). Indian J Entomol 56:370–380

Khan M (2013) Potential of liquid larval diets for mass rearing of Queensland fruit fly, *Bactrocera tryoni* (Froggatt) (Diptera: Tephritidae). Aust J Entomol 52:268–276

Khan L, Haq M, Inayatullah C, Mohsan A (1993) Biology and behaviour of melon fruit fly, *Dacus cucurbitae* Coquillet. (Diptera: Tephritidae). Pak J Zool 25:203–208

Khan M, Hossain MA, Shakil AK, Islam MS, Chang CL (2011) Development of liquid larval diet with modified rearing system for *Bactrocera dorsalis* (Diptera: Tephritidae) for the application of sterile insect technique. ARPN J Agric Biol Sci 6:52–57

Khan M, Mahin AA, Pramanik MK, Akter H (2014) Identification of gut bacterial community and their effect on the fecundity of pumpkin fly, *Bactrocera tau* (Walker). J Entomol 11:68–77

Koul VK, Bhagat KC (1994) Biology of melon fly, *Bactrocera* (*Dacus*) *cucurbitae* Coquillet (Diptera: Tephritidae) on bottle gourd. Pest Manag Econ Zool 2:123–125

Krainacker DA, Carey JR, Vargas RI (1987) Effect of larval host on life history traits of the Mediterranean fruit fly, *Ceratitis capitata*. Oecologia 73:583–590

Langar AG, Sahito HA, Talpur MA (2013) Biology and population of melon fruit fly on musk melon and Indian squash Intl. J Farm Allied Sci 2:42–47

Leppla NC, Ashley TR (1989) Quality control in insect mass production: a review and model. Bull Entomol Soc Am 35:33–44

Manousis T, Ellar DJ (1988) *Dacus oleae* microbial symbionts. Microbiol Sci 5:149–152

Manrakhan A, Lux SA (2006) Contribution of natural food sources to reproductive behaviour, fecundity and longevity of *Ceratitis cosyra*, *C. fasciventris* and *C. capitata* (Diptera: Tephritidae). Bull Entomol Res 96:259–268

Marchini D, Rosetto M, Dallai R, Marri L (2002) Bacteria associated with the oesophageal bulb of the medfly *Ceratitis capitata* (Diptera: Tephritidae). Curr Microbiol 44:120–124

Mir SH, Dar SA, Mir GM, Ahmad SB (2014) Biology of *Bactrocera cucurbitae* (Diptera: Tephritidae) on cucumber. Fla Entomol 97:753–758

Narayanan ES, Batra HN (1960) Fruit flies and their control. A monograph by ICAR, New Delhi. 66 pp

Nash WJ, Chapman T (2014) Effect of dietary components on larval life history characteristics in the Medfly (*Ceratitis capitata*: Diptera, Tephritidae). PLoS One 9:e86029

Nestel D, Nemny-Lavy E, Zilberg L, Weiss M, Akiva R, Gazit Y (2004) The fruit fly PUB: a phagostimulation unit bioassay system to quantitatively measure ingestion of baits by individual flies. J Appl Entomol 128:576–582

Orankanok W, Chinvinijkul S, Sawatwangkhoung A, Pinkaew S, Orankanok S (2013) Methyl eugenol and pre-release diet improve mating performance of young *Bactrocera dorsalis* and *Bactrocera correcta* males. J Appl Entomol 137:200–209

Papadopoulos NT, Katsoyannos BI, Kouloussis NA, Economopoulos AP, Carey JR (1998) Effect of adult age, food, and time of the day on sexual calling incidence of wild and mass reared *Ceratitis capitata* males. Entomol Exp et Appl 89:175–182

Papadopoulos NT, Katsoyannos BI, Carey JR (2002) Demographic parameters of the Mediterranean fruit fly (Diptera: Tephritidae) reared in apples. Anns Entomol Soc Am 95:564–569

Pascacio-Villafa C, Williams T, Sivinski J, Birke A, Aluja M (2015) Costly nutritious diets do not necessarily translate into better performance of artificially reared fruit flies (Diptera: Tephritidae). J Econ Entomol Adv 108:53–59 Access published January

Petri L (1910) Untersuchung uber die darmbakterien der olivenfliege. Zentbl Bakteriolog 26:357–367

Placido-Silva MC, Da Silva Neto AM, Zucoloto FS, Joachim-Bravo IS (2006) Effects of different protein concentrations on longevity and feeding behavior of two adult populations of *Ceratitis capitata* Wiedemann (Diptera: Tephritidae). Neotrop Entomol 35:747–752. doi:10.1590/S1519-566x2006000600004

Quilici S, Schmitt C, Vidal J, Franck A, Deguine JP (2013) Adult diet and exposure to semiochemicals influence male mating success in *Ceratitis rosa* (Diptera: Tephritidae). J Appl Entomol 137:142–153

Rauf I, Ahmad N, Rashdi MSMS, Ismail M, Khan MH (2013) Laboratory studies on ovipositional preference of the peach fruit fly *Bactrocera zonata* (Saunders) (Diptera: Tephiritidae) for different host fruits. Afr J Agric Res 8:1300–1303

Renjhen PL (1949) On the morphology of the immature stage of *Dacus* (Strumeta) *cucurbitae* (Cog.) (The melon fruit fly) with notes on its biology. Indian J Entomol 11:83–100

Rivnay E (1950) The Mediterranean fruit fly in Israel. Bull Entomol Res 41:321–341

Rossler Y (1975) Reproductive differences between laboratory-reared and field-collected populations of the Mediterranean fruit fly *Ceratitis capitata*. J Econ Entomol 68:987–991

Sarwar M, Hamed M, Rasool B, Yousaf M, Hussain M (2013) Host preference and performance of fruit flies *Bactrocera zonata* (Saunders) and *Bactrocera cucurbitae* (Coquillett) (Diptera: Tephritidae) for various fruits and vegetables. Int J Sci Res Environ Sci 1:188–194

Schroeder WJ, Miyabara RY, Chambers DL (1971) A fluid larval medium for rearing the melon fly. J Econ Entomol 64:1221–1223

Shelly TE, Kennelly S (2002) Influence of male diet on male mating success and longevity and female remating in the Mediterranean fruit fly (Diptera: Tephritidae) under laboratory conditions. Fla Entomol 85:572–579

Shelly TE, McInnis DO (2003) Influence of adult diet on the mating success and survival of male Mediterranean fruit flies (Diptera: Tephritidae) from two mass rearing strains on field caged host trees. Fla Entomol 86:340–344

Shelly TE, Kennelly SS, McInnis DO (2002) Effect of adult diet on signalling activity, mate attraction, and mating success in male Mediterranean fruit flies (Diptera: Tephritidae). Fla Entomol 85:150–155

Shelly TE, Edu J, Pahio E (2005) Influence of diet and methyl eugenol on the mating success of males of the Oriental fruit fly, *Bactrocera dorsalis* (Diptera: Tephritidae). Fla Entomol 88:307–313

Shivarkar DT, Dumbre RB (1985) Bionomics and chemical control of melon fly. J Maharashtra Agric Univ 10:298–300

Shivayya V, Ashok-Kumar CT, Chakravar-Thy AK (2007) Biology of melon fly, *Bactrocera cucurbitae* on different food sources. Indian J Plant Prot 35:25–28

Shoukry A, Hafez M (1979) Studies on the biology of the Mediterranean fruit fly *Ceratitis capitata*. Entomol Exp et Appl 26:33–39

Singh OP, Teotia TPS (1970) A simple method of mass culturing melon fruit fly, *Dacus cucurbitae* (Cog.). Indian J Entomol 32:28–31

Steiner LF, Mitchell S (1966) Tephritid fruit flies. In: Smith CN (ed) Insect colonization and mass production. Academic Press, New York, pp 555–583

Tanaka N, Steiner F, Ohinata K, Okamoto R (1965) Low-cost larval rearing medium for mass production of oriental and Mediterranean fruit flies. J Econ Entomol 62:967–968

Taylor PW, Kaspi R, Yuval B (2000) Copula duration and sperm storage in Mediterranean fruit flies from a wild population. Phys Entomol 25:94–99

Teal PEA, Pereira R, Segura DF, Haq I, Go'mez-Simuta Y, Robinson AS, Hendrichs J (2013) Methoprene and protein supplements accelerate reproductive development and improve mating success of male Tephritid flies. J Appl Entomol 137:91–98

Tsiropoulos GJ (1977) Survival and reproduction of *Dacus oleae* on chemically defined diets. Zeitsch Angewan Entomol 84:192–197

Tsiropoulos GJ (1980) Major nutritional requirements of adult *Dacus oleae*. Anns Entomol Soc Am 73:251–253

Tsiropoulos GJ (1981) Effect of varying the dietary nitrogen to carbohydrate ratio upon the biological performance of adult *Dacus oleae*. Arch Int Physiol Biochim 89:101–105

Tsiropoulos GJ (1982) Effect of vitamins overdosing on adult *Dacus oleae*. In: Proceedings international symposium on fruit flies of economic importance CEC/IOBC, Athens, Nov 1992, pp 85–90

Tsiropoulos GJ (1989) Biosynthetic activity of the microflora associated with the olive fruit fly, *Dacus oleae*. International Colloquium on Microbiology in Poikilotherms, Paris, July 1989

Tsiropoulos GJ (1992) Feeding and dietary requirements of the tephritid fruit flies. In: Anderson TE, Leppla NC (eds) Advances in insect rearing for research and pest management. Westview Press, Boulder, pp 93–118

Tsitsipis JA (1977) An improved method for the mass rearing of the olive fruit fly, *Dacus oleae* (Gmel.) (Diptera, Tephritidae). Z Angew Entomol 83:419–426

Tsitsipis JA (1983) Changes of a wild ecotype of the olive fruit fly during adaptation to lab rearing. In: Cavalloro R (ed) Proceedings of the 1st international symposium on fruit flies of economic importance, Athens, Greece, pp 416–422

Vargas RI (1989) Mass production of fruit flies. In: Robinson AS, Hooper G (eds) Fruit flies: their biology, natural enemies and control, vol 3B. Elsevier, Amsterdam, pp 141–151

Vargas RI, Carey JR (1989) Comparison of demographic parameters for wild and laboratory adapted Mediterranean fruit fly (Diptera: Tephritidae). Anns Entomol Soc Am 82:55–59

Vargas RI, Nishida T, Beardsley JW (1983) Distribution and abundance of *Dacus dorsalis* in native and exotic forest environments on Kauai. Environ Entomol 12:1185–1189

Vargas RI, Stark JD, Uchida GK, Purcell M (1993) Opiine parasitoids (Hymenoptera: Braconidae) of oriental fruit fly (Diptera: Tephritidae) on Kauai island, Hawaii: islandwide relative abundance and parasitism rates in wild and orchard guava habitats. Environ Entomol 21:246–253

Via S (1986) Genetic covariance between oviposition preference and larval performance in an insect herbivore. Evolution 40:778–785

Vishva NS (2005) Studies on biology, behaviour and management of melon fruit fly, *Bactrocera cucurbitae* (Coq.) (Diptera: Tephritidae), M.Sc thesis submitted to U.A.S. Banglore, pp 122

Walker GP, Tora VE, Hamacek EL, Allwood AJ (1997) Laboratory-rearing techniques for Tephritid fruit flies in the South Pacific. In: Management of fruit flies in the Pacific. ACIAR, Canberra, pp 145–152

Waseem MA, Naganagoud A, Sagar D, Abdul-Kareem M (2012) Biology of melon fly, *Bactrocera cucurbitae* (Coquillett) on cucumber. Bioinfolet 9:232–239

Whittier TS, Nam FY, Shelly TE, Kaneshiro KY (1994) Male courtship success and female discrimination in the Mediterranean fruit fly (Diptera:Tephritidae). J Insect Behav 7:159–170

Yuval B, Kaspi R, Shloush S, Warburg MS (1998) Nutritional reserves regulate male participation in Mediterranean fruit fly leks. Ecol Entomol 23:211–215

Yuval B, Kaspi R, Field SA, Blay S, Taylor P (2002) Effects of post-teneral nutrition on reproductive success of male Mediterranean fruit flies. Fla Entomol 85:165–170

Yuval B, Maor M, Levy K, Kaspi R, Taylor P, Shelly T (2007) Breakfast of champions or kiss of death? Survival and sexual performance of protein-fed, sterile Mediterranean fruit flies (Diptera: Tephritidae). Fla Entomol 90:115–122

Yuval B, Ben-Ami E, Behar A, Ben-Yosef M, Jurkevitch E (2010) The Mediterranean fruit fly and its bacteria – potential for improving sterile insect technique operations. J Appl Entomol 137:1–4
Zografou EN, Tsiropoulos GJ, Margaritis LH (1998) Survival, fecundity and fertility of *Bactrocera oleae*, as affected by amino acid analogues. Entomol Exp Appl 87:125–132
Zucoloto FS (1993) Acceptability of different Brazilian fruits to *Ceratitis capitata* (Diptera, Tephritidae) and fly performance on each species. Braz J Med Biol Res 26:291–298

Samira A. Mohamed works in the African fruit fly programme at the International Centre of Insect Physiology and Ecology (*icipe*), Nairobi, Kenya and is a senior scientist and co-ordinator of the German BMZ funded *Tuta* IPM project. She has over 20 years of research experience focusing on the development of IPM strategies for suppression of horticultural crop pests. Her main area of interest is in classical biological control of invasive pests within the context of IPM. Samira was instrumental in importing in to Africa, and later releasing, two-efficient co-evolved parasitoid species (***Fopius arisanus*** and ***Diachasmimorpha longicaudata***) from Hawaii. So far, she has authored or co-authored over 40 peer-reviewed publications and several book chapters.

Fathiya M. Khamis Fathiya works in the African fruit fly programme and the Arthropod Pathology Unit at the International Centre of Insect Physiology and Ecology (*icipe*), Nairobi, Kenya and she is co-ordinator of the *icipe* Biovision Foundation (Switzerland). Her main focus is to employ molecular tools for the identification and delineation of species, inference of genetic variability and population genetic structures of pests as a prelude for the development of effective management strategies. She has worked on tephritid fruit flies and other pests of economic importance, both invasive and non-invasive. She has authored or co-authored over 15 peer-reviewed publications in this field.

Chrysantus Mbi Tanga is an Agricultural Entomologist/Ecologist working in the Plant Health Theme at the International Centre of Insect Physiology and Ecology (*icipe*), Nairobi, Kenya. He obtained his PhD from the University of Pretoria, South Africa and is currently a postdoctoral research fellow at *icipe*. His area of research interest is sustainable integrated pest management (IPM) that is based on the use of biopesticides, baiting techniques, orchard sanitation and classical biological control for the management of arthropod pests in horticulture. His research activities involve the promotion and up-scaling of the dissemination of fruit fly IPM amongst smallholder growers in Africa. He also develops temperature-based phenology models for African fruit flies and other crop pests to improve implementation of IPM strategies for their management. He also has research activities that are focused on filling critical gaps in knowledge aimed to exploit the commercial potential of insects in the production of affordable, high-quality protein for the poultry, fish and pig industries. He seeks to create an effective agribusiness model by ensuring high nutrition and microbial safety of insect-based protein products through research as well as by creating awareness and identifying market opportunities for the technology using participatory approaches that involves farmer groups, with a particular focus on the involvement of women and young adults in Africa.

Chapter 11
The Ontological Modelling of Fruit Fly Control and Management Knowledge

Caroline C. Kiptoo, Aurona Gerber, and Alta Van der Merwe

Abstract Fruit fly control and management in Africa has been the topic of several scientific investigations resulting in diverse sources of knowledge on the topic. Despite the existence of this knowledge, frequently it is not readily accessible to all targeted beneficiaries; this can be due to, for example, the remote locations of farms and the complexity of the knowledge. However, recent technological developments such as web technologies and networking allow for the engagement and participation of stakeholder groups in the acquisition and dissemination of knowledge and these technologies can also be applied to fruit fly knowledge. In order to facilitate this stakeholder participation in fruit fly knowledge sharing, the relevant domain knowledge needs to be available in a format that can support stakeholder engagement, preferably through the Web. Fruit fly knowledge has not been modelled in this manner and this paper reports on an investigation to model and capture the relevant domain knowledge using ontologies. The objective of this work is thus the development of the domain ontology and its evaluation using a prototype stakeholder participation system for fruit fly control and management that was capable of utilising the ontology. We describe our findings on the use of ontology technologies for representation of fruit fly knowledge, the fruit fly ontology developed, as well as a prototype Web-based system that uses the ontology as a source of knowledge.

Keywords Ontological modelling • Ontology-driven • Taxonomic key

---

C.C. Kiptoo (✉) • A. Van der Merwe
Department of Informatics, University of Pretoria, Pretoria, South Africa
e-mail: cckiptoo@gmail.com

A. Gerber
Department of Informatics, University of Pretoria, Pretoria, South Africa

CAIR, CSIR Meraka Institute, Pretoria, South Africa

1 Introduction

Fruit flies (Diptera: Tephritidae) are one of the most important pests affecting fruit and vegetable production worldwide (Allwood and Drew 1997; Badii et al. 2015). The spread of fruit flies, particularly exotic invasive species, is as a result of both natural processes and human activities. Natural processes depend on the traits of the species e.g. availability of hosts plants, mating patterns, and survival patterns in different environments amongst others (Malacrida et al. 2007; Vargas et al. 2000). Human activities that encourage spread include transportation of infected fruit by travelers, and trade between countries. Losses due to fruit flies result from direct damage to fruit, reduction in quality and quantity of fruit and loss of markets due either to the quarantine restrictions of the importing countries or because the importer's Maximum Residue Limits (MRL) for pesticides are exceeded (de Bon et al. 2014; Dominiak and Ekman 2013; Ekesi et al. 2005; Manrakhan et al. 2013). Production of high quality fruits and vegetables that meet the required MRL and quarantine measures is therefore a prerequisite for targeting lucrative export markets.

At the request of growers, regional authorities and national authorities, the *icipe*-led African Fruit Fly Programme (AFFP) was established. The broad objective of AFFP is to help stakeholders in the horticulture industry to effectively manage fruit flies and build the capacity of agricultural officers, extension workers, quarantine personnel and growers (Ekesi 2010). The AFFP programme has undertaken different research activities and has developed effective management packages for growers across Africa. Key outcomes of these initiatives include the development of knowledge on the identification of fruit fly species (Billah et al. 2007), on attractants that can be used for monitoring different fruit fly species (Manrakhan 2007; Nagaraja et al. 2014), on the species distribution across Africa (Ekesi and Muchugu 2007) and on the host plant relationships of different fruit fly species (Ekesi and Muchugu 2007; Rwomushana et al. 2008). However, the accessibility of this knowledge to farmers in Africa is still limited due to inadequate resources, complexity of the knowledge, the nature of farming and lack of capacity to engage experts on a continuous basis.

The emergence of Web 2.0 and the social web has created new opportunities for online collaboration in different domains. In ecology and environmental sciences, this type of online collaboration has been used to support citizen science projects such as the eBird project and BioBlitz projects (Bonney et al. 2009; Delaney et al. 2008; Karns et al. 2006; Lewington and West 2008; Lundmark 2003; Silvertown 2009; Sullivan et al. 2009). By harnessing these technologies, we aim to create a platform to support online collaboration between scientists and citizens on the management of fruit flies. Such a platform requires modelling approaches that adequately represent the available expert knowledge to drive the collaboration and we found that the use of ontological modelling was most appropriate. Here we present the development of an ontology that represents the key knowledge necessary for management of fruit flies. We have focused on species with the greatest economic importance in Africa, specifically 30 species from the genera *Ceratitis, Dacus, Bactrocera* and *Trirhithrum*. The scope of knowledge that was modelled included the morphological features of the different taxonomic groups, the attractants used to lure different species and the sets of species supported by different host plants (Billah et al. 2007; Rwomushana et al. 2008; Manrakhan 2007; Ekesi and Muchugu 2007).

2 What are Ontologies and How Are They Developed?

The term ontology has its origin in philosophy (Kunne et al. 1982) and in philosophy, an ontology is defined as "a branch of philosophy that deals with the science of what is, of the kinds and structures of objects, properties, events, processes and relations in every area of reality" (Smith 2003). This concept was adopted by Computer Science where an ontology is used to refer to an information object that contains formally symbolized knowledge. Gruber (1993) defined an ontology as "a body of formally represented knowledge which is based on a conceptualization" and a conceptualization as "an abstract simplified view of the world that we wish to represent for some purpose". The desired facts on the target set of objects (universe of discourse) is represented in a declarative formalism using computational logic and the relationships between these sets of objects are modelled. In Guarino et al. (2009) an ontology is described as something used to embody the structure of a system. In Horridge et al. (2009) an ontology is seen to be a formalized representation of knowledge consisting of classes, properties and individuals. In this work we adopt the definition of Gruber (1993) where an ontology consists of formally represented knowledge based on a conceptualization.

In Noy and McGuinness (2001), the advantages of using ontologies over other forms of knowledge representation are presented. In our research, the need to bring together stakeholders comprised of experts and non-experts was one of the reasons that motivated the use of ontology to represent knowledge. The gap between experts and novices can be effectively reduced by using ontologies because people and software who adopt a formalism have a common understanding of the represented facts and, therefore, software can be used easily to aid humans in answering questions. In addition, in ontological modelling assumptions are made explicit and, therefore, those who adopt the ontology are aware of all the assumptions.

Building ontologies is arguably still a craft skill. Different methodologies have been proposed, most of them derived from project experiences (Iqbal et al. 2013). Some methodologies are comprehensive while others address specific aspects of ontology development (López 1999). Comprehensive methodologies can be categorized into two types: stage-based models and evolving-prototype models (Jones et al. 1998). Stage-based models, as the name suggests, have step by step processes with clear inputs and outputs at each stage. They are more suitable for building ontologies where the full requirements are clear from the onset. Evolving-prototype methodologies create an initial prototype ontology which is then evolved over time ideally creating improved versions at every iteration. Methodologies of this type are ideal for problems where the requirements are not clear at the initial stages and can emerge and become clearer after some iterative improvements on the initial version.

Noy and McGuinness (2001) argued that there is no correct way to create domain ontologies and approaches depend on the targeted application. A methodology for design and evaluation of ontologies is presented in Grüninger and Fox (1995). The steps include: document motivating scenarios and clearly establishing why existing ontologies cannot meet the needs at hand; developing informal competency questions that the ontology must answer; enumerating first order logic terminology by identifying objects, attributes and relations; creating formal competency questions from the informal questions and the formal terminology; defining first order logic axioms that capture the relationship between the objects and answer the formal competency questions; and developing completeness theorems that guide when the answers to the competency questions are complete. Noy and McGuinness (2001) present a seven step process for building domain ontologies: determine the domain and scope of the ontology, consider reusing existing ontologies, enumerate important terms in the ontology, define the classes and the class hierarchy, define the properties of classes – slots, define the facets of the slots, and create instances. They use terminology from the Web Ontology Language (OWL) to explain their methodology and, although it is slightly different in the steps, the general approach is closely similar to the methodology of Grüninger and Fox (1995).

Construction and maintenance of formal ontologies is done using ontology representation languages. In 2001, a working group called Web Ontology (WebOnt) Working Group was formed by the World Wide Web Consortium (W3C) and their mandate was to make a new ontology markup language for the Semantic Web, called OWL. The second edition, OWL 2, is now recommended as a standard by W3C and to promote interoperability of the web (Corcho et al. 2003). As result of these standardization efforts, tools that ease creation of OWL ontologies have been created (Corcho et al. 2003). An example is Protégé, which is open-source software developed at Stanford University. Protégé allows interactive creation and editing of ontologies in various formats. Protégé comes with core functionalities that can be expanded by the importation of available plugins (Sivakumar and Arivoli 2011). Protégé is available as a stand-alone version that can be installed in individual machines, and as a web version, WebProtege, which can be installed on a web server and allows users to share, create and edit ontologies through a web browser (Gennari et al. 2003; Rubin et al. 2007).

The use of ontologies in representations of biological knowledge is not entirely new. Examples of ontology use within the biological sciences include the highly cited Gene ontology (GO) which represents knowledge on molecular functions, biological processes and cell components (Ashburner et al. 2000; Bard and Rhee 2004); the Plant ontology, which links plant anatomy, morphology, growth and development to plant genomics data; and Mouse gross anatomy ontology representing knowledge on the anatomy of adult mice. More biological ontologies are hosted at the Open Biological Ontologies (OBO) website, hosted by the Berkeley Bioinformatics Open Source Project.

Two ontologies that are closely related to our own study described here are the Hymenoptera Anatomy Ontology (HAO) presented in Yoder et al. (2010) and the Morphology of Afrotropical Bees Ontology (ABO) presented in Gerber et al. (2014). These both model knowledge about organisms from the same class as fruit flies. The HAO represents the anatomy of members of the family Hymenoptera, and its objective was to address the challenge of language discrepancies in anatomical terminology. The ABO uses concepts from the HAO ontology to represent knowledge about the morphological features of different taxonomic groups of bees and a model for modelling morphological features of different taxonomic groups is presented (Gerber et al. 2014).

3 Development of a Fruit Fly Ontology

The construction of the fruit fly ontology was done following the guidelines developed by Grüninger and Fox (1995), Horridge et al. (2009) and Noy and McGuinness (2001).

3.1 Materials and Tools

3.1.1 Materials

The Materials used for this research were:

1. Fruit fly taxonomic key (Billah et al. 2007).
2. Data on the set of host plants used by different fruit fly species (Ekesi and Muchugu 2007).
3. Sets of species attracted by different lures (Manrakhan 2007).

3.2 Tools

The tools used in this research were:

4. Protégé, a graphical editor for building ontologies and knowledge bases (Gennari et al. 2003; Noy et al. 2003).
5. Fact++ reasoner (Tsarkov and Horrocks 2006).
6. Java programming language and JSP (Java Server Pages).

3.3 Construction of the Ontology

In this section, the construction of the ontology is presented. To improve reading, the class names are italicized. The naming convention proposed by Noy and McGuinness (2001) was used to name the concepts in the ontology. The construction of the ontology was done using Protégé and built in Fact++ reasoner. The process began with defining the scope of the ontology. The scope was defined by specifying the competency questions that the ontology must answer and the questions are as listed in Box 11.1.

> **Box 11.1: Competency Questions**
> 1. Which species have a given set of taxonomic features
>   (e.g. which species have patterned wings, dark brown femora and three black spots on the scutellum?)
> 2. Which set of host plants can a given species attack
>   (e.g. which hosts can be attacked by *Bactrocera latifrons* (Hendel)?)
> 3. Which set of species can a given lure attract
>   (e.g. which species are attracted by Trimedlure?).

In the different methodologies, it is recommended that existing ontologies should be used if they can serve the needs of the target project. Modelling of the ontology therefore proceeded with the identification of ontologies that could be re-used. We found that the TAXRANK ontology (http://www.phenoscape.org/wiki/Taxonomic_Rank_Vocabulary) was suitable for the association of our taxonomic groupings with the biological taxon information. Use of the TAXRANK ontology is illustrated later in this section. The other knowledge on identification features, host plants and lures were modelled in the fruit fly ontology presented here and, for easy reading, the names of the concepts are italicized. Class names begin with a capital letter, while properties begin with lower case letters. Top concepts and their sub concepts consisting of classes and properties were identified. The top classes included: *BodyPart* which contained the anatomical parts of the organism; *Feature* which consisted of general characteristics such as colour, shape and texture; *DiagnosticFeature* which described compound features and had three sub-classes (*MorphDiagnosticFeature, HostDiagnosticFeature* and *AttractantDiagnosticFeature*); *Organism* which contained the taxonomic groupings of the fruit flies and host plants. The top properties included: *hasDiagnosticFeature* which was used to associate taxonomic groupings with diagnostic features and had three sub properties (*hasHostDiagnosticFeature, hasAttractantDiagnosticFeature, hasMorphDiagnosticFeature*); *hasPart* which was used to describe the parts of a body segment. The *hasTaxaRank* object property was used to associate the taxonomic groupings in the fruit fly key with the biological taxon defined in the TAXRANK ontology. The top concepts and top object properties are as shown in Fig. 11.1.

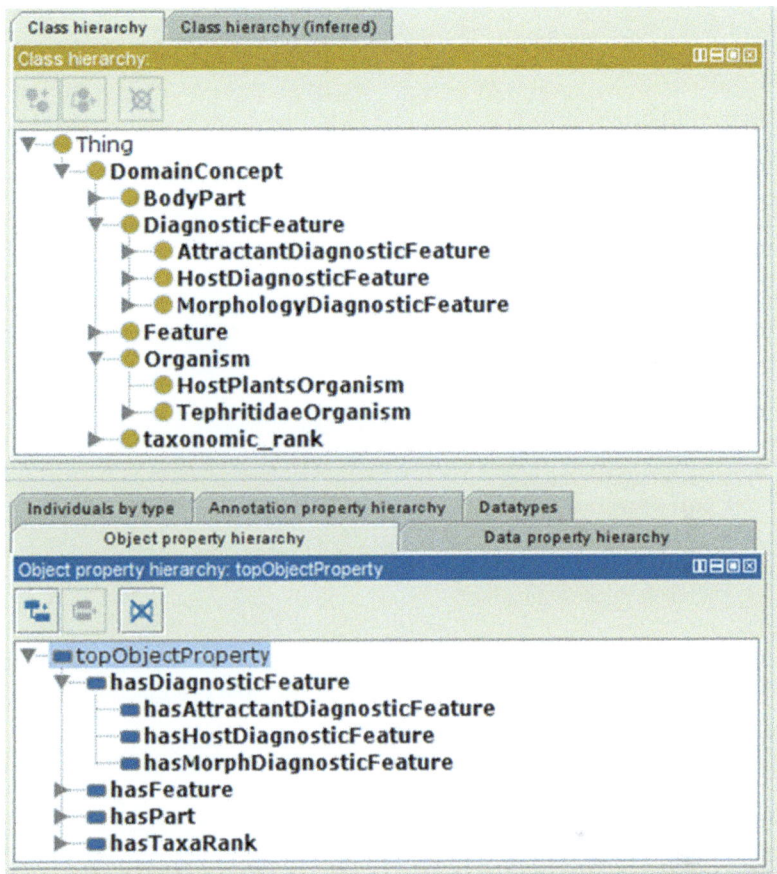

Fig. 11.1 Top concepts in fruit fly ontology

The *BodyPart* class was modelled as sub-classes consisting of the organism body parts and any relationship between parts was inferred using the *hasPart* object property.

The *Feature* class was modelled as sub classes consisting of features such as colour, shape, size and texture.

The *Morphological Diagnostic feature* was modelled using the Gerber et al. (2014) model, where the concept *MorphDiagnosticFeature* was defined as a *BodyPart* that has a feature *Feature*.

MorphDiagnosticFeature = BodyPart and (*hasFeature* some *Feature*)

For instance, to define a feature of a leg that is yellow in colour, the diagnostic feature was represented using the body part class '*LegBodyPart*' and feature class '*YellowColourFeature*' and was modelled as:

(*LegBodyPart* and (*hasColour* some *YellowFeature*))

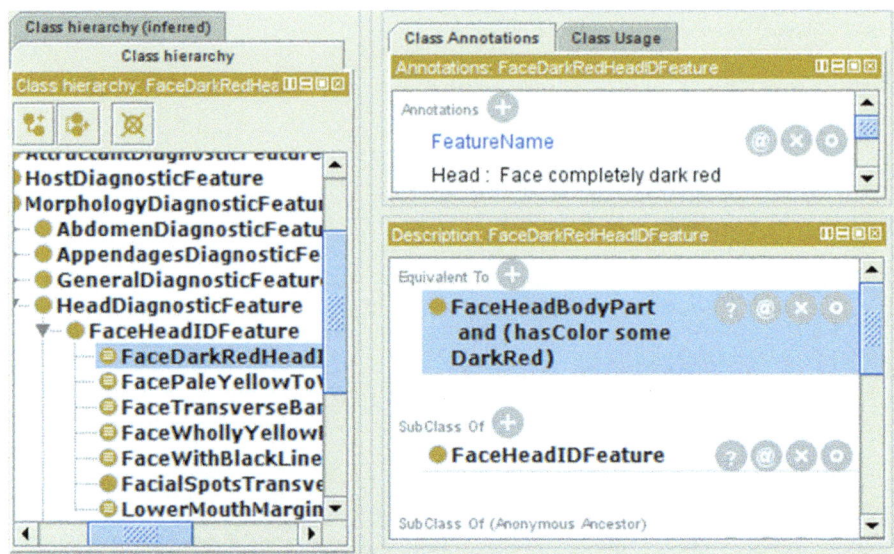

Fig. 11.2 Face dark red diagnostic feature

Note that *hasColour* is a sub property of the property *hasFeature*. As another example the modelling of the diagnostic feature for a dark red face is shown in Fig. 11.2.

Modelling of attractants and host plants was done by defining them as subclasses of *AttractantDiagnosticFeature* and *HostDiagnosticFeature* respectively.

After the diagnostic features were modelled, the next step was to associate them with the taxonomic grouping of the organism through the *hasDiagnosticFeature* object property using the relevant sub property. The diagnostic features *MorphDiagnosticFeature*, *HostDiagnosticFeature* and *AttractantDiagnosticFeature* were associated with the taxonomic group using *hasMorphDiagnosticFeature*, *hasHostDiagnosticFeature* and *hasAttractantDiagnosticFeature*, respectively. For example, a taxonomic grouping *TG* that has a set of morphological diagnostic features *MDF1*, *MDF2*…. *MDFn* is modelled as:

TG and (*hasMorphDiagnosticFeature* some *MDF1*) and (*hasMorphDiagnosticFeature* some *MDF2*) …and (*hasMorphDiagnosticFeature* some *MDFn*)

To associate attractants and host plants to the taxonomic groupings, the same modelling structure was used. A complete example of modelling all three categories of diagnostic features is described below. The taxonomic group *TG* that has yellow femora, is attracted to protein bait and can be hosted by the custard apple, *Annona muricata* L., was modelled as:

TG and (*hasMorphDiagnosticFeature* some *FemurAllFemoraYellowLegIDFeature*) and (*hasAttractantDiagnosticFeature* some *ProteinBaitAttractantDFeature*) and (*hasHostDiagnosticFeature* some *AnnonaMuricata*)

11 The Ontological Modelling of Fruit Fly Control and Management Knowledge

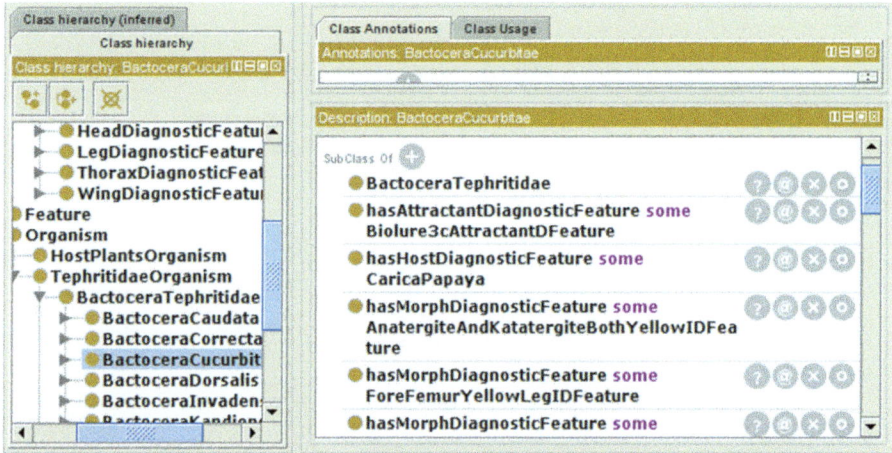

Fig. 11.3 Association of taxonomic groupings with morphology, host plant and attractant

A further example showing the association of a morphological feature, a host and an attractant with a taxonomic grouping is presented in Fig. 11.3.

Associating our taxonomic groupings with biological taxa was done by importing the TaxaRank ontology into our ontology and associating our taxonomic groupings with the appropriate taxon in TaxaRank. The association was done using the *hasTaxaRank* object property. This association will enable other applications using the TaxaRank ontology to process our data. For example, associating a taxonomic group *TG* with *subfamily* taxon is as shown below:

TG and (*hasTaxaRank* some *subfamily*)

The development of the ontology was iterative across the presented activities until the ontology was completed. The ontology in its current form has captured knowledge on the simplified taxonomic key and the set of lures for different species. The host plant for the different species have not been captured fully yet, but the basic structure exists. According to Protégé metrics the ontology has 1181 classes and 4600 axioms.

3.4 Evaluating the Ontology

Evaluation of the ontology was done based on the competency questions that guided its development (see Box 11.1). The evaluation was done using the DL Query tool within Protégé and through development of a prototype application that used the ontology as a knowledge source.

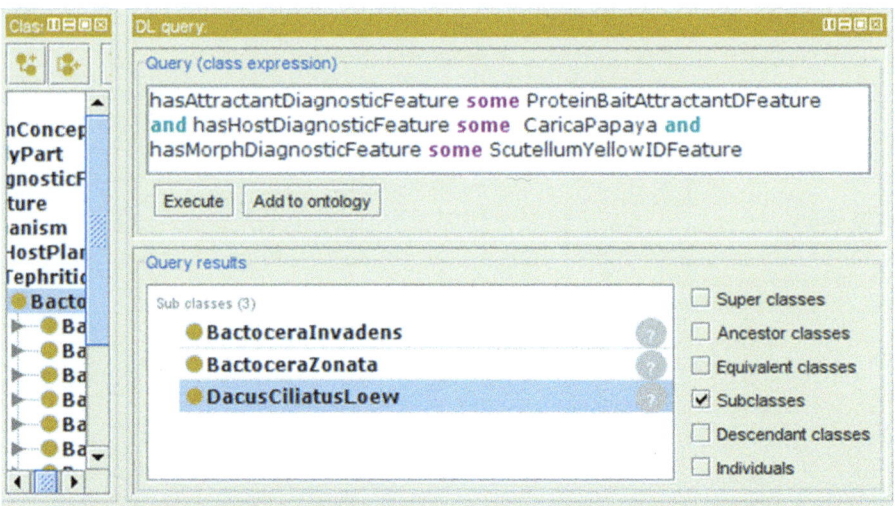

Fig. 11.4 Species that match query criteria

3.4.1 Evaluation Using DL Query

The Protégé DL Query tool was used in conjunction with the integrated Fact++ reasoner to evaluate the ontology. The evaluation was done using the competency questions and was found to give correct answers. For example, the species that have a yellow scutellum, are attracted to protein bait and can be hosted by papaya, *Carica papaya* L., are extracted by the query as shown in Fig. 11.4.

3.4.2 Evaluation Using Application Prototype

An important quality of a domain ontology is to meet the requirements of the application it was intended for (Noy and McGuinness 2001). In this section we present a prototype of an application developed for the purposes of accessing fruit fly knowledge for practical purposes. The fundamental requirements of the application include provision of a taxonomic key for the stakeholders, provision of guidance on baits to use based on the fly species being targeted and the host plants that could be affected by different fly species. The application was developed to evaluate whether the ontology could meet these requirements.

The Prototyping approach (Canning 1981) was adopted in the development of the application and Java, JSP and html were used for programming the application. All the tools were supported by the Fact++ reasoner. The application consisted of a multi-entry key for species identification, tools for querying hosts that a given species could attack, and also species that could be attracted by a selected attractant. The key allowed selection of features from an observation and querying for the

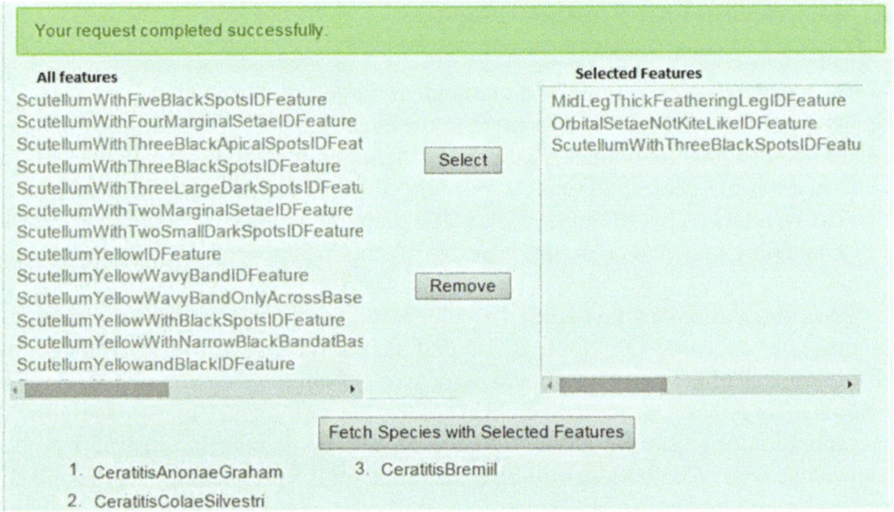

11 The Ontological Modelling of Fruit Fly Control and Management Knowledge

Fig. 11.5 Key sample prototype

Fig. 11.6 Species hosted by selected host plant prototype

species that have all the selected features. The basic interface for the key is as shown in Fig. 11.5.

The tool for querying species that affect a particular host plant consists of an interface where the user can select the host plant name and search for the fly species that the host can support. An example of fruit fly species hosted by cashew, *Anacardium occidentale* L., was queried as shown in Fig. 11.6. Querying for fly species attracted to a given attractant uses a similar interface. By selecting an attractant, it is possible to view the fly species that it can lure. All these tools incorporate the services of a reasoner and therefore use both explicit and implicit facts represented in the ontology to answer questions.

4 Discussion and Further Work

In this chapter, we have presented ontological modeling of knowledge on fruit fly biology and management. The targeted knowledge includes that which can be reasoned upon to provide guidance on management options, including knowledge on identification of species, attractants that lure different species and the host plants that can be attacked by different species. The outcome of this modelling is an ontology containing knowledge on thirty species of fruit flies of economic importance in Africa.

The contributions of this study include, amongst other things, mechanisms to incorporate reasoning into applications that support user access to knowledge on fruit flies. Use of the ontology should reduce the difficulties that non-experts find in understanding these areas of knowledge since the ontological knowledge allows incorporation of reasoning services to provide answers to questions asked over the knowledge base. Another use will be to facilitate tools for collecting user feedback in a structured manner since the ontology exists as a reusable artifact, thus making it possible to collect useful feedback to enhance scientific research. A prototype application that uses the ontology as discussed above is presented and the tools that were developed in the prototype include an ontology-based multi-entry taxonomic key, querying of fruit flies that are attracted to a given attractant and querying of those that are hosted by a given plant species.

It is envisaged that the ontology will be included in the development of an online Web-based platform that supports online collaboration between citizens and scientists in fruit fly biology and management. The platform will use the ontology to enable citizen access to expert knowledge. In future we intend to extend the ontology to capture knowledge on survival environments for the different species of fruit flies. We also intend to develop tools that facilitate creation and editing of this kind of ontology by biologists. This will facilitate knowledge sharing amongst stakeholders without the need to engage specialists in ontological modelling.

References

Allwood AJ, Drew RAI (1997) Management of fruit flies in the Pacific. A regional symposium, Nadi, Fiji 28–31 October 1996. In: Management of fruit flies in the Pacific. A regional symposium, Nadi, Fiji 28–31 October 1996. Australian Centre for International Agricultural Research (Australia)

Ashburner M, Ball CA, Blake JA, Botstein D, Butler H, Cherry JM, Davis AP, Dolinski K, Dwight SS, Eppig JT, Harris MA (2000) Gene ontology: tool for the unification of biology. Nat Genet 25:25–29

Badii KB, Billah MK, Afreh-Nuamah K, Obeng-Ofori D, Nyarko G (2015) Review of the pest status, economic impact and management of fruit-infesting flies (Diptera: Tephritidae) in Africa. Afr J Agric Res 10:1488–1498

Bard JB, Rhee SY (2004) Ontologies in biology: design, applications and future challenges. Nat Rev Genet 5:213–222

Billah MK, Mansell MW, De Meyer M, Goergen G (2007) Fruit fly taxonomy and identification. In: Billah MK, Ekesi S (eds) A field guide to the management of economically important tephritid fruit flies in Africa, 2nd edn. icipe Science Press, H1-H32

Bonney R, Cooper CB, Dickinson J, Kelling S, Phillips T, Rosenberg KV, Shirk J (2009) Citizen science: a developing tool for expanding science knowledge and scientific literacy. Bioscience 59:977–984

Canning RG (1981) Developing systems by prototyping. EDP Anal 19:1–14

Corcho O, Fernández-López M, Gómez-Pérez A (2003) Methodologies, tools and languages for building ontologies. Where is their meeting point? Data Knowl Eng 46:41–64

de Bon H, Huat J, Parrot L, Sinzogan A, Martin T, Malézieux E, Vayssières J-F (2014) Pesticide risks from fruit and vegetable pest management by small farmers in sub-Saharan Africa. A review. Agron Sust Devel 34:723–736

Delaney DG, Sperling CD, Adams CS, Leung B (2008) Marine invasive species: validation of citizen science and implications for national monitoring networks. Biol Invasions 10:117–128

Dominiak BC, Ekman JH (2013) The rise and demise of control options for fruit fly in Australia. Crop Prot 51:57–67

Ekesi S (2010) Combating fruit flies in eastern and southern africa (COFESA): elements of a strategy and action plan for a regional cooperation program. International Centre of Insect Physiology and Ecology, Nairobi/Kenya, pp 9–11

Ekesi S, Muchugu E (2007) Tephritid fruit flies in Africa – fact sheet of some economically important species. In: Billah MK, Ekesi S (eds) A field guide to the management of economically important tephritid fruit flies in Africa, 2nd edn. icipe Science Press, B1-B20

Ekesi S, Maniania NK, Mohamed SA, Lux SA (2005) Effect of soil application of different formulations of *Metarhizium anisopliae* on African tephritid fruit flies and their associated endoparasitoids. Biol Contr 35:83–91

Gennari JH, Musen MA, Fergerson RW, Grosso WE, Crubézy M, Eriksson H, Noy NF, Tu SW (2003) The evolution of Protégé: an environment for knowledge-based systems development. Int J Hum Comput Stud 58:89–123

Gerber A, Eardley C, Morar N (2014) An ontology-based taxonomic key for Afrotropical bees. In: Garbacz P, Kutz O (eds) Formal ontology in information systems, vol 267. Ios Press, Amsterdam, pp 277–288

Gruber TR (1993) A translation approach to portable ontology specifications. Knowl Acquisi 5:199–220

Grüninger M, Fox MS (1995) Methodology for the design and evaluation of ontologies. In: Proceedings of the workshop on basic ontological issues in knowledge sharing. IJCAI-95

Guarino N, Oberle D, Staab S (2009) What is an ontology? In: Staab S, Studer R (eds) Handbook on ontologies. Springer, Berlin, pp 1–17

Horridge M, Jupp S, Moulton G, Rector A, Stevens R, Wroe C. (2009) A practical guide to building OWL ontologies using protégé 4 and CO-ODE tools edition1. 2. *The University of Manchester* 107

Iqbal R, Murad MAA, Mustapha A, Sharef NM (2013) An analysis of ontology engineering methodologies: a literature review. Res J Appl Sci Eng Technol 6:2993–3000

Jones D, Bench-Capon T, Visser P (1998) Methodologies for ontology development. In: Proceedings of IT&KNOWS conference, XV IFIP world computer congress

Karns DR, Ruch DD, Brodman RD, Jackson MT, Rothrock PE, Scott PE, Simon TP, Whitaker JO Jr (2006) Results of a short-term BioBlitz of the aquatic and terrestrial habitats of Otter Creek, Vigo County, Indiana. Proc Indiana Acad Sci 115:82–88

Kunne W, Mulligan K, Null G, Simons P, Simons R, Smith B, Willard D (1982) Parts and moments: studies in logic and formal ontology. Retrieved from http://ontology.buffalo.edu/smith/book/P&M/

Lewington RJ, West CJ (2008) Otari Bioblitz: detailing vascular plants, mosses and liverworts. Wellington Bot Soc Bull 51:5–23

López FM (1999) Overview of methodologies for building ontologies. In: Proceedings of IJCAI99's workshop on ontologies and problem solving methods: lessons learned and future trends, Stockholm, Sweden, pp 4.1–4.13

Lundmark C (2003) BioBlitz: getting into backyard biodiversity. Bioscience 53:329

Malacrida AR, Gomulski LM, Bonizzoni M, Bertin S, Gasperi G, Guglielmino CR (2007) Globalization and fruitfly invasion and expansion: the medfly paradigm. Genetica 131:1–9. http://doi.org/10.1007/s10709-006-9117-2

Manrakhan A (2007) Fruit fly monitoring – purpose tools and methodology. In: Billah MK, Ekesi S (eds) A field guide to the management of economically important tephritid fruit flies in Africa, 2nd edn. *icipe* Science Press, C1-C17

Manrakhan A, Kotze C, Daneel J-H, Stephen PR, Beck RR (2013) Investigating a replacement for malathion in bait sprays for fruit fly control in South African citrus orchards. Crop Prot 43:45–53

Nagaraja KS, Gowda VN, Basavarajeshwari SLG, Shabarish PR, Husainnaik M (2014) Effect of protein food baits in attracting fruit flies in mango orchard. Trends Biosci 7:1971–1973

Noy NF, McGuinness DL (2001) Ontology development 101: a guide to creating your first ontology. Stanford knowledge systems laboratory technical report KSL-01-05 and Stanford medical informatics technical report SMI-2001-0880

Noy NF, Crubézy M, Fergerson RW, Knublauch H, Tu SW, Vendetti J, Musen MA (2003) Protégé-2000: an open-source ontology-development and knowledge-acquisition environment. AMIA Annu Symp Proc 953:953

Rubin DL, Noy NF, Musen MA (2007) Protege: a tool for managing and using terminology in radiology applications. J Digit Imag 20:34–46

Rwomushana I, Ekesi S, Gordon I, Ogol CKPO (2008) Host plants and host plant preference studies for *Bactrocera invadens* (Diptera: Tephritidae) in Kenya, a new invasive fruit fly species in Africa. Anns Entomol Soc Amer 101:331–340

Silvertown J (2009) A new dawn for citizen science. TREE 24:467–471

Sivakumar R, Arivoli PV (2011) Ontology visualization protégé tools: a review. Int J Adv Inf Technol 1:1–11. doi:10.5121/ijait.2011.1401

Smith B (2003) Ontology. In: Floridi L (ed) Blackwell companion to philosophy, information and computers. Blackwell, Oxford, pp 155–166

Sullivan BL, Wood CL, Iliff MJ, Bonney RE, Fink D, Kelling S (2009) eBird: A citizen-based bird observation network in the biological sciences. Biol Conserv 142:2282–2292

Tsarkov D, Horrocks I (2006) FaCT++ description logic reasoner: system description. Automated Reasoning. Retrieved from http://link.springer.com/chapter/10.1007/11814771_26

Vargas RI, Walsh WA, Kanehisa D, Stark JD, Nishida T (2000) Comparative demography of three Hawaiian fruit flies (Diptera: Tephritidae) at alternating temperatures. Anns Entomol Soc Amer 93:75–81

Yoder MJ, Miko I, Seltmann KC, Bertone MA, Deans AR (2010) A gross anatomy ontology for Hymenoptera. PLoS One 5:1–8. doi:10.1271/journal.pone.0015991

Caroline Chepkoech Kiptoo is currently a doctoral student at the Department of Informatics, School of Information Technology within the University of Pretoria, South Africa. Her research motivation is in applied computing, particularly the practical application of computing technologies to address challenges in day to day activities. Ms. Kiptoo has experience in information systems development, database systems and development of web based applications. Her research interests are in knowledge technologies, semantic technologies and ontological modelling. She is also interested in crowd computing and more specifically crowdsourcing techniques. Her current practical application areas include biodiversity informatics, knowledge management and knowledge transfer.

Aurona Gerber is an Associate Professor in the Department of Informatics in the Engineering, Build Environment and Information Technology (EBIT) faculty of the University of Pretoria. Her primary interest is research in to the development and use of models and modelling languages (including formal ontologies) within all aspects of Information Systems including Enterprise Architecture and Engineering. She is also interested in IS research methodologies, specifically Design Science Research. She has published more than 50 accredited publications as well as supervised several Masters and Doctoral students to completion. She is current Chapter Chair of the IEEE SMC South African Chapter, as well as Co-chair and founder of the International Technical Committee of the IEEE SMC on Enterprise Architecture and Engineering (EAE TC). She is a member of the AIS and a board member the SA Chapter of the AIS (AISSAC). She is also a board member of SAICSIT (the South African Institute for Computer Scientists and Information Technologists).

Alta Van der Merwe is currently the Head of Department in the Informatics Department within the School of Information Technology, University of Pretoria. She focuses on the design of socio-technical solutions with research activities in Enterprise Architecture, Data Science and different theories supporting the successful use of technology in the organization. Her interest also includes the design of systems using innovative and new approaches such as Crowd Sourcing and Content Awareness. Prof van der Merwe is serving her second term as president of the South African Institute of Computer Scientists and Information Technologists. She is the founder and past chair of the South African IEEE SMCS Chapter, specialist editor of the SAIEE journal (Software Engineering track) and co-founder and past chair of the Enterprise Architecture Research Forum (EARF). On the international level she was involved in the proposal and acceptance of the IEEE SMC Enterprise Engineering and Enterprise Architecture Technical Committee, where she still serves as co-chair.

Part II
Pre-harvest and Post-harvest Management Measures

Chapter 12
Detection and Monitoring of Fruit Flies in Africa

Aruna Manrakhan

Abstract The production and trade of fresh fruit is currently increasing in Africa, as is the movement of people into and within the region. This increases the risk of new fruit fly invasions. The increase in production of fresh fruit also requires for more effective management of established insect pests like fruit flies in order to maximise yield and facilitate trade. It is imperative, therefore, that effective fruit fly detection and monitoring systems are developed, set up and maintained in Africa in order to protect and expand the fresh fruit sector which brings income and employment to the region. Effective trapping systems have been developed for many fruit fly pests and they enable the early detection and monitoring of these pests. However, for a number of important established fruit fly pests in Africa, notably in the *Dacus* group, trapping systems are yet to be developed and optimised. Moreover, new recently developed fruit fly attractants have yet to be tested on African species.

Keywords Bactrocera • Zeugodacus • Dacus • Ceratitis • Traps • Attractants

1 Introduction

Fruit fly populations can be characterised by surveys using traps and attractants (IAEA 2003; Cunningham 1989b). These surveys have two purposes: detection of the presence of new fruit fly species in an area and monitoring of the fluctuations in populations of established fruit fly species (IAEA 2003; Cunningham 1989b).

Olfactory attractants form the basis of most fruit fly detection and monitoring systems (Light and Jang 1996). The olfactory attractants are contained in traps with different retention systems: liquid, insecticide, sticky insert (IAEA 2003). A number of fruit fly attractants have been discovered over the years and a thorough understanding of fruit fly behaviour in response to them, has led to the development of effective trapping systems for some species (Economopoulos and Haniotakis 1994). Many fruit fly attractants are commercially available. Trapping procedures using

---

A. Manrakhan (✉)
Citrus Research International, PO Box 28, Nelspruit 1200, South Africa
e-mail: aruna@cri.co.za

commercially available attractants and traps for fruit fly detection and monitoring have been harmonised in published guidelines (IAEA 2003; IPPC 2008; Manrakhan 2006).

Although attractants and traps are commercially available and trapping procedures have been established for many species, fruit fly trapping surveys at producer, national and regional levels are not systematically implemented in many African countries. Trapping surveys are costly investigations. Due to funding constraints it is difficult for most African countries to implement extensive detection and monitoring surveys. However, strategic detection networks should still be established and maintained to protect individual countries and indeed the region from invasions of exotic fruit fly species in the future (Papadopoulos 2014). Moreover, there should be more transparency concerning fruit fly interceptions within the region so that regional approaches can be taken to effectively deal with invasive fruit fly species (Papadopoulos 2014). Currently in the African region, new fruit fly interceptions are reported through bilateral meetings and on the International Plant Protection Portal. Often though, detections of new fruit fly pests are not reported immediately, and the status of actions against new fruit fly species are not regularly updated, leaving neighbouring countries uninformed of the current pest status in the region and at risk of invasion. The lack of effective national and regional detection surveys has resulted in the introduction, spread and establishment of four exotic fruit fly species in Africa over the last 200 years. With increasing trade and movement of people in the African region, there is an increased risk of introduction and spread of more exotic fruit fly species. Detection surveys are therefore required nationally and regionally in Africa. Furthermore, for development of competitive fresh fruit export sectors in Africa, monitoring surveys of established fruit fly pests are essential at the level of the producers as well as at the national level, in order to implement timely and effective control actions and ensure export of fruit-fly-free fresh fruit consignments.

Here I review the tools available for fruit fly detection and monitoring and the current status of trapping surveys in Africa. Future prospects in the establishment of detection and monitoring systems in Africa are discussed.

2 Tools Used in Fruit Fly Trapping Surveys

There are currently four exotic, invasive fruit fly species within the African region: the Oriental fruit fly, *Bactrocera dorsalis* (Hendel) (previously reported in Africa as *Bactrocera invadens* Drew, Tsuruta & White); the solanum fruit fly, *Bactrocera latifrons* (Hendel); the melon fly, *Zeugodacus cucurbitae* (Coquillet) (previously known as *Bactrocera cucurbitae* (Coquillett)); and the peach fruit fly, *Bactrocera zonata* (Saunders) (Ekesi and Muchugu 2006). To date, only *B. dorsalis* is widely distributed within the region (Ekesi and Muchugu 2006), making trapping efforts for early detection of the other three species essential in currently pest-free countries to prevent further spread. Other potentially invasive species that have currently not been reported in Africa should also be included in detection surveys and these include: other members of the *B. dorsalis* complex (Clarke et al. 2005), other exotic *Bactrocera* species such as the Queensland fruit fly *Bactrocera tryoni* (Froggatt) and species in the genus *Anastrepha* (White and Elson-Harris 1994; Norrbom and Foote 1989). Furthermore, it would also be worthwhile to include indigenous *Ceratitis* and *Dacus* species that currently have a limited distribution in Africa in detection programmes as it is likely that their distribution could expand in the future with increasing movement of people and trade within the region.

For fruit fly species already present within a country, it is important that their populations are monitored within commercial fruit production areas for timely control actions. Monitoring of fruit flies should, ideally, be area-wide, covering all habitat types within a fruit-producing region to underpin more effective area-wide management actions.

There are known olfactory attractants for many fruit fly species of which some are present in African countries, but others are under exclusion. Olfactory attractants can be categorised as male lures, food-odour attractants and pheromones. Other categories of attractants such as fruit volatiles are still being investigated but could, in the future, play a role in detection and monitoring systems in Africa. Male lures and pheromones are species-specific and are usually preferred in detection programmes due to their specificity. For those species with no known male lures or pheromones, food-odour attractants are usually recommended. Different types of traps are recommended for use with different categories of attractants (IAEA 2003). Depending on the type of attractant, some traps are more effective than others. However, the cost, availability and handling efficacy of traps, as well as the environment being sampled often dictate the choice of traps. The efficacy of fruit fly detection systems has been evaluated for only a few species of economic importance and remains to be evaluated for other important fruit fly pests in Africa under African field conditions.

2.1 Olfactory Attractants

2.1.1 Male Lures

Dacine Flies (Bactrocera, Zeugodacus, Dacus *species*)

Populations of many species in the subtribe Dacina (*Bactrocera* species, *Zeugodacus* species, *Dacus* species) can be surveyed using species-specific male lures (Drew 1974; Drew and Hooper 1981; White 2006), most of which are commercially available. Ever since the mid 1980s there has been extensive research on new and improved male lures; most recent investigations have provided promising results on the efficacy of some of these new male lures, thereby expanding the toolkit of male lures for detection and monitoring of Dacine flies. The availability and cost of these new male lures are likely to influence their use in the African region.

Male lures for Dacine fruit fly pests are either compounds that naturally occur in many plant species or synthetic analogues of naturally occurring compounds (Cunningham 1989a), and they are categorised in two groups: methyl eugenol (4-allyl-1,2-dimethoxy-benzene) and cue-lure (4-(p-acetoxyphenyl)-2-butanone) (Drew 1974; Drew and Hooper 1981). The functional significance of male lures in this group of fruit flies has been widely debated (see review by (Raghu 2004)). There is strong support for the hypothesis that the response of Dacine flies to male lures is a mechanism of sexual selection and this is based on evidence of an increased reproductive success in Dacine flies following exposure to male lures (Kumaran et al. 2013; Shelly 2000; Shelly et al. 2003).

Males of most *Bactrocera* species are attracted to methyl eugenol (Cunningham 1989a; Tan et al. 2014; White 2006). Due to the potential carcinogenicity of methyl eugenol, there have been studies to identify alternative compounds (Khrimian et al. 1994; Mitchell et al. 1985). Fluorination of methyl eugenol does reduce its toxicity (Liquido et al. 1998), but also reduces its attractiveness as a bait in the field (Khrimian et al. 2009). Subsequently human exposure to methyl eugenol was deemed to be too low to be of concern (Tan et al. 2014), and as such the search for further alternatives to methyl eugenol have not been prioritised. In the African region, methyl eugenol has to be imported and is often unaffordable for resource-poor farmers. For this reason researchers in Senegal evaluated local products as alternatives to methyl eugenol (Ndiaye et al. 2008). They compared the attractiveness of methyl eugenol, ground nutmeg and a local beauty cream to adult male *B. dorsalis*. They found that methyl eugenol was the most attractive product compared to the two other products but that there was some degree of attraction to ground nutmeg which they recommended for use if methyl eugenol was unavailable (Ndiaye et al. 2008). Further investigations on natural sources with high methyl eugenol content are required in order to increase the use of methyl eugenol-based products as monitoring and control tools for resource-poor farmers in Africa.

Bactrocera latifrons is attracted to methyl eugenol (Flath et al. 1994). However during screening of a range of essential oils, aroma formulations and synthetic compounds, *B. latifrons* males were more attracted to α-ionol, a compound used in the perfume and flavour industry, than to methyl eugenol (Flath et al. 1994). Subsequently, cade oil and eugenol were found to separately synergise the attractiveness of α-ionol to males of *B. latifrons* (McQuate and Peck 2001; McQuate et al. 2004). For monitoring of *B. latifrons* in Africa, the combination of α-ionol and cade oil is currently recommended (Manrakhan 2006). A combination of α-ionol and eugenol could also become an option for this pest species. The synergist eugenol was reported to have less of a 'smoky' smell than the cade oil and would possibly enhance handling of the attractant (McQuate et al. 2004).

Males of other important *Bactrocera*, *Zeugodacus* and some *Dacus* species are attracted to cue lure (Cunningham 1989a; Tan et al. 2014; White 2006). Amongst the *Dacus* species, response to cue lure is known only for particular subgenera with no reported male lures for the other subgenera which include important pest species (White 2006). Cue lure is a synthetic compound whilst its analogue and break down product, raspberry ketone (4-(4-hydroxyphenyl)-2-butanone), is known to occur in a number of plants (Cunningham 1989a). Males of *Z. cucurbitae* are attracted to raspberry ketone (Cunningham 1989a). In field tests done in the early 2000s, the formate ester of raspberry ketone (i.e. raspberry ketone formate) was more attractive to *Z. cucurbitae* than cue lure (Oliver et al. 2002). However, in more recent field investigations, conflicting results were found for *Z. cucurbitae*. In one trial, raspberry ketone formate outperformed cue lure in terms of catches of *Z. cucurbitae* when exposed as a liquid lure (1 g per wick) or as a plug (Jang et al. 2007a). In a separate trial cue lure was more attractive to *Z. cucurbitae* than raspberry ketone formate (Shelly et al. 2012). The raspberry ketone formate, recently referred to as melolure, was also inferior in attracting *B. tryoni* compared with cue lure (Dominiak et al. 2015). However Dominiak et al. (2015) reported increased capture rates of *Dacus* spp in melolure-baited traps compared with cue lure-baited traps. These conflicting results indicate the need for further evaluation of raspberry ketone formate under contrasting field conditions, and further evaluations on other 'cue-lure responding' flies.

Vertlure is one of several male attractants for Dacine flies that is in a category of its own (Hancock 1989). Vertlure is specific to the jointed pumpkin fly, *Dacus vertebratus* Bezzi, an important pest of cucurbits in Southern Africa (Hancock 1989). To date there are no reports of other *Dacus* species being attracted to this lure.

A novel and promising male attractant for Dacine fruit flies is zingerone (4-hydroxy, 3-methoxyphenyl-2-butanone) which was identified from the floral components of a wild orchid (Tan and Nishida 2000). The discovery of this attractant followed observations of visits of 'methyl eugenol-responding flies' and 'cue lure-responding flies' to the flowers of this wild orchid (Tan and Nishida 2000). In subsequent studies done in northern Australia, zingerone was attractive to *Bactrocera jarvisi* (Tryon), and other Dacine species, some of which were not responsive to other known male lures (Fay 2012; Royer 2015). In a recent study conducted in South Africa, zingerone was attractive to a few *Dacus* species including the pumpkin fly, *Dacus frontalis* Becker, which is an important pest of water melon (Manrakhan et al. unpublished data). Further evaluation of this new attractant is warranted in Africa in order to determine the response of African Dacine to this lure.

Other new compounds in the phenylpropanoid and phenylbutanoid groups (such as isoeugenol, methyl isoeugenol, dihydroeugenol and 2-methoxy-4-propyl phenol) also attract Dacine flies that were previously non-responsive to methyl eugenol and cue lure in northern Australia (Royer 2015; Fay 2010). It would be important to trial these same compounds in Africa in order to determine the full range of Dacine flies that may be responsive.

Ceratitis *species*

Terpinyl acetate and trimedlure are the most commonly used male lures for pest fruit flies in the *Ceratitis* group (Cunningham 1989a; Ripley and Hepburn 1935; Beroza et al. 1961). Some early studies conducted in South Africa showed greater responses of the Natal fly, *Ceratitis rosa* Karsch and the Mediterranean fruit fly (medfly), *Ceratitis capitata* (Wiedemann) to trimedlure than to terpinyl acetate (Georgala 1964). Currently in South Africa, the male lure capilure, which is a blend of trimedlure and extenders (Leonhardt et al. 1984), is recommended for monitoring *C. capitata* and *C. rosa* (Manrakhan and Grout 2010). Studies comparing the relative attractiveness of trimedlure and capilure to *C. capitata* have produced varied results; one study showed a better performance for capilure compared with trimedlure (Hill 1987) while other studies showing the reverse (Baker et al. 1988; Rice et al. 1984; Shelly 2013). In trapping studies done in avocado orchards in South Africa (Grove et al. 1998), the mango fly, *Ceratitis cosyra* (Walker), was non-responsive to capilure-baited traps. Male *C. cosyra* were instead attracted to traps baited with β-caryophyllene (Grove et al. 1998). In mark-release-recapture studies conducted by Grout et al. (2011), capilure was less attractive to *C. rosa* than to *C. capitata* and the authors confirmed the lack of response of *C. cosyra* to capilure. In those same studies *C. cosyra* responded to a β-caryophyllene-based lure commercially sold as 'Ceratitis lure' in South Africa (Grout et al. 2011). Interestingly, Grout et al. (2011) also reported higher catches of *C. capitata* and *C. rosa* in traps baited with 'Ceratitis lure' than capilure. A new promising male lure for *Ceratitis* species is enriched ginger root oil (EGO) which attracted a wide range of *Ceratitis* species in studies done in Tanzania (Mwatawala et al. 2013). The EGO lure is a commercially available product in South Africa and in recent trapping studies was a superior male lure for *Ceratitis* flies compared with trimedlure (Manrakhan et al., unpublished data). In South Africa, an EGO-based product known as 'Last FF' is registered for control of *Ceratitis* species such as *C. capitata* and *C. rosa*. Further evaluation of EGO lures and the 'Ceratitis lure' is required across the African region in order to establish the responsiveness of all *Ceratitis* species. Studies done in Hawaii showed that the EGO lure was less effective than trimedlure in attracting *C. capitata* (Shelly 2013; Shelly and Pahio 2002). A new promising lure for *C. capitata* called ceralure B1 (an analogue of trimedlure) is currently being evaluated and has shown promise in some recent field trials (Jang et al. 2010). Ceralure B1 is yet to be evaluated in the African region.

2.1.2 Food-Odour Attractants

Many fruit fly species require both sugar and protein during their adult life for survival and reproduction. The protein requirements of fruit fly pests, and subsequently their attraction to protein sources, have been exploited in the development of food-odour attractants for fruit fly monitoring and control. The history and development of food-based attractants were recently reviewed by Epsky et al. (2014). Currently, there are two main types of food-odour attractants being used for fruit fly detection and monitoring: (1) liquid protein hydrolysates and (2) synthetic lures that contain synthetic versions of the main volatile components found in protein hydrolysate lures. Food-odour attractants are not species specific and are generally female biased. Protein hydrolysates, similar to the ones developed by the pioneers in this field such as McPhail (1939) and Gow (1954), are still in use today. The protein hydrolysates developed are from hydrolysis (using alkalis, acids or enzymes) of proteinaceous compounds such as corn protein, animal protein and yeast. Protein hydrolysates are usually available in liquid form and require dilution with water before use in traps. There are, however, practical problems in the use of liquid lures. Liquid lures are cumbersome to carry to the field and difficult to handle during trap placement and servicing. Putrefaction of liquid protein baits inside traps is problematic as it leads to decomposition of the fruit flies captured, rendering identification impossible. Lopez and Becerril (1967) investigated the addition of borax to liquid protein hydrolysates to prevent decomposition of captured flies. Addition of borax at a rate of 2 % to protein hydrolysates prevented putrefaction but led to a reduction in the number of flies captured (Lopez and Becerril 1967). In the late 1960s, Lopez et al. (1968) developed pelleted lures in order to improve transport and handling of the liquid protein hydrolysates. However, pelleted lures were slightly less effective than freshly-diluted protein lures and this lower efficacy was attributed to the fact that the pellets dissolved slowly (Lopez et al. 1968). The role of ammonia in the attraction of flies to protein hydrolysates was suggested by McPhail (1939) in the late 1930s and later proven by Bateman and Morton (1981). The latter authors also showed that, at high concentrations, ammonia was repellent. Since solutions of ammonium salts were more attractive to flies at high pH, Bateman and Morton (1981) suggested that other volatiles may also be released at high pH. There were then further investigations in to other possible volatile components of the protein hydrolysates that may have been luring fruit flies (Buttery et al. 1983; Morton and Bateman 1981). Synthetic mixtures matching the attractiveness of protein hydrolysates were then investigated with the aim of finding alternatives to the bulky protein hydrolysate liquid lures that were difficult to handle. Wakabayashi and Cunningham (1991) found a nine component synthetic mixture that was as attractive as protein hydrolysate to *Z. cucurbitae*; they described four of the components as being most important in attraction: ammonium bicarbonate, linoleic acid, putrescine, pyrrolidine. Heath et al. (1997) investigated the efficacy of synthetic mixtures in luring *C. capitata* and found that, for this particular species, the addition of methyl-ammonia derivatives to a mixture of ammonium acetate and putrescine increased the capture of the pest. Work in the 1990s led to the development, recommendation and use of what we now know as 'three-component lures' and 'two-component lures' as food-odour attractants for fruit flies (IPPC 2008). These synthetic food lures have the advantages of easy handling, long field life and the possibility of use in dry traps. The major disadvantages of synthetic lures are cost and availability. In Africa, these have to be imported and are likely to be unaffordable for large-scale government-funded detection programmes and for resource-poor farmers. Moving on from sophisticated synthetic mixtures there has been a line of research in many parts of the world, including Africa, on the development of low-cost protein baits. In Australia and the Pacific islands, the use of beer waste as an alternative low-cost fruit fly bait was explored in the late 1990s (Lloyd and Drew 1996). Lloyd and Drew (1996) described a simple method of converting waste yeast slurry from breweries into protein bait for fruit flies using a batch process involving the enzyme papain for digestion of the waste yeast cells. In Mauritius, waste brewer's yeast was shown to be a promising equivalent to local protein baits in terms of control efficacy (Sookar et al. 2002). In Nigeria, waste yeast slurry from breweries that was autolysed by heat was as attractive to fruit flies as imported protein hydrolysate (Umeh and Garcia 2008). Other promising low-cost fruit fly protein baits that were explored include human urine and chicken faeces, which were attractive to *Anastrepha* species (Piñero et al. 2003; Aluja and Pinero 2004). Other lines of investigation evaluated bacterial cultures and bacterial filtrates as attractants for fruit flies, but, to date, these have not been commercially produced and used (Reddy et al. 2014; Jang and Nishijima 1990). Food-odour attractants elicit different responses from different fruit fly species. For *C. capitata* and most *Anastrepha* species, synthetic lure mixtures were more effective than protein hydrolysates (Thomas et al. 2001; Robacker 1995; Katsoyannos et al. 1999; Holler et al. 2006; Epsky et al. 2011; Broughton and De Lima 2002). In contrast, for Dacine flies (*Bactrocera* spp. and *Zeugodacus* spp), comparative studies on the relative performance of synthetic mixtures and protein hydrolsates were variable (Wakabayashi and Cunningham 1991; Leblanc et al. 2010; Ekesi et al. 2014; Cornelius et al. 2000). For *B. dorsalis*, the dominant pest species in many African countries, liquid protein hydrolysates were more effective than synthetic mixtures (Leblanc et al. 2010; Ekesi et al. 2014; Cornelius et al. 2000). Responses of some fruit fly species to protein hydrolysates were increased with additives such as propylene glycol (antifreeze) and ammonium acetate (Thomas et al. 2001; Pinero et al. 2015). Depending on the type of protein hydrolysate, Duyck et al. (2004) improved efficacy by acidification or alkalinisation. It must be emphasized that not all congeneric fruit fly species have the same level of response to protein baits. In studies with *Anastrepha* species, the West Indian fruit fly *Anastrepha obliqua* (Macquart), was more responsive to Biolure (synthetic food-based attractant) and Nulure (protein hydrolysate) compared with the Mexican fruit fly, *Anastrepha ludens* (Loew) (Diaz-Fleischer et al. 2009). Differences in responses to baits between congeneric species were linked to differences in the life history traits of the species concerned (Diaz-Fleischer et al. 2009). Similarly in studies on *Ceratitis* species, *C. capitata* responded more strongly to protein baits than other *Ceratitis* species such as *C. rosa* and *C. cosyra* (Manrakhan and Kotze 2011). The methods for determining bait efficacy have recently been debated by Mangan and Thomas (Mangan and Thomas 2014) who suggested that baits should not only be evaluated by the magnitude of the catches, but also by the frequency of zero catches over time. In other words, a good bait should also be the one with the lowest frequency of zero captures. In their recent study, the authors showed that the performance of olfactory attractants varied over time (Mangan and Thomas 2014).

2.1.3 Pheromones

For most fruit fly species, except the olive fruit fly, *Bactrocera oleae* (Rossi), and the papaya fruit fly, *Toxotrypana curvicauda* Gerstaecker, there are no commercial pheromones available or recommended as attractants for use in detection and monitoring programmes (IPPC 2008;

IAEA 2003). The identification of a female sex pheromone produced by *B. oleae* led to the subsequent synthesis and commercial development of the olive fruit fly pheromone referred to as spiroketal (1,7-dioxaspiro [5,5] undecane) (Mazomenos 1989). In recent studies, Yokoyama et al. (2006) found that responses of *B. oleae* to pheromone-baited traps can be greatly improved when combined with food-odour attractants. For *T. curvicauda*, which is non-responsive to food-odour attractants, the synthetic male sex pheromone (2-6-methylvinyl pyrazine abbreviated as MVP and commonly referred to as papaya fruit fly pheromone) was effective when incorporated within a visual stimulus such as a fruit model (Landolt et al. 1988). The non-availability of pheromones for monitoring of other fruit fly pests is due to the lack of a detailed understanding of pheromone-mediated attraction during courtship and the current availability of effective lures for some important fruit fly pests (Tan et al. 2014). However, for fruit fly species that respond poorly or are non-responsive to commercially available olfactory attractants, an understanding of the mating behaviour and pheromone mediated courtship behaviour would be important. If pheromone mediated responses are found to be long range, it would then be important to explore the synthesis of these chemicals for further development and commercialisation.

2.1.4 Other Olfactory Attractants such as Fruit Volatiles

For some fruit fly species that respond poorly or are non-responsive to male lures and food-odour attractants, the potential use of plant volatiles has been investigated (see review by Quilici et al. [2014]) Recent research studies conducted on the lesser pumpkin fly, *Dacus ciliatus* (Loew), and *Dacus demmerezi* (Bezzi) have indicated that these species are poor responders to protein-based attractants (Deguine et al. 2012). Current trapping studies being conducted in the Limpopo and Mpumalanga provinces in South Africa also show that very few *Dacus* flies are collected in traps baited with food odour attractants, although the same species are trapped within the same sites in high numbers in male lure traps (Manrakhan et al. unpublished data). The use of fruit volatiles as attractants for *Dacus* species should be explored further. For *B. dorsalis*, which does not respond strongly to currently available synthetic food odour attractants, the responses of the pest to host and non-host odours were also studied (Jang et al. 1997; Kimbokota et al. 2013; Biasazin et al. 2014). Jang et al. (1997) showed that mated *B. dorsalis* females were highly attracted to volatile components of leaves from a non-host plant commonly known as Panax (*Polyscias guilfoylei* [Bull]). Volatiles from host fruit, in particular mature fruit, were also highly attractive to *B. dorsalis* (Kimbokota et al. 2013; Biasazin et al. 2014).

2.1.5 Lure Formulations and Combinations

Male lures are commercially available in different formulations: liquid, polymeric plug, laminate and wafers (IPPC 2008). The formulations differ in their field longevity (IPPC 2008). Liquid lures are generally dispensed on dental rolls or cotton wicks (IAEA 2003). Some earlier studies have shown that evaporation rates of male lures from cotton wicks were fairly constant. Cotton wicks impregnated with trimedlure were, however, ineffective after 2–4 weeks (Leonhardt et al. 1989) while those impregnated with methyl eugenol remained effective for at least 9 weeks (Qureshi et al. 1992). Polymeric plugs containing male lures had slower release rates than other dispensers such as mesoporous dispensers and wafers (Suckling et al. 2008; Domingues-Ruiz et al. 2008). Release rates of polymeric plugs are temperature dependent with higher release rates at higher temperatures (Domingues-Ruiz et al. 2008). Larger fruit fly catches were achieved when using wafers and mesoporous dispensers compared with polymeric plugs (Suckling et al. 2008; Domingues-Ruiz et al. 2008).

The effect of mixtures of different male lures has been investigated by a number of researchers with the justification that it could lead to savings in cost and time by monitoring more species of fruit flies at a time. In studies conducted by Vargas et al. (2000) and Shelly et al. (2004), mixtures of methyl eugenol and cue lure led to a decrease in catches of *B. dorsalis* compared with methyl eugenol on its own. While Vargas et al. (2000) found that mixtures of methyl eugenol and cue lure did not significantly change catches of *Z. cucurbitae* compared with cue lure on its own, Shelly et al. (2004) found that methyl eugenol had a synergistic effect when mixed with cue lure for *Z. cucurbitae*. In some more recent studies by Vargas et al. (2010), wafers containing methyl eugenol and raspberry ketone were as effective as either methyl eugenol or cue lure on their own for capturing *B. dorsalis* and *Z. cucurbitae*, respectively. Shelly et al. (2012) found that wafers containing mixtures of trimedlure, methyl eugenol and raspberry ketone were as effective as each separate lure used on its own. Such 'trilure' wafers could certainly reduce the time and costs associated with fruit fly monitoring.

With respect to food-odour attractants, the three-component and two-component lures are generally available as two and three separate dispensers, respectively – each containing a particular component. The use of three-component or two-component lures entails opening up all the required dispensers before placement in traps. The time involved in preparation and servicing of traps requiring two or three separate membrane dispensers could be considerably shortened if each lures were combined into one dispenser. In studies conducted on *C. capitata* and the guava fruit fly, *Anastrepha suspensa* (Loew), single dispensers of the three- and two-component lures were as effective as lures comprising separate dispensers (Jang et al. 2007b; Holler et al. 2009). In trials done in citrus orchards in South Africa, the single dispenser of the three-component lure was less effective than the lure comprised of three separate dispensers for females of *C. rosa* and *C. capitata*, but not for *C. cosyra* (Manrakhan et al., unpublished data).

Combinations of male lures and food-odour lures have also been tested; results from studies on combinations of trimedlure or ceralure with synthetic food lures have shown that catches of female *C. capitata* were reduced when the two groups of lures were combined compared with catches using synthetic food lures only (Toth et al. 2004; Broughton and De Lima 2002). In contrast, there were no changes in catches of male *C. capitata* when male lures were combined with synthetic food lures (Toth et al. 2004; Broughton and De Lima 2002). Liquido et al. (1993) found that male catches in traps baited with trimedlure and ammonia (ammonium bicarbonate solution) were significantly higher than catches in traps baited with trimedlure alone.

There are, therefore, positive and negative effects of combining lures depending on the target pest groups and the survey objectives. As such, the effects of lure combinations should be properly assessed before implementation in fruit fly detection and monitoring programmes.

2.2 Trap Types for Use with Olfactory Attractants

Trap types for fruit flies can be classified into two main categories, dry traps and wet traps, depending on their retention systems (IPPC 2008). Some traps such as modified and plastic McPhail-type traps can be used as either dry or wet traps. Sticky inserts or insecticidal strips are generally the retention systems used in dry traps. With wet traps, a liquid medium (liquid protein bait or water) is the retention system with flies being killed or retained by drowning in the liquid. A number of commercially available dry and wet traps are available for use with olfactory fruit fly attractants (IPPC 2008). The widely used dry traps with sticky inserts include the Cook and Cunningham trap, Jackson/Delta trap and ChamP trap. Widely used dry traps with insecticidal strips include the Lynfield trap and the Steiner trap. The most widely used wet trap is the McPhail trap. Male lures are mostly used with dry traps while food-odour attractants can be used in either dry or wet traps depending on the type of attractants, trapping environment and trapping methodology.

Amongst the dry traps, bucket type traps and modified McPhail type traps were more effective than Jackson traps when baited with male lures (Katsoyannos 1994; Cowley 1992; Cowley et al. 1990). Prokopy et al. (1996) studied the behaviour of *C. capitata* males arriving and entering Jackson traps and Nadel-Harris bucket traps under field cage conditions; they found no significant differences in the numbers of males arriving and entering either of these trap types. Jackson traps were deemed to be more suitable for early detection as opposed to regular monitoring programmes because the capture surface of the traps is limited (Uchida et al. 1996). When using male lures in regular monitoring programme for established fruit fly pests, bucket traps would be suitable (Uchida et al. 1996). There were no significant effects of entrance hole size or trap colour of the bucket traps baited with trimedlure on catches of *C. capitata* (Uchida et al. 1996). In contrast, for *B. dorsalis*, the colour of methyl-eugenol baited bucket traps did influence catches of males; white and yellow bucket traps attracted more males than green, red or black bucket traps (Stark and Vargas 1992). In a recent study using methyl eugenol or cue lure, targeting *B. dorsalis* and *Z. cucurbitae* respectively, toxicant-free bucket traps containing holes with a one way entrance design where flies could easily enter traps and not escape were more effective than toxicant-containing bucket traps with similar sized entrance holes (Jang 2011).

With food-odour attractants, trapping efficiency was also influenced by trap type; McPhail-type traps, Probodelt and multilure traps were the most effective for a range of different fruit fly species (Robacker and Czokajlo 2005; Navarro-Llopis et al. 2008; Heath et al. 1995; Hall et al. 2005; Katsoyannos and Papadopoulos 2004). When using synthetic food lures, the colour of dry traps influenced capture rates of *C. capitata*; *C. capitata* females were more attracted to green traps compared with orange traps (Epsky et al. 1995).

Research has also been ongoing on the development of inexpensive traps for resource poor farmers. In a study on *A. ludens*, inexpensive plastic bottles with 10 mm lateral holes were as effective as costly McPhail-type traps, in particular when used with the effective liquid-food bait, ceratrap, which is enzymatic hydrolysed protein from pig intestinal mucosa (Lasa et al. 2015). Lasa et al. (2014) recently argued that, under field conditions, the type of lure was more critical for trap efficacy than trap type and design.

2.3 Efficacy of Trapping Systems

Despite the availability of effective olfactory attractants and traps, information on the efficacy of trapping systems for early detection of fruit fly pests is, to date, still limited. The efficacy of male-lure-based trapping systems has been determined for only a few major pest species, namely *C. capitata*, *B. dorsalis* and *Z. cucurbitae* (Shelly and Edu 2010; Shelly et al. 2010; Lance and Gates 1994; Cunningham and Couey 1986). Estimates of detection probability have suggested that, for trimedlure-based traps, a density of 10 traps per 2.6 km^2 would only be able to detect *C. capitata* infestations consisting of more than 1000 individuals, whilst for methyl eugenol and cue lure-baited traps densities of 5 traps per 2.6 km^2 would be able to detect *B. dorsalis* and *Z. cucurbitae* infestations of 50 and 350 individuals respectively (Shelly et al. 2010; Lance and Gates 1994; Cunningham and Couey 1986). Studies have shown that congeneric species responding to a particular male lure might have different sensitivity to the same male lure (Wee et al. 2002; Grout et al. 2011). As such generic trapping density recommendations for early detection using a particular male lure might not be optimal for all pests within a particular lure category.

For species that do not respond to existing male lures and pheromones, food-odour attractants are usually used. Efficacy of trapping grids based on food-odour lures have not been determined for many pest species. Calkins et al. (1984) estimated that McPhail traps baited with liquid protein lures would only be able to detect a low population of *A. suspensa* if the traps are placed at a density of at least 83 traps per ha. More recently the effective sampling range of traps baited with synthetic food lures was estimated as ≈ 30 m for *A. suspensa* and *C. capitata* (Kendra et al. 2010; Epsky et al. 2010). Trapping grids based on food-odour attractants would therefore be not as sensitive as those based on male lures.

3 Current Status of Fruit Fly Detection and Monitoring in Africa

3.1 Detection Surveys of Exotic Fruit Flies in Africa

The discovery in Africa of the exotic invasive species, *B. dorsalis*, and its subsequent rapid spread to several African countries (Lux et al. 2003; Drew et al. 2005) was a wake-up call to the risk of exotic fruit fly invasions for many countries in the African region. This led to a number of nationally and internationally-funded fruit fly detection surveys in many countries across the region. However, these detection surveys were often short term, possibly due to financial constraints and lack of resources (materials and labour). The discontinuity in detection surveys is worrying and leaves countries in the African region at risk. Early detection of an invasive pest is one of the key factors in the successful eradication of the pest (Liebhold and Tobin 2008).

There are few examples of successful national fruit fly early detection programmes in Africa. In South Africa, the initiation of a Plant Health Early Warning Systems Division and the forging of an industry and government participatory forum led to the implementation of a national fruit fly surveillance programme and relevant control actions that did eventually delay the introduction of *B. dorsalis* into South Africa (Manrakhan et al. 2015; Manrakhan et al. 2009; Barnes and Venter 2008). In Mauritius, early detection of *B. dorsalis* led to two separate and successful eradication campaigns of *B. dorsalis* between 1996 and 1997 (Seewooruthun et al. 2000; Sookar et al. 2008) and more recently between 2013 and 2014 (Sookar P., personal communication).

With increasing trade in fresh fruit within the African region, a regional early detection programme would be more advantageous (Ekesi 2010). There are a few ongoing regional fruit fly detection programmes in Africa, notably in the Indian Ocean region and in West Africa. There is, however, a real need to initiate similar programmes in other parts of Africa: East Africa, North Africa and Southern Africa. Ideally, collaboration and sharing of knowledge between the programmes should be fostered.

3.2 Monitoring of Established Fruit Fly Pests

Effective control of established fruit fly pests in fruit-producing areas requires knowledge of the fruit fly present and their population dynamics. This can be achieved by monitoring surveys in both commercial and non-commercial orchards within fruit producing areas. Currently fruit fly monitoring surveys are not done systematically in all fruit producing areas within the African region. In South Africa, it is recommended that monitoring of fruit fly pests in citrus starts at least 2 months before the earliest harvest dates (Manrakhan and Grout 2010). The recommendation is for traps to be checked weekly and on the basis of catches, decisions are made on the intensity of control actions. Thresholds for specific trap and lure combinations have been developed for particular fruit fly pests in the citrus industry (Grout et al. 2011). Adherence to these threshold levels has led to very few fruit fly export market interceptions in fresh fruit consignments from South Africa. It is important that threshold levels for fruit flies in recommended trap lure systems is established for different fruit producing areas within the region in order to provide guidance to fruit growers on the necessary control actions to achieve fruit fly free consignments.

4 Future Prospects

Although trapping systems have been developed for most major fruit fly species in and outside of Africa, the trapping systems for some important *Dacus* pest species are yet to be developed and optimised. Effective olfactory attractants should be sought for those *Dacus* species that do not respond to commercially-available lures. For fruit fly pests that respond to known lures, new improved attractants have been developed, but are yet to be tested on African fruit fly species.

With regards to exotic fruit fly pests, plant health organisations in each country within the African region should prioritise the establishment and maintenance of detection-focussed trapping systems and should encourage participation of the fruit industry to ensure sustainability. Ideally there should be sharing, co-ordination, collaboration and transparency between countries to enable a more regional approach to fruit fly detection in Africa.

Acknowledgements I would like to thank Dr Sunday Ekesi for his invitation in writing about detection and monitoring of fruit flies in Africa. This book chapter was funded by the International Centre of Insect Physiology and Ecology (*icipe*), Nairobi, Kenya.

References

Aluja M, Pinero J (2004) Testing human urine as a low-tech bait for *Anastrepha* spp. (Diptera: Tephritidae) in small guava, mango, sapodilla and grapefruit orchards. Fla Entomol 87:41–50

Baker PS, Hendrichs J, Liedo P (1988) Improvement of attractant dispensing systems for the Mediterranean fruit fly (Diptera: Tephritidae) sterile release program in Chiapas, Mexico. J Econ Entomol 81:1068–1072

Barnes BN, Venter J-H (2008) The South African fruit fly action plan- area-wide suppression and exotic species surveillance. Paper presented at the 7th international symposium on fruit flies of economic importance Salvador, Bahia, Brazil, 10–15 September 2006

Bateman MA, Morton TC (1981) The importance of ammonia in proteinaceous attractants for fruit flies (Family: Tephritidae). Aus J Agric Res 32:883–903

Beroza M, Green N, Gertler SI, Steiner LF, Miyashita DH (1961) New attractants for the Mediterranean fruit fly. Agric & Food Chem 9:361–365

Biasazin TD, Karlsson MF, Hillbur Y, Seyoum E, Dekker T (2014) Identification of host blends that attract the African invasive fruit fly, *Bactrocera invadens*. J Chem Ecol 40:966–976

Broughton S, De Lima CP (2002) Field evaluation of female attractants for monitoring *Ceratitis capitata* (Diptera: Tephritidae) under a range of climatic conditions and population levels in Western Australia. J Econ Entomol 95:507–512

Buttery RG, Ling LC, Teranishi R, Mon TR (1983) Insect attractants: volatiles of hydrolyzed protein insect baits. J Agric & Food Chem 31:689–692

Calkins CO, Schroeder WJ, Chambers DL (1984) Probability of detecting Carribean fruit fly, *Anastrepha suspensa* (Loew) (Diptera: Tephritidae), populations with McPhail traps. J Econ Entomol 77:198–201

Clarke AR, Armstrong KF, Carmichael AE, Milne JR, Raghu S, Roderick GK, Yeates DK (2005) Invasive phytophagous pests arising through a recent tropical evolutionary radiation: the *Bactrocera dorsalis* complex of fruit flies. Ann Rev Entomol 50:293–319

Cornelius ML, Nergel L, Duan JJ, Messing RH (2000) Responses of female Oriental fruit flies (Diptera:Tephritidae) to protein and fruit odors in field cage and open field tests. Environ Entomol 29:14–19

Cowley JM (1992) A new system of fruit fly surveillance trapping in New Zealand. NZ Entomol 13:81–84

Cowley JM, Page FD, Nimmo PR, Cowley DR (1990) Comparison of the effectiveness of two traps for *Bactrocera tryoni* (Diptera: Tephritidae) and implications for quarantine surveillance systems. J Aus Entomol Soc 29:171–176

Cunningham RT (1989a) Parapheromones. In: Robinson AS, Hooper G (eds) Fruit flies, their biology, natural enemies and control, vol 3A. Elsevier, Amsterdam, pp 221–229

Cunningham RT (1989b) Population detection. In: Robinson AS, Hooper G (eds) Fruit flies: their biology, natural enemies and control, vol 3B. Elsevier, Amsterdam, pp 169–173

Cunningham RT, Couey HM (1986) Mediterranean fruit fly (Diptera: Tephritidae): distance/response curves to trimedlure to measure trapping efficiency. Environ Entomol 15:71–74

Deguine J-P, Douraguia E, Atiama-Nurbel T, Chiroleu F, Quilici S (2012) Cage study of spinosad-based bait efficacy on *Bactrocera cucurbitae*, *Dacus ciliatus*, and *Dacus demmerezi* (Diptera: Tephritidae) in Reunion island. J Econ Entomol 105:1358–1365

Diaz-Fleischer F, Arredondo J, Flores S, Montoya P, Aluja M (2009) There is no magic fruit fly trap: multiple biological factors influence the response of adult *Anastrepha ludens* and *Anastrepha obliqua* (Diptera: Tephritidae) individuals to multilure traps baited with Biolure or Nulure. J Econ Entomol 102:86–94

Domingues-Ruiz J, Sanchis J, Navarro-Llopis V, Primo J (2008) A new long-life trimedlure dispenser for Mediterranean fruit fly. J Econ Entomol 101(4):1325–1330

Dominiak BC, Campbell AJ, Jang EB, Ramsey A, Fanson BG (2015) Field evaluation of melolure, a formate analogue of cuelure, and reassessment of fruit fly species trapped in Sydney, New South Wales. Aus J Econ Entomol 108:1176–1181

Drew RAI (1974) The responses of fruit fly species (Diptera: Tephritidae) in the south pacific area to male attractants. J Aus Entomol Soc 13:267–270

Drew RAI, Hooper GHS (1981) The responses of fruit fly species (Diptera: Tephritidae) in Australia to various attractants. J Aus Entomol Soc 20:201–205

Drew RAI, Tsuruta K, White IM (2005) A new species of pest fruit fly (Diptera: Tephritidae) from Sri Lanka and Africa. Afri Entomol 13:149–154

Duyck PF, Rousse P, Ryckewaert P, Fabre F, Quilici S (2004) Influence of adding borax and modifying pH on effectiveness of food attractants for melon fly (Diptera: Tephritidae). J Econ Entomol 97:1137–1141

Economopoulos AP, Haniotakis GE (1994) Advances in attractant and trapping technologies for Tephritids. In: Calkins CO, Klassen W, Liedo P (eds) Fruit flies and the sterile insect technique. CRC Press, Boca Raton, pp 57–66

Ekesi S (2010) Combating fruit flies in Eastern and Southern Africa (COFESA): elements of a strategy and action plan for a regional cooperation program. The World Bank

Ekesi S, Muchugu E (2006) Tephritid fruit flies in Africa- fact sheets of some economically important species. In: Ekesi S, Billah MK (eds) A field guide to the management of economically important Tephritid fruit flies in Africa, 2nd edn. icipe Science Press, Nairobi, pp B-1–B-19

Ekesi S, Mohamed S, Tanga CM (2014) Comparison of food-based attractants for *Bactrocera invadens* (Diptera: Tephritidae) and evaluation of mazoferm-spinosad bait spray for field suppression in mango. J Econ Entomol 107:299–309

Epsky ND, Heath RR, Guzman A, Meyer WL (1995) Visual cue and chemical cue interactions in a dry trap with food-based synthetic attractant for *Ceratitis capitata* and *Anastrepha ludens* (Diptera: Tephritidae). Environ Entomol 24:1387–1395

Epsky ND, Espinoza HR, Kendra PE, Abernathy R, Midgarden D, Heath RR (2010) Effective sampling range of a synthetic protein-based attractant for Ceratitis capitata (Diptera: Tephritidae). J Econ Entomol 103:1886–1895

Epsky ND, Kendra PE, Pena J, Heath RR (2011) Comparison of synthetic food-based lures and liquid protein baits for capture of *Anastrepha suspensa* (Diptera: Tephritidae) adults. Fla Entomol 94:180–185

Epsky ND, Kendra PE, Schnell EQ (2014) History and development of food-based attractants. In: Shelly TE, Epsky ND, Jang EB, Reyes-Flores J, Vargas R (eds) Trapping and the detection, control and regulation of Tephritid fruit flies: lures, area-wide programs and trade implications. Springer Science+Business, Dordrecht, pp 75–118

Fay HAC (2010) Exploring structure-activity relationships in the phenylpropanoids to procure new male lures for non-responsive *Bactrocera* and *Dacus*. In: Sabater-Munoz B, Navarro Llopis V, Urbaneja A (eds) 8th international symposium on fruit flies of economic importance, Valencia, 2010. Laimprenta CG, pp 270–280

Fay HAC (2012) A highly effective and selective male lure for *Bactrocera jarvisi* (Tryon) (Diptera: Tephritidae). Aus J Entomol 51:189–197

Flath RA, Cunningham RT, Liquido NJ, McGovern TP (1994) α-Ionol as attractant for trapping *Bactrocera latifrons* (Diptera: Tephritidae). J Econ Entomol 87:1470–1476

Georgala MB (1964) The response of the males of *Pterandus rosa* (Karsch) and *Ceratitis capitata* (Wied.) to synthetic lures (Diptera: Trypetidae). J Entomol Soc South Afr 27:67–73

Gow PL (1954) Proteinaceous bait for the Oriental fruit fly. J Econ Entomol 47:153–160

Grout TG, Daneel JH, Ware AB, Beck RR (2011) A comparison of monitoring systems used for *Ceratitis* species (Diptera: Tephritidae) in South Africa. Crop Prot 30:617–622

Grove T, De Beer MS, Dreyer S, Steyn WP (1998) Monitoring fruit flies in avocado orchards. S Afr Avocado Growers' Assoc Yearb 21:80–82

Hall DG, Burns RE, Jenkins CC, Hibbard KL, Harris DL, Sivinski JM, Nigg HN (2005) Field comparison of chemical attractants and traps for Caribbean fruit fly (Diptera: Tephritidae) in Florida citrus. J Econ Entomol 98:1641–1647

Hancock DL (1989) Pest status; southern Africa. In: Robinson AS, Hooper G (eds) Fruit flies; their biology, natural enemies and control, vol 3A, World crop pests. Elsevier, Amsterdam, pp 51–58

Heath RR, Epsky ND, Guzman A, Dueben BD, Manukian A, Meyer WL (1995) Development of a dry plastic insect trap with food-based synthetic attractant for the Mediterranean and Mexican fruit flies (Diptera: Tephritidae). J Econ Entomol 88:1307–1315

Heath RR, Epsky ND, Dueben BD, Rizzo J, Jeronimo F (1997) Adding methyl-substituted ammonia derivatives to a food-based synthetic attractant on capture of Mediterranean and Mexican fruit flies (Diptera: Tephritidae). J Econ Entomol 90:1584–1589

Hill AR (1987) Comparison between trimedlure and capilure®- attractants for male *Ceratitis capitata* (Wiedemann) (Diptera: Tephritidae). J Aus Entomol Soc 26:35–36

Holler T, Sivinski J, Jenkins C, Fraser S (2006) A comparison of yeast hydrolysate and synthetic food attractants for capture of *Anastrepha suspensa* (Diptera: Tephritidae). Fla Entomol 89:419–420

Holler TC, Peebles M, Young A, Whiteman L, Olson S, Sivinski J (2009) Efficacy of the suterra biolure individual female fruit fly attractant packages vs. the unipak version. Fla Entomol 92:667–669

IAEA (2003) Trapping guidelines for area-wide fruit fly programmes. IAEA, Vienna, 48 pp

IPPC (2008) Annex 1 to ISPM No. 26 (Establishment of pest free areas for fruit flies (Tephritidae)). Fruit fly trapping. International Plant Protection Portal. p 25

Jang EB (2011) Effectiveness of plastic matrix lures and traps against *Bactrocera dorsalis* and *Bactrocera cucurbitae* in Hawaii. J Appl Entomol 135:456–466

Jang EB, Nishijima KA (1990) Identification and attractancy of bacteria associated with *Dacus dorsalis* (Diptera: Tephritidae). Environ Entomol 19:1726–1731

Jang EB, Carvalho LA, Stark JD (1997) Attraction of female Oriental fruit fly, *Bactrocera dorsalis*, to volatile semiochemicals from leaves and extracts of a nonhost plant, Panax (*Polyscias guilfoylei*) in laboratory and olfactometer assays. J Chem Ecol 23:1389–1401

Jang EB, Casana-Giner V, Oliver JE (2007a) Field captures of wild melon fly (Diptera: Tephritidae) with an improved male attractant, raspberry ketone formate. J Econ Entomol 100:1124–1128

Jang EB, Holler TC, Moses AL, Salvato MH, Fraser S (2007b) Evaluation of a single-matrix food attractant tephritid fruit fly bait dispenser for use in fedderal trap detection programs. Proc Hawaiian Entomol Soc 39:1–8

Jang EB, Khrimian A, Holler T (2010) Field response of Mediterranean fruit flies to ceralure B1 relative to the most active isomer and commercial formulation of trimedlure. J Econ Entomol 103:1586–1593

Katsoyannos BI (1994) Evaluation of Mediterranean fruit-fly traps for use in sterile-insect-technique programmes. J Appl Entomol 118:442–452

Katsoyannos BI, Papadopoulos NK (2004) Evaluation of synthetic female attractants against *Ceratitis capitata* (Diptera: Tephritidae) in sticky coated spheres and McPhail type traps. J Econ Entomol 97:21–26

Katsoyannos BI, Heath RR, Papadopoulos NK, Epsky ND, Hendrichs J (1999) Field evaluation of Mediterranean fruit fly (Diptera:Tephritidae) female selective attractants for use in monitoring programs. J Econ Entomol 92:583–589

Kendra PE, Epsky ND, Heath RR (2010) Effective sampling range of food-based attractants for female *Anastrepha suspensa* (Diptera: Tephritidae). J Econ Entomol 103:533–540

Khrimian AP, DeMilo AB, Waters RM, Liquido NJ, Nicholson JM (1994) Monofluoro analogs of eugenol methyl ether as novel attractants for the Oriental fruit fly. J Org Chem 59:8034–8039

Khrimian A, Siderhurst MS, McQuate GT, Liquido NJ, Nagata J, Carvalho L, Guzman F, Jang EB (2009) Ring-fluorinated analog of methyl eugenol: attractiveness to and metabolism in the Oriental fruit fly, *Bactrocera dorsalis* (Hendel). J Chem Ecol 35:209–218

Kimbokota F, Njagi PGN, Torto B, Ekesi S, Hassanali A (2013) Responses of *Bactrocera invadens* (Diptera: Tephritidae) to volatile emissions of fruits from three hosts. J Biol Agric & Healthcare 3:53–60

Kumaran N, Balagawi S, Schutze MK, Clarke AR (2013) Evolution of lure response in tephritid fruit flies: phytochemicals as drivers of sexual selection. Anim Behav 85:781–789

Lance DR, Gates DB (1994) Sensitivity of detection trapping systems for Mediterranean fruit flies (Diptera: Tephritidae) in Southern California. J Econ Entomol 87:1377–1383

Landolt PJ, Heath RR, Agee HR, Tumlinson JH, Calkins CO (1988) Sex pheromone-based trapping system for papaya fruit fly (Diptera: Tephritidae). J Econ Entomol 81:1163–1169

Lasa R, Velazquez OE, Ortega R, Acosta E (2014) Efficacy of commercial traps and food odor attractants for mass trapping of *Anastrepha ludens* (Diptera: Tephritidae). J Econ Entomol 107:198–205

Lasa R, Herrera F, Miranda E, Gomez E, Antonio S, Aluja M (2015) Economic and highly effective trap-lure combination to monitor the Mexican fruit fly (Diptera: Tephritidae) at the orchard level. J Econ Entomol 108:1637–1645

Leblanc L, Vargas RI, Rubinoff D (2010) Captures of pest fruit flies (Diptera: Tephritidae) and nontarget insects in Biolure and Torula yeast traps in Hawaii. Environ Entomol 39:1626–1630

Leonhardt BA, Rice RE, Harte EM, Cunningham RT (1984) Evaluation of dispensers containing Trimedlure, the attractant for the Mediterranean fruit fly (Diptera: Tephritidae). J Econ Entomol 77:744–749

Leonhardt BA, Cunningham RT, Rice RE, Harte EM, Hendrichs J (1989) Design, effectiveness, and performance criteria of dispenser formulations of trimedlure, an attractant of the Mediterranean fruit fly (Diptera: Tephritidae). J Econ Entomol 82:860–867

Liebhold AM, Tobin PC (2008) Population ecology of insect invasions and their management. Ann Rev Entomol 53:387–408

Light DM, Jang EB (1996) Olfactory semiochemicals of Tephritids. In: McPheron BA, Steck GJ (eds) Fruit fly pests: a world assessment of their biology and management. St Lucie Press, Delray Beach, pp 73–90

Liquido NJ, Teranishi R, Kint S (1993) Increasing the efficiency of catching Mediterranean fruit fly (Diptera: Tephritidae) males in Trimedlure-baited traps with ammonia. J Econ Entomol 86:1700–1705

Liquido NJ, Khrimian AP, DeMilo AB, McQuate GT (1998) Monofluoro analogues of methyl eugenol: new attractants for males of *Bactrocera dorsalis* (Hendel) (Dipt., Tephritidae). J Appl Entomol 122:259–264

Lloyd AC, Drew RAI (1996) Modification and testing of brewery waste yeast as a protein source for fruit fly bait. In: Allwood AJ, Drew RAI (eds) Regional management of fruit flies in the Pacific, Nadi, Fiji. ACIAR, pp 192–198

Lopez FD, Becerril OH (1967) Sodium borate inhibits decomposition of two protein hydrolysates attractive to the Mexican fruit fly. J Econ Entomol 60:137–140

Lopez FD, Spishakoff LM, Becerill H (1968) Pelletized lures for trapping the Mexican fruit fly. J Econ Entomol 61:316–317

Lux SA, Ekesi S, Dimbi S, Mohamed S, Billah M (2003) Mango-infesting fruit flies in Africa: perspectives and limitations of biological approaches to their management. In: Neuenschwander P, Borgemeister C, Langewald J (eds) Biological Control in IPM systems in Africa. CAB International, Oxon, pp 277–293

Mangan RL, Thomas DB (2014) Comparison of torula yeast and various grape juice products as attractants for Mexican fruit fly (Diptera: Tephritidae). J Econ Entomol 107:591–600

Manrakhan A (2006) Fruit fly monitoring – purpose, tools and methodology. In: Ekesi S, Billah MK (eds) A field guide to the management of economically important Tephritid fruit flies in Africa. icipe Science press, Nairobi, pp c1–c17

Manrakhan A, Grout TG (2010) Fruit fly. In: Grout TG (ed) Integrated production guidelines for export citrus. Integrated pest and disease management, vol 3. Citrus Research International, Nelspruit, pp 1–7

Manrakhan A, Kotze C (2011) Attraction of *Ceratitis capitata*, *C. rosa* and *C. cosyra* (Diptera: Tephritidae) to proteinaceous baits. J Appl Entomol 135:98–105

Manrakhan A, Grout T, Hattingh V (2009) Combating the African invader fly *Bactrocera invadens*. SA Fruit J 8:57–61

Manrakhan A, Venter JH, Hattingh V (2015) The progressive invasion of *Bactrocera dorsalis* (Diptera: Tephritidae) in South Africa. Biol Invasions 17:2803–2809

Mazomenos BE (1989) *Dacus oleae*. In: Robinson AS, Hooper G (eds) Fruit flies: their biology, natural enemies and control, vol 3A. Elsevier, Amsterdam, pp 169–178

McPhail M (1939) Protein lures for fruitflies. J Econ Entomol 32:758–761

McQuate GT, Peck SL (2001) Enhancement of attraction of alpha-ionol to male *Bactrocera latifrons* (Diptera: Tephritidae) by addition of a synergist, cade oil. J Econ Entomol 94:39–46

McQuate GT, Keum YS, Sylva CD, Li QX, Jang EB (2004) Active ingredients in cade oil that synergize attractiveness of α-Ionol to male *Bactrocera latifrons* (Diptera: Tephritidae). J Econ Entomol 97:862–870

Mitchell WC, Metcalf RL, Metcalf ER, Mitchell S (1985) Candidate substitutes for methyl eugenol as attractants for the area-wide monitoring and control of the Oriental fruit fly, *Dacus dorsalis* Hendel (Diptera: Tephritidae). Environ Entomol 14:176–181

Morton TC, Bateman MA (1981) Chemical studies on proteinaceous attractants for fruit flies, including the identification of volatile constituents. Aus J Agric Res 32:905–916

Mwatawala MW, Virgilio M, Quilici S, Dominic M, De Meyer M (2013) Field evaluation of the relative attractiveness of enriched ginger root (EGO)lure and trimedlure for African *Ceratitis* species (Diptera: Tephritidae). J Appl Entomol 137:392–397

Navarro-Llopis V, Alfaro F, Dominguez J, Sanchis J, Primo J (2008) Evaluation of traps and lures for mass trapping of Mediterranean fruit fly in citrus groves. J Econ Entomol 101:126–131

Ndiaye M, Dieng EO, Delhove G (2008) Population dynamics and on-farm fruit fly integrated pest management in mango orchards in the natural areas of Niayes in Senegal. Pest Man Hort Ecosyst 14:1–8

Norrbom AL, Foote RH (1989) The taxonomy and zoogeography of the genus *Anastrepha* (Diptera: Tephritidae). In: Robinson AS, Hooper G (eds) Fruit flies: their biology, natural enemies and control, vol 3A. Elsevier, Amsterdam, pp 15–26

Oliver JE, Casana-Giner V, Jang EB, McQuate GT, Carvalho L (2002) Improved attractants for the melon fly, *Bactrocera cucurbitae*. In: Barnes BN (ed) 6th international fruit fly symposium. Isteg Scientific Publications, Stellenbosch, pp 283–290

Papadopoulos NT (2014) Fruit fly invasion: historical, biological, economic aspects and management. In: Shelly T, Epsky N, Jang EB, Reyes-Flores J, Vargas R (eds) Trapping and the detection, control and regulation of Tephritid fruit flies: lures, area-wide programs and trade implications. Springer Science+Business media, Dordrecht, pp 219–252

Piñero J, Aluja M, Vazquez A, Equiiiua M, Varon J (2003) Human urine and chicken feces as fruit fly (Diptera: Tephritidae) attractants for resource-poor fruit growers. J Econ Entomol 96:334–340

Pinero JC, Souder SK, Smith TR, Fox AJ, Vargas RI (2015) Ammonium acetate enhances the attractiveness of a variety of protein-based baits to female *Ceratitis capitata* (Diptera: Tephritidae). J Econ Entomol 108:694–700

Prokopy RJ, Resilva SS, Vargas RI (1996) Post alighting behaviour of *Ceratitis capitata* (Diptera: Tephritidae) on odor-baited traps. Fla Entomol 79:422–428

Quilici S, Atiama-Nurbel T, Brevault T (2014) Plant odors as fruit fly attractants. In: Shelly T, Epsky N, Jang EB, Reyes-Flores J, Vargas R (eds) Trapping and the detection, control and regulation of Tephritid fruit flies: lures, area-wide programs and trade implications. Springer Science +Business, Dordrecht, pp 119–144

Qureshi ZA, Siddiqui QH, Hussain T (1992) Field evaluation of various dispensers for methyl eugenol, an attractant of *Dacus zonatus* (Saund.) (Dipt., Tephritidae). J Appl Entomol 113:365–367

Raghu S (2004) Functional significance of phytochemical lures to dacine fruit flies (Diptera: Tephritidae): an ecological and evolutionary synthesis. Bull Entomol Res 94:385–399

Reddy K, Sharma K, Singh S (2014) Attractancy potential of culturable bacteria from the gut of peach fruit fly, *Bactrocera zonata* (Saunders). Phytoparasitica 42:691–698

Rice RE, Cunningham RT, Leonhardt BA (1984) Weathering and efficacy of trimedlure dispensers for attraction of Mediterranean fruit flies (Diptera: Tephritidae). J Econ Entomol 77:750–756

Ripley LB, Hepburn GA (1935) Olfactory attractants for male fruit flies. Entomol Mem, Dep Agric, Union of S Afr 9:3–17

Robacker DC (1995) Attractiveness of a mixture of ammonia, methylamine and putrescine to Mexican fruit flies (Diptera: Tephritidae) in a citrus orchard. Fla Entomol 78:571–578

Robacker DC, Czokajlo D (2005) Efficacy of two synthetic food-odor lures for Mexican fruit flies (Diptera: Tephritidae) is determined by trap type. J Econ Entomol 98:1517–1523

Royer JE (2015) Responses of fruit flies (Tephritidae: Dacinae) to novel male male attractants in north Queensland, Australia, an improved male lures for some pest species. Aus Entomol 54:411–426

Seewooruthun SI, Permalloo S, Gungah S, Soonnoo AR, Alleck M (2000) Eradication of an exotic fruit fly from Mauritius. In: Tan KH (ed) Area-wide control of fruit flies and other insect pests. Penerbit Universiti Sains Malaysia, Penang, pp 389–394

Shelly TE (2000) Fecundity of female Oriental fruit flies (Diptera: Tephritidae): effects of methyl-eugenol fed and multiple mates. Anns Entomol Soc Am 93:559–564

Shelly TE (2013) Detection of male Mediterranean fruit flies (Diptera: Tephritidae): Performance of trimedlure relative to capilure and enriched ginger root oil. Proc Hawaiian Entomol Soc 45:1–7

Shelly TE, Edu J (2010) Mark-release-recapture of males of *Bactrocera cucurbitae* and *B. dorsalis* (Diptera: Tephritidae) in two residential areas of Honolulu. J Asia Pac Entomol 13:131–137

Shelly TE, Pahio E (2002) Relative attractiveness of enriched ginger root oil and trimedlure to male Mediterranean fruit flies (Diptera: Tephritidae). Fla Entomol 85:545–551

Shelly TE, Edu J, Pahio E (2003) Influence of diet and methyl eugenol on the mating success of males of the Oriental fruit fly, *Bactrocera dorsalis* (Diptera: Tephritidae). Fla Entomol 88:307–313

Shelly TE, Pahio E, Edu J (2004) Synergistic and inhibitory interactions between methyl eugenol and cue lure influence trap catch of male fruit flies, *Bactrocera dorsalis* (Hendel) and *B. cucurbitae* (Diptera: Tephritidae). Fla Entomol 87:481–486

Shelly T, Nishimoto J, Diaz A, Leathers J, War M, Shoemaker R, Al-Zubaidy M, Joseph D (2010) Capture probability of released males of two *Bactrocera* species (Diptera: Tephritidae) in detection traps in California. J Econ Entomol 103:2042–2051

Shelly TE, Kurashima RS, Nishimoto JI (2012) Field capture of male melon flies, *Bactrocera cucurbitae* (Coquillett), in Jackson traps baited with cue-lure versus raspberry ketone formate in Hawaii. Proc Hawaiian Entomol Soc 44:63–70

Sookar P, Facknath S, Permalloo S, Seewooruthun SI (2002) Evaluation of modified waste brewer's yeast as a protein source for the control of the melon fly, *Bactrocera cucurbitae* (Coquillett). In: Barnes B (ed) 6th international fruit fly symposium. Isteg Scientific Publications, Stellenbosch, pp 295–299

Sookar P, Permalloo S, Gungah S, Alleck M, Seewooruthun SI, Soonnoo AR (2008) An area wide control of fruit flies in Mauritius. In: Sugayama RL, Zucchi RA, Ovruski SM, Sivinski J (eds) 7th international symposium on fruit flies of economic importance, Salvador. vol 261–269. Biofabrica Moscamed Brasil

Stark JD, Vargas RI (1992) Differential response of male Oriental fruit fly (Diptera: Tephritidae) to colored traps baited with methyleugenol. J Econ Entomol 85:808–812

Suckling DM, Jang EB, Holder P, Carvalho L, Stephens AEA (2008) Evaluation of lure dispensers for fruit fly surveillance in New Zealand. Pest Man Sci 64:848–856

Tan KH, Nishida R (2000) Mutual reproductive benefits between a wild orchid, *Bulbophyllum patens*, and *Bactrocera* fruit flies via a floral synomone. J Chem Ecol 26:533–546

Tan KH, Nishida R, Jang EB, Shelly TE (2014) Pheromones, male lures and trapping of Tephritid fruit flies. In: Shelly T, Epsky N, Jang EB, Reyes-Flores J, Vargas R (eds) Trapping and the detection, control and regulation of Tephritid fruit flies: lures, area-wide programs and trade implications. Springer Science + Business, Dordrecht, pp 15–75

Thomas DB, Holler TC, Heath RR, Salinas EJ, Moses AL (2001) Trap-lure combinations for surveillance of *Anastrepha* fruit flies (Diptera: Tephritidae). Fla Entomol 84:344–351

Toth M, Nobili P, Tabilio R, Ujvary I (2004) Interference between male-targeted and female-targeted lures of the Mediterranean fruit fly *Ceratitis capitata* (Dipt., Tephritidae) in Italy. J Appl Entomol 128:64–69

Uchida GK, Walsh WA, Encarnacion C, Vargas RI, Stark JD, Beardsley JW, Mc Innis DO (1996) Design and relative efficiency of Mediterranean fruit fly (Diptera: Tephritidae) bucket traps. J Econ Entomol 89:1137–1142

Umeh VC, Garcia LE (2008) Monitoring and managing *Ceratitis* spp. complex of sweet orange varieties using locally made protein bait of brewery waste. Fruits 63:209–217

Vargas RI, Stark JD, Kido M, Ketter HM, Whitehand LC (2000) Methyl eugenol and cue-lure traps for suppression of male Oriental fruit flies and melon flies (Diptera: Tephritidae) in Hawaii: effects of lure mixtures and weathering. J Econ Entomol 93:81–87

Vargas RI, Mau RFL, Stark JD, Pinero JC, Leblanc L, Souder SK (2010) Evaluation of methyl eugenol and cue-lure traps with solid lure and insecticide dispensers for fruit fly monitoring and male annihilation in the Hawaii areawide pest management program. J Econ Entomol 103:409–415

Wakabayashi N, Cunningham RT (1991) Four-component synthetic food bait for attracting both sexes of the melon fly (Diptera: Tephritidae). J Econ Entomol 84:1672–1676

Wee S-L, Hee AK-W, Tan KH (2002) Comparative sensitivity to and consumption of methyl eugenol in three *Bactrocera dorsalis* (Diptera: Tephritidae) complex sibling species. Chemoecology 12:193–197

White IM (2006) Taxonomy of the Dacina (Diptera: Tephritidae) of Africa and the Middle East. Afr Entomol Mem 2:156

White IM, Elson-Harris MM (1994) Fruit flies of economic significance: their identification and bionomics. CAB International, Oxon, 601 pp

Yokoyama VY, Miller GT, Stewart-Leslie J, Rice RE, Phillipps PA (2006) Olive fruit fly (Diptera: Tephritidae) populations in relation to region, trap type, season, and availability of fruit. J Econ Entomol 99:2072–2079

Aruna Manrakhan is a research entomologist and co-ordinator of the fruit fly research programme at Citrus Research International, South Africa. Soon after completing her MSc degree in Environmental Technology at Imperial College, University of London, in 1996, Aruna began working on fruit flies in Mauritius as part of the Indian Ocean Regional Fruit Fly Programme led by Professor John Mumford from Imperial College. In 2000, she joined *icipe*, Kenya, and completed her PhD on fruit fly feeding behaviour supervised by Dr Slawomir Lux, who contributed to the 'African Fruit Fly Initiative', now known as the 'African Fruit Fly Programme'. In 2006, Aruna began a post-doctoral fellowship with Dr Pia Addison at Stellenbosch University, South Africa where she focussed on the ecology and management of *Ceratitis capitata* and *Ceratitis rosa*. Aruna's current research interests are the behaviour, monitoring and control of fruit fly pests that affect the citrus industry in Southern Africa. Aruna has also been involved in the South African national surveillance and eradication programme targeting *Bactrocera dorsalis*. Her work has been published in several peer-reviewed journals.

Chapter 13
Baiting and Male Annihilation Techniques for Fruit Fly Suppression in Africa

Sunday Ekesi

Abstract The use of protein baits and the male annihilation technique (MAT) is gaining importance in Africa as one of the major components of fruit fly suppression strategies. A variety of commercial food baits including GF-120, Nulure, Mazoferm, Hymlure, Questlure, Biolure and Fruit fly Mania are available locally for use by growers although the need for identification and registration of more efficient baits cannot be overemphasized. Spray mixtures of these protein hydrolysates and killing agents applied as spot sprays commonly reduce fruit fly populations by 80–90 %. Based on bait spray costs, yield data and monetary gains, a cost-benefit ratio of 1:9.1 has been reported, which is generally acceptable for smallholder and large-scale fruit producers alike. Research has also seen formulation of liquid protein baits into solid bait stations and they are now availability as commercial products for management of fruit flies on the continent. A deployment strategy of M3 bait stations that decreases the number of units required from the perimeter to the centre of the orchard was found to effectively maintain low fruit fly populations below the threshold level before, during and after citrus harvest. The male annihilation technique (MAT) involves the deployment of high densities of trapping stations consisting of a male attractant combined with a killing agent. MAT is being promoted across Africa as a component of IPM strategies against different fruit fly species. Several male attractants such as methyl eugenol, cuelure, trimedlure, EGOlure, ceralure, terpinyl acetate can be used with an appropriate toxicant. In suppression trials using methyl eugenol or terpinyl acetate plus malathion for the suppression of *Bactrocera dorsalis* or *Ceratitis cosyra*, respectively, up to 60–96 % reduction in puparia/kg fruits was achieved. Due to the quarantine nature of the complex of fruit flies inhabiting Africa and zero tolerance of fruit flies in sensitive quarantine markets, the use of a combination of management techniques is being advocated and a combined application of protein baiting and MAT have been reported to further reduce fruit infestation to 1.1–3.1 %. To access sensitive quarantine markets, postharvest treatment of the fruits is encouraged to remove any remaining insects in the produce.

S. Ekesi (✉)
Plant Health Theme, International Centre of Insect Physiology & Ecology (*icipe*),
PO Box 30772–00100, Nairobi, Kenya
e-mail: sekesi@icipe.org

1 Background

Fruit fly population suppression mainly relies on the use of food baits (hydrolysed proteins or their ammonium-based mimics) combined with a killing agent and applied in localized spots. Food baits were first recommended for used against the Mediterranean fruit fly (medfly), *Ceratitis capitata* (Wiedemann) in South Africa in 1908–1909 (Roessler 1989). The baits used included carbohydrate and fermenting substances such as sugars, molasses and syrups. Experiments done against the cucumber fruit fly, *Dacus vertebratus* Bezzi, reported 95% protection of cucumber fruits from damage by the pest (Gunn 1916). A spray mixture of protein hydrolysate bait, sugar and parathion insecticide was later introduced for the management of the oriental fruit fly, *Bactrocera dorsalis* (Hendel), in Hawaii (Steiner 1952). Thereafter, the combination of malathion and protein hydrolysate bait sprays was instrumental in the successful eradication of *C. capitata* in Florida in 1956 (Steiner et al. 1961). Ammonia is believed to be the major olfactory factor in protein hydrolysate and amino acids provide the phagostimulatory factors (Roessler 1989). Several insecticides including malathion, spinosad, fipronil and cypermethrin have been tested and used in combination with baits with varying degrees of efficacy (Roessler 1989; Keiser 1989).

2 Mechanisms Underpinning the Use of Baiting Technique for Fruit Fly Suppression

Adult female fruit flies lay their eggs just under the fruit's skin; eggs and larvae develop inside the fruits and are, therefore, somewhat protected from the direct application of most management tools except systemic insecticides. However, the use of systemic insecticides is associated with residues and consumer sensitivity precludes their use on many fresh fruits. Mature fruit fly larvae emerge from the fruits and pupate on the ground. Soil treatment with insecticides is possible but can be damaging to ground fauna and flora (Roessler 1989). Although ground treatments with biopesticides are recommended (see Maniania and Ekesi et al. 2016), they must be integrated with other management measures within the context of integrated pest management (IPM). The adult stage is therefore a key developmental stage against which management is directed. Females fruit flies require proteinaceous food to mature their eggs (Hagen and Finney 1950) and, given the existence of a pre-oviposition period in most species (Back and Pemberton 1918), a mixture of protein bait and a killing agent targeted at pre-ovipositing females should ensure that oviposition on fruit is prevented. In fact, the IAEA (2009) recommends that the use of baiting technique should be considered as one of the primary components of fruit fly suppression strategies because it has the following advantages: (1) it is less harmful to beneficial organisms, the environment and the operator; (2) the cost is substantially lower because less insecticide and bait are required compared with cover sprays; (3) less time is required for application and thus there is less demand for labour; (4) cheap and simple spraying equipment can be used and; (5) application can be directed away from fruits to minimize the problems of residues in fruits. Despite these advantages, baiting techniques should not be used as a stand-alone approach because their effectiveness can be less than desirable under high pest population pressure. These techniques must therefore be integrated with all the other management methods available including the male annihilation technique (MAT), use of biopesticides, releases of natural enemies (parasitoids), ant technology and field sanitation (Ekesi et al. 2016).

3 Some Commonly Used Protein Baits and Application Techniques

Protein baits attract both male and female fruit flies and are usually not species specific. They are available in both liquid and dry synthetic forms but the use of liquid baits are known to have limited field life, require weekly renewal, are difficult to handle and also attract non-target species (Ekesi and Billah 2007). Recent efforts have therefore concentrated on the development of potent and selective dry protein attractants and trapping systems for fruit flies that can be used for population monitoring, mass trapping and suppression. In general, the field longevity of liquid protein baits is usually 1–2 weeks requiring repeated application. However, dried synthetic lures can last 4–6 weeks. A number of protein baits are available commercially in Africa including GF-120, NuLure, Mazoferm, HymLure, Biolure, Prolure, Questlure and Fruit fly Mania. There is, however, the need for more efficient products to be registered on the continent.

3.1 Bait Sprays

Application of bait spray aims largely at attracting (and killing) female flies before they lay eggs on fruits. The bait attracts the fruit flies from a distance to the spot of application, where the flies feed on the bait, ingest the toxicant and later die. For fruit trees, once the bait has been mixed with a pesticide, it is normally sprayed on to one or more (depending on the bait) 1 m^2 areas, or 'spots' in the canopy of each tree in the orchard (Vayssières et al. 2009; Ekesi et al. 2011, 2014). The sprays are preferably applied to the lower surface of leaves to enhance persistence and reduce the chances of wash-off by rain. Bait is applied on a weekly basis beginning when fruits are, for example, about 1.3 cm (1/2″) in size in the case of mango, and when monitoring data indicates that fruit flies are already present; applications continue until the very end of harvest. Specialized application methods based on spot treatments ensures relative safety to beneficial and non-target organisms, and it is also compatible with other fruit fly management methods such as classical biological control and the use of biopesticides (see Maniania and Ekesi 2016; Mohamed et al. 2016).

Bait spray application technique for control of native *Dacus* species and the exotic melon fly *Zeugodacus cucurbitae* (Coquillett) that mainly damage vegetable crops (e.g. cucurbits and peppers) is different from fruit flies that attack fruits tree crops. Cucurbitaceous and some solanaceous fruit flies generally inhabit the vegetation around the edges of the field and female flies only enter the fields to lay eggs. To suppress the population of these fruit flies, bait sprays are applied to the border or trap plants where the fruit flies roost, rather than directly on the cucurbit plants (Prokopy et al. 2003, 2004). Phytotoxicity of some baits to delicate cucurbit leaves is another reason for the bait application to border vegetation. Important alternative host plants that these flies roost in, and that could be utilized as trap crops on the border include sorghum, castor bean, cassava and corn. Three to six weeks before planting the vegetable crops, appropriate alternative host plants (e.g. maize, sorghum, castor bean) that are favoured by the target fruit flies, should be planted around the borders of the vegetable crop to be protected. Flies are monitored with relevant attractants (male lure or food baits) in or on the border of crop fields and food attractants mixed with a toxicant are applied to the non-host border plant when the first flowers appear on the vegetable crop.

3.2 Bait Stations

A bait station (BS) is a distinct device that combines a lure or bait with a toxicant that attracts and kills fruit flies without retaining them in the device. This method of fruit fly management is an alternative to liquid bait sprays. BS could be (i) a retrievable device, (ii) a biodegradable device that can remain in the field for a specific time period or, (iii) a device that is based on direct application to a substrate (IAEA 2009). Attractants used in BS can include both male-specific lures and baits that attract both male and female flies. Recommended killing agents can be contact or ingestable pesticides, or an entomopathogen (e.g. a *Metarhizium anisopliae* (Metchnikoff) Sorokin-based biopesticide) and can be fast acting or slow acting and even utilized through autodissemination. BS has evolved as a response to the need for cheaper, more efficient and selective systems for management of fruit flies. Several BS devices that encompass male and female attractants, corn cobs, killing bags, biodegradable wax-coated cardboard are now available commercially for use globally, including in Africa (Ware et al. 2003; Heath et al. 2007; Moreno and Mangan 2002; Epsky et al. 2012; Manrakhan and Daneel 2013; Navarro-Llopis et al. 2012, 2015). IAEA (2009) identified the following desirable characteristics as crucial for a BS: (1) the ability to target and suppress female populations; (2) low cost of the device including the attractant and the killing agent, (3) attracted flies should not be trapped and retained; (4) attractants should be long lasting with long residual toxicity of the killing agent thus minimizing servicing or replacement time and less labour; (5) ease of use, disposable and/or biodegradable; (6) highly selective i.e., no negative non-target effects, and; (7) a positive cost-benefit relationship. An additional feature that enhances the response of flies to an odour source is the presence of additional visual cues that act synergistically with the odour.

4 Evaluation of Protein Baits for the Management of Fruit Flies in Africa

4.1 Relative Attractiveness of Different Baits

In the Nguruman division, Kajiado district of Kenya, Rwomushana (2008) evaluated, on mango, attraction of the invasive species *B. dorsalis* to Nulure, torula yeast, corn steep-water and a local yeast-based attractant between October and December 2006 and between November 2007 and January 2008. Three trap types were also compared: Multilure, Easy and Lynfield traps. All the food baits evaluated were attractive to *B. dorsalis* but the level of attraction varied. Results showed that the Multilure trap baited with torula yeast or Nulure were the most attractive trap-bait combinations with total captures of 19.7–30.3 flies/trap/day (FTD) and 10.54–22.97 FTD, respectively. All attractants caught more females than males except for corn steep-water. The highest proportion of females captured was 61% in the torula yeast/Lynfield trap combination during the 2007/2008 season.

Mwatawala et al. (2006) studied the diversity of fruit flies in four main agroecological zones of the Morogoro region, Tanzania using five different attractants. The authors reported that the protein-baited traps attracted the highest diversity of fruit flies (21 different species) in comparison with the more specific parapheromones (methyl eugenol attracted three species; Cue lure attracted 12 species; and Trimedlure attracted nine species). The synthetic three-component food attractant, Biolure, attracted eight species.

In an FAO-IAEA coordinated research programme, the response of *C. capitata* and *Ceratitis fasciventris* Bezzi to varying combinations of the constituents of the Biolure protein bait (putrescine [PT], trimethylamine [TMA] and ammonium acetate [AA])) on coffee was evaluated in a series of different experiments (Ekesi et al. 2007). The treatment combinations included PT/TMA/AA, TMA/AA, PT/TMA, PT/TMA/AA, AA alone and Nulure and they were evaluated in plastic Multilure traps either as wet or dry forms. In the first trial, wet PT/TMA/AA, TMA/AA and PT/TMA captured significantly more female flies than the other treatments. In the second experiment, both the wet and dry PT/TMA/AA and wet TMA/AA were superior to the other treatments in capturing females. Amongst the different three-component treatments, female *C. capitata* accounted for 67–72% of the total captures while in the Nulure trapping system they accounted for 72–74% in both trials.

On mango in Kenya, Ekesi et al. (2014) evaluated catches of *B. dorsalis* in Multilure traps baited with six commercial food-based attractants: Mazoferm E802, torula yeast, GF-120, Hymlure, Biolure and Nulure. In 2007, Mazoferm E802 and torula yeast were the most effective attractants and captured 2.4–2.6 times more females and 3.4–4.0 times more males than the standard Nulure. All attractants captured more females than males (ranging from 63 to 74%). In 2008, Mazoferm E802 was the most effective bait capturing 5.6 and 9.1 times more females and males, respectively, than the standard Nulure. Amongst all the attractants, in both years, Nulure captured the greatest proportion of females: 74% compared with 51–68% for the other attractants.

4.2 Field Suppression of Fruit Flies Using Protein Baits in Africa

4.2.1 Suppression Using Liquid Protein Baits

In Benin, Vayssières et al. (2009) assessed the effectiveness of GF-120 fruit fly bait containing spinosad for controlling the mango fruit fly, *Ceratitis cosyra* (Walker), and *B. dorsalis* over two seasons (2006 and 2007). Spot sprays were used to apply limited quantities (~0.07–0.09 l) of the mixture to 1 m^2 of the canopy. The results showed that GF-120 provided an 81% reduction in the number of pupae per kilogram of fruit after 7 weeks of weekly application of bait sprays in 2006 and an 89% reduction after 10 weeks of weekly applications in 2007.

In trials conducted over two seasons on mango (2006/2007 and 2008/2008), Ekesi et al. (2011) reported mango fruit infestation ranging from 28 to 30% by *B. dorsalis* in orchards receiving six applications of GF-120 bait spray compared to 52–60% fruit infestation in an untreated mango orchard. Additional trials by Ekesi et al. (2014) on mango assessed the application of Mazoferm E802 in combination with spinosad as a bait spray in mango orchards in 2008 and 2009. Results showed that weekly application of the bait spray reduced *B. dorsalis* catches relative to the control by 87% within 4 weeks and 90% within 8 weeks. At harvest, the proportion of fruit infested was significantly lower in the treated orchards (8%) compared to the control orchards (59%). Estimated mango yield was significantly higher in orchards receiving the bait sprays (12,487 kg/ha) compared to control orchards (3606 kg/ha). Based on bait spray costs, yield data and monetary gains, a cost-benefit ratio of 1:9 was realized which is acceptable for growers. In 2009 the experiment was repeated with similar results.

In the Morogoro region of Tanzania, Mwatawala et al. (2015) evaluated the effectiveness of three IPM programmes against *B. dorsalis* over five mango seasons. Spot applications of molasses bait (formulated with crude extracts of *Derris elliptica* Benth, water and waste brewer's yeast) was compared with cover sprays of the insecticide dimethoate/lambda cyhalothrin (Karate 5 EC) and spot applications of Success bait containing spinosad in combination with mass trapping using methyl eugenol. In all treatments orchard sanitation and early harvesting of fruits were included as standard practice. Results showed that the lowest incidence of *B. dorsalis* was recorded in fruits harvested from orchards where the Success bait treatment/ mass trapping strategy was employed, while the highest incidence was recorded in fruits from orchards assigned to the molasses bait treatment. Incidences ranged from 9 to 25%, 2–15.56% and 4–17.5% in mangoes harvested from molasses bait, Success bait/mass trapping and insecticide-treated orchards, respectively. Fruits harvested from orchards under the Success bait/mass trapping treatment had the lowest *B. dorsalis* infestation rates. Overall infestation rates ranged from 0.58 to 2.52%, 0.004–1.7%, and 0.4–2.03% in mango from orchards under the Success bait/mass trapping, molasses bait and dimethoate/lambda cyhalothrin treatments, respectively. The number of flies trapped per week (FTW) ranged from 26.0 to 84.7, 14.5–186.9, and 13.7–150.0 in orchards under the molasses bait, Success bait/mass trapping and insecticide treatments, respectively. Cumulatively, all these studies attest to the role of baiting techniques in the suppression of fruit flies on the African continent.

4.2.2 Suppression Using Protein Bait Stations

In South Africa, the M3 Fruit Fly Bait Station has been registered for use in fruit fly control since 2000 and is widely used by horticultural growers for the management of a variety of fruit fly species (Ware et al. 2003; Manrakhan and Daneel 2013). The use of the M3 bait station comes with the advantages of zero residues on fruits, a single application per season and the fact that it is environmentally friendly. The recommended application rates of M3 bait stations vary with different fruit types but generally range from 300–400 stations per ha, depending on the susceptibility of the fruits to fruit fly damage. Although M3 bait stations represent an alternative to fruit fly bait sprays, their widespread use in South Africa has been somewhat impeded by the cost factor of this control technique. As such, ways to improve the cost-competitiveness of bait stations by lowering the number of units required per hectare requires investigation. Manrakhan et al. (2010) evaluated the use of M3 bait stations with declining numbers from the perimeter to the centre of citrus orchards, leaving a core area untreated and bringing the total density of the stations down from the standard rate of 416 units per ha, to 324 units per ha (a decrease of 22.1%). Such a deployment strategy was found to be effective in sustaining fruit fly populations below the threshold level. Furthermore, the treatments also completely prevented fruit fly damage to the fruits of sweet orange, *Citrus sinensis* (L.) Osbeck.

In South Africa, Manrakhan and Daneel (2013) evaluated the performance of M3 bait stations and weekly applications of GF-120 in navel orange (*C. sinensis*) orchards for the control of *C. capitata*. Both treatments were effective in keeping *C. capitata* populations below the threshold of 1 female/trap until the start of the harvest, when the orchard was treated with protein baits. No fruit fly infestation was recorded in 2400 orange fruit sampled across the treated blocks.

Most recently however, Manrakhan et al. (2015) noted phytotoxicity of GF-120 on the fruit of 'Nadorcott' mandarins, *Citrus reticulata* Blanco, which were at the green and colour-break stages; the incidence of burn increased with increasing concentrations of GF-120. The authors recommended alternative methods of GF-120 application other than ground-based canopy sprays for 'Nadorcott' mandarin, especially if baiting was to start early before full colour development of the fruit. In this regard, the use of bait stations that minimize direct contact with fruits should minimize the risk of phytotoxicity on fruit of 'Nadorcott' mandarin at the green and colour break stages.

5 Male Annihilation Technique (MAT)

The term male annihilation was coined by Steiner and Lee (1955) following the realization that populations of fruit flies could be controlled using male lures. The male annihilation technique (MAT) is a control strategy that involves the deployment of high density trapping stations consisting of a male attractant combined with an insecticide. The aim is to reduce male fruit fly populations to a low level such that mating does not occur (in the case of eradication) or is reduced to low levels (in the case of suppression). Compressed fibre boards, coconut husks and string or cord soaked in the attractant/insecticide mixture are examples of MAT 'block's. These blocks are distributed at the rate of 250 per km^2 and the entire application is repeated every 6–8 weeks. Cotton wicks soaked in the attractant/insecticide mixture and placed in Lynfield traps is also a common practice. MAT is currently being promoted across Africa as a component of IPM for *Bactrocera, Ceratitis* and *Dacus* fruit fly pests. MAT can be used both for eradication and for population suppression (Cunningham 1989a). However, in this review, we discuss the use of MAT within the context of suppression as opposed to eradication.

5.1 Key Attractants Used for MAT

Male attractants such as methyl eugenol (ME), cue-lure, trimedlure, EGOlure, ceralure, vertlure and terpinyl acetate can be used with an appropriate toxicant such as malathion, fipronil, spinosad or deltamethrin and deployed in orchards and on vegetable farms, preferably at the periphery of the farms for population suppression (Cunningham 1989a; Vargas et al. 2010; Mwatawala et al. 2013). Although a variety of male lures are available for use in MAT, the three predominant male attractants that have been largely exploited for MAT are ME, cue-lure and trimedlure. Limited trials have also been undertaken in Africa to evaluate the possibility of using terpinyl acetate for MAT.

5.1.1 Methyl Eugenol (ME)

ME, i.e. 4-allyl-1, 2-dimethoxybenzene-carboxylate, is believed to be the most powerful of all fruit fly male lures. Tan and Nishida (2012) reported that 450 plant species belonging to 80 families possessed ME as a constituent component and/or as a component of their floral fragrance. The role of ME as a plant kairomone in dacine fruit fly ecology has been discussed (Cunningham 1989b; Metcalf 1990; Metcalf and Metcalf 1992). ME is reported to act as a precursor or booster of components of the male fruit fly sex pheromone and is produced in the rectal gland of some *Bactrocera* species (Nishda et al. 1988; 1990; 1993). Males of several species of the *B. dorsalis* complex (e.g. the carambola fruit fly, *Bactrocera carambolae* Drew and Hancock; *Bactrocera caryeae* (Kapoor); *B. dorsalis*; *Bactrocera kandiensis* Drew and Hancock; *Bactrocera occipitalis* [Bezzi]) are all attracted to ME. Consumption of ME by male flies enhances mating competitiveness in: *B dorsalis* (Shelly and Dewire 1994, 2000; Tan and Nishida 1996, 1998); *B. carambolae* (Wee et al. 2007); the guava fruit fly, *Bactrocera correcta* (Bezzi) (Orankanok et al. 2009); and the peach fruit fly, *Bactrocera zonata* (Saunders) (Quilici et al. 2004; Sookar et al. 2009). In male *B. dorsalis* the strength of the attraction to ME and pharmacophagous feeding (i.e. consumption of non–nutritive and non–essential chemicals) has been described as truly intriguing. According to Cunningham (1989b) "if pure liquid methyl eugenol is offered, the males will drink it until they fill their crops and die". As a result, ME is recommended for various suppression programmes for this group of fruit flies.

5.1.2 Cue-lure

Cue-lure (C-L), i.e. 4-(*p*-aceacetoxyphenyl)-2-butanone, has never been isolated as a natural product but quickly hydrolyzes to form raspberry ketone (RK), i.e. 4-(*p*-hydroxyphenyl)-2-butanone, which is known from plants (Schinz and Seidel 1961; Metcalf and Metcalf 1992; Cunningham 1989b). There are over 80 fruit fly species that respond to C-L. According to Drew (1974), attraction to C-L is reported from several taxons of the tribe Dacini, the best known species on the African continent is probably *Z. cucurbitae*. Recent developments in the Hawaiian fruit fly IPM programme recommends the use of C-L in MAT for *Z. cucurbitae* suppression (Vargas et al. 2010, 2012) and this requires evaluation in Africa.

5.1.3 Trimedlure

Trimedlure (TML), i.e. t Butyl-4 (or 5)-chloro-2-methyl cyclohexane carboxylate, is the standard male attractant for *C. capitata* and is used in many detection programmes. However, it can also be used for suppression or control through MAT (Vargas et al. 2012). Despite its wide acceptance, TML is not a particularly powerful attractant, especially compared with ME. Ceralure, an iodinated analogue of TML, was shown to be 4–9 times more attractive to male *C. capitata* than TML (Jang et al. 2003, 2005), but its synthesis on a commercial scale is not yet cost effective (Jang et al. 2010). In an effort to consolidate 'attract & kill' techniques in to solid single or double-lure insecticidal formulations, Vargas et al. (2012) successfully evaluated solid Mallet TMR wafers (=TML+ME+RK) and Mallet CMR wafers (= ceralure+ME+RK+benzyl acetate) impregnated with DDVP insecticide (2,2- dichlorovinyl dimethyl phosphate) in traps as potential combined detection and MAT devices for management of *C. capitata*, *B. cucurbitae* and *B. dorsalis*. They recommended the solid Mallet TMR wafer as the safer system that was also more convenient to handle and could be used in the place of several individual lure and trap systems, potentially reducing costs of large survey and detection programmes.

5.1.4 Terpinyl Acetate

Terpinyl acetate (TA), i.e. 2-(4-trimethyl-1-cyclohex-3-enyl) propan-2-yl acetate, was first reported to be attractive to the Natal fruit fly, *Ceratitis rosa* Karsch (Ripley and Hepburn 1935), and 12 related species in the same subfamily including *C. cosyra* and *C. capitata*. The authors reported that responses to this male lure followed recognized taxonomic lines with only species from related genera, i.e. *Ceratitis*, *Pterandrus*, *Pardalaspis* and *Pinacochlaeta* responding to the attractant.

5.2 MAT for Suppression of Fruit Flies in Africa

Hanna et al. (2008) tested MAT devices charged with 3 ml of ME or TA plus 1 ml of malathion for the suppression of *B. dorsalis* and *C. cosyra*, respectively, in Benin. Results showed that during the first 2 weeks of trapping, populations of *B. dorsalis* and a complex of *Ceratitis* spp. were larger in MAT orchards compared with control orchards. However, the trend was reversed in subsequent sampling dates when *B. dorsalis* populations increased steeply in the control orchards, while populations in the MAT orchards remained relatively flat and at levels of less than 10% of those in the control orchards. For the *Ceratitis* spp., the situation was also reversed in subsequent sampling dates with about a 65% reduction in peak population density achieved in the MAT orchards compared with control orchards. Up to 60% reductions in puparia/kg fruits were reported from the MAT orchards compared with the controls.

Ndlela et al. (2016) used ME laced with deltamethrin for field suppression of *B. dorsalis* on mango for two seasons in coastal Kenya. Following the application of MAT, the total FTD was significantly lower in MAT-treated orchards (0.1 and 2.7 FTD for season 1 and 2, respectively) compared with the control orchards (18.6 and 21.5 FTD for season 1 and 2, respectively). This represented a reduction in *B. dor-*

salis population of 99.5 % in both seasons. The number of puparia/kg of mango fruit were 17 and 24 fold smaller in MAT-treated orchards compared with control orchards for the two consecutive seasons representing over a 90 % reduction.

6 Combined Application of Bait Sprays or Bait Stations with MAT

The complexes of fruit flies that growers in Africa have to deal with are mostly of quarantine significance (Ekesi et al. 2016). Based on experience in similar agro-ecologies in Latin America and the South Pacific, management of any fruit fly complex is unlikely to be successful if based on a single management technique (Aluja et al. 1996; Allwood and Drew 1997). An IPM approach offers the best method to improve the economies of production systems by reducing yield losses and enabling growers to comply with stringent quality assessments made by the export market (Lux et al. 2003). In this regard, the combined use of baiting techniques and MAT is encouraged (Ekesi et al. 2016). Grout and Stephen (2013) compared the catches of different fruit fly species in citrus orchards after 4 weeks of treatment with either: (1) MAT + M3 baits stations at 400 units/ha; (2) MAT + 2 % Prolure bait mixed with mercaptothion; (3) MAT + 7 % Mazoferm with mercaptothion; (4) MAT alone and; (5) an untreated border 0.5 km distant from the treated orchards. The M3 bait stations were comprised of protein bait and apha-cypermethrin at 2 g/l while the MAT blocks were made of soft-board blocks impregnated with ME and mercaptothion 500 EC in a 3:1 ratio. Mean catches of *B. dorsalis* and *C. capitata* sampled using three-component Biolure traps were smallest in the MAT + M3 bait station treatment (1.2 flies/trap/week [FTW]) and the MAT + Prolure treatment (1.7 FTW) compared with the untreated border (8.6 FTW). *Ceratitis cosyra* was the dominant insect in the orchards and there were 12.8 FTW in the MAT + M3 bait station treatment and 10.8 FTW in the MAT + Prolure treatment. The largest catches were in the border field (74.1 FTW). The authors concluded that commercial control of *B. dorsalis* was possible when MAT is used in combination with protein baits for field suppression.

On mango, studies done at Nguruman, Kajiado South, Kenya over two mango seasons (October-December, 2009 and 2010) also demonstrated the benefits of combined applications of MAT and protein baits sprays for suppression of *B. dorsalis*. Four distinct localities or blocks that were ~4.6–5.2 km apart were identified at the experimental location. At each location, four mango orchards (0.3–0.8 ha) were selected and the following treatments were applied: (1) MAT + Mazoferm bait spray; (2) MAT alone; (3) bait spray alone; (4) untreated control. MAT was achieved using cotton wicks impregnated with ME and placed in Lynfied traps; traps were recharged after 4 weeks. Mazoferm was applied weekly for 8 weeks at a rate of 400 ml active ingredient ha^{-1} with spinosad at 100 g active ingredient ha^{-1} and at an output volume of 5 l ha^{-1} (approx. 50 ml tree^{-1}). The experiment, therefore, con-

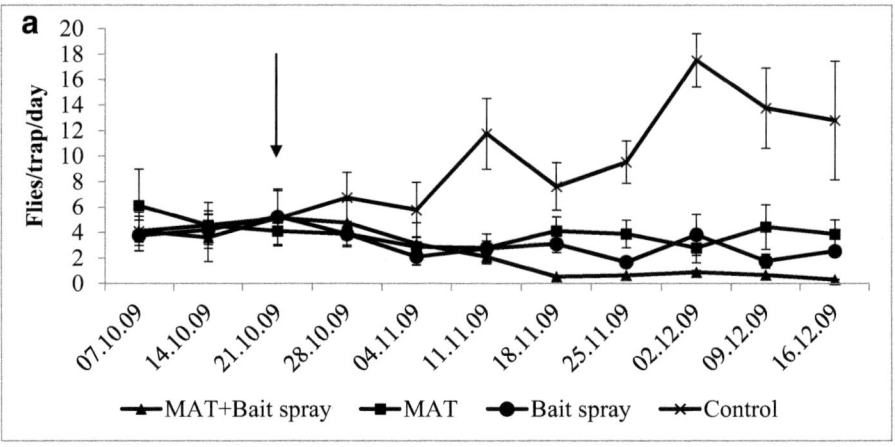

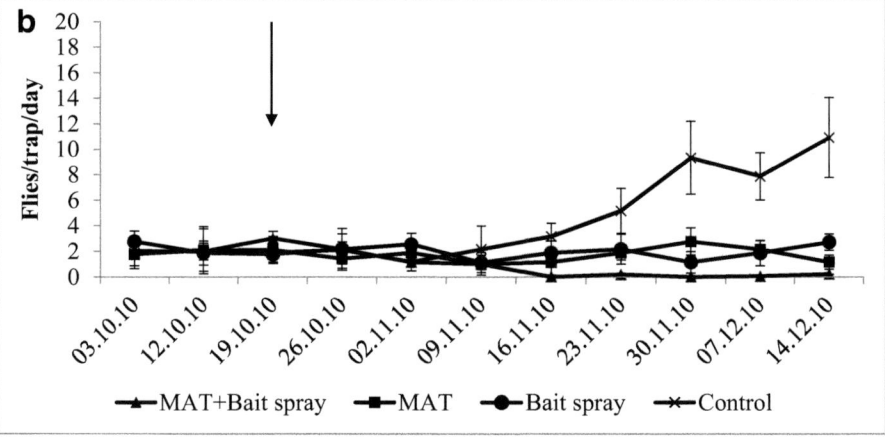

Fig. 13.1 Mean number of *B. dorsalis* captured per trap per day following application of bait spray and male annihilation techniques (MAT) during the 2009 (**a**) and 2010 (**b**) mango seasons. *Arrow* indicates date of first deployment of MAT and bait spray. Error *bars* denote SE

sisted of four treatments replicated four times. All orchards were comprised of a mixture of mango varieties namely Boribo, Apple, Tommy Atkins and Van Dyke, in rows of 10 m apart (100 trees per ha).

The temporal trend of *B. dorsalis* (the predominant fruit fly species during the experiment) was monitored using 2% Nulure in Multilure traps. During the 2009 season and prior to the application of MAT blocks and bait sprays, mean fruit fly populations ranged from 3.11 to 5.14 FTD in the MAT and bait spray treatment, 3.87–6.12 FTD in the MAT treatment, 3.76–5.23 FTD in the bait spray treatment and 4.09–6.76 FTD in orchards assigned to control treatment (Fig. 13.1a). Following application of the treatments, major reductions in populations were observed in the treated orchards: MAT and bait spray (0.32–2.08 FTD), MAT alone (2.78–4.44 FTD), bait spray alone (1.67–3.82) and control (7.62–17.51 FTD) (Fig. 13.1a).

During the 2010 season fruit fly population was generally lower than in the 2009 mango season. Pre-treatment infestation levels were 1.98–2.99 FTD, 1.43–2.11 FTD, 1.76–2.75 FTD and 1.87–2.11 FTD in the MAT+bait spray, MAT alone, bait spray alone, and control orchards, respectively (Fig. 13.1b). Following the application of suppression methods, *B. dorsalis* populations were massively reduced in the MAT+bait spray treatment (0.00–1.01 FTD). In orchards where MAT and bait sprays were applied individually, fruit fly populations were 0.98–2.76 and 1.12–2.72 FTD, respectively. In the control treatments, *B. dorsalis* populations were generally high and continued to increase throughout the season (2.17–10.92 FTD).

During the 2009 mango season the proportion of mango fruits infested (%) was lowest in the MAT+bait spray treatment (3.1%) followed by the bait spray alone treatment (9.1%) and the MAT alone treatment (16.3%). Fruit infestation in the control was 40.1% (Fig. 13.2a). In the 2010 mango season, mango fruit infestation was 1.1%, 8.2%, 7.1% and 32.6% in the MAT+bait spray, MAT alone, bait spray alone and control treatments, respectively, at harvest (Fig. 13.2b). Although the individual use of bait spray and MAT has a demonstrable effect as seen in previous studies (Peck and McQuate 2000; Yee 2006; Mangan and Moreno 2007; Vayssières et al. 2009; Piñero et al. 2009), our results shows that a combined application of protein bait spray and MAT within the context of IPM further reduced fruit fly population pressure and fruit infestation compared with applications of individual components within the management package.

7 Conclusions

Clearly, the use of baiting and MAT offers one of the greatest opportunities for fruit fly suppression in Africa. Although individual use of control methods is unlikely to achieve total control of fruit fly populations, especially under high population pressure, the combined application of both techniques can significantly impact on the complex of fruit fly species that smallholders have to grapple with on the continent. The use of bait spray or BS in combination with MAT ensures that less insecticide is applied with minimal impact on humans, the environment and biodiversity; there is huge potential for zero residues on fruits, reduced labour costs for the grower and this is strongly recommended for smallholder and large-scale producers of fruits and vegetables across Africa. However, without postharvest treatment to provide quarantine security, export of fruits and vegetables attacked by fruit flies will still be limited by quarantine restrictions. To complement the pre-harvest management measures described here, postharvest treatment should be applied to the produce before export (see Grout 2016).

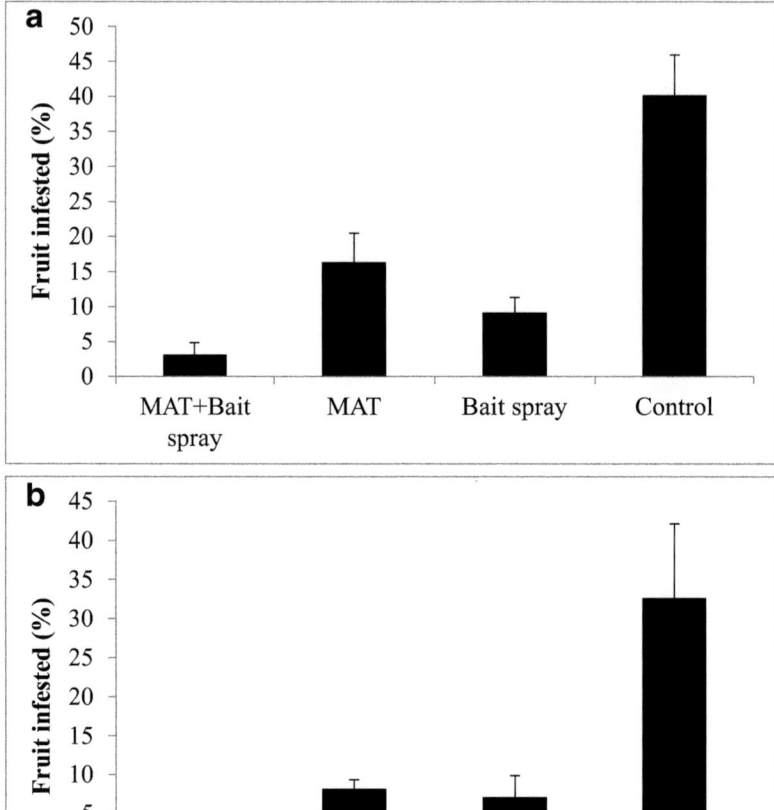

Fig. 13.2 Mean percentage of mango fruits infested by *B. dorsalis* following application of bait spray and male annihilation techniques (MAT) during the 2009 (**a**) and 2010 (**b**) mango seasons. Error *bars* denote SE

Acknowledgements We sincerely thank GIZ/BMZ, Biovision Foundation, IAEA, FAO and DFID for their funding to the Mango IPM project of *icipe*.

References

Allwood AJ, Drew RAI. (1997) Management of fruit flies in the Pacific. A regional symposium, Nadi, Fiji 28–31 October 1996. ACIAR Proceedings No. 76. p 267

Aluja M, Celedonio-Hurtado H, Liedo P, Cabrera M, Castillo F, Guillen J, Rios E (1996) Seasonal population fluctuations and ecological implications for management of *Anastrepha* fruit flies

(Diptera: Tephritidae) in commercial mango orchards in southern Mexico. J Econ Entomol 89:654–667

Back EA, Pemberton CE (1918) The Mediterranean fruit fly in Hawaii. USDA Bull 538, p 118

Cunningham RT (1989a) Male annihilation. In: Robinson AS, Hooper G (eds) Fruit flies, their biology, natural enemies and control, vol 3B. Elsevier, Amsterdam, pp 345–351

Cunningham RT (1989b) Parapheromones. In: Robinson AS, Hooper G (eds) Fruit flies, their biology, natural enemies and control, vol 3B. Elsevier, Amsterdam, pp 221–229

Drew RAI (1974) The responses of fruit fly species (Diptera: Tephritidae) in the South Pacific area to male attractants. J Aus Entomol Soc 13:267–270

Ekesi S, Billah MK (2007) A field guide to the management of economically important Tephritid fruit flies in Africa. ICIPE Science Press, Nairobi, 90 p

Ekesi S, Lux SA, Billah MK (2007) Field comparison of food-based synthetic attractants and traps for African Tephritid fruit flies. In: Development of improved attractants and their integration into fruit fly SIT management programmes. Proceedings of a final research coordination meeting, IAEA-TECDOC-1574, 5–7 May 2005, pp 205–222. International Atomic Energy Agency, Vienna, Austria

Ekesi S, Maniania NK, Mohamed SA (2011) Efficacy of soil application of *Metarhizium anisopliae* and the use of GF-120 spinosad bait spray for suppression of *Bactrocera invadens* (Diptera: Tephritidae) in mango orchards. Biocontr Sci Technol 21:299–316

Ekesi S, Mohamed SA, Tanga CM (2014) Comparison of food-based attractants for *Bactrocera invadens* (Diptera: Tephritidae) and evaluation of mazoferm–spinosad bait spray for field suppression in mango. J Econ Entomol 107:299–309

Ekesi S, De Meyer M, Mohamed SA, Virgilio M, Borgemeister C (2016) Taxonomy, ecology and management of native and exotic fruit fly species in Africa. Ann Rev Entomol. doi:10.1146/annurev-ento-010715-023603

Epsky ND, Midgarden D, Rendon P, Villatoro D, Heath RR (2012) Efficacy of wax matrix bait stations for Mediterranean fruit flies (Diptera: Tephritidae). J Econ Entomol 105:471–479

Grout T (2016) Cold and heat treatment technologies for post-harvest control of fruit flies in Africa. In: Ekesi S, Mohamed SA, De Meyer M (eds) Fruit fly research and development in Africa – towards a sustainable management strategy to improve horticulture. Springer, Cham. doi:10.1007/978-3-319-43226-7_20

Grout TG, Stephen PR (2013) Controlling *Bactrocera invadens* by using protein bait and male annihilation. SA Fruit J 7(8):61–65

Gunn D (1916) The cucumber and vegetable marrow fly (*Dacus vertebratus*). Report of the Division of Entomology, Department of Agriculture, Union of South Africa, Pretoria

Hagen KS, Finney GL (1950) A food supplement for effectively increasing the fecundity of certain tephritid species. J Econ Entomol 43:735

Hanna R, Gnanvossou D, Grout T (2008) Male annihilation technique (MAT) in eliminating *B. invadens* in northern Bénin. In: Fighting fruit and vegetable flies regionally in Western Africa. COLEACP/CIRAD, Information Letter 7:3

Heath RR, Burns RE, Kendra PE, Mangan R (2007) Fruit fly trapping and control – past, present and future. In: Development of improved attractants and their integration into fruit fly SIT management programmes. Proceedings of a final research coordination meeting, IAEA-TECDOC-1574, 5–7 May 2005, pp 7–10. International Atomic Energy Agency, Vienna, Austria

[IAEA] International Atomic Energy Agency (2009) Development of bait stations for fruit fly suppression in support of SIT. Report and recommendations of the consultants group meeting organized by the Joint FAO/IAEA Division of Nuclear Techniques in Food and Agriculture, Mazatlán, Mexico, 30 October –1 November 2008

Jang EB, Holler T, Cristofaro M, Lux S, Raw AS, Moses AL, Carvalho LA (2003) Improved attractants for area-wide detection and control of Mediterranean fruit fly, *Ceratitis capitata* (Weidemann): response of sterile and wild flies to (−) enantiomer of ceralure B1. J Econ Entomol 96:1719–1723

Jang EB, Khrimian A, Holler T, Casana-Giner V, Lux S, Carvalho LA (2005) Field response of Mediterranean fruit fly (Diptera: Tephritidae) to ceralure B1: evaluations of enantiomer B1 ratios on fly captures. J Econ Entomol 98:1139–1143

Jang EB, Khrimian A, Holler T (2010) Field response of Mediterranean fruit flies to ceralure B1 relative to most active isomer and commercial formulation of trimedlure. J Econ Entomol 103:1586–1593

Keiser I (1989) Insecticide resistance status. In: Robinson AS, Hooper G (eds) Fruit flies, their biology, natural enemies and control, vol 3B. Elsevier, Amsterdam, pp 337–344

Lux SA, Ekesi S, Dimbi S, Mohamed SA, Billah MK (2003) Mango infesting fruit flies in Africa – perspectives and limitations of biological approaches to their management. In: Neuenschwander P, Borgemeister C, Langewald J (eds) Biological control in integrated pest management systems in Africa. CAB International, Wallingford, pp 277–293

Mangan RL, Moreno D (2007) Development of bait stations for fruit fly population suppression. J Econ Entomol 100:440–450

Maniania NK, Ekesi S (2016) Development and application of mycoinsecticides for the management of fruit flies in Africa. In: Ekesi S, Mohamed SA, De Meyer M (eds) Fruit fly research and development in Africa – towards a sustainable management strategy to improve horticulture. Springer, Cham. doi:10.1007/978-3-319-43226-7_15

Manrakhan A, Daneel J-H (2013) Efficacy of M3 bait stations and GF-120 for the control of fruit flies in a citrus orchard in Mpumalanga. SA Fruit J 4(5):54–57

Manrakhan A, Daneel J-H, Grout T (2010) Testing a perimeter baiting strategy for fruit fly control using M3 bait station. SA Fruit J 8(9):57–60

Manrakhan A, Stephen PR, Cronje PJR (2015) Phytotoxic effect of GF-120 NF fruit fly bait on fruit of mandarin (*Citrus reticulata* Blanco cv. Nadorcott): influence of bait characteristics and fruit maturity stage. Crop Prot 78:48–53

Metcalf RL (1990) Chemical ecology of Dacinae fruit flies (Diptera: Tephritidae). Anns Entomol Soc Am 83:1017–1030

Metcalf RL, Metcalf ER (1992) Fruit flies of the family Tephritidae. In: Metcalf RL, Metcalf ER (eds) Plant kairomones in insect ecology and control. Chapman Hall, New York, pp 109–152

Mohamed SA, Ramadan MM, Ekesi SE (2016) In and out of Africa: parasitoids used for biological of fruit flies. In: Ekesi S, Mohamed SA, De Meyer M (eds) Fruit fly research and development in Africa – towards a sustainable management strategy to improve horticulture. Springer, Cham. doi:10.1007/978-3-319-43226-7_16

Moreno DS, Mangan RL (2002) Bait matrix for novel toxicants for use in control of fruit flies (Diptera: Tephritidae). In: Hallman G, Schwalbe CP (eds) Invasive arthropods in agriculture. Science Publishers Inc., Enfield, pp 333–362

Mwatawala MW, De Meyer M, Makundi RH, Maerere AP (2006) Biodiversity of fruit flies (Diptera: Tephritidae) at orchards in different agro-ecological zones of the Morogoro region, Tanzania. Fruits 61:321–332

Mwatawala M, Virgilio M, Quilici S, Dominic M, De Meyer M (2013) Field evaluation of the relative attractiveness of enriched ginger root oil (EGO)lure and trimedlure for African *Ceratitis* species (Diptera: Tephritidae. J Appl Entomol 137:392–397

Mwatawala MW, Mziray H, Malebo H, De Meyer M (2015) Guiding farmers' choice for an integrated pest management program against the invasive *Bactrocera dorsalis* Hendel (Diptera: Tephritidae) in mango orchards in Tanzania. Crop Prot 76:103–107

Navarro-Llopis C, Primo J, Vacas S (2012) Efficacy of attract-and-kill devices for the control of *Ceratitis capitata*. Pest Man Sci. doi:10.1002/ps.3393

Navarro-Llopis V, Primo J, Vacas S (2015) Bait station devices can improve mass trapping performance for the control of the Mediterranean fruit fly. Pest Manag Sci 71:923–927

Ndlela S, Mohamed SA, Ndegwa, PN, Ong'amo GO, Ekesi S (2016) Male annihilation technique using methyl eugenol laced with a toxicant for field suppression of *Bactrocera dorsalis* (Hendel) (Diptera: Tephritidae) on mango in Kenya. Afr Entomol (in press)

Nishda R, Tan KH, Serit M, Lajis NH, Sukari AM, Takahashi S, Fukami H (1988) Accumulation of phenylpropanoids in the rectal glands of male Oriental fruit fly, *Dacus dorsalis*. Experientia 44:534–536

Nishida R, Tan KH, Takahashi S, Fukami H (1990) Volatile components of male rectal glands of the melon fly, *Dacus cucurbitae* Coquillett (Diptera: Tephritidae). Appl Entomol Zool 25:105–112

Nishida R, Iwahashi I, Tan KH (1993) Accumulation of *Dendrobium* (Orchidaceae) flower fragrance in the rectal glands by males of the melon fly, *Dacus cucurbitae* (Tephritidae). J Chem Ecol 19:713–722

Orankanok W, Chinvinijkul S, Sawatwangkhoungm A, Pinkaew S, Orankanok S (2009) Application of chemical supplements to enhance *Bactrocera dorsalis* and *B. correcta* sterile males performance in Thailand. Fourth FAO/IAEA research coordination meetings on 'improving sterile male performance in fruit fly SIT programmes'

Peck SL, McQuate GT (2000) Field tests of environmentally friendly malathion replacements to suppress wild Mediterranean fruit fly (Diptera: Tephritidae) populations. J Econ Entomol 93:280–289

Piñero JC, Mau RFL, Vargas RI (2009) Managing Oriental fruit fly (Diptera: Tephritidae), with spinosad-based protein bait sprays and sanitation in papaya orchards in Hawaii. J Econ Entomol 102:1123–1132

Prokopy RJ, Miller NW, Piñero JC, Barry JD, Tran LC, Oride L, Vargas RI (2003) Effectiveness of GF-120 fruit fly bait spray applied to border area plants for control of melon flies. J Econ Entomol 96:1485–1493

Prokopy RJ, Miller NW, Piñero JC, Oride L, Chaney N, Revis H, Vargas RI (2004) How effective is GF-120 fruit fly bait spray applied to border area sorghum plants for control of melon flies (Diptera: Tephritidae)? Fla Entomol 87:354–360

Quilici S, Duyck PF, Franck A (2004) Preliminary experiments on the influence of exposure to methyl eugenol on mating success of males in the peach fruit fly, *Bactrocera zonata*. 1st RCM on improving sterile male performance in fruit fly SIT, Antigua, Guatemala

Ripley LB, Hepburn GA (1935) Olfactory attractants for male fruit flies. Department of Agriculture, Union of South Africa. Entomol Mem 9, pp 17

Roessler Y (1989) Insecticidal bait and cover sprays. In: Robinson AS, Hoopher G (eds) World crop pests: fruit flies: their biology, natural enemies and control. Elsevier, Amsterdam, pp 329–336

Rwomushana I (2008) Bioecology of the invasive fruit fly *Bactrocera invadens* (Diptera: Tephritidae) and its interaction with indigenous mango infesting fruit fly species, PhD dissertation, Kenyatta University, Department of Zoology

Schinz H, Seidel CF (1961) Nachtrag zu der arbeit Nr. 194 im. Hel Chim Acta 40:1829

Shelly TE, Dewire A-LM (1994) Chemically mediated mating success in male Oriental fruit flies (Diptera: Tephritidae). Anns Entomol Soc Am 87:375–382

Shelly TE, Dewire A-LM (2000) Flower-feeding affects mating performance in male Oriental fruit flies (Diptera: Tephritidae). Ecol Entomol 25:109–114

Sookar P, Alleck M, Ahseek N, Khayrattee FB, Permalloo S (2009) Improving male reproductive performance of *Bactrocera zonata* and *Bactrocera cucurbitae*. Fourth FAO/IAEA research coordination meetings on 'improving sterile male performance in fruit fly SIT programmes'

Steiner LF (1952) Fruit fly control in Hawaii with poison-bait sprays containing protein hydrolysate. J Econ Entomol 45:838–843

Steiner LF, Lee RKS (1955) Large-area tests of a male annihilation method for oriental fruit fly control. J Econ Entomol 48:311–317

Steiner LF, Rohwer GG, Ayers EL, Christenson LD (1961) The role of attractants in the recent Mediterranean fruit fly eradication program in Florida. J Econ Entomol 54:30–35

Tan KH, Nishida R (1996) Sex pheromone and mating competition after methyl eugenol consumption in *Bactrocera dorsalis* complex. In: McPheron BA, Steck GJ (eds) Fruit fly pests – a world assessment of their biology and management. St. Lucie Press, Delray Beach, pp 147–153

Tan KH, Nishida R (1998) Ecological significance of a male attractant in the defence and mating strategies of the fruit fly pest, *Bactrocera papayae*. Entomol Exp Applic 89:155–158

Tan KH, Nishida R (2012) Methyl eugenol: its occurrence, distribution, and role in nature, especially in relation to insect behavior and pollination. J Ins Sci 12:56. Available online: insectscience.org/12.56

Vargas RI, Shelly TE, Leblanc L, Piñero JC (2010) Recent advances in methyl eugenol and cuelure technologies for fruit fly detection, monitoring, and control in Hawaii. In: Gerald L (ed) Vitamins and hormones, vol 83. Academic, Burlington, pp 575–596

Vargas RI, Souder SK, Mackey B, Cook P, Morse JG, Stark JD (2012) Field trials of solid triple lure (Trimedlure, methyl eugenol, raspberry ketone, and DDVP) dispensers for detection and male annihilation of *Ceratitis capitata*, *Bactrocera dorsalis*, and *Bactrocera cucurbitae* (Diptera: Tephritidae) in Hawaii. J Econ Entomol 105:1557–1565

Vayssières J-F, Sinzogan A, Korie S, Ouagoussounon I, Thomas-Odjo AS (2009) Effectiveness of spinosad bait sprays (GF-120) in controlling mango-infesting fruit flies (Diptera: Tephritidae) in Benin. J Econ Entomol 102:515–521

Ware T, Richards G, Daneel J-H (2003) The M3 bait station, a novel method of fruit fly control. SA Fruit J 1:44–47

Wee SL, Tan KH, Nishida R (2007) Pharmacophagy of methyl eugenol by males enhances sexual selection of *Bactrocera carambolae* (Diptera: Tephritidae). J Chem Ecol 33:1272–1282

Yee WL (2006) GF-120, Nu-Lure, and Mazoferm effects on feeding responses and infestations of western cherry fruit fly (Diptera: Tephritidae). J Agric Urban Entomol 23:125–140

Sunday Ekesi is an Entomologist at the International Centre of Insect Physiology and Ecology (*icipe*), Nairobi, Kenya. He heads the Plant Health Theme at *icipe*. Sunday is a professional scientist, research leader and manager with extensive knowledge and experience in sustainable agriculture (microbial control, biological control, habitat management/conservation, managing pesticide use, IPM) and biodiversity in Africa and internationally. Sunday has been leading a continent-wide initiative to control African fruit flies that threaten production and export of fruits and vegetables. The initiative is being done in close collaboration with IITA, University of Bremen, Max Planck Institute for Chemical Ecology together with NARS, private sectors and ARI partners in Africa, Asia, Europe and the USA and focuses on the development of an IPM strategy that encompasses baiting and male annihilation techniques, classical biological control, use of biopesticides, ant technology, field sanitation and postharvest treatment for quarantine fruit flies. The aim is to develop a cost effective and sustainable technology for control of fruit flies on the African continent that is compliant with standards for export markets while also meeting the requirements of domestic urban markets. Sunday has broad perspectives on global agricultural research and development issues, with first-hand experience of the challenges and opportunities in working with smallholder farmers, extension agents, research organizations and the private sector to improve food and nutritional security. He sits on various international advisory and consultancy panels for the FAO, IAEA, WB and regional and national projects on fruit fly, arthropod pests and climate change-related issues. Sunday is a Fellow of the African Academy of Sciences (FAAS).

Chapter 14
Waste Brewer's Yeast as an Alternative Source of Protein for Use as a Bait in the Management of Tephritid Fruit Flies

Sunday Ekesi and Chrysantus M. Tanga

Abstract Yeast, and products of yeast, are composed of proteins, carbohydrates and minerals that fruit flies can utilize for development. They are being considered widely as alternative sources of protein for baits used for fruit fly suppression. In this chapter, we describe how we used techniques developed in Australia to developed fruit fly protein baits from waste brewer's yeast (WBY) from breweries in Kenya, Tanzania and Uganda. In attraction and feeding response studies, protein originating from waste yeasts compared favourably with commercially available protein baits used for fruit fly suppression. In field evaluation trials, the total number of oriental fruit fly, *Bactrocera dorsalis*, captured in Kenyan (112.1 flies/trap/day [FTD]) and Ugandan (109.8 FTD) WBY was superior to the standard NuLure (56.8 FTD) although Torula yeast gave the highest catch of 132.4 FTD. In field suppression trials, mango fruit infestation by *B. dorsalis* in treatments receiving protein bait originating from Kenyan WBY in combination with other management methods incurred 6.7–20.0 % fruit damage compared to 65.6 % fruit damage in the untreated control. The results suggest that locally developed protein baits made from WBY offers a suitable alternative to expensive imported food baits for the management of fruit flies in Africa. On the basis of these results, a local commercial protein bait production plant using WBY is being established in Kenya inspired by a similar facility in Mauritius.

Keywords Protein bait • Attraction • Feeding • Suppression

S. Ekesi (✉) • C.M. Tanga
Plant Health Theme, International Centre of Insect Physiology & Ecology (*icipe*),
PO Box 30772-00100 Nairobi, Kenya
e-mail: sekesi@icipe.org

© Springer International Publishing Switzerland 2016
S. Ekesi et al. (eds.), *Fruit Fly Research and Development in Africa - Towards a Sustainable Management Strategy to Improve Horticulture*,
DOI 10.1007/978-3-319-43226-7_14

1 Introduction

The process of yeast-catalyzed chemical conversion of sugars into ethyl alcohol and carbon dioxide has led to the production of alcoholic beverages including wine, beer or cider and other products that are subsequently distilled into brandy, whisky or vodka (Menezes et al. 2016). Associated with the brewing process is the production of large quantities of by-products usually referred to as waste yeast or spent grain (Hernandez-Pinerua and Lewis 1975). In fact, waste brewer's yeast (WBY) falls within the category of environmental hazard and disposal methods increase the energy costs for the factory (Grieve 1979; Peel 1999; Mirzaei-Aghsaghal and Maheri-Sis 2008). Despite environmental concerns, the use of brewery waste streams, especially WBY for various industrial products has been recognized for decades (Henry 1990; Lloyd and Drew 1997; Fillaudeau et al. 2006; Ferreira et al. 2010). Indeed, there is a continuous and increasing demand for brewer's grains, spent hops and other waste materials, some of which were previously considered of little or no value. Although dried WBY and spent hops are unsuitable for human consumption, they are known to be valuable as cattle feed (Grieve 1979; Tripathi and Karim 2011), and hops are also be used as a manure when mixed with other materials. WBY is reported to contain: a variety of enzymes, most of them capable of scientific or industrial application; nucleic acids, which are of great therapeutic value; vitamins; and proteins, carbohydrates and minerals that can be used as a food source by animals including fruit flies (Lloyd and Drew 1997; Sookar et al. 2002; Ferreira et al. 2010).

Lloyd and Drew (1997) pioneered the development of autolyzed protein bait from WBY for fruit fly control. The authors used pasteurized waste yeast from a Brisbane brewery in Australia and from the Royal Brewery in Tonga to produce a protein bait that could be used in fruit fly management strategies. Since this report several other studies have assessed the utility of WBY for fruit fly control with varying degrees of success (Seewooruthun et al. 1998, 2000; Chinajariyawong et al. 2003; Sookar et al. 2002, 2006; Umeh and Garcia 2008). In this chapter, we describe our attempt to develop protein baits from three different brewery sources in Kenya, Tanzania and Uganda using the methodology outlined by Lloyd and Drew (1997) with some minor modifications to the procedure. The attraction and feeding responses of the oriental fruit fly *Bactrocera dorsalis* (Hendel) to protein baits produced from the three sources of waste yeasts were evaluated in the laboratory and in field cages. The most effective yeasts were compared with commercially available protein baits for their attraction to *B. dorsalis*. Fruit fly suppression trials targeted at *B. dorsalis* on mango were also undertaken using a formulated protein bait product and are reported in this chapter.

2 Protein Bait Production from Waste Brewer's Yeast

Waste yeast was obtained from three breweries: Kenya Breweries Limited, Tanzania Breweries Limited and Uganda Breweries Limited. The solid content of the start-up materials ranged from 11 to 16%, the alcohol content from 5.8 to 6.2% and the pH from 5.2 to 6.1. Production methodology followed the procedure of Lloyd and Drew (1997) with slight modifications. Briefly, yeast slurry was boiled using aluminum pots in a water bath set at 120 °C with intermittent manual stirring to minimize burning and to remove as much of the alcohol as possible. Boiling continued until the material had reached 40% solid state. The concentrated yeasts were then digested with 0.4% papain for 24 h in a water bath held at 70 °C. Following digestion, 0.4% methyl p-hydrozybenzoate (Nipagin) was added as a preservative and the baits were evaluated directly for attractiveness to fruit flies, feeding response of fruit flies and field suppression, without further formulation.

3 Attraction and Feeding Responses of *Bactrocera dorsalis* to Commercial Protein Baits Compared with Protein Baits Made from Waste Brewer's Yeast: Field Cage Studies

A field cage trial were done at *icipe*'s Thomas Odhiambo Centre, Mbita, Kenya, which is located on the shores of Lake Victoria. Four field cages (5 m tall and 5 m dia.) made of nylon screen were set up under a shaded tree and experimental procedures were similar to that described by Vargas et al. (2002; Vargas and Prokopy 2006), one for each WBY bait (Kenyan, Ugandan and Tanzanian) and one for the water control. Briefly, 2–4 potted mango trees were placed in each cage to provide roosting sites and a shady canopy for the insects. To ensure insects were not attracted to bait because they were thirsty, the cages and trees were sprayed with water before initiation of the trials. After 30 min, 250 males and 250 females were released in to each cage. Test bait substances (20 droplets, ca 10 µl) were applied to mango leaf strips cut to fit into Petri dishes and covered with a screened lid. The dishes with the bait (or water in the control) were hung randomly around the perimeter of the tree canopy. Every 5 min, an observer walked around the canopy and recorded the number of male and female flies that had landed on the top of the screened Petri dishes. Both sexes of *B. dorsalis* were attracted to the different food baits although significantly more females were attracted than males (Fig. 14.1). Amongst the protein baits from the different waste yeast sources, the Kenyan and Ugandan WBY attracted more females compared with the Tanzanian WBY (Fig. 14.1). The lowest fly counts were observed in the water control. In a subsequent similar trials comparing the WBY baits from Kenya and Uganda with commercial food attractants our results showed that both the WBY baits were as effective at attracting female *B. dorsalis* as corn steep liquor, GF-120, Mazoferm and Buminal (Fig. 14.2). However, *B. dorsalis* females responded more to NuLure and Torula yeast than either the

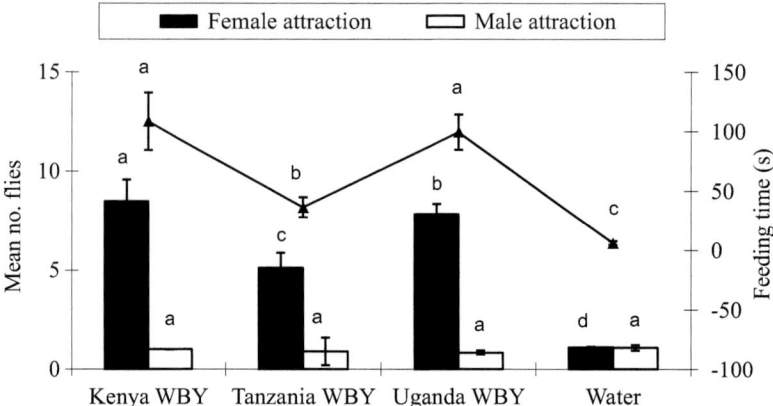

Fig. 14.1 Attraction of *Bactrocera dorsalis* to protein bait originating from waste brewer's yeast (WBY) and their feeding responses. For each parameter and sex bars and lines bearing the same letter do not differ significantly from each other using Tukey's (HSD) test. *Bars* denote SE

WBY baits from Kenya and Uganda. Overall, the WBY baits were as good as the commercially available protein baits tested in our studies. In a laboratory experiment, Lloyd and Drew (1997) showed that the relative attractiveness of WBY baits to the Queensland fruit fly, *Bactrocera tryoni* (Froggatt), ranged from ~0.6–0.9 compared with Mauri's bait, the commercial standard. The relative attractiveness of different dilutions of WBY baits (available commercially as Tongalure®) to *Bactrocera facialis* (Coquillett) were also high (0.66–1.97) at dilutions ranging from 1:5–1:50. Vargas et al. (2002) reported that the type of protein used influenced the attraction of the Mediterranean fruit fly (medfly), *Ceratitis capitata* (Wiedemann), to baits. The authors showed that *C. capitata* responded more to Provesta protein than the standard NuLure. Similar observations have also been reported for the melon fly, *Zeugodacus cucurbitae* (Coquillett) (= *B. cucurbitae*), and *B. dorsalis* (Vargas and Prokopy 2006).

Feeding assays were done using only female *B. dorsalis*. Mango leaf disks (5 cm × 5 cm) were placed in glass Petri dishes and treated with one drop of bait placed at the middle of the leaf. Each leaf disk was then transferred to a Perspex cage (30 cm × 30 cm × 30 cm). Female *B. dorsalis* (14 d old) were released individually on to the leaf surface and observed; the experiment ended after 600 s if the insect remained on the leaf all the time, or when the insect flew away or crawled off from the leaf disk. The time spent feeding was recorded and the total feeding time calculated. Results showed that the total feeding time varied significantly depending on the bait (Fig. 14.1). Female *B. dorsalis* fed for longer on the Kenyan and Ugandan WBY baits compared with the Tanzanian WBY bait. Female *B. dorsalis* spent the shortest time feeding in the water control. When we compared our WBY baits with commercial food attractants, feeding time was longest on NuLure and Kenyan WBY compared with the Ugandan WBY and other commercial baits (Fig. 14.2). Feeding times were similar on Corn steep liquor, GF-120, Torula yeast, Mazoferm and

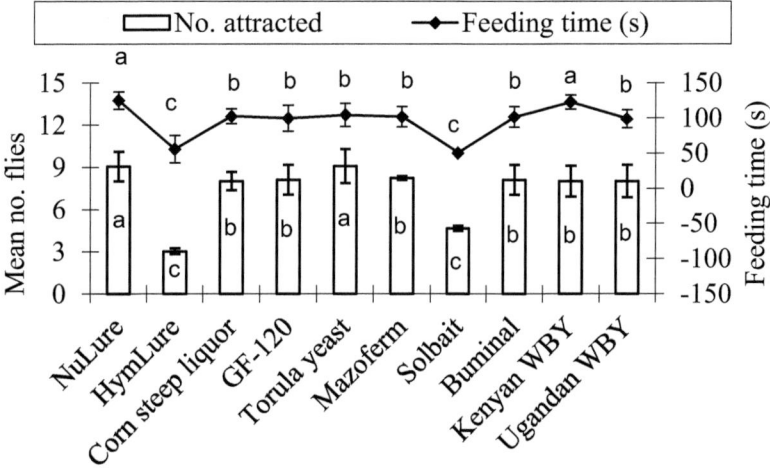

Fig. 14.2 Attraction and feeding response of *Bactrocera dorsalis* to protein baits made from waste brewer's yeast (WBY) in comparison with commercially available food attractants. For each parameter and sex, bars and lines bearing the same letter do not differ significantly from each other using Tukey's (HSD) test. *Bars* denote SE

Ugandan WBY (Fig. 14.2). The shortest feeding times were observed for HymLure and Solbait. Clearly, the type of protein in the bait also influences feeding and food attractants must be examined in the context of both attraction and feeding response (Vargas et al. 2002; Vargas and Prokopy 2006).

4 Attraction of *Bactrocera dorsalis* to Commercial Protein Baits Compared with Protein Baits made from Waste Brewer's Yeast: Open Field Studies

Larger scale protein bait evaluations were done at Nguruman, Kenya in an orchard that had a mixture of different mango varieties (Boribo, Tommy Atkins, Kent, Apple and Ngowe). Seven protein baits were evaluated between October 8, 2012 and November 28, 2012: (1) Protein bait from Kenyan WBY at a rate of 9 %, (2) Protein bait from Ugandan WBY at a rate of 9 %, (3) Torula yeast (ISCA Technologies, Riverside, CA, USA) at the rate of three pellets (4.78 g/pellet) per 1000 ml of water, (4) GF-120® (Dow AgroSciences, Indianapolis, IN, USA) at the on-the-label rate of 18.2 %, (5) Hymlure® (Savoury Food Industries Ltd, Johannesburg, South Africa) at the on-the-label rate of 4 %, (6) Biolure®, a three-component synthetic lure containing ammonium acetate, trimethylamine and putrescine (Suttera LLC, Bend, OR, USA) as a wet trap also containing 0.01 % Triton X-100 and (7) Nulure® (Miller Chemical & Fertilizer Corporation, Hanover, PA, USA) at a rate of 9 %. Each bait was placed in a Multi-lure trap® (Better World Manufacturing, Fresno, CA, USA)

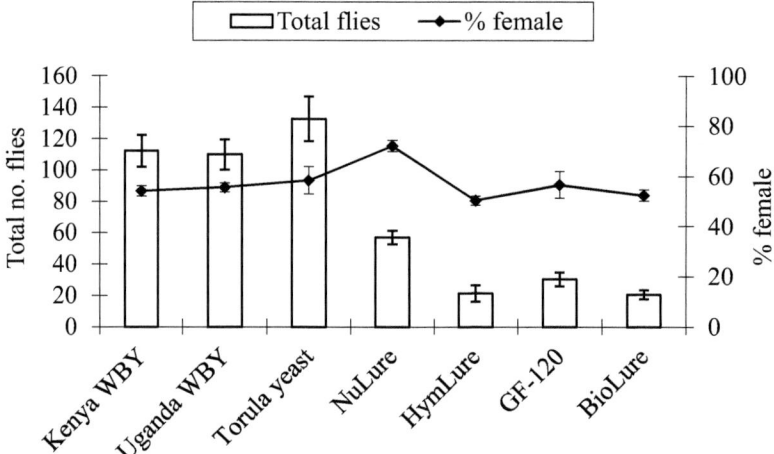

Fig. 14.3 Field responses of *Bactrocera dorsalis* to protein bait made from waste brewer's yeast (WBY) and to commercial food attractants in Nguruman, Kenya. *Bars* denote SE

for evaluation. In all the traps, 3 % borax was added to preserve trapped flies. Traps were placed on randomly selected trees of the various cultivars in the orchard and their position within each block was rotated sequentially every week, at the time they were checked for the presence of *B. dorsalis*. Apart from Biolure®, all baits were replaced weekly at the time of trap checking. Biolure® was replaced every 4 weeks. At each weekly check, flies were removed from the trap and the number and sex of *B. dorsalis* captured was recorded and a daily catch rate estimated. Results showed that the total number of *B. dorsalis* captured was greatest in traps baited with Torula yeast (132.4 flies/trap/day [FTD]) followed by the Kenyan WBY bait (112.1 FTD) and the Ugandan WBY bait (109.8 FTD) (Fig. 14.3). Fly catches in the standard NuLure-baited traps were 56.8 FTD. The percentage of female catches in all the food attractants ranged from 50.4 % in HymLure-baited traps to 72.1 % in NuLure-baited traps. Although Torula yeast was the best attractant, the use of this food bait in fruit fly management has been restricted to detection and monitoring, largely because its high pH may lead to phytotoxicity problems when applied directly to plants as part of field suppression strategies. Sookar et al. (2006) evaluated the catches of the peach fruit fly, *Bactrocera zonata* (Saunders), the Natal fruit fly, *Ceratitis rosa* Karsch and *C. capitata* using various combinations of Biolure (ammonium acetate [AA], putrescine [PT] and trimethylalamine [TMA]) and WBY and reported that traps baited with AA + PT + TMA with water/Triton as a retention device in multilure traps and the WBY treatment captured significantly more female flies compared with other treatments and that females accounted for more than 75 % of the catches. In Nigeria, Umeh and Garcia (2008) compared the performance of WBY with commercial protein hydrolyate for monitoring *Ceratitis* spp. on citrus and reported that average catches of fruit flies ranged from 8.3 to 10.4 flies/trap in WBY-baited traps compared with 6.2–12.2 flies/trap in protein hydrolysate-baited traps. Our observations in Kenya are in agreement with the previous studies and suggest that the two protein sources from Kenyan and Ugandan WBY offer excellent option for fruit fly monitoring and management in Africa.

5 Field Suppression of *Bactrocera dorsalis* Using Protein Baits Developed from Kenyan Waste Brewer's Yeast

Suppression trials were done in Meru county, Kenya (Fig. 14.4) between December 7, 2013 and April 4, 2014. Four sub-Counties in the major mango production areas of the County were selected as study sites and a census of mango growers in the sub-Counties was made with the assistance of County agricultural extension workers and an agribusiness NGO promoter (TechnoServe). Three of the four sub-Counties (Central Imenti, North Imenti and South Imenti) were assigned as fruit fly management areas and the remaining sub-County (Tigania West) served as a control area; this division was largely based on the farmers' methods of fruit fly management at the time which ranged from smoking the trees to drive away fruit flies to broad-spectrum cover sprays of different pesticides.

Seven treatments with varying combinations of tools from the fruit fly management package were applied to a total of 1071 mango orchards (153 farms per treatment) and the control treatment was assigned to 152 farms totaling 1223 farms used in total. The treatments included: [1] releases of two parasitoids *Fopius arisanus* (Sonan) and *Diachasmimorpha longicaudata* (Ashmead) (P), orchard sanitation (OS) and male annihilation technique (MAT); [2] P, OS and application of the Kenyan WBY bait laced with spinosad (= DuduLure®) (FB); [3] P, OS and soil application of a biopesticide based on *Metarhizium anisopliae* (Metchnikoff) Sorokin (= Met 69®) (BIOP); [4] P, OS, MAT and FB; [5] P, OS, FB and BIOP; (6) P, OS, MAT and BIOP; (7) P, OS, MAT, FB and BIOP; and [8] control. The socio-economic impact of these interventions have been recently reported by Muriithi et al. (2016) and details of how each treatment was applied can be found in this publication.

A sub-set of 40 orchards (five farms per treatment) were randomly selected from the 1223 farms (eight treatments) for monitoring of fruit fly populations. Also at harvest 50 fruits were sampled from each orchard at colour break and transported to the laboratory. Fruit samples from each treatment were incubated separately in plastic containers with sterile sand to facilitate pupation. The number of infested fruits, pupae recovered and adult emergence per treatment were recorded. The bulk of the insects that emerged from infested mango fruits were *B. dorsalis*, perhaps not surprising given that the native *Ceratitis* species have been displaced (Ekesi et al. 2009). After 8–12 weeks of treatment (depending on the mango variety), results showed that average post-treatment catches of *B. dorsalis* ranged from 84.9 flies/trap/week in orchards where FB had been applied in combination with P, OS, MAT and BIOP to 337.3 flies/trap/week in orchards assigned to P, OS and MAT (Fig. 14.5).

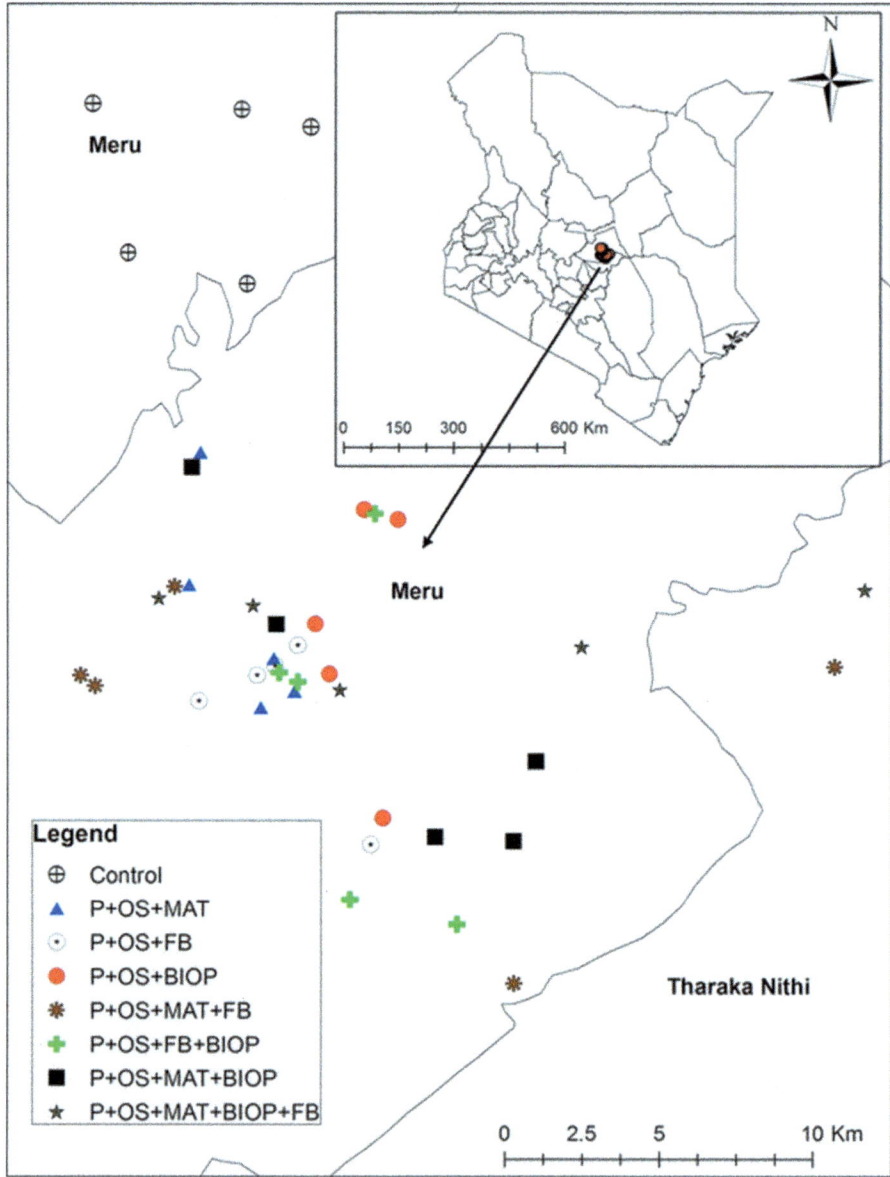

Fig. 14.4 Map showing locations of farms used for the field suppression trials in Meru County, Kenya. *P* parasitoids, *OS* orchard sanitation, *MAT* male annihilation technique, *FB* food bait and *BIOP* biopesticide

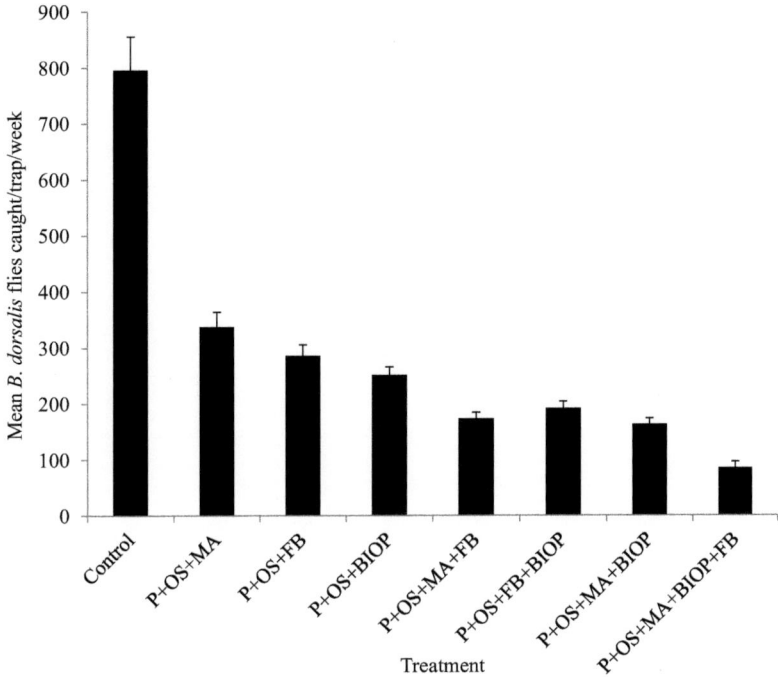

Fig. 14.5 Average post-treatment catches of *Bactrocera dorsalis*/trap/week following application of different management methods at Meru County, Kenya in 2014. *P* parasitoids, *OS* orchard sanitation, *MAT* male annihilation technique, *FB* food bait and *BIOP* biopesticide. Bars denote SE

The highest fruit fly catches (796.1 flies/trap/week) was observed in the control orchards. In locations where the fruit fly IPM toolbox was implemented, mango fruit infestation by fruit flies ranged from 7% in the treatment using all the tools in the package (P, OS, MAT, BIOP and FB) to 30% in orchards assigned to P, OS and BIOP (Fig. 14.6). Fruit infestation by fruit flies in treatments incorporating FB did not exceed 23%. In the control orchards, mango fruit damage reached 66% (Fig. 14.6). In Tonga, weekly treatment of chilli with protein baits developed from WBY reduced fruit damage due to *B. facialis* from 90 to 7% while damage in untreated control plots increased from 27 to 100% (Heimoana et al. 1997). Following the success story in the Pacific Islands, locally produced neutralized WBY was used as a source of protein bait for the eradication of *B. dorsalis* in Mauritius (Seewooruthun et al. 1998). Although good results were obtained with the bait based on the success of the eradication campaign (Seewooruthun et al. 2000), the neutralized WBY was observed to clog and damage the nozzles of knapsack sprayers and the product was also phytotoxic to young leaves of pawpaw and cucurbits. Sookar et al. (2002) modified the WBY product by digesting the yeast slurry with papain (0.8%), raw pawpaw or pineapple juice (4% v/v) and tested the new products for the control of *Z. cucurbitae* on ridged gourd, *Luffa acutangulata* (L.) Roxb. After 12 weeks of bait application, fruit infestation by *Z. cucurbitae* was reported to be 1% in treatment

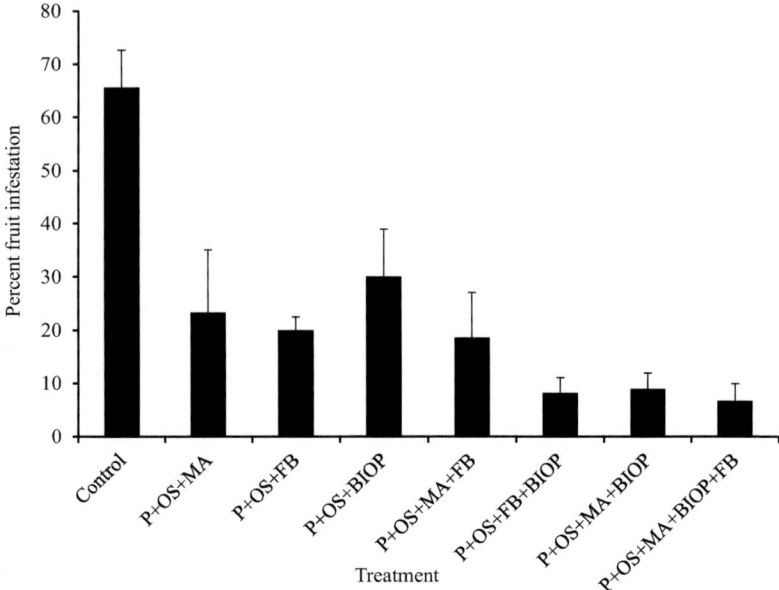

Fig. 14.6 The proportion of mango fruits infested (%) following application of different combinations of treatments to suppress *Bactrocera dorsalis* in Meru County, Kenya in 2014. *P* parasitoids, *OS*, orchard sanitation, *MAT* male annihilation technique, *FB* food bait and *BIOP* biopesticide. *Bars* denote SE

using WBY autolysed with papain, 2% in the treatment using WBY autolysed with pawpaw, 6% in treatment using WBY autolysed with pineapple, 7% in the boiled/unautolysed treatment, 2% in the standard protein hydrolysate treatment and 100% infestation in the untreated control plots. No bait exhibited phytotoxicity effects on the plants. In Tanzania, Mwatawala et al. (2015) formulated WBY in molasses, water and an extract of *Derris elliptica* Benth. (insecticidal plant) and reported that mango fruit infestation in orchards treated with the WBY ranged from 0.004 to 1.7% compared with 0.58–2.52% in orchards treated with Success® bait spray and 0.4–2.03 in the treatment receiving a cover spray of dimethoate/Karate. Generally, the results obtained from the suppression trials in Kenya are in agreement with other studies and demonstrate that protein baits from WBY offer cheap readily available alternatives to expensive imported protein hydrolysate for the management of fruit flies in Africa.

6 Conclusions and Recommendations for Future Research

Organic waste that originates from the brewing process including excess yeast, spent hops and grains have been the subject of various investigations for their potential use as industrial byproducts. Because WBY is a rich source of protein,

Fig. 14.7 Waste brewer's yeast-based protein bait facility in Réduit, Mauritius

carbohydrates, vitamins and minerals, and since these nutrient are essential for the development of fruit flies, it is being widely exploited as a source of protein for use as baits in the management of fruit flies. In most Sub-Saharan African (SSA) countries, the majority of the protein baits used for fruit fly management have to be imported from Europe or North America at a very high price. The cost of the imported protein baits also limits their use, especially for smallholder growers. The ability of various African countries to produce bait from locally available raw materials such as WBY could make a difference in reducing the cost and enhancing the sustainability of fruit fly monitoring and management programmes. Although inconsistent and inadequate performance of brewery wastes has been cited as one of the drawbacks of using WBY in fruit fly suppression, the results presented here show that the use of protein baits made from WBY holds great promise in ameliorating fruit fly problems in Africa. The technology, adapted from Australia (Lloyd and Drew 1997), is gradually gaining importance as a real-time solution to the cost of baiting techniques for fruit fly suppression. Based on this knowledge, the government of Mauritius has recently funded the establishment of a 1000 l capacity facility in Réduit (Fig. 14.7.) providing affordable protein baits based on WBY to over 15,000 fruit and vegetable growers across the country. Under the *icipe*-led project on 'Establishment of a Pilot Commercial Processing Plant for Food Bait Production for the Management of Fruit Flies in Kenya' funded by GIZ/BMZ, a similar bait production plant is being established in Kenya in collaboration with Kenya Biologic Ltd (http://www.worldagroforestry.org/itaacc/projects/food-bait). Currently, bulk waste yeast materials from various lager beers have been used by researchers for the production of protein bait. However, it is unknown how the raw materials, yeast types and various other substrates that are combined in the brewing process influence the quality of the proteins that are autolyzed for fruit fly management and these are potential areas of research that require attention. Research should also focus on appropriate formulation of the baits, addressing shelf life and training of entrepreneurs that are willing to take up business ventures in the production of food baits for fruit fly management on the African continent.

Acknowledgements The German Federal Ministry for Economic Cooperation and Development (BMZ) provided funding for research on the Mango IPM and Fruit Fly Food Bait projects at *icipe*. We thank the mango growers at Nguruman and Meru, Kenya for providing the experimental orchards for all the trials. We also greatly appreciate the efforts by Mr P. Agola and Mr E.F. Mlato for bait production, data collection and processing.

References

Chinajariyawong A, Kritsaneepaiboon S, Drew RAI (2003) Efficacy of protein bait sprays in controlling fruit flies (Diptera: Tephritidae) infesting angled luffa and bitter gourd in Thailand. Raffles Bull Zool 51:7–15

Ekesi S, Billah MK, Nderitu PW, Lux SA, Rwomushana I (2009) Evidence for competitive displacement of the mango fruit fly, *Ceratitis cosyra* by the invasive fruit fly, *Bactrocera invadens* (Diptera: Tephritidae) on mango and mechanisms contributing to the displacement. J Econ Entomol 102:981–991

Ferreira I, Pinho O, Viera E, Tavarela J (2010) Brewer's *Saccharomyces* yeast biomass: characteristics and potential applications. Trends Food Sci Technol 21:77–84

Fillaudeau L, Blanpain-Avet P, Daugin G (2006) Water, wastewater, and waste management in brewing industries. J Clean Prod 14:463–471

Grieve DG (1979) Feed intake and growth of cattle fed liquid brewer's yeast. Can J Anim Sci 59:89–94

Heimoana V, Nemeye P, Drew RAI (1997) Assessment of protein bait sprays for the control of fruit flies in chilli and capsicum crops in Tonga. In: Allood AJ, Drew RAI (eds) Management of fruit flies in the Pacific. Nadi, Fuji. 28–31 October 1996. ACIAR Proceedings No.76, pp 179–182

Henry WA (1990) Dried distillery grains compared with oats. Feeds and feeding, 2nd edn. Morrison Publishing Co., Ithaca, 421 pp

Hernandez-Pinerua JR, Lewis MJ (1975) Disposal of excess brewer's yeast by recycling to the brewerhouse. J Inst Brew 81:476–483

Lloyd A, Drew RAI (1997) Modification and testing of brewery waste yeast as a protein source for fruit fly bait. In: Allwood AJ, Drew RAI (eds) Fruit fly management in the Pacific. ACIAR Proceedings No. 76, pp 192–198

Menezes AGT, Menenzes EGT, Alves JGLF, Rodrigues LF, Cardoso M (2016) Vodka production from potato (*Solanum tuberosum* L.) using three *Saccharomyces cerevisiae* isolates. J Inst Brew 122:76–83

Mirzaei-Aghsaghal A, Maheri-Sis N (2008) Nutritive value of some agro-industrial by-products for ruminants. World J Zool 3:40–46

Muriithi B, Affognon HD, Diiro GM, Kingori SW, Tanga CM, Nderitu PW, Mohamed SA, Ekesi S (2016) Impact assessment of Integrated Pest Management (IPM) strategy for suppression of mango-infesting fruit flies in Kenya. Crop Prot 81:20–29

Mwatawala MW, Mziray H, Malebo H, De Meyer M (2015) Guiding farmers' choice for an integrated pest management program against the invasive *Bactrocera dorsalis* Hendel (Diptera: Tephritidae) in mango orchards in Tanzania. Crop Prot 76:103–107

Peel R (1999) Ecological sustainability in the brewing industry. J Inst Brew 105:14–22

Seewooruthun SI, Sookar P, Permalloo S, Joomaye A, Alleck A, Gungah B, Soonnoo AR (1998) An attempt at the eradication of the oriental fruit fly, *Bactrocera dorsalis* (Hendel) from Mauritius. In: Lalouette JA, Bachraz DY, Sukurdeep N, Seebaluck BD (eds) Proceedings of the 2nd annual meeting of agricultural scientists. Food and Agricultural Research Council, Réduit Mauritius, pp 135–144

Seewooruthun SI, Permalloo S, Sookar P, Soonnoo AR (2000) The oriental fruit fly, *Bactrocera dorsalis* eradicated from Mauritius. In: Price NS, Seewooruthun I (eds) Proceedings of the Indian Ocean Commission, Regional fruit fly symposium, Flic en Flac, Mauritius, 5th–9th June 2000. Quatre Bornes. Indian Ocean Commission, pp 207–210

Sookar P, Facknath S, Permalloo S, Seewoorothum SI (2002) Evaluation of modified waste brewer's yeast as a protein source for the control of the melon fly, *Bactrocera cucurbitae* (Coquillet). In: Barnes BN (ed) Proceedings of 6th international fruit fly symposium, 6–10 May 2002, Stellenbosch, South Africa, pp 295–299

Sookar P, Permalloo S, Alleck M, Seewooruthun SI (2006) Development of improved attractants and their integration into fruit fly management programmes. Fruit flies of economic importance: from basic to applied knowledge, pp 71–79. In: Proceedings of the 7th international symposium on fruit flies of economic importance. 10–15 September 2006, Salvador, Brazil

Tripathi M, Karim S (2011) Effect of yeast cultures supplementation on live weight change, rumen fermentation, ciliate protozoa population, microbial hydrolytic enzymes status and slaughtering performance of growing lamb. Livest Sci 135:17–25

Umeh VC, Garcia LE (2008) Monitoring and managing *Ceratitis* spp. complex of sweet orange varieties using locally made protein bait of brewery waste. Fruits 63:209–217

Vargas RI, Miller NW, Prokopy RJ (2002) Attraction and feeding responses of Mediterranean fruit fly and a natural enemy to protein laced with novel toxinx, Phloxine B and spinosad. Entomol Exp Appl 102:273–282

Vargas RI, Prokopy R (2006) Attraction and feeding responses of melon flies and oriental fruit flies (Diptera: Tephritidae) to various protein baits with and without toxicants. Proc Hawaiian Entomol Soc 38:49–60

Sunday Ekesi is an Entomologist at the International Centre of Insect Physiology and Ecology (*icipe*), Nairobi, Kenya. He heads the Plant Health Theme at *icipe*. Sunday is a professional scientist, research leader and manager with extensive knowledge and experience in sustainable agriculture (microbial control, biological control, habitat management/conservation, managing pesticide use, IPM) and biodiversity in Africa and internationally. Sunday has been leading a continent-wide initiative to control African fruit flies that threaten production and export of fruits and vegetables. The initiative is being done in close collaboration with IITA, University of Bremen, Max Planck Institute for Chemical Ecology together with NARS, private sectors and ARI partners in Africa, Asia, Europe and the USA and focuses on the development of an IPM strategy that encompasses baiting and male annihilation techniques, classical biological control, use of biopesticides, ant technology, field sanitation and postharvest treatment for quarantine fruit flies. The aim is to develop a cost effective and sustainable technology for control of fruit flies on the African continent that is compliant with standards for export markets while also meeting the requirements of domestic urban markets. Sunday has broad perspectives on global agricultural research and development issues, with first-hand experience of the challenges and opportunities in working with smallholder farmers, extension agents, research organizations and the private sector to improve food and nutritional security. He sits on various international advisory and consultancy panels for the FAO, IAEA, WB and regional and national projects on fruit fly, arthropod pests and climate change-related issues. Sunday is a Fellow of the African Academy of Sciences (FAAS).

Chrysantus Mbi Tanga is an Agricultural Entomologist/Ecologist working in the Plant Health Theme at the International Centre of Insect Physiology and Ecology (*icipe*), Nairobi, Kenya. He obtained his PhD from the University of Pretoria, South Africa and is currently a postdoctoral research fellow at *icipe*. His area of research interest is sustainable integrated pest management (IPM) that is based on the use of biopesticides, baiting techniques, orchard sanitation and classical biological control for the management of arthropod pests in horticulture. His research activities involve the promotion and up-scaling of the dissemination of fruit fly IPM amongst smallholder growers in Africa. He also develops temperature-based phenology models for African fruit flies and other crop pests to improve implementation of IPM strategies for their management. He also has research activities that are focused on filling critical gaps in knowledge aimed to exploit the commercial potential of insects in the production of affordable, high-quality protein for the poultry, fish and pig industries. He seeks to create an effective agribusiness model by ensuring high nutrition and microbial safety of insect-based protein products through research as well as by creating awareness and identifying market opportunities for the technology using participatory approaches that involves farmer groups, with a particular focus on the involvement of women and youths in Africa.

Chapter 15
Development and Application of Mycoinsecticides for the Management of Fruit Flies in Africa

Jean N.K. Maniania and Sunday Ekesi

Abstract In many African countries, the management of fruit flies relies heavily on applications of synthetic chemical insecticides that have detrimental effects on producers and consumers, the environment and often disrupt the activity of natural enemies. Entomopathogenic fungi (EPF) which infect their hosts through the cuticle are being considered as alternatives to synthetic chemical insecticides and are being developed as mycoinsecticides. Since the life cycle of fruit flies takes place in the soil as pupating larvae and pupae, and in the canopy as adults, a two-pronged approach is being considered for their control by EPF. Prophylactic applications of mycoinsecticides to the soil before fruit set and infestation would be the ideal strategy for control of the soil-dwelling stages of fruit flies. However, a number of challenges need to be addressed because of the complex interactions amongst mycoinsecticides, the insect host, and the biotic and abiotic factors that occur in the soil. The second approach, autodissemination, is based on both the behaviour of adult flies, which are attracted to semiochemicals, and the unique characteristics of EPF that allows them to be transmitted horizontally. Examples of the development of EPF as mycoinsecticides and their use for the management of fruit flies in Africa are provided in this review. EPF fungi as components within an IPM approach, can provide sustainable control of fruit flies.

Keywords Entomopathogenic fungi • Virulence • Biocontrol • Formulation • Biotic and abiotic factors • Autodissemination • Attractant

---

J.N.K. Maniania • S. Ekesi (✉)
Plant Health Theme, International Centre of Insect Physiology & Ecology (*icipe*),
PO Box 30772-00100, Nairobi, Kenya
e-mail: sekesi@icipe.org

© Springer International Publishing Switzerland 2016
S. Ekesi et al. (eds.), *Fruit Fly Research and Development in Africa - Towards a Sustainable Management Strategy to Improve Horticulture*,
DOI 10.1007/978-3-319-43226-7_15

1 Background

There is worldwide support for reducing chemical insecticide usage for the management of crop pests due to the associated problems of environmental contamination and adverse effects on non-target organisms (Grant et al. 2010). The demand for contaminant-free foods amongst other drivers has necessitated the development of alternative strategies for managing pests, including fruit flies. In many African countries, fruit fly management still relies heavily on applications of synthetic chemical insecticides. The majority of fruit and vegetable growers, who are smallholders with a limited knowledge of pesticides, often apply cocktails of synthetic chemical insecticides (Williamson et al. 2008). This practice substantially increases production costs, increases health risks for producers and consumers, and often disrupts the activity of natural enemies that would otherwise contribute to fruit fly population regulation (Ekesi et al. 2007). Therefore, exploring non-chemical alternatives for control of fruit flies is fundamental to realizing sustainable fruit and vegetable production, especially amongst smallholder farmers on the African continent.

There is considerable interest in the exploitation of naturally occurring organisms, including fungi, viruses, protozoa and bacteria for the control of crop pests. Amongst the various entomopathogens that attack fruit flies, fungi have received more attention as candidates for the control of these pests, than all the other groups of pathogens combined. Entomopathogenic fungi (EPF) are unique amongst insect pathogens in that they infect their host directly through the external cuticle without the need to be ingested which is a particular advantage for control of piercing and sucking insect pests (Burges 2007; McCoy et al. 2009; Lacey et al. 2011). The infection process includes attachment of the spore, or conidium, to the cuticle, germination, penetration of the cuticle and interactions with the host's immune response. EPF must overcome and/or avoid the host immune defenses and proliferate within the insect haemocoel in order to obtain nutrients. Host death occurs as a result of physical damage and loss of normal function following colonization of tissues and organs. After death, growth may continue in the mycelial phase with invasion of the vital organs; ultimately the fungus emerges to the exterior of the insect and sporulates to produce infective conidia for horizontal transmission (Roberts and Humber 1981). EPF inhabit a diverse array of geographic, climatic and agro-ecological zones (see Lacey et al. 2015).

EPF have several characteristics that make them ideal alternatives to synthetic chemical insecticides. They are relatively host specific with minimal effects on non-target beneficial organisms such as bees, earthworms and Collembola which provide key ecosystem services, and natural enemies such as parasitic wasps and predators (Goettel et al. 2001; Zimmermann 2007a, b; O'Callaghan and Brownbridge 2009; Garrido-Jurado et al. 2011a). Hence they are compatible with other biological options within an IPM programme. EPF have low mammalian toxicity (Zimmermann 2007a, b). Their production is easy and cheap and does not require high input technology (Prior 1988) and so their use is feasible within low input agriculture. In recent years, stable products, called mycoinsecticides, that are easily mixed for spraying and that are cost competitive with chemical insecticides have been developed and are now available in different parts of world including Africa (Shah and Goettel 1999; Faria and Wraight 2007; Lacey et al. 2015; Gwynn and Maniania 2010; www.RealIPM.com). In Europe and North America several commercial products based on *Metarhizium anisopliae* (Metchnikoff) Sorokin, *Beauveria bassiana* (Bals-Criv.) Vuill., *Isaria* species and *Lecanicillium* species have been registered for control of a variety of insect pests. They include BotaniGard® and Naturalis-L® (*B. bassiana*-based products), Met52® (a *M. anisopliae*-based product), Preferal® (an *Isaria javanica* (Friedrichs & Bally)-based product) and Vertalec® (a *Lecanicillium longisporum* Zare and Gams-based product) (Kabaluk et al. 2010; Ravensberg 2011; Jandricic et al. 2014). In recent times, similar products have become available on the African continent including Met 69® (a *M. anisopliae icipe* 69-based product), Met 62® (a *M. anisopliae icipe* 62-based product), Met 78® (a *M. anisopliae icipe* 78-based product) (www.RealIPM.com) and Green Muscle® (a *Metarhizium acridum* (Driver & Milner) Bisch., Rehner & Humber-based product) (Lomer et al. 1999). The present chapter focuses on the Phylum Ascomycota because of its many desirable traits and because most of the currently commercialized mycoinsecticides are primarily based on species from this group (Faria and Wraight 2007).

Attack of fruit flies by EPF is relatively uncommon in nature (Ekesi et al. 2007). However, experimental infections induced under controlled conditions allow the pathogenic activity of EPF and the susceptibility of different target species to be evaluated (Hall and Papierok 1982). Subsequently, several EPF including *B. bassiana*, *M. anisopliae*, *Isaria fumosorosea* Wize and *Lecanicillium* sp. have been exploited for biological control of fruit flies (Ekesi et al. 2016). Development of these fungi has followed an 'industrial' pathway with mass production systems devised to provide large quantities of inoculum that can be formulated and repeatedly applied as sprays or granules (Feng et al. 1994; Shah and Pell 2003; Chandler et al. 2008). These products are regularly applied to regulate target pest populations and their development and commercialization has generally been similar to that of synthetic pesticides.

2 Screening and Selection of Potent Isolates

There are many steps in developing EPF as microbial control agents (Zimmermann 1986), amongst which selection of the best isolate is the most important (Soper and Ward 1981; Heale 1988). Virulent isolates generally express high levels of conidium-bound proteases, efficiently produce and release exoenzymes during cuticular penetration, and also produce toxins to overcome the host immune system during colonization of the internal organs (Vey et al. 2001; Ortiz-Urquiza et al. 2013; Khan et al. 2012). Several EPF are ubiquitous inhabitants of the soil worldwide (Hummel et al. 2002; Keller et al. 2003; Jaronski 2010; Meyling and Eilenberg 2007). Because fruit flies also spend part of their life cycle in contact with soil as pupating larvae and pupae, application of mycoinsecticides to the soil has been recommended for management of fruit flies (Ekesi et al. 2007). In this regard, several isolates of *M. anisopliae* and *B. bassiana* have been evaluated in the laboratory and in the field against native *Ceratitis* species: the Mediterranean fruit fly (medfly), *Ceratitis capitata* (Wiedemann); the mango fruit fly, *Ceratitis cosyra* (Walker); *Ceratitis fasciventris* (Bezzi) and the Natal fruit fly, *Ceratitis rosa* Karsch) and the exotic oriental fruit fly, *Bactrocera dorsalis* (Hendel). For example, seven fungal isolates (*M. anisopliae icipe* 18, 20, 32, 60 and 69 and *B. bassiana icipe* 44 and 82) caused significantly higher mortality of *C. capitata* pupae than other Kenyan isolates (Ekesi et al. 2002). All these isolates also significantly reduced the lifespan of surviving adults that emerged following treatment as late third-instar larvae. With the exception of *icipe* 32, the other four isolates of *M. anisopliae* were equally pathogenic to *C. fasciventris*. Dose-mortality response assay revealed that the regression lines of *icipe* 18 and 20 were steeper with lower LC_{50} values when compared with *icipe* 60 and 69 against *C. capitata*, *C. fasciventris* and *C. cosyra*. When tested against pupae of different ages, adult emergence increased with increasing pupal age; this was correlated with decreases in mortality of pupae with age and was similar in the three species of fruit flies (Ekesi et al. 2002). Further studies by Ouna (2012) also demonstrated the pathogenicity of several isolates of EPF to *B. dorsalis*. Exposure of third-instar larvae to soil treated with conidial suspensions of *M. anisopliae* isolates *icipe* 95 and 18 (1×10^9 conidia ml^{-1}) reduced adult emergence to 8 and 10%, respectively, compared with 82% emergence in the control.

In South Africa, Goble et al. (2011) evaluated the pathogenicity of 15 isolates of *B. bassiana*, five isolates of *M. anisopliae* and one isolate of *Metarhizium flavoviride* Gams and Roszypal against larval stages of *C. rosa* and *C. capitata*. At a concentration of 1×10^7 conidia ml^{-1}, the fungal isolates caused significantly higher mortality in the adults of both species than in the pupae. Further, mycosis in both adult and pupae did not differ significantly. In Morocco, Imoulan et al. (2011) evaluated 118 Moroccan isolates of *B. bassiana* against *C. capitata* and also assessed their thermo-tolerance. The authors reported that 86% of the isolates were pathogenic to pupae and 55% of these were tolerant to temperature stress (i.e. 45 °C for 2 h). However, they concluded that only 60% of the *B. bassiana* isolates evaluated could be considered as virulent with the potential for use in biological control of *C. capitata*. In Mauritius, Sookar et al. (2008) demonstrated that seven isolates of *M. anisopliae*, five isolates of *B. bassiana* and two isolates of *I. fumosorosea* applied at the concentration of 1×10^6 conidia ml^{-1} caused mortalities of 12–98% in the peach fruit fly, *Bactrocera zonata* (Saunders) and 2–94% in the melon fly, *Zeugodacus cucurbitae* (Coquillett) within 5 days. On the basis of these results, it was suggested that a prophylactic application of these isolates as mycoinsecticides before fruit set and potential infestation, that was targeted at pupating larvae, would be the ideal strategy for control of the soil-dwelling stages of fruit flies.

Autodissemination of mycoinsecticides is being widely advocated for the management of adult fruit flies (Ekesi et al. 2007; Toledo et al. 2007; Navarro-Llopis et al. 2015). This strategy consists on attracting insects to an autoinoculator where they are contaminated/infected with a pathogen before returning to the environment to disseminate the pathogen to conspecifics (Vega et al. 2007). In the laboratory, Dimbi et al. (2003a) designed and evaluated a simple autoinoculator in the form of a cylindrical plastic tube in which fruit flies could be contaminated with dry conidia of EPF. Velvet carpet material that was impregnated with conidia covered the inside of the cylindrical plastic tube (95×48 mm with white nylon netting over one end) and then more dry conidia were evenly spread over the velvet and adult fruit flies allowed to walk on the velvet for 3 min. Mortality of *C. capitata*, *C. fasciventris* and *C. cosyra* following exposure to different isolates of *B. bassiana* and *M. anisopliae* ranged from 7–100%, 11–100% and 72–78%, respectively, within 4 days of contamination. Six isolates of *M. anisopliae* (*icipe* 18, 20, 32, 40, 41 and 62) were reported to be highly pathogenic to fruit flies when the flies were contaminated in this way (Dimbi et al. 2003a). Using a similar contamination technique, Ouna (2012) evaluated several fungal isolates against adult *B. dorsalis* in the laboratory. Fly mortality ranged from 46 to 100% within 4 days of contamination, compared with 2% mortality in the control. Under laboratory conditions in Egypt, Mahmoud (2009) evaluated the pathogenicity of three commercial mycoinsecticide-based products: Bio-Power (*B. bassiana*), Bio-Magic (*M. anisopliae*) and Bio-Catch (*Lecanicillium* sp.) (T. Stanes & Company Ltd.), India, against adult olive fruit flies, *Bactrocera oleae* (Rossi); two methods of inoculation were compared (contact and oral). Oral applications resulted in higher mortalities than contact inoculations. Moreover, the *Lecanicillium* sp. was most virulent than the other products regardless of the method of application. LT_{50} values were shorter following oral application than following contact inoculation in all treatments.

3 Factors Affecting Infectivity and Efficacy

Insect susceptibility to EPF can be affected by a number of factors, including environmental variables, traits of the host population and traits of the fungus (Benz 1987; Fuxa and Tanada 1987; McCoy et al. 2000; Inglis et al. 2001). Of the various environmental parameters that affect EPF, temperature, humidity and solar radiation are probably the most important (Inglis et al. 2001). Variability in temperature tolerance is common amongst species and amongst isolates within a species. Therefore, it is important to establish temperature optima for candidate isolates being developed as mycoinsecticides. Ekesi et al. (2003) assessed the infectivity of four isolates of *M. anisopliae* against pupae of *C. capitata* treated as late third-instar larvae in unsterilized soil in the laboratory, under controlled temperature and moisture conditions. Between 20 °C and 30 °C, mortality of pupae was highest at water potentials of −0.1 and −0.01 mega Pascal (MPa) and lowest at water potentials of −0.0055 and −0.0035 MPa for all isolates. In wetter soil, isolates *icipe* 20 and 60 caused significantly higher mortality than *icipe* 18 and 69. The survival of conidia in drier soil (−0.1 MPa) was not adversely affected at all temperatures. However, in wet soil (0.0035 MPa) there was a drastic reduction in survival of *icipe* 18 and 69 at 25 °C and 30 °C while *icipe* 20 and 60 were unaffected for 14 days after inoculation at all temperatures. Evaluation of isolate *icipe* 20 against three other fruit fly species (*C. cosyra, C. rosa* and *C. fasciventris*) showed a significant reduction in adult emergence and higher pupal mortality in *C. cosyra* and *C. fasciventris* than in *C. rosa* at combinations of 15 °C or 20 °C and −0.1 MPa or −0.0035 MPa. However, at higher temperatures and the same moisture levels, all three fruit fly species were equally susceptible. The authors concluded that survival and infectivity of EPF in soil seems to be governed by a complex interaction that occurs between soil temperature and moisture in addition to pathogen cycling and multiplication (on dead, infected insects) in the soil. It is probable that a balance between microbial degradation and replenishment of inoculum of virulent isolates occurs through fluctuations and intricate interactions between temperature and moisture levels. Using a multiple logistic regression model for mycosis, Garrido-Jurado et al. (2011b) studied the relationships between temperature, moisture and infectivity of two isolates of *B. bassiana* (Bb-1333, EABb 01/33-Su) and two isolates of *M. anisopliae* (EAMa 01/58-Su, EAMa 01/158-Su) against last-instar larvae of *C. capitata* in sterile soil under controlled conditions. *Beauveria bassiana* isolates were generally less virulent than *M. anisopliae* isolates. For all isolates, lower rates of infection were achieved under extreme moisture conditions (1 % and 17 % wt.:wt.) than under drier conditions. For the most virulent isolate, *M. anisopliae* EAMa 01/58-Su, the highest rates of infection were observed at intermediate temperatures. Conversely there was a direct relationship between temperature and infection rate for *M. anisopliae* isolate EAMa 01/158-Su. Infection rates of both *B. bassiana* isolates displayed a parabolic relationship with moisture, and infection was only achieved at low temperatures (15–20.1 °C). The authors concluded that each species and each isolate are independent biological entities with different responses to environmental conditions; therefore, EPF for use as mycoinsecticides should be matched to the range of temperatures and humidities in the environment where they will be applied.

Dimbi et al. (2004) demonstrated that the mortality caused by six fungal isolates against adults of *C. capitata, C. fasciventris* and *C. cosyra* varied with temperature, isolate and fruit fly species. Fungal isolates were more effective at 25, 30 and 35 °C (48.8–100 %) than at 20 °C (5.0–26.4 %). The LT_{90} values decreased with increasing temperature up to the optimum temperature of 30 °C.

Amongst the attributes of the host, host species, host age, host developmental stage and host sex can affect susceptibility to EPF (Ferron 1985; Feng et al. 1985; Maniania and Odulaja 1998). Dimbi et al. (2003b) assessed the effect of host age and sex on the susceptibility of *C. capitata*, *C. cosyra* and *C. fasciventris* to *M. anisopliae* in the laboratory. Three adult host ages, 0 (<1 day-old), 7-day-old and 14-day-old, were compared. Differences were observed in the level of susceptibility amongst the different host species but age accounted for the most variability in mortality, while sex accounted for the least variability. Of the three host ages tested, the 0- and 7-day-old flies were more susceptible to infection than the 14-day-old flies. Females of *C. cosyra* and *C. fasciventris* were more susceptible to infection than the males. Mean lethal time-mortality values generally indicated that the speed of kill was faster in younger flies than in older flies. Overall, LT_{95} values ranged from 3.9 to 4.9 days in the 0-day-old flies, 4.3–6.1 days in the 7-day-old flies and

4.6–6.1 days in 14-day-old flies. The implication of these results is that management strategies using mycoinsecticides should be targeted at young females seeking protein baits to mature their eggs before they cause significant damage to fruits.

Mycoinsecticides must be appropriately formulated to be effective in controlling the target pest. Several aspects should be considered including ease of field application against target insects within their habitats, enhancement of shelf-life, and environmental persistence after application (Feng et al. 1994). In general, selection of ingredients for formulation is crucial as they should not interfere with the infection process and, at best, should enhance fungal viability, virulence, transmission and persistence (Soper and Ward 1981; Feng et al. 1994). Ekesi et al. (2005) evaluated the persistence and infectivity of three formulations (aqueous, oil/aqueous [50:50] and granular) of *M. anisopliae* against pupating larvae of *C. capitata, C. fasciventris* and *C. cosyra* in field cage experiments. Compared with the untreated control, all formulations of the fungus and the chemical insecticide diazinon (used as a standard for comparison), significantly reduced adult emergence from the soil. Exposure of pupating larvae to treated soil samples collected from the field 183 and 366 days after treatment showed that the three formulations were more effective than diazinon in reducing adult emergence. By 668 days after soil inoculation, the granular formulation was, however, more effective than the aqueous and oil/aqueous formulations of conidia, achieving 37, 42 and 54 % reduction in emergence of *C. capitata, C. fasciventris* and *C. cosyra*, respectively. The granules used in this study consisted of pumice/maize on which the *M. anisopliae* had been grown.

Studies by Ekesi et al. (2005) also showed that the density of conidia in soil samples immediately after inoculation was between 1.9 and 3.0×10^5 colony forming units (cfu) g^{-1} of dry weight of soil (enumerated on agar plates). The density was relatively stable in the soil for a period of 183 days after treatment, but thereafter decreased substantially over time in the aqueous and oil/aqueous formulations. Compared with the other formulations, high levels of conidia (4.9–9.5×10^4 cfu g^{-1}) could still be recovered from the soil treated with the granular formulation between 448 to 668 days after treatment. Soil temperatures of 30 °C and above can be detrimental to the survival of EPF (Li and Holdom 1994; Quintela and McCoy 1998) although it can vary depending on isolate (McCoy et al. 2000). Under the tree canopies where the experiments were conducted, soil temperatures never exceeded 26 °C and this evidently benefited the persistence of the fungus in the soil. Low temperatures during the study period may have also played a significant role in slowing down the loss of viability of the fungus in the soil. Additionally, despite three seasons of rainfall, Ekesi et al. (2005) were able to recover large numbers of *M. anisopliae* colonies similar to densities reported elsewhere (Zimmermann 1982; Reinecke et al. 1990; Vänninen et al. 2000) from all the fungus-treated soils over the sampling period. Since pupation of most fruit fly species occurs at a depth not exceeding 7 cm (Ruiz 1945; Hennessey 1994), infection by the fungus should be maximized in this region of the soil. Soil properties can affect the availability, movement and virulence of EPF applied against soil pests (Duniway and McCoy 1990). For example, Quintela and McCoy (1998) observed that movement of 2-d-old larvae of the citrus root weevil, *Diaprepes abbreviatus* L., in sandy soil was not affected by soil moisture ranging from 2 to 12 %, but they did not penetrate soil with <1 % moisture. In contrast, movement of 2-d-old larvae was retarded as moisture and soil depth increased. The availability of *M. anisopliae* conidia in soil differing in pH, texture, organic matter and carbonate content tended to be lower in sandy soil than in clay soil and was not influenced by ionic strength (Garrido-Jurado et al. 2011c). Neither soil texture nor ionic strength affected the infectivity of conidia of the fungus to *C. capitata* pupae.

Fruit fly population suppression by autodissemination can only be successful through horizontal transmission, mainly during mating and lek formation by male flies. If fungal contamination with dry conidia impacts negatively on mating behaviour, population suppression will be compromised. Dimbi et al. (2009) conducted bioassays in the laboratory to investigate the effect of inoculation with *M. anisopliae* on mating behaviour of *C. cosyra, C. fasciventris* and *C. capitata*. In all three species, contamination with dry conidia resulted in a significant delay in the commencement of calling and mating in males due to the substantial amount of time they spent grooming to remove the conidia lodged on their bodies. For fungus-treated male flies, calling and mating began 70–80 min after exposing them to untreated females. However, when females were treated with dry conidia, calling and mating began within 15–16 min. When the grooming period was over, both treated and untreated males began normal pre-mating activities such as calling and wing vibration. Fungus-treated males competed equally with untreated males for virgin female flies from day 0–2 post-inoculation. There were, however, significant differences on day 3, with untreated males of the three fruit fly species having higher percentages of pairing than fungus-treated males. There was no significant difference in the duration of pairing of fungus-treated males and untreated males of the three fruit fly species at day 0, 1 and 2 post-inoculation. However, on day 3 post-inoculation, there was a significant difference in the duration of mating with fungus-treated males having the lowest duration of mating. There was no significant difference in the percentage of matings between fungus-treated male/female flies and the untreated control flies at day 0, 1 and 2 days post-inoculation. Recently, Thaochan and Ngampongsai (2015) also investigated the effects of inoculation with *Metarhizium guizhouense* Chen & Guo on mating propensity and mating competitiveness of *Z. cucurbitae*. On day 4 post-inoculation, the *M. guizhouense*-treated male flies had significantly reduced mating propensity and mating competitiveness, while the treated female flies had reduced mating propensity on day 4 and reduced mating competitiveness on day 5. The mating propensity and competitiveness of treated male and female flies then further declined until death. Kaplan–Meier survival analysis of treated male and female flies gave average survival times (AST) of 6.2 ± 0.2 and 5.4 ± 0.3 days in the mating propensity assay, and about 5.0 ± 0.1 and 4.4 ± 0.2 days in the mating competitiveness assay. Overall, these studies demonstrated that, under laboratory conditions, exposure to fungal infection does not adversely affect the mating competitiveness of fruit fly species until the third day following infection when mortality due to infection began; despite this initial mating behaviour such as calling, can be delayed.

The effectiveness of horizontal transmission can be lost if females preferentially select healthy males over fungal-contaminated or infected males. Dimbi et al. (2013), however, demonstrated horizontal transmission of *M. anisopliae* conidia from treated to untreated adult *C. capitata*, *C. cosyra* and *C. fasciventris*. In all the three fruit fly species, males and females exposed to *M. anisopliae* conidia became infected and exhibited 100% mortality within 5–6 days. Treated (donor) males of all species maintained for 24 h with recipient untreated females were able to transmit infection to those females, resulting in mortalities ranging from 71 to 83% within 15 days. When the process was reversed and female flies were the donors, they were also able to transmit infection to healthy male recipients, resulting in mortalities of 85–100%. The experiments showed that repeated fly-to-fly transfer of fatal doses of inoculum was possible from one donor male to a series of females during mating. The fact that flies mate more than once will obviously increase the chances for contaminated/infected flies to transmit infection to several mates before they die.

In Mauritius, Sookar et al. (2014) demonstrated that when fungus-treated male *B. zonata* or *Z. cucurbitae* were maintained with untreated female flies, they were able to transmit infection to those females, resulting in high mortalities. Similarly, fungus-infected female flies maintained with untreated males also transmitted infection to those males, also resulting in high mortalities.

Sexually mature flies may lay 300–1000 eggs, depending on species. Any effect of fungal infection on adult fecundity and fertility is therefore critical in any fruit fly suppression campaign. Peak oviposition in most species occurs during the initial 10–15 days of egg laying and then gradually declines. As young sexually-mature gravid female fruit flies are more susceptible to fungal infection than older flies (Dimbi et al. 2003b), then any control strategy that targets this age group and reduces the level of oviposition in field populations should contribute to overall management of the insect. Although infection by *M. anisopliae* did not affect the likelihood that eggs would hatch, Dimbi et al. (2003b) recorded a drastic reduction in fecundity of infected *C. capitata*, *C. cosyra* and *C. fasciventris* in addition to their shorter lifespan due to fungus-induced mortality. Sookar et al. (2014) reported that infection by *M. anisopliae* also resulted in a reduction in the number of eggs produced by *Z. cucurbitae*.

4 Field Suppression

As outlined above the use of mycoinsecticides for suppression of fruit flies can either be exploited to target pupating larvae and pupae in the soil, or adult flies through the autodissemination technique. In either case, the approach must be implemented within the context of IPM in combination with other techniques.

4.1 Soil Inoculation

Ekesi et al. (2011) conducted trials over two seasons in mango orchards at Nthagaiya, Kenya, to evaluate the efficacy of soil inoculation with *M. anisopliae* either alone or in combination with GF-120 spinosad bait sprays for suppression of *B. dorsalis*. In the 2006/2007 season, average post-treatment samples showed that *B. dorsalis* catches from the control orchards were four times higher than the number of flies captured in the plots receiving *M. anisopliae* + GF-120. Fruit infestation was 16, 45 and 60 % in the *M. anisopliae* + GF-120, *M. anisopliae* alone and control orchards, respectively. In the 2007/2008 season, average *B. dorsalis* post-treatment samples in the control orchards were seven times higher than the treatment with *M. anisopliae* + GF-120 and fruit infestation was 11, 38 and 52 % in the orchards assigned to *M. anisopliae* + GF-120, *M. anisopliae* alone and control treatments, respectively. The number of conidia in samples of soil from the treated orchards (enumerated on agar) showed initial densities of $1.1–2.1 \times 10^5$ cfu g^{-1} of dry weight of soil but decreased to $1.0–1.4 \times 10^3$ cfu g^{-1} at the end of the experimental period. These experiments clearly demonstrated that the combined use of a soil application of *M. anisopliae* and GF-120 spinosad bait spray is an effective IPM strategy for field suppression of *B. dorsalis* in mango orchards.

4.2 Autodissemination of Entomopathogenic Fungi Using Autoinoculators Baited with Attractants

A number of autodissemination devices have been developed in Africa for the control of fruit flies (Fig. 15.1). Autodissemination of conidia of *M. anisopliae* from autoinoculators baited with attractants (Fig. 15.1a) was evaluated for the management of *C. cosyra* in field trials at Nguruman, Rift Valley Province, Kenya, over two seasons (October to January 2000/2001 and 2001/2002) (Ekesi et al. 2007). Three treatments were compared: (i) autodissemination of dry conidia of *M. anisopliae* presented on maize cobs dipped in molasses, and using a food bait as the attractant (ii) malathion (Malathion 50®) bait spray and (iii) untreated control. The dry conidia were applied at the rate of 0.8–1.0 g of conidia/autoinoculation device ($\sim 4–8 \times 10^{10}$ conidia g^{-1}) and evenly spread on the surface of each maize cob. Cobs were wrapped with a piece of cheesecloth to hold the conidia in place and thereafter suspended from the lid of the autoinoculation device in the uppermost chamber. One hundred (100) ml of food bait (waste brewers' yeast, 43 ml/litre) was added to each device to attract fruit flies. Compared to the untreated controls, application of the *M. anisopliae*-based mycoinsecticide significantly reduced fruit fly population densities compared with the control. More female flies were captured in monitoring traps than males in both seasons. No fruit infestation data were collected due to farmer interference with the trials. Despite the lack of fruit infestation data, this trial

Fig. 15.1 Pictures of autodissemination devices developed in Africa. (**a**) Autoinoculator made from a 750 ml plastic mineral water bottle (Dimbi et al. 2003a); (**b**) autoinoculator made from a Lynfield trap. Inside it is coated with velvet on which conidia are applied (Ekesi et al. unpublished); (**c**) Real IPM-made autodissemination device composed of three parts (**d**): one side of the compartment is coated with velvet on to which conidia are applied and the other compartment serves as a receptacle for the bait; a screen separates the two compartments to prevent flies coming in to direct contact with the bait (Real IPM, unpublished)

provided the first concrete evidence of the potential for autodissemination in the management of fruits flies.

Ekesi and Maniania (2015) conducted a field experiment in the 2011 mango season in Malindi, Kenya, to evaluate the use of an autodissemination strategy with *M. anisopliae* isolate *icipe* 69, for the suppression of *B. dorsalis* and its impact on fruit infestation and yield. Velvet-coated Lynfield traps were used as autoinoculators (Fig. 15.1b) loaded with 0.5 g of dry conidia and laced with methyl eugenol on cotton-wicks (as the attractant). The autoinoculators were deployed at the rate of 30 traps ha^{-1} in orchards assigned to the autodissemination treatment. Control orchards were left untreated. During the experimental period, the average pre-treatment fruit fly population density was 4.2 flies/trap/day (FTD) in the control treatment and 6.8 FTD in orchards assigned to the autodissemination treatment (Fig. 15.2). After 8 weeks, average post-treatment fly catches were 56.2 FTD and 1.4 FTD in the control and the autodissemination treatments, respectively (Fig. 15.2), indicating a significant impact of the treatment on *B. dorsalis* populations. Overall, the use of autodissemination reduced *B. dorsalis* catches relative to the control by 94.6 % within 8 weeks. Samples of field-collected adult *B. dorsalis* that subsequently died due to *M. anisopliae* ranged from 54.2 % 6 weeks post-treatment to 76.5 % 10 weeks after initiation of the experiment (Fig. 15.3). A small percentage of flies (10.8 %) were infected by *M. anisopliae* in the control plot despite it being 1.4 km away from the treatment plots. This was attributed to fly migration, especially because the attractant used (methyl eugenol) can attract insects over distances in excess of 500 m. The fungus persisted for 3–4 weeks (80 % conidial germination) in the autoinoculator but declined significantly thereafter. Fruit infestation by *B. dorsalis* was assessed both at fruit maturity (~7 weeks after treatment) and at the fruit ripening stage (12 weeks after treatment). In orchards assigned to the control, fruit infesta-

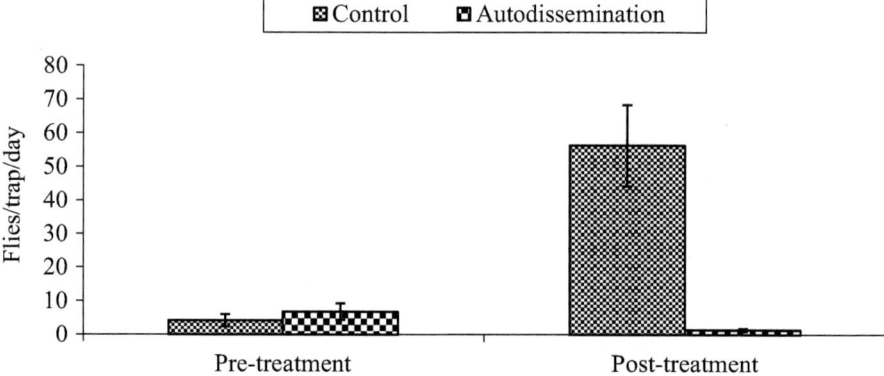

Fig. 15.2 Average pre- and post-treatment catches of adult *Bactrocera dorsalis* following deployment of methyl eugenol-baited autoinoculators containing a *Metarhizium anisopliae*-based mycoinsecticide on mango in Malindi in 2011. Error *bars* denote SE

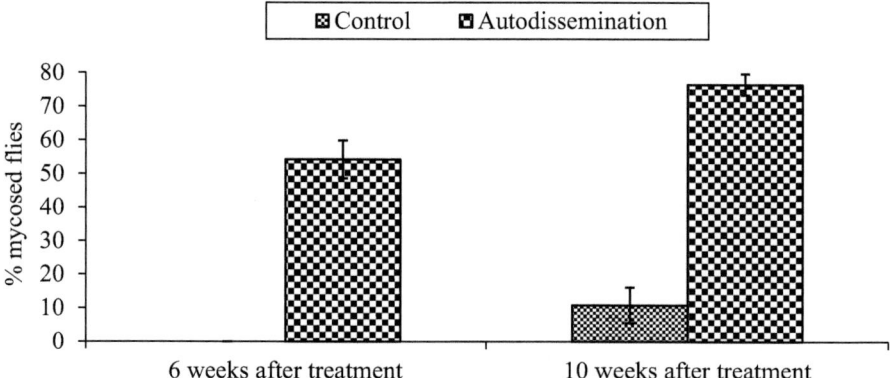

Fig. 15.3 Percentage of adult *Bactrocera dorsalis* killed by fungal infection following deployment of methyl eugenol-baited autoinoculators containing a *Metarhizium anisopliae*-based mycoinsecticide on mango in Malindi in 2011. Error *bars* denote SE

tion at maturity was 48.2 % while infestation was 12.6 % in the orchards assigned to the autodissemination treatment (Fig. 15.4). At the fruit ripening stage, infestation was 61.7 % and 6.4 % in the control and autodissemination treatment, respectively (Fig. 15.4). Based on the results of these experiments and other trials across Africa, *M. anisopliae* isolate *icipe* 69 (although originally developed for control of thrips and mealybugs) (Ekesi et al. 1998, 1999), is now registered as Metarhizium 69® (www.realipm.org) for fruit fly control across many African countries using newly developed autodissemination devices (Fig. 15.1c). Although the product has proven to be an effective tool for suppressing fruit flies, it is recommended that it is used within the context of IPM for fruit flies as this offers the best strategy for achieving economies in the production system by reducing yield losses and enabling growers

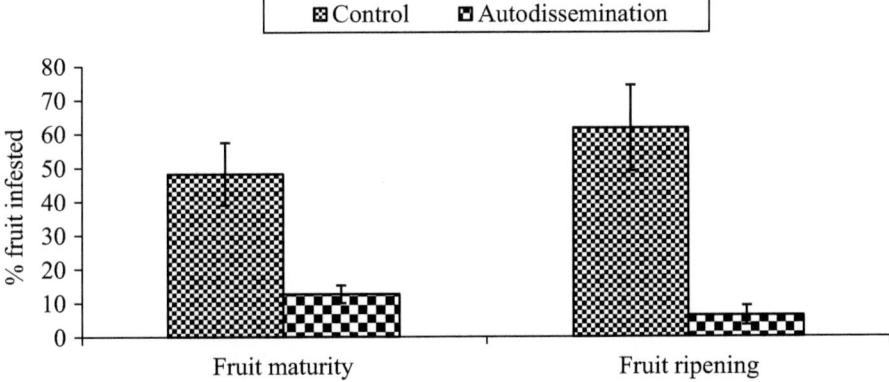

Fig. 15.4 Percentage of mango fruits infested by *Bactrocera dorsalis* following deployment of methyl eugenol-baited autoinoculators containing a *Metarhizium anisopliae*-based mycoinsecticideon mango in Malindi in 2011. Error *bars* denote SE

to comply with the stringent quality standards required by the export market (Aluja et al. 1996; Allwood 1997; Lux et al. 2003).

Acknowledgements The research activities were partly funded by grants from IFAD, GIZ/BMZ, Biovision, the EU, DFID and DAAD to the African Fruit Fly Programme of *icipe*.

References

Allwood AJ (1997) Control strategies for fruit flies (Family Tephritidae) in the South Pacific. In: Allwood AJ, Drew RAI (eds) Management of fruit flies in the Pacific. A regional symposium, Nadi, Fiji 28–31 October 1996. ACIAR Proceedings No. 76, Brown, Prior and Anderson, Melbourne, 171

Aluja M, Celedonio-Hurtado H, Liedo P, Cabrera M, Castillo F, Guillen J, Rios E (1996) Seasonal population fluctuations and ecological implications for management of *Anastrepha* fruit flies (Diptera: Tephritidae) in commercial mango orchards in Southern Mexico. J Econ Entomol 89:654–667

Benz G (1987) Environment. In: Fuxa JR, Tanada Y (eds) Epizootiology of insect diseases. Wiley, New York, pp 177–214

Burges HD (2007) Techniques for testing microbials for control arthropod pests in greenhouses. In: Lacey LA, Kaya HK (eds) Field manual of techniques in invertebrate pathology: application and evaluation of pathogens for control of insects and other invertebrate pests, 2nd edn. Springer, Dordrecht, pp 463–479

Chandler D, Davidson G, Grant WP, Greaves J, Tatchell GM (2008) Microbial biopesticides for integrated crop management: an assessment of environmental and regulatory sustainability. Trends Food Sci Technol 19:275–283

Dimbi S, Maniania NK, Lux SA, Ekesi S, Mueke JK (2003a) Pathogenicity of *Metarhizium anisopliae* (Metsch.) Sorokin and *Beauveria bassiana* (Balsamo) Vuillemin, to three adult fruit fly species: *Ceratitis capitata* (Weidemann), *C. rosa* var. *fasciventris* Karsch and *C. cosyra* (Walker) (Diptera: Tephritidae). Mycopathologia 156:375–382

Dimbi S, Maniania NK, Lux SA, Ekesi S, Mueke JK (2003b) Host species, age and sex as factors affecting the susceptibility of the African Tephritid fruit fly species, *Ceratitis capitata, C. cosyra* and *C. fasciventris* to infection by *Metarhizium anisopliae*. J Pest Sci 76:113–117

Dimbi S, Maniania NK, Lux SA, Mueke JK (2004) Effect of constant temperatures on germination, radial growth and virulence of *Metarhizium anisopliae* to three species of African tephritid fruit flies. BioControl 49:83–94

Dimbi S, Maniania NK, Ekesi S (2009) Effect of *Metarhizium anisopliae* inoculation on the mating behavior of three 3 species of African Tephritid fruit flies, *Ceratitis capitata, Ceratitis cosyra* and *Ceratitis fasciventris*. Biol Control 50:111–116

Dimbi S, Maniania NK, Ekesi S (2013) Horizontal transmission of *Metarhizium anisopliae* in fruit flies and effect of fungal infection on egg laying and fertility. Insects 4:206–216

Duniway JM, McCoy CW (1990) Influence of the soil environment on insect and plant pathogens. In: Baker RR, Dunn PE (eds) New directions in biological control: alternatives for suppressing agricultural pests and diseases. Alan R Liss, New York, pp 473–477

Ekesi S, Maniania NK (2015) Autodissemination strategy for field suppression of *Bactrocera dorsalis* (Diptera: Tephritidae) using *Metarhizium anisopliae*-based biopesticide on mango. International Congress on invertebrate pathology and microbial control and the 48th annual meeting of the society for invertebrate pathology, Vancouver, Canada, August 9–13th, 2015

Ekesi S, Maniania NK, Ampong-Nyarko K, Onu I (1998) Potential of the entomopathogenic fungus, *Metarhizium anisopliae* for the control of legume flower thrips, *Megalurothrips sjostedti* (Trybom) on cowpea in Kenya. Crop Protect 17:661–668

Ekesi S, Maniania NK, Ampong-Nyarko K, Onu I (1999) Effect of intercropping cowpea with maize on the performance of *Metarhizium anisopliae* against the legume flower thrips, *Megalurothrips sjostedti*, and some predators. Environ Entomol 28:1154–1161

Ekesi S, Maniania NK, Lux SA (2002) Mortality in three economically important African tephritid fruit fly puparia and adults caused by the entomopathogenic fungi, *Metarhizium anisopliae* and *Beauveria bassiana*. Biocontrol Sci Technol 12:7–17

Ekesi S, Maniania NK, Lux SA (2003) Effect of soil temperature and moisture on survival and infectivity of *Metarhizium anosopliae* to four tephritid fruit fly puparia. J Invertebr Pathol 83:157–167

Ekesi S, Maniania NK, Mohamed SA, Lux SA (2005) Effect of soil application of different formulations of *Metarhizium anisopliae* on African tephritid fruit flies and their associated endoparasitoids. Biol Control 35:83–91

Ekesi S, Dimbi S, Maniania NK (2007) The role of entomopathogenic fungi in the integrated management of fruit flies (Diptera: Tephritidae) with emphasis on species occurring in Africa. In: Ekesi S, Maniania NK (eds) Use of entomopathogenic fungi in biological pest management. Research Signpost, Kerala, pp 239–274

Ekesi S, Maniania NK, Mohamed SA (2011) Efficacy of soil application of *Metarhizium anisopliae* and the use of GF-120 spinosad bait spray for suppression of *Bactrocera invadens* (Diptera: Tephritidae) in mango orchards. Biocontrol Sci Technol 21:299–316

Ekesi S, De Meyer M, Mohamed SM, Virgilio M, Borgemeister C (2016) Taxonomy, ecology, and management of native and exotic fruit fly species in Africa. Annu Rev Entomol 61:219–238

Faria MR, Wraight SP (2007) Mycoinsecticides and mycoacaricides: a comprehensive list with worldwide coverage and international classification of formulation types. Biol Control 43:237–256

Feng Z, Carruthers RI, Roberts DW, Robson DS (1985) Age-specific dose-mortality effects of *Beauveria bassiana* (Deuteromycotina: Hyphomycetes) on the European corn borer *Ostrinia nubilialis* (Lepidoptera: Pyralidae). J Invertebr Pathol 46:259–264

Feng MG, Poprawski TJ, Khachatourians GG (1994) Entomopathogenic fungus *Beauveria bassiana* for insect control: current status. Biocontrol Sci Technol 4:3–34

Ferron P (1985) Fungal control. In: Kerkut GA, Gilbert LI (eds) Comprehensive insect physiology, biochemistry and pharmacology, vol 12. Pergamon Press, Oxford, pp 313–346

Fuxa JR, Tanada Y (1987) Epizootiology of insect diseases. Wiley, New York, 555 pp

Garrido-Jurado I, Ruano F, Campos M, Quesada-Moraga E (2011a) Effects of soil treatments with entomopathogenic fungi on soil dwelling non-target arthropods at a commercial olive orchard. Biol Control 59:239–244

Garrido-Jurado I, Valverde-García P, Quesada-Moraga E (2011b) Use of a multiple logistic regression model to determine the effects of soil moisture and temperature on the virulence of entomopathogenic fungi against pre-imaginal Mediterranean fruit fly *Ceratitis capitata*. Biol Control 59:366–372

Garrido-Jurado I, Torrent J, Barrón V, Corpas A, Quesada-Moraga E (2011c) Soil properties affect the availability, movement, and virulence of entomopathogenic fungi conidia against puparia of *Ceratitis capitata* (Diptera: Tephritidae). Biol Control 58:277–285

Goble TA, Dames JF, Hill MP, Moore SD (2011) Investigation of native isolates of entomopathogenic fungi for the biological control of three citrus pests from citrus soils in the Eastern Cape Province, South Africa. Biocontrol Sci Technol 21:1193–1211

Goettel MS, Hajek AE, Siegel JP, Evans HC (2001) Safety of fungal biocontrol agents. In: Jackson C, Magan N, Butt TM (eds) Fungi as biocontrol agents. CAB International, Oxon, pp 347–375

Grant WP, Chandler D, Bailey A, Greaves J, Tatchell M, Prince G (2010) Biopesticides: pest management and regulation. CABI, Wallingford, 238 pp

Gwynn RL, Maniania JNK (2010) Africa with special reference to Kenya. In: Kabaluk JT, Svircev AM, Goettel MS, Woo SG (eds) The use and regulation of microbial pesticides in representative jurisdictions worldwide, IOBC Global, pp 1–6

Hall RA, Papierok B (1982) Fungi as biological control agents of arthropods of agricultural and medical importance. Parasitology 84:205–240

Heale JB (1988) The potential impact of fungal genetics and molecular biology on biological control, with particular reference to entomopathogens. In: Burges MN (ed) Fungi in biological control system. Manchester University Press, Manchester/New York, pp 211–234

Hennessey MK (1994) Depth of pupation of Caribbean fruit fly (Diptera: Tephritidae) in soils in the laboratory. Environ Entomol 23:1119–1123

Hummel RL, Walegenbach JF, Barbercheck ME, Kennedy GG, Hoyt GD, Arellano C (2002) Effects of production practices on soil-borne entomopathogens in western North Carolina vegetable systems. Environ Entomol 31:84–91

Imoulan A, Alaoui A, El Meziane A (2011) Natural occurrence of soil-borne entomopathogenic fungi in the Moroccan endemic forest of *Argania spinosa* and their pathogenicity to *Ceratitis capitata*. World J Microbiol Biotechnol. DOI:10.1007/s11274-011-0735-1

Inglis GD, Goettel MS, Butt TM, Strasser H (2001) Use of hyphomycetous fungi for managing insect pests. In: Jackson C, Magan N, Butt TM (eds) Fungi as biocontrol agents. CAB International, Oxon, pp 23–69

Jandricic SE, Filotas M, Sanderson JP, Wraight SP (2014) Pathogenicity of conidia-based preparations of entomopathogenic fungi against the greenhouse pest aphids *Myzus persicae*, *Aphis gossypii*, and *Aulacorthum solani* (Hemiptera: Aphididae). J Invertebr Pathol 118:34–46

Jaronski ST (2010) Ecological factors in the inundative use of fungal entomopathogens. Biocontrol 55:159–185

Kabaluk JT, Svircev AM, Goettel MS and Woo SG (eds) (2010) The use and regulation of microbial pesticides in representative jurisdictions worldwide. IOBC Glob, p 99. Available online through www.IOBC-Global.org

Keller S, Kessler P, Schweizer C (2003) Distribution of insect pathogenic soil fungi in Switzerland with special reference to *Beauveria brongniartii* and *Metarhizium anisopliae*. BioControl 48:307–319

Khan S, Guo L, Maimaiti Y, Mijit M, Qiu D (2012) Entomopathogenic fungi as microbial biocontrol agent. Mol Plant Breed 3:63–79

Lacey LA, Liu T-X, Buchman JL, Munyaneza JE, Goolsby JA, Horton DR (2011) Entomopathogenic fungi (Hypocreales) for control of potato psyllid, *Bactericera cockerelli* (Šulc) (Hemiptera: Triozidae) in an area endemic for zebra chip disease of potato. Biol Control 36:271–278

Lacey LA, Grzywacz D, Shapiro-Ilan DI, Frutos R, Brownbridge M, Goettel MS (2015) Insect pathogens as biological control agents: back to the future. J Invertebr Pathol 132:1–41

Li DP, Holdom DG (1994) Effects of pesticides on growth and sporulation of *Metarhizium anisopliae* (Deuteromycotina: Hyphomycetes). J Invertebr Pathol 63:209–211

Lomer CJ, Bateman RP, Dent D, De Groote H, Douro-Kpindou O-K, Kooyman C, Langewald J, Ouambama Z, Peveling R, Thomas M (1999) Development of strategies for the incorporation of biological pesticides into the integrated management of locusts and grasshoppers. Agric Forest Entomol 1:71–88

Lux SA, Ekesi S, Dimbi S, Mohamed S, Billah M (2003) Mango infesting fruit flies in Africa – perspectives and limitations of biological approaches to their management. In: Neuenschwander P, Borgemeister C, Langewald J (eds) Biological control in IPM systems in Africa. CABI, Wallingford, pp 277–293

Mahmoud MF (2009) Pathogenicity of three commercial products of entomopathogenic fungi, *Beauveria bassiana*, *Metarhizium anisopliae* and *Lecanicillium lecanii* against adults of olive fly, *Bactrocera oleae* (Gmelin) (Diptera: Tephritidae) in the laboratory. Plant Protect Sci 45:98–102

Maniania NK, Odulaja A (1998) Effect of species, age, and sex of tsetse on response to infection by *Metarhizium anisopliae*. BioControl 43:311–323

McCoy CW, Quintela ED, De Faria M (2000) Environmental persistence of entomopathogenic fungi. In: Baur ME, Fuxa JR (eds) Factors Affecting the survival of entomopathogens. Lousiana State University Agricultural Center, Southern Cooperative Series Bulletin 400. http://www.agrctr.lsu.edu/s265/default.htm

McCoy CW, Samson RA, Boucias DG, Osborne LS. Pena JE, Buss LJ (2009) Pathogens infecting insects and mites of citrus. LLC Friends of Microbes, Winter Park, FL, USA, 193 p

Meyling NV, Eilenberg J (2007) Ecology of the entomopathogenic fungi *Beauveria bassiana* and *Metarhizium anisopliae* in temperate agroecosystems: potential for conservation biological control. Biol Control 43:145–155

Navarro-Llopis V, Ayala I, Sanchis J, Primo J, Moya P (2015) Field efficacy of a *Metarhizium anisopliae*-based attractant-contaminant device to control *Ceratitis capitata* (Diptera: Tephritidae). J Econ Entomol 108:1570–1578

O'Callaghan M, Brownbridge M (2009) Environmental impacts of microbial control agents used for control of invasive insects. In: Hajek AE, Glare TR, O'Callaghan M (eds) Use of microbes for control and eradication of invasive arthropods. Springer, Dordrecht, pp 305–327

Ortiz-Urquiza A, Keyhani NO, Quesada-Moraga E (2013) Culture conditions affect virulence and production of insect toxic proteins in the entomopathogenic fungus *Metarhizium anisopliae*. Biocontrol Sci Biotechnol 23:1199–1212

Ouna EA (2012) Entomopathogenicity of Hyphomycetes fungi to the invasive fruit flies *Bactrocera invadens*. MSc dissertation, Kenyatta University, Kenya

Prior C (1988) Biological pesticides for low external-input agriculture. Biocontrol News Info 10:17N–22N

Quintela ED, McCoy CW (1998) Synergistic effect of imidacloprid and two entomopathogenic fungi on the behavior and survival of larvae of *Diaprepes abbreviatus* (Coleoptera: Curculionidae) in soil. J Econ Entomol 91:110–122

Ravensberg WJ (2011) A roadmap to the successful development and commercialization of microbial pest control products for control of arthropods, vol 10, Progress in biological control. Springer, Dordrecht

Reinecke P, Andersch W, Stenzel K, Hartwig J (1990) BIO 1020, a new microbial insecticide for use in horticultural crops. Brighton Crop Protect Conf Pests Dis 1:49–84

Roberts DW, Humber RA (1981) Entomopathogenic fungi. In: Roberts DW, Aist JR (eds) Infection processes of fungi. Rockfeller Fdn, New York, pp 1–12

Ruiz CA (1945) Estudio sistematico-biologico de las especies de mayor importancia economica. III. (Diptera). Madrid

Shah, PA Goettel MS (1999) Directory of microbial control products and services. Microbial Control Division, Society for Invertebrate Pathology, Gainsville, Florida

Shah PA, Pell JK (2003) Entomopathogenic fungi as biological control agents. Appl Microbiol Biotechnol 61:413–423

Sookar P, Bhagwant S, Ouna EA (2008) Isolation of entomopathogenic fungi from the soil and their pathogenicity to two fruit fly species (Diptera: Tephritidae). J Appl Entomol 132:778–788

Sookar P, Bhagwant S, Khayrattee FB, Chooneea Y, Ekesi S (2014) Mating compatibility of wild and sterile melon flies, *Bactrocera cucurbitae* (Diptera: Tephritidae) treated with entomopathogenic fungi. J Appl Entomol 138:409–417

Soper RS, Ward MG (1981) Production, formulation and application of fungi for insect control. In: Papavizas GC (ed) Beltsville symposia in agricultural research: biological control in crop production. Allanheld, Osmun Publishers, Totowa, pp 161–180

Thaochan N, Ngampongsai A (2015) Effects of autodisseminated *Metarhizium guizhouense* PSUM02 on mating propensity and mating competitiveness of *Bactrocera cucurbitae* (Diptera: Tephritidae). Biocontr Sci Technol 25:629–644

Toledo J, Campos SE, Flores S, Liedo P, Barrera JF, Villaseñor A, Montoya P (2007) Horizontal transmission of *Beauveria bassiana* in *Anastrepha ludens* (Diptera: Tephritidae), under laboratory and field-cage conditions. J Econ Entomol 100:291–297

Vänninen I, Tyni-Juslin J, Hokkanen H (2000) Persistence of augmented *Metarhizium anisopliae* and *Beauveria bassiana* in Finnish agricultural soils. Biocontrol 45:201–222

Vega FE, Dowd PF, Lacey LA, Pell JK, Jackson DM, Klein MG (2007) Dissemination of beneficial microbial agents by insects. In: Lacey LA, Kaya HK (eds) Field manual of techniques in invertebrate pathology. Springer, Dordrecht, pp 127–146

Vey A, Hoagland RE, Butt TM (2001) Toxic metabolites of fungal biocontrol agents. In: Jackson C, Magan N, Butt TM (eds) Fungi as biocontrol agents. CAB International, Oxon, pp 310–346

Williamson S, Ball A, Pretty J (2008) Trends in pesticide use and drivers for safer pest management in four African countries. Crop Prot 27:1327–1334

Zimmermann G (1982) Effect of high temperatures and artificial sunlight on the viability of conidia of *Metarhizium anisopliae*. J Invertebr Pathol 40:36–40

Zimmermann G (1986) Insect pathogenic fungi as pest control agents. In: Franz HG (ed) International symposium of the Akademic der Wissens Chaften und der literature: biological plant and health protection. Fortschritte der Zoologie Bd, 15–17 November 1984, Gustav Fisher Verlag, Stuttgart, Germany, pp 217–231

Zimmermann G (2007a) Review on safety of the entomopathogenic fungi *Beauveria bassiana* and *Beauveria brongniartii*. Biocontrol Sci Technol 17:553–596

Zimmermann G (2007b) Review on safety of the entomopathogenic fungus *Metarhizium anisopliae*. Biocontrol Sci Technol 17:879–920

Jean N.K. Maniania is an Insect Pathologist who gained his PhD from the Université Paris VI (Pierre and Marie Curie), France. For the first 7 years of his carrier he worked as Resident Fellow with the French National Institute for Agricultural Research (INRA, La Minière, France) and also as Research Associate with the European Parasite Laboratory (EPL), a United States Department of Agriculture, Agricultural Research Service (USDA/ARS) overseas laboratory based in France. He carried out extensive surveys in France and Spain for isolation of insect pathogens from insects that had become major pests in the USA in order to develop them as biological control agents. Dr Maniania joined *icipe* in 1989 as an insect pathologist during which time he conducted research on the development of insect pathogens, particularly entomopathogenic fungi (EPF) as microbial control agents of arthropod crop pests (stemborers, termites, thrips, mites, leafminer flies, locusts) and disease vectors (tsetse flies, sandflies, ticks). His other research interests include the development of delivery systems to autodisseminate EPF amongst insect pests and disease vectors in the field, the use of EPF as endophytes, as well as understanding tritrophic interactions plant/insect pathogen/insect. He provided the first attempt to control tsetse flies with EPF in the field using an autodissemination device. He has also led research activities associated with the introduction of Green Muscle® (*M. acridum* IMI330189) into Madagascar for the control of Malagasy migratory locust, *Locusta migratoria capito*, which resulted in its registration there in July 2009. During his 33-year career as an insect pathologist, Dr. Maniania has supervised 22 PhD and 17 MSc students. He has written/co-authored over 100 papers in refereed journals, co-edited one book and authored/co-authored ten book chapters. As a result of his collaborative efforts, he and Dr Ekesi have developed four mycoinsecticides which are commercialized by a private company. Dr Maniania is a prominent member of many professional societies including the Society for Invertebrate Pathology (SIP), Group of Experts of UNEP/FAO/IPM Facilities on Termite Biology and Control, Fungus Consortium on Malaria. In December 2015 he retired from *icipe* and is now based in Canada where he works as a consultant in microbial control and insect pathology.

Sunday Ekesi is an Entomologist at the International Centre of Insect Physiology and Ecology (*icipe*), Nairobi, Kenya. He heads the Plant Health Theme at *icipe*. Sunday is a professional scientist, research leader and manager with extensive knowledge and experience in sustainable agriculture (microbial control, biological control, habitat management/conservation, managing pesticide use, IPM) and biodiversity in Africa and internationally. Sunday has been leading a continent-wide initiative to control African fruit flies that threaten production and export of fruits and vegetables. The initiative is being done in close collaboration with IITA, University of Bremen, Max Planck Institute for Chemical Ecology together with NARS, private sectors and ARI partners in Africa, Asia, Europe and the USA and focuses on the development of an IPM strategy that encompasses baiting and male annihilation techniques, classical biological control, use of biopesticides, ant technology, field sanitation and postharvest treatment for quarantine fruit flies. The aim is to develop a cost effective and sustainable technology for control of fruit flies on the African continent that is compliant with standards for export markets while also meeting the requirements of domestic urban markets. Sunday has broad perspectives on global agricultural research and development issues, with first-hand experience of the challenges and opportunities in working with smallholder farmers, extension agents, research organizations and the private sector to improve food and nutritional security. He sits on various international advisory and consultancy panels for the FAO, IAEA, WB and regional and national projects on fruit fly, arthropod pests and climate change-related issues. Sunday is a Fellow of the African Academy of Sciences (FAAS).

Chapter 16
In and Out of Africa: Parasitoids Used for Biological Control of Fruit Flies

Samira A. Mohamed, Mohsen M. Ramadan, and Sunday Ekesi

Abstract This chapter is a demonstration of the wealth of African natural resources and their contribution to biological control of tephritid fruit flies (Diptera: Tephritidae). Africa is the native region of more than 900 species of fruit flies, many of which are significant agricultural pests. Highly diverse assemblages of indigenous hymenopteran parasitoid species have evolved with these fruit flies, which makes Africa a valuable source of parasitoids for use in classical biological control of fruit flies around the world. Interest in the use of parasitoids for biological control has recently increased due to advances in mass rearing techniques for exotic and native parasitoid species alongside the need to reduce synthetic insecticide use. Here we review the diversity of indigenous African parasitoid species and their role in classical biological control of fruit flies in other parts of the world; we also discuss their contribution to the management of native fruit flies in Africa. Likewise, the prospects and potential for using exotic parasitoids for management of newly-established invasive fruit flies in Africa is discussed, particularly for *Batrocera zonata* (Saunders), *Bactrocera dorsalis* (Hendel), *Bactrocera latifrons* (Hendel) and *Zeugodacus cucurbitae* (Coquillett). We cover the introduction and spread of exotic parasitoid species released in Africa for biological control of invasive fruit flies. The rich diversity of indigenous parasitoids of African fruit flies continues to be unraveled as more new species are discovered and recognized as potential biological control agents for fruit fly management.

Keywords Indigenous parasitoids • Exotic parasitoids • Exploration • Introduction

---

S.A. Mohamed (✉) • S. Ekesi
Plant Health Theme, International Centre of Insect Physiology & Ecology (*icipe*),
PO Box 30772-00100 Nairobi, Kenya
e-mail: sfaris@icipe.org

M.M. Ramadan
State of Hawaii Department of Agriculture, Division of Plant Industry,
Plant Pest Control Branch, 1428 South King Street, Honolulu, HI 96814, USA

© Springer International Publishing Switzerland 2016
S. Ekesi et al. (eds.), *Fruit Fly Research and Development in Africa - Towards a Sustainable Management Strategy to Improve Horticulture*,
DOI 10.1007/978-3-319-43226-7_16

1 Introduction

Management of tephritid fruit flies requires an holistic IPM approach of which biological control is one of the essential components. Hymenopteran parasitoids are considered to be well suited to biological control of fruit flies because they are generally more host specific compared with predators and entomopathogens. For successful development endoparasitoids must deal with the host immune response and ectoparasitoids must deal with host mobility; for these reasons they are highly co-evolved with their particular hosts. Moreover, parasitoids are able to locate and attack the concealed immature stages of fruit flies inside fruits of both wild and cultivated plants.

Although the history of fruit fly biological control dates back to the beginning of the last century (Silvestri 1914a, b; Clausen 1978), it has recently received increasing attention (Wharton 1989; Knipling 1992; Headrick and Goeden 1996; Sivinski 1996; Purcell 1998). This has been facilitated by technological advances and ease in transportation of parasitoid consignments across the globe. Ovruski et al. (2000) attributed the renewed interest in using parasitoids for fruit fly biological control to the advances made in mass rearing techniques for exotic and native parasitoid species and their tephritid hosts. Increasing pressure to reduce the use of synthetic insecticides and the current drive towards conservation of biodiversity through the use of ecologically acceptable pest management tactics have made classical and augmentative biological control a desirable method to reduce fruit fly populations.

In almost all the published literature on biological control of fruit flies, Africa is highlighted as a source of parasitoids for use in classical biological control of fruit flies that are invasive pests elsewhere in the world; there is also a high species richness of fruit fly parasitoids in Africa (Silvestri 1914a, b, 1915; Clausen et al. 1965; Greathead 1976; Clausen 1978; Neuenschwander 1982; Wharton 1989 and reference there in; Waterhouse 1993; Mkize et al. 2008). In this chapter, we have compiled information on the diversity of indigenous African parasitoid species that attack fruit flies and their role in classical biological control in other parts of the world. Additionally, we highlight the contribution of these parasitoids in management of native fruit flies in Africa. Parasitoid species used for classical biological control of alien fruit flies that have invaded and become established in Africa are also reviewed in this chapter including four newly established Asian fruit flies: the peach fruit fly, *Batrocera zonata* (Saunders); the oriental fruit fly, *Bactrocera dorsalis* (Hendel); the solanaceous fruit fly, *Bactrocera latifrons* (Hendel); and the melon fly, *Zeugodacus cucurbitae* (Coquillett).

2 Diversity of the Indigenous Parasitoids of African Fruit Flies

Africa is the native range of several genera and more than 1000 species of fruit flies in the subfamily Dacinae (Diptera: Tephritidae), many of which are of significant agricultural importance as pests of commercial fruits and vegetables in sub-Saharan and Afrotropical regions (White and Elson-Harris 1992; Thompson 1998; De Meyer and Ekesi 2016). It is not surprising that a highly diverse assemblage of native hymenopteran parasitoid species have evolved with these fruit flies. However, much of our knowledge on the species composition of indigenous African parasitoids of tephritids is derived from the information generated during foreign explorations for natural enemies of African fruit flies that had invaded and become pests in other parts of the world, namely the Mediterranean fruit fly (medfly), *Ceratitis capitata* (Wiedemann), and the olive fruit fly, *Bactrocera oleae* (Rossi) (White and Elson-Harris 1992; CABI 2016).

A comprehensive record of indigenous African fruit fly parasitoids was first documented as early as 1912 by the prominent Italian entomologist Filippo Silvestri during his exploration for natural enemies in the West Coast of Africa (between 1912 and 1913) and Australia for use in biological control in the State of Hawaii (Territory of Hawaii at that time; Silvestri (1914a, b, 1915). He reported a high diversity of hymenopteran parasitoid species attacking fruit flies (*Ceratitis* species were attacked by ten species of parasitoids and *Dacus* species were attacked by seven parasitoid species) in the families Braconidae, Eulophidae, Chalcididae and Diapriidae from West Africa and South Africa (Table 16.1). However, the members of the family Braconidae (14 species), particularly in the subfamily Opiinae, were the most numerous in his collection. Additional information on the African parasitoid fauna is also reported from surveys by the earlier Hawaiian explorers e.g. D.T. Fullaway 1914; J.C. Bridwell 1914; F.A. Bianchi and N.L.H. Krauss 1936–1937 (reported in Bianchi and Krauss 1936) in Kenya; R.H. Van Zwaluwenburg in West Africa 1936; J.M. McGough 1949 in Kenya, Congo, Uganda and South Africa; F.E. Skinner 1948 in Kenya, Congo and South Africa; D.W. Clancy 1951 in Congo (reported by Clausen et al. 1965; Greathead 1976; Clausen 1978; Wharton 1989; Waterhouse 1993; Ovruski and Fidalgo 1994). In Hawaii the parasitoids collected were mass reared and introduced into many countries around the world for biological control of invasive fruit flies, where they subsequently became established (Table 16.1).

In contrast, invasions of the African continent by exotic fruit flies in the genus *Bactrocera* prompted many scientists in Africa to carry out inventories of the indigenous parasitoid species as a prerequisite prior to introduction of coevolved natural enemies from the native region of the exotic pest. Records from the indigenous parasitoid species inventories can be found in Appiah (2012) and Vayssières et al. (2011, 2012). Also Fischer and Madl (2008) provided a review for the Opiinae parasitoids of the Malagasy sub-region, most of which are of unknown biology or attack other non-tephritid hosts.

Table 16.1 Hymenopteran parasitoids (grouped by taxon family, subfamily and known biology) originating from Africa and reported to attack frugivorous African Tephritidae

Valid names of parasitoids[a]	Family and subfamily	Mode of parasitism	Country and region of origin	Host records	Distribution outside Africa when used in tephritid biological control programmes	References for distribution and host association
Dirhinus giffardii Silvestri	Chalcididae, Dirhininae	Idiobiont solitary ectophagous pupal parasitoid	Cape Verde, Egypt, Kenya, Nigeria, West Africa	*Ceratitis capitata* *Ceratitis rosa* *Dacus demmerezi* *Dacus frontalis* *Trirhithromyia cyanescens*	Australia 1956 for *Bactrocera tryoni* via Hawaii. Bolivia 1971 for *C. capitata*. China released for Tephritidae. Greece 1962 for *C. capitata* via Israel. Hawaiian Islands 1913 for *C. capitata*, 1950 for *Bactrocera dorsalis*. Israel 1956 for *C. capitata*. Italy 1913 for *B. oleae*. Mexico 1955 for *Anastrepha ludens* via Hawaii. Puerto Rico 1935 for *Anastrepha suspensa* via Hawaii. Samoa 1935 for *Bactrocera passiflorae* via Fiji.	3, 5, 10, 19, 24, 27, 31, 35
Dirhinus ehrhorni Silvestri	"	Idiobiont solitary ectophagous pupal parasitoid	Nigeria	*Ceratitis capitata* *Ceratitis giffardi* *Ceratitis* sp.	Hawaiian Islands for *C. capitata*	27, 31
Dirhinus sp.	"	Idiobiont solitary ectophagous pupal parasitoid	Africa	*Dacus bivittatus*		19
Coptera magnificus (Nixon)	Diapriidae, Diapriinae	unknown	Kenya	*Ceratitis contramedia*	Hawaii 1947–1952 for *B. dorsalis*, imported but not released.	41

Coptera silvestrii (Kieffer)	"	Idiobiont solitary ectophagous pupal parasitoid	Benin, Ghana, Guinea, Kenya, Mozambique, Niger, Nigeria, Senegal, South Africa, Zululand, Uganda, West Africa	*Bactrocera oleae*	Hawaiian Islands 1913 for *C. capitata*. Hawaii 1947–1952 for *B. dorsalis*, imported but not released.	**4, 12, 13, 19, 22, 27, 31, 37**
				Ceratitis anonae		
				Ceratitis capitata		
				Ceratitis colae		
				Ceratitis contramedia	Italy 1913 for *Bactrocera oleae*.	
				Ceratitis cosyra		
				Ceratitis giffardi		
				Ceratitis punctata		
				Ceratitis rosa		
				Ceratitis simi		
				Dacus bivittatus		
				Dacus ciliatus		
				Dacus sp.		
				Trirhithrum coffeae		
				Trirhithrum nigerrimum		
Coptera robustior Silvestri	"	Pupal parasitoid	Guinea, Kenya, Nigeria, South Africa	*Ceratitis capitata*		34, 37
				Ceratitis punctata		
Coptera sp.	"	Idiobiont pupal endophagous parasitoid	Kenya	*Ceratitis capitata*		34
Trichopria capensis Kieffer	"	Pupal parasitoid	South Africa	*Ceratitis capitata*	Hawaii 1913 for *C. capitata*	27
					Israel 1951 for *C. capitata*	
Trichopria sp.	"		Congo	*Ceratitis anonae*	Hawaii 1947–1952 for *B. dorsalis*, imported but not released.	30, 41
				Dacus ciliatus		
Meraporus graminicola Walker	Pteromalidae, Pteromalinae	Pupal parasitoid	Ethiopia	*Bactrocera oleae*		31

(continued)

Table 16.1 (continued)

Valid names of parasitoids[a]	Family and subfamily	Mode of parasitism	Country and region of origin	Host records	Distribution outside Africa when used in tephritid biological control programmes	References for distribution and host association
Pachycrepoideus vindemmiae (Rondani)	"	Idiobiont solitary ectophagous pupal parasitoid	Congo, Kenya, Morocco	*Bactrocera oleae* *Ceratitis capitata* *Ceratitis rosa* *Ceratitis* sp. *Dacus ciliatus* *Dacus demmerezi* *Trirhithromyia cyanescens*	Argentina 1960 for *Anastrepha fraterculus* and *C. capitata* via Hawaii, 1961 via Mexico and 1986 via Cost Rica. Bolivia for *Anastrepha* sp. Brazil for *Anastrepha* sp. Costa Rica 1955 for *C. capitata* and *Anastrepha* sp. El Salvador for *Anastrepha* sp. France, Italy, Spain, and USA for *Drosophila suzukii*. Florida for *Anastrepha* sp. via Hawaii. Hawaiian Islands 1947–1952 for *C. capitata* and *B. dorsalis*. Mexico for *Anastrepha* sp. Nicaragua 1955 for *C. capitata* via Hawaii. New Zealand released, not established. Peru for *Anastrepha* sp. Puerto Rico 1935 for *Anastrepha suspensa* via Hawaii. La Réunion. Syria 2014 for *C. capitata* (inadvertently)	9, 19, 31
Cyrtoptyx latipes (Rondani)	"	Pupal parasitoid (also attack larvae)	Egypt, Eritrea	*Bactrocera oleae*	Crete, Cyprus, Greece, India (inadvertently). Italy for *C. capitata* and *B. oleae*.	8, 20

Sphegigaster sp	"	Pupal parasitoid	Tanzania	Ceratitis rosa	19, 31
				Ceratitis sp.	
Cyrtoptyx latipes Rondani	"	Unknown	Egypt, Eritrea	Bactrocera oleae	Greece 1975–1980 for B. oleae	8
Mesopolobus modestus (Silvestri)	"	Unknown	Eritrea, Ethiopia	Bactrocera oleae	Italy (inadvertently)	26, 28
Pteromalus semotus Walker	"	Idiobiont solitary ectophagous larval parasitoid	Egypt, Canary Islands, Cape Verde, South Africa	Bactrocera oleae	Cosmopolitan (inadvertently)	20
Halticoptera daci Silvestri	Pteromalidae, Miscogastrinae	Koinobiont endophagous larval parasitoid	Eritrea, Ethiopia, South Africa	Bactrocera oleae	Italy 1914 for B. oleae	19, 20, 21, 26, 28
Spalangia afra Silvestri.	Pteromalidae, Spalanginae	Idiobiont solitary ectophagous pupal parasitoid	East Africa, Kenya, Nigeria, Tanzania, West Africa	Ceratitis anonae	Hawaii 1947–1952 for B. dorsalis, imported but not released	1, 19, 27, 41
				Ceratitis coleae		
				Ceratitis sp.		
				Dacus bivittatus		
				Dacus ciliatus		
				Pardalaspis cyanescens		

(continued)

Table 16.1 (continued)

Valid names of parasitoids[a]	Family and subfamily	Mode of parasitism	Country and region of origin	Host records	Distribution outside Africa when used in tephritid biological control programmes	References for distribution and host association
Spalangia cameroni Perkins	"	"	Canary Islands, Malawi, Mauritius, Madagascar, Morocco, Senegal, Somalia, South Africa, Tanzania	Ceratitis capitata	Cosmopolitan (inadvertently).	42
				Dacus sp.	Fiji 1935 for B. passiflorae and B. xanthodes via Hawaii.	
Spalangia simplex Perkins	"	"	Congo, Mali, South Africa, Uganda	Ceratitis anonae?		45
				Ceratitis cosyra		
				Ceratitis ditissima?		
				Ceratitis fasciventris?		
				Ceratitis quinaria?		
				Ceratitis silvestrii?		
Tachinaephagus zealandicus Ashmead	Encyrtidae, Encyrtinae	Pupal parasitoid (also reported on larvae)	Congo, South Africa, Uganda	Tephritidae		20, 30
				Bactrocera oleae		
Pnigalio agraules (Walker)	Eulophidae, Eulophinae	Larval and pupal parasitoid	Egypt	Bactrocera oleae	Austria. England. France.	8
					Greece 1975–1980 for B. oleae.	
					Italy. Spain. Turkey	
Eulophus sp.	"	Larval ectophagous parasitoid (?)	Libya	Bactrocera oleae		31

Species	Family/subfamily	Biology	Countries	Host	Notes	References
Zagrammosoma variegatum (Masi)	"	Larval parasitoid	Cameroon, Eritrea, Ethiopia	Bactrocera oleae Euleia heraclei	Italy 1915 for B. oleae.	20, 21, 26, 28, 31
Asecodes notandus (Silvestri)	Eulophidae, Entedoninae	Egg-larval parasitoid (?)	Eritrea, Ethiopia	Bactrocera oleae		19, 26, 28, 31
Entedon atrocyanea (Silvestri)	"	Unknown	Eritrea, Ethiopia	Bactrocera oleae		19, 28, 31
Entedon viridis Silvestri	"	Larval parasitoid	Eritrea, Ethiopia	Bactrocera oleae		19, 26, 28, 31
Neochrysocharis formosus (Westwood)	"	larval parasitoid	Eritrea, Ethiopia, South Africa	Bactrocera oleae	Italy 1915 for B. oleae.	18, 19, 20, 21, 26, 28, 31
Euderus cavasolae (Silvestri)	Eulophidae, Entiinae	Larval parasitoid	Eritrea, Ethiopia	Bactrocera oleae		28
Tetrastichus giffardii Silvestri	Eulophidae, Tetrartichinae	Koinobiont gregarious endophagous larval (egg-larval?) parasitoid	Benin, Cameroon, Congo, Ghana, Kenya, Nigeria, Tanzania, Uganda, Victoria	Ceratitis antistictica Ceratitis capitata Ceratitis colae Ceratitis fasciventris Ceratitis giffardi Ceratitis sticica Dacus bivitatus Dacus ciliatus Dacus humeralis Trirhithrum coffeae	Hawaiian Islands 1950 on Oahu and Kauai for B. dorsalis and C. capitata. Mexico	12, 13, 19, 25, 27, 31, 35, 43

(continued)

Table 16.1 (continued)

Valid names of parasitoids[a]	Family and subfamily	Mode of parasitism	Country and region of origin	Host records	Distribution outside Africa when used in tephritid biological control programmes	References for distribution and host association
Terrastichus giffardianus Silvestri	"	Koinobiont gregarious endophagous larval parasitoid	Benin, Cameroon, Egypt, Kenya, Nigeria, Reunion Sierra Leone, Tanzania, South Africa	*Ceratitis anonae*	Australia 1932, 1956 for *Bactrocera tryoni* via Hawaii.	**4, 5, 19, 31, 42**
				Ceratitis capitata		
				Ceratitis rosa	Argentina 1947 for *C. capitata* via Brazil. Brazil 1937 for *C. capitata* and *Anastrepha* sp. via Hawaii. Cook Island 1938 for *B. melanotus* via Fiji. Costa Rica and Nicaragua 1955 for *C. capitata* via Hawaii. Fiji 1935, 1959, 1960 for *Bactrocera xanthodes* and *B. passiflorae* via Hawaii. Hawaiian Islands 1914 for *C. capitata*, and 1947–1952 for *B. dorsalis*. Greece for *B. oleae*. Italy 1916 for *C. capitata* via Hawaii. New Caledonia 1936 for *Bactrocera umbrosa* and *Bactrocera psidii* via Fiji.	
				Ceratitis cosyra		
				Dacus bivittatus		
				Dacus ciliatus		
				Dacus demmerezi		
				Trirhithromyia cyanescens, Trirhithrum queritum		
Terrastichus maculifer Silvestri	"	Larval parasitoid (?)	Eritrea, Ethiopia	*Bactrocera oleae*	La Réunion. Western Samoa 1935, 1938 for *B. passiflorae* via Fiji. USA 1931 for *Rhagoletis* sp. via Hawaii. Puerto Rico 1935 for *A. suspensa* via Hawaii. Spain. Vanuatu.	**26, 28, 31**

Tetrastichus oxyurus Silvestri	"	Larval endophagous parasitoid	Kenya, Nigeria, West Africa	*Carpophthoromyia tritea*	Hawaiian Islands	27
Tetrastichus sp.	"	larval gregarious endophagous parasitoid	South Africa, West Africa	*Ceratitis anonae* *Bactrocera oleae*	Hawaiian Islands 1936 for *C. capitata* and *Bactrocera cucurbitae*.	20, 21, 43
Syntomosphyrum sp.	"	Larval parasitoid	Uganda	*Trirhithrum coffeae*		12, 13
Macroneura sp.	Eupelmidae, Eupelminae	Unknown	Egypt	*Bactrocera oleae*		8
Eupelmus urozonus Dalman	"	Idiobiont solitary ectophagous larval and pupal parasitoid	Algeria, Egypt, Libya, South Africa	*Bactrocera oleae*	Greece 1975–1980 for *B. oleae*	8, 20, 21, 31
Eupelmus afer Silvestri	"	Idiobiont solitary ectophagous larval or pupal parasitoid.	Eritrea, Ethiopia, South Africa	*Bactrocera oleae*	Italy 1915 for *B. oleae*.	20, 21, 26, 28
Eupelmus spermophilus Silvestri	"	Idiobiont solitary ectophagous larval or pupal parasitoid	Eritrea, Ethiopia, South Africa	*Bactrocera oleae*		28
Eupelmus sp.	"	Unknown	Tanzania	*Ceratitis* sp.		31
Eurytoma martelli Domenichini	Eurytomidae, Eurytomiae	larval or pupal ectophagous parasitoid	Egypt, North Africa	*Bactrocera oleae*	Greece 1975–1980 for *B. oleae*	8, 19
Eurytoma sp.	"	larval or pupal ectophagous parasitoid	Libya, Egypt	*Bactrocera oleae*		8, 31

(continued)

Table 16.1 (continued)

Valid names of parasitoids[a]	Family and subfamily	Mode of parasitism	Country and region of origin	Host records	Distribution outside Africa when used in tephritid biological control programmes	References for distribution and host association
Allocerellus inquirendus Silvestri	Encyrtidae, Tetracneminae	Unknown	Eritrea, Ethiopia	*Bactrocera oleae*		
Microdontomerus sp.	Torymidae, Toryminae	"	South Africa	*Bactrocera oleae*		20, 21
Aganaspis sp.	Figitidae, Eucoilinae	"	Central African Republic, Congo, Kenya, Reunion, South Africa, Tanzania	Fruit-infesting Tephritidae		40
Eucoila sp.	"	"	Mauritius	*Ceratitis capitata*		23
Ealata clava Quinlan	"	"	Cameroon, Congo, Kenya, Mauritius, Principe, South Africa, Uganda	Fruit-infesting Tephritidae	Taiwan (inadvertently)	40
Ealata marica Quinlan	"	"	Congo			40
Ealata saba Quinlan	"	"	Congo, Nigeria, South Africa, Uganda, Zimbabwe			40
Ganaspis kilimandjaroi (Kieffer)	"	"	Tanzania			40
Ganaspis mahensis Kieffer	"	"	Seychelles			40
Ganaspis ruandana (Benoit)	"	"	Rwanda			40

Asobara sp.	Braconidae, Alysiinae	Unknown	Mali	*Ceratitis cosyra?* *Ceratitis fasciventris?* *Ceratitis silvestrii?*	45
Triaspis daci (Szépligeti)	Braconidae, Brachistinae	"	Congo, Ethiopia, South Africa	*Bactrocera oleae*		19, 26, 28, 31
Bracon celer Szépligeti	Braconidae, Braconinae	Idiobiont solitary ectophagous larval parasitoid	Cape Verde Island, Eritrea, Ethiopia, Kenya Namibia, South Africa	*Bactrocera oleae* *Ceratitis capitata* *Trirhithrum nigerrimum*	Hawaii 1947–1952 for *B. dorsalis*, imported but not released. California 2003–2007 for *B. oleae*. Israel 2011 on *B. oleae* (inadvertently). Italy 1915 for *B. oleae*.	19, 20, 21, 26, 27, 28, 34, 41
Microbracon sp.	"	Unknown	Tanzania	*Ceratitis* sp.		31
Diachasmimorpha brevistyli (Paoli)	Braconidae, Opiinae	Koinobiont solitary endophagous larval parasitoid	Somalia	*Dacus ciliatus*		31
Diachasmimorpha carinata (Szépligeti)	"	Koinobiont solitary endophagous larval parasitoid	Cameroon, Cape Verde, Congo, Guinea, Kenya, Senegal, Sierra Leone, Tanzania, Zaire	*Ceratitis anonae* *Ceratitis contramedia* *Ceratitis ditissima* *Ceratitis giffardi* *Ceratitis punctate* *Dacus bivitatus* *Dacus ciliatus*	Hawaiian Islands 1936 for *C. capitata* and *Bactrocera cucurbitae*, released as *Hedylus giffardii* (Silvestri). Hawaii 1947–1952 for *B. dorsalis*, imported but not released.	11, 19, 27, 35, 39, 41

(continued)

Table 16.1 (continued)

Valid names of parasitoids[a]	Family and subfamily	Mode of parasitism	Country and region of origin	Host records	Distribution outside Africa when used in tephritid biological control programmes	References for distribution and host association
Diachasmimorpha fullawayi (Silvistri)	"	Koinobiont solitary endophagous larval parasitoid	Cameroons, Congo, Guinea, Kenya, Nigeria, Senegal, West Africa, Reunion, Sierra Leone, South Africa, Togo, Uganda, Zaire	*Bactrocera amplexa* *Carpophthoromyia tritea* *Ceratitis anonae* *Ceratitis capitata* *Ceratitis cosyrae* *Ceratitis giffardi* *Ceratitis punctata* *Ceratitis tritea* *Ceratitis* sp. *Dacus bivitatus* *Dacus* sp. *Trirhithrum coffeae*	Australia 1932 for *B. tryoni* via Hawaii. Brazil for *C. capitata*. Fiji for *C. capitata*. Hawaiian Islands 1914 for *C. capitata*. Hawaii 1947–1952 for *B. dorsalis*, imported but not released. Mauritius on *Ceratitis* spp (new to fauna), Puerto Rico for Tephritidae 1941. La Réunion. Spain.	**3, 19, 27, 29, 35, 39, 41, 44**
Diachasmimorpha insignis (Granger)	"	Koinobiont solitary endophagous larval parasitoid	Madagascar	*Ceratitis* sp.		35
Fopius bevisi (Brues)	"	Koinobiont solitary endophagous larval parasitoid	Kenya, South Africa	*Ceratitis capitata* *Dacus ciliatus* *Trirhithrum queritum*	Hawaiian Islands 1949 for *C. capitata* and *B. dorsalis*, imported but not released.	**6, 35**

Fopius caudatus (Szépligeti)	"	Koinobiont solitary endophagous egg-larval parasitoid	Benin, Cameroon, Congo, Mali, Guinea, Kenya, Nigeria, Senegal, Sierra Leone, Victoria, West Africa, Zaire	*Carpophthoromyi atritea* *Ceratitis anonae* *Ceratitis antisticticta* *Ceratitis capitata* *Ceratitis giffardi* *Ceratitis tritea* *Dacus bivittatus* *Dacus ciliatus* *Dacus humeralis* *Dacus momordicae* *Trirhithrum coffeae* *Trirhithrum nigerrimum*	Guatemala for *C. capitata*. Hawaiian Islands 1936 for *C. capitata*. Hawaii 1947–1952 for *B. dorsalis*, imported but not released. Hawaii (quarantine facility via Kenya 1996–2004 for *C. capitata*)	**19, 27, 29,** 33, 34, **35, 39,** 41
Fopius ceratitivorus Wharton	"	Koinobiont solitary endophagous egg-larval parasitoid	Kenya	*Ceratitis capitata* *Ceratitis rosa* *Trirhithrum coffeae*	Australia. Guatemala 2003 for *C. capitata*. Hawaii (2004 cultured for release on *C. capitata* via Kenya and Guatemala, pending release). Spain. Puerto Rico. Israel 2011 for *C. capitata* and *B. oleae*.	33, 34, 35
Fopius desideratus (Bridwell)	"	Koinobiont solitary endophagous larval (? egg-larval) parasitoid	Cameroon, Congo, Nigeria, Senegal, Togo, Uganda	*Ceratitis anonae* *Ceratitis capitata* *Ceratitis* sp. *Dacus bivittatus* *Dacus* sp. *Trirhithrum coffeae*	Hawaii 1947–1952 for *B. dorsalis*, imported but not released.	**2, 12, 13, 19,** 29, 31, **35, 39,** 41

(continued)

Table 16.1 (continued)

Valid names of parasitoids[a]	Family and subfamily	Mode of parasitism	Country and region of origin	Host records	Distribution outside Africa when used in tephritid biological control programmes	References for distribution and host association
Fopius niger (Szépligeti)	"	Koinobiont solitary endophagous larval parasitoid	Cameroon, Kenya, Tanzania	*Dacus humeralis*		35
Fopius okekai Kimani-Njogu & Wharton	"	Koinobiont solitary endophagous larval parasitoid	Kenya	*Trirhithrum inscriptum* *Trirhithrum nigrum*		16
Fopius ototomoanus (Fullaway)	"	Koinobiont solitary endophagous larval parasitoid	Cameroon	*Dacus* sp.		35
Fopius silvestrii (Wharton)	"	Koinobiont solitary endophagous larval parasitoid	Cameroon, Senegal, Western Kenya	*Ceratitis anonae* *Ceratitis capitata* *Ceratitis cosyra,* *Ceratitis fasciventris* *Ceratitis flexuosa* *Dacus bivittatus*		39, 43
Opius sp.	"	Unknown	Cameroon, West Africa	*Ceratitis cosyra* *Trirhithrum coffeae*		30
Opius sp.	"	"	Tanzania	*Ceratitis* sp.	Hawaii 1947–1953 for *B. dorsalis*	4, 31
Pseudorhinoplus fuscipennis (Szépligeti)	"	Koinobiont solitary endophagous larval parasitoid	Cameroon, Congo, Uganda	*Ceratitis anonae* *Ceratitis ditissima* *Trirhithrum coffeae*	Hawaii 1947–1952 for *B. dorsalis*, imported but not released.	40

Psyttalia concolor (Szépligeti)	"	Koinobiont solitary endophagous larval parasitoid	Algeria, Benin, Cape Verde, Congo, Eritrea, Kenya, Libya, Madagascar, Morocco, Senegal, South Africa, Tunisia	Bactrocera oleae	Australia for B. tryoni 1932, 1933 via Hawaii. Bermuda. Bolivia 1969 for C. capitata, Anastrepha spp. Bulgaria.	3, 7, 8, 10, 14, 17, 19, 21, 27, 31, 35, 39, 41, 42
				Capparimyia savastani		
				Carpomya incompleta		
				Ceratitis capitata	California 2005 for B. oleae.	
				Ceratitis cosyra,	Cook Island 1927 for B. melanotus via Hawaii. Costa Rica 1956 for C. capitata, and Anastrepha spp. Crete.	
				Ceratitis colae,		
				Dacus brevistylus		
				Dacus ciliatus		
				Dacus frontalis	El Salvador 1971 for C. capitata.	
				Trirhithromyia cyanescens	Fiji 1935 for B. passiflorae and B. xanthodes via Hawaii. Florida 1977–1979 for Anastrepha suspensa.	
				Trirhithrum nigrum	Finland (inadvertently on Euphranta connexa). France 1919, 1931, 1958 for B. oleae. Greece 1954 for B. oleae via France. Hawaii 1947–1952 for B. dorsalis, imported but not released. Hawaii (quarantine facility 1996–2004 for C. capitata). Greece 1968 for B. oleae via France. Guam. Guatemala mass reared. Italy 1914, 1917 from Libya, 1918, 1923, 1934, for B. oleae. Israel. Jordan. Lebanon. New Caledonia 1966 for B. psidii, and B. frenchi via France. Pakistan. Peru. Puerto Rico 1935, 1936, 1941 for A. suspensa.and A. obliqua. La Réunion. Spain 1923 for C. capitata and B. oleae. Turkey. Yugoslavia.	

(continued)

Table 16.1 (continued)

Valid names of parasitoids[a]	Family and subfamily	Mode of parasitism	Country and region of origin	Host records	Distribution outside Africa when used in tephritid biological control programmes	References for distribution and host association
Psyttalia cosyrae (Wilkinson)	"	Koinobiont solitary endophagous larval parasitoid	Congo, Kenya, Nigeria, North Africa, Reunion, Senegal, Sierra Leone, South Africa, Tanzania, Uganda, Zaire	*Ceratitis capitata* *Ceratitis cosyra* *Trirhithrum coffeae*		13, 19, 31, 35, 39
Psyttalia dacicida (Silvestri)	"	Koinobiont solitary endophagous larval parasitoid	Eritrea, Ethiopia, Kenya, South Africa	*Bactrocera oleae* Tephritidae	Italy (inadvertently)	19, 20, 21, 26, 28, 31, 35
Psyttalia dexter (Silvestri)	"	Koinobiont solitary endophagous larval parasitoid	Senegal	*Dacus longistylus*		35
Psyttalia distinguenda (Granger)	"	Koinobiont solitary endophagous larval parasitoid	Madagascar, Mascarenes, Mauritius, Reunion	*Ceratitis capitata* *Ceratitis rosa*		33, 34
Psyttalia halidayi Wharton	"	Koinobiont solitary endophagous larval parasitoid	Kenya	*Ceratitis rosa*		36

Psyttalia humilis (Silvestri)	"	Koinobiont solitary endophagous larval parasitoid	Egypt, Kenya, Namibia, South Africa	*Bactrocera oleae*	Australia 1932 for *B. tryoni*. Bermuda 1926 for *C. capitata* via Hawaii. California 1932 for *Rhagoletis* sp. via Hawaii.	3, 18, 19, 27, 31, 35, 41, 42
				Ceratitis capitata	California 2003–2007 for *B. oleae* via France. Cook Islands.	
					Fiji 1935 for *B. xanthodes* and *B. passiflorae* via Hawaii. Guatemala. Hawaiian Islands (Oahu and Maui Islands 1913 for *C. capitata*).	
					Hawaii 1947–1952 for *B. dorsalis*, imported but not released. Israel 1926 for *C. capitata*. Puerto Rico for Tephritidae 1940. Spain 1932 for *C. capitata* via Hawaii.	
Psyttalia inconsueta (Silvestri)	"	Koinobiont solitary endophagous larval parasitoid	Nigeria	*Carpophthoromyia tritea*	Hawaiian Islands	27
Psyttalia insignipennis (Granger)	"	Koinobiont solitary endophagous larval parasitoid	Madagascar, Mauritius, Reunion	*Ceratitis capitata*		23, 33
				Ceratitis catoirii		
				Neoceratitis cyanescens		
				Trirhithromyia cyanescens		
Psyttalia lounsburyi (Silvestri)	"	Koinobiont solitary endophagous larval parasitoid	Kenya, Namibia, South Africa, Transvaal	*Ceratitis capitata*	California 2003–2007 for *B. oleae* via France. France 2007 for *B. oleae*. Hawaii 1947–1952 for *B. dorsalis*, imported but not released. Hawaii (quarantine facility 1996 via Kenya for *C. capitata*)	20, 21, 26, 27, 28, 35, 41
				Bactrocera oleae		

(continued)

Table 16.1 (continued)

Valid names of parasitoids[a]	Family and subfamily	Mode of parasitism	Country and region of origin	Host records	Distribution outside Africa when used in tephritid biological control programmes	References for distribution and host association
Psyttalia masneri Wharton	"	Koinobiont solitary endophagous larval parasitoid	Kenya	*Taomyia marshalli*		36
Psyttalia perproximus (Silvestri).	"	Koinobiont solitary endophagous larval parasitoid	Benin, Cameroon, Ghana, Kenya, Mali, Nigeria, Sierra Leone, South Africa, Tanzania, Togo	*Ceratitis capitata*	Hawaiian Islands 1913, 1936 for *C. capitata*. Hawaii 1947–1952 for *B. dorsalis*, imported but not released.	**19, 27, 30, 31, 35, 41**
				Ceratitis colae		
				Ceratitis cosyra		
				Ceratitis flexuosa		
				Ceratitis giffardi		
				Ceratitis pedestris		
				Ceratitis punctata		
				Dacus bivittatus		
				Dacus ciliatus		
				Dacus sp.		
				Trirhithrum nigerrimum		
				Trirhithrum nigrum		
				Trirhithrum senex		
				Trirhithrum teres		
Psyttalia phaeostigma (Wilkinson)	"	Koinobiont solitary endophagous larval parasitoid	Cameroon, Congo, East Africa, Kenya, Madagascar, Mauritius, Mayotte, Reunion, South Africa	*Ceratitis anonae*	Hawaiian Islands 1951 for *Bactrocera dorsalis*. Hawaii 1996 quarantine facility for *B. cucurbitae*. Mauritius for Tephritidae 1934.	**19, 31, 35, 38**
				Ceratitis capitata		
				Ceratitis catoirii		
				Dacus ciliatus		
				Dacus bivittatus	Mayotte (new to fauna).	
				Dacus demmerezi	Madagascar (new to fauna).	
				Trirhithrum queritum	La Réunion.	

Psyttalia ponerophaga (Silvestri)	"	Koinobiont solitary endophagous larval parasitoid	Reunion (Pakistan origin?)	Bactrocera oleae	California 2003–2007 for B. oleae via Pakistan. 15
Psyttalia sanctamariana (Fischer)	"	Unknown	Madagascar, Mauritius, Reunion	Spathulina acroleuca (Schiner)	38
Psyttalia subsulcata (Granger)	"	"	Madagascar, Reunion	Spathulina acroleuca (Schiner	38
Psyttalia sp.	"	"	Kenya	Ceratitis anonae	43
				Ceratitis lexuosa	
				Ceratitis fasciventris	
				Ceratitis rosa	
Rhynchosteres mandibularis Kimani-Njogu and Wharton	"	Larval parasitoid	Kenya	Trirhithrum sp.	16
Rhynchosteres clypeatus (Bridwell)	"	Koinobiont solitary endophagous larval parasitoid	Africa, Nigeria	Ceratitis sp.	19
Sternaulopius bisternaulicus Fischer	"	Unknown	Cameroon, Congo, Kenya	Ceratitis sp.	44
				Trirhithrum sp.	
Sternaulopius sp.	"	"	Kenya	Ceratitis flexuosa	43
				Ceratitis fasciventris	

(continued)

Table 16.1 (continued)

Valid names of parasitoids[a]	Family and subfamily	Mode of parasitism	Country and region of origin	Host records	Distribution outside Africa when used in tephritid biological control programmes	References for distribution and host association
Utetes africanus (Szépligeti)	"	Koinobiont solitary endophagous larval parasitoid	Eritrea, Ethiopia, Kenya, Namibia, Senegal, South Africa	*Bactrocera oleae* *Ceratitis capitata* *Ceratitis rosa* *Trirhithrum coffeae*	California 1990s, 2006 for *B. oleae*. Hawaii 1996–2004 for *C. capitata* and 1947–1952 for *B. dorsalis*, imported but not released. Italy 1910, 1915, 1917 for *B. oleae*.	**19, 20, 21, 26, 27, 28, 31, 35, 41**
Isurgus sp.	Ichneumonidae, Tersilochinae	Unknown (? ectophagous)	Tanzania	*Ceratitis capitata* *Ceratitis* sp. *Ceratitis. rosa* *Trirhithrum nigerrimum*		19, 31

[a]Accepted names of Chalcidoids and Braconids were revised according to Noyes, J.S. 2012, Chalcidoidea Universal Database, World Wide Web electronic publication http://www.nhm.ac.uk/chalcidoids

van Noort, S. 2015. Afrotropical Waspweb Database of Braconidae http://www.waspweb.org/Ichneumonoidea/Braconidae/; and Wharton Lab Database http://mx.speciesfile.org/projects/8/public/public_content/show/1318?content_template_id=88

(?) Indicates uncertain interpretation or host information not based on rearing or experiments, e.g. when parasitoids emerged from fruit infested by more than one tephritid species

References cited in table: (1) Boucek 1963; (2) Bridwell 1918; (3) Clausen 1956; (4) Clausen et al. 1965; (5) Cochereau 1970; (6) Daiber 1966; (7) Delucchi 1957; (8) El-Heneidy et al. 2001; (9) Etienne 1973; (10) Fry 1987; (11) Ghani 1972; (12) Greathead 1972; (13) Greathead 1976; (14) Kapatos et al. 1977; (15) Hoelmer et al. 2011; (16) Kimani-Njogu and Wharton 2002; (17) Kimani-Njogu et al. 2000; (18) Monaco 1978; (19) Narayanan and Chawla 1962; (20) Neuenschwander 1982; (21) Neuenschwander et al. 1983; (22) Nixon 1930; (23) Orian and Moutia 1960; (24) Rivnay 1968; (25) Rogg and Camacho 2000; (26) Silvestri 1914a; (27) Silvestri 1914b; (28) Silvestri 1915; (29) Steck et al. 1986; (30) Stibick 2004; (31) Thompson 1943; (32) Wharton 1999a; (33) Wharton 1999b; (34) Wharton et al. 2000; (35) Wharton and Gilstrap 1983; (36) Wharton 2009; (37) Yoder and Wharton 2002; (38) Fischer and Madl 2008; (39) Vayssières et al. 2012; (40) van Noort et al. 2015; (41) Gilstrap and Hart 1987; (42) Waterhouse 1993; (43) Copeland et al. 2006; (44) Wharton 2006; (45) Vayssières et al. 2002

The rich diversity of the African tephritid parasitoid fauna continues to be unravelled as more new species are described and careful studies on their biology and host specificity are made. For example, *Fopius ceratitivorus* Wharton was first described by Wharton in 1999 and recognized as an important egg-larval parasitoid of *C. capitata* (Wharton 1999a); *Fopius okekai* and *Rhynchosteres mandibularis* were described in 2002 (Kimani-Njogu and Wharton 2002). More recently, two new Kenyan species have been described: *Psyttalia halidayi* Wharton (from the Natal fruit fly, *Ceratitis rosa* Karsch) and *Psyttalia masneri* Wharton (from an uncommon tephritid, *Taomyia marshalli* Bezzi, in cornstalk dracaena, *Dracaena fragrans* [L.] Ker Gawl) (Wharton 2009). In general, coffee, *Coffea arabica* L. and wild olive, *Olea europaea* ssp. *cuspidate* (Wall. ex G. Don) Cif, the closest relative to cultivated olives, supported the greatest diversity of parasitoid fauna (Clausen et al. 1965; Greathead 1972; Steck et al. 1986; Wharton et al. 2000; Copeland et al. 2004; Hoelmer et al. 2004, 2011).

It is important to note that some taxa reported in these early records have undergone several taxonomic revisions and changes in nomenclature (Fischer 1972, 1977, 1987; Wharton 1983, 1987; Wharton and Gilstrap 1983). Lists of synonyms and previously used combinations have been produced for the Braconidae and Opiinae (Wharton 1989) and for the superfamily Chalcidoidea (Noyes 2012).

3 Contribution of Indigenous Parasitoids to Fruit Fly Management

The level of parasitism achieved by indigenous parasitoid species in various fruit fly species on cultivated fruits is variable but generally quite low (<5 %) (Steck et al. 1986; Lux et al. 2003; Vayssières et al. 2012). For example, Vayssières et al. (2012) reported combined parasitism by seven parasitoid species of various wild and cultivated crops to be just 2.4 %. These observations may not entirely reflect the field situation as some parasitized larvae might have already left the sampled fruits to pupate in the soil, thus escaping observation (Lux et al. 2003). Also, unripe fruits collected during the surveys are likely to yield fewer larval parasitoids than ripe fruits, especially of *Psyttalia* species which have short ovipositors and prefer mature larvae close to the surface of ripe and fallen fruits. Wong and Ramadan (1987) working in Maui Island, Hawaii reported 19 % parasitism of *C. capitata* and *B. dorsalis* larvae in green fruit samples compared with 43 % in ripe and fallen fruits. Similar relationships between fruit ripeness and rates of parasitism have been reported for *Psyttalia fletcheri* (Silvestri) (Purcell and Messing 1996).

Of all the cultivated crops, coffee not only supported the highest diversity of parasitoids attacking fruit flies, but also high levels of parasitism. Steck et al. (1986) recorded a combined percent parasitism by *Psyttalia perproximus* (Silvestri), *Fopius caudatus* (Szépligeti) and *Fopius caudatus* auc C, *Diachasmimorpha fullawayi* (Silvestri), *Fopius desideratus* (Bridwell) and an undescribed species of *Opius* that ranged between 10 and 56%; the average was 35% parasitism in a research plantation and 17% parasitism in a commercial plantation. This could be because coffee has a relatively small fruits compared with mango, *Mangifera indica* L., guava, *Psidium guajava* L., and papaya, *Carica papaya* L.. Opiine larval parasitoids do not enter the infested fruits to locate fruit fly larvae and their success is, therefore, limited by the length of their ovipositor and the size of the fruit. Moreover, in coffee ripe fruits remain on the tree allowing for full larval exposure to parasitoids.

Other tephritid host plants that support high levels of parasitism are members of the family Oleaceae, e.g. *Olea europaea* ssp. cuspidata (Wall. ex G. Don). During the 1999–2003 survey for insects associated with fruits of indigenous species of Oleaceae in Kenya, the rates of parasitization of *B. oleae* by *Psyttalia lounsburyi* (Silvestri) alone exceeded 30% in some of the collections (Copeland et al. 2004). In a recent study by Mkize et al. (2008) on wild olives in the Eastern Cape Province, South Africa, the combined percent parasitism of *B. oleae* and *Bactrocera biguttula* (Bezzi) by *Psyttalia concolor* (Szépligeti), *P. lounsburyi*, *Utetes africanus* (Szépligeti) and *Bracon celer* Szépligeti, was in some instances as high as 83%, leading to very low infestation levels (1–8%). The authors indicated that these parasitoids were more closely associated with *B. oleae* as the number of *B. oleae* recovered was far smaller than the number of *B. biguttula* recovered. They also argued that fruit flies might not have become economic pests of commercial olives in the Eastern Cape due to the activity of these natural enemies. In Egypt, El-Heneidy et al. (2001) reported parasitism rates for *P. concolor* and *Pnigalio agraules* (Walker) (= *Pnigalio mediterraneus* (F.)), attacking *B. oleae* of 39% and 11%, repectively.

The performance of native parasitoids on different fruit fly species has been evaluated under laboratory conditions; high to moderate rates of parasitism were achieved in some host species. For example, Mohamed et al. (2003) reported parasitism rates of 37 and 46% by *Psyttalia cosyrae* (Wilkinson) in *C. capitata* and the mango fruit fly, *Ceratitis cosyra* (Walker), respectively. In a different study the same authors, reported parasitism rates by *P. concolor* of 46 and 28% in *C. capitata* and *C. cosyra*, respectively (Mohamed et al. 2007). Both parasitoid species were unable to develop on the *C. rosa*, *Ceratitis fasciventris* (Bezzi), *Ceratitis anonae* (Graham) and *Z. cucurbitae* (Mohamed et al. 2003, 2007) (Fig. 16.1). In contrast, the Eulophid *Tetrastichus giffardii* Silvestri achieved parasitism rates of 44.3 and 41.8% on *C. capitata* and the lesser pumpkin fly, *Dacus ciliatus* Loew, respectively. Although members of the genus *Tetrastichus* are known to be rather generalist parasitoids, *T. giffardii* achieved zero parasitism on all members of the *Ceratitis* FAR group (*C. fasciventris*, *C. annonae* and *C. rosa*) as well as on the exotic *Bactrocera* species (*Z. cucurbitae* and *B. dorsalis*) (Fig. 16.1).

Although the role of pupal parasitoids in biological control of fruit flies cannot be denied, no systematic studies to evaluate their impact on fruit fly populations have been made, and hence no accurate statistics are available on their role as biological control agents. They are not host specific and may also attack nontarget Diptera in the suborder Cyclorhapha (e.g. Agromyzidae, Drosophilidae, Muscidae). Also they are difficult to evaluate in the field as they need to be collected by sifting

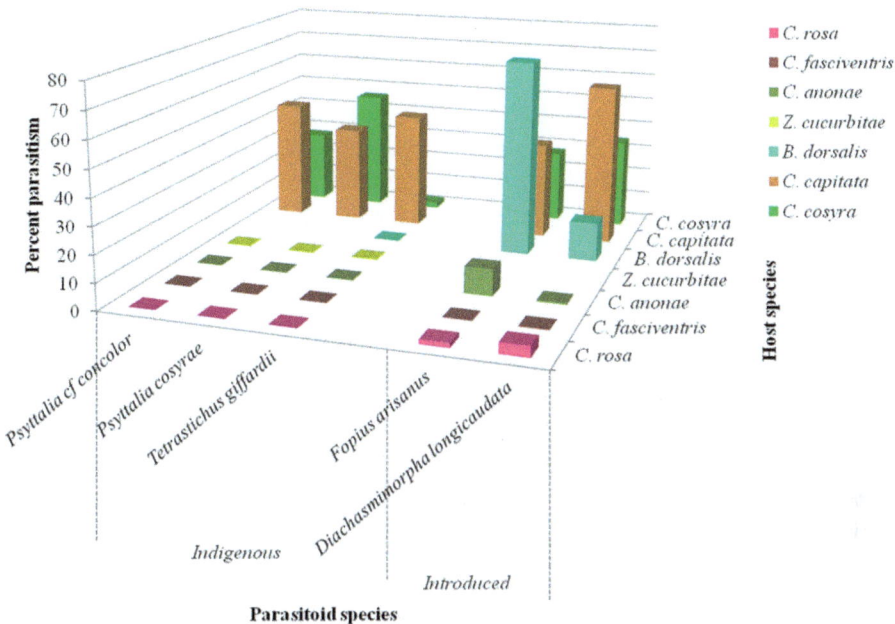

Fig. 16.1 Performance of indigenous and introduced parasitoid species on key native and invasive fruit flies in Africa

the soil to retrieve fruit fly pupae, compared with collecting and incubating fruits to evaluate parasitoid species attacking the egg and larval stages of their hosts (M.M. Ramadan unpublished data; Wang and Messing 2004a, b).

4 Exploration for Fruit Fly Parasitoid Species in Africa for Introduction Elsewhere

Numerous species of hymenopteran fruit fly parasitoids have been recorded from native African tephritids since Silvestri's famous survey in 1912 (Table 16.1). The table includes parasitoid species reared from fruit-infesting Tephritidae but excludes parasitoids specialized on tephritids infesting flowerheads (e.g. the African *Psyttalia vittator* group), stem and gall forming tephritids, various African opiines from agromyzid leafminers (e.g. *Opius importatus* Fischer and *Opius phaseoli* Fischer imported from Africa into Hawaii in 1969), and seed feeders (e.g. *Psyttalia sanctamariana* [Fischer] reared from the seed tephritid and *Spathulina acroleuca* [Schiner]). Parasitoids without confirmed host records, doubtful hosts, or doubtful identifications (e.g. *Psyttalia insignipennis* [Granger] from Madagascar and Singapore), are not reported here.

The African fruit fly species, *C. capitata*, has invaded and become established in many parts of the world including Western Australia and the Hawaiian Islands from as early as 1897 and 1910, respectively (Froggatt 1909; Compere 1912; both cited in Headrick and Goeden 1996). Being an alien pest, and lacking resident parasitoids in these countries, it continued to cause massive yield losses on various types of fruit. This prompted searches for efficient natural enemies of this devastating pest. The first classical biological control attempt was directed against *C. capitata* by George Compere when he was hired by the government of Western Australia between 1902 and 1907 to search for natural enemies of *C. capitata* (Wharton 1989). However, Compere was unable to determine the native range of *C. capitata*, and hence the parasitoids that he introduced to Australia from Brazil and India never established in *C. capitata* populations. A decade later, following the accidental introduction and establishment of *C. capitata* in Hawaii (then the Territory of Hawaii), Filippo Silvestri travelled to Africa and Australia, on behalf of the Hawaiian Board of Agriculture and Forestry, to search for efficient natural enemies of *C. capitata* (Silvestri 1914a, b). He identified 21 species of African hymenopteran parasitoids as having potential as biological control agents of *C. capitata*; he made collections from fruit infested with ten *Ceratitis* species and seven *Dacus* species. However, few parasitoids survived his long steamship trip and he returned to Hawaii with only *Dirhinus giffardii* Silvestri, *Coptera silvestrii* (Kieffer), *Psyttalia humilis* (Silvestri) and *Psyttalia perproximus* (Silvestri) from Africa, and *Diachasmimorpha tryoni* (Cameron) from Australia.

Silvestri returned from Hawaii to Italy in 1913 with some *D. giffardii* and *C. silvestrii* for biological control of *B. oleae*. A year later, he travelled back to East Africa (Eritrea), this time in search of more parasitoids for classical biological control of *B. oleae* in his homeland of Italy. He found 14 species attacking *B. oleae*, ten of which were reared and released in Italy although none became established. Fullaway, travelled to Nigeria in 1914 to re-collect parasitoid species that had not survived Silvestri's expedition and he returned with *Tetrastichus giffardianus* Silvestri and *Diachasmimorpha fullawayi* (Silvestri), which were then released and established in Hawaii (Fullaway 1914).

Although Silvestri and Fullaway collected many parasitoid species belonging to different genera and families, only a few survived the long voyage to Hawaii. Amongst those that survived, four species were released and established of which three were from Africa. These were, *P. humilis* from South Africa and, *D. fullawayi* and *D. giffardii* both from West Africa. The two former species are koinobiont larval parasitoids while the latter is an idiobiont pupal parasitoid. Two decades after introduction in to Hawaii the combined parasitism rates achieved by *P. humilis* and another introduced Australian parasitoid, *D. tryoni* in *C. capitata* populations ranged from 46 to 94 % (Willard and Mason 1937). The two parasitoid species achieved approximately equal levels of parasitism in *C. capitata* populations. As a result, *C. capitata* infestations were significantly reduced on coffee and, to a lesser extent, on other fruits; success was not so good against *C. capitata* in large sized fruits such as mangoes (http://paroffit.org/public/site/paroffit/home). Subsequently, *P. humilis* was mass reared and redistributed from Hawaii to several other countries with tephritid fruit fly problems (Table 16.1). However, this parasitoid has not been recorded in Hawaii since 1933, even in recent surveys (M.M. Ramadan unpublished data) and is thought to be extinct there (http://paroffit.org/public/site/paroffit/home). Similarly, although it did establish after introduction, *D. fullawayi* has only rarely been recorded in Hawaii since 1949 (Bess 1953; Bess et al. 1961). From Hawaii, *D. fullawayi* and *P. humilis* were also introduced into Spain, Puerto Rico and Australia, without success (Table 16.1). Following its introduction into Hawaii, *D. giffardii* became established in *C. capitata* populations; it was later introduced from Hawaii into Australia in 1956, Mexico in 1955, Puerto Rico in 1935 and Bolivia in 1971 (Bennett and Squire 1972), and Israel in 1956 for biological control of *C. capitata* and other resident tephritids (Table 16.1).

During a separate expedition at around the same time, the gregarious parasitoid, *T. giffardianus* was also introduced into Hawaii from West Africa by D.T. Fullaway and J.C. Bridwell in 1914, where it became established (Clausen et al. 1965). Subsequently, this species was mass-reared and redistributed from Hawaii to the Pacific Islands and Latin American countries. For example, it was imported into Brazil in 1937 where it established (Ovruski and Schliserman 2012), and from there it was also imported into Argentina in 1947 (Flávio et al. 2013) (Table 16.1).

Africa was also targeted in world-wide surveys for parasitoids made during the Hawaiian biological control campaign against *B. dorsalis*, in the 1950s. Import of African fruit flies into Hawaii (from South Africa in 1949, from Kenya in 1949–1950, from Congo in 1950–1951, and from Cameroon in 1951) with the purpose of collecting any parasitoids that emerged, was comprised of 571,995 pupae from 26 different tephritid species (Clausen et al. 1965). At least 22 different parasitoid species were recovered from these shipments, propagated and evaluated for their ability to develop on, *B. dorsalis*, *Z. cucurbitae* and *C. capitata*. Only six parasitoid species were released (*D. giffardii*, *T. giffardii*, *T. giffardianus*, *Fopius bevisi* (Brues), *Psyttalia phaeostigma* (Wilkinson) and an *Opius* sp. (Clausen et al. 1965).

Within the framework of a USDA grant (2001–2004) through the Texas A&M University entitled 'Facilitating Identification and Suppression of African Fruit-infesting Tephritidae (Diptera): Invasive Species That Threaten U.S. Fruit and Vegetable Production' the recently described parasitoid species, *Fopius ceratitivorus* Wharton and a related species, *Fopius caudatus* (Szépligeti) were imported from Kenya into the USDA-APHIS/MOSCAMED quarantine facility in Guatemala (Lopez et al. 2003), and from Guatemala into Hawaii. They where both evaluated for potential effects on non-target hosts and found not to parasitize eggs or larvae of the non-target tephritids, *Procecidochares alani* Steyskal, a biological control agent of the invasive weed, *Ageratina riparia* (Regel), and the native Hawaiian tephritid *Trupanea dubautia* (Bryan) found in the flowerheads of the endemic shrub, *Dubautia raillardioides* Hillebr. (Bokonon-Ganta et al. 2007; Wang et al. 2004). Under the same initiative *P. phaeostigma* and *P. halidayi* were, respectively, sent to St. Helena for control of *D. ciliatus* (2000–2001) and La Réunion (2000–2001) for control of *C. rosa* (S.A. Mohamed unpublished data). However, no follow up on their release and establishment has been made.

Psyttalia concolor a parasitoid of North African origin that is similar to the South African *P. humilis*, was initially imported from Tunisia (Monastero 1931; Silvestri 1939), and then released in Italy in 1913 for control of *B. oleae*, where it only became established at low densities. Since then, biological control of *B. oleae* in southern European countries has been almost exclusively based on importation and repeated releases of *P. concolor* (Raspi 1995; Raspi and Loni 1994). This parasitoid also parasitizes *C. capitata* in the Mediterranean basin.

In Israel, classical biological control targeting *C. capitata* and *B. oleae* has a relatively long history (Argov and Gazit 2008 and references therein). Between 2002 and 2004 four parasitoid species were imported from Hawaii and released against *C. capitata*. Two of these parasitoid species, the egg-larval parasitoid, *F. ceratitivorus* and the larval parasitoid, *P. concolor* were originally from Kenya. Of the African parasitoid species, *F. ceratitivorus* has shown signs of long-term establishment in Israel (Argov and Gazit 2008). A few years later (2009–2010), two other African parasitoid species were imported in to Israel, this time targeting *B. oleae*. These were *P. lounsburyi* (from Kenya and South Africa), and *Psyttalia* sp. nr. *concolor* (also called *P. humilis*) (from Namibia). A total of 37,000 and 97,000 wasps of the former and the later species, respectively were released in Israeli olive groves.

In 1998 *B. oleae* was detected in Californian olive groves (Rice et al. 2003). On the recommendation of earlier explorers highlighting the high diversity of *B. oleae*-associated parasitioids in Africa (e.g. Silvestri 1914a, b; Neuenschwander 1982), more expeditions across Africa were made to study these parasitoid species further. The parasitoid, *P. concolor*, was obtained from tephritid fruit flies infesting coffee in Kenya, reared on *C. capitata* in Guatemala by USDA-APHIS, PPQ, and then imported and released in Californian olive groves for biological control of *B. oleae*. Following this further exploration was attempted, this time for parasitoids that were more specific to *B. oleae* on wild African olives. Robert Copeland, an American entomologist based at *icipe*, Nairobi, Kenya, was contracted by USDA-APHIS to search for parasitoids attacking *B. oleae* in Kenya. He collected *P. concolor*, *P. lounsburyi* and *Utetes africanus* for importation into California via the USDA-ARS European Biological Control Laboratory (EBCL) in Montferrier, Montpellier, France (Copeland et al. 2004). This was followed by more expeditions to Kenya, South Africa, Namibia, La Réunion and Morocco. During these expeditions, *P. lounsburyi*, *P. humilis*, *P. concolor*, *Bracon* spp. and *U. africanus* were reared from wild olives and shipped to California for release via France (Hoelmer et al. 2011).

In Central America, African parasitoids were also the main focus for classical biological control of *C. capitata*. For example, in Costa Rica two African parasitoids, *D. giffardii* and *P. concolor* were introduced following the invasion by *C. capitata* in 1955 (Purcell 1998). A further six African parasitoid species were obtained by Gary Steck during his exploration for natural enemies of *C. capitata* in Togo and Cameroon between 1980 and 1982 (Steck et al. 1986). Following mass-rearing in Guatemala, *F. ceratitivorus* from Kenya was released on a large scale against *C. capitata* in the coffee-growing highlands along the Mexican borders (Sivinski and Aluja 2012). Detailed information regarding African parasitoid introductions for classical biological control of tephritid fruit flies in other countries is given in Table 16.1.

5 Introduction of Exotic Parasitoid Species into Africa for Biological Control of Invasive Fruit Flies

The first, though unsuccessful, attempt at classical biological control of exotic, invasive fruit flies in Africa was done in 1905. During this period, Charles Lounsbury and Claude Fuller, entomologists from South Africa, travelled to South America (Sao Paulo and Bahia, Brazil) to collect natural enemies for control of *C. capitata* in South Africa because, at the time, the native range of *C. capitata* was unknown (Lounsbury 1905 as cited in Ovruski et al. 2000). They collected the braconid, *Opius trimaculatus* Spinola and another unidentified parasitoid, from fruits infested by *Anastrepha fraterculus* (Wiedemann) and *Anastrepha serpentina* (Wiedemann) (Table 16.2). According to Wharton and Gilstrap (1983) this braconid could have been a misidentification of *Opius bellus* Gahan, *Utetes anastrephae* (Viereck), or a *Doryctobracon* sp. *Opius trimaculatus* was an important species to collect as field parasitism rates ranged from 7% in large guava fruits to 38% in the smaller fruits of Surinam cherry, *Eugenia uniflora* L. Because of the length of the trip from Brazil to South Africa via England, none of the imported braconid parasitoids survived the journey. Three years later, from a laboratory-reared colony in Australia, G. Compere sent to South Africa 20,000 *Aceratoneuromyia indica* (Silvestri) parasitoids, which he had initially collected from India during his expedition for natural enemies of *C. capitata* in Western Australia (Table 16.2). However, this parasitoid never became established in South Africa (Clausen 1956). Other failed attempts included the introduction of *Diachasmimorpha longicaudata* (Ashmead), *Opius* sp., *Psyttalia incisi* (Silvestri) and *P. phaeostigma* into Mauritius; and *D. tryoni* into both Mauritius and La Réunion (Fischer and Madl 2008).

Apart from the initiatives already mentioned, and despite the fact that Africa has been invaded by four Asian *Bactrocera* species (see De Meyer and Ekesi 2016), for which the first records date back to the 1930s (White and Elson-Harris 1992), classical biological control programmes for invasive fruit flies in Africa have not been taken up in the same way as in other continents that have been invaded by exotic species. For example, in Hawaii where *C. capitata* and three species in the genus *Bactrocera* have become established as key pests of fruits and vegetables, several expeditions were undertaken to various parts of the world in search of co-evolved natural enemies of these pests for introduction in to Hawaii. This resulted in the most successful classical biological control programme ever undertaken against tephrtids fruit flies (Wharton 1989; Purcell 1998).

In Africa, the earliest record of successful classical biological control of an exotic fruit fly species was in 1995, when *P. fletcheri* was introduced from Hawaii for biological control of *Z. cucurbitae* on the island of La Réunion (Quilici et al. 2004) (Table 16.2). The parasitoid is currently well established on the island though rates of parasitism of *Z. cucurbitae* are quite variable ranging from 1 to 75% on bitter gourd, *Momordica charantia* L (Cucurbitaceae) (Quilici et al. 2008). This was followed by introduction of another parasitoid species, the egg-larval parasitoid *Fopius arisanus* (Sonan) for biological control of another alien invasive pest,

Table 16.2 Hymenopteran larval and egg-larval parasitoids introduced into Africa for classical biological control against invasive fruit flies

Parasitoid species	Family and subfamily	Country/region of origin (where reared and where exported to)	Country and year of introduction	Target hosts	Status	References
Aceratoneuromyia indica (Silvestri)	Eulophidae, Tetrastichinae	India, Malaysia, Sri Lanka (mass reared, redistribution via Australia 1908). Southeast Asia (mass reared on *C. capitata*, redistribution via Hawaii 1957).	Egypt, Mauritius (1957–1959), Reunion (1972), South Africa (1908), Tunisia	*Ceratitis capitata* *Ceratitis rosa*	Established on Mauritius, La Réunion South Africa, Tunisia	**2, 5, 6, 8**
Aganaspis daci (Weld)	Figitidae, Eucoilinae	Southeast Asia (mass reared in Hawaii on *Bactrocera dorsalis* and *B. latifrons*. Strain ex. *C. capitata* from Greece, redistribution via Hawaii).	Egypt 2009 via Hawaii Reunion 1975 via Malaysia	*Bactrocera zonata* *Ceratitis capitata*	Egypt, established in *C. capitata* and *B. zonata* populations 2010, ongoing evaluations.	**9**, M.M. Ramadan unpublished data
Fopius arisanus (Sonan)	Braconidae Opiinae	Indo Malayan, Asia (mass reared on *Bactrocera dorsalis*, redistribution via Hawaii).	Benin (2009). Botswana (2015?). Cameroon (2009?). Comoros (2015). Egypt (2009). Kenya (2006, 2010). Madagascar (1993). Mauritius (1957–1959). Morocco. Mozambique (2009, 2012). Namibia (2015). Reunion (1965, 1972, 1995, 2003, 2009). Senegal (2012). Tanzania (2010). Togo (2009). Uganda. Zambia (2015). Zimbabwe (2015).	*Bactrocera dorsalis* *Bactrocera zonata* *Ceratitis capitata* *Ceratitis catoirii*. *Ceratitis rosa* *Dacus ciliatus*, *Dacus demmerezi*, *Neoceratitis cyanescens*	Established in all except Egypt	**2, 3, 7, 11**

Species		Origin	Location (year)	Host	Status	References
Fopius vandenboschi (Fullaway)	"	Indo Malayan, Asia (mass reared on *Bactrocera dorsalis*, redistribution via Hawaii).	Egypt (2009)	*Ceratitis capitata*	Pending release in Egypt	M.M. Ramadan unpublished data
				Bactrocera zonata		
Diachasmimorpha longicaudata (Ashmead)	"	Southeast Asia (mass reared on *Bactrocera dorsalis*, redistribution via Hawaii).	Kenya (2006, 2010). Mauritius (1957–1959). Egypt (2009). Cape Verde. Madagascar. Morocco. Mozambique (2009). Reunion (1971). Tanzania (2010). Zambia	*Bactrocera dorsalis*	Established in all except Mauritius	**6, 2, 5, 8, 11**, MMR
				Bactrocera zonata		
				Ceratitis capitata	Pending release in Egypt	
				Ceratitis rosa		
Diachasmimorpha tryoni (Cameron)	"	Queensland, Australia to Hawaii (mass reared on *Ceratitis capitata*, redistribution via Hawaii).	1957? Algeria, Canary Islands, Cape Verde. Egypt (2009). Madagascar. Mauritius. Reunion (1995). South Africa, Zambia	*Bactrocera zonata*	Egypt not established.	2, 5, 11
				Ceratitis capitata	Established in Mauritius on *Ceratitis* sp.	
Diachasmimorpha kraussii (Fullaway)	"	Queensland Australia to Hawaii (mass reared on *Bactrocera latifrons*, redistribution via Hawaii).	Egypt (2009)	*Bactrocera zonata*	Pending release in Egypt	M.M. Ramadan unpublished data
Psyttalia fletcheri (Silvestri)	"	Indo Malayan region, mass reared in Hawaii ex. strain from Java and India 1916, redistributed 1995–1997, 2009 via Hawaii.	Egypt (2009). La Réunion (1995)	*Ceratitis capitata*	Established	9, 10
				Bactrocera cucurbitae	Pending release in Egypt	
				Dacus ciliatus		
				Dacus demmerezi		
				Neoceratitis cyanescens		

(continued)

Table 16.2 (continued)

Parasitoid species	Family and subfamily	Country/region of origin (where reared and where exported to)	Country and year of introduction	Target hosts	Status	References
Psyttalia incisi (Silvestri)	"	Indo Malayan region, mass reared in Hawaii on *Bactrocera dorsalis*, redistributed via Hawaii.	Egypt (2009).	*Bactrocera zonata*	Not established	M.M. Ramadan unpublished data
			Mauritius	*Ceratitis rosa* ?	Pending release in Egypt	
Phaedrotoma trimaculata (Spinola)	"	Brazil reared on *Anastrepha* spp., Chile and Argentina ex. *Drosophila flavopilosa* (Drosophilidae).	South Africa (1905)	*Ceratitis capitata*	Not established	1, 2, 4
Opius sp.	"	Southeast Asia to Hawaii for *Bactrocera dorsalis*, redistribution via Hawaii.	Mauritius.	*Ceratitis capitata*	Not established	2
			Morocco (1954)	*Bactrocera oleae*		

(?) Indicates uncertain interpretation, or host information not based on rearing or experiments, e.g. when parasitoids emerged from fruit infested by more than one tephritid species

References cited in table: (1) van Achterberg and Salvo 1997; (2) Clausen 1978; (3) Ekesi and Billah 2006; (4) Fischer 1971; (5) Greathead and Greathead 1992; (6) Noyes 2012; (7) Mohamed et al. 2008; (8) Orian and Moutia 1960; (9) Quilici et al. 2004; (10) Quilici et al. 2000; (11) CABI 2016

B. zonata, on the same island (Rousse et al. 2006). A survey conducted on Indian almond, *Terminalia catappa* L., on which *B. zonata* is the dominant species, found that the level of parasitism on this host-fruit could reach 70–80 % (Quilici et al. 2008).

The most prominent fruit fly classical biological control programme in Africa to date was directed against *B. dorsalis* after it proved to be lacking resident parasitoid species capable of regulating its populations; all indigenous parasitoid species evaluated failed to form new associations with this pest due to its strong immune system, resulting in encapsulation and melanization of parasitoid eggs (Mohamed et al. 2006; S.A. Mohamed unpublished data). For example, two solitary larval parasitoids, *P. cosyrae* and *P. phaeostigma* and one gregarious parasitoid, *T. giffardii* were evaluated. *Bactrocera dorsalis* was readily accepted as a potential host by adult female *T. giffardii* and to a lesser extent by females of the two *Psyttalia* species. However, all eggs of the two *Psyttalia* species and nearly all the eggs of *T. giffardii* were encapsulated within larvae of *B. dorsalis* (Mohamed et al. 2006; S.A. Mohamed unpublished data). None of the *T. giffardii* progeny that escaped encapsulation were able to complete development to the adult stage. Furthermore, 34,430 kg of various host fruits of *B. dorsalis* were sampled in East Africa (Rwomushana et al. 2008) and West Africa (Vayssières et al. 2012; R. Hanna unpublished data), but not a single parasitoid species was recovered, confirming the fact that the indigenous African parasitoids were unable to parasitize *B. dorsalis*. These findings paved the way for identification and introduction of efficient parasitoids that had a shared history and origin with *B. dorsalis*. In this regard, the subsequent and logical approach was exploration for co-evolved parasitoid species in the pest's presumed native range of Sri Lanka. Three expeditions were made between 2005 and 2008 by scientists from the International Centre of Insect Physiology and Ecology (*icipe*), Nairobi, Kenya, the International Institute of Tropical Agriculture (IITA) and the University of Bremen, Germany, in collaboration with staff from the Horticultural Crop Research and Development Institute (HORDI), Peradeniya, Sri Lanka within the framework of the Mango IPM BMZ-funded project. Eight parasitoid species from different guilds (one egg-larval, five larval and two pupal) including *F. arisanus*, *D. longicaudata* and *P. fletcheri* were recovered from the sampled fruits and evaluated in the laboratory against target hosts (Billah et al. 2008; S.A. Mohamed unpublished data). Despite this, none were introduced into Africa due to issues relating to the Convention on Biological Diversity (CBD) to which Sri Lanka is a signatory. Thereafter, contacts were made between scientists on the *icipe*-led African Fruit Fly Programme and scientists at the USDA-ARS Pacific Basin Agricultural Research Center at Hilo, Hawaii and the University of Hawaii at Manoa. This led to introduction of the egg-larval parasitoid *F. arisanus* and the larval parasitoid, *D. longicaudata* into Africa (Table 16.2). These parasitoid species had been credited with outstanding success in the biological control of *B. dorsalis* following its invasion and establishment in Hawaii in 1944/1945 (Fullaway 1949). The two parasitoid species were imported into the *icipe* quarantine facility in 2006, following the FAO code of conduct for the importation and release of exotic biological control agents (IPPC 2005), and were later released in Kenya in 2008, in Tanzania in 2010, and in

Mozambique in 2012. *Fopius arisanus* was also released in the Comoros Islands in 2015. In Western Africa and under the umbrella of the same collaborative project, IITA released *F. arisanus* in Benin, Cameroon and Togo from a colony initially obtained from *icipe* in 2006 and subsequently maintained by IITA at Yaoundé, Cameroon and Cotonou, Benin. A detailed account of the release, establishment and spread of this parasitoid in Benin is given in Gnanvossou et al. (2016).

The post release assessment of colonization of these parasitoid species so far indicates that *F. arisanus* has established in all the countries where it was released but to varying degrees; the rates of *B. dorsalis* parasitism achieved on cultivated fruits was 33–40% in Kenya at the Northern Coast region of Kilifi (elevation > 400 masl) (Ekesi et al. 2010, 2016; S. Ndlela, unpublished data). On wild host fruit rates of *B. dorsalis* parasitism reached 46.5% on bush mango, *Irvingia gabonensis* (Aubry-Lecomte), in Benin (Gnanvossou et al. 2016). While establishment of *D. longicaudata* has been reported only in Kenya, at Embu in the Eastern Province (elevation range of 694M–1509 masl) and the Coast region (elevation < 400masl) with parasitism rates of up to 17% and 15.4%, respectively. Under a separate initiative, yet still targeting *B. dorsalis*, USDA-APHIS in collaboration with the Senegalese Plant Protection Department introduced *F. arisanus* into Senegal from Hawaii (Vargas et al. 2016). Between 2013 and 2014 14 shipments of 66,000 parasitoids were received in Senegal and released in the Casamance region (Vargas et al. 2016). This resulted in 20–30% parasitism of *B. dorsalis*. The authors indicated that additional parasitoid shipments were sent from Hawaii and released in other regions of Senegal to improve control during the mango fruiting season (Vargas et al. 2016).

In southern Africa, within the framework of the BONAZAZI FAO-funded project for suppression of *B. dorsalis*, both *F. arisanus* and *D. longicaudata* have recently been introduced into Botswana, Namibia, Zambia and Zimbabwe. However, a post evaluation survey to evaluate their establishment will only be undertaken during the 2016/2017 mango fruiting season.

In North Africa there has been a control programme targeted at another exotic invasive species, *B. zonata*. This species was first detected in 1997 and has since become widespread over most of the Egyptian governorates causing serious damage to many fruit crops. The Agricultural Research Centre (ARC), Giza, Egypt in collaboration with the State of Hawaii Department of Agriculture, have imported five parasitoid species from Hawaii for evaluation and release in Egypt (El-Heneidy and Ramadan 2010). These are *Aganaspis daci* (Weld), *F. arisanus*, *D. kraussii*, *D. tryoni* and *D. longicaudata* (Table 16.2). The five species were evaluated in the laboratory against *B. zonata*. Surprisingly, *F. arisanus*, which achieved high rates of parasitism on *B. zonata* in La Réunion, performed poorly on the same host in Egypt (El-Heneidy personal communication).

Following the promising performance in the laboratory evaluation of *A. daci* against *B. zonata*, this parasitoid has been released in the El-Arish district, North Sinai Governorate, during the guava season of 2010, and was recovered 1 month after release. Post-release assessment in the El-Arish district indicated 9.7% parasitism. Further studies on its natural dispersal and effectiveness in suppressing *B. zonata* and other tephritid fruit fly populations in Egypt, are still in progress

(El-Heneidy unpublished data). This parasitoid is an important candidate for *B. zonata* control, especially in large sized fruits (mango, peach, and guava), as it uses an ingress and sting strategy (i.e it enters the fruits to parasitize the larvae). All opiines use only drill and sting strategies; therefore, their accessibility to the host inside the fruit can be limited by the length of their ovipositors.

Currently, efforts are underway to introduce *F. arisanus* and *D. longicaudata* from *icipe* into Sudan for control of *B. zonata* and Ethiopia and South Africa for control of *B. dorsalis*.

6 Prospects and Potential Use of Parasitoids for Fruit Fly Management in Africa

Since the turn of last century, considerable advances have been made in both classical and augmentative biological control of fruit flies. However, this has not progressed at the same pace in Africa.

In general, parasitoids are unlikely to provide complete control of tephrid fruit flies because they act in a density dependant manner. Furthermore, the majority of susceptible produce is high-value fruit, making the damage threshold extremely low to ensure that the consumers' zero tolerance to blemished fruits is achieved. Nevertheless, parasitoids can significantly reduce fruit fly populations when used within the framework of an area-wide IPM approach. This is evidenced by the outstanding success of biological control programmes using parasitoids against the same and/or related tephritid fruit fly species in other parts of the world. Undeniably, the outcome of *B. dorsalis* and *C. capitata* control in Hawaii, and *B. dorsalis, B. kirki* and *B. tryoni* control in French Polynesia using *F. arisanus* and *D. longicaudata* (Vargas et al. 2007) are good examples of success that can be achieved and could be replicated in Africa. Indeed, the earlier explorers such as Silvestri (1914a, b) and van Zwaluwenburg (1937) indicated that *C. capitata* was rare in West Africa; the former author attributed the paucity of *C. capitata* in West Africa to the role of parasitoids. Also Steck et al. (1986) stated that *C. capitata* was of no economic importance in Central and West Africa due to the action of natural enemies. Similar observations of low infestation levels on olives in the Eastern Cape, South Africa have also been attributed to the action of parasitoids (Hancock 1989; Mkize et al. 2008).

Although *Bactrocera invadens* (as *B. dorsalis* was initially called in Africa) was recently synonymized with *Bactrocera dorsalis sensu stricto* (Schutze et al. 2015) populations in the native range could still be phenotypically different to populations in Africa with respect to their susceptibility to parasitoids; for example, African populations of *B. dorsalis* performed differently compared with the Hawaiian population where there are no reports of host immunity to *D. longicaudata* and *F. arisanus* (Mohamed et al. 2006, 2008). Therefore, more expeditions to the pest's area of origin are needed in Southeast Asia to evaluate the parasitoid species that did not establish in Hawaii during the *B. dorsalis* biological control programme in 1950s. Although *Z. cucurbitae* was presumed to have invaded Africa in the 1930s, no parasitoid species were introduced for its control. Considering that *Z. cucurbitae* mounts a strong immune response against almost all African parasitoid species and only *P. fletcheri* from its native range is capable of overcoming its immune system, it would be worthwhile to source this parastoid species from its native range and release it in Africa. Indeed, this parastoid species has been imported and released for classical biological control in several countries with promising results. For example, the release of *P. fletcheri* in Hawaii resulted in up to 29.8 and 96.9 % rates of parasitism of *Z. cucurbitae* on cucumber and wild bitter gourd, respectively (Willard 1920). Other parasitoids that are promising candidates for classical biological control of *Z. cucurbitae* need to be considered for importation into Africa and include four opiine parasitoids: *Diachasmimorpha albobalteata* (Cameron) from North Borneo, *Diachasmimorpha dacusii* (Cameron) from North India, *Diachasmimorpha hageni* (Fullaway) from Fiji and *Fopius skinneri* (Fullaway) from Thailand. *Fopius skinneri* should be considerd due to its tendency to parasitize tephritid larvae in cucurbits rather than other fruits (Waterhouse 1993). The larval-pupal parasitoid, *A. daci*, introduced into Hawaii from Queensland, Australia and Malaysia in 1949, has been reported as a primary parasitoid of *Z. cucurbitae* as has an *Aceratoneuromyia* sp. from northern Thailand (Ramadan and Messing 2003). However, a strain of *A. daci* from Greece was unable to develop in *Z. cucurbitae* (M.M. Ramadan unpublished data).

Although *B. latifrons* is of less economic importance than some species it can be a serious pest on solanaceous crops in the absence of natural enemies. Its management in Africa would greatly benefit from introduction of a co-evolved and efficient exotic parasitoid species from its native range. Laboratory experiments showed that most of the parasitoid species that attack *B. dorsalis* and *C. capitata* can survive in *B. latifrons*. *Diachasmimorpha kraussii* was released in Hawaii after it was successfully reared on *B. latifrons*, but subsequently it was rarely recovered from *B. latifrons* in wild fruits in the field. Exploration for parasitoids attracted to infested solanaceous fruits in the Indo-Malaysian region is required.

The introduction of *F. arisanus* for biological control of *B. zonata* resulted in mixed outcomes. This also calls for exploration and evaluation of more efficient parasitoid species from its native range. Such expeditions should aim at finding parasitoid species attacking both egg and larval stages of *B. zonata* to maximize the chances of pest suppression. Moreover, *A. daci* which has been promising for *B. zonata* control in Egypt should be evaluated further as a potential candidate for classical biological control of *B. zonta* in other African countries that are affected.

The native fruit fly, *C. rosa*, and its close relatives in the FAR complex, were immune to all the indigenous solitary and gregarious parasitoid species evaluated (Mohamed et al. 2003, 2006, 2007); furthermore, the two introduced parasitoids, *F. arisanus* and *D. longicaudata*, performed very poorly on *Ceratitis* species in the FAR complex (Mohamed et al. 2008, 2010). For these reasons a search for efficient parasitoids against these pests is urgently needed. Fortunately, the recently described *P. halidayi* was reared from field-collected *C. rosa* developing in fruits of

Lettowianthus stellatus Diels in coastal Kenya (Wharton 1996b) and its efficiency against *C. rosa* was further confirmed in laboratory studies (S.A. Mohamed unpublished data). Therefore, this parasitoid is a promising candidate that could be developed for biological control of *C. rosa* in mainland Africa; it could also be introduced for classical biological control in La Réunion and Mauritius where *C. rosa* has invaded. There is also a need for further research to identify parasitoid species that can overcome the immune response and develop successfully in *C. fasciventris* and *C. anonae* which cause significant yield losses in many tropical fruits (White and Elson-Harris 1992; Copeland et al. 2006).

Augmentation of parasitoid populations should also be considered to boost the efficiency of introduced parasitoids. In the same way, the role of native parasitoids in controlling native fruit flies could be enhanced by augmentative releases. This calls for involvement of the private sector in mass rearing of these parasitoids.

Parasitoid conservation, whether introduced or indigenous, is a fundamental pillar in ensuring the success of biological control programmes. It is, therefore, essential to make fruit and vegetable growers in Africa more aware of how to conserve parasitoids by using more eco-friendly management approaches rather than expensive blanket cover sprays of insecticide. Additionally, growers should be encouraged to practice habitat management that provides refuges and food sources for parasitoids in the areas surrounding orchards and gardens. Finally, the role of pupal parasitoids, particularly for biological control of native species should not be overlooked.

Acknowledgements Funding for the *icipe* fruit fly activities came from GIZ/BMZ, Biovision, the EU and DFID.

References

Appiah EF (2012) Evaluation of introduced parasitoids against *Bactrocera invadens* and their interaction with indigenous natural enemies (Ph.D. thesis). University of Ghana

Argov Y, Gazit Y (2008) Biological control of the Mediterranean fruit fly in Israel: Introduction and establishment of natural enemies. Biol Cont 46:502–507

Bennett FD, Squire FA (1972) Investigations on the biological control of some insect pests in Bolivia PANS (Pest Articles and News Summaries) 18:459–467

Bess HA (1953) Status of *Ceratitis capitata* in Hawaii following the introduction of *Dacus dorsalis* and its parasites. Proc Hawaii Entomol Soc 15:221–233

Bess HA, van den Bosch R, Haramoto RH (1961) Fruit fly parasites and their activities in Hawaii. Proc Hawaii Entomol Soc 17:367–378

Bianchi FA, Krauss NH (1936) Report on the United States Department of Agriculture East-African fruit fly expedition of 1935–1936. United States Department of Agriculture, Washington, DC

Billah MK, Ekesi S, Hanna R, Goergen G, Bandara KANP (2008) Exploration for natural enemies of *Bactrocera invadens* in Sri Lanka for classical biological control of the pest in Africa. First meeting of TEAM. Palma of Mallorca, Spain. 7–8th April 2008

Bokonon-Ganta AH, Ramadan MM, Messing RH (2007) Reproductive biology of *Fopius ceratitivorus* (Hymenoptera: Braconidae), an egg–larval parasitoid of the Mediterranean fruit fly, *Ceratitis capitata* (Diptera: Tephritidae). Biol Cont 41:361–367

Boucek Z (1963) A taxonomic study in *Spalangia*. Acta Entomol Mus Nat Prague 35:430–512

Bridwell JC (1918) Descriptions of new species of hymenopterous parasites of muscoid Diptera with notes on their habits. Proc Hawaiian Entomol Soc 4:166–179

CABI (2016) Invasive species compendium. CAB International, Wallingford. Available:http://cabi.org/isc/datasheet. (January 2016)

Clausen CP (1956) Biological control of fruit flies. J Econ Entomol 49:176–178

Clausen CP (1978) Tephritidae (Trypetidae, Trupaneidae), pp 320–335. In: Clausen CP (ed) Introduced parasites and predators of arthropod pests and weeds: a world *review*. United States Department of Agriculture, Agriculture Handbook 480. U.S. Government Printing Office, Washington, DC, 545 p.

Clausen CP, Clancy DW, Chock QC (1965) Biological control of the oriental fruit fly (*Dacus dorsalis* Hendel) and other fruit flies in Hawaii. US Dept Agric Tech Bull 1322:1–102

Cochereau P (1970) Les mouches des fruits et leurs parasites dans la zone Indo-Australo-Pacifique et particulierement en ouvelle-Caledonie. Cah ORSTOM Ser Bio 12:15–50

Copeland RS, White IM, Okumu M, Machera P, Wharton RA (2004) Insects associated with fruits of the Oleaceae (Asteridae, Lamiales) in Kenya, with special reference to the Tephritidae (Diptera). Bishop Mus Bull Entomol 12:135–164

Copeland RS, Wharton RA, Luke Q, De Meyer M, Lux S, Zenz N, Machera P, Okumu M (2006) Geographic distribution, host fruit, and parasitoids of African fruit fly pests *Ceratitis anonae*, *Ceratitis cosyra*, *Ceratitis fasciventris*, and *Ceratitis rosa* (Diptera: Tephritidae) in Kenya. Anns Entomol Soc Am 99:261–278

Daiber CC (1966) Notes on two pumpkin fly species and some of their host plants. S Afr J Agri Sc 9:863–876

De Meyer M, Ekesi S (2016) Exotic invasive fruit flies (Diptera: Tephritidae): In and out of Africa. In: Ekesi S et al (eds) Fruit fly research and development in Africa – towards a sustainable management strategy to improve horticulture. Springer International Publishing, Cham. doi:10.1007/978-3-319-43226-7_7

Delucchi V (1957) Les parasites de la mouche del olives. Entomophaga 2:107–118

Ekesi S, Billah MK (eds) (2006) A field guide to the management of economically important tephritid fruit flies in Africa. The International Centre of Insect Physiology and Ecology (ICIPE). ICIPE Science Press-Nairobi, Nairobi

Ekesi S, Mohamed S, Hanna R (2010) Rid fruits and vegetables in Africa of notorious fruit flies. Technical Innov. Brief No. 4. CGIAR SP-IPM. http://www.spipm.cgiar.org/c/document_library/get_file?p_l_id=17830&folderId=18484&name=DLFE-882.pdf

Ekesi S, De Meyer M, Mohamed SA, Virgilio M, Borgemeister C (2016) Taxonomy, ecology and management of native and exotic fruit fly species in Africa. Ann Rev Entomol 61:219–238

El-Heneidy AH, Ramadan MM (2010) *Bacterocera zonata* (Saunders) status and its natural enemies in Egypt. 8th international symposium on fruit flies of economic importance, Valencia, Spain, 26 Sept–10 Oct 2010

El-Heneidy AH, Omar AH, El-Sherif H et al (2001) Survey and seasonal abundance of the parasitoids of the olive fruit fly, *Bactrocera* (*Dacus*) *oleae* Gmel. (Diptera: Trypetidae). Egypt Arab J Plant Prot 19:80–85

Etienne J (1973) Lutte biologique et apercu sur les etudes entomologiques diverses effectuees ces dernieres annees a la Réunion. L'Agronomie Tropicale 6–7:683–687

Fischer M (1971) Index of entomophagous insects. Hymenoptera Braconidae. World Opiinae. Le François, Paris, 189 pp

Fischer M (1972) Hymenoptera Braconidae (Opiinae I). Das Tierreich 91:1–620

Fischer M (1977) Hymenoptera Braconidae (Opiinae II-Amerika). Das Tierreich 96:1–1001

Fischer M (1987) Hymenoptera: Opiinae III – athiopische, orientalische, australische und ozeanische region. Das Tierreich 104:1–734

Fischer M, Madl M (2008) Review of the Opiinae of the Malagasy Subregion (Hymenoptera: Braconidae). Linzer Biolog Beitr 40:1467–1489

Flávio R, Garcia M, Ricalde MP (2013) Augmentative biological control using parasitoids for fruit fly management in Brazil. Insects 4:55–70. doi:10.3390/insects4010055

Fry JM (1987) Natural enemy databank. A catalogue of natural enemies of arthropods derived from records in the CIBC Natural Enemy Databank. CAB International, Oxon, pp 34–35

Fullaway DT (1914) Report of the work of the insectary pp 143–151. In: Report of the board of commissioners of agriculture and forestry of the territory of Hawaii, Honolulu Star Bulletin LTD, Honolulu, pp 246 https://books.google.com/books?id=YVZJAAAAMAAJ&dq

Fullaway DT (1949) *Dacus dorsalis* Hendel in Hawaii. Proc Hawaii Entomol Soc 13:351–355

Ghani MA (1972) Final report: studies on the ecology of some important species of fruit flies and their natural enemies in West Pakistan. Pakistan Station, CIBC, Rawalpindi, 1–42

Gilstrap FE, Hart WG (1987) Biological control of the Mediterranean fruit fly in the United States and Central America. US Dep Agric Agric Res Serv 56:68

Gnanvossou D, Hanna R, Bokonon-Ganta AH, Ekesi S, Mohamed SA (2016) Release, establishment and spread of the natural enemy Fopius arisanus (Hymenoptera: Braconidae) for control of the invasive oriental fruit fly Bactrocera dorsalis (Diptera: Tephritidae) in Benin, West Africa. In: Ekesi S et al (eds) Fruit fly research and development in Africa – towards a sustainable management strategy to improve horticulture. Springer International Publishing, Cham. doi:10.1007/978-3-319-43226-7_26.

Greathead DJ (1972) Notes on coffee fruit-flies and their parasites at Kawanda. Commonwealth Inst. Biol Cont CAB 13:11–18

Greathead DJ (1976) Report on a survey for natural enemies of olive pests in Kenya and Ethiopia, September–December 1975. Commonwealth Institute of Biological Control, Delemont, Switzerland. 24p.

Greathead DJ, Greathead AH (1992) Biological control of insect pests by insect parasitoids and predators: the BIOCAT database. Biocontrol News Inf 13:61N–68N

Hancock DL (1989) Pest status: southern Africa. In: Robinson A, Hooper G (eds) World crop pests: fruit flies—their biology, natural enemies and control, 3A:51–58. Elsevier, Amsterdam

Headrick DH, Goeden RD (1996) Issues concerning the eradication or establishment and biological control of the Mediterranean fruit fly, *Ceratitis capitata* (Wiedemann) (Diptera: Tephritidae), in California. Biol Cont 6:412–421

Hoelmer KA, Kirk A, Wharton R, Pickett CH (2004) Foreign exploration for parasitoids of the olive fruit fly, *Bactrocera oleae*. In: Woods D (ed) Biological control program annual summary, 2003. California Department of Food & Agriculture, Sacramento, pp 12–14

Hoelmer KA, Kirk AA, Pickett CH, Daane KM, Johnson MW (2011) Prospects for improving biological control of olive fruit fly, *Bactrocera oleae* (Diptera: Tephritidae), with introduced parasitoids (Hymenoptera). Biocont Sci Tech 21:1005–1025. doi:10.1080/09583157.2011.594951

IPPC (2005) International plant protection convention: international standards for phytosanitary measures No. 3 – Guidelines for the export, shipment, import and release of biological control agents and other beneficial organisms. International Plant Protection Convention, Rome, 32 pp

Kapatos E, Fletcher BS, Pappas S, Laudeho Y (1977) The release of *Opius concolor* and *O. concolor* var. *siculus* against the spring generation of *Dacus oleae* on Corfu. Entomophaga 22:265–270

Kimani-Njogu SW, Wharton RA (2002) Two new species of Opiinae (Hymenoptera: Braconidae) attacking fruit-infesting Tephritidae (Diptera) in western Kenya. Proc Entomol Soc Wash 104:79–90

Kimani-Njogu SW, Trostle MK, Wharton RA, Woolley JB, Raspi A (2000) Biosystematics of the *Psyttalia concolor* species complex (Hymenoptera: Braconidae: Opiinae): the identity of populations attacking *Ceratitis capitata* (Diptera: Tephritidae) in coffee in Kenya. Biol Cont 20:167–174

Knipling EE (1992) Principles of insect parasitism analyzed from new perspectives. Practical implications for regulating insect populations by biological means. US Dep Agric Agric Res Serv Agric Handb 693:1–335

Lopez M, Sivinski J, Rendon P, Holler T, Bloem K, Copeland R, Trostle M, Aluja M (2003) Colonization of *Fopius ceratitivorus*, a newly discovered African egg-pupal parasitoid (Hymenoptera: Braconidae) of *Ceratitis capitata* (Diptera: Tephritidae). Fl Entomol 86:53–60

Lux SA, Ekesi S, Dimbi S, Mohamed S, Billah M (2003) Mango infesting fruit flies in Africa – perspectives and limitations of biological approaches to their management. In: Neuenschwander P, Borgemeister C, Langewald J (eds) Biological control in integrated pest management systems in Africa, pp 277–293. CAB International, Wallingford, 414 pp. ISBN: 0-85199-639-6

Mkize N, Hoelmer KA, Villet MH (2008) A survey of fruit-feeding insects and their parasitoids occurring on wild olives, *Olea europaea* ssp. cuspidata, in the Eastern Cape of South Africa. Biocont Sci Tech 18:991–1004

Mohamed SA, Overholt WA, Wharton RA, Lux SA, Eltoum EM (2003) Host specificity of Psyttalia cosyrae (Hymenoptera: Braconidae) and the effect of different host species on parasitoid fitness. Biol Control 28:155–163

Mohamed SA, Wharton RA, von Mérey G, Schulthess F (2006) Acceptance and suitability of different host stages of *Ceratitis capitata* (Wiedemann) (Diptera: Tephritidae) and seven other tephritid fruit fly species to *Tetrastichus giffardii* Silvestri (Hymenoptera: Eulophidae). Biol Cont 39:262–271

Mohamed SA, Overholt WA, Lux SA, Wharton RA, Eltoum EM (2007) Acceptability and suitability of six fruit fly species (Diptera: Tephritidae) for Kenyan strains of *Psyttalia concolor* Szépligeti (Hymenoptera: Braconidae). Biocont Sci Tech 17:247–259

Mohamed SA, Ekesi S, Hanna R (2008) Evaluation of the impact of *Diachasmimorpha longicaudata* on *Bactrocera invadens* and five African fruit fly species. J Appl Entomol 132:789–797

Mohamed SA, Ekesi S, Hanna R (2010) Old and new host-parasitoid associations: parasitization of the invasive fruit fly *Bactrocera invadens* (Diptera: Tephritidae) and five other African fruit fly species by *Fopius arisanus*, an Asian opiine parasitoid. Biocont Sci Tech 10:183–196

Monaco R (1978) Note sui parassiti del *Dacus oleae* Gmel. (Dipt. Tephritidae) in Sud Africa. Proceedings XI Congresso Nazionale Italiano di Entomologia, Portici-Sorrento 11:303–310

Monastero S (1931) Un nuovo parassita endofago della mosca delle olive trovato in Altavilla Milcia (Sicilia) Atti della Reale Accademia della Scienze. Lettere e Belle Arti di Palermo 16:195–201

Narayanan ES, Chawla SS (1962) Parasites of fruit fly pests of the world. Beitrage zur Entomologie 12:437–476

Neuenschwander P (1982) Searching for parasitoids of *Dacus oleae* in South Africa. J Appl Entomol 94:509–522

Neuenschwander P, Bigler F, Delucchi V, Michelakis SE (1983) Natural enemies of preimaginal stages of *Dacus oleae* Gmel. (Dipt., Tephritidae) in Western Crete. Bionomics and phenologies. Boll Lab Entomol Agrar Filippo Silvestri 40:3–32

Nixon GEJ (1930) The Ethiopian representatives of the Genus *Galesus*, with descriptions of new species. Ann Ma N Hist Ser 10(6):399–414

Noyes JS (2012) Universal chalcidoidea database. Natural History Museum, London. http://www.nhm.ac.uk/our-science/data/chalcidoids/

Orian AJE, Moutia LA (1960) Fruit flies of economic importance in Mauritius. Revue Agricole et Sucriere de L'Ile Maurice 39:142–150

Ovruski SM, Fidalgo P (1994) Use of parasitoids (Hym.) in the control of fruit flies (Dip.: Tephritidae) in Argentina: bibliographic review (1937–1991). Bull IOBC West Palearctic Reg Sec 17:84–92

Ovruski SM, Schliserman P (2012) Biological control of tephritid fruit flies in Argentina: historical review, current status, and future trends for developing a parasitoid mass-release program. Insects 3:870–888. doi:10.3390/insects3030870

Ovruski SM, Aluja M, Sivinski J, Wharton RA (2000) Hymenopteran parasitoids on fruit-infesting Tephritidae (Diptera) in Latin America and the southern United States: diversity, distribution, taxonomic status and their use in fruit fly biological control. Int Pest Manag Rev 5:81–107

Purcell MF (1998) Contribution of biological control to integrated pest management of tephritid fruit flies in the tropics and subtropics. Int Pest Manag Rev 3:63–83

Purcell MF, Messing RH (1996) Ripeness effects of three vegetable crops on abundance of augmentatively released *Psyttalia fletcheri* (Hym.: Braconidae): improved sampling and release methods. Entomophaga 41:105–116

Quilici S, Brevault T, Hurtrel B (2000) Major research achievements in Réunion within the Indian Ocean. In: Price NS, Seewooruthun I (eds) Proceedings of the Indian Ocean commission, regional fruit fly symposium, Flic en Flac, 5–9 June 2000

Quilici S, Hurtrel B, Messing RH, Montagneux B, Barbet A, Gourdon F, Malvotti A, Simon A (2004) Successful acclimatization of *Psyttalia fletcheri* (Braconidae: Opiinae) for biological control of the melon fly, *Bactrocera cucurbitae* (Diptera: Tephritidae), on Réunion Island. In: Barnes BN (ed) Proceedings of the 6th international symposium on fruit flies of economic importance, Stellenbosch, 6–10 May 2002, Isteg Scientific Publication, Irene, pp 457–459

Quilici S, Rousse P, Deguine JP, Simiand C, Franck A, Gourdon F, Mangine T, Harris E (2008) Successful acclimatization of the ovo-pupal parasitoid *Fopius arisanus* in Réunion island for the biological control of the peach fruit fly, *Bactrocera zonata*. First Meeting of TEAM. Palma of Mallorca, Spain. 7–8 April 2008

Ramadan MM, Messing RH (2003) A survey for potential biocontrol agents of *Bactrocera cucurbitae* (Diptera: Tephritidae) in Thailand. Proc Hawaii Entomol Soc 36:115–122

Raspi A (1995) Lotta biological in olivicoltura pp 483–495. In: Atti del convegno su: Techniche, norme e qualita' in olivicolture. Potenza (Italia), 15–17 Dicembre 1993

Raspi A, Loni A (1994) Alcune note sull'allevamento di *Opius concolor* Szépl. (Hymenoptera Braconidae) e su recent tentativi d'introduzione della specie in Toscana ed in Liguria. Frustula Entomol 17:135–145

Rice RE, Phillips PA, Stewart-Leslie J, Sibbett GS (2003) Olive fruit fly populations measured in central and southern California. Calif Agric 57:122–127

Rivnay E (1968) Biological control of pests in Israel (a review 1905–1965). Isr J Entomol 3:1–156

Rogg H, Camacho E (2000) History of fruit flies and their control in Bolivia. Santa Cruz de la Sierra, SANINET-IICA, Bolivia, 9 pp

Rousse R, Gourdon F, Quilici S (2006) Host specificity of the egg pupal parasitoid *Fopius arisanus* (Hymenoptera: Braconidae) in La Réunion. Biol Cont 37:284–290

Rwomushana I, Ekesi S, Gordon I, Ogol CKPO (2008) Host plants and host plant preference studies for *Bactrocera invadens* (Diptera: Tephritidae) in Kenya, a new invasive fruit fly species in Africa. Anns Entomol Soc Am 101:331–340

Schutze MK, Aketarawong N, Amornsak W, Armstrong KF, Augustinos AA, Barr NB, Bo W, Bourtzis K, Boykin L, Cáceres CE, Cameron SL, Chapman TA, Chinvinijkul S, Chomič A, De Meyer M, Drosopoulou E, Englezou A, Ekesi S, Gariou-Papalexiou A, Geib SM, Hailstones DL, Hasanuzzaman M, Haymer DS, Hee AKW, Hendrichs J, Jessup AJ, Ji Q, Khamis FM, Krosch MN, Leblanc L, Mahmood K, Malacrida AR, Mavragani-Tsipidou P, Mwatawala MW, Nishida R, Ono H, Reyes J, Rubinoff DZ, San Jose M, Shelly TE, Srikachar S, Tan K, Thanaphum S, Haq IU, Vijaysegaran S, Wee S, Yesmin F, Zacharopoulou A, Clarke AR (2015) Synonymization of key pest species within the *Bactrocera dorsalis* species complex (Diptera: Tephritidae): taxonomic changes based on a review of 20 years of integrative morphological, molecular, cytogenetic, behavioural and chemoecological data. Systemat Entomol 40:456–471. doi:10.1111/syen.12113

Silvestri F (1914a) Viaggio in Eritrea per cercare parassiti della mosca dell olive. Bollettino del Laboratorio di Zoologia Generale e Agraria della R. Scuola Superiore d'Agricoltura, Portici 9:186–226

Silvestri F (1914b) Report of an expedition to Africa in search of the natural enemies of fruit flies (Trypaneidae) with descriptions, observations and biological notes Territory of Hawaii Board of Agriculture and Forestry. Div Entomol Bull 3:1–146

Silvestri F (1915) 'Contributo alla Conoscenza degli Insetti dell'Olivo dell'Eritrea e dell' Africa Meridionale', Bollettino del Laboratorio di zoologia generale e agrarian. Portici 9:240–334

Silvestri F (1939) La lotta biologica contro le mosche dei frutti della famiglia Trypetidae. Verhandlungen der VII Internationaler Kongress der Entomologie (Berlin) 4:2396–2418

Sivinski JM (1996) The past and potential of biological control of fruit flies, pp 369–375. In: McPheron BA, Steck GJ (eds) Fruit fly pests: a world assessment of their biology and management. St. Lucie Press, Delray Beach, 586 p

Sivinski J, Aluja M (2012) The roles of parasitoid foraging for hosts, food and mates in the augmentative control of Tephritidae. Insects 3:668–691. doi:10.3390/insects3030668

Steck GJ, Gilstrap FE, Wharton RA, Hart WG (1986) Braconid parasitoids of Tephritidae (Diptera) infesting coffee and other fruits in west-central Africa. Entomophaga 31:59–67

Stibick JNL (2004) Natural enemies of true fruit flies (Tephritidae). United States Department of Agriculture APHIS. PPQ. 86 pp

Thompson WR (1943) A catalogue of the parasites and predators of insect pests. Parasites of the dermaptera and diptera. Imperial Agricultural Bureaux, Slough, 2:99

Thompson FC (1998) Fruit fly expert identification system and systematic information database, the international journal of the North American dipterists' society, vol 9. Backhuys Publishers, Leiden, p 524

van Achterberg C, Salvo A (1997) Reared Opiinae (Hymenoptera: Braconidae) from Argentina. Zool Med Leiden 71:189–214

van Noort S, Buffington M, Forshage M (2015) Afrotropical Cynipoidea (Hymenoptera). ZooKeys 493:1–176

van Zwaluwenburg RH (1937) West African notes. Hawaiian Plant Rec 41:57–83

Vargas RI, Leblanc L, Putoa R, Eitam A (2007) Impact of introduction of *Bactrocera dorsalis* (Diptera: Tephritidae) and classical biological control releases of *Fopius arisanus* (Hymenoptera: Braconidae) on economically important fruit flies in French Polynesia. J Econ Entomol 100:670–679

Vargas R, Badji K, Mckenney M, Leblanc L, Dieng EO (2016) Releases of *Fopius arisanus* against *Bactrocera dorsalis* (Hendel) in French Polynesia and Senegal. 3rd international symposium of TEAM (Working group on tephritids of Europe, Africa and the Middle East), 11–14 April 2016, Stellenbosch, South Africa

Vayssières J-F, Wharton RA, Delvare G, Sanogo F (2002) Diversity and pest control potential of hymenopteran parasitoids of *Ceratitis* spp. on mangos in Mali. In: Proceedings of 6th international fruit fly symposium, Stellenbosch, South Africa pp 461–464

Vayssières J-F, Wharton R, Adandonon A, Sinzogan A (2011) Preliminary inventory of parasitoids associated with fruit flies in mangoes, guavas, cashew, pepper and wild fruit crops in Benin. Biocontrol 56:35–43

Vayssières J-F, Adandonon A, N'Diaye O, Sinzogan A, Kooymann C, Badji K, Rey J-Y, Wharton RA (2012) Native parasitoids associated with fruit flies (Diptera: Tephritidae) in cultivated and wild fruit crops in Casamance, Senegal. Afr Entomol 20:308–315

Wang X-G, Messing RH (2004a) The ectoparasitic pupal parasitoid, *Pachycrepoideus vindemmiae* (Hymenoptera: Pteromalidae), attacks other primary tephritid fruit fly parasitoids: host expansion and potential non-target impact. Biol Contr 31:227–236

Wang X-G, Messing RH (2004b) Potential interactions between pupal and egg- or larval-pupal parasitoids of tephritid fruit flies. Environ Entomol 33:1313–1320

Wang XG, Bokonon-Ganta AH, Ramadan MM, Messing RH (2004) Egg-larval opiine parasitoids (Hym., Braconidae) of tephritid fruit fly pests do not attack the flowerhead-feeder Trupanea dubautiae (Dip., Tephritidae). J Appl Entomol 128:716–722

Waterhouse DF (1993) Biological control: pacific prospects – Supplement 2. ACIAR Monogr 20:138

Wharton RA (1983) Variation in *Opius hirtus* Fischer and discussion of *Desmiostoma* Foerster (Hymenoptera, Braconidae). Proc Entomol Soc Washington 85:327–330

Wharton RA (1987) Changes in nomenclature and classification of some opiine Braconidae (Hymenoptera). Proc Entomol Soc Washington 89:61–73

Wharton RA (1989) Classical biological control of fruit-infesting Tephritidae. In: Robinson AS, Hooper G (eds) Fruit flies, their biology, natural enemies and control, vol 3B. Elsevier, Amsterdam

Wharton RA (1999a) A review of the old world genus *Fopius*, with description of two new species reared from fruit-infesting Tephritidae. J Hymen Res 8:48–64

Wharton RA (1999b) The status of two species of *Psyttalia* (Hymenoptera: Braconidae: Opiinae) reared from fruit-infesting Tephritidae (Diptera) on the Indian Ocean Island of Réunion and Mauritius. Afric Entomol 7:85–90

Wharton RA (2006) The species of *Sternaulopius* Fischer (Hymenoptera: Braconidae, Opiinae) and the braconid sternaulus. J Hymen Res 15:317–347

Wharton RA (2009) Two new species of *Psyttalia* Walker (Hymenoptera, Braconidae, Opiinae) reared from fruit-infesting tephritid (Diptera) hosts in Kenya. ZooKeys 20:349–377

Wharton RA, Gilstrap FE (1983) Key to and status of opiine braconid (Hymenoptera) parasitoids used in biological control of *Ceratitis* and *Dacus* s. l. (Diptera: Tephritidae). Anns Entomol Soc Am 76:721–742

Wharton RA, Trostle MK, Messing RH, Copeland RS, Kimani-Njogu SW, Lux S, Overholt WA, Mohamed S, Sivinski J (2000) Parasitoids of medfly, *Ceratitis capitata,* and related tephritids in Kenyan coffee: a predominantly koinobiont assemblage. Bull Entomol Res 90:517–526

White IM, Elson-Harris MM (1992) Fruit flies of economic significance: their identification and bionomics. International Institute of Entomology, London, xii+601 p

Willard HF (1920) *Opius fletcheri* as a parasite of the melon fly in Hawaii. J Agr Res 20:423–438

Willard HF, Mason AC (1937) Parasitization of the Mediterranean fruit fly in Hawaii, 1914–33. US Dep Agric Circ 439:1–17

Wong TTY, Ramadan MM (1987) Parasitization of the Mediterranean and oriental fruit flies (Diptera: Tephritidae) in the Kula area of Maui, Hawaii. J Econ Entomol 80:77–80

Yoder MJ, Wharton RA (2002) Nomenclature of African Psilini (Hymenoptera: Diapriidae) and status of *Coptera robustior*, a parasitoid of Mediterranean fruit fly (Diptera, Tephrididae). Can Entomol 134:561–576

Samira A. Mohamed works in the African fruit fly programme at the International Centre of Insect Physiology and Ecology (*icipe*), Nairobi, Kenya and is a senior scientist and co-ordinator of the German BMZ funded *Tuta* IPM project. She has over 20 years of research experience focusing on the development of IPM strategies for suppression of horticultural crop pests. Her main area of interest is in classical biological control of invasive pests within the context of IPM. Samira was instrumental in importing in to Africa and later releasing two-efficient co-evolved parasitoid species (*Fopius arisanus* and *Diachasmimorpha longicaudata*) from Hawaii. So far, she has authored or co-authored over 40 peer-reviewed publications and several book chapters.

M.M. Ramadan is an exploratory entomologist for the State of Hawaii Department of Agriculture, Plant Pest Control Branch, USA. He is responsible for planning and conducting explorations in foreign countries for natural enemies of insect pests and weeds targeted by the branch for suppression. Dr Ramadan earned his Doctorate and Masters degrees in entomology from the University of Hawaii at Manoa, and a Bachelor degree in entomology from Alexandria University, Egypt. He has conducted research and published extensively in areas related to classical biological control, insect behaviour, biology, evaluation of biocontrol agents and mass rearing of parasitoids for augmentative biological control programmes.

Sunday Ekesi is an Entomologist at the International Centre of Insect Physiology and Ecology (*icipe*), Nairobi, Kenya. He heads the Plant Health Theme at *icipe* and is a member of the Senior management team. Sunday is a professional scientist, research leader and manager with extensive knowledge and experience in sustainable agriculture (microbial control, biological control, habitat management/conservation, managing pesticide use, IPM) and biodiversity in Africa and internationally. Sunday has been leading a continent-wide initiative to control African fruit flies that threaten production and export of fruits and vegetables. The initiative is being done in close collaboration with IITA, University of Bremen, Max Planck Institute for Chemical Ecology together with NARS, private sectors and ARI partners in Africa, Asia, Europe and the USA and focuses on the development of an IPM strategy that encompasses baiting and male annihilation techniques, classical biological control, use of biopesticides, ant technology, field sanitation and postharvest treatment for quarantine fruit flies. The aim is to develop a cost effective and sustainable technology for control of fruit flies on the African continent that is compliant with standards for export markets while also meeting the requirements of domestic urban markets. Sunday has broad perspectives on global agricultural research and development issues, with first-hand experience of the challenges and opportunities in working with smallholder farmers, extension agents, research organizations and the private sector to improve food and nutritional security. He sits on various international advisory and consultancy panels for the FAO, IAEA, WB and regional and national projects on fruit fly, arthropod pests and climate change-related issues. Sunday is a Fellow of the African Academy of Sciences (FAAS).

Chapter 17
From Behavioural Studies to Field Application: Improving Biological Control Strategies by Integrating Laboratory Results into Field Experiments

Katharina Merkel, Valentina Migani, Sunday Ekesi, and Thomas S. Hoffmeister

Abstract Biological control and integrated pest management (IPM) of tephritid fruit flies has repeatedly made use of parasitoids as natural enemies to suppress fly populations. Parasitoids, however, are selected to maximize their individual reproductive success, and thus do not necessarily maximize pest suppression at the population level. Furthermore, more than one parasitoid species within a pest-natural enemy assemblage might be available as a potential control agent. This calls for a thorough understanding of behavioural processes in pest-natural enemy interactions to select the best single species, or multiple species, to achieve pest suppression at the population level. We make a case for the importance of laboratory studies in informing field application, while acknowledging that they cannot replicate all the complexity present within an ecological community. Thus, there is still the need for integrating laboratory-based research with field application. Therefore, manipulative field studies are needed to determine whether the insights from laboratory results hold true in a more complex system. We describe how laboratory results can

---

K. Merkel (✉) • V. Migani
FB2 Biology/Chemistry, Population and Evolutionary Ecology Group, University of Bremen, Leobener Str. NW2, D-28359 Bremen, Germany

International Centre of Insect Physiology & Ecology (*icipe*),
PO Box 30772-00100, Nairobi, Kenya

School of Earth, Environmental & Biological Sciences,
Queensland University of Technology, Gardens Point, Brisbane, QLD 4001, Australia
e-mail: katharina.merkel@qut.edu.au

S. Ekesi
Plant Health Theme, International Centre of Insect Physiology & Ecology (*icipe*),
PO Box 30772-00100, Nairobi, Kenya

T.S. Hoffmeister
FB2 Biology/Chemistry, Population and Evolutionary Ecology Group, University of Bremen, Leobener Str. NW2, D-28359 Bremen, Germany

be used to inform field studies and predict the likelihood of success of natural enemy releases. We use the example of the natural enemy-*Bactrocera dorsalis* system in Africa. We show that behavioural ecology offers a powerful tool to understand species interactions on the basis of individual decisions. We further discuss how these findings can be exploited in an agricultural context to improve the control effort. Finally, we describe the need for comprehensive field studies based on the behavioural observations made in the laboratory.

Keywords Invasion biology • Intra-specific interaction • Inter-specific interaction • *Bactrocera dorsalis* • *Fopius arisanus*

1 Introduction

In the field of ecology, theory and practice are often closely connected (Murdoch 1994; Doak and Mills 1994; Jervis 2005). Ecological research aims at explaining the complexity of natural systems by investigating at the level of its individual components. Investigations on individual organisms/species that compose natural communities can uncover the mechanisms that structure these communities and facilitate the development of adequate theory to make predictions, which may help to solve applied problems (Beddington et al. 1978). Species that impact human economy have been given special focus and studies on biological traits of agricultural pests and beneficial organisms are widely published (e.g. Waage and Hassell 1982). Many studies aim to understand the population dynamics of economically important species, including the influence of varying biotic and abiotic conditions that impact on population growth and spread (Symondson et al. 2002; Kausrud et al. 2012). Therefore, it is not surprising that linkages exist between biological control and the ecological theory of several disciplines including population dynamics (Murdoch et al. 1985; May and Hassell 1988), invasion biology (Fagan et al. 2002) and coevolution (Roitberg 2000). Applied ecology may raise research questions that can be answered using manipulative experiments, the results of which can contribute to theory development (Kareiva 1996). Furthermore, many theoretical approaches in ecology can help solve applied problems. In fact, population dynamics and the underlying mechanisms for it are not only of interest to population ecologists and theoreticians, but make an important contribution to addressing applied problems, such as increasing the efficacy of pest management and biological control strategies (Kidd and Jervis 2005) (Fig. 17.1).

Biological control is based upon the fact that the population growth of pest species is frequently limited by their natural enemies such as predators, parasitoids and pathogens. Biological control can be classified in to three major types, i.e. classical, augmentative and conservation biological control (see Box 17.1). The issues addressed hereafter have particular relevance to classical biological control but can be in part transferred to augmentative and conservation biological control.

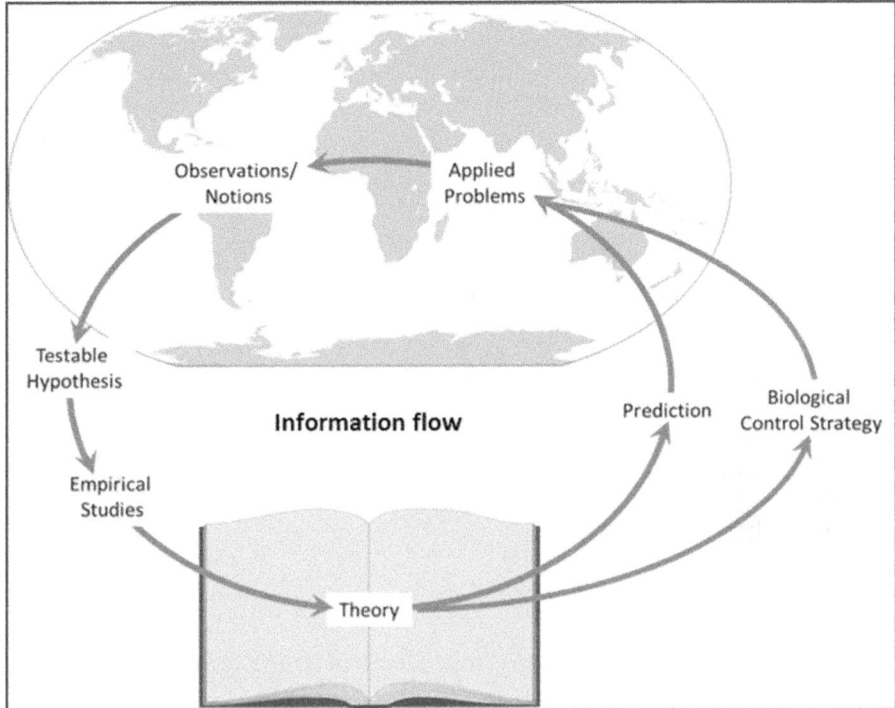

Fig. 17.1 Information flow between ecological theory and applied problems

Box 17.1. Biological Control
Biological (pest) control refers to all human activities that involve the use of organisms to reduce pest populations and/or maintain pest populations at low densities.

This practice is considered environmentally friendly, as it exploits natural mechanisms, such as predation and diseases, in order to reduce damage caused by agricultural pests. Biological control can be classified in to different categories according to the methodologies used for the practice. The most important and well known are (i) classical, (ii) augmentative and (iii) conservation biological control.

The term **classical** (or "importation") **biological control** is mostly used to describe the introduction of exotic natural enemies into target areas to control exotic pests. The success of this strategy relies, amongst others, on the ability of the introduced natural enemy to establish in the new environment. Once a biological control agent establishes any impact on pest populations can be permanent, potentially maintaining pest populations at low levels. While the

(continued)

permanent effect makes this strategy economically friendly, this advantage can at the same time become a disadvantage as any negative effects of the introduction are mostly irreversible.

Augmentative biological control utilizes one or several releases of natural enemies (either introduced or native) to increase the population of the enemy in order to achieve a sufficient level of pest control. In contrast to classical biological control, augmentative releases do not aim to establish a population of the released enemy. It can employ inoculative (seasonal) or inundative (mass) releases. While classical biological control may only be effective after several generations post establishment, augmentative biological control acts on a shorter time scale.

Conservation biological control takes advantage of the natural enemies already living in the environment in order to control and suppress a target pest. Thus, the environment is modified to increase the effectiveness of the already existing natural enemies, making it a cost-effective strategy.

1.1 A Short History of Biological Control and Its Risks

The use of natural enemies to control organisms detrimental to agriculture dates back to ancient times. The first reports originate from 304 A.D. when the Chinese used the predatory ant *Oecophylla smaragdina* (Fabricius) to control pests of citrus (reviewed in Huang and Yang 1987). However, biological control did not receive much attention until much later when the vedalia beetle, *Rodolia cardinalis* (Mulsant), was successfully introduced into California for classical biological control of the invasive cottony cushion scale, *Icerya purchasi* Maskell (Caltagirone 1981; Howarth 1983, 1991). Encouraged by this success and the popular assumption at the time that the introduction of exotic natural enemies to control exotic invasive pests was entirely 'environmentally safe' followed a period of trial and error in classical biological control attempts (De Clercq et al. 2011). The active exchange of potential natural enemies across the globe resulted in the worldwide release of thousands of arthropod species as control agents for agricultural pest insects (van Lenteren et al. 2006a). However, despite the fact that these natural enemies often proved to be environmentally safe and very powerful pest control agents, there were cases in which the control agent itself became a pest. Some species intentionally released for beneficial purposes established with undesirable negative effects on the indigenous ecosystem (Simberloff and Stiling 1996; De Clercq et al. 2011; Simberloff 2012). This was due to a lack of behavioural and ecological knowledge about the species concerned and/or ignorance of the potentially deleterious effects of releasing polyphagous control agents (e.g. many predators) rather than highly specialized control agents (e.g. many parasitoids). For example, the cane toad, *Rhinella marina* (Linnaeus), was introduced into Queensland, Northern Australia as a control agent of the greyback cane beetle, *Dermolepida albohirtum* (Waterhouse) and French's canegrub, *Lepidiota frenchi* Blackburn; however, it is now considered a pest as, due to its toxicity, it endangers native Australian species that feed on frogs (Lampo and De Leo 1998; Griffiths and McKay 2007; Doody et al. 2009). The fallacy that release of exotic natural enemies could have no harmful effects has led to many careless introductions (Howarth 1991), particularly during the early stages of classical biological control development. There are also examples highlighting the fact that, even with a good understanding of the ecology of a species, it is not always possible to predict all possible side effects that could result from their introduction. For example, when the parasitoid, *Diachasmimorpha tryoni* (Cameron), was introduced into Hawaii in 1913 to control populations of the Mediterranean fruit fly (medfly), *Ceratitis capitata* (Wiedemann) (Pemberton 1964), it could not have been predicted that, many years later, it would undergo a host shift and begin attacking larvae of the lantana gall fly, *Eutreta xanthochaeta* Aldrich (Bess and Haramoto 1972). Although, *C. capitata* and *E. xanthochaeta* belong to the same family, they inhabit very different ecological niches (Duan et al. 2000). Therefore, any introduction of exotic species, accidentally or intentionally, may have consequences on the indigenous ecosystem and thus classical biological control, where an exotic enemy is released, cannot be considered free of risk. For this reason, classical biological control agents should be selected to ensure that they pose a minimal risk to the existing ecosystem, while simultaneously having a high potential to suppress the targeted pest population. Trust in the safety of exotic biological control agents has suffered from past mistakes (Messing and Wright 2006) and success in the future greatly depends on a shift to knowledge-based introductions. Research questions should be particularly geared to the characteristics of the agent as well as the affected ecosystem (Louda et al. 2003). Pest control efforts relying on the use of exotic species in new environments must be carefully planned and based on a detailed knowledge of both potential efficiency and possible side effects (van Lenteren et al. 2006a).

1.2 Invasion Biology and Classical Biological Control

Reviewing classical biological control shows how pest management strategies can be improved by the integration of ecological knowledge. Since the potential impact of alien species has been recognized, modern scientific research has tried to reveal the mechanisms of biological invasions (Kenis et al. 2009). Several life-history traits, such as phenotypic plasticity, favour future reproductive success and have been associated with successful invaders (Chown et al. 2007; Davidson et al. 2011; Sol et al. 2012). Research on invasion biology includes many studies of former biological control agents, because the spread of organisms for biological control can be better traced than accidental introductions. Fagan et al. (2002) reviewed various aspects of invasion biology to identify attributes that would inform the practice of classical biological control. Hastings (2000) used records of parasitoid spread to verify the usefulness of using simplified models to predict the rate of spread. The term 'planned invasion' has been used for biological control (van Driesche 2012), clearly indicating that species invasion and species introduction are just different perspectives of the same biological processes. One challenge for classical biological control projects is to identify species that establish well but do not spread beyond the target region, thereby limiting effects on non-target populations. Both establishment and spread are closely connected and thus, to achieve one but not the other, requires a thorough understanding of the mechanisms that underpin dispersal. While an enemy with a very low dispersal rate may not be able to keep up with the pest population, an enemy with a very high dispersal rate may not establish or it may migrate from the target area (Fagan et al. 2002). Behavioural experiments can help to elucidate the mechanisms that drive natural enemy dispersal.

1.3 The Influence of Intra- and Interspecific Interactions on Foraging Decisions

One important aspect of the foraging decisions that herbivores and their natural enemies make is the decision on when to leave a resource patch currently being exploited and disperse to other potential patches. The decision, how much time to allocate towards individual patches and when to leave may be influenced by the presence of other individuals, either conspecifics or heterospecifics. Competition for shared resources can result in either the reduction or prolongation of the time an individual is willing to spend foraging within a patch. The parasitoid *Fopius arisanus* (Sonan), for example, spends less time foraging in previously exploited patches (Wang and Messing 2003). Competition may also play a role for the pest species as shown in the example of *C. capitata* females, where females spend less time in trees bearing infested fruits than in trees bearing uninfested fruits (Papaj et al. 1989). While patch-leaving decisions are often studied on a small scale, such as within-tree dispersal, population ecologists have also identified intra- and interspecific competition as two of the major reasons that cause an organism to disperse on the landscape scale (Poethke and Hovestadt 2002; Mashanova et al. 2008).

1.4 The Role of Multi-trophic Interactions in Biological Control Systems

When generalist and specialist predators or parasitoids are present within the same trophic level of a foodweb, they usually interfere with each other through apparent competition (Holt 1977) or intraguild predation (Polis and Holt 1992). One or both of these two phenomena often occur within guilds of biological control agents and their effects can differ according to the system. For example, Snyder et al. (2006) showed that increased predator diversity was beneficial for the control of two phloem-feeding aphids on collards. In another study, biological control of aphids infesting cereals benefited from a combination of predators and parasitoids: while the parasitoids attacked aphids present in the upper parts of the wheat, the predators fed on those aphids infesting the lower parts. Moreover, ground dwelling predators could feed on aphids that dropped to the ground in order to escape from parasitoid attack (Schmidt et al. 2003). While in some cases the use of a combination of different natural enemies improves pest suppression, in other cases a single biological control agent is more effective than a combination of different species (Messelink et al. 2012). An example of the latter has been reported for control of the indian mealmoth, *Plodia interpunctella* Hübner; integration of both an anthocorid predatory bug, *Xylocoris flavipes* (Reuters) and the ectoparasitoid *Bracon hebetor* Say was unsuccessful because the anthocorid bug fed on larvae of the ectoparasitoid, reducing its effectiveness (Press et al. 1974). Another study on an aphid-parasitoid-generalist predator community revealed that the predators involved in the system were mainly feeding on the adult parasitoids used to control the aphid population, rather than on the aphids themselves (Traugott et al. 2012). Likewise, the presence of a generalist predator, *Pterostichus melanarius* (Illiger) disrupted the effectiveness of the aphid parasitoid, *Aphidius ervi* (Haliday), against aphid pests on alfalfa plants (Snyder and Ives 2001). By manipulating the densities of all three species, Snyder and Ives (2001) showed that the aphid population increased in the presence of a generalist carabid predator; the carabid was actually predating not only on the adult parasitoids, but also parasitoid pupae, thus affecting more than one life stage of the parasitoid.

While we have discussed direct trophic effects, where a predator directly consumes the control agent, there are also cases in which the mere presence of the predator may have a negative impact on the density of other natural enemies, even if consumption does not occur (Peacor and Werner 1997; Lima 1998). Predator presence may, in fact, induce other natural enemies to adopt anti-predator strategies in order to avoid or reduce predation risk (Preisser et al. 2005). Theory and experimental data suggest that responses to potential intraguild predation should be dynamic, based on the cost of potential predator encounters and the benefits a forager might currently accrue from the patch (Roitberg et al. 2010). These so called 'trait-mediated interactions' can be as influential as direct consumption of other natural enemies for herbivore population regulation. For example, *Pauesia silvestris* Stary, a specialized endoparasitoid of conifer aphids in the genus *Cinara*, alters its foraging strategies in the presence of *Formica polyctena* Förster workers. While this parasitoid prefers foraging on *C. pinea* on the bark of *Pinus silvestris* it switches to foraging on needles of the trees, parasitizing a less preferred host, *C. pini*, in order to reduce its own risk of predation by *Formica* workers patrolling the bark. Host switching behaviour leads to a fitness reduction in *P. silvestris* when using *C. pini* as a host (Völkl and Kroupa 1997). In some cases, predator presence is mediated by chemical cues that can be detected by any potential prey (including other natural enemies), and influences prey choice and behaviour. Even though predation does not involve consumption here, the changes induced are costly in terms of reproductive success and fitness of the parasitoids (Gonthier 2012).

Besides direct and indirect interactions of parasitoids and predators at the same trophic level, indirect multitrophic interactions across trophic levels may influence the foraging behaviour of parasitoids. In the same way that herbivores may demonstrate dispersal behaviour in response to parasitoid cues (Kunert et al. 2005), Höller and colleagues (1994) have suggested that cues of secondary parasitoids lead to dispersal in primary aphid parasitoids.

1.5 A Plea for Behavioural Ecological Lab Studies

As discussed above, multiple natural enemy assemblages including specialists and generalists may, in some cases, lead to lower pest densities compared with situations where a single species of natural enemy is used (Vance-Chalcraft et al. 2007; Messelink et al. 2012). Thus, the complexity of interactions within ecological communities must be considered in order to develop effective and successful biological control strategies (Weisser 2003; Snyder et al. 2006; Straub and Snyder 2006). However, as the complexity of interactions within community assemblages can be difficult to elucidate under field conditions, we argue for the use of laboratory-based experiments under controlled conditions in order to understand the forces that structure pest – natural enemy communities, particularly when considering the release of an exotic natural enemy species for classical biological control. In this respect, behavioural ecology offers a powerful investigative approach. Behavioural ecology takes into account the dependence of reaction norms to the information status of the candidate species involved, as well as their physiological state. It helps us to understand what biological control agents could achieve with respect to pest population suppression and why this is not always achieved in the field; this may help to develop theoretical models that can be used as predictive tools to solve applied problems.

2 Biological Control of *Bactrocera dorsalis* in Africa

2.1 The Invasion of **Bactrocera dorsalis** *into Africa*

The arrival and spread of the oriental fruit fly, *Bactrocera dorsalis* (Hendel), in sub-Saharan Africa is a very good example of an invasive species to which the principles mentioned above can be applied. The first record of *B. dorsalis* in Africa dates back to 2003 when it was first described from the Kenyan coast, as *Bactrocera invadens* Drew, Tsuruta & White (Drew et al. 2005; Schutze et al. 2015). *Bactrocera dorsalis* belongs to a problematic subgroup of highly polyphagous fruit fly pests in the family Tephritidae (Christenson and Foote 1960). Tephritid fruit flies have been accidently introduced into many parts of the world where they have become invasive and resulted in severe economic problems (Duyck et al. 2004). The rapid spread of *B. dorsalis* across the African continent increased interest in this species and the fear of further unintentional spread. The successful establishment of *B. dorsalis* and its ability to invade new environments can be partially linked to its extreme polyphagy (Fletcher 1987); it has been found developing in numerous host plant species from many different families. Another aspect that contributes to the pest status of *B. dorsalis* is its high lifetime fecundity (Fletcher 1987), leading to rapidly growing populations and hence high infestation levels on crops. Considering the characteristics of this species it is not surprising that the detection of *B. dorsalis* in Africa resulted in an immediate call for measures to limit crop damage. Since chemical control is both economically unfeasible for many producers and can be environmentally unsafe, the integration of biological control strategies into sustainable management plans is desirable. However, given the biological traits of *B. dorsalis*, control of this species by natural enemies presents an ambitious task: (1) its extreme polyphagy calls for control agents that are able to control the pest in a range of different environments; (2) its high fecundity leads to large egg batches in both mature and immature fruits (Migani et al. 2014), requiring efficient agents that are able to negatively impact the growth rate of the fly population and; (3) the resistance of *B. dorsalis* to native parasitoid species attacking African species of *Ceratitis* fruit flies (Mohamed et al. 2006; Vayssieres et al. 2011) demands the use of a classical biological control approach and the introduction of exotic parasitoid species that have the ability to overcome host immunity.

2.2 Classical Biological Control: The Exotic Parasitoid **Fopius arisanus**

The success of the egg-prepupal parasitoid *Fopius arisanus* (Sonan) as a classical biological control agent of *B. dorsalis* and *C. capitata* in Hawaii (Haramoto and Bess 1970) made it the most promising candidate for biological control programmes in Africa. In Hawaii *F. arisanus* not only established itself after introduction, but also became the most abundant parasitoid species and was presumed to be maintaining the local fly populations at low levels (Haramoto and Bess 1970; Vargas et al. 2007). Although the main use of *F. arisanus* was for control of *B. dorsalis*, its host range is much broader. *Fopius arisanus* has been reported parasitizing several species in the genus *Bactrocera*, and it also attacks *C. capitata* and *Anastrepha ludens* (Loew) (Rousse et al. 2005). However, *F. arisanus* has not been reported attacking any non-frugivorous tephritids, so far (Duan and Messing 1997; Rousse et al. 2005). While the risk of unwanted non-target effects on potential hosts of *F. arisanus* could be considered as small, negative effects on other parasitoid species due to interspecific competition may pose a higher risk. Being an egg-prepupal parasitoid *F. arisanus* may outcompete other parasitoid species that attack the same host (Wang and Messing 2002; Wang et al. 2003). For this reason, we argue that releases in new areas should always be preceded by thorough laboratory investigations on possible non-target effects, particularly host range studies (van Lenteren et al. 2006b).

The first pre-release evaluation should identify the species that are most likely to be negatively affected (van Lenteren et al. 2006a), and determination of the processes that cause negative impacts, e.g. species that could be potential hosts or species that could be affected by competition. Furthermore, when conducting experiments to estimate non-target effects researchers should be aware of the associated technical difficulties, i.e., while demonstrating a negative effect might be relatively easy, it is difficult to categorically say that there is no negative effect (Hoffmeister et al. 2006). Establishing that a potential control agent only poses a low risk to the ecosystem in the target area is not the end of the evaluation. The next step should be to estimate the effectiveness of the potential control agent, which requires a demonstration of successful establishment followed by a significant level of pest suppression. In the case of *F. arisanus*, there was success in some control programmes while in others it failed to establish or only caused low parasitism rates in the targeted pest species (Rousse et al. 2005). This may partly be explained by the fact that *F. arisanus* has been introduced into a range of different habitats that differed in both the composition of plants available as food sources for tephritid flies and the fruit fly species that were present. While *F. arisanus* was effective in suppressing populations of *B. dorsalis* and *C. capitata* (Haramoto and Bess 1970), its impact on populations of *Bactrocera kirki* (Froggatt); the Queensland fruit fly, *Bactrocera tryoni* (Froggatt); and the Pacific fruit fly, *Bactrocera xanthodes* (Broun), were minimal (Quimio and Walter 2001; Purcell 1998). Secondly, the parasitism rates achieved by *F. arisanus* infesting fruit flies on different fruiting plants varied greatly irrespective of the fruit fly species present. In 2009 (7 years after its introduction into Tahiti) *F. arisanus* achieved more than 50 % parasitism of fruit flies on guava (*Psidium guajava* L.), Polynesian chestnut (*Inocarpus fagifer* (Parkinson)) and tropical almond (*Terminalia catappa* L.). In contrast, the rates of parasitism in mango (*Mangifera indica* L.) remained below 30 % (Vargas et al. 2012). By comparing the different systems in which *F. arisanus* was introduced in the past suggests that the parasitoid has great potential for suppressing fly populations but the unsuccessful attempts also highlight that it is not yet possible to predict, with certainty, which target regions and ecosystems are most likely to support establishment of the parasitoid. The complexity of ecosystems seems to make identification of the causes that lead to an unsuccessful introduction impossible to determine from post-release observations. Therefore, we stress the importance of manipulative small-scale experiments to identify the processes that impact on establishment and effectiveness, as an important foundation for larger-scale studies.

In any classical biological control campaign against *B. dorsalis* it is important to identify target areas for release that are most likely to support parasitoid establishment and reproduction. One might assume that parasitoid releases in areas of high fruit fly infestation should be more successful than releases into areas of low fruit fly infestation, due to the greater host availability for parasitoid reproduction, i.e. the more hosts present, the higher the chances that parasitoid females will reproduce successfully. However, high infestation levels may also correspond to high host aggregation. For prey species aggregation can be advantageous as it dilutes the risk of attack by predators (Hamilton 1971) or parasitoids in cases of inverse density dependence (Lessells 1985). Thus, increased host densities may offer a 'refuge' for the pest to escape parasitism. Clearly there is only one way to determine what actually happens and that is to conduct behavioural assays in the laboratory to measure parasitism at different host densities. Manipulating the density of fruit fly eggs within the fruit and observing the rules used by female parasitoids to attack fly eggs as a function of fly egg density, and by analysing the rate of parasitism, we found evidence for an inverse density-dependent functional response in *F. arisanus* whereby flies benefitted from a density dependent refuge (Merkel, Ekesi, Migani, Hoffmeister, unpublished data). As a consequence, it may be better to release the parasitoid in areas of low fly density or, in areas of high fly density, try reducing fly densities prior to release, arguing that *F. arisanus* might be more effective at these densities.

As a next logical step there should be consideration for the release strategy to be adopted in order to increase the chances of parasitoid population establishment. Commonly parasitoids are mass released from a single release point. The theory behind this was that by inundating the target area with parasitoids, the chances of establishment would increase. However, point mass releases create high parasitoid densities within a small area, and this leads to parasitoids competing for host resources. If there are no investigations into parasitoid behaviour under competition, it is impossible to estimate how competitive interactions between females might influence establishment and success of a biological control agent in the release area. Parasitoids may compete directly or indirectly for the available resources resulting in decreased parasitism efficiency, i.e. a reduction in the per capita parasitism rate. Laboratory studies on the effect of varying densities of *F. arisanus* suggest that, at high densities, females suffer from severe mutual interference especially when the host distribution is aggregated (Merkel, Ekesi, Migani, Hoffmeister, unpublished data). Behavioural observations further suggest that the high level of interference measured was, at least partly, caused by an increased tendency for females to leave the patch. This implies that point mass releases, particularly in environments with highly aggregated host populations, leads to a high dispersal rate, which may in turn reduce the agent's density in the target region below the minimum density necessary for establishment. While locally the reduced efficiency of parasitoids due to interference has a negative impact on the suppression of the pest population, dispersal may provide suppression of pest population beyond the release area. In the case of *B. dorsalis* the latter may be desired as it may encourage the parasitoid to follow the fruit fly populations as they spread. The dispersal behaviour of control agents has been related to their successful establishment and to their ability to follow the pest populations. Heimpel and Asplen (2011) proposed that an intermediate level of dispersal is optimal for successful biological control. Thus, we suggest that biological control practitioners should consider releasing intermediate numbers of parasitoids at several locations in order to maximize the probability of local establishment. Behavioural observations of parasitoids further indicated that the decision to stay or to leave a host patch when encountering conspecifics, was dependent on the previous investment in that patch. This mechanism may ensure that only part of the parasitoid population disperses while the other remains. Thus, future studies should address the effect of experience on the dispersal behaviour of *F. arisanus*.

2.3 Weaver Ants as Biological Control Agents

Given the discussion above, we may conclude that *F. arisanus* alone may not achieve sufficient control of *B. dorsalis* in Africa and hence additional management approaches must be considered to develop a strong IPM strategy. Despite the use of bait sprays or pesticides, the suppression of fruit fly populations within mango orchards could be enhanced by augmentation of the African weaver ant, *Oecophylla longinoda* (Latreille). *Oecophylla longinoda* are known to reduce fruit fly infestation levels in mango orchards (van Mele et al. 2007; Sinzogan et al. 2008). Moreover, the cues from this ant species deter the flies from ovipositing on the mango fruit (Adandonon et al. 2009; van Mele et al. 2009). Thus, *O. longinoda* seem a suitable potential biological control agent against fruit flies on mango. However, we have to consider that, as generalist predators of other insects (Hölldobler and Wilson 1990; Dejean 1990), *O. longinoda* might potentially interfere with specialist fruit fly parasitoids via intraguild predation. Experiments with *O. longinoda* and *F. arisanus* in screen houses showed a decrease in parasitization rates of *B. dorsalis* by *F. arisanus* compared with when *O. longinoda* was absent (Migani, Ekesi, Merkel, Hoffmeister, unpublished data). Thus, we need further research to determine whether *F. arisanus* and *O. longinoda* can be used together for successful fruit fly control. Initially we need to determine whether the effects of *O. longinoda* on *F. arisanus* are mostly direct or indirect. Behavioural bioassays in which foraging female *F. arisanus* were observed in the presence of *O. longinoda* suggested that the ants were interfering with parasitoid foraging; the ant workers were, in fact, chasing the parasitoids away from the fruit rather than killing them (Migani, Ekesi, Merkel, Hoffmeister, unpublished data). It is known that prey surviving predator attacks may learn to avoid predation, adopting changes in behaviour and defence strategies (Völkl 2001). Thus behavioural responses may be induced in the parasitoids to increase predator avoidance.

However, in order to decide whether *O. longinoda* and parasitoids could be compatible, the relative effect of *O. longinoda* on *F. arisanus* and on *B. dorsalis* must be considered. If intra-guild predation occurs, there is evidence suggesting that, within terrestrial ecosystems, a single control agent will be more effective than a combination of multiple control agents (Rosenheim et al. 1995; Snyder and Ives 2001; Vance-Chalcraft et al. 2007). However, overall biological control effectiveness can also be greater through implementation of multiple natural enemies (Messelink et al. 2012). Thus, behavioural bioassays are needed to determine the relative degree by which *O. longinoda* interferes with both the pest and the parasitoid. By looking at the effect of *O. longinoda* on patch leaving probability of *B. dorsalis* and *F. arisa-*

nus, respectively, we found that the ants had a greater impact on *F. arisanus* than on the fruit flies, making the ants a potentially disruptive agent for parasitoid pest control effectiveness (Migani, Ekesi, Merkel, Hoffmeister, unpublished data).

In summary, we have shown how behavioural and laboratory-based bioassays could help to identify aspects of pest-natural enemy interactions that could be important for the effectiveness of biological control. Furthermore, investigating and designing experiments that include multiple-species assemblages of natural enemies are crucial to understand whether interactions between natural enemies could interfere with pest suppression. This knowledge should be used to significantly improve the strategies adopted when planning successful biological control programmes.

3 Conclusions

Using concepts of behavioural ecology as an approach for studies on biological control at the laboratory and greenhouse level offers insights into the processes that may be important drivers of natural enemy – *B. dorsalis* interactions in the field. Based on what we have learned in this system so far we believe that the knowledge gained may guide further important studies at the population level under semi-field conditions and post release in the field.

On the basis of our own studies and earlier studies we can identify some important questions that, as yet, remain unanswered in the mango – *B. dorsalis* – *F. arisanus* – *O. longinoda* system. Firstly, our results on parasitoid interference suggest that it is necessary to test different spatial release strategies under field conditions. Mass point releases may lead to strong interference and, from what is known in other systems, this may in turn lead to parasitoid dispersal rather than local establishment. The alternative would be to make releases at numerous points in an orchard leading to an even distribution of parasitoids across the crop. Clearly, field experiments analysing post-release parasitism rates in relation to release strategy would be needed to confirm this hypothesis. Secondly, we found inverse density dependent parasitism by *F. arisanus* due to a refuge effect caused by aggregated fly eggs. Two angles of research are needed to better understand the system and allow manipulation. Quantifying control effectiveness at the population level in the field when flies occur at different densities at the time of parasitoid release is needed. It seems intuitive to assume that the parasitoid might be effective at low densities, but there may be a switch point after which control cannot be achieved due to flies profiting from the refuge. Given that this is true, other control strategies within an IPM approach would be necessary to reduce the fly populations to moderate densities prior to parasitoid release. Studies are needed to determine whether *O. longinoda*, with their harassment of ovipositing flies, increase the usage of previous oviposition sites by flies in their attempts to escape harassment by ants, in turn further increasing the aggregation effect on fly eggs that leads to the refuge against *F. arisanus*. If this is the case, then the fly density at which *F. arisanus* might provide successful control would need to be even lower. Further, the combined use of *F. arisanus* and *O. longinoda* needs to be analysed under semi-field and field conditions. At this point our results suggest that their combined use might be of limited effectiveness because the presence of the ants reduced the time that *F. arisanus* females spent on the fruit, which in turn resulted in a reduced rate of parasitism. Subsequent experiments should clarify whether, under more natural conditions, parasitoid efficiency against flies might be greater.

To conclude, behavioural ecology offers useful tools to highlight important ecological and behavioural aspects that should be considered to better understand the pest-natural enemy system and to plan and design experiments at the field level. The knowledge gained from these experiments may in turn benefit pest management practitioners by improving the success of biological control programmes.

Acknowledgements The chapter has benefited greatly from the intellectual support of all team members of the African Fruit Fly Programme at *icipe*, Nairobi and the Population and Evolutionary Ecology Group at the University of Bremen. This study was financially supported by the BMZ (project no.: 06.7860.7-00100 and 06.7860.9-00100).

References

Adandonon A, Vayssieres JF, Sinzogan A, Van Mele P (2009) Density of pheromone sources of the weaver ant *Oecophylla longinoda* affects oviposition behaviour and damage by mango fruit flies (Diptera: Tephritidae). Int J Pest Man 55:285–292. doi:10.1080/09670870902878418

Beddington JR, Free CA, Lawton JH (1978) Characteristics of successful natural enemies in models of biological control of insect pests. Nature 273:513–519

Bess HA, Haramoto FH (1972) Biological control of pamakani, *Eupatorium adenophorum*, in Hawaii by a tephritid gall fly, *Procecidochares utilis*. 3. Status of weed, fly and parasites of fly in 1966–71 versus 1950–57. Proc Hawaiian Entomol Soc 21:165–178

Caltagirone LE (1981) Landmark examples in classical biological control. Ann Rev Entomol 26:213–232

Chown SL, Slabber S, McGeoch MA, Janion C, Leinaas HP (2007) Phenotypic plasticity mediates climate change responses among invasive and indigenous arthropods. Proc Roy Soc Ser B 274:2531–2537. doi:10.2307/25249362

Christenson LD, Foote RH (1960) Biology of fruit flies. Ann Rev Entomol 5:171–192

Davidson AM, Jennions M, Nicotra AB (2011) Do invasive species show higher phenotypic plasticity than native species, and if so, is it adaptive? A meta-analysis. Ecol Lett 14:419–431. doi:10.1111/j.1461-0248.2011.01596.x

De Clercq P, Mason P, Babendreier D (2011) Benefits and risks of exotic biological control agents. BioControl 56:681–698. doi:10.1007/s10526-011-9372-8

Dejean A (1990) Circadian-rhythm of *Oecophylla longinoda* in relation to territoriality and predatory behaviour. Phys Entomol 15:393–403

Doak DF, Mills LS (1994) A useful role for theory in conservation. Ecology 75:615–626. doi:10.2307/1941720

Doody JS, Green B, Rhind D, Castellano CM, Sims R, Robinson T (2009) Population-level declines in Australian predators caused by an invasive species. Anim Conserv 12:46–53. doi:10.1111/j.1469-1795.2008.00219.x

Drew RAI, Tsuruta K, White IM (2005) A new species of pest fruit fly (Diptera: Tephritidae: Dacinae) from Sri Lanka and Africa. Afr Entomol 13:149–154

Duan JJ, Messing RH (1997) Biological control of fruit flies in Hawaii: factors affecting non-target risk analysis. Agric Hum Values 14:227–236

Duan JJ, Messing RH, Dukas R (2000) Host selection of *Diachasmimorpha tryoni* (Hymenoptera: Braconidae): comparative response to fruit-infesting and gall-forming tephritid flies. Env Entomol 29:838–845

Duyck PF, David P, Quilici S (2004) A review of relationships between interspecific competition and invasions in fruit flies (Diptera: Tephritidae). Ecol Entomol 29:511–520. doi:10.1111/j.0307-6946.2004.00638.x

Fagan WF, Lewis MA, Neubert MG, Van Den Driessche P (2002) Invasion theory and biological control. Ecol Lett 5:148–157. doi:10.1046/j.1461-0248.2002.0_285.x

Fletcher BS (1987) The biology of dacine fruit flies. Ann Rev Entomol 32:115–144. doi:10.1146/annurev.ento.32.1.115

Gonthier DJ (2012) Do herbivores eavesdrop on ant chemical communication to avoid predation? Plos One 7:e28703. doi:10.1371/journal.pone.0028703

Griffiths AD, McKay JL (2007) Cane toads reduce the abundance and site occupancy of Merten's water monitor (*Varanus mertensi*). Wildl Res 34:609–615. doi:10.1071/wr07024

Hamilton WD (1971) Geometry for selfish herd. J Theor Biol 31:295–311. doi:10.1016/0022-5193(71)90189-5

Haramoto FH, Bess HA (1970) Recent studies on the abundance of the oriental and mediterranean fruit flies and the status of their parasites. Proc Hawaiian Entomol Soc 20:551–566

Hastings A (2000) Parasitoid spread: lessons for and from invasion biology. In: Ives AR, Hochberg ME (eds) Parasitoid population biology. Princeton University Press, Princeton, pp 70–82

Heimpel GE, Asplen MK (2011) A 'Goldilocks' hypothesis for dispersal of biological control agents. BioControl 56:441–450

Hoffmeister TS, Babendreier D, Wajnberg É (2006) Statistical tools to improve the quality of experiments and data analysis for assessing non-target effect. In: Bigler F, Babendreier D, Kuhlmann U (eds) Environmental impact of invertebrates for biological control of arthropods: methods and risk assessment. CAB International, Wallingford, pp 222–240

Hölldobler B, Wilson EO (1990) The ants. Springer, Berlin

Höller C, Micha SG, Schulz S, Francke W, Pickett JA (1994) Enemy-induced dispersal in a parasitic wasp. Experientia 50:182–185

Holt RD (1977) Predation, apparent competition, and structure of prey communities. Theor Pop Biol 12:197–229. doi:10.1016/0040-5809(77)90042-9

Howarth FG (1983) Classical biocontrol: panacea or Pandora's box. Proc Hawaiian Entomol Soc 24:239–244

Howarth FG (1991) Environmental impacts of classical biological control. Ann Rev Entomol 36:485–509. doi:10.1146/annurev.en.36.010191.002413

Huang HT, Yang P (1987) The ancient cultured citrus ant. Bioscience 37:665–671. doi:10.2307/1310713

Jervis MA (ed) (2005) Insects as natural enemies: a practical perspective. Springer, Dordrecht

Kareiva P (1996) Special feature: contributions of ecology to biological control. Ecology 77:1963–1964. doi:10.2307/2265692

Kausrud K, Økland B, Skarpaas O, Gregoire JC, Erbilgin N, Stenseth NC (2012) Population dynamics in changing environments: the case of an eruptive forest pest species. Biol Rev 87:34–51. doi:10.1111/j.1469-185X.2011.00183.x

Kenis M, Auger-Rozenberg M-A, Roques A, Timms L, Péré C, Cock MJW, Settele J, Augustin S, Lopez-Vaamonde C (2009) Ecological effects of invasive alien insects. In: Langor D, Sweeney J (eds) Ecological impacts of non-native invertebrates and fungi on terrestrial ecosystems. Springer, Dordrecht, pp 21–45. doi:10.1007/978-1-4020-9680-8_3

Kidd NA, Jervis MA (2005) Population dynamics. In: Insects as natural enemies: a practical perspective. Springer, Dordrecht, pp 435–523

Kunert G, Otto S, Röse USR, Gershenzon J, Weisser WW (2005) Alarm pheromone mediates production of winged dispersal morphs in aphids. Ecol Lett 8:596–603. doi:10.1111/j.1461-0248.2005.00754.x

Lampo M, De Leo GA (1998) The invasion ecology of the toad *bufo marinus*: from South America to Australia. Ecol Appl 8:388–396. doi:10.2307/2641079

Lessells CM (1985) Parasitoid foraging: should parasitism be density dependent? JAE 54:27–41

Lima SL (1998) Nonlethal effects in the ecology of predator-prey interactions – What are the ecological effects of anti-predator decision-making? Bioscience 48:25–34. doi:10.2307/1313225

Louda SM, Pemberton RW, Johnson MT, Follett PA (2003) Nontarget effects – The Achilles' Heel of biological control? Retrospective analyses to reduce risk associated with biocontrol introductions. Ann Rev Entomol 48:365–396. doi:10.1146/annurev.ento.48.060402.102800

Mashanova A, Gange AC, Jansen VAA (2008) Density-dependent dispersal may explain the mid-season crash in some aphid populations. Pop Ecol 50:285–292. doi:10.1007/s10144-008-0087-3

May RM, Hassell MP (1988) Population dynamics and biological control. Philos Trans R Soc B 318:129–169

Messelink GJ, Sabelis MW, Janssen A (2012) Generalist predators, food web complexities and biological pest control. In: Sonia S (ed) Greenhouse crops, integrated pest management and pest control – current and future tactics. InTech, pp 191–214. doi:10.5772/30835

Messing RH, Wright MG (2006) Biological control of invasive species: solution or pollution? Front Ecol Environ 4:132–140. doi:10.2307/3868683

Migani V, Ekesi S, Hoffmeister TS (2014) Physiology vs. environment: what drives oviposition decisions in mango fruit flies (*Bactrocera invadens* and *Ceratitis cosyra*)? J Appl Entomol 138:395–402. doi:10.1111/jen.12038

Mohamed SA, Wharton RA, von Merey G, Schulthess F (2006) Acceptance and suitability of different host stages of *Ceratitis capitata* (Wiedemann) (Diptera: Tephritidae) and seven other tephritid fruit fly species to *Tetrastichus giffardii* Silvestri (Hymenoptera: Eulophidae). Biol Contr 39:262–271. doi:10.1016/j.biocontrol.2006.08.016

Murdoch WW (1994) Population regulation in theory and practice. Ecology 75:271–287. doi:10.2307/1939533

Murdoch WW, Chesson J, Chesson PL (1985) Biological control in theory and practice. Am Nat 125:344–366. doi:10.1086/284347

Papaj DR, Roitberg BD, Opp SB (1989) Serial effects of host infestation on egg allocation by the mediterranean fruit fly: a rule of thumb and its functional significance. J Anim Ecol 58:955–970. doi:10.2307/5135

Peacor SD, Werner EE (1997) Trait-mediated indirect interactions in a simple aquatic food web. Ecology 78:1146–1156. doi:10.2307/2265865

Pemberton C (1964) Highlights in the history of entomology in Hawaii 1778–1963. Pac Insects 6:689–729

Poethke HJ, Hovestadt T (2002) Evolution of density- and patch-size-dependent dispersal rates. Proc Roy Soc Ser B 269:637–645. doi:10.1098/rspb.2001.1936

Polis GA, Holt RD (1992) Intraguild predation – The dynamics of complex trophic interactions. TREE 7:151–154. doi:10.1016/0169-5347(92)90208-s

Preisser EL, Bolnick DI, Benard MF (2005) Scared to death? The effects of intimidation and consumption in predator-prey interactions. Ecology 86:501–509. doi:10.1890/04-0719

Press JW, Flaherty BR, Arbogast RT (1974) Interactions among *Plodia interpunctella*, *Bracon hebetor*, and *Xylocoris flavipes*. Environ Entomol 3:183–184. doi:10.1093/ee/3.1.183

Purcell MF (1998) Contribution of biological control to integrated pest management of tephritid fruit flies in the tropics and subtropics. Int Pest Man Rev 3:63–83. doi:10.1023/a:1009647429498

Quimio GM, Walter GH (2001) Host preference and host suitability in an egg-pupal fruit fly parasitoid, *Fopius arisanus* (Sonan) (Hym., Braconidae). J Appl Entomol 125:135–140. doi:10.1046/j.1439-0418.2001.00514.x

Roitberg BD (2000) Threats, flies, and protocol gaps: can evolutionary ecology save biological control? In: Hochberg ME, Ives AR (eds) Parasitoid population biology. Princeton University Press, Princeton, pp 254–265

Roitberg BD, Zimmermann K, Hoffmeister TS (2010) Dynamic response to danger in a parasitoid wasp. Behav Ecol Sociobiol 64:627–637. doi:10.1007/s00265-009-0880-9

Rosenheim JA, Kaya HK, Ehler LE, Marois JJ, Jaffee BA (1995) Intraguild predation among biological-control agents: theory and evidence. Biol Contr 5:303–335. doi:10.1006/bcon.1995.1038

Rousse P, Harris EJ, Quilici S (2005) *Fopius arisanus*, an egg–pupal parasitoid of Tephritidae. Overview. Biocontrol News Inf 26:59N–69N

Schmidt MH, Lauer A, Purtauf T, Thies C, Schaefer M, Tscharntke T (2003) Relative importance of predators and parasitoids for cereal aphid control. Proc R Soc Lond B 270:1905–1909. doi:10.1098/rspb.2003.2469

Schutze MK, Aketarawong N, Amornsak W, Armstrong KF, Augustinos AA, Barr N, Bo W, Bourtzis K, Boykin LM, Caceres C, Cameron SL, Chapman TA, Chinvinijkul S, Chomic A, De Meyer M, Drosopoulou E, Englezou A, Ekesi S, Gariou-Papalexiou A, Geib SM, Hailstones D, Hasanuzzaman M, Haymer D, Hee AKW, Hendrichs J, Jessup A, Ji QG, Khamis FM, Krosch MN, Leblanc L, Mahmood K, Malacrida AR, Mavragani-Tsipidou P, Mwatawala M, Nishida R, Ono H, Reyes J, Rubinoff D, San Jose M, Shelly TE, Srikachar S, Tan KH, Thanaphum S, Haq I, Vijaysegaran S, Wee SL, Yesmin F, Zacharopoulou A, Clarke AR (2015) Synonymization of key pest species within the *Bactrocera dorsalis* species complex (Diptera: Tephritidae): taxonomic changes based on a review of 20 years of integrative morphological, molecular, cytogenetic, behavioural and chemoecological data. Syst Entomol 40:456–471. doi:10.1111/syen.12113

Simberloff D (2012) Risks of biological control for conservation purposes. BioControl 57:263–276. doi:10.1007/s10526-011-9392-4

Simberloff D, Stiling P (1996) Risks of species introduced for biological control. Biol Conserv 78:185–192. doi: http://dx.doi.org/10.1016/0006-3207(96)00027-4

Sinzogan AAC, Van Mele P, Vayssieres JF (2008) Implications of on-farm research for local knowledge related to fruit flies and the weaver ant *Oecophylla longinoda* in mango production. Int J Pest Manag 54:241–246. doi:10.1080/09670870802014940

Snyder WE, Ives AR (2001) Generalist predators disrupt biological control by a specialist parasitoid. Ecology 82:705–716. doi:10.1890/0012-9658(2001)082[0705:gpdbcb]2.0.co;2

Snyder WE, Snyder GB, Finke DL, Straub CS (2006) Predator biodiversity strengthens herbivore suppression. Ecol Lett 9:789–796. doi:10.1111/j.1461-0248.2006.00922.x

Sol D, Maspons J, Vall-llosera M, Bartomeus I, García-Peña GE, Piñol J, Freckleton RP (2012) Unraveling the life history of successful invaders. Science 337:580–583. doi:10.1126/science.1221523

Straub CS, Snyder WE (2006) Species identity dominates the relationship between predator biodiversity and herbivore suppression. Ecology 87:277–282. doi:10.1890/05-0599

Symondson WOC, Sunderland KD, Greenstone MH (2002) Can generalist predators be effective biocontrol agents? Ann Rev Entomol 47:561–594. doi:10.1146/annurev.ento.47.091201.145240

Traugott M, Bell JR, Raso L, Sint D, Symondson WOC (2012) Generalist predators disrupt parasitoid aphid control by direct and coincidental intraguild predation. Bull Entomol Res 102:239–247. doi:10.1017/s0007485311000551

van Driesche RG (2012) The role of biological control in wildlands. BioControl 57:131–137. doi:10.1007/s10526-011-9432-0

van Lenteren JC, Bale J, Bigler F, Hokkanen HMT, Loomans AJM (2006a) Assessing risks of releasing exotic biological control agents of arthropod pests. Ann Rev Entomol 51:609–634. doi:10.1146/annurev.ento.51.110104.151129

van Lenteren JC, Cock MJW, Hoffmeister TS, Sands DPA (2006b) Host specificity in arthropod biological control, methods for testing and interpretation of the data. In: Bigler F, Babendreier D, Kuhlmann U (eds) Environmental impact of invertebrates for biological control of arthropods: methods and risk assessment. CAB International, Wallingford, pp 38–63 van Mele P, Vayssieres JF, Van Tellingen E, Vrolijks J (2007) Effects of an African weaver ant, *Oecophylla longinoda*, in controlling mango fruit flies (Diptera: Tephritidae) in Benin. J Econ Entomol 100:695–701. doi:10.1603/0022-0493(2007)100[695:eoaawa]2.0.co;2

van Mele P, Vayssieres JF, Adandonon A, Sinzogan A (2009) Ant cues affect the oviposition behaviour of fruit flies (Diptera: Tephritidae) in Africa. Phys Entomol 34:256–261. doi:10.1111/j.1365-3032.2009.00685.x

Vance-Chalcraft HD, Rosenheim JA, Vonesh JR, Osenberg CW, Sih A (2007) The influence of intraguild predation on prey suppression and prey release: a meta-analysis. Ecology 88:2689–2696. doi:10.1890/06-1869.1

Vargas RI, Leblanc L, Putoa R, Eitam A (2007) Impact of introduction of *Bactrocera dorsalis* (Diptera: Tephritidae) and classical biological control releases of *Fopius arisanus* (Hymenoptera: Braconidae) on economically important fruit flies in French Polynesia. J Econ Entomol 100:670–679. doi:10.1603/0022-0493(2007)100[670:ioiobd]2.0.co;2

Vargas RI, Leblanc L, Putoa R, Piñero JC (2012) Population dynamics of three *Bactrocera* spp. fruit flies (Diptera: Tephritidae) and two introduced natural enemies, *Fopius arisanus* (Sonan) and *Diachasmimorpha longicaudata* (Ashmead) (Hymenoptera: Braconidae), after an invasion by *Bactrocera dorsalis* (Hendel) in Tahiti. Biol Control 60:199–206. doi:10.1016/j.biocontrol.2011.10.012

Vayssieres JF, Wharton R, Adandonon A, Sinzogan A (2011) Preliminary inventory of parasitoids associated with fruit flies in mangoes, guavas, cashew pepper and wild fruit crops in Benin. BioControl 56:35–43. doi:10.1007/s10526-010-9313-y

Völkl W (2001) Parasitoid learning during interactions with ants: how to deal with an aggressive antagonist. Behav Ecol Sociobiol 49:135–144. doi:10.1007/s002650000285

Völkl W, Kroupa AS (1997) Effects of adult mortality risks on parasitoid foraging tactics. Anim Behav 54:349–359. doi:10.1006/anbe.1996.0462

Waage JK, Hassell MP (1982) Parasitoids as biological-control agents – a fundamental approach. Parasitology 84:241–268

Wang XG, Messing RH (2002) Newly imported larval parasitoids pose minimal competitive risk to extant egg-larval parasitoid of tephritid fruit flies in Hawaii. Bull Entomol Res 92:423–429. doi:10.1079/ber2002181

Wang XG, Messing RH (2003) Foraging behavior and patch time allocation by *Fopius arisanus* (Hymenoptera: Braconidae), an egg-larval parasitoid of tephritid fruit flies. J Insect Behav 16:593–612. doi:10.1023/B:Joir.0000007698.01714.56

Wang XG, Messing RH, Bautista RC (2003) Competitive superiority of early acting species: a case study of opiine fruit fly parasitoids. Biocontrol Sci Technol 13:391–402. doi:10.1080/0958315031000104514

Weisser WW (2003) Additive effects of pea aphid natural enemies despite intraguild predation. In: Soares AO, Ventura MA, Garcia V, Hemptinne J-L (eds) 8th international symposium on ecology of aphidophaga: biology, ecology and behaviour of aphidophagous insects, Ponta Delgada, 2003. University of the Azores, Azores, pp 11–15

Katharina Merkel is a postdoctoral researcher at Queensland University of Technology (QUT). She has substantial experience in ecological and behavioural studies on honeybees and tephritid fruit flies. Behavioural ecology and how it is linked to population dynamics – especially in insects – represents her primary field of interest.

Valentina Migani is currently registered as a PhD student at Bremen University (Germany). She developed her research at the International Institute for Insect Physiology and Ecology (*icipe*) in Kenya, investigating the role of a generalist predator, the weaver ant ***Oecophylla longinoda***, as biological control agent of mango-infesting fruit flies and the impact of the ants on the exotic parasitoids introduced to control the fruit fly infestations.

Sunday Ekesi is an Entomologist at the International Centre of Insect Physiology and Ecology (*icipe*), Nairobi, Kenya. He heads the Plant Health Theme at *icipe*. Sunday is a professional scientist, research leader and manager with extensive knowledge and experience in sustainable agriculture (microbial control, biological control, habitat management/conservation, managing pesticide use, IPM) and biodiversity in Africa and internationally. Sunday has been leading a continent-wide initiative to control African fruit flies that threaten production and export of fruits and vegetables. The initiative is being done in close collaboration with IITA, University of Bremen, Max Planck Institute for Chemical Ecology together with NARS, private sectors and ARI partners in Africa, Asia, Europe and the USA and focuses on the development of an IPM strategy that encompasses baiting and male annihilation techniques, classical biological control, use of biopesticides, ant technology, field sanitation and postharvest treatment for quarantine fruit flies. The aim is to develop a cost effective and sustainable technology for control of fruit flies on the African continent that is compliant with standards for export markets while also meeting the requirements of domestic urban markets. Sunday has broad perspectives on global agricultural research and development issues, with first-hand experience of the challenges and opportunities in working with smallholder farmers, extension agents, research organizations and the private sector to improve food and nutritional security. He sits on various international advisory and consultancy panels for the FAO, IAEA, WB and regional and national projects on fruit fly, arthropod pests and climate change-related issues. Sunday is a Fellow of the African Academy of Sciences (FAAS).

Thomas S. Hoffmeister is a reseacher and academic teacher with more than 20 years of experience. He studies the evolutionary ecology of insect communication; the behavioural ecology, information use and foraging decisions of insects (mainly parasitoids); the life-history and spatial ecology of insects; and contributes to the study of biological control using a combined experimental and theoretical approach. After his post-doc at Simon Fraser University (B.C., Canada) he has lead a research group at the University of Kiel (Germany) and received a professorship position at the University of Bremen (Germany) in 2004. School of Earth, Environmental and Biological Sciences, Queensland University of Technology, Gardens Point, Brisbane, QLD 4001, Australia

Chapter 18
The Use of Weaver Ants in the Management of Fruit Flies in Africa

Jean-François Vayssières, Joachim Offenberg, Antonio Sinzogan, Appolinaire Adandonon, Rosine Wargui, Florence Anato, Hermance Y. Houngbo, Issa Ouagoussounon, Lamine Diamé, Serge Quilici*, Jean-Yves Rey, Georg Goergen, Marc De Meyer, and Paul Van Mele

Abstract Generalist predators such as the weaver ant, *Oecophylla longinoda* (Latreille), play an important role as biological control agents in West African orchards and, by extension, also in forest and savanna ecosystems within sub-Saharan Africa. These weaver ants are one of the most effective and efficient predators of arthropods in perennial tropical tree crops; their presence also acts as a deterrent to insect herbivores, particularly tephritid female fruit flies, due to the semiochemicals they produce. Emerging African markets for organic and sustainably-managed fruits and nuts have encouraged an interest in the use of weaver ants. Protection of tropical forests and savannas is ecologically and environmentally crucial and also essential for the protection of *O. longinoda*.

---

*Author was deceased at the time of publication.

J.-F. Vayssières (✉)
UPR HortSys, CIRAD Persyst, 34398 Montpellier, France

IITA, Biological Control Unit for Africa, IITA, 08BP 0932 Cotonou, Benin
e-mail: jean-francois.vayssieres@cirad.fr

J. Offenberg
Department of Bioscience, Aarhus University, Vejlsoevej 25, 8600 Silkeborg, Denmark

A. Sinzogan • R. Wargui • F. Anato • H.Y. Houngbo • I. Ouagoussounon
FSA, UAC, Université d'Abomey Calavi, 03BP 2819 Cotonou, Benin

A. Adandonon
ENSTA – Kétou, Université d'Agriculture de Kétou, BP 43 Kétou, Benin

L. Diamé
ISRA/CDH, BP 3120 Dakar, Senegal

Université Cheikh Anta Diop, BP 5005 Dakar, Senegal

© Springer International Publishing Switzerland 2016
S. Ekesi et al. (eds.), *Fruit Fly Research and Development in Africa - Towards a Sustainable Management Strategy to Improve Horticulture*,
DOI 10.1007/978-3-319-43226-7_18

Keywords *Oecophylla longinoda* • Tephritidae • Trophic interactions • Ant cues • Conservation biological control

1 Introduction

Generalist predators, such as the weaver ant, *Oecophylla longinoda* (Latreille) (Hymenoptera: Formicidae: Formicinae), play an important role in orchard, forest and savanna ecosystems in sub-Saharan Africa (Leston 1973). Weaver ants are highly effective and efficient in controlling arthropod pests in perennial crops due to their tireless predatory activities (Dejean 1991). The presence of weaver ants also deters the activity of insect herbivores such as fruit flies (Diptera: Tephritidae) who recognize and avoid semiochemicals produced by the ants (Adandonon et al. 2009). Under some conditions fruit fly populations can be controlled by biological control agents such as parasitoids and generalist predators like *Oecophylla* species (Van Mele and Cuc 1999).

To determine the conditions necessary to enhance pest regulation by natural enemies, it is necessary to better understand the exact nature of their interactions across all trophic levels in the food web (Dejean et al. 2007). Interactions affecting the abundance of organisms may be direct or indirect and can have cascading effects across several trophic levels (Table 18.1) in the ecosystem before they influence fruit productivity and quality (Vayssières 2012). A detailed understanding of the ecology and behaviour of natural enemies is, therefore, essential if we are to encourage their activity against fruit flies within integrated management strategies (Quilici and Rousse 2012).

J.-Y. Rey
UPR HortSys, CIRAD Persyst, 34398 Montpellier, France
ISRA/CDH, BP 3120 Dakar, Senegal

G. Goergen
IITA, Biological Control Unit for Africa, IITA, 08BP 0932 Cotonou, Benin

M. De Meyer
Biology Department, Royal Museum for Central Africa, Leuvensesteenweg 13, B3080, Tervuren, Belgium

P. Van Mele
Agro-Insight, 3670 Meeuwen-Gruitrode, Belgium

Table 18.1 Food web in mango orchards in Benin

Primary level	Secondary level	Tertiary level		
Main fruit-hosts	Tephritidae	Insect parasitoids	Insect predators	Arachnid predators
Mangifera indica L. (mango)	*Bactrocera dorsalis* (Hendel)	*Pachycrepoideus vindemiae* (Rondani)	**Oecophylla longinoda** (Latreille) *Crematogaster* spp. *Odontomachus* spp. *Pachycondula* spp. *Camponotus* spp. *Dorylus* spp.	Salticidae Araneidae Thomisidae Sparassidae Oxyopidae
Terminalia catappa L. (tropical almond)		*Fopius arisanus* (Sonan)		
Mango	*Ceratitis cosyra* (Walker)	*Fopius caudatus* (Széligeti)		
Mango		*Psyttalia cosyrae* (Wilkinson)		
Mango		*Psyttalia concolor* Széligeti		
Mango		*Diachasmimorpha fullawayi* (Silvestri)		
Mango		*Tetrastichus giffardianus* Silvestri		
Sarcocephalus latifolius (Smith) Bruce (African peach)		*F. caudatus*		
Psidium guajava L. (common guava)		*F. caudatus*		
Annona senegalensis Pers. (wild custard apple)		*F. caudatus*		
Wild custard apple		*Psyttalia perproxima* (Silvestrii)		
Anacardium occidentale L. (cashew)		*P. vindemiae*		
Mango	*Ceratitis silvestrii* Bezzi	*F. caudatus*		
Mango		*P. cosyrae*		
Vitellaria paradoxa L. (shea butter tree)		*P. cosyrae*		
Shea butter tree		*F. caudatus*		
Mango	*Ceratitis quinaria* (Bezzi)	*P. perproxima*		
Shea butter tree		*P. perproxima*		
Shea butter tree		*F. caudatus*		
Capsicum annuum L. (chilli)	*Ceratitis capitata* (Wiedemann)	*D. fullawayi*		
Chilli	*Ceratitis fasciventris* (Bezzi)	*T. giffardianus*		
Mango	*Ceratitis ditissima* (Munro)	*T. giffardianus*		
Citrus sinensis L. (sweet orange)	*Ceratitis anonae* Graham	*Spalangia simplex* Perkins		
Mango		*T. giffardianus*		

2 Fruit Flies in Sub-Saharan Africa

African fruit producers throughout sub-Saharan Africa, and particularly in certain regions and countries, are confronted with a number of closely connected problems: (a) severe deterioration of fruit quality due to fruit flies (White and Elson-Harris 1992; De Meyer et al. 2007); (b) inadequate post-harvest fruit fly control methods (Van Melle and Buschmann 2013); (c) invasive plant diseases such as *Xanthomonas citri* pv. *mangiferaeindicae* (Patel et al.) comb. nov., which causes mango bacterial canker across Ghana and Benin (Pruvost et al. 2011a, b; Zombré et al. 2015) and *Phaeoramularia angolensis* (De Carv. & Mendes) P.M. Kirk which causes citrus fruit and leaf spot in West Africa (Vayssières 1995); (d) over-production in national markets leading to wastage and low prices (Vayssières et al. 2009a); (e) inadequate selection of appropriate mango (*Mangifera indica* L.) varieties (Vayssières et al. 2008); and (f) under-development and/or under-utilisation of export markets (Van Melle and Buschmann 2013).

In sub-Saharan Africa, many fruit fly species attack agricultural crops. Amongst them, six are considered of the greatest economic importance: the mango fruit fly, *Ceratitis cosyra* (Walker); the Mediterranean fruit fly (medfly), *Ceratitis capitata* (Wiedemann); the Natal fruit fly, *Ceratitis rosa* Karsch; and three exotic invasive species, the peach fruit fly, *Bactrocera zonata* (Saunders) that has been present in Sudan for several years (Salah et al. 2012); the melon fruit fly, *Zeugodacus cucurbitae* (Coquillett) (De Meyer et al. 2015); and *Bactrocera invadens* Drew Tsuruta & White. The latter has recently been placed in synonymy with the oriental fruit fly *Bactrocera dorsalis* (Hendel) (Schutze et al. 2014a, b).

Bactrocera dorsalis, originating from South-East Asia, was found for the first time in Africa in Kenya in 2003 (Lux et al. 2003) and then in Tanzania (Mwatawala et al. 2004). This invasive species was subsequently reported in Sudan (Luckman 2004), Senegal (Vayssières 2004), then in Côte d'Ivoire (Hala et al. 2006), Cameroon (Ndzana Abanda et al. 2008), Nigeria (Umeh et al. 2008), and subsequently in other African countries (Vayssières et al. 2010a). *Bactrocera dorsalis* causes extensive economic losses to horticultural crops throughout sub-Saharan Africa and is especially harmful to the mango value chain in East and West Africa (Ekesi et al. 2006; Vayssières et al. 2009a), increasing the already considerable damage caused by native fruit flies. With its high reproductive rate (Ekesi et al. 2006; Salum et al. 2013; Gomina et al. 2014), a large host plant range (De Meyer et al. 2007; Rwomushana et al. 2008; Mwatawala et al. 2009a; Goergen et al. 2011) and high mobility (Vayssières et al. 2009b), this species is a major pest of economic significance.

In Benin several fruit value chains of commercial interest are severely jeopardized by *B. dorsalis*, such as mango (Vayssières et al. 2009a), citrus (Vayssières et al. 2010b) and guava (Vayssières et al. 2010c). According to Ekesi et al. (2009) and Salum et al. (2013) exploitative competition through larval scrambling for resources and interference competition accompanied by the aggressive adult behaviour of *B. dorsalis* compared with native species, are important displacement mechanisms. As a result, *B. dorsalis* has displaced *C. cosyra* in mango agro-ecosystems and out-competed *C. capitata* in citrus plantations (Vayssières et al. 2010b). *Bactrocera dorsalis* has changed the landscape of pest fruit flies in Africa.

Studies on fruit fly control have been conducted in: Southern Africa (Labuschagne et al. 1996; De Meyer 2001, 2005; White 2006; Correia et al. 2008; Grové et al. 2009; De Villiers et al. 2013; José and Santos 2013; Hill and Terblanche 2014; Manrakhan et al. 2015); in East Africa (De Meyer 2001, 2005; Lux et al. 2003; Ekesi et al. 2006, 2009; Mwatawala et al. 2006a, b, 2009a, b; Copeland et al. 2006; White 2006; Rwomushana et al. 2009; Geurts et al. 2012, 2014; Salum et al. 2013); in central Africa (De Meyer 2001, 2005; Ndzana Abanda et al. 2008; Ngamo et al. 2010; Virgilio et al. 2011; Mayamba et al. 2014); and in West Africa (Vayssières and Kalabane 2000; Vayssières et al. 2004, 2005, 2008, 2009a, b, 2010a, b, c, 2011a, 2012, 2014, 2015a; De Meyer 2001, 2005; Hala et al. 2006; Ndiaye et al. 2008, 2012, 2015; Umeh et al. 2008; Amevoin et al. 2009; Appiah et al. 2009; N'Dépo et al. 2009, 2010, 2013; Ouedraogo 2011; Ouedraogo et al. 2011; Zakari et al. 2012; Gomina et al. 2014).

3 Importance of Natural Enemies, Particularly Predators, in Controlling Fruit Flies

Biological control is the use, by introduction, augmentation and/or conservation of beneficial organisms to control harmful organisms. In the case of fruit fly pests, the main biological control agents known to date are parasitoid wasps (Hymenoptera: Braconidae, Eulophidae, Pteromalidae), ant predators (Hymenoptera: Formicidae) and a fungus called *Metarhizium anisopliae sensu lato* (Metschn.) Sorokin (Ekesi et al. 2003).

Releases of parasitoids, mainly Braconidae, have shown promise for the area-wide management of tephritid pests, either as part of classical biological control programmes against exotic species, or augmentation programmes against native species (Wharton 1989; Sivinski et al. 1996). In several West African countries, we found that immature stages of fruit flies were parasitized by a wide range of parasitoid species. Only a few of them are appropriate for biological control (Sivinski et al. 1996; Quilici and Rousse 2012). To date, nearly all the actual or potential biological control agents are egg or larval parasitoids, frequently from the family Braconidae, subfamily Opiinae. The Opiinae are the most abundant and species-rich group of tephritid parasitoids and are frequently used in biological control or as part of integrated pest management programmes. The most common hosts are known for almost half of the currently recognized parasitoid species (Wharton 1997; Rugman-Jones et al. 2009) and they are all solitary, koinobiont endoparasitoids of Tephritidae, developing inside immature flies and killing them in the process. In response to the success achieved in Hawaii and elsewhere in the biological control of several *Bactrocera* species of economical importance (Vargas et al. 1993, 2007), the braconid parasitoid *Fopius arisanus* (Sonan) has been introduced into several African countries including Benin, Kenya, Senegal and Togo (Goergen, personal communication). While it appears that repeated releases have resulted in parasitoid establishment in these countries, their overall impact on fruit fly populations under afrotropical climatic conditions remains to be assessed. An initial small scale study in Senegal to evaluate impacts on fruit fly populations has shown that mangoes in the control orchard were 5–6 times more heavily infested with fruit flies than mangoes in the orchards where *F. arisanus* was released (Ndiaye et al. 2015).

The braconid genus *Psyttalia* also bears some potential but more fundamental studies are required (Billah et al. 2008). Other families of tephritid parasitoids, such as the Eulophidae (koinobionts) and the Pteromalidae (idiobionts), are less frequently used in biological control of tephritid pests. Use of the fungus *M. anisopliae* is another effective way of fruit fly control as developed and successfully tested by *icipe* in East Africa (Ekesi et al. 2005, 2007, 2011).

Species of weaver ant in the genus *Oecophylla* Smith, 1860 (Hymenoptera Formicidae) are a key element within conservation biological control approaches targeted at tephritids in several regions (Peng and Christian 2006; Van Mele et al. 2007). In conservation approaches, the biological control agents concerned are not introduced or augmented but naturally occurring individuals are encouraged through cultural practices (Vanderplank 1960). The Asian weaver ant, *Oecophylla smaragdina* (Fabricius), has provided protection of citrus trees against tephritids in Southern China since the fourth century AD (Huang and Yang 1987; Barzman et al. 1996). More recent reports have demonstrated that *O. smaragdina* provided protection against fruit flies in Asian and Australian cashew (*Anacardium occidentale* L.) and mango orchards (Peng et al. 1995; Van Mele and Cuc 2000; Peng and Christian 2006). The weaver ant, *Oecophylla longinoda* (Latreille), a close relative of *O. smaragdina*, is distributed across sub-saharan Africa but its potential for control of tephritids has received less attention until recently. Since 2005 exploratory studies on this species have been ongoing in Benin, West Africa.

3.1 Distribution of the Genus Oecophylla

To date the genus *Oecophylla* has been recorded throughout forested regions of the Old World tropics (Fig. 18.1). There are only two *Oecophylla* species globally, both of which are exclusively arboreal: the African species *O. longinoda* (Fig. 18.2) and the Asian species *O. smaragdina* (Fig. 18.3a). The African species is widespread throughout sub-Saharan Africa between the sixteen-degrees-north latitude and the twenty-degrees-south latitude (Lokkers 1986). However, these general data require fine-tuning for each sub-Saharan country. The Asian species extends from Southern Asia, including India, to northern tropical Australia including many tropical western Pacific islands (Cole and Jones 1948). Both species are similar in their ecological and morphological traits with powerful mandibles (Fig. 18.3b). *Oecophylla smaragdina* can be distinguished by its very slender petiole, its very prominent stigmata, and its ventral surface which is nearly straight or very feebly convex in profile; in contrast, *O. longinoda* has a stouter and higher petiole (Fig. 18.2), with

18 The Use of Weaver Ants in the Management of Fruit Flies in Africa

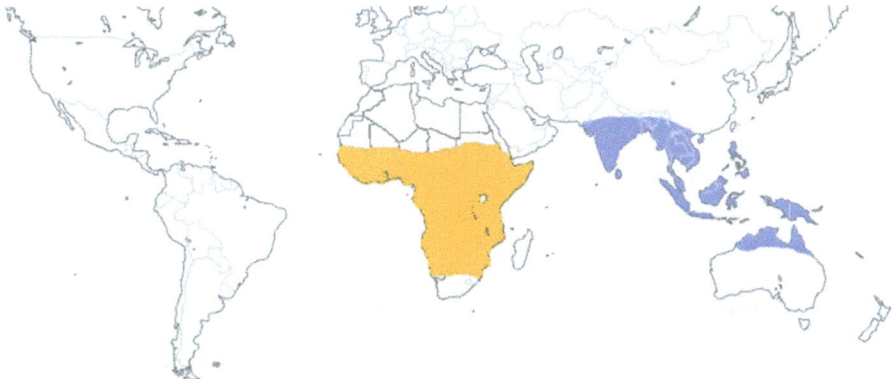

Fig. 18.1 Worldwide distribution of the genus *Oecophylla* species (*O. longinoda* in *orange*, *O. smaragdina* in *blue*)

Fig. 18.2 *Oecophylla longinoda* (Georg Goergen credit)

Fig. 18.3 (**a**) *O. smaragdina*, (**b**) *O. smaragdina* detail of mandibles (Kim Aaen credit)

the stigmata not prominent from above, and the ventral surface strongly convex in profile (Wheeler 1922).

The critical parameters that positively affect the distribution of *Oecophylla* species are (i) perennial vegetation, (b) temperatures above 18 °C, (c) high annual rainfall (above 500 mm) (Lokkers 1990), (d) low altitude (in relation to the temperature threshold), but also (e) conservation measures that enhance densities in plantations. Even though these critical thresholds, especially rainfall and temperature, can help to explain delimitation of *O. longinoda* in the African Sudano-Sahelian zone, further field work is necessary to determine the exact limits for the presence of the weaver ant in this zone. The distribution of *O. longinoda* remains unclear with regard to its northern limit within the south-Sahelian zone, and it has been suggested that climate change may impact on this (Vayssières 2012). Under favourable conditions the ant can survive in the Sahelian zone, as was observed in mango plantations at the fruit station of Kaedi (Mauritania 16.1353 N–13.5826E) from 1987 to 1991 (Vayssières et al. 1991); however, it seems that *O. longinoda* is currently no longer present in this location.

3.2 Key Traits of the African Weaver Ant, O. longinoda

The genus *Oecophylla* belongs to the sub-family Formicinae, an ant taxon lacking a functional sting. The painful 'sting' reported by fruit growers is actually the ant's bite, which it inflicts with its powerful mandibles. *Oecophylla longinoda* is a large ant reaching up to 9 mm in length; the species is characterized by populous colonies and the ability to build large and polydomous nests with a highly developed intra- and interspecific territoriality (Crozier et al. 2009) that allows populations to be distributed in a mosaic pattern in tree canopies (Majer 1972; Dejean et al. 1994; Blüthgen and Stork 2007).

Oecophylla longinoda is endemic to sub-Saharan Africa and frequently found in forested wild vegetation and unsprayed orchards of: mango, *Mangifera indica* L. (Fig. 18.4a); *Citrus* species; (Fig. 18.4b); cashew, *Anacardium occidentale* L. (Fig. 18.4c); guava, *Psidium guajava* L. (Fig. 18.4d); custard apple, *Annona muricata* L.; cocoa, *Theobroma cacao* L.; coconut, *Cocos nucifera* L.; coffee, *Coffea* x *arabusta* Capot & Aké Assi; and oil palm, *Elaeis guineensis* Jacq. amongst others. Many tropical trees, shrubs (Fig. 18.5a, b) and lianas can host this common arboreal ant species. Detailed accounts of the ecology of *O. longinoda* (Wheeler 1910; Ledoux 1950; Way 1954a), its behaviour (Wheeler 1910; Ledoux 1950; Chauvin 1952; Way 1954b; Sudd 1963; Wilson 1971; Hölldobler and Wilson 1977), and its role in plant protection (Way 1953; Vanderplank 1960; Sudd 1963; Dejean 1991; Varela Barros 1992; Peng et al. 1995; Offenberg 2015 amongst others) are available.

Each *O. longinoda* colony maintains its territory and coordinates its activities using a highly developed chemical communication system in combination with visual and tactile cues (Hölldobler and Wilson 1978). They secrete various pheromones from their heads but also from sternal and rectal glands (Hölldobler and

Fig. 18.4 Nests of *O. longinoda* on fruit trees: (**a**) mango, (**b**) grapefruit, (**c**) cashew, (**d**) guava (Jean-François Vayssières credit)

Fig. 18.5 Nest of *O. longinoda* on shrubs: (**a**) *Antidesma venosum*, (**b**) *Sarcocephalus latifolius* (Jean-François Vayssières credit)

Fig. 18.6 Different life stages of *O. longinoda* on mango (Jean-François Vayssières credit) (**a**) queen, (**b**) males with small and large workers

Wilson 1990). With their large colonies and their ability to build nests almost anywhere (with the exception of many urbanized areas), *O. longinoda* populations are able to closely control their environment. In African savannahs and forests, the construction of numerous communal silk nests by each colony has probably facilitated the success of these arboreal ants (Hölldobler and Wilson 1990). Different life stages and forms can be distinguished in a colony and include the queen (Fig. 18.6a), males, small and large workers (Fig. 18.6b), pupae and larvae. Way (1954a) observed one colony of *O. longinoda* inhabiting 151 nests, scattered throughout eight coconut trees and four clove trees (*Syzygium aromaticum* (L.) Merrill & Perry) covering an area of 800 m². He estimated that this colony contained 480,000 worker ants and 280,000 brood. Similarly, in an assessment of the abundance of *O. smaragdina* in Australian mango plantations, it was found that the biomass of colonies was up to almost 3 kg per colony and that the average number of worker ants per occupied mango tree was up to 60,000 individuals (Pinkalski et al. 2015). Such dense worker populations may afford intensive patrolling in the canopies of host trees.

3.3 Colony Establishment and Reproduction in O. longinoda

Various authors have revealed the importance of different patterns of vegetation on *O. longinoda* distribution (Leston 1973; Room 1971; Majer 1972; Way 1963). Many types of vegetation can support *O. longinoda*, and vegetation manipulation can induce changes in the ant-mosaic. *Oecophylla longinoda* requires thick vegetation, especially perennial species, usually with an interconnected canopy to provide nesting sites and foraging areas (Taylor and Adedoyin 1978). Nests of *O. longinoda* can be found in different positions, at different heights in the tree, and on many tree species; they can utilize many plants in a wide range of tropical habitats (Dejean et al. 1999). In West Africa, the nests of *O. longinoda* are commonly found in (i) fruit plantations (such as mango, citrus and cashew), (ii) wild hosts (trees, shrubs and lianas) around fruit plantations, and (iii) in dry as well as humid forests (Vayssières 2012). In Benin, *O. longinoda* nests were recorded on 34 tree species belonging to 21 families (Vayssières 2012) with new records continuously being added.

Oecophylla longinoda ants are highly organized when building their nests (Ledoux 1950; Chauvin 1952; Way 1954a; Hölldobler and Wilson 1977). In Benin leaves of some mango cultivars (e.g. Keitt) which are very long and narrow, were first pulled together edge to edge by the large workers building chains between the leaves by attaching to each other; by shortening these chains the leaves were forced together (J.-F. Vayssières, personal observation). Then the leaves were joined together with silk produced by the last larval instars (both females and males) that were held as a tool by the worker ants (Fig. 18.7) (Hölldobler and Wilson 1977). Only the larvae of *Oecophylla* species can produce silk. In the nest, some of the walls and galleries are made entirely of silk.

Winged male and female sexual stages are produced during the rainy season, which is also when nuptial flights take place (Way 1954b; Rwegasira et al. 2015). Both haplometrotic colony founding (by a single queen) and pleometrotic colony founding (by multiple queens) have been recorded (Ledoux 1950; Vanderplank 1960). The alate (winged) queens that leave the colony during the nuptial flight become founders of new future colonies. Such new queens can be attracted to artificial nests (Fig. 18.8) where they detach their wings, seek shelter and settle during their founding stage (Ouagoussounon et al. 2013, 2015). Queens caught in artificial nests can be used to establish new colonies for farmers. In order to favour the development of large *O. longinoda* populations, transplantation of pupae may be used to shorten the time needed to produce mature *O. longinoda* colonies in ant nurseries (Ouagoussounon et al. 2013). These methods could be used to implement biological

Fig. 18.7 Construction of a nest using the last larval instar (Jean-François Vayssières credit)

Fig. 18.8 Artificial nest on mango (Jean-François Vayssières credit)

control of fruit flies by *O. longinoda* throughout West and sub-Saharan Africa (Ouagoussounon et al. 2013).

Once established, there are many constraints that threaten the viability of *O. longinoda* populations. Its main abiotic constraint is the harsh environment, especially long and dry winds such as the harmattan (dusty trade wind in western Africa, blowing from the Sahara into the Gulf of Guinea during the dry season) combined with long drought periods with low temperatures (December-January). Other major constraints are (i) limited food supply, (ii) unsuccessful mating, (iii) the use of insecticides by farmers, (iv) and bush fires (Vayssières 2012).

3.4 Host-Plants and Associated Trophobionts Exploited by O. longinoda

Oecophylla longinoda are mostly found on plants that also support heteropteran symbionts, also known as trophobionts. These trophobionts excrete sugar-rich honeydew, which the ants feed on. Since *O. longinoda* obtain rich sugar and amino acid food sources via these interactions, in return they protect the trophobionts from predators and parasitoids (Way 1963). Trophobionts are a key component in tropical foodwebs that link ants with plants. The presence of trophobionts improves the establishment of *O. longinoda* colonies first by providing the ants with sugar and secondly by serving as alternative prey (i.e. a protein source) when the predatory activities of the ants are unsuccessful. This is particularly the case during the dry season when, in the Sudan zone of Benin, *O. longinoda* also obtain nutritional ressources by harvesting seeds (Fig. 18.9a, b) and plant debris (Fig. 18.9c) (Vayssières et al. 2015b). There are several species of trophobionts that *O. longinoda* colonies tend for food and they come from the families: Coccidae (Fig. 18.10a), Pseudococcidae, Stictococcidae, Membracidae and Tettigometridae (Fig. 18.10b) (Bluthgen et al. 2004). Across Benin, approximately 20 species of Coccoidea (Table 18.2) have been reported interacting mutualistically with *O. longinoda* (Vayssières 2012). In the three agroecological zones of Benin the most common species associated with *O. longinoda* on mango is the coccid *Udinia catori* (Green) (Fig. 18.10a) (Vayssières 2012).

Fig. 18.9 Plant material harvested and carried back to the nest along the trunk of a mango tree (Jean-François Vayssières credit) (**a**) Harvested seed; (**b**) Detail of seed being carried to the nest; (**c**) Harvested plant debris

Fig. 18.10 *O. longinoda* attending: (**a**) *Udinia catori* (Hem.: Coccidae) on mango and (**b**) *Hilda* sp. (Hem.: Tettigometridae) on *Albizia glaberrima* (Jean-François Vayssières credit)

Table 18.2 Main host plants of *Oecophylla longinoda* and their associated trophobionts in Benin

	Host-plant species of *Oecophylla longinoda*		Associated trophobionts		
	Family	Name	Name	Family	Position on the tree
1	Anacardiaceae	*Mangifera indica* L. (mango)	*Udinia catori* (Green)	Coccidae	on stem (flush) and on fruits
2		*Anacardium occidentale* L. (cashew)	*Parasaissetia nigra* (Nietner)	Coccidae	on stem (flush)
3	Annonaceae	*Monodera tenuifolia* Benth.	*Stictoccocus intermedius* (Newstead)	Stictococcidae	on stem (flush)
4	Apocynaceae	*Holarrhena floribunda* (G. Don) Durand & Schinz	*U. catori*	Coccidae	on stem (flush)
5	Caesalpinioideae	*Isoberlinia doka* Craib & Stapf	*Hilda funesta* (Stal)	Tettigometridae	on stem (flush)
		I. doka	*Coccus hesperidum* (Linnaeus)	Coccidae	on leaves (inside the nest)
6		*Senna siamea* (Lam.) Irwin & Barneby	*Tylococcus westwoodi* Strickland	Pseudococcidae	on stem (flush)
7	Celestraceae	*Maytenus senegalensis* (Lam.) Exell	*Udinia farqharsoni* (Newstead)	Coccidae	on leaves (inside the nest)
8	Chrysobalanaceae	*Maranthes polyandra* (Benth.) Prance	*U. farqharsoni*	Coccidae	on white flowers
9		*Maranthes robusta* (Oliv.) Prance ex F. White	*U. farqharsoni*	Coccidae	on stem (flush)
10	Combretaceae	*Combretum nigricans* Lepr. Ex Guill. & Perr.	*Parasaissetia* sp.	Coccidae	on stem (flush)
		C. nigricans	*Coccus* sp.	Coccidae	on stem (flush)
11	Ebenaceae	*Diospyros mespiliformis* Hochst. Ex A. Rich.	*U. catori*	Coccidae	on stem (flush)
		D. mespiliformis	*U. farqharsoni*	Coccidae	on stem (flush)

(continued)

Table 18.2 (continued)

	Host-plant species of *Oecophylla longinoda*		Associated trophobionts		
	Family	Name	Name	Family	Position on the tree
12	Euphorbiaceae	*Alchornea cordifolia* [Schumah. & Thonn.] Müll. Arg.	*T. westwoodi*	Pseudococcidae	on leaves (inside the nest)
		A. cordifolia	*P. nigra*	Coccidae	on stem (flush)
		A. cordifolia	*U. catori*	Coccidae	on leaves (inside the nest)
13	Loganiaceae	*Anthocleista nobilis* G. Don	*Saissetia* sp.	Coccidae	on stem (flush)
14		*Strychnos spinosa* Lam.	*U. farqharsoni*	Coccidae	on stem (flush)
15	Meliaceae	*Khaya senegalensis* (Desr.) A. Juss.	*U. catori*	Coccidae	on leaves (inside the nest)
		K. senegalensis	*U. farqharsoni*	Coccidae	on stem (flush)
		K. senegalensis	*U. catori*	Coccidae	on stem (flush)
16	Mimosoideae	*Acacia auriculiformis* A. Cunn. Ex Benth.	*Coccus* sp.	Coccidae	on stem (flush)
17		*Albizia glaberrima* (Schumach. & Thonn.) Benth.	*Oxyrhachis tarandus* Fab.	Membracidae	on stem (flush)
18	Moraceae	*Ficus sur* Forssk.	*U. catori*	Coccidae	on fruits
		F. sur	*H. funesta*	Tettigometridae	on fruits
19		*Ficus vallis-choudae* Del.	*Hilda undata* (Walker)	Tettigometridae	on fruits
20	Myrtaceae	*Psidium guajava* L. (common guava)	*P. nigra*	Coccidae	on stem (flush)
		common guava	*U. catori*	Coccidae	on leaves (inside the nest)
21		*Syzygium guineense* (Willd.) DC.	*U. catori*	Coccidae	on stem (flush)
22	Ochnaceae	*Lophira lanceolata* Van Tiegh. ex Kay	*U. catori*	Coccidae	on leaves (inside the nest)
		L. lanceolata	*P. nigra*	Coccidae	on stem (flush)
23	Rubiaceae	*Aidia genipiflora* (DC.) Dandy	*U. farqharsoni*	Coccidae	on leaves (inside the nest)

(continued)

Table 18.2 (continued)

	Host-plant species of *Oecophylla longinoda*		Associated trophobionts		Position on the tree
	Family	Name	Name	Family	
24		*Gardenia erubescens* Stapf & Hutch.	*P. nigra*	Coccidae	on fruits
		G. erubescens	*Saissetia privigna* De Lotto	Coccidae	on fruits
		G. erubescens	*Planococcus kenyae* (Le Pelley)	Pseudococcidae	on fruits
25		*Psydrax horizontalis* (Schumach. & Thonn.) Bridson	*U. farqharsoni*	Coccidae	on stem of this liana
26		*Sarcocephalus latifolius* (Smilth) Bruce	*P. nigra*	Coccidae	on leaves (inside the nest)
		S. latifolius	*H. funesta*	Tettigometridae	on stem (flush)
		S. latifolius	*H. undata*	Tettigometridae	on stem (flush)
		S. latifolius	*Parasaissetia* sp.	Coccidae	on stem (flush)
27	Rutaceae	*Citrus limon* (L.) Burm. F. (lemon)	*Coccus hesperidum* (Linnaeus)	Coccidae	on stem (flush)
28		*Citrus sinensis* (L.) Osbeck (sweet orange)	*C. hesperidum*	Coccidae	on stem (flush)
29	Sapindaceae	*Paullinia pinnata* L.	*T. westwoodi*	Pseudococcidae	on leaves (inside the nest)
30		*Blighia unijugata* Haker	*P. nigra*	Coccidae	on stem (flush)
31		*Lecaniodiscus cupanioides* Planch. Ex Benth.	*C. hesperidum*	Coccidae	on stem (flush)
32	Sapotaceae	*Vitellaria paradoxa* Gaertn. F. (shea butter tree)	*U. catori*	Coccidae	on leaves (inside the nest)
		shea butter tree	*Parasaissetia* sp.	Coccidae	on stem (flush)
33	Simaroubaceae	*Hannoa undulata* (Guill. & Perr.) Planch.	*U. catori*	Coccidae	on stem (flush)
34	Verbenaceae	*Vitex doniana* Sweet	*U. catori*	Coccidae	on leaves (inside the nest)

3.5 Circadian Rhythms and the Importance of Ant Density on the Biological Control Potential of *O.* longinoda

The ability of *O. longinoda* to protect plants against pests is related to their activity patterns, i.e. the times at which they forage for prey but also on other associated activities, such as patrolling, during which time ant-derived semiochemicals are deposited on leaves, branches and fruits; these semiochemical cues (ant cues) have the potential to act as deterrents to fruit fly pests. In the Democratic Republic of Congo, early studies investigating the activity patterns of *O. longinoda* were inconclusive (Dejean 1990), although it was shown that when *O. longinoda* hunted by sight, light intensity was crucial and other parameters such as temperature and humidity also played an important role (Dejean 1986). Study of the circadian activity patterns of *O. longinoda* and the influence of particular ecological factors on this activity is of great interest because it affords an opportunity to understand what drives the spatiotemporal distribution of *O. longinoda* in trees and in orchards. In Benin, preliminary results indicate that circadian activity of *O. longinoda* is continuous (Fig. 18.11) although diurnal activity is greater than nocturnal activity (Vayssières et al. 2011b). In the South Sudan Beninese zone, there seems to be no

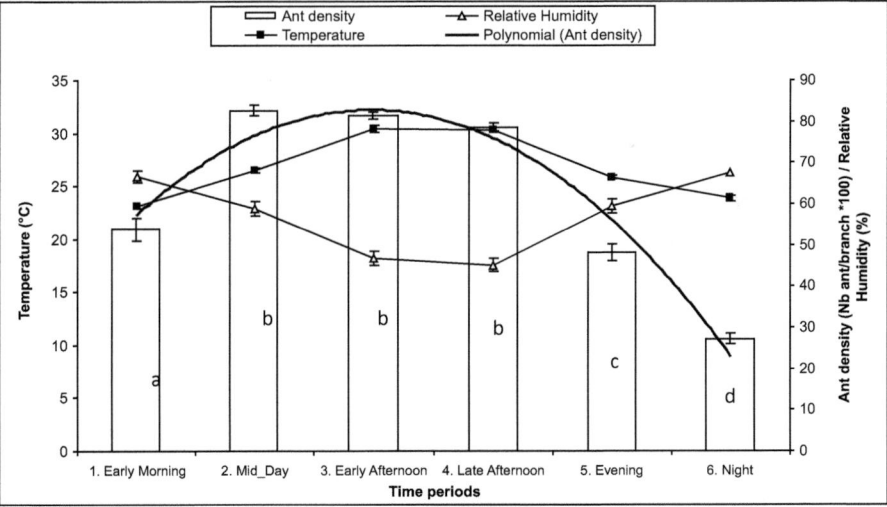

Fig. 18.11 Mean diurnal rhythms of activity of *Oecophylla longinoda* outside the nest in relation to variation in temperature and humidity (Each observation for each of the different factors (ant density, temperature and humidity) is the mean ± SE of observations made hourly from 7:30 AM to 6:30 AM the following day, carried out on a fortnightly basis for 2 years and pooled according to the following time periods: 1. Early Morning (5:30 AM to 8:30 AM); 2. Mid-Day (9:30 AM to 11:30 AM); 3. Early Afternoon (12:30 PM to 2:30 PM); 4. Late Afternoon (3:30 PM to 6:30 PM); 5. Evening (7:30 PM to 10:30 PM); 6. Night (11:30 PM to 4:30 AM). Bar charts with a different letter are significantly different ($P<0.05$). The polynomial curve line is the software-derived trend for the bar chart (Vayssières et al. 2011b))

difference in *O. longinoda* activity between dry, rainy and harmattan seasons (Vayssières et al. 2011b). However, this may be different in the North Sudan and South Sahelian zones and requires further study.

Oecophylla longinoda density is crucial for biological control. The effectiveness of both *Oecophylla* species in controlling insect pests was positively correlated with high ant abundance on host trees (Van Mele et al. 2007; Peng and Christian 2008). For this reason, *O. longinoda* abundance is a key factor that should be regularly monitored to determine the optimal density for biological control. Various methods used to assess ant densities have been tested in Benin including three methods based on the number of ant trails on the main branches of a tree (called the Peng 1, Peng 2 and Offenberg indices) and one method based on the number of ant nests per tree. According to Wargui et al. (2015), nest counting is not recommended, whereas the Peng 1 index can track dynamics at low ant abundances and the Peng 2 and Offenberg indices can be used in most situations. The number of nests fluctuated widely suggested that their number was unlikely to reflect ant abundance, but rather the influence of tree phenology on nest-building behaviour (Wargui et al. 2015).

4 *Oecophylla smaragdina*, as a Biological Control Agent in Southern Asia Through to Northern Australia

In a 1992 review the two species, *O. smaragdina* and *O. longinoda*, were reported to effectively protect eight crops against different insect pests in many countries (Way and Khoo 1992). In a more recent review covering studies made between 2004 and 2014, the two species were shown to protect nine different crops against a number of different pests in eight different countries (Offenberg 2015). By combining previous reviews Peng and Christian (2004) were able to show that the two *Oecophylla* species controlled more than 50 different pest species in more than 12 different crops. However, the biological control potential of *Oecophylla* species is far better known in Asia than in Africa; research on *O. smaragdina* and the use of this ant for biological control is much more advanced in South East Asia and Australia than research on *O. longinoda* in Africa. Having said that, the use of *O. longinoda* in Africa is currently increasing, particularly in organic cashew, cocoa and mango production (Van Mele and Vayssières 2007a). Of particular relevance, the prey of *O. longinoda* includes fruit flies and these ants are increasingly being recognized as successful endemic natural enemies of fruit flies in mango (Peng and Christian 2006; Van Mele et al. 2007; Vayssières et al. 2013).

Oecophylla smaragdina is one of the most ancient biological control agents used against insect pests on citrus (Groff and Howard 1925). Old Chinese records report that *O. smaragdina* nests were being gathered, sold and established in citrus trees to control insect pests over 1700 years ago (Chen 1962). Chinese growers in the Canton area are still using *O. smaragdina* to control *Tesseratoma papillosa* (Drury) (Hemiptera Tessaratomidae) on lychees (*Litchi chinensis* Sonn.) (Jianzhong 1990). In the Mekong Delta of Vietnam, there is also a long tradition of *O. smaragdina* husbandry (Barzman et al. 1996; Van Mele and Cuc 2000). In Australia, *O. smaragdina* is currently used in biological control programmes against pests on cashew (Peng et al. 1995), mango (Peng and Christian 2005a) and mahogany, *Swietenia macophylla* King (Peng et al. 2011). In South-East Asian citrus plantations, *O. smaragdina* effectively controls many insects including key pests such as citrus bugs, *Rhynchocoris humeralis* (Thunb.), the aphids *Toxoptera aurantii* (Boyer de Fonscolombe) and *Toxoptera citricidus* (Kirkaldy), leaf-miners *Phyllocnistis citrella* Stainton and the weevil *Hypomeces squamosus* (Fabricius) (Groff and Howard 1925; Yang 1982; Huang and Yang 1987; Barzman et al. 1996; Van Mele and Van Lenteren 2002; Offenberg et al. 2013).

In Australian mango plantations, *O. smaragdina* effectively controls thrips *Selenothrips rubrocinctus* (Giard), leafhoppers *Idioscopus nitidulus* (Walker), the fruit fly, *Bactrocera jarvisi* (Tryon), mango seed weevils *Sternochetus mangiferae* (Fabricius) and mango bugs *Campylomma austrina* Malipatil (Peng and Christian 2004, 2005b, 2006, 2007, 2008). However, it has also been reported that *O. smaragdina* was unable to protect mango (Thai variety) against the leafhopper *Ideoscopus clypealis* (Lethierry) in Thailand (Offenberg et al. 2013). In Australian cashew plantations *O. smaragdina* effectively controls key pests such as the cashew bug *Helopeltis pernicialis* Stonedahl Malipatil & Houston, the fruit-spotting bug, *Amblypelta lutescens lutescens* Distant and also the moth *Anigraea ochrobasis* (Hampson) (Peng et al. 1997a, b; Peng and Christian 2005c).

5 *Oecophylla longinoda*, as a Biological Control Agent in Sub-Saharan Africa

In sub-Saharan Africa, the history of using *O. longinoda* to control insect pests is relatively new compared with South East Asia. There are reports of *O. longinoda* significantly reducing damage in cocoa due to *Distantiella theobroma* (Distant) (Room 1971; Majer 1972), and in coconut due to *Pseudotheraptus wayi* Brown (Simmonds 1924; Way 1951, 1953; Vanderplank 1960). Studies in Ghana have also shown that the presence of *O. longinoda* reduces the incidence of two serious diseases of cocoa that are transmitted by a mirid bug and they suggest that this was because *O. longinoda* workers were efficient in capturing phytophagous insects, especially the cacao mirid bugs (Leston 1973).

Recently *O. longinoda* has become more widely accepted as a precious biological control tool in Ghana, Guinea, Tanzania, Senegal and Benin (Ativor et al. 2012; Olotu et al. 2013; Diamé et al. 2015; Anato et al. 2015). In Ghana, Dwomoh et al. (2009) demonstrated that *O. longinoda* controlled sap-sucking bugs in cashew plamtations, even though the ants caused some disturbance to farmers during harvest. Studies in Benin demonstrated that the presence of *O. longinoda* reduced the incidence of pests such as cashew bugs (Fig. 18.12a) (Anato et al. 2015) and mango seed weevils (Fig. 18.12b) in orchards (Vayssières et al. unpublished data).

In Benin, *O. longinoda* was shown to be of economic significance in protecting mango orchards against fruit flies (Van Mele et al. 2007) even though predation of tephritid adults is, globally, rarely observed (Fig. 18.13a, b, c, d). The repulsive effect of 'ant-cues' left on the fruits are far more important than the predation issue (Vayssières et al. 2013). Unfortunately, about 50 % of the mango pickers considered *O. longinoda* as a nuisance (Sinzogan et al. 2008) and this percentage was even

Fig. 18.12 *O. longinoda* capturing: (**a**) *Sternochetus mangiferae* (Col.: Curculionidae) on a mango (Department of Atlantique, Benin) and (**b**) *Pseudotheraptus devastans* (Hem.: Coreidae) on cashew flowers (Department of Borgou, Benin) (Jean-François Vayssières credit)

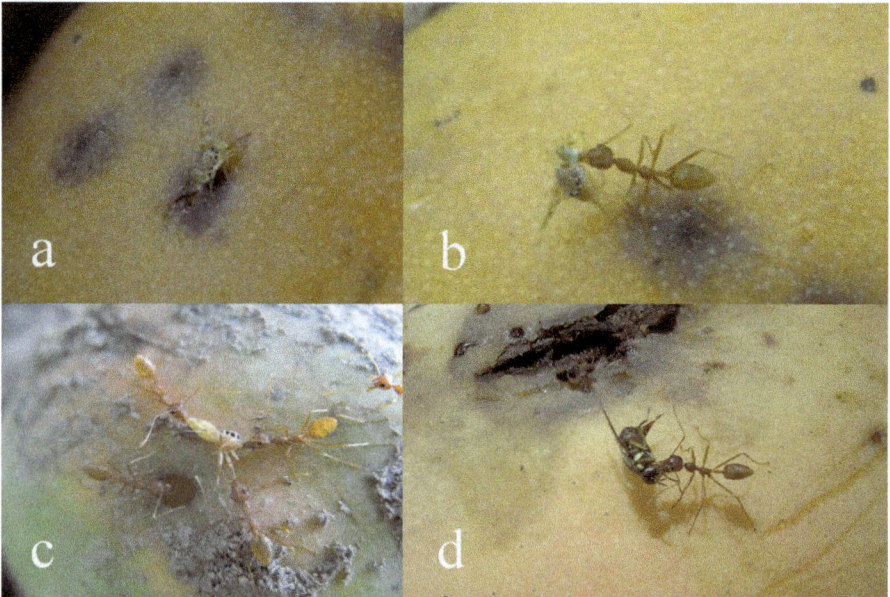

Fig. 18.13 *Ceratitis cosyra* males (**a**, **b**), female (**c**), and *B. dorsalis* female (**d**) captured... on mango by *O. longinoda* (Department of Borgou, Benin) (Jean-François Vayssières credit)

higher in some other West African countries such as Côte-d'Ivoire, Mali and Senegal among others (Diamé et al. 2015). Between 2009 and 2012, the West African Fruit Fly Initiative (WAFFI: a regional control programme to develop and promote sustainable area-wide integrated fruit fly management in West Africa), began to reverse this trend using regional participatory approaches. Adverse perceptions have now been revised and there is currently a more positive attitude towards *O. longinoda* in Ghana and Benin (Ouagoussounon et al. 2015). Consequently, *O. longinoda* has been included as an integral component within Integrated Pest Management (IPM) strategies against fruit pests in some West and East African countries (Dwomoh et al. 2008; Vayssières et al. 2011b; Seguni et al. 2011; Abdulla et al. 2015).

After an initial article demonstrating a key role for *O. longinoda* in mango orchards in Benin (Van Mele et al. 2007), subsequent field and laboratory studies have elucidated the tritrophic interactions between mangoes, the two fruit fly species *B. dorsalis* and *C. cosyra*, and *O. longinoda*. Laboratory experiments showed that: (a) fruit flies landed significantly more often on fruits from ant-free trees than on fruits from ant-colonized trees (Fig. 18.14); (b) time spent on mangoes was significantly lower for fruit on which ants had previously foraged and deposited ant cues (Fig. 18.15); (c) ant cues inhibited fruit fly oviposition behaviour (Fig. 18.16); (d) the concentration of ant-cues was significantly negatively correlated with the number of fruit fly pupae collected per kg fruit (Tables 18.3 and 18.4) as was the distance from ant-nests (Fig. 18.17) (Adandonon et al. 2009; Van Mele et al. 2009a; Vayssières et al. 2013). However, field observations revealed that there was no difference in damage to fruit collected at different distances from ant nests, suggesting that physical or visual mechanisms could complement the deterrent effect of ant

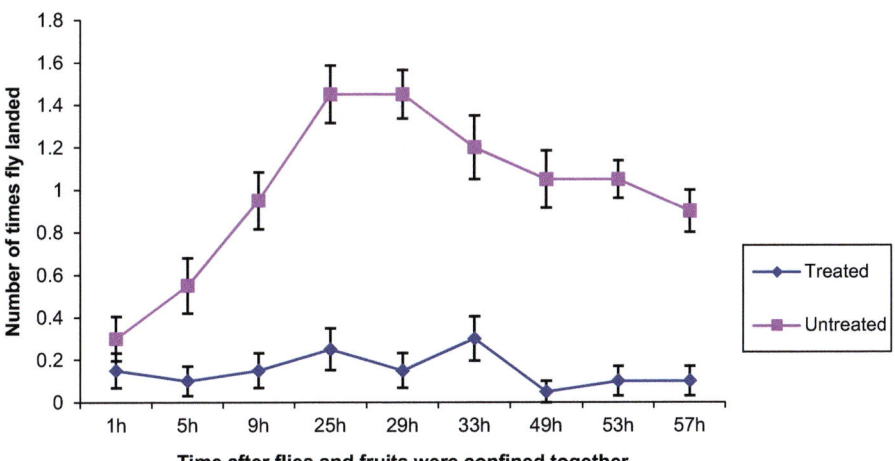

Fig. 18.14 Number of occasions that a fruitfly landed on mangoes in relation to ant treatment (mangoes that were exposed or unexposed to *O. longinoda*). Interactions between treatment and fruit fly species, *B. dorsalis* and *C. cosyra,* were not significant ($P=0.05$), hence data were pooled. Data are the mean ± SE

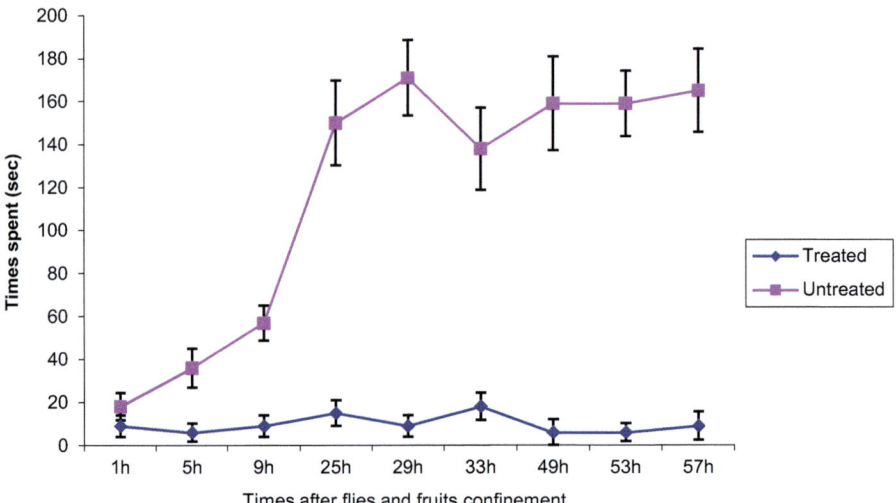

Fig. 18.15 Time spent, per fruit fly, walking on mangoes in relation to ant treatment (mangoes that were exposed or unexposed to *O. longinoda*). Interactions between treatment and fruit fly species, *B. dorsalis* and *C. cosyra*, were not significant ($P = 0.05$). Data are the mean ± SE

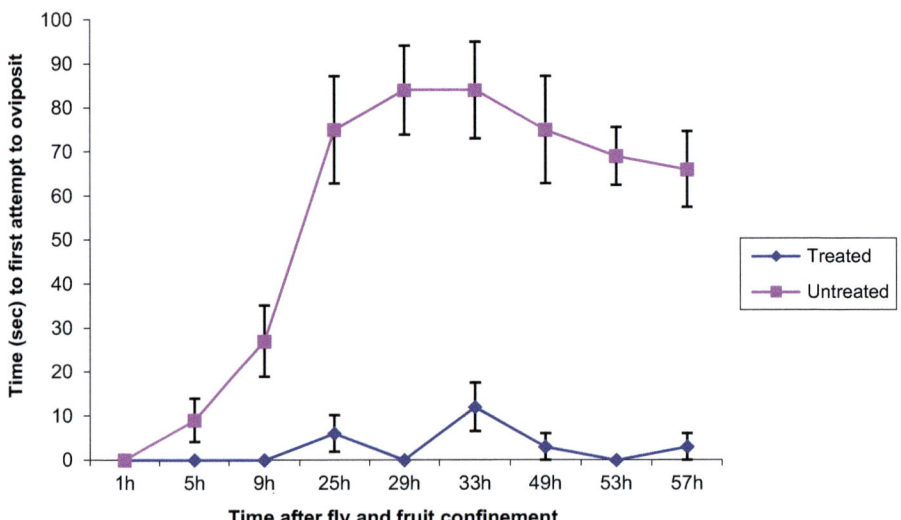

Fig. 18.16 Time to first fruit fly oviposition attempt to oviposit on mangoes in relation to ant treatment (mangoes that were exposed or unexposed to *O. longinoda*). Interactions between treatment and fruit fly species, *B. dorsalis* and *C. cosyra*, were not significant ($P = 0.05$). Data are the mean ± SE

Table 18.3 Number of fruit fly pupae present in mangoes in relation to the density of ant pheromone sources in a choice test (n=20)

	Number of pupae (*B. dorsalis* and *C. cosyra*) per kg of mango fruits[c]	
Treatment[a]	Initial[b]	Final
d<1 m	0	8.73±1.74 a
1 m<d<3 m	0	25.41±2.16 b
No ants	0	53.61±4.17 c

[a]Mango fruits were collected from fruits with *Oecophylla* within 1 m and 1–3 m distance from ant nests, as well as from trees without *Oecophylla*. The three treatments were put in the same cage and offered to tephritids for 72 h oviposition

[b]Fruits unconfined to tephritids were incubated for pupae emergence to test initial field infestation (Initial) while pupae emerging from 72 h fly-oviposited fruits was considered as greenhouse infestation (Final)

[c]Interaction between treatment and fly species was not significant ($P=0.05$). Each value is a mean (± SE) of 20 replicates. In the same column, values followed by a different letter are significantly different ($P<0.05$) according to Student-Newman Keuls test (Adandonon et al. 2009)

Table 18.4 Fruit fly damage of ant-marked and unmarked fruits

	Number of pupae per kg of mango fruits		
Treatment	Initial[a]	Final	
		B. dorsalis	*C. cosyra*
Oecophylla -marked	0	9.65±1.33 c	10.28±1.34 c
Unmarked	0	65.04±3.19 a	44.88±3.32 b

[a]Number of pupae from mangoes collected in the orchard and incubated for initial infestation. Each value is a mean (± SE) of 10 replicates. In the same column, values followed by a different letter are significantly different ($P<0.05$) according to GLM using Student-Newman Keuls test (Adandonon et al. 2009)

cues to fruit flies (Adandonon et al. 2009). Oviposition behaviour seemed to be independent of fruit fly species when ant cues were present, especially after fruit flies have landed on mangoes marked with ant cues (Vayssières et al. 2013). Based on the above studies it can be concluded that the presence of *O. longinoda* in mango orchards reduces the damage caused by tephritid fruit flies by: (i) rare predation of adult fruit flies (Fig. 18.13a, b); (ii) quite frequent predation of third instar larvae (Fig. 18.18); and especially via, (iii) the deterrent effect of ant cues deposited by *O. longinoda* on fruit and other parts of the tree. In Benin, early findings also suggest that ant cues protect fruits of *Citrus sinensis* (L.) Osbeck and *Citrus reticulata* Blanco from *B. dorsalis* oviposition (Wargui 2010). These studies now need to be repeated on other fruit tree species such as *Citrus paradisi* Macfad., which are also attacked by *B. dorsalis* throughout western Africa.

Because fruit flies are prey for *O. longinoda* it is not surprising that they detect and avoid ant-related semiochemicals, as shown for other herbivores that are preyed upon by ants (Offenberg et al. 2004; Adandonon et al. 2009; Van Mele et al. 2009a; Offenberg 2014). All major groups of arthropods (insects, millipedes and spiders) are captured by *O. longinoda* within their host trees as well as in secondary territories (Fig. 18.19a, b, c, d). Regular weekly monitoring of prey capture and food scaveng-

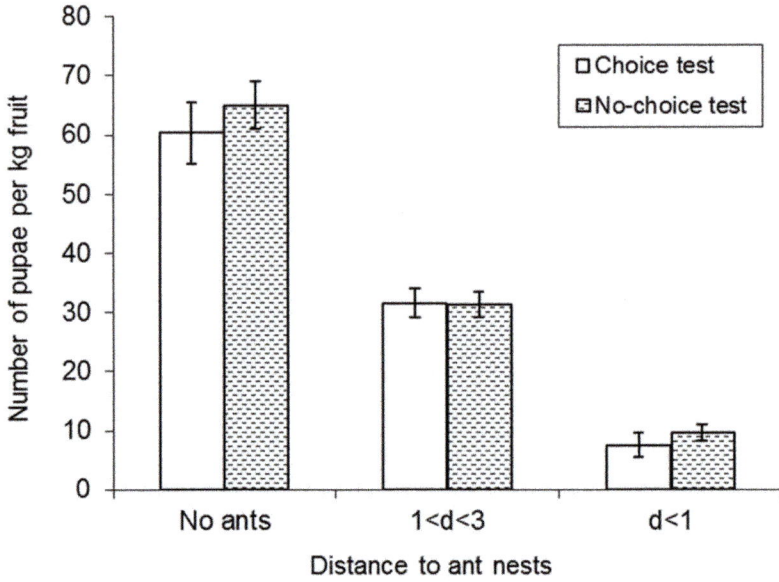

Fig. 18.17 Mean+SE number of *B. dorsalis* pupae per kg fruit in relation to the density of ant pheromone sources. Mango fruits were collected from trees with *O. longinoda* within 1 m and 1–3 m distance from ant nests, as well as from trees without *O. longinoda*. In the choice test, fruit flies were offered the three mangoes at the same time, whereas in the no-choice test, flies were offered fruit from one treatment at a time

Fig. 18.18 Tephritid larvae captured on a mango (cv Eldon) by *O. longinoda* (Department of Borgou, Benin) (Jean-François Vayssières credit)

Fig. 18.19 *O. longinoda* capturing: (**a**) *Pachycondyla* sp. (Hym.: Formicidae) on a mango tree, (**b**) *Dysdercus* sp. (Hem.: Pyrrhocoreidae) on a mango (also with a *Dorylus* sp. to the left), (**c**) Nitidulidae (Col.) on a mango, (**d**) larva of *Euschmidtia* sp. (Orth.: Euschmidtiidae) under a mango tree (Jean-François Vayssières credit)

ing activities of *O. longinoda* in a large mango orchard in Benin over two consecutive years (2009–2010) have shown that, in both years, there were similar patterns of insect groups and plant debris recovered in nests (Table 18.5). During this study a total of 241 species of insects were recovered from ant nests; this included 61 species associated specifically with mango of which 48 were pest species representing 78.7 % of the species associated with mango (Vayssières et al. 2015b).

Oecophylla longinoda may not only protect crops against invertebrate pests. A study of the perceptions of mango and cashew growers in Guinea showed that more than half of the growers said that *O. longinoda* helped protect their orchards against theft as people are afraid to be bitten by the ants (Van Mele et al. 2009b). Growers also indicated that *O. longinoda* deterred snakes and regularly mentioned that *O. longinoda* reduced damage by fruit-eating bats; some farmers said that bats disliked the smell of *O. longinoda* (Van Mele et al. 2009b). Another advantage is the improvement of physicochemical, microbiological and organoleptic properties of mangoes picked in Benin from trees with *O. longinoda* nests compared with fruit picked from trees without *O. longinoda* (Houngbo 2011).

Table 18.5 Relationship between the mango fruit developmental stages and the insect groups and plant debris recovered from *Oecophylla longinoda* nests (from Vayssières et al. 2015b)

Fruit stage	N	Hymeno Mean±S.E	Hemiptera Mean±S.E	Isoptera Mean±S.E	Coleoptera Mean±S.E	Diptera Mean±S.E	Other_insects Mean±S.E	Debris Mean±S.E	TOTAL_Contents Mean±S.E	Temp Mean±S.E	RH Mean±S.E
Flowering	11	36.3±7.0 b	0.3±0.2 c	0.0±0.0 c	0.2±0.2 d	0.6±0.2 c	25.6±8.2 cd	39.0±10.0 a	102.0±10.5 bc	28.4±0.5 b	26.0±2.7 d
Fruit growing	18	78.1±5.3 a	0.9±0.4 b	0.0±0.0 c	1.8±0.5 c	0.6±0.2 c	16.9±4.0 d	6.0±1.2 b	104.4±7.3 bc	31.3±0.5 a	40.0±2.9 c
Pre-maturity	6	27.7±8.2 b	6.7±1.6 a	3.0±2.6 ab	4.0±1.5 bc	4.2±1.4 b	34.5±11.4 bc	0.8±0.4 c	80.8±15.9 c	31.2±1.2 a	54.8±6.4 bc
Maturity	27	15.4±2.6 c	4.7±1.1 a	6.7±2.0 a	26.2±4.4 a	28.6±3.5 a	71.4±9.0 a	0.1±0.1 c	153.0±9.9 a	28.2±0.4 b	71.7±1.7 a
Without fruit	42	13.2±2.3 c	2.1±0.4 b	2.9±0.9 b	4.6±0.9 b	2.0±0.4 b	58.6±7.0 ab	52.1±8.4 a	135.5±7.7 ab	27.6±0.2 b	60.0±2.8 b
F-value (4, 53)		26,07	6,33	4,85	21,05	56,37	18,59	11,03	5,00	11,48	18,05
P-value		<0.0001	0,0003	0,0021	<0.0001	<0.0001	<0.0001	<0.0001	0,0017	<0.0001	<0.0001

Linear correlations between different six groups of insect species, plant debris and also between insect species – debris and abiotic factors (temperature and relative humidity) in relation to different mango tree stages. In the same column, values followed by a different letter are significantly different ($P<0.05$) according to ANOVA (Vayssières et al. 2015b)

Effects of *O. longinoda* may not only be positive as they may have negative effects on other beneficial species. An important issue to consider is the potential predation and disturbance of tephritid parasitoids (Hymenoptera: Braconidae) by *O. longinoda*. This has long been a point of discussion by colleagues involved in fruit fly biological control using parasitoids. However, during our own field studies (2005–2012) in all the agroecological zones of Benin, we never saw any predation of parasitoids (*Fopius* species, *Psyttalia* species, *Diachasmimorpha* species) by *O. longinoda* (Vayssières et al., unpublished data). Adult parasitoids were often observed foraging on the same mango fruits as ant workers without the ants attacking them. In confined conditions, however, the results could be quite different: presence of *O. longinoda* inhibiting the activity of *Fopius arisanus* (Sonan) has been observed (Appiah et al. 2014). According to Aluja and Birke (2003), however, valid conclusions should only be deduced from experiments carried out under natural conditions (vs confined conditions). Furthermore, Peng and Christian (2013) showed that *O. longinoda*, in the field, either benefited or had no impact on natural enemy diversity and abundance. They argued that ants improved the health of their host trees and that this enhanced subsequent conditions thereby attracting more arthropods, including natural enemies of pests.

Other ant species than *O. longinoda* may also have potential as biological control agents in orchards. Throughout the world, examples of benefits from ants as predators of fruit pests have been demonstrated. In Brazil, for example, *Pheidole* species are important predators of *Anastrepha* species (Fernandes et al. 2012) and in south Morocco, the ant *Monomorium subopacum* Mayr was by far the most efficient predator of larvae of the fruit fly *C. capitata*, under argan trees (*Argania spinosa* (L.) Skeels) (El Keroumia et al. 2010). Also the ant *Wasmania auropunctata* Roger has been observed to protect cocoa from pests in Cameroon (Bruneau De Miré 1969) and the ant *Dolichoderus thoracicus* (Smith) has been observed to protect sapodilla, *Manilkara sapota* (L.) from pests in Vietnam (Van Mele and Cuc 2001).

6 Research Gaps

6.1 Main Research Interests

Some of the most important research topics that must be covered in order to develop and improve the use of *O. longinoda* in tropical African agriculture are: (i) characterizing the genetic structure of different *O. longinoda* populations in Africa and their agroecological diversity because "some ant populations seem to be more efficient predators than others"; (ii) characterizing the semiochemical deposits made by ants and quantifying the persistence of their deterrent properties against fruit flies and other pests; (iii) developing methods to conserve and increase *O. longinoda* populations when conditions are suboptimal; (iv) assessing the feasibility of *O. longinoda* as biological control agents in more crops; (v) elucidating the mechanisms behind the improvement of physicochemical, microbiological and organoleptic properties of mangoes picked from mango trees with *O. longinoda*; and lastly (vi) mapping the main factors influencing the activity patterns of *O. longinoda* in relation to the different agroecological zones. The latter point is especially important for West Africa where WAFFI is encouraging the use of *O. longinoda* within IPM strategies.

6.2 Additional Research Interests

Apart from their biological control activities, *O. smaragdina* are also used in South East Asia as edible insect protein and as a healthy human food (Offenberg et al. 2010; Offenberg 2011; Van Huis et al. 2013; Van Itterbeeck 2014), and also in traditional medicine (Chen and Alue 1994; Oudhia 2002). These services should also be investigated for *O. longinoda* throughout sub-Saharan Africa.

7 Conclusion

Since 2005 we have suggested that the use of *O. longinoda* colonies for biological control seems well suited for perennial cropping systems in Benin because these generalist predators are constantly available, widespread, effective and efficient against many tree pests, and self-regenerating (Vayssières 2007; Ouagoussounon et al. 2013; Anato et al. 2015; Wargui et al. 2015; Vayssières et al. 2015b). In this review we have shown that *O. longinoda* can enhance both the quality and quantity of fruit production in mango (Van Mele et al. 2007) and cashew orchards (Anato et al. 2015) and preliminary experiments in Benin have shown that similar effects may also be achieved in citrus orchards (Wargui 2010). Furthermore, emerging African markets for organic and sustainably-managed fruit and nut production may promote further interest in the use of *O. longinoda* (Van Mele and Vayssières 2007b). Other African countries such as Ghana (Dwomoh et al. 2009), Senegal (Diamé et al. 2015) and Tanzania (Abdulla et al. 2015; Kirkegaard et al. 2015) are also increasing research on the applied aspects of *O. longinoda* and are providing interesting new results (Table 18.6).

Long term trends suggest that growing attention is being paid to biological control and particularly the use of *O. longinoda* (Table 18.7). Following successful biological control campaigns against the cassava mealybug, *Phenococcus manihoti* Matile-Ferrero and the mango mealybug, *Rastrococcus invadens* Williams in Africa in the 1980s (Herren and Neuenschwander 1991; Neuenschwander 2003), confidence in this technology has increased resulting in new investment into classical biological control. Accordingly, records of publications on biological control increased fourfold between the early 1980s and the early 2010s (Table 18.7). Of these 16–25 % have dealt with predators. Although ants are one of the most abun-

Table 18.6 Synthesis of research in the last 40 years on *Oecophylla* species as beneficial predators

Crop	Ant species	Country	Pest	Pest reduction	Damage reduction	Yield increase	Quality improved	Relation with orchard type	Comparison to alternatives	Cost-benefit analysis	Reference
Mango	*O. longinoda*	Senegal	*Bactrocera dorsalis* (Hendel)	+	+	+		+			Diamé et al. 2015
	O. longinoda	Benin	*B. dorsalis, Ceratitis cosyra* (Walker)	+	+	+					VanMele et al. 2007
	O. longinoda	Benin	*B. dorsalis, C. cosyra*	+	+						Adandonon et al. 2009; Van Mele et al. 2009a, b; Vayssières et al. 2013
	O. longinoda	Benin	*B. dorsalis, C. cosyra*	+	+		+				Houngbo 2011
	O. smaragdina	Thailand	*Idioscopus clypealis* (Lethierry)	–	–	–			–	–	Offenberg et al. 2013
	O. smaragdina	Australia	*Bactrocera jarvisi* (Tryon)	+	+	+			+		Peng and Christian 2006
	O. smaragdina	Australia	*Sternochetus mangiferae* (Fabricius)	+	+	+			+		Peng and Christian 2007
	O. smaragdina	Australia	several mango pests	+	+	+			+	+	Peng and Christian 2005b
	O. smaragdina	Australia	*Campylomma austrina* Malipatil	+	+	+			+		Peng and Christian 2008
	O. smaragdina	Australia	Mealybugs	–	–						Peng and Christian 2005b
	O. smaragdina	India	several mango pests (*B. dorsalis* included)	+	+	+					Bharti and Silla 2011

(continued)

Table 18.6 (continued)

Crop	Ant species	Country	Pest	Pest reduction	Damage reduction	Yield increase	Quality improved	Relation with orchard type	Comparison to alternatives	Cost-benefit analysis	Reference
Citrus	O. longinoda	Ghana	B. dorsalis, Ceratitis ditissima (Munro)	+	+	+			+		Ativor et al. 2012
	O. longinoda	Benin	B. dorsalis	+	+						Wargui 2010
	O. smaragdina	Thailand	Hypomeces squamosus (Fabricius)	+	+	n			+	n	Offenberg et al. 2013
	O. smaragdina	Vietnam	pest species not specified	+	+	+			+	+	Offenberg et al. 2013
	O. smaragdina	India	several citrus pests	+	+	+					Bharti and Silla 2011
	O. smaragdina	Vietnam	several citrus pests						+	+	Van Mele and Cuc 2000
	O. smaragdina	China	Tessaratoma papillosa (Drury) & other Heteroptera	+	+	+					Huang and Yang 1987
	O. smaragdina	China	Rhynchoris humeralis (Thnb.)	+	+	+					Yang 1982

18 The Use of Weaver Ants in the Management of Fruit Flies in Africa

Crop	Ant sp.	Country	Pests						Reference
Cashew	*O. longinoda*	Benin	*Anoplocnemis curvipes* (Fabricius); *Helopeltis schoutedeni* Reuter; *Mirperus jaculus* (Thnb.); *Pseudotheraptus devastans* (Distant); *Tupalus fasciatus* (Dallas)	+	+	+	+		Anato et al. 2015
	O. longinoda	Ghana	*A. curvipes*; *H. schoutedeni*; *P. devastans*	+	+	+		+	Dwomoh et al. 2009
	O. longinoda	Ghana	*A. curvipes*; *H. schoutedeni*; *P. devastans*	+	+	+		+	Aidoo 2009
	O. longinoda	Tanzania	*Helopeltis* spp.; *Pseudotheraptus wayi* Brown	+	+		+		Olotu et al. 2013
	O. smaragdina	Australia	*Helopeltis pernicialis* Stonedahl; *Solenothrips rubrocinctus* (Giard) and other cashew pests	+	+	+	+	+	Peng et al. 1997a, b; Peng and Christian 2004

Table 18.7 Changes in the number of publications related to biological control and particularly the use of weaver ants over time (CAB abstracts)

Keywords	1965–1969	1970–1974	1975–1979	1980–1984	1985–1989	1990–1994	1995–1999	2000–2004	2005–2009	2010–2014
Biological control	186	6575	13,134	16,404	23,479	27,065	27,393	38,430	53,892	71,175
Predator	59	1665	3424	3902	5776	6197	6548	7759	9754	11,879
Ant & biological control	2	106	234	311	394	377	373	499	686	774
Both *Oecophylla* species	0	15	25	18	20	23	40	54	62	88
Oecophylla longinoda	0	7	15	6	3	10	13	9	15	19
Oecophylla smaragdina	0	8	10	12	17	13	27	45	47	69

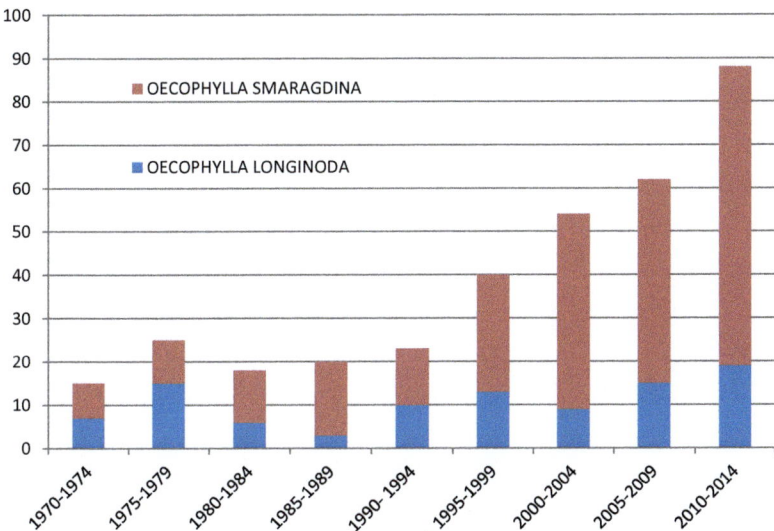

Fig. 18.20 Number of publications relating to *O. longinoda* and *O. smaragdina* in the CAB Abstracts database

dant arthropod groups, only about 6 % of publications centered on predators address the role of ants. The relative proportion of records dealing with ants in conjunction with biological control remained rather stable; about 10 % of publications on predators are devoted to *Oecophylla* species, predominantly *O. smaragdina* (Fig. 18.20). All these data came from CAB Abstracts database subsets and were adjusted for potential overlap between biological control and ecology.

Oecophylla longinoda could offer a valuable and substantial contribution to sustainable biological control and IPM of fruit flies. The knowledge is incomplete and seems to be restricted, for the moment, to Ghana, Tanzania and Benin. Recommendations should be provided to growers throughout sub-Saharan Africa who want to use *O. longinoda* as a tool in conservation biological control. According to Offenberg (2015), both *O. longinoda* and *O. smaragdina* provide examples of documented and efficient conservation biological control. To better exploit their biological control activities, protection of *O. longinoda* colonies in fruit plantations in the African savannahs and forests should be promoted in stakeholder awareness campaings across sub-Saharan Africa. Moreover, considering all the additional benefits (physicochemical, microbiological, organoleptic, food, medicinal properties) this should really be a primary goal.

Lastly, we would like to draw attention to the tremendous deforestation of tropical forests and savannas, which inexorably leads to irreversible losses in biodiversity (Ahrends et al. 2010; Norris et al. 2010; Sodhi et al. 2009). Protection of tropical forests and savannas is crucial to halt biodiversity losses, but also, more specifically, to protect *O. longinoda*, which can provide so many services to mankind.

Acknowledgements The authors thank the growers in Benin for their encouragement, interest and assistance in protecting *O. longinoda*. We thank Manuele Tamo for his continuous support and Brian Taylor, Jean-François Germain and Aristide Adomou for taxonomic identification of Formicidae, Coccidae and tree species respectively. This research was part of the West African Fruit Fly Initiative (WAFFI) and helped to promote *O. longinoda* as a key element of conservation biological control activities within the IPM packages proposed to growers. Thanks are due to WAFFI staff in Parakou and Cotonou. This Research for Development was supported by donors, particularly WB-EU, STDF, CFH and DANIDA to whom the authors are very grateful. Thanks also to IITA and CIRAD for their support.

Dedication This chapter is dedicated to our friend and colleague, Serge Quilici, who always encouraged our activities with the greatest enthusiasm, generosity and practical assistance.

References

Abdulla N, Rwegasira G, Jensen K-M, Mwatawala M, Offenberg J (2015) Effect of supplementary feeding of *Oecophylla longinoda* on their abundance and predatory activities against cashew insect pests. Biocontr Sci & Technol. doi:10.1080/09583157.2015.1057476

Adandonon A, Vayssières J-F, Sinzogan A, Van Mele P (2009) Density of pheromone sources of the weaver ant *Oecophylla longinoda* (Hymenoptera: Formicidae) affects oviposition behaviour and damage by mango fruit flies (Diptera Tephritidae). Int J Pest Man 55:85–292

Ahrends A, Burgess N, Milledge S, Bulling M, Fisher B, Smart J, Clarke G, Mhoro B, Lewis S (2010) Predictable waves of sequential forest degradation and biodiversity loss spreading from an African city. PNAS 107:14556–14561

Aidoo KS (2009) Boosting cashew production in Ghana. Bees Dev 91:8–9

Aluja M, Birke A (2003) Panorama general sobre los principios éticos aplicables a la investigación científica y educación superior. En: Aluja M, Birke A (eds) El Papel de la Ética en la Investigación Científica y la Educación Superior, (Capítulo 3). Academia Mexicana de Ciencias, México D.F, pp 35–75

Amevoin K, Sanbena B, Nuto Y, Gomina M, De Meyer M, Glitho IA (2009) Les mouches des fruits (Diptera: Tephritidae) au Togo: Inventaire, prévalence et dynamique des populations dans la zone urbaine de Lomé. Int J Biol Chem Sci 3:912–920

Anato F, Wargui R, Sinzogan A, Offenberg J, Adandonon A, Kossou D, Vayssières J-F (2015) Reducing losses inflicted by insect pests on cashew, using weaver ants as efficient biological control agent. Agr For Entomol. doi: 10.1111/afe.12105

Appiah EF, Afreh-Nuamah K, Obeng-Ofori D (2009) Abundance and distribution of the Mediterranean fruit fly *Ceratitis capitata* (Diptera: Tephritidae), in Late Valencia citrus orchards in Ghana. Int J Trop Ins Sci. doi:10.1017/S1742758409351036

Appiah EF, Ekesi S, Afreh-Nuamah K, Obeng-Ofori D, Mohamed S (2014) African weaver ant-produced semiochemicals impact on foraging behavior and parasitism by the Opiine parasitoid, *Fopius arisanus* on *Bactrocera invadens* (Diptera: Tephritidae). Biocontrol 79:49–57

Ativor IN, Afreh-Nuamah K, Billah M, Obeng-Ofori D (2012) Weaver ant *Oecophylla longinoda* (Latreille) (Hymenoptera Formicidae) activity reduces fruit fly damage in citrus orchards. J Agr Sci Technol A 2:449–458

Barzman MZ, Mills NJ, Cuc NTT (1996) Traditional knowledge and rationale for weaver ant husbandry in the Mekong delta of Vietnam. Agr Hum Values 13:1–9

Bharti H, Silla S (2011) Notes on life history of *Oecophylla smaragdina* and its potential as biological control agent. Halteres 3:57–64

Billah M, Kimani-Njogu S, Wharton RA, Woolley J, Masiga D (2008) Comparison of five allopatric fruit fly parasitoid populations (*Psyttalia* species) (Hymenoptera: Braconidae) from coffee fields using morphometric and molecular methods. Bull Entomol Res 98:63–75

Blüthgen N, Stork NE (2007) Ant mosaics in a tropical rainforest in Australia and elsewhere: a critical review. Austral Ecol 32:93–104

Bluthgen N, Stork NE, Fiedler K (2004) Bottom-up control and co-occurrence in complex communities: honeydew and nectar determine a rainforest ant mosaic. Oikos 106:344–358

Bruneau De Miré P (1969) Une fourmi utilisée au Cameroun dans la lutte contre les mirides du cacaoyer *Wasmania auropunctata* Roger. Café Cacao Thé 13:209–212

Chauvin R (1952) Sur la reconstruction du nid chez les fourmis oecophylles. Behaviour 4:190–201

Chen S (1962) The oldest practice of biological control: the culture and efficacy of *Oecophylla smaragdina* Fabr. in orange orchards. Acta Entomol Sin 11:401–407

Chen Y, Alue RD (1994) Ants used as food and medicine in China. Food Insects Newsl 7:1–10

Cole AC, Jones JW (1948) A study of the weaver ant, *Oecophylla smaragdina*. Am Mid Nat 39:641–651

Copeland RS, Wharton RA, Luke Q, De Meyer M, Lux S, Zenz N, Machera P, Okumu M (2006) Geographic distribution, host fruit, and parasitoids of African fruit fly pests *Ceratitis anonae*, *Ceratitis cosyra*, *Ceratitis fasciventris*, and *Ceratitis rosa* (Diptera: Tephritidae) in Kenya. Ann Entomol Soc Am 99:261–278

Correia AR, Rego J, Olmi M (2008) A pest of significant economic importance detected for the first time in Mozambique: *Bactrocera invadens* Drew, Tsuruta & White (Diptera: Tephritidae: Dacinae). Boll Zool Agrarian di Bachicoltura Ser U 40:9–13

Crozier RH, Newey PS, Schlüns H, Robson SK (2009) A masterpiece of evolution *Oecophylla* weaver ants (Hymenoptera: Formicidae). Myrmecol News 13:57–71

De Meyer M (2001) Distribution patterns and host-plant relationships within the genus *Ceratitis* MacLeay (Diptera Tephritidae) in Africa. Cimbebasia 17:219–228

De Meyer M (2005) Phylogenetic relationships within the fruit fly genus *Ceratitis* MacLeay (Diptera Tephritidae) derived from morphological and host plant evidence. Ins Syst Evol 36:459–479

De Meyer M, Mohamed M, White IM (2007) Invasive fruit fly pests in Africa. A diagnostic tool and information references for the four Asian species of fruit fly that have become accidentally established as pests in Africa, including the Indian Ocean Islands. Available online at http://www.africamuseum.be/fruitfly/AfroAsia.htm

De Meyer M, Delatte H, Mwatawala M, Quilici S, Vayssières J-F, Virgilio M (2015) A review of the current knowledge on *Zeugodacus cucurbitae* (Diptera: Tephritidae) in Africa. Zookeys 540:539–557

De Villiers M, Hattingh V, Kriticos D (2013) Combining field phenological observations with distribution data to model the potential distribution of the fruit fly *Ceratitis rosa* Karsch (Diptera: Tephritidae). Bull Entomol Res 103:60–73

Dejean A (1986) Predation of the ant *Oecophylla longinoda* in Zaïre. 37th Annual AIBS. Meeting. University of Massachussets, Amherst. Assoc Trop Biol p 10

Dejean A (1990) Circadian rhythm of *Oecophylla longinoda* in relation with territoriality and predatory behaviour. Physiol Entomol 15:393–403

Dejean A (1991) Adaptation d'*Oecophylla longinoda* (Formicidae-Formicinae) aux variations spatio-temporelles de proies. Entomophaga 36:29–54

Dejean A, Akoa A, Djieto-Lordon C, Lenoir A (1994) Mosaic ant territories in African secondary rain forest (Hymenoptera: Formicidae). Sociobiol 23:275–292

Dejean A, Corbara B, Orivel J (1999) The arboreal ant mosaic in two Atlantic rain forests. Selbyana 20:133–145

Dejean A, Corbara B, Orivel J, Leponce M (2007) Rainforest canopy ants: the implications of territoriality and predatory behavior. Funct Ecosyst Comm 1:105–120

Diamé L, Grechi I, Rey J-Y, Sané CAB, Diatta P, Vayssières JF, Yasmine A, De Bon H, Diarra K (2015) Influence of *Oecophylla longinoda* Latreille (Hymenoptera: Formicidae) on mango infestation by *Bactrocera dorsalis* (Hendel) (Diptera: Tephritidae) in relation to Senegalese orchard design and management practices. Afr Entomol 23:294–305

Dwomoh EA, Afun JV, Ackonor JB, Agene VN (2009) Investigations on *Oecophylla longinoda* Latreille (Hymenoptera: Formicidae) as a biocontrol agent in the protection of cashew plantations. Pest Man Sci 65:41–46

Ekesi S, Maniania NK, Lux SA (2003) Effect of soil temperature and moisture on survival and infectivity of *Metarhizium anisopliae* to four tephritid fruit fly puparia. J Invertebr Pathol 83:157–167

Ekesi S, Maniania N, Mohamed S, Lux S (2005) Effect of soil application of different formulations of *Metarhizium anisopliae* on African tephritid fruit flies and their associated endoparasitoids. Biol Con 35:83–91

Ekesi S, Nderitu P, Rwomushana I (2006) Field infestation, life history and demographic parameters of the fruit fly *Bactrocera invadens* (Diptera: Tephritidae) in Africa. Bull Entomol Res 96:379–386

Ekesi S, Dimbi S, Maniania N (2007) The role of entomopathogenic fungi in the integrated management of tephritid fruit flies (Diptera: Tephritidae) with emphasis on species ccurring in Africa. In: Ekesi S, Maniania NK (eds) Use of entomopathogenic fungi in biological pest management. Research Sign Post, Kerala, pp 239–274

Ekesi S, Billah M, Nderitu P, Lux S, Rwomushana I (2009) Evidence for competitive displacement of *Ceratitis cosyra* by the invasive fruit fly *Bactrocera invadens* (Diptera: Tephritidae) on mango and mechanisms contributing to the displacement. J Econ Entomol 102:981–991

Ekesi S, Maniania N, Mohamed S (2011) Efficacy of soil application of *Metarhizium anisopliae* and the use of GF-120 spinosad bait spray for suppression of *Bactrocera invadens* (Diptera: Tephritidae) in mango orchards. Biocontrol Sci Tech 21:299–316

El Keroumia A, Naamania K, Dahbib A, Luquec I, Carvajalc A, Cerdac X, Boulayc R (2010) Effect of ant predation and abiotic factors on the mortality of medfly larvae, *Ceratitis capitata*, in the Argan forest of Western Morocco. Biocon Sci Technol 20:751–762

Fernandes WD, Sant'Ana MV, Raizer J, Lange D (2012) Predation of fruit fly larvae *Anastrepha* (Diptera: Tephritidae) by ants in grove. Psyche. doi:10.1155/2012/108389

Geurts K, Mwatawala M, De Meyer M (2012) Indigenous and invasive fruit fly diversity along an altitudinal transect in Eastern central Tanzania. J Ins Sci 12:1536–2442

Geurts K, Mwatawala M, De Meyer M (2014) Dominance of an invasive fruit fly species, *Bactrocera invadens*, along an altitudinal transect in Morogoro (Eastern central Tanzania). Bull Entomol Res. doi:10.1017/S0007485313000722

Goergen G, Vayssières J-F, Gnanvossou D, Tindo M (2011) *Bactrocera invadens* (Diptera Tephritidae), a new invasive fruit fly pest for the Afrotropical region: host plant range and distribution in west and central Africa. Environ Entomol 40:844–854

Gomina M, Mondedji A, Nyamador W, Vayssières J-F, Amevouin K, Glitho AI (2014) Development and demographic parameters of *Bactrocera invadens* (Diptera: Tephritidae) in Guinean climatic zone of Togo. Int J Nat Sci Res 2:263–277

Groff WG, Howard CW (1925) The cultured citrus ant of South China. Ling Agric Rev 2:108–114

Grové T, De Beer MS, Joubert PH (2009) Monitoring fruit flies in mango orchards in South Africa and determining the time of infestation. Acta Hortic 820:589–596

Hala NF, Quilici S, Gnago A, N'depo OR, Kouassi K, Allou K (2006) Status of fruit flies (Diptera: Tephritidae) in Côte d'Ivoire and implications for mango exports. In: Proceeding of 7th international symposium on fruit flies of economic importance, September 10–15, 2006, Salvadore, pp 233–239

Herren HR, Neuenschwander P (1991) Biological control of cassava pests in Africa. Ann Entomol Soc Am 36:257–283

Hill MP, Terblanche JS (2014) Niche overlap of congeneric invaders supports a single-species hypothesis and provides insight into future invasion risk: implications for global management of the *Bactrocera dorsalis* complex. Plos One. doi:10.1371/journal.pone.0090121

Hölldobler B, Wilson EO (1977) Weaver ants: social establishment and maintenance of territory. Science 195:900–902

Hölldobler B, Wilson EO (1978) The multiple recruitment systems of the African weaver ant *Oecophylla longinoda* (Latreille). Behav Ecol Sociobiol 3:19–60

Hölldobler B, Wilson EO (1990) The ants. Harvard University Press, Cambridge, MA, p 732

Houngbo H (2011) Evaluation de la qualité nutritionnelle, microbiologique et organoleptique de la mangue (*Mangifera indica*) protégée par les fourmis rouges (*Oecophylla longinoda*). Mémoire du Master en Normes, Contrôle de Qualité et Technologie Alimentaire. Universite D'Abomey-Calavi, Cotonou, p 64

Huang HT, Yang P (1987) The ancient cultural citrus ant, a tropical ant is used to control insect pests in southern China. Bioscience 37:665–671

Jianzhong B (1990) Progress of the biological control research and application in China. In: Proceedings of the first Asia-Pacific conference Entomol., Chiang Mai, Thailand, 8–13 November 1989, pp 493–503. Entomological and Zoological Associations of Thailand, Bangkok. pp 855

José D, Santos L (2013) Assessment of invasive fruit fly infestation and damage in Cabo Delgado province, northern Mozambique. Afr Crop Sci J 21:21–28

Kirkegaard N, Offenberg J, Msogoya TJ, Grout BW (2015) Indigenous weaver ants and fruit fly control in Tanzanian smallholder mango production. Acta Hortic 1111:355–362

Labuschagne TI, Brink T, Steyn WP, De Beer MS (1996) Fruit flies attacking mangoes – their importance and post-harvest control. Suid-Afrikaanse Mango Kwekersvereninging Jaarboek 16:17–19

Ledoux A (1950) Recherche sur la biologie de la fourmi fileuse (*Oecophylla longinoda* Latreille). Ann Sci Nat 12:313–461

Leston D (1973) The ant mosaic, tropical tree crops and the limiting of pests and diseases. Pest Art News Sum 19:311–341

Lokkers C (1986) The distribution of the weaver ant, *Oecophylla smaragdina*, in northern Australia. Aust J Zool 34:683–687

Lokkers C (1990) Colony dynamics of the green tree ant (*Oecophylla smaragdina* Fab.) in a seasonal tropical climate. PhD thesis, James Cook University of North Queensland, pp 301

Luckman G (2004) A helping hand for Sudan. Bull Aus Quarantine Insp Serv 05, 04.

Lux SA, Copeland RS, White IM, Manrakhan A, Billah M (2003) A new invasive fruit fly species from the *Bactrocera dorsalis* (Hendel) group detected in East Africa. Insect Sci Appl 23:355–361

Majer JD (1972) The ant-mosaic in Ghana cocoa farms. Bull Entomol Res 62:151–160

Manrakhan A, Venter J, Hattingh V (2015) The progressive invasion of *Bactrocera dorsalis* (Diptera Tephritidae) in South Africa. Biol Inv. doi 10.1007/s10530-015-0923-2

Mayamba A, Nankinga CK, Isabirye B, Akol AM (2014) Seasonal population fluctuations of *Bactrocera invadens* (Diptera: Tephritidae) in relation to mango phenology in the Lake Victoria Crescent, Uganda. Fruits 69:473–480

Mwatawala MW, White IM, Maerere AP, Senkendo FJ, De Meyer M (2004) A new invasive *Bactrocera* species (Diptera Tephritidae) in Tanzania. Afr Entomol 12:154–156

Mwatawala MW, De Meyer M, Makundi R, Maerere A (2006a) Seasonality and host utilization of the invasive fruit fly, *Bactrocera invadens* (Diptera Tephritidae) in central Tanzania. J Appl Entomol 130:530–537

Mwatawala MW, De Meyer M, Makundi R, Maerere A (2006b) Biodiversity of fruit flies (Diptera: Tephritidae) in orchards in different agro-ecological zones of the Morogoro region, Tanzania. Fruits 61:321–332

Mwatawala MW, De Meyer M, Makundi R, Maerere A (2009a) Host range and distribution of fruit infesting pestiferous fruit flies (Diptera Tephritidae) in selected areas of Central Tanzania. Bull Entomol Res 99:629–641

Mwatawala MW, De Meyer M, Makundi R, Maerere A (2009b) An overview of *Bactrocera* (Diptera: Tephritidae) invasions and their speculated dominancy over native fruit fly species in Tanzania. J Entomol 6:18–27

N'Dépo OR, Hala NF, Allou K, Aboua LR, Kouassi KP, Vayssières J-F, De Meyer M (2009) Abondance des mouches des fruits dans les zones de productions fruitières de Côte d'Ivoire: dynamique des populations de *Bactrocera invadens* (Diptera Tephritidae). Fruits 64:313–324

N'Dépo OR, Hala NF, Gnago A, Allou K, Kouassi KP, Vayssières J-F, De Meyer M (2010) Inventory of fruit flies of three agroecological areas and host plants associated to the new species Bactrocera (Bactrocera) invadens Drew et al., 2005 (Diptera Tephritidae) in Côte d'Ivoire. Eur J Sci Res 46:62–72

N'Dépo OR, Hala NF, Adopo A, Coulibaly F, Kouassi K, Vayssières J-F, De Meyer M (2013) Fruit flies (Diptera Tephritidae) populations dynamic in mangoes production zone of Côte d'Ivoire. Agric Sci Res J 3:352–363

Ndiaye M, Dieng EO, Delhove G (2008) Population dynamics and on-farm fruit fly integrated pest management in mango orchards in the natural area of Niayes (Senegal). Pest Manag Hortic Ecosyst 14:1–8

Ndiaye O, Vayssières J-F, Rey J-Y, Ndiaye S, Diedhiou P, Ba CT, Diatta P (2012) Seasonality and range of fruit fly (Diptera Tephritidae) host plants in Niayes and the Thiès plateau (Senegal). Fruits 67:311–331

Ndiaye O, Ndiaye S, Djiba S, Ba CT, Vaughan L, Rey J-Y, Vayssières J-F (2015) Preliminary surveys after release of the fruit fly parasitoid *Fopius arisanus* Sonan (Hymenoptera Braconidae) in mango production systems in Casamance (Senegal). Fruits 70:1–9

Ndzana Abanda FX, Quilici S, Vayssières J-F, Kouodiekong L, Woin N (2008) Inventaire des espèces de mouches des fruits sur goyave dans la région de Yaoundé au Cameroun. Fruits 63:19–26

Neuenschwander P (2003) Biological control of cassava and mango mealybugs in Africa. In: Neuenschwander P, Borgemeister C, Langewald J (eds) Biological control in integrated pest management systems in Africa. CABI Publishing, Wallingford, pp 45–59

Ngamo L, Ladang D, Vayssières J-F, Lyannaz J-P (2010) Diversité des espèces de mouches des fruits (Diptera Tephritidae) dans un verger mixte dans la localité de Malang (Ngaoundéré, Cameroun). Int J Biol Chem Sci 4:1425–1434

Norris K, Asase A, Collen B, Gockowksi J, Mason J, Phalan B, Wadea A (2010) Biodiversity in a forest-agriculture mosaic – The changing face of West African rainforests. Biol Conserv 143:2341–2350

Offenberg J (2011) *Oecophylla smaragdina* food conversion efficiency: prospects for ant farming. J Appl Entomol 135:575–581

Offenberg J (2014) Pest repelling properties of ant pheromones. IOBC-WPRS Bull 99:173–176

Offenberg J (2015) Ants as tool in sustainable agriculture. J Appl Ecol doi: 10.1111/1365-2664.12496

Offenberg J, Nielsen M, Mc Intosh D, Havanon S, Aksornkoae S (2004) Evidence that insect herbivores are deterred by ant pheromones. Biol Lett 271:434–435

Offenberg J, Cuc TTN, Wiwatwitaya D (2010) Sustainable weaver ant (*Oecophylla smaragdina*) farming: harvest yields and effects on worker ant density. Asian Myrmecol 3:55–62

Offenberg J, Cuc TTN, Wiwatwitaya D (2013) The effectiveness of weaver ant (*Oecophylla smaragdina*) biocontrol in Southeast Asian citrus and mango. Asian Myrmecol 5:139–149

Olotu MI, Plessis H, Seguni Z, Maniania NK (2013) Efficacy of the African weaver ant *Oecophylla longinoda* (Hymenoptera: Formicidae) in the control of *Helopeltis* spp. (Hemiptera: Miridae) and *Pseudotheraptus wayi* (Hemiptera: Coreidae) in cashew crop in Tanzania. Pest Man Sci 69:911–918

Ouagoussounon I, Sinzogan A, Offenberg J, Adandonon A, Vayssières J-F, Kossou D (2013) Pupae transplantation to boost early colony growth in the weaver ant *Oecophylla longinoda* Latreille (Hymenoptera: Formicidae). Sociobiol 60:374–379

Ouagoussounon I, Offenberg J, Sinzogan A, Adandonon A, Kossou D, Vayssières J-F (2015) Founding weaver ant queens increase production and nanitic worker size. Springer Plus. doi:10.1186/2193-1801-4-6

Oudhia P (2002) Traditional medicinal knowledge about red ant *Oecophylla smaragdina* (Fab.) [Hymenoptera: Formicidae] in Chhattisgarh, India. Ins Environ 8:114–115

Ouedraogo S (2011) Dynamique spatio-temporelle des mouches des fruits (Diptera Tephritidae) en fonction des facteurs biotiques et abiotiques dans les vergers de manguiers de l'Ouest du Burkina. Thèse de Doctorat, Paris Est, p 184

Ouedraogo S, Vayssières J-F, Dabiré R, Rouland-Lefèvre C (2011) Biodiversité des mouches des fruits (Diptera Tephritidae) en vergers de manguiers de l'Ouest du Burkina Faso: structure et comparaison des communautés de différents sites. Fruits 66:393–404

Peng R, Christian K (2004) The weaver ant, *Oecophylla smaragdina* (F.) (Hymenoptera: Formicidae), an effective biological control agent of the red-banded thrips, *Selenothrips rubrocinctus* (Thysanoptera: Thripidae) in mango crops in the Northern Territory of Australia. Int J Pest Man 50:107–114

Peng R, Christian K (2005a) Integrated pest management in mango orchards in the Northern Territory Australia, using the weaver ant, *Oecophylla smaragdina*, (Hymenoptera Formicidae) a key element. Int J Pest Man 51:149–155

Peng R, Christian K (2005b) Integrated pest management for mango orchards using green ants as a major component. A manual for conventional and organic mango growers in Australia. School of Science and Primary Industries Charles Darwin University, Darwin, p 53

Peng R, Christian K (2005c) Ecology of the fruit spotting bug, *Amblypelta lutescens lutescens* distant (Hemiptera: Coreidae) in cashew plantations, with particular reference to the potential for its biological control. Aus J Entomol 44:45–51

Peng R, Christian K (2006) Effective control of Jarvi's fruit fly, *Bactrocera jarvisi* (Diptera Tephritidae), by the weaver ant, *Oecophylla smaragdina* (Hymenoptera: Formicidae) in mango orchards in the Northern Territory of Australia. Int J Pest Man 52:275–282

Peng R, Christian K (2007) The effect of the weaver ant, *Oecophylla smaragdina* (Hymenoptera: Formicidae), on the mango seed-weevil, *Sternochetus mangiferae* (Coleoptera: Curculionidae) in mango orchards in the Northern Territory of Australia. Int J Pest Man 53:15–24

Peng R, Christian K (2008) The dimpling bug, *Campylomma austrina* Malipatil (Hemiptera: Miridae): the damage and its relationship with ants in mango orchards in the Northern Territory of Australia. Int J Pest Man 54:173–179

Peng R, Christian K (2013) Do weaver ants affect arthropod diversity and the natural- enemy-to-pest ratio in horticultural systems? J Appl Entomol 137:711–720

Peng R, Christian K, Gibb K (1995) The effect of the green ant, *Oecophylla smaragdina* (Hymenoptera Formicidae), on insect pests of cashew trees in Australia. Bull Entomol Res 85:279–284

Peng R, Christian K, Gibb K (1997a) Control threshold analysis for the tea mosquito bug, *Helopeltis pernicialis* (Hemiptera Miridae) and preliminary results concerning the efficiency of control by the green ant, *Oecophylla smaragdina* (F.) (Hymenoptera: Formicidae), in northern Australia. Int J Pest Man 43:233–237

Peng R, Christian K, Gibb K (1997b) Distribution of the green ant, *Oecophylla smaragdina* (F.) (Hymenoptera: Formicidae), in relative to native vegetation and the insect pests in cashew plantations in Australia. Int J Pest Man 43:203–211

Peng R, Christian K, Reilly D (2011) Weaver ants, *Oecophylla smaragdina* (Hymenoptera: Formicidae), as biocontrol agents on African mahogany trees, *Kaya senegalensis* (Sapindales: Meliaceae), in the Northern Territory of Australia. Int J Pest Man 56:363–370

Pinkalski C, Damgaard C, Jensen KM, Gislum R, Peng R, Offenberg J (2015) Non-destructive biomass estimation of *Oecophylla smaragdina* colonies: a model species for the ecological impact of ants. Ins Cons Div. doi:10.1111/icad.12126

Pruvost O, Boyer C, Vital K, Vernière C, Gagnevin L, de Bruno AL, Rey J-Y (2011a) First report in Ghana of *Xanthomonas citri* pv. *mangiferaeindicae* causing mango bacterial canker on *Mangifera indica* L. Plant Dis 95:6

Pruvost O, Boyer C, Vital K, Vernière C, Gagnevin L, Somda I (2011b) First report in Burkina Faso of *Xanthomonas citri* pv. *mangiferaeindicae* causing bacterial canker on *Mangifera indica* L. Plant Dis 95:10

Quilici S, Rousse P (2012) Location of host and host habitat by fruit fly parasitoids. Insects 3:1220–1235

Room PM (1971) The relative distributions of ant species in Ghana's cocoa farms. JAE 40:735–741

Rugman-Jones P, Wharton RA, van Noort T, Stouthamer R (2009) Molecular differentiation of the *Psyttalia concolor* (Szépligeti) species complex (Hymenoptera: Braconidae) associated with olive fly, *Bactrocera oleae* (Rossi) (Diptera: Tephritidae), in Africa. Biol Cont 49:17–26

Rwegasira RG, Mwatawala M, Rwegasira GM, Offenberg J (2015) Occurrence of sexuals of African weaver ant (*Oecophylla longinoda* Latreille) (Hymenoptera: Formicidae) under a bimodal rainfall pattern in eastern Tanzania. Bull Entomol Res 105:182–186

Rwomushana I, Ekesi S, Gordon I, Ogol C (2008) Host plant and host plant preference studies for *Bactrocera invadens* (Diptera: Tephritidae) in Kenya, a new invasive fruit fly species in Africa. Ann Entomol Soc Am 101:331–340

Rwomushana I, Ekesi S, Ogol C, Gordon I (2009) Mechanisms contributing to the competitive success of the invasive fruit fly *Bactrocera invadens* over the indigenous mango fruit fly, *Ceratitis cosyra*: the role of temperature and resource pre-emption. Entomol Exp Appl 133:27–37

Salah FE, Abdelgader H, De Villiers M (2012) The occurrence of the peach fruit fly, *Bactrocera zonata* (Saunders) (Tephritidae), in Sudan. pp 128, In TEAM 2nd international meeting: biological invasions of tephritidae: ecological and economic impacts, 3–6 July 2012, Kolymbari

Salum JK, Mwatawala M, Kusolwa P, De Meyer M (2013) Demographic parameters of the two main fruit fly (Diptera Tephritidae) species attacking mango in Central Tanzania. J Appl Entomol. doi:10.1111/jen.12044

Schutze MK, Mahmood K, Pavasovic A, Bo W et al (2014a) One and the same: integrative taxonomic evidence that the African Invasive Fruit Fly *Bactrocera invadens* (Diptera: Tephritidae), is the same species as the Oriental Fruit Fly *Bactrocera dorsalis*. Syst Entomol. doi:10.1111/syen.12114

Schutze MK, Aketarawong N, Amornsak W, Armstrong K et al (2014b) Synonymization of key pest species within the *Bactrocera dorsalis* species complex (Diptera: Tephritidae): taxonomic changes based on a review of 20 years of integrative morphological, molecular, cytogenetic, behavioural and chemoecological data. Syst Entomol. doi:10.1111/syen.12113

Seguni ZS, Way MJ, Van Mele P (2011) The effect of ground vegetation management on competition between the ants *Oecophylla longinoda* and *Pheidole megacephala*. Crop Prot 30:713–717

Simmonds HW (1924) Mission to New Guinea, Bismarck Solomons and New Hebrides. Council Papua Fidji 2:13

Sinzogan A, Van Mele P, Vayssières J-F (2008) Implications of on-farm research for local knowledge related to fruit flies and the weaver ant *Oecophylla longinoda* in mango production. Int J Pest Man 54:241–246

Sivinski JM, Calkins C, Baranowski R, Harris D, Brambila J, Dıaz J, Burns R, Holler T, Dodson G (1996) Suppression of a Caribbean fruit fly (*Anastrepha suspensa* (Loew) Diptera: Tephritidae) populations through augmented releases of the parasitoid *Diachasmimorpha longicaudata* (Ashmead) (Hymenoptera: Braconidae). Biol Cont 6:177–185

Sodhi N, Koh L, Clements R, Wanger T, Hill J, Hamer K, Clough Y, Tscharntke T, Posa M, Lee T (2009) Conserving Southeast Asian forest biodiversity in human-modified landscapes. Biol Conserv 143:2375–2384

Sudd JH (1963) How insects work in groups. Discovery, London 24:15–19

Taylor B, Adedoyin SF (1978) The abundance and interspecific relations of common ant species (Hymenoptera Formicidae) on cocoa farms in Western Nigeria. Bull Entomol Res 8:105–121

Umeh V, Garcia L, De Meyer M (2008) Fruit flies of Citrus in Nigeria: diversity, relative abundance and spread in major producing areas. Fruits 63:145–153

Van Huis A, van Itterbeeck J, Klunder H, Mertens E, Alloran E, Muir G, Vantomme P (2013) Edible insects: future prospects for food and feed security. Wageningen UR; FAO forestry paper n° = 171, pp 201

Van Itterbeeck J (2014) Prospects of semi-cultivating the edible weaver ant *Oecophylla smaragdina*. PhD thesis. Wageningen University, Wageningen, pp 111

Van Mele P, Cuc NTT (1999) Predatory ants in orchards in the Mekong Delta of Vietnam. Biological control in the tropics: towards efficient biodiversity and bioresource management for effective biological control. In: Wright and Hong (eds) Proceedings of the international symposium on biological control in the tropics, CAB International Serdang, pp 118–122

Van Mele P, Cuc NTT (2000) Evolution and status of *Oecophylla smaragdina* as a pest control agent in citrus in the Mekong Delta. Int J Pest Man 46:295–301

Van Mele P, Cuc NTT (2001) Farmers perception and practices in use of *Dolichoderus thoracicus* (Smith) (Hymenoptera Formicidae) for biological control of pests of sapodilla. Biol Cont 20:23–29

Van Mele P, Van Lenteren JC (2002) Survey of current crop management practices in a mixed-ricefield landscape, Mekong Delta, Vietnam—potential of habitat manipulation for improved control of citrus leafminer and citrus red mite. Agric Ecosyst Environ 88:35–48

Van Mele P, Vayssières J-F (2007a) West Africa's mango farmers have allies in the trees. Biocont News Inf 28:56–58

Van Mele P, Vayssières J-F (2007b) Weaver ants help farmers to capture organic markets. Alt Pest News 75:9–11

Van Mele P, Vayssières J-F, Van Tellingen E, Vrolijks J (2007) Effects of an African weaver ant, *Oecophylla longinoda*, in controlling mango fruit flies (Diptera Tephritidae) in Benin. J Econ Entomol 100:695–701

Van Mele P, Vayssières J-F, Adandonon A, Sinzogan A (2009a) Ant cues affect the oviposition behaviour of fruit flies (Diptera Tephritidae) in Africa. Physiol Entomol 34:256–261

Van Mele P, Camara K, Vayssières J-F (2009b) Thieves, bats and fruit flies: local ecological knowledge on the weaver ant *Oecophylla longinoda* in relation to three "invisible" intruders in orchards in Guinea. Int J Pest Man 55:57–61

Van Melle C, Buschmann S (2013) Comparative analysis of mango value chain models in Benin, Burkina Faso and Ghana. In: Elbehri A (ed) Rebuilding West Africa's food potential. FAO/IFAD, Rome, pp 317–345

Vanderplank FL (1960) The bionomics and ecology of the red ant tree, *Oecophylla* sp., and its relationship to the coconut bug *Pseudotheraptus wayi* Brown (Coreidae). JAE 29:15–33

Varela Barros AM (1992) Role of *Oecophylla longinoda* (Formicidae) in control of *Pseudotheraptus wayi* (Coreidae) on coconuts in Tanzania. PhD thesis. Imperial College, Silwood Park, pp 231

Vargas RI, Stark J, Uchida G, Purcell M (1993) Opiine parasitoids (Hymenoptera: Braconidae) of Oriental fruit fly (Diptera: Tephritidae) on Kauai Island, Hawaii: islandwide relative abundance and parasitism rates in wild and orchard guava habitats. Environ Entomol 22:246–253

Vargas RI, Leblanc L, Putoa R, Eitam A (2007) Impact of introduction of *Bactrocera dorsalis* (Diptera: Tephritidae) and classical biological control releases of *Fopius arisanus* (Hymenoptera: Braconidae) on economically important fruit flies in French Polynesia. J Econ Entomol 100:670–679

Vayssières J-F (1995) Rapport d'activités du conseiller auprès du chef de la filière fruits au Centre de Recherche Agronomique de Foulaya (Projet FAC-IRAG). MIN COOP-IRAG, Conakry, p 96

Vayssières J-F (2004) Rapport de mission sur les Tephritidae au Sénégal du 11 au 20 décembre 2004. Consultation du CIRAD pour l'Union Européenne. Bruxelles, Belgique

Vayssières J-F (2007) L'IITA dans la lutte contre les mouches des fruits du manguier. Information letter of COLEACP and CIRAD: fighting fruit flies regionally in sub-Saharan Africa 1:3

Vayssières J-F (2012) Tri-trophic relations in food web structure of tropical horticultural crops with ant and fruit fly complexes. DHDR memoir. Paris: University of Paris Est, pp 158

Vayssières J-F, Kalabane S (2000) Inventory and fluctuation of the catches of Diptera Tephritidae associated with mangoes in coastal Guinea. Fruits 55:259–270

Vayssières J-F, Mangassouba BA, Diop MA (1991) Projet Recherche et Développement en zone de palmeraie (Kouroudjel-Assaba et Kaedi-Gorgol). MIN COOP-CNRA, Nouakchott, p 88

Vayssières J-F, Sanogo F, Noussourou M (2004) Inventaire des espèces de mouches des fruits (Diptera Tephritidae) inféodées au manguier au Mali et essais de lutte raisonnée. Fruits 59:3–16

Vayssières J-F, Goergen G, Lokossou O, Dossa P, Akponon C (2005) A new *Bactrocera* species detected in Benin among mango fruit flies (Diptera Tephritidae) species. Fruits 60:371–377

Vayssières J-F, Korie K, Coulibaly O, Temple L, Boueyi S (2008) The mango tree in central and northern Benin: cultivar inventory, yield assessment, infested stages and loss due to fruit flies (Diptera Tephritidae). Fruits 63:335–348

Vayssières J-F, Korie S, Ayegnon D (2009a) Correlation of fruit fly (Diptera Tephritidae) infestation of major mango cultivars in Borgou (Benin) with abiotic and biotic factors and assessment of damages. Crop Prot 28:477–488

Vayssières J-F, Sinzogan A, Ouagoussounon I, Korie S, Thomas-Odjo A (2009b) Effectiveness of spinosad bait sprays (GF-120) in controlling mango-infesting fruit flies (Diptera Tephritidae) in Benin. J Econ Entomol 102:515–521

Vayssières J-F, Sinzogan A, Adandonon A (2010a) New developments in the fruit fly issue in West Africa. Information letter of COLEACP and CIRAD: fighting fruit flies regionally in sub-Saharan Africa 11:3

Vayssières J-F, Sinzogan A, Adandonon A, Korie S (2010b) Diversity of fruit fly species (Diptera Tephritidae) associated with citrus crops (Rutaceae) in southern Benin in 2008–2009. Int J Biol Chem Scie 4:1881–1897

Vayssières J-F, Adandonon A, Sinzogan A, Ayegnon D, Ouagoussounon I, Modjibou S (2010c) Main wild fruit trees of Guineo-Sudanian zones of Benin: inventory, period of production and losses due to fruit flies. Global Sci Book 4:42–46

Vayssières J-F, Wharton R, Adandonon A, Sinzogan A (2011a) Preliminary inventory of parasitoids (Hymenoptera) associated with fruit flies in mangoes, guavas, cashew, pepper and wild fruit crops in Benin. Biocontrol 56:35–43

Vayssières J-F, Sinzogan A, Korie S, Adandonon A, Worou S (2011b) Field observational studies on circadian activity pattern of *Oecophylla longinoda* (Latreille) (Hymenoptera Formicidae) in relation to abiotic factors and mango cultivars. Int J Biol Chem Sci 5:790–802

Vayssières J-F, Adandonon A, N'Diaye O, Sinzogan A, Kooymann C, Badji K, Rey J-Y, Wharton R (2012) Native parasitoids associated with fruit flies (Diptera Tephritidae) in cultivated and wild fruit crops in Casamance (Senegal) during the season 2010. Afr Entomol 20:308–315

Vayssières J-F, Sinzogan A, Adandonon A, Van Mele P, Korie S (2013) Ovipositional behavior of two mango fruit fly species in relation to *Oecophylla* cues as compared to natural conditions without ant cues. Int J Biol Chem Sci 7:447–456

Vayssières J-F, Sinzogan A, Adandonon A, Rey J-Y, Dieng E, Camara K, Sangaré M, Ouédraogo S, Hala N, Sidibé A, Keita Y, Gogovor G, Korie S, Coulibaly O, Kikissagbé C, Tossou A, Billah M, Biney K, Nobime O, Diatta P, N'Dépo R, Noussourou M, Traoré L, Saizonou S, Tamo M (2014) Annual population dynamics of mango fruit flies (Diptera Tephritidae) in West

Africa: socio-economic aspects, host phenology and implications for management. Fruits 69:207–222

Vayssières J-F, De Meyer M, Ouagoussounon I, Sinzogan A, Adandonon A, Korie S, Wargui R, Anato F, Houngbo H, Didier C, De Bon H, Goergen G (2015a) Seasonal abundance of mango fruit flies (Diptera: Tephritidae) and ecological implications for their management in mango and cashew orchards in Benin (Centre & North). J Econ Entomol. doi:10.1093/jee/tov143

Vayssières J-F, Ouagoussounon I, Adandonon A, Sinzogan A, Korie S, Todjihoundé R, Alassane S, Wargui R, Anato F, Goergen G (2015b) Seasonal pattern in food gathering of the weaver ant *Oecophylla longinoda* (Hymenoptera: Formicidae) in mango orchards in Benin. Biocon Sci Technol. doi:10.1080/09583157.2015.1048425

Virgilio M, Backeljau T, Emeleme R, Juakali J, De Meyer M (2011) A quantitative comparison of frugivorous tephritids (Diptera: Tephritidae) in tropical forests and rural areas of the Democratic Republic of Congo. Bull Entomol Res 101:591–597

Wargui R (2010) Pheromonal effect of the weaver ant *Oecophylla longinoda* (Hymenoptera: Formicidae) on citrus fruit flies (Diptera Tephritidae). IITA report. pp 14

Wargui R, Offenberg J, Sinzogan A, Adandonon A, Kossou D, Vayssières J-F (2015) Comparing different methods to assess weaver ant abundance in plantation trees. Asian Myrmecol 7:1–12

Way MJ (1951) An insect pest of coconuts and its relationship to certain ant species. Nature 168:302

Way MJ (1953) The relationship between certain ant species with particular reference to the biological control of the coreid, *Theraptus* sp. Bull Entomol Res 44:669–691

Way MJ (1954a) Studies on the life history and ecology of the ant *Oecophylla longinoda* Latreille. Bull Entomol Res 45:93–112

Way MJ (1954b) Studies on the association of the ant *Oecophylla longinoda* Latreille (Formicidae) with the scale insect *Saissetia zanzibarensis* Williams (Coccidae). Bull Entomol Res 45:113–134

Way MJ (1963) Mutualism between ants and honeydew producing Homoptera. Ann Rev Entomol 8:307–344

Way MJ, Khoo KC (1992) Role of ants in pest management. Ann Rev Entomol 37:479–503

Wharton RA (1989) Classical biological control of fruit infesting Tephritidae. In: Robinson AS, Hooper G (eds) Fruit flies their biology, natural enemies and control, vol 3B. Elsevier, Amsterdam, pp 303–313

Wharton RA (1997) Generic relationships of opiine Braconidae (Hymenoptera) parasitic on fruit-infesting Tephritidae (Diptera). Contrib Am Entomol Inst 30:1–53

Wheeler WM (1910) Ants: their structure, development and behavior. Columbia University Press, New York, p 663

Wheeler WM (1922) Ants of the American museum congo expedition. A contribution to the myrmecology of Africa. Bull Am Mus Nat Hist 45:1–1139

White IM (2006) Taxonomy of the Dacina (Diptera Tephritidae) of Africa and the Middle East. Afr Entomol Memoir 2:1–156

White IM, Elson-Harris M (1992) Fruit flies of significance: their identification and bionomics. CABI, ACIAR, Redwood Press, Wallingford, p 601

Wilson EO (1971) The insect societies. Belknap Press of Harvard University Press, Cambridge, p 548

Yang P (1982) Biology of the yellow citrus ant, *Oecophylla smaragdina*, and its utilization against insect pests. Acta Scie Nat Université Sunyatseni 3:102–105

Zakari O, Ratnadass A, Vayssières J-F, Nikiema A, Fatondji D, Salha H, Aboubacar K, Ryckewaert P, Pasternak D (2012) GF-120 effects on fruit fly species (Diptera Tephritidae) in Sahelian agroforestry-based horticultural cropping systems. Fruits 67:333–339

Zombré C, Sangara P, Pruvost O, Boyer C, Vernière C, Adandonon A, Vayssières J-F, Ahohuendo BC (2015) First report of *Xanthomonas citri* pv. *mangiferaeindicae* causing mango bacterial canker on *Mangifera indica* in Benin. Plant Dis. doi:10.1094/PDIS-04-15-0392-PDN

J. -F. Vayssières is a senior entomologist attached to the Centre de Coopération Internationale en Recherche Agronomique pour le Développement (CIRAD) in Montpellier (France) specializing in agricultural entomology. Previously he worked in La Réunion, French Guiana and West Africa, including Benin where he was based at the International Institute of Tropical Agriculture IITA for 11 years. His research interests include ecology of pest-natural enemy complexes on tropical fruits (food web structure) and biodiversity issues. He has been involved in regional IPM projects focused on fruit fly management such as the 'Regional control program against the Carambola Fruit Fly' in South America. In West Africa he was one of the principal investigators of the projects 'Fight against fruit flies in West Africa' (WB-EU), 'West African fruit fly fighting initiative' (WB-WTO) and 'Mise au point et diffusion de méthodes alternatives et durables de lutte contre les mouches des fruit en Afrique de l'Ouest' (UEMOA) (also known as WAFFI). In Benin, he participated in the project 'Increasing value of African mango and cashew production' (DANIDA) and was involved in the project 'Fruit fly control technologies and dissemination' (CORAF).

J. Offenberg is senior researcher at Aarhus University in Denmark, specializing in ant-plant interactions. Since 2001 he has been studying tropical weaver ants and how they can be used in agriculture, including pest control, nutrient dynamics between ants and plants, and the utilization of ant larvae as a source of edible protein. Currently these research topics are also being addressed with temperate ants in temperate agriculture. He has published more than 50 peer reviewed scientific papers on ants and their interactions with plants and works as an associated editor for the journal, Asian Myrmecology.

A. Sinzogan is a *Beta-gamma* scientist (Beta stand for biological sciences and gamma for social sciences) working in the area of agricultural crop protection research. He has more than 10-years experience as a field researcher and university lecturer, a good understanding of what is required for managing team research and a good scientific publication record. Current research interests include using innovative agricultural research to provide viable methods for moving towards sustainable integrated pest management (IPM). This involves linking various technologies (developed jointly with farmers) to parallel capacity building, thereby ensuring long lasting benefits at both the farm and community level. The crops concerned are mango and cotton and the control methods involved are IPM with natural pesticides (neem in conjunction with GF120 lures) and natural enemies (weaver ants).

A. Adandonon is an Assistant Professor at the University of Agriculture of Kétou, specializing in Crop Protection and with a focus on plant pathology and agricultural entomology. For the past 10 years his research interests have not only been on the incidence and integrated management and reduction of fruit fly damage in West Africa but also on pests and diseases of other crops including citrus, *Jatropha*, soybean, cowpea and mango. Through the West African Fruit Fly Initiative project he conducted applied research on the development of fruit fly control technologies in eight West African countries. He currently coordinates, at the national level, a technology dissemination project on control of fruit flies in the Republic of Benin. He has published a number of papers and book chapters on fruit flies. He is a reviewer for many journals including the Journal of Applied Biosciences, Crop Protection, European Journal of Plant Pathology, Global Research Journal of Microbiology, Journal of Agricultural Sciences and African Plant Protection.

R. Wargui is a research entomologist specializing in crop science, agricultural entomology, and with an emphasis on Integrated Pest Management on perennial tree crops. She graduated in January 2016 at the University of Abomey-Calavi, Benin. Previously, she worked as a research assistant at the International Institute of Tropical Agriculture (IITA) on the West African Fruit Fly Initiative (WAFFI). Her research focuses on the use of weaver ants, *Oecophylla longinoda* Latreille (Hymenoptera: Formicidae), for biological control of insect pests associated with economically important fruit crop species in Benin.

F. Anato is a young entomologist who has worked with various insect species since 2006. Her main field of expertise is biological control using parasitoïds and predators. She has worked on rice stem borer, termites and fruit flies at AfricaRice and the International Institute of Tropical Agriculture (IITA), Benin. For her PhD degree, she studied biological control of pests on mango and cashew using weaver ants as part of the project 'Increasing value of African mango and cashew' based in the University of Abomey-Calavi, Benin.

H.Y. Houngbo is currently a PhD student working in food technology and nutrition. She graduated with her Masters degree in food quality control and technology from the University of Abomey Calavi, Benin in November 2012. Previously, she worked as a research assistant on nutritional, microbiological and chemical quality of mango fruits protected by weaver ants, as part of the project 'Increasing value of African mango and cashew' based in the University of Africa Calavi, Benin. Her work focuses on possible impacts on fruit quality of the use of weaver ants for biological control of fruits flies in mango orchards.

I. Ouagoussounon is a young entomologist specializing in crop science and agricultural entomology. He graduated in January 2016 at the University of Abomey-Calavi, Benin. Previously, he worked as a research assistant at the International Institute of Tropical Agriculture (IITA) on the West African Fruit Fly Initiative (WAFFI) and, in Belgium, on aphid glycosyl hydrolase inhibitors. His research focuses on strategies to improve rearing methods for colonies of the weaver ant, *Oecophylla longinoda* Latreille (Hymenoptera: Formicidae), for use in biological control of fruit flies in Benin.

L. Diamé received his PhD in Ecology and Ecosystems Management from Cheikh Anta Diop University, Dakar, Senegal, and previously a Masters degree in Management of Horticultural Agroecosystems from the same university. His main field of expertise is with the ant fauna of Sahelian ecosystems and their use in biological control, with a particular emphasis on use of the weaver ant, *Oecophylla longinoda*, for control of fruit flies. His research includes characterization of Senegalese agroecosystems in relation to fruit fly populations.

S. Quilici* was a senior entomologist attached to the Centre de Coopération Internationale en Recherche Agronomique pour le Développement (CIRAD) in Montpellier (France) specializing in agricultural entomology. Based in La Réunion, he was a world renowned expert on fruit flies and contributed significantly to improve the scientific knowledge on the biology and ecology of this important taxonomic group. He was a longstanding member of the international committee of the International Congress of Fruit Flies and the'Conférence Internationale Francophone d'Entomologie' (CIFE). He was also an international expert working for Ministère de l'Agriculture, de l'Agro-alimentaire et de la Forêt (MAAF), the International Atomic Energy Agency (IAEA) and the United States Department for Agriculture (USDA). Serge Quilici has always been very concerned by the practical application of his results, so he engaged early in his carrer with promoting integrated pest management, especially within the International Organisation of Biological Control (IOBC). His profound entomological knowledge allowed him to develop bio-technical methods for improving trapping systems and biological control approaches by successfully establishing hymenopteran parasitoids. As a passionate naturalist, he was personaly engaged in several committees of the Ministry of Environment and was also active in nature conservation.

J.-Y. Rey is a senior agronomist attached to the Centre de Coopération Internationale en Recherche Agronomique pour le Développement (CIRAD) in Montpellier, France. He graduated from the Ecole Nationale Supérieure d'Horticulture de Versailles and did his PhD at the University of Montpellier II. Although currently based in Senegal at the Institut Sénégalais de Recherche Agricole, he has worked in various countries in West and Central Africa on tropical fruit crops for more than 45 years. His research interests concern the improvement of sustainable fruit tree-based systems, particularly the relationships between the design and management of orchards and the effects of pests and diseases.

G. Goergen is a research entomologist at the International Institute of Tropical Agriculture. He joined in 1987 to work on a biological control programme against the cassava mealybug under the framework of a PhD fellowship. Over a 20-year period based at the IITA station Cotonou, Benin, he has established an institutional taxonomic capacity building programme to study arthropods of agricultural importance with a particular focus on biological control of pests introduced into tropical Africa. His professional experience and research interests are in the field of integrated pest management, habitat management and biodiversity of afrotropical insects including Tephritid flies.

M. De Meyer is an entomologist attached to the Royal Museum for Central Africa in Tervuren (Belgium). Previously he worked in Botswana and Kenya for a number of years. His main field of expertise is the taxonomy and systematics of selected groups of Diptera such as Pipunculidae, Syrphidae and Tephritidae, with emphasis on the Afrotropical fauna. His research includes different aspects, from taxonomic revisions to studies on evolutionary trends and speciation events.

P. Van Mele is an agricultural scientist who obtained his PhD in 2000 from Wageningen University on fruit farmers' knowledge and innovations with regard to integrated pest management. He currently runs his own business, Agro-Insight (Belgium), making training videos for farmers in developing countries. He is co-founder of the international NGO Access Agriculture that enables south-south exchange of and access to quality audio-visual training materials to support smallholder farmers.

Chapter 19
Sterile Insect Technique (SIT) for Fruit Fly Control – The South African Experience

Brian N. Barnes

Abstract Development and implementation of the Sterile Insect Technique (SIT) in South Africa began in the mid 1990s with Phase 1 to evaluate the feasibility of eradicating Mediterranean fruit fly (medfly), *Ceratitis capitata* (Wiedemann), in export table grapes in the Hex River Valley, Western Cape Province. Since then the SIT programme has progressed through a further two phases. During these phases the scale and cost-effectiveness of the production of high-quality sterile male flies for release has improved significantly. Additional areas with different fruit crops have been incorporated and the relative efficacy of aerial and ground releases evaluated. Overall, *C. capitata* SIT in South Africa has been more successful in some areas than in others, but the programme continues to evolve based on the many valuable lessons that have been learnt. For SIT to be effective there must be sustained funding, no compromise in quality, and good management, communication and training for the staff involved. There must be buy in from the growers to ensure that SIT runs alongside other management strategies and orchard sanitation. Rather than eradication, the programme focus is now population suppression in a limited number of production areas. In the long term we aim to increase the area under SIT, returning to aerial releases of sterile *C. capitata*, creating areas of low fruit fly prevalence, and possibly fruit fly-free areas. This could lead to a sustainable international fruit market without the need for fruit fly trade restrictions.

Keywords *Ceratitis capitata* • Mass-rearing • Sterile male release • Population suppression

---

B.N. Barnes (✉)
ARC Infruitec-Nietvoorbij (Retired), Stellenbosch 7600, South Africa

FruitFly Africa (Pty) Ltd (Technical Adviser), Stellenbosch 7600, South Africa

PO Box 5092, 7135 Helderberg, South Africa
e-mail: bnb303@gmail.com

1 Background and Basics of the Sterile Insect Technique (SIT)

The Sterile Insect Technique (SIT) uses releases of radiation-sterilised insects as part of environmentally-compatible and area-wide Integrated Pest Management (IPM). It was pioneered in the USA, when it was implemented against New World screwworm, *Cochliomya hominivorax* (Coquerel) (Diptera: Calliphoridae) in the 1950s. Since then, it has been refined in many ways, and its use extended to include at least 15 other pests, most of them lepidopterans and tephritids (Klassen and Curtis 2005). Subsequent research and development of SIT, and its promotion worldwide, has been led by the Joint Food & Agriculture Organisation of the United Nations/International Atomic Energy Agency (FAO/IAEA) Programme on Nuclear Techniques in Food and Agriculture, based in Vienna, Austria. SIT is now used on all six continents (Dyck et al. 2005). Development and improvement of SIT is on-going worldwide, and its use against other pests is still under development, e.g. mosquitoes (Klassen and Curtis 2005) and sugarcane borers in southern Africa (Conlong 2007; Conlong and Rutherford 2009; Potgieter et al. 2013).

SIT involves mass-rearing the target species, sterilizing them (where possible, only the males) with ionising radiation, and releasing these sterilized insects into the target area in their millions every week (Klassen 2005). Sterilized males mate with fertile 'wild-type' females, which subsequently lay only infertile eggs. Provided particular cultural practices, such as management of alternative non-commercial host plants (including pest control, plant removal or stripping of unripe fruit) and on-farm sanitation are carried out simultaneously, the size of successive generations of the pest is thus systematically reduced (e.g. Barnes et al. 2004; Conlong and Rutherford 2009). An important principle for the success of SIT is achieving a sufficient 'over-flooding ratio' of sterile males to wild-type females in the field. The released sterile males must sufficiently outnumber the wild-type females in the release area to ensure a sufficient proportion of the population are sterile, thereby overcoming the natural rate of increase of wild-type females. This overflooding ratio varies from species to species (Klassen 2005) but for fruit flies, it is generally regarded as 80:1.

The use of SIT offers various strategic options; pest eradication, suppression, containment and prevention (Hendrichs et al. 2005). Eradication is often perceived to be the desired objective, particularly for additional export trade benefits; however, the degree of difficulty and the cost of achieving pest-free status are very high. As a result, using SIT to create 'areas of low pest prevalence' (ALPPs) within a 'systems approach' that guarantees pest-free agricultural produce, is viewed as a more realistic goal (Klassen 2005; Hendrichs et al. 2005). Achieving ALPP status is quicker, less complex and management intensive, and therefore less expensive than an eradication programme, and can still allow export trade benefits (Hendrichs et al. 2005).

Worldwide, SIT is currently practiced on an industrial and area-wide scale against at least ten pest species (Klassen and Curtis 2005). For fruit fly control alone, at least 20 area-wide IPM programmes use SIT as one of the pest suppression tools (Enkerlin 2005). However, before considering initiating an SIT programme a

> **Box 19.1: Basic Principles of SIT** Some basic principles of SIT are given below. The degree of success in adhering to these principles is likely to determine whether an SIT programme can be used successfully against a target pest.
>
> *Cost-effective mass rearing of the target pest*: Some insect species are more difficult and more expensive to rear than others. The cost-effectiveness of mass rearing will play a role in whether or not an SIT programme can be used against a candidate pest.
>
> *Isolation of target area*: Because of the threat of reinfestation of an SIT area by the target pest from surrounding areas, geographical isolation (e.g. islands) or topographical isolation (e.g. cultivation in valleys) is required. Host crop isolation (where commercial plantings are surrounded for many kilometres by non-host species or desert areas) could also make SIT feasible.
>
> *Area-wide implementation*: SIT works best in an area-wide situation (e.g. thousands of hectares). It is not a technique usually suited to individual farms or small areas of less than a few hundred hectares.
>
> *An integrated approach*: SIT is not a 'stand-alone' technology – it needs to be integrated simultaneously with other pest management techniques.
>
> *Adequate funding for infrastructure and resources*: SIT is an expensive undertaking, needing significant resources and very good management. Broad-based, multi-organisational funding for the programme is preferable, ideally involving funding by national and provincial governments, and by the beneficiaries (e.g. fruit growers).
>
> *Beneficiary and community buy in*: An SIT programme cannot simply be introduced into an area and implemented on a whim without an understanding of what will be involved, and acceptance/active cooperation by the beneficiaries in that area. Similarly, members of the community need to understand what will be taking place in their area.
>
> *Promises and expectations*: Those responsible for implementing the SIT programme need to guard against unrealistic assurances in terms of deliverables from the programme. Beneficiaries and the communities involved must likewise be realistic about what they can expect from the programme, and the time frame involved.

number of basic principles need to be considered (Dyck et al. 2005; Barnes 2007) (Box 19.1). A detailed discussion of the principles, development and implementation of SIT is given by Dyck et al. (2005).

2 Fruit Flies of Economic Importance in South Africa

Three species of tephritid fruit flies have long been pests of economic importance in South Africa: the Mediterranean fruit fly (medfly), *Ceratitis capitata* (Wiedemann); the Natal fruit fly (*Ceratitis rosa* Karsch); and the mango (marula) fruit fly, *Ceratitis cosyra* (Walker) (Blomefield et al. 2015; Grové et al. 2015). Now there is a fourth species, the oriental fruit fly, *Bactrocera dorsalis* (Hendel), which was first recorded in South Africa in 2010 having systematically spread throughout Africa since it was first detected in 2003 (Manrakhan et al. 2011).

Ceratitis capitata and *C. rosa* are fairly widespread across South Africa. *Ceratitis rosa* is more predominant in the cooler and wetter maritime areas, and absent or in very low numbers in the arid and hotter regions. In contrast *C. capitata* is more abundant in the warmer, drier inland areas (Barnes 2006; De Villiers et al. 2012). Both species infest all types of commercial fruit and many wild berries. *Ceratitis cosyra* occurs only in the north-eastern parts of the country where a sub-tropical climate prevails, causing damage mostly to citrus and sub-tropical fruits (Prinsloo and Uys 2015). *Bactrocera dorsalis* was first recorded in the far north of South Africa, where an initial eradication programme was successful; however, it has since been reported from parts of the Limpopo, Mpumalanga, North West, Gauteng and Kwa-Zulu Natal provinces. At the time of writing it is absent from the Free State, Northern Cape, Eastern Cape and Western Cape Provinces (HortGro Science 2015). *Bactrocera dorsalis* has a preference for mango, banana and citrus (Manrakhan et al. 2011).

The export of deciduous fruits and table grapes is an industry of significant economic importance in South Africa. Nearly 118 million cartons are exported annually, with gross annual export earnings of approximately US$ 940 million. About 80,000 ha are under cultivation, with the Western Cape Province and parts of the Northern and Eastern Cape Provinces being the largest and most important production regions (OABS 2013). Most fruit is grown in valley systems or, in the Northern Cape, surrounded by desert; due to their relative isolation, some of these regions are ideal for area-wide pest management such as SIT. *Ceratitis capitata* and *C. rosa* both occur in the Western Cape and Eastern Cape Provinces, while only *C. capitata* occurs in the Northern Cape. In 1997 it was calculated that crop losses and control costs due to fruit flies in the Western Cape exceeded R20 million (then ~ US$4 million) per annum (Mumford and Tween 1997). In 2015 this figure was estimated to be at least R90 million (currently ~ US$7.5 million) per annum (N. Baard, pers. comm.).

3 The *Ceratitis capitata* SIT Programme – Phase 1, 1996–2003

In 1996 the IAEA approached the Agricultural Research Council (ARC) Infruitec-Nietvoorbij Institute for Fruit, Vine and Wine in Stellenbosch with a view to collaboration in an area-wide project to investigate the feasibility of using SIT to eradicate or suppress *C. capitata* in the Western Cape Province. As *C. capitata* is a pest of international quarantine importance, South Africa already complied with strict phytosanitary measures imposed against this pest by a number of countries (e.g. USA, EU, Japan, China and South Korea) for the trade of deciduous fruits, citrus and fresh table grapes (Venter 2012; Barnes et al. 2015). However, creation of fruit fly-free areas, or even areas of low fruit fly prevalence, would facilitate greater export opportunities for South Africa. Accordingly, an IAEA-funded project, administered and driven by ARC Infruitec-Nietvoorbij, began in 1997.

3.1 Hex River Valley Study

After a visit to the Western Cape Province by SIT experts from the IAEA, the site chosen for a feasibility study was the Hex River Valley, some 100 km from Stellenbosch, where approximately 4500 ha of table grapes were under cultivation. This is the largest single area of table grapes in the Western Cape Province. The Hex River Valley is long and narrow (about 20 km×5 km), surrounded by mountains, with a single town (De Doorns) and a single major national road running through it (Fig. 19.1). At the start of the project 90–95% of the cultivated area was planted with table grapes, with the remaining area comprising small areas of citrus. About 140 growers farmed in the valley, with the majority of the crop being exported. The many farm gardens and associated labourers' dwellings, together with the small town, harboured a great many alternative fruit fly host plants throughout the valley. Further details are provided by Barnes et al. (2004).

This site was chosen for a number of reasons:

- its geographic isolation, with little chance of natural reinvasion by *C. capitata*.
- a large area of, predominantly, one crop – table grapes, an important and lucrative export fruit crop in South Africa.
- *Ceratitis capitata* is a key pest, and the predominent fruit fly species in the valley.
- there was buy-in into the proposed SIT project by all of the growers in the valley.
- the project was supported by the Hex River Valley Table Grape Association, which oversees the interests of all table grape growers in the valley.

Two nearby production areas were also included in the feasibility study. The De Wet area (350 ha of table grapes for export), immediately outside the western end of the Hex River Valley, was included as a buffer area to reduce the likelihood of fertile wild-type fruit flies moving into the Hex River Valley. The Brandwacht area (275 ha of table grapes for export), approximately 10 km further west, was also included because table grapes from this area were routinely sent through to the Hex River Valley for cooling and subsequent transport to the Cape Town docks. Barnes and Venter (2008) reported only very low numbers of *C. rosa* in a few localities in the Hex River Valley. Furthermore, in a subsequent 2-year trapping survey the fruit fly species composition was 100% *C. capitata* in the Hex River Valley and 99.5% *C. capitata*/0.5% *C. rosa* in the Brandwacht area (Manrakhan and Addison 2014).

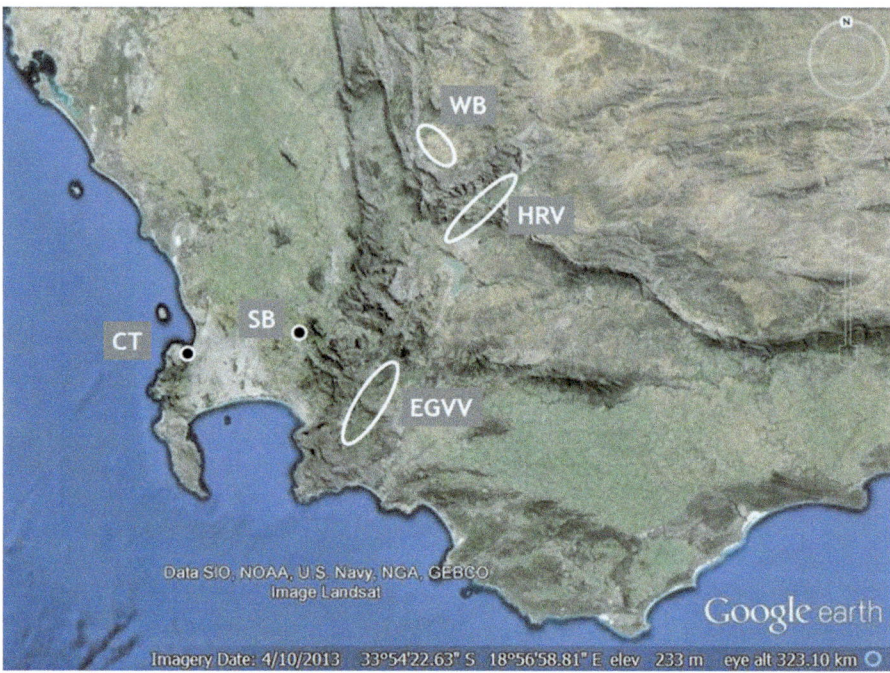

Fig. 19.1 Map of the south-western area of the Western Cape Province. CT=Cape Town, SB=Stellenbosch, EGVV=Elgin, Grabouw, Vyeboom & Villiersdorp SIT area, HRV=Hex River Valley SIT area, WB=Warm Bokkeveld SIT area

All subsequent mention of the Hex River Valley includes all three areas (the Hex River Valley *per se*, the De Wet area and the Brandwacht area [approximately 5000 ha in total]).

3.1.1 Project Funding and Management

The IAEA committed substantial and sustained funding for the feasibility project, mainly in the form of equipment and technology transfer. At the local level, a formal SIT Partnership Agreement was formed between the ARC in Stellenbosch and the governing body of the deciduous fruit industry, the Deciduous Fruit Producers Trust (DFPT, now HortGro). This provided additional funding on a 50:50 basis, for project management, the rearing and sterilization of *C. capitata* and the coordination of field activities. The cost of producing sterile male flies in this early phase of the programme was high. This was due to poor economies of scale in a facility that could only produce 8 million sterile flies per week; this was then the equivalent of approximately US$1200 per million sterile males, excluding the cost of delivery, release and monitoring. Hex River Valley growers paid a subsidized price for the sterile flies as they could not afford to pay the full cost; the balance was covered by the SIT Partnership in the interests of keeping the pilot project running (Barnes 2007).

Additional funding was raised through a voluntary levy on all table grape cartons for export from the area. This provided for all field operations, including the aerial release of the sterile flies, hiring of personnel, trapping and laboratory equipment, and partially contributed to the cost of rearing the sterile *C. capitata* released in the area. At this stage of the programme the project was not supported financially or managerially by government, and relied solely on funding from the sources mentioned above (Barnes et al. 2004; Barnes 2007).

Overall management of the project was the responsibility of the Pest Management Division of ARC Infruitec-Nietvoorbij. Nevertheless, by virtue of the financial contribution made by the Hex River Valley growers, they retained a significant degree of independence in decision making; for example, the growers controlled the number of sterile flies released per hectare per week, the management of trapping and baiting programmes, and host plant management.

3.1.2 Attempts at Fruit Fly Population Reduction Prior to SIT, 1997[1]–1999

In 1997[1] a 3-year fruit fly population reduction programme was initiated in the Hex River Valley, with the intention of decreasing fruit fly populations to very low levels over two to three successive seasons, prior to initiation of SIT and release of sterile *C. capitata* males. The strategy involved: coordinated applications of fruit fly baits; vineyard sanitation; and management of alternative host plants beyond the two or three organophosphate cover sprays normally used against fruit flies. These actions were coordinated by an SIT field manager based in the Hex River Valley, but in practice were carried out by the growers. Standard protein and mercaptothion insecticide bait mixtures were used, and applied to all vineyards weekly during summer and fortnightly during winter according to the manufacturer's recommendations. The field manager attempted (with limited success) to coordinate vineyard sanitation and management of alternative host plants throughout the valley. Fruit fly populations in the area were monitored throughout this period using a network of 360 Jackson traps (delta traps with a disposable sticky liner on the base) baited with trimedlure, and deployed at a density of one trap per 25 ha block. Traps were inspected once a week on the same day, all trapped insects removed, and the numbers and species of fruit flies recorded. The trimedlure bait plug was replaced every 4–6 weeks, and the sticky liner replaced when the glue started to become ineffective

---

[1] In the southern hemisphere the deciduous fruit growing-season extends from early spring (August/September) of one year to late-autumn/winter (June/July) of the next. In this chapter a growing season and related pest management activities and outcomes are referenced as the second year of any season, e.g. the season 1996/97 is referred to as 1997.

in retaining insect visitors. An effort was made to move each trap to another location within the block every 4–6 weeks, although this was not rigidly adhered to (Barnes et al. 2004).

3.1.3 Mass Rearing and Sterilization of *C. capitata*

Mass-rearing of *C. capitata* began at ARC Infruitec-Nietvoorbij, Stellenbosch, in 1998, in a small pilot facility comprising old insect breeding rooms converted for the purpose. Staff at the Stellenbosch rearing facility were trained by experts in the mass rearing of genetic sexing strains (GSS) of *C. capitata*, both on-site and at the IAEA Laboratories at Seibersdorf, Austria. GSS strains were developed at the Insect Pest Control Laboratory of the IAEA in Seibersdorf, specifically for the production of pure 'male-only' streams of *C. capitata* (Franz 2005); the use of male-only strains in SIT programmes increases efficacy and cost-effectiveness since only males are produced (Hendrichs et al. 1995). Initially the GSS Vienna 7–97 strain was used but this strain was genetically unstable and, between 1999 and 2001, was superceeded at the rearing facility by a series of improved strains: Vienna 7/Mix 99, Vienna 7/Tol 2000, Vienna 7/Mix 2000 and Vienna 7-D53/Mix 2001. Finally, in 2003, a new generation GSS, designated Vienna-8, was made available by the IAEA, and incorporated into the Stellenbosch *C. capitata* rearing facility. This strain has a very high degree of rearing stability and is currently used by most *C. capitata* mass rearing facilities supporting area-wide integrated SIT programmes around the world (Franz 2005).

All pupae from the male-only stream destined for release were sterilised with gamma radiation doses of between 90 and 110 Gy in the ARC Infruitec-Nietvoorbij Co^{60} irradiator. Resulting sterile pupae were dyed with Day-Glo fluorescent dye, packed in paper bags and stored in fibreglass plastic adult release containers (PARC boxes, U.S. Plastic Corp., Lima, Ohio; seven bags per box) in an eclosion room at 25 °C and 65 % RH; each PARC box held approximately 29,000 adult flies. Food for the flies was provided in the form of agar cakes placed on gauze vents on top of each box. Five to 7 days after eclosion the PARC boxes were placed in a refrigerated truck at 4 °C and transported approximately 130 km to either the Worcester airfield for aerial releases, or directly to the three ground release areas.

The initial production target for sterile males was based on the requirements for the Hex River Valley project – 5 million per week (i.e. 500 sterile males/ha/wk) during the growing season, and 1 million per week (100 sterile males/ha/wk) during winter/early spring. These release levels were set by the Hex River Valley SIT team but, due to expenditure limitations, were below the level of 1000 sterile males/ha/wk recommended by international SIT experts (Barnes et al. 2004; Barnes 2007).

The rearing facility faced many challenges during this early phase of the SIT programme. The small building in which the insects were mass reared was not designed for the purpose, and cramped space compromised rearing efficiency. Budget restrictions precluded the procurement of rearing equipment with the necessary high specification. The rearing staff were initially inexperienced and needed time to become familiar with the processes and procedures. High humidity in the rearing rooms caused rusting of equipment and door frames, and resulted in high levels of fungal contamination on walls and ceilings; the latter served as a source of contamination for the artificial larval diet. Larvae of the vinegar fly, *Drosophila melanogaster* Meigen (Diptera: Drosophilidae), sometimes colonised the larval diet and, at one stage, ant invasions seriously threatened the entire adult colony. At this stage there was no formal quality monitoring and management system in place.

These factors compromised the production of sufficient numbers of high quality sterile males and, as a result, the number produced seldom exceeded 1 million per week for the first 16 months of operation. For a number of months during this period it was, therefore, necessary to import sterile males from the El Pino *C. capitata* rearing facility in Guatemala and *C. capitata* GSS eggs from the IAEA's Seibersdorf Laboratories in Austria, to supplement local production of sterile males. Local production only began to approach target levels by the end of 2001 (Barnes et al. 2004).

3.1.4 Sterile *C. capitata* Release Strategy

Three years of population reduction efforts in the Hex River Valley (1997–1999) did not yield the desired results, and wild-type *C. capitata* populations remained high (Table 19.1). Nevertheless, a decision was made to begin aerial releases of sterile males in the spring (October) of 1999, at a stage in the season when wild-type *C. capitata* populations were very low. All Jackson traps were replaced with 222 McPhail traps baited with a three-component food lure (one trap/49 ha) to monitor wild-type females and sterile males, as suggested for evaluation of SIT programmes by IAEA (2003) and Vreysen (2005). These dry, bucket-type traps each contained a 1-cm square block of plastic impregnated with dichlorvos to kill flies entering the trap. Dead fruit flies were removed weekly and the number and species recorded. The lure and dichlorvos block were replaced every 4–6 weeks.

Two release strategies were adopted at different times of the year. Between spring and autumn (October to May inclusive), 500 sterile males/ha/wk were released by air (5 million/wk over the whole target area). Twice-weekly releases were made from a Cessna 207 aircraft flying at an altitude of 700 m above ground level; the plane was fitted with a special chilled fly release machine. Different flight paths, separated laterally by 1.4 km, were flown on each release occasion to achieve better distribution of sterile flies over time. The starting point for releases was also varied on each release occasion for the same reason. The Cessna aircraft was on permanent hire for the duration of the aerial releases (winter period included), which was paid for by the Hex River Valley growers. Between winter and early spring (June to September inclusive), when vineyards were inhospitable to fruit flies and weather conditions were often unsuitable for aerial releases, all releases were made on the ground. Release sites were distributed amongst approximately 240 farm gardens and labourers' settlements in the Hex River Valley, and also in the

Table 19.1 Annual mean and peak numbers of wild-type *C. capitata* (flies/trap/day – FTD) in the Hex Valley before and after releases of sterile males during three phases of SIT implementation in three areas in the Western Cape Province

Year	Annual mean and peak FTD					
	Hex River Valley		EGVV		Warm Bokkeveld	
	Mean	Peak	Mean	Peak	Mean	Peak
Phase 1						
	Pre-SIT					
1997	0.9	4.1	–	–	–	–
1998	1.0	5.9	–	–	–	–
1999	1.0	4.0	–	–	–	–
	Aerial release		–	–	–	–
2000	0.3	1.2	Pre-SIT		–	–
2001	0.2	1.8	1.2	7.3	–	–
2002	0.4	2.8	1.3	6.2	–	–
2003	0.1	0.3	0.8	5.2	–	–
Phase 2						
	Ground release[a]		Ground release[a]		Ground release[b]	
2004	0.2	1.0	–	–	–	–
2005	1.0	6.0	0.6	2.6	–	–
2006	0.6	3.2	0.2	1.0	–	–
2007	2.7	24.8	0.2	0.7	–	–
2008	3.8	24.7	0.4	2.4	–	–
2009	2.1	14.2	0.4	1.7	–	–
2010	4.5	31.4	0.4	1.6	–	–
Phase 3						
2011	2.1	6.3	0.5	2.2	0.6	3.9
2012	1.9	16.5	0.2	1.4	0.3	2.8
2013	2.8	21.9	0.1	0.7	0.2	2.8
2014	1.5	11.9	0.2	1.1	0.1	0.7
2015[c]	3.2	11.8	0.2	1.4	0.2	1.2

[a]Sterile flies released in farm and urban gardens
[b]Between 2010 and 2014, sterile males were released only in urban gardens; from 2015, they were also released in farm gardens
[c]To end May

town; the focus was on fruit fly 'hotspots' including all backyards, around any neglected fruit fly host plants and any trees with the potential to provide shelter. The aim was to release 100 sterile males/garden ha/wk.

From October 1999 until October 2001, the summer and winter sterile male release targets were seldom achieved due to initial production problems. However, from 2002 this was rectified following the introduction of a Quality Management System in the rearing facility (Barnes et al. 2004).

3.1.5 Sterile/Wild-Type Fly Identification

All fruit flies found in the traps were identified in the SIT laboratory in the Hex River Valley, and all *C. capitata* classified as either 'sterile' or 'wild-type' on the basis of the presence (sterile) or absence (wild-type) of fluorescent dye as detected under a high intensity ultra-violet lamp (Barnes et al. 2004). Where the dye was not immediately evident, the heads of the flies were crushed, slide-mounted and further checked for the presence of dye under a compound microscope equipped with an ultra-violet light. All flies determined not to have dye present were classified as wild-type flies. During this phase no differentiation of the sexes was made due to time constraints.

3.1.6 Quality Management and Infrastructure Improvements

In 2001 a grant from the Western Cape Department of Agriculture enabled the construction of a new adult room and quality control laboratory equipped with the necessary high specification equipment. Up to this point no structured system existed to monitor, record, analyse and quickly rectify any rearing problems that could lead to the production of poor quality insects for release. In 2002 a comprehensive Quality Management System (QMS) was compiled, comprising comprehensive Instructions and Standard Operating Procedures (SOPs).

3.1.7 Wild-Type *C. capitata* Outbreak in Hex River Valley, 2001/2002

In December 2001, growers in the Hex River Valley complained of increased *C. capitata* infestation in table grapes which they, and SIT monitors in the valley, attributed to either (i) *C. capitata* males not being adequately irradiated in the Stellenbosch facility, and/or (ii) the presence of fertile *C. capitata* females in deliveries of sterile males from the rearing facility. The SIT programme was blamed for the resulting crop losses. As grapes were being packed for export at this time, it was imperative to determine the origin of the fertile female flies that had caused the infestation, and to take corrective measures.

Three investigations were made. The first was a DNA analysis of *C. capitata* specimens from infested grapes collected in the Hex River Valley using known Vienna 8 GSS genetic markers (Barnes et al. 2006). *Ceratitis capitata*-infested grapes were collected from the areas reporting infestation in the Hex River Valley, and maintained on dry vermiculite at ARC Infruitec-Nietvoorbij. The 137 pupae emerging from this material were couriered to the IAEA Laboratories at Seibersdorf, Vienna, where 88 adults emerged. These adults were subjected to mitochondrial DNA analyses to determine the presence or absence of the genetic markers present in the GSS. A further 58 flies from the GSS colony from the Stellenbosch rearing

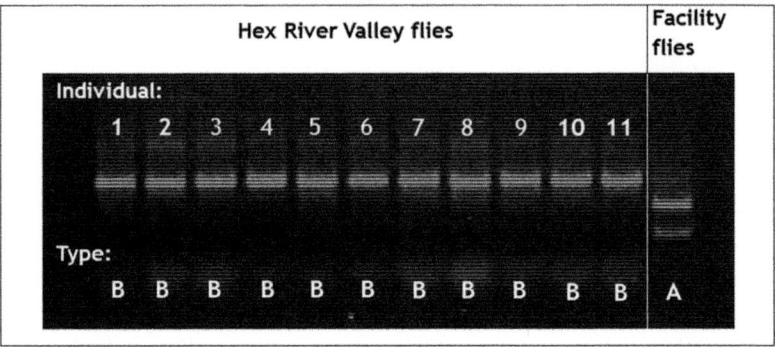

Fig. 19.2 An example of the results obtained from DNA analysis of 11 *C. capitata* adults recovered from infested grapes from the Hex River Valley; the band patterns from wild-type flies from the Hex River Valley are distinctly different to those from representative GSS flies from the rearing facility. All wild-type flies recovered from the Hex River Valley were type B, and all facility flies were type A

facility were also analysed. Results demonstrated that none of the *C. capitata* from the Hex River Valley infestations had the AAAA or the AAAB (type A) genetic markers present in the GSS from the rearing facility (Fig. 19.2). Secondly, mating trials in field and laboratory cages at ARC Infruitec-Nietvoorbij were made with wild-type females and sterile males. Unmated wild-type female *C. capitata* (fertile) were placed in cages (either in the field or the laboratory) with males that had been irradiated (sterile), and uninfested fruit. The fruit was later examined for eggs and all eggs found were monitored for fertility. No fertile eggs were laid from sterile male/wild-type female matings. Finally an audit of the irradiation procedure at the ARC Infruitec-Nietvoorbij rearing facility was made. The irradiation records for all pupae sent to the Hex River Valley were checked and an independent consultant engaged to check the calibration of the irradiator. This revealed no irregular procedures and demonstrated that all batches of pupae sent to the Hex River Valley for release had received the designated radiation dose. The independent consultant found that the calibration of the irradiator was correct and appropriate doses were being delivered.

In conclusion all male *C. capitata* sent to the Hex River Valley had been exposed to the correct radiation dose and were sterile. From this evidence it was concluded beyond doubt that the Hex River Valley infestation did not originate from the rearing facility, but from poor vineyard sanitation and poor management of alternative host plants by growers. At the time this was a serious problem in the Hex River Valley (Barnes et al. 2006), and remains problematic to this day.

3.2 Outcomes from Phase 1, 1996–2003

3.2.1 *Ceratitis capitata* Quality and Production Improvement

Construction of the new adult rearing building and quality control room substantially improved adult rearing and quality control, and eased the pressure on the old rearing building, which was still used for larval rearing and pupal maturation. Fly quality and production volumes were also substantially increased following the implementation of the QMS into the rearing procedure. By the end of December 2002 production of sterile males for release reached and sometimes exceeded target levels (Barnes et al. 2004).

3.2.2 Wild-Type Fly Populations

It was evident that the baiting programme in the Hex River Valley was unsuccessful in reducing *C. capitata* populations prior to release of sterile males in October 1999; wild-type fly populations were between 4 and 6 flies/trap/day (FTD) (Table 19.1). This was attributed to (i) the relatively weak protein component (2%) of the legally-permitted bait solutions being inadequate for attracting and killing fruit flies, (ii) inadequate area-wide coordination of bait applications in the release area, and (iii) poor vineyard sanitation and poor management of alternative fruit fly host plants by growers.

The effect of sterile *C. capitata* releases only became evident in the second year of releases (2000). This was probably because the number of sterile males released in the facility's first year of operation (1999) was below the target level for much of the time. However, between 2000 and 2003, wild-type fly populations decreased substantially, with means between 0.1 and 0.4 FTD, and peaks between 0.4 and 2.8 FTD (Table 19.1; Fig. 19.3) (Barnes et al. 2004, 2015).

Population levels before and after sterile male releases were not directly comparable because the types of traps used were different. Nevertheless, McPhail traps with three-component lures were considered at least as effective as Jackson traps in trapping *C. capitata* (D. Moreno, pers. comm.). It was therefore concluded that aerial releases led to substantial reductions in *C. capitata* populations in the Hex River Valley.

3.2.3 Sterile *C. capitata* Release Strategy

The Cessna aircraft used for aerial releases was specially adapted to carry the release machine, and could not be used for other purposes. An aerial release took about 3 h a week, and for the rest of the time it remained unused at the airfield. By the end of 2003 neither the Hex River Valley growers nor the SIT Partnership could afford

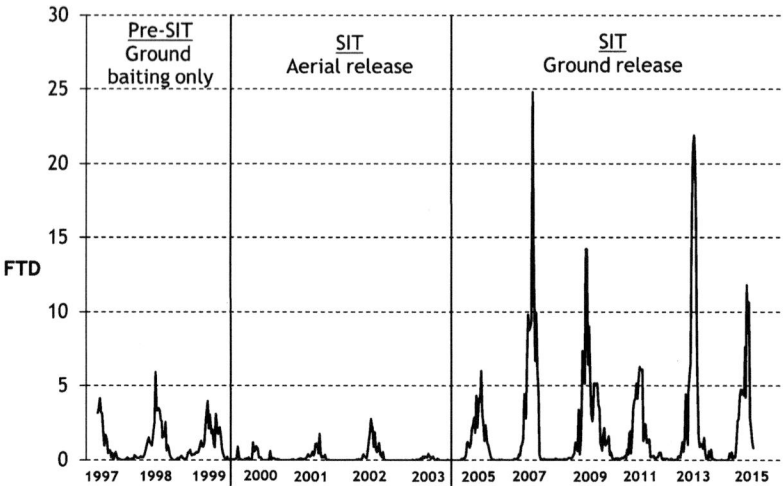

Fig. 19.3 Seasonal fluctuations in wild-type *C. capitata* populations (flies/trap/day – FTD) in the Hex River Valley before (1997–1999) and after (2000–2003) aerial releases of sterile males, and during subsequent ground releases (2005–2015). Data for ground releases are given only every alternate year from 2005 to illustrate the trend

this cost. As a result it was decided that 'blanket' aerial releases of sterile *C. capitata* over the full 5000 ha would be replaced with ground releases throughout the year, focussing on strategic areas for release (see next section).

4 The *C. capitata* SIT Programme – Phase 2, 2003–2010

4.1 Privatisation, Changing Strategies and Expansion

4.1.1 Privatisation

Without long-term funding from the government, and with the inability of the Hex River Valley growers to pay the high production costs of sterile *C. capitata* in a small facility with poor economies of scale, funding for the SIT programme through the SIT Partnership became economically unsustainable. As a result, the programme was privatised in 2003 under the name SIT Africa (Pty) Ltd. (now FruitFly Africa [Pty] Ltd.). This company manages all *C. capitata* monitoring activities, rearing, irradiation, sterile fly delivery and releases in the areas covered by the SIT programme. It is a non-profit-making organization owned by the deciduous fruit, table grape, dried fruit and canning fruit industry organizations, and the Agricultural Research Council.

4.1.2 Ground Release Strategy, Population Reduction and Monitoring

The ground release strategy was developed in 2003 in conjunction with Dr Gerardo Ortiz, Mexico, an international expert on the topic. This strategy involves releasing sterile *C. capitata* on the ground directly from the paper bags in which they emerged at the rearing facility (±4500 flies/bag). The focal sites for all ground releases has remained the same throughout the programme; they include the many farm gardens and workers' settlements throughout the Hex River Valley, any fruit fly 'hotspots' in vineyards and orchards, and any other hotspots outside these areas (e.g. neglected fruit fly host plants). The rationale behind targeting these sites rather than commercial plantings was the fact that from approximately August to December, populations in commercial plantings are extremely low or even absent, making releases there ineffective. Commercial plantings are unattractive to fruit flies at this time because they offer little or no shelter or food for the adults, and there is little or no ripe fruit for female oviposition and larval development. In contrast, during this period fruit flies are concentrated in gardens on farms and in urban areas where ornamental trees and alternative fruit fly host plants commonly occur, providing them with shelter and food for continued breeding through the colder months (Barnes and Venter 2008). By concentrating ground releases in these 'hotspot' areas in conjunction with other fruit fly management tactics such as bait sprays and bait stations, the sterile males are targetted at localised breeding populations of *C. capitata* with the potential to reduce these populations to very low levels during the winter months. By limiting breeding through the winter, the number of wild-type flies that migrate later (in summer) to the ripening fruit of commercial plantings, should also be substantially reduced. By 2015 there were approximately 650 ground release sites in all three release areas.

Attempts to improve vineyard sanitation and pest management of alternative fruit fly host plants were continued, and the use of M3® bait stations extended. For budgetary reasons the use of these bait stations was limited to farm gardens and other known breeding sites (e.g. neglected host trees) during winter (June to September), and the town of De Doorns (in the Hex River Valley) where permission could be obtained from the residents. The number of bait stations per tree depended on the type of host tree and its potential as a fruit fly host. There was also a summer and winter programme of monitoring using Chempack® bucket traps baited with a three-component food lure. For the winter programme (June to October; the specific dates varying from area to area) most bucket traps were deployed in home and farm gardens where the overwintering *C. capitata* populations were breeding. In October, most traps were moved into commercial vineyards and orchards, at least 100 m away from any garden, to detect *C. capitata* migrating into commercial plantings from their overwintering breeding sites. The density of traps in commercial plantings was one trap per 20 ha.

4.1.3 New Release Areas

In 2004 two new areas were incorporated into the ground release programme: the Berg River area in the vicinity of Riebeek Kasteel (mostly table grapes) and the Elgin-Grabouw, Vyeboom and Villiersdorp area, known as the EGVV (mixed fruit farming; apples, pears, stonefruit and wine grapes; Fig. 19.1). Both new areas were each comprised of three separate and isolated sub-areas. All releases were made in farm gardens and urban areas.

4.1.4 Aerial Fruit Fly Bait Applications

In January 2008, the first aerial applications of fruit fly bait by fixed-wing aircraft were made in the Hex River Valley. The bait was GF-120™ NF Naturalyte (Dow AgroSciences), which contained 0.24 g/L spinosad, an ecologically-compatible and organically-approved insecticide; it was applied at a rate of between 1.0 and 1.2 L GF-120™ NF Naturalyte per ha, as stipulated by the product label. Two aerial bait applications were made in 2009, and from 2010 four or five applications at intervals of between 2 and 3 weeks were made each year. In addition to these applications, all growers were advised to apply their own routine ground-based fruit fly baits as described under Sect. 3.1. However, this advice was not always followed.

All aerial bait application programmes were financed by the growers in the respective SIT areas; this meant that they determined the application dates and intervals between applications. Growers always timed their applications to reduce *C. capitata* infestation (and therefore associated crop losses) in vineyards shortly before harvest. At this time (summer) *C. capitata* populations in commercial fruit plantings would have already been well established and growing rapidly as reproduction rates are highest in summer. This was despite advice from experts to start applications much earlier in the season, as part of a strategic *C. capitata* population reduction programme.

4.1.5 Sterile *C. capitata* Production and Release

Production targets for sterile males produced by the facility varied from 7 million per week during the winter months (June to August) to 15 million per week during the spring and summer months. The target numbers released per garden varied over time and depended on the size and nature of the gardens, but averaged between 8000 per garden in winter and 24,000 per garden in summer. There were issues in some urban areas that were a problem, e.g. access to private property; a belief by some residents that dyed pupal cases left behind in gardens were a hazard to human health; and the perception of some residents that the sterile fruit flies around and in the home were a nuisance.

4.2 Outcomes from Phase 2, 2003–2010

4.2.1 Release Areas

The SIT programme in the Berg River area which began in 2004 had to be terminated in 2007 due to inadequate support for the programme by growers in the area; in particular an unsustainable funding mechanism for the area and objections by some of the urban communities to standard SIT practices, had resulted in fragmented releases (Barnes 2007). The Hex River Valley and EGVV programmes continued throughout this phase, with weekly releases of sterile male *C. capitata* throughout the year in farm and urban gardens, supplemented by other fruit fly management practices.

4.2.2 Aerial Fruit Fly Bait Applications

Ceratitis capitata population levels during aerial baiting programmes in the Hex River Valley decreased significantly in the 8–10 week application period, but this was only a short-term effect. After the last application of the season populations always increased rapidly, reaching seasonal peak levels within a few weeks (Barnes et al. 2015). This was not unexpected; applications were made during the season when *C. capitata* populations increase most rapidly, and intervals between applications were sometimes too long. However, the application dates and intervals were decided by the growers, and were aimed at late-season crop protection rather than early-season population reduction.

The contribution of these late-season aerial bait applications to lowering the mean annual and peak *C. capitata* populations (and thus the long-term success of SIT) is regarded as being small. Controlling populations in infestation loci, from where they migrate and re-establish damaging populations in commercial crops, is best achieved by starting early in the season (e.g. spring). This is because the wild-type population density is at its lowest in spring following natural winter mortality (Klassen 2005). Results in the Hex River Valley from 2011 to 2015 (Table 19.1; Fig. 19.3) illustrate that limited numbers of aerial bait applications applied in summer over a maximum of two *C. capitata* generations, is inadequate for area-wide population suppression.

4.2.3 Wild-Type Fruit Fly Populations

Hex River Valley: After aerial releases were replaced by ground releases in 2004, there was an increase in wild-type *C. capitata* populations. From annual means of 0.1–0.4 FTD and peaks of 0.4–2.8 with aerial releases, the annual mean (and peak) increased to 1.0 FTD (6.0 FTD) in 2005. From 2007 to the end of Phase 2 in 2010, wild-type populations increased even more dramatically, reaching an annual mean

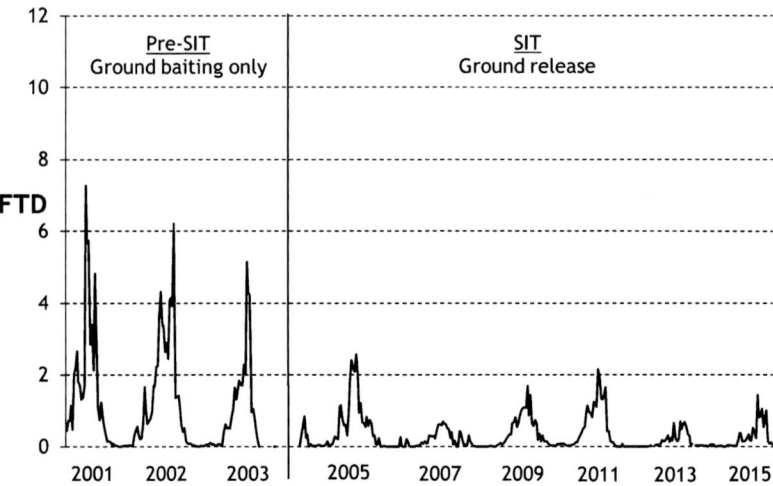

Fig. 19.4 Seasonal fluctuations in wild-type *C. capitata* populations (flies/trap/day – FTD) in the EGVV area before (2001–2003) and after (2005–2015) ground releases of sterile males. Data for ground releases are given only every alternate year from 2005 to illustrate the trend

(and peak) of 4.5 FTD (31.4 FTD) in 2010 (Table 19.1; Fig. 19.3). These are average FTDs over the entire 5000 ha, but *C. capitata* populations were not large throughout the entire area. In most cases a disproportionately small number of traps (and farms) had higher than average *C. capitata* populations. This was due to poor sanitation and other fruit fly management problems on those farms, which contributed to the high FTDs for the whole area. To illustrate this, wild-type fly trap data on all farms (208 traps on 162 farms) in the Hex Valley sub-zone of the greater Hex River Valley SIT area was analyzed for a 16-week period from February to May 2009; this is the period in the season when fruit fly populations are at their highest. A mean of 68 % (range 41–100 %) of the total number of wild-type flies trapped during this period came from only 10 % of the traps in the area. Of the 162 farms in the valley, nine occured between ten and 25 times in the top 10 % list, and 20 occurred between five and 25 times in the top 10 % list (B.N. Barnes, unpublished data).

Despite the overall poor response of the wild-type *C. capitata* populations to SIT, the growers persevered; they realized that ground based fruit fly baiting was not very effective, and that routine aerial bait applications through the year would be too expensive. Nevertheless, there remained an unwillingness to address the sanitation problems associated with table grape production that prevailed in the area.

EGVV: Following the introduction in 2004 of ground releases of sterile males in conjunction with other management practices such as bait applications and orchard sanitation, wild-type *C. capitata* populations in this area decreased. Between 2005 and 2010 annual means fell to between 0.2 and 0.5 FTD with peaks between 0.7 and 2.6 FTD. This represented a considerable reduction in *C. capitata* populations compared with pre-SIT years (Table 19.1; Fig. 19.4). Unlike the situation in the Hex

River Valley, growers in the EGVV area generally practised good sanitation and other integrated pest management interventions; they were also, as a group, very satisfied with the SIT programme.

5 The *C. capitata* SIT Programme – Phase 3, Post-2010

5.1 Improved Funding

In 2012 FruitFly Africa obtained approval for a statutory SIT levy for all deciduous fruit growers. In a further development, a 50:50 Public-Private Partnership was established between FruitFly Africa and the national Department of Agriculture, Forestry and Fisheries (DAFF) whereby DAFF contributes 50% of the costs associated with monitoring *C. capitata* populations; identification of sterile and wild-type flies caught in the monitoring traps; and for production, delivery and release of sterile *C. capitata* in the respective SIT areas (Barnes et al. 2015). With this additional funding, FruitFly Africa built a new larval and pupal rearing facility adjoining the adult rearing room. The entire rearing process now takes place under one roof. Other *C. capitata* management costs such as bait applications (ground and aerial), use of bait stations, orchard and vineyard sanitation and management of alternative host plants, is borne by the growers.

5.2 Improved Sterile C. capitata *Production and Releases*

Between 2012 and 2014, improvements to *C. capitata* rearing procedures were implemented following laboratory trials and quality control evaluation. These improvements are described below and helped facilitate an increase in production in the first quarter of 2014; currently 45 million sterile males are produced per week.

- *Larval 'starter packs'*: Eggs for each 5-kg tray of larval diet are allowed to hatch on a 1-kg block of larval diet, where they develop more uniformly in the confined space for 2 days before the block is replaced in the tray for completion of larval development (Fig. 19.5).
- *Egg brushing of adult cages*: Most eggs laid through the screen walls of the adult cages fall immediately into water baths below; however, some stick to the screens or fall into the water later at which stage they are often desiccated and of poor quality. The screens are now lightly brushed downwards twice daily to dislodge any eggs sticking to the screens so that more eggs land sooner in the water baths for improved egg quality.
- *Lighting in the adult room*: Additional light units were installed on walls and pillars in the adult room to improve vertical light distribution and facilitate optimum oviposition.

Fig. 19.5 A 1-kg starter pack of larval diet containing 2-day-old *C. capitata* larvae being placed into a larval rearing tray for further larval development

- *Aeration of eggs during egg maturation*: After egg collection from the water baths, the eggs are further matured by oxygenating them in water in plastic containers by bubbling air through the water. More constant and uniform aeration in all bottles was achieved by replacing the aquarium air pumps (one per bottle) with an air supply for each bottle from a single line fitted with an airflow regulator.
- *Crowding in adult cages*: The number of adults in the cages was reduced by 20 %, from 4400 to 3600 per cage. Although this reduced the number of eggs harvested from each cage, adult crowding stress was reduced.

Due to improvements in production it was possible to increase the number of sterile males released in the Hex River Valley and EGVV SIT areas in 2014 (also in a new area – see Sect. 5.3). In addition to routine releases in all farm gardens, additional sterile flies were released to supplement those already released in urban areas and other hotspots such as neglected fruit trees and alternative fruit fly host plants. By mid-2014 between 9000 and 38,000 sterile flies were being released per hectare of gardens, per week, depending on the *C. capitata* population pressure rating in the area. These ratings are subjectively based on mean annual population levels recorded during the previous season: low = up to 0.5 FTD; moderate = 0.6–1.0 FTD; high = above 1.0 FTD. Ground releases were also made in urban residential areas.

5.3 New Release Area

In 2010 the Warm Bokkeveld area, comprising the fruit production areas of Ceres and Prince Alfred's Hamlet, was brought into the SIT programme (Fig. 19.1). This valley lies on the other side of a range of mountains west of the Hex River Valley; predominantly pears and stonefruit are cultivated on 75 farms over an area of approximately 4600 ha. In a 2-year survey of this area the fruit fly species composition was found to be 100 % *C. capitata* (Manrakhan and Addison 2014). Until 2014, only weekly ground releases of sterile males were made in the two urban areas of Ceres and Prince Alfred's Hamlet. From October 2014 releases were also made in all farm gardens over the entire area, in a similar way to the programmes in the Hex River Valley and the EGVV.

5.4 Increased Use of Aerial Bait Applications and Other Population Reduction Measures

In 2010 the EGVV and Warm Bokkeveld areas followed the Hex River Valley example and began aerial applications of GF-120™ NF Naturalyte; between one and four applications were made annually with 2-year intervals. Application rates were the same as in the Hex River Valley. Growers were also recommended to apply their own routine ground-based fruit fly baits, but as in the Hex River Valley, this advice was not always followed by all growers. From 2012 helicopters replaced fixed-wing aircraft for bait application.

From 2011, M3® 'attract-and-kill' bait stations were deployed during the winter months in all farm gardens and in strategic locations where there was easy access to private property, and in areas known to harbour alternative fruit fly host plants. From mid-2014 the number of these bait stations was doubled, and their use in farm gardens continued throughout the year. The number per garden varied by area, according to fruit fly pressure.

5.5 Aerial Release Pilot Project

Ground releases of sterile *C. capitata* are unlikely to be the optimum long-term solution to *C. capitata* management in deciduous fruit-growing areas, as the nature of the releases precludes treating large areas of commercial fruit plantings. With a view to resuming an aerial release programme such as that carried out in the Hex River Valley in the early 2000s, a pilot area-wide aerial release programme was initiated in the spring (October) of 2014. The release area comprised 2200 ha of

Fig. 19.6 Gyroplane used in pilot aerial releases of sterile male *C. capitata*. The insulated box of pre-chilled flies is visible behind the pilot, as is the venturi release tube below the fuselage

apples, pears and peaches in the Warm Bokkeveld fruit production area. This area was selected as it was relatively isolated within the rest of the Warm Bokkeveld area, and historically had the lowest *C. capitata* populations in the area; during the previous 3 years, the mean seasonal population size was 0.02 FTD, compared with a mean of 0.12 FTD over the same period for the entire Warm Bokkeveld area (Barnes et al. 2015).

The release aircraft was a gyroplane (Sycamore Mk I Eagle tandem); despite its smaller payload, it was more economical to operate than a fixed-wing aircraft. The insulated release machine was designed locally and built specifically for the gyroplane with advice from Mr Roberto Angulo Kladt, Mubarqui Group, Mexico, an expert on this type of release technology; it had a capacity of 4.5 million pre-chilled flies, which were released through a venturi tube (Fig. 19.6), and was calibrated to release 2000 sterile flies per hectare.

Each week between the last week of October 2014 and the end of April 2015 (27 consecutive weeks), 4–5-day-old sterile male *C. capitata* were transported 120 km by road to the release area. The flies were then chilled to approximately 5 °C in a nearby cold-room facility, and loaded into an insulated box placed over the auger of the release machine in the gyroplane. Between 2000 and 2500 sterile flies were released per hectare each week in two release flights. Releases were made at an average altitude of 300 m above ground level and an average speed of 120 km/h. The *C. capitata* population (wild-type and sterile males) was monitored using 64 Chempac ® bucket traps baited with Biolure ® Fruit Fly lure (a food-based lure attracting both sexes), and 17 Chempac Yellow Delta sticky traps baited with

Chempac Fruit Fly lure (trimedlure, a male attractant) to monitor the distribution of sterile males. Traps were emptied weekly and lures and sticky pads were replaced every 6 weeks.

5.6 New Monitoring Areas

During this phase FruitFly Africa extended coordinated area-wide fruit fly monitoring programmes to two new areas, with a view to evaluating them for future SIT releases. These areas included approximately 4000 ha of export table grapes in the Lower Orange River area of the Northern Cape Province (since 2014; ~ 800 km from Stellenbosch; only *C. capitata* present), and about 2700 ha of apples and pears in the Langkloof area of the Eastern Cape Province (since 2012; ~ 600 km from Stellenbosch; *C. capitata* and *C. rosa* present).

5.7 Outcomes from Phase 3, Post-2010

5.7.1 Wild-Type Fruit Fly Populations

Hex River Valley Between 2011 and 2013 trap catches fluctuated between means of 1.9 and 2.8 FTD with peaks between 6.3 and 21.9 FTD. Improved population management in gardens and an increase in the number of sterile male *C. capitata* released in 2014 and 2015 did not, however, result in population reductions to the degree anticipated (Barnes et al. 2004, 2015; Table 19.1; Fig. 19.3). We believe that this reflects the ineffective timing of the aerial bait application programme, and the on-going difficulties in this area with vineyard sanitation and management of alternative host plants.

EGVV Between 2011 and 2015 *C. capitata* populations remained very low throughout this area. Highest populations were recorded in 2011 with an annual mean (and peak) of 0.4 (2.2) FTD. From 2012 to 2015, populations continued to decrease further, to annual means (and peaks) of 0.1–0.2 FTD (0.7–1.4) (Barnes et al. 2015; L. Hoofd pers. comm; Table 19.1; Fig. 19.4).

Warm Bokkeveld Prior to 2011 there was no reliable area-wide *C. capitata* population data available for this area. From 2011, population levels in this area steadily decreased from an annual mean (and peak) of 0.6 FTD (3.9) to between 0.1 and 0.2 FTD (0.7 and 1.2) in 2014 and 2015, respectively (Barnes et al. 2015, Table 19.1). This was attributed to greater grower awareness of, and compliance with, orchard sanitation and management of alternative fruit fly host plants, which began when the SIT programme was initiated in 2010. The effect of ground releases of sterile male *C. capitata* in all gardens in 2014/15 was not evident.

5.7.2 Aerial Release Pilot Project

For the majority of the duration of the pilot project, 4.5–5.5 million sterile flies (2000–2500/ha) were released weekly (27 consecutive weeks). Adverse weather conditions for the gyroplane resulted in sub-optimum releases on a few occasions, mostly in the first 2 months of the project before the expected increase in wild-type *C. capitata* populations began in late summer (February to April). Due to incomplete datasets for the first weeks of release, the results discussed below are all from the last 18 consecutive release weeks from January to April inclusive; this coincides with the period of the season when *C. capitata* populations are at their highest.

Wild-type *C. capitata* populations in the release area were retained at an exceptionally low level during this period of the pilot project. Based on weekly trap catches, the mean population level of wild-type *C. capitata* was 0.01 FTD, compared with 0.02 and 0.04 FTD for the same period in 2014 and 2013 respectively. Most wild-type flies were recorded in very localised zones within the area; only 48 % of the traps recorded one or more *C. capitata*, with 84 % of flies being trapped in only 20 % of the traps, and 35 % of the flies in just 2.5 % of traps. In the two preceding years 70 % (2014) and 100 % (2013) of the traps recorded catches. In most cases the sources of the wild-type flies could be traced to poor fruit fly management practices by a limited number of growers (N. Baard, unpub. data).

The distribution of sterile males during the pilot project was regarded as satisfactory, considering that the release method and release machine were being tested for the first time. Sterile males were recorded in an average of 70 % of the trimedlure traps per week, although during the entire trial every trap did capture sterile males. The mean sterile male:wild-type male ratio over the 18-week analysis period was approximately 42:1, and the mean number of sterile flies trapped was 1.05 FTD (N. Baard, unpub. data).

Following the success of the pilot project, gyroplane releases are scheduled to be extended during the 2015/2016 season to include the entire Warm Bokkeveld production area (releases over ~8500 ha) and the Hex River Valley area (also ~8500 ha), using a larger gyroplane with a newly-designed release machine with a capacity of 10 million pre-chilled flies per flight.

6 SIT and *Ceratitis rosa*

Ceratitis capitata is the only species of fruit fly in many of the fruit production areas in the Northern and Western Cape Provinces. However, Manrakhan and Addison (2014) found that, in some fruit production areas of the Western Province, *C. rosa* accounted for between 1 % and 51 % of the fruit fly population; in the more maritime areas such as around Stellenbosch and Elgin, the numbers were even higher. In an unpublished two-season study by the author in Somerset West (20 km from Stellenbosch, 5 km from the sea), *C. rosa* accounted for 87 % of the total fruit fly population. Substantial numbers of *C. rosa* have also been recorded in the Langkloof area of the Eastern Cape Province (S. Vosloo pers. comm.).

In the areas where *C. capitata* is the only fruit fly present, the creation of either a fruit fly-free area or an area of low fruit fly prevalence is possible using SIT. In areas where only a very low prevalence of *C. rosa* occurs, this may also be achievable if a concerted programme of *C. rosa* baiting and management of alternative host plants is undertaken in combination with SIT for *C. capitata*. However, where higher populations of *C. rosa* occur, fruit fly suppression/eradication with SIT will only be achievable if the technique can be developed and refined for *C. rosa* as well.

An effective artificial larval diet has been developed for *C. rosa*, based on a modified diet for *C. capitata*, but it remains to be evaluated for large-scale mass rearing (Barnes 2010). Unlike *C. capitata* females, female *C. rosa* do not oviposit through gauze screens. A cost-effective and reasonably practical egg collection system was developed for this species using cylindrical plastic jars (Barnes 2010), but again requires further evaluation for large-scale mass rearing. However, the greatest limiting factor for SIT for *C. rosa* is the fact that there is no genetic sexing strain for this species; as a result, both sterile males and sterile females would need to be released. This would be problematic in commercial fruit production as released sterile females would still puncture the fruit during attempted oviposition. Furthermore, even though no fertile eggs would be laid, the punctures would allow infestation by secondary pests and pathogens (e.g. vinegar fly and botrytis). For these reasons, SIT for fruit flies in the Western Cape is being focussed on those areas where *C. capitata* predominates.

7 Overall Success of SIT and Lessons Learnt

As a suppression tool against *C. capitata* in South Africa, SIT has had more success in some areas than in others (Table 19.1; Figs. 19.2 and 19.3). It has been very successful in the EGVV area, and far less successful in the Hex River Valley area. Some of the factors that have compromised the success of SIT in the Hex River Valley have already been mentioned here, and are discussed in detail by Barnes (2007). A further factor is climate; long-term data show that the Hex River Valley has higher average maximum temperatures than the EGVV area (Barnes et al. 2015), conditions which favour development of *C. capitata* (Nyamukondiwa et al. 2013). However, other contributing factors include the specific nature of the cultivation and harvesting of table grapes and the methods used for area-wide pest management, which vary between the two areas (Box 19.2).

Box 19.2: Factors Affecting Fruit Fly Management in the Hex River Valley and EGVV The following factors influenced the relative success of SIT in the Hex River Valley and the EGVV areas (Barnes et al. 2015, I. Sutherland, pers. comm.):

In the Hex River Valley:

Table grape cultivars: *Ceratitis capitata* development and survival in berries of table grape cultivars grown up to the end of the 1990s was relatively poor; berries were more juicy and most larvae drowned in the juice before maturing. Newer cultivars bred subsequently are crisper and firmer and more suitable for the completion of *C. capitata* larval development. Further, in contrast to earlier years many more table grape cultivars are now grown, and they are grown over a longer growing season. Different cultivars ripen in quick succession, often simultaneously, with the result that growers often neglect proper vineyard sanitation.

Ground fruit fly baiting: This is largely ineffectual in table grape vineyards for a variety of reasons.

Harvesting process: The process by which table grape cultivars are harvested (particularly late-season grapes) is very detrimental to fruit fly management. At optimal ripeness, pickers harvest all export quality bunches, leaving behind lower grade bunches. At this point growers allow fresh-fruit market agents to harvest any remaining bunches suitable for the local market. These agents have no regard for vineyard sanitation and discard on the ground, or leave on the vines, any bunches that are not suitable for their purposes. The grower generally takes no further responsibility for the grapes that remain in the vineyards. The longer these remain in the vineyard, the more suitable they become for *C. capitata* infestation, seriously compromising *C. capitata* management.

Number of fruit packers: All growers in the valley have their own packhouses and pack their own grapes. Their management decisions are thus likely to be based on self interest rather than an interest in the Hex River Valley grower community as a whole.

Table grape management: In general, the level of *C. capitata* monitoring, management of alternative host plants and sanitation in the Hex River Valley is at a lower level than in other fruit production areas in the Western Cape (e.g. the EGVV area). This is extremely deleterious to effective *C. capitata* management, and has a very negative impact on SIT.

In contrast, in the EGVV area (mixed fruit farming):

- Growers have been using intensive pest monitoring programmes successfully for many years, and are more compliant with supplementary fruit fly management actions such as orchard sanitation and management of alternative host plants

(continued)

Box 19.2: (continued)
- At harvest, fruit pickers generally 'clean-pick' apple trees in one operation, leaving behind very few fruit that could support later fruit fly infestation
- Nearly all growers send their fruit to a packing cooperative which packs fruit for a number of growers. These cooperatives provide technical advisers who assist growers in making sound pest management decisions that are in the broader interest of the local fruit-growing community.

These factors complement the area-wide *C. capitata* SIT programme, and have helped to make it more successful in the EGVV area than in the Hex River Valley.

Many lessons have been learnt during the *C. capitata* SIT programme, and they are summarised below. Some of these factors could be considered obvious by those who are well acquainted with the technique, and be well established as principles in an SIT programme. Nevertheless, the significance and implications of these elements became more clear as SIT for *C. capitata* evolved in South Africa.

- Broad-based, multi-organisational and sustainable funding must be available from the start of the SIT programme.
- The area for the SIT programme, especially its initiation, must be carefully selected. Besides geographic or topographic isolation, the target pest should already be well managed by conventional methods, with growers who are progressive in their pest management outlook.
- There must be buy-in to the project by all fruit growers in the selected area(s).
- SIT cannot be introduced into any area without a desire for it from the growers. A 'push-pull' policy by the stakeholders in the programme is advisable: while the SIT technologists should 'push' (i.e. advocate) SIT where such a programme is appropriate, there must also be a 'pull' (i.e. receptiveness) on the part of the growers.
- Good public relations and communication between SIT service providers and growers is crucial.
- Mechanisms whereby growers financially contribute to the programme should be in place before the programme starts, and be consistent and fair between all production areas.
- Delivery assurances by the implementers of the SIT programme, as well as expectations by the beneficiaries, need to be realistic.
- A logical and practical expansion plan from any pilot projects should exist before the start of the full programme.
- There should be no compromise on quality – whether of infrastructure, equipment, or the insects produced.
- A good quality management system must be in place in the rearing facility, and include regular audits (internal and external) of procedures, processes and performance. An automatic electronic message system in the rearing facility should warn key staff members of any equipment deviation from specified performance.

- Rearing facility and field staff should be carefully selected, well-trained, enthusiastic and observant. The staff component should be stable to avoid rapid turnover and retraining. Audits (internal and external) of procedures and performance should be made regularly.
- Ground releases of sterile *C. capitata* are not a long-term solution to population suppression. Releases should be by air if at all possible.
- Effective management of alternative fruit fly host plants and orchard/vineyard sanitation at farm level is crucial to the success of an SIT programme.
- Good supervision in the rearing facility and the field is essential. Training of facility and field staff should be an on-going process.
- Programme managers should keep abreast of the latest international developments in the field of SIT, and make good use of knowledge and input from international SIT experts.
- Databases of both rearing facility and field operations must be well maintained.
- Research and development should be on-going, and all cost-effective and affordable improvements in procedures and processes should be implemented.

Acknowledgements I thank N. Baard, J. Claassen, R. Du Preez, L. Hoofd and I. Sutherland of FruitFly Africa (Pty) Ltd., for access to unpublished data on fruit fly populations and for their input to various aspects of the SIT programme. I also acknowledge the major role played by the Joint FAO/IAEA Division of Nuclear Techniques in Food and Agriculture, Vienna, Austria, its staff members in the Insect Pest Control Section and many other SIT experts around the world who are involved in the development of SIT in South Africa; this invaluable support continues today.

References

Barnes BN (2006) Fruit flies on wine grapes – infestation success, cultivar effects and impact on area-wide control. Wynboer Tech Yearb 2006. Wineland Media, Suider Paarl, South Africa, pp 44–46

Barnes BN (2007) Privatizing the SIT: a conflict between business and technology? In: Vreysen MJB, Robinson AS, Hendrichs J (eds) Area-Wide Control of Insect Pests: from Research to Field Implementation. Springer, Dordrecht, pp 449–456

Barnes BN (2010) Improvements in mass-rearing Natal fruit fly, *Ceratitis rosa* (Karsch), to facilitate its inclusion in a fruit fly SIT programme. S Afr Fruit J 9:16–21

Barnes BN, Venter JH (2008) The South African fruit fly action plan – area-wide suppression and exotic species surveillance. In: Sugayama RL, Zucchi RA, Ovruski SM, Savinski J (eds) Fruit flies of economic importance: from basic to applied knowledge. Proceedings of the 7th international symposium on fruit flies of economic importance, Salvador, Bahia, Brazil 2006. Biofábrica Moscamed, Brazil, pp 271–283

Barnes BN, Eyles DK, Franz G (2004) South Africa's fruit fly SIT programme – the Hex River Valley pilot project and beyond. In: Barnes BN (ed) Proceedings of the 6th international symposium on fruit flies of economic importance, Stellenbosch, South Africa 2002. Isteg Scientific Publications, Irene, South Africa, pp 131–141

Barnes BN, Targovska A, Franz G (2006) Origin of Mediterranean fruit fly, *Ceratitis capitata* (Wiedemann), outbreak determined by DNA analysis. Afr Entomol 14:205–209

Barnes BN, Hofmeyr JH, Groenewald S, Conlong DE, Wohlfarter M (2015) The sterile insect technique in agricultural crops in South Africa: a metamorphosis… but will it fly? Afr Entomol 23:1–18

Blomefield TL, Barnes BN, Giliomee JH (2015) Peach and nectarine. In: Prinsloo GL, Uys VM (eds) Insects of Cultivated Plants and Natural Pastures in Southern Africa. Entomological Society of Southern Africa, Hatfield, pp 320–339

Conlong DE (2007) Prospects for the sterile insect production and quality control of *Eldana saccharina* and *Chilo sacchariphagus* (Lepidoptera: Pyralidae, Crambidae): stalk borers threatening sugarcane industries in south-eastern Africa and the Mascarene islands. In: Van Lenteren JC, De Clercq P, Johnson M (eds) Proceedings of the 11th meeting of the Working Group on Arthropod Mass Rearing and Quality Control, Montreal, Canada 2007. Bulletin IOBC Global No. 3, pp 30–33

Conlong DE, Rutherford SR (2009) Conventional and new biological and habitat interventions for integrated pest management systems: review and case studies using *Eldana saccharina* Walker (Lepidoptera: Pyralidae). In: Peshin R, Dhawan AK (eds) Integrated pest management: innovation-development process. Springer, Dordrecht, pp 241–261

De Villiers M, Hattingh V, Kriticos DJ (2012) Combining field phenological observations with distribution data to model the potential distribution of the fruit fly *Ceratitis rosa* Karsch (Diptera: Tephritidae). Bull Entomol Res. doi:10.1017/S0007485312000454

Dyck VA, Hendrichs J, Robinson AS (eds) (2005) Sterile insect technique. Principles and practice in area-wide pest management. Springer, Dordrecht

Enkerlin WR (2005) Impact of fruit fly control programmes using the sterile insect technique. In: Dyck VA, Hendrichs J, Robinson AS (eds) Sterile insect technique. Principles and practice in area-wide integrated pest management. Springer, Dordrecht, pp 651–676

Franz G (2005) Genetic sexing strains in Mediterranean fruit fly, an example for other species amenable to large scale rearing as required for the sterile insect technique. In: Dyck VA, Hendrichs J, Robinson AS (eds) Sterile insect technique. Principles and practice in area-wide integrated pest management. Springer, Dordrecht, pp 427–451

Grové T, De Villiers EA, Daneel MS (2015) Mango. In: Prinsloo GL, Uys VM (eds) Insects of cultivated plants and natural pastures in Southern Africa. Entomological Society of Southern Africa, Hatfield, pp 574–588

Hendrichs J, Franz G, Rendon P (1995) Increased effectiveness and applicability of the sterile insect technique through male-only releases for control of Mediterranean fruit flies during fruiting seasons. J Appl Entomol 119:371–377

Hendrichs J, Vreysen MJB, Enkerlin WR, Cayol JP (2005) Strategic options in using sterile insects for area-wide integrated pest management. In: Dyck VA, Hendrichs J, Robinson AS (eds) Sterile insect technique. Principles and practice in area-wide integrated pest management. Springer, Dordrecht, pp 563–600

HortGro Science (2015) Name change of the invasive fruit fly and an update on its pest status in South Africa. Fresh Notes 107 (20 February)

(IAEA) International Atomic Energy Agency (2003) Trapping guidelines for area-wide fruit fly programmes, IAEA/FAO-TG/FFP. IAEA, Vienna

Klassen W (2005) Area-wide integrated pest management and the sterile insect technique. In: Dyck VA, Hendrichs J, Robinson AS (eds) Sterile insect technique. Principles and practice in area-wide integrated pest management. Springer, Dordrecht, pp 39–68

Klassen W, Curtis CF (2005) History of the sterile insect technique. In: Dyck VA, Hendrichs J, Robinson AS (eds) Sterile insect technique. Principles and practice in area-wide integrated pest management. Springer, Dordrecht, pp 3–36

Manrakhan A, Addison P (2014) Assessment of fruit fly (Diptera: Tephritidae) management practices in deciduous fruit growing areas in South Africa. Pest Man Sci 70:651–660. doi:10.1002/ps.3604

Manrakhan A, Hattingh V, Venter J-H, Holtzhausen M (2011) Eradication of *Bactrocera invadens* (Diptera: Tephritidae) in Limpopo Province, South Africa. Afr Entomol 19:650–659

Mumford J, Tween G (1997) Economic feasibility study for the control of the Mediterranean fruit and Natal fruit fly in the Western Cape Province. Expert Mission Report RU-7135, International Atomic Energy Agency, Vienna

Nyamukondiwa C, Weldon CW, Chown SL, Le Roux PC, Terblanche JS (2013) Thermal biology, population fluctuations and implications of temperature extremes for the management of two globally significant insect pests. J Insect Physiol 59:1199–1211

Potgieter L, Van Vuuren JH, Conlong DE (2013) A reaction–diffusion model for the control of *Eldana saccharina* Walker in sugarcane using the sterile insect technique. Ecol Model 250:319–328

Prinsloo G, Uys V (eds) (2015) Insects of cultivated plants and natural pastures in Southern Africa. Entomological Society of Southern Africa, Hatfield

(OABS) Optimal Agricultural Business Systems (2013) Key deciduous fruit statistics. Hortgro, Paarl

Venter JH (2012) Pest risk assessment for regulatory control of *Bactrocera invadens* (Diptera: Tephritidae) in the Musina area (Limpopo province). M. Env. Sc. dissertation, North-West University, Potchefstroom, South Africa

Vreysen MJB (2005) Monitoring sterile and wild insects in area-wide integrated pest management programmes. In: Dyck VA, Hendrichs J, Robinson AS (eds) Sterile insect technique. Principles and practice in area-wide integrated pest management. Springer, Dordrecht, pp 325–361

Brian N. Barnes majored in Entomology at university, ultimately being awarded a PhD (Agric.) degree by Stellenbosch University in 1987. For the greatest part of his career he carried out research on the development of pest management systems for the deciduous fruit industry in the Western Cape Province, based at the ARC Infruitec-Nietvoorbij Institute for Fruit, Vine & Wine. In 1992 he was promoted to Manager of the Pest Management Division, and in 2001 became involved full-time with coordinating the fruit fly SIT programme. During his career he has published numerous articles on fruit pest management in peer-reviewed and fruit-industry journals, and in agricultural magazines. He has written a number of book chapters and compiled a number of pest management manuals. He also served as Chairman of the International Fruit Fly Steering Committee from 2002 to 2010, and was a founding member of the Working Group on Tephritids of Europe, Africa and the Middle East (TEAM). He retired in 2009 and has since served as the Technical Advisor for FruitFly Africa.

Chapter 20
Cold and Heat Treatment Technologies for Post-harvest Control of Fruit Flies in Africa

Tim G. Grout

Abstract Research on cold treatments for post-harvest control of tephritid fruit flies began in South Africa more than 100 years ago. Once commercial cold treatments were accepted by the USA for deciduous fruit exported from Africa, research turned to evaluating the same treatments on citrus. Experiments in shipments to New Zealand gave promising results in the late 1940s and a treatment of <0 °C for 12 days was accepted for exports of some types of citrus from Africa to Japan in 1970. In 1956 and again in 2001, it was shown that the Natal fruit fly, *Ceratitis rosa*, was as susceptible as the targeted medfly, *Ceratitis capitata*, to cold treatments. After the arrival in Africa of the oriental fruit fly, *Bactrocera dorsalis*, further cold treatment research showed that this species was similar in its cold tolerance to *C. capitata* and commercial cold treatments were developed for both citrus and avocado. Post-harvest heat treatments for tephritids in tropical fruit did not receive much attention before the arrival of *B. dorsalis*. Hot water immersion is the only heat treatment that has been evaluated and shown to be effective against *B. dorsalis*, *C. capitata* and the mango fruit fly, *Ceratitis cosyra*, in mango. These results confirmed previous studies from other countries and so this relatively simple treatment may permit widespread exports of mango from Africa in the future.

Keywords *Bactrocera dorsalis* • *Ceratitis capitata* • *Ceratitis rosa* • *Ceratitis cosyra* • Quarantine pests • Disinfestation

1 Introduction

As countries in Africa seek to export more of their horticultural produce (Smith 2010; Viviers et al. 2014), so importing countries usually demand guarantees that this produce will not contain viable fruit flies (Tephritidae). In Africa, fruit is grown under a range of different climatic conditions that can be broadly categorised as

T.G. Grout (✉)
Citrus Research International, P O Box 28, 1200 Nelspruit, South Africa
e-mail: tg@cri.co.za

Mediterranean, subtropical and tropical. The conditions that are optimal for growing particular fruits also determine which post-harvest cold or heat treatments are most suitable for those fruits. Deciduous fruits tolerate cold treatments well and are often stored for long periods of time at low temperatures, so this is the treatment of choice for this type of fruit (Pryke and Pringle 2008). Subtropical fruits such as *Citrus* species and avocado, *Persea americana* Mill., generally tolerate cold treatments better than heat treatments, but can be damaged if the cold treatment temperature is too low or the treatment period too long. Tropical fruits tolerate heat treatments better than cold treatments and, as these treatments are much quicker than cold treatments, they are the preferred means of controlling fruit fly in tropical fruit such as mango, *Mangifera indica* L.

Other post-harvest disinfestation treatments besides low or high temperatures are in use or under development in some countries, but are not widely used in Africa. These include ionizing radiation; for control of tephritids a suggested generic dose of 150 Gy is gaining acceptance in some countries (Hallman and Loaharanu 2002). However, there are only a limited number of irradiation sources available in Africa and it is logistically difficult to disassemble and then reassemble pallets of fruit cartons if required to prevent over-exposure of the outer layer of cartons due to the dose uniformity ratio (Hallman 2000; Viljoen 2011). Furthermore, consumers in many importing countries do not currently accept irradiated fruit, thereby limiting market opportunities (Ferrier 2010; Loaharanu and Ahmed 1991). Fumigation of fruit with methyl bromide either prior to export or on arrival in an importing country is being phased out, although a few countries still make use of it (Phillips et al. 2015; Shamilov 2012). The most promising alternative fumigant against internal pests like fruit fly larvae is phosphine, although this can require a fumigation period of 48 h, compared with 2 h for methyl bromide, to achieve the same level of control (Williams et al. 2000). Outside Africa, phosphine is increasingly used commercially so it may become available in Africa in the near future. The use of a 'controlled atmosphere', where oxygen levels are lowered and carbon dioxide levels may be increased, is sometimes suggested as an alternative means of post-harvest disinfestation but much more research is required before commercial treatments are likely to become available (Pryke and Pringle 2008); attempts to control the Natal fruit fly, *Ceratitis rosa* Karsch, and the mango (marula) fruit fly, *Ceratitis cosyra* (Walker), using this technique were ineffective (Grové et al. 2000).

2 Cold Treatments

In South Africa, research on cold storage for the post-harvest control of the medfly, *Ceratitis capitata* (Wiedemann), in deciduous fruit began in the early 1900s and was reviewed by Nel (1936) in a paper that also evaluated various temperature-time treatment combinations and included observations from commercial shipments of infested grapes to the UK. These experiments showed that a temperature of −0.6 °C for 9 days, 1.1 °C for 12 days or a precooling temperature of 1.1 °C for 4 days followed by 12 at 2.8 °C were effective in controlling *C. capitata* and that applying these treatments during shipping could allow for the export of deciduous fruit from South Africa to the UK (Nel 1936). In 1937, the Bureau of Entomology and Plant Quarantine in the USDA published notices 463 and 464 which permitted the importation of fruit from South Africa after treatment at a temperature of 1.1 °C or below for a period of at least 12 days, provided that it also met the precooling requirements (Myburgh 1956). Subsequent studies showed that *C. rosa* was equally susceptible to the above cold treatments used for *C. capitata* (Myburgh 1956). At this time, research focus shifted to the control of Tortricidae such as the false codling moth, *Thaumatotibia leucotreta* (Meyrick), (Myburgh 1965), and later the oriental fruit moth, *Grapholita molesta* (Busck) (Dustan 1963; Blomefield and Geertsema 1990), in deciduous fruits, because larvae of these lepidopterans were known to be more cold tolerant than fruit flies (Pryke and Pringle 2008). Thus, for export of deciduous fruit to various countries outside of Africa, the commercial cold treatments developed to control these lepidopterans were more than sufficient to also control fruit flies.

Citrus is less tolerant to cold storage than deciduous fruit and low temperature for an extended period of time can result in unsightly chilling injury in the peel. With citrus there is therefore an incentive to use as short a treatment as possible or at as high a temperature as possible (Henriod et al. 2005). African research on a cold treatment to control *C. capitata* in citrus began in the 1930s at the Low Temperature Research Laboratory in Cape Town. In 1948, the South African Co-operative Citrus Exchange was asked by the New Zealand government to export a trial shipment of citrus to New Zealand using the treatment schedules approved for export of deciduous fruit to the USA (Boyes and Ginsburg 1969). The citrus arrived in good condition encouraging further research which found that oranges harvested in the coldest weather during mid-winter were more susceptible to chilling injury during postharvest cold treatments, than oranges harvested in autumn or early spring (Boyes and Ginsburg 1969). Subsequent research to develop a cold treatment for exports of citrus to Japan found that a treatment of <0 °C for 12 days was effective against *C. capitata*; this was accepted by the Japanese authorities (Anonymous 1970) and is still in use today for export of oranges (*Citrus sinensis* [L.] Osbeck), grapefruit (*Citrus paradisi* Macfad.) and lemons (*Citrus limon* [L.] Burm.f.) to Japan. However, this cold treatment was implemented at the dock-side where it caused logistical delays and it was only approved for use in-transit on ships in 1995 (McGlashan 1995) when on-board cooling systems were considered reliable enough. Ware et al. (2001) demonstrated that this treatment was equally effective against *C. rosa*. Since then research by Citrus Research International (CRI) has convinced the Japanese authorities to also accept mandarins, *Citrus reticulata* Blanco, after treatment at temperatures below 0 °C for 14 days (Ware et al. 2005; Hattingh and Carstens 2008). Using temperatures below 0 °C (as described above) often results in chilling injury to lemon. However, an alternative cold treatment of <1.4 °C for 16 days was equally effective against *C. capitata* and also caused less damage to lemons (Grout et al. 2011a).

The invasive oriental fruit fly, *Bactrocera dorsalis* (Hendel), previously known as *Bacterocera invadens* Drew, Tsuruta and White (Drew et al. 2005), has now spread throughout most of Africa (Goergen et al. 2011; Lux et al. 2003) leading to collaborative research between South African and Kenyan scientists to develop post-harvest cold treatments for citrus and avocado, targeted against this fruit fly. This research was done at the International Centre of Insect Physiology and Ecology (*icipe*) north of Nairobi, Kenya using techniques developed at Citrus Research International in Nelspruit, South Africa. The cold room used at *icipe* was small and the ambient temperature was high so it took around 2 days for the internal fruit temperature to drop to the required treatment level. However, the results showed that *B. dorsalis* was killed in citrus treated at a temperature of ≤0.9 °C for 16 days (Grout et al. 2011b) and in avocado, which is more susceptible to chilling injury, at a temperature of ≤1.5 °C for 18 days (Ware et al. 2012). Complementary research conducted in Europe on diet-reared third instar larvae of both *C. capitata* and *B. dorsalis* (two strains, one originating from Africa and the other from Asia) found no significant difference in their susceptibility to cold; mortality of all strains was similar at a temperature of 0.94 °C ± 0.65 °C for 8–11 days (Hallman et al. 2011). These data were sufficient to convince the USDA that the cold treatment schedules they imposed for control of *T. leucotreta* in citrus fruit would also be suitable for control of fruit flies in citrus from Africa (Hallman et al. 2011). Further research by Hallman et al. (2013) found that larvae of both *C. capitata* and *B. dorsalis* from Africa developing within oranges were equally susceptible to cold temperature treatments of 0.94 °C ± 0.01 °C for 9 days. Together these studies gave importing countries the assurance that *B. dorsalis* from Africa could be controlled using established cold treatment schedules already in place for a range of other commodities. In addition, now that strains of *B. dorsalis* from Africa and Asia are known to be the same (Schutze et al. 2013, 2015), cold treatment schedules established for control of the Asian strain in mandarins from Taiwan intended for export to Japan and South Korea (≤1.0 °C for 14 days; Dohino et al. 2016) can also be adopted by other countries importing fruit from Africa.

All South African research on cold treatments for control of *C. capitata* and *C. rosa* followed the approaches of Nel (1936), and others before him, who evaluated treatment efficacy based on larval mortality, also known as the larval endpoint. The pros and cons of this approach have been discussed by Grout et al. (2011a). The larval endpoint approach evaluates larval mortality in fruit after treatment, which is what government inspectors do when evaluating samples of fruit received at ports of entry. When samples of fruit are cut open by inspectors at the harbour there is no time or space to hold a consignment of fruit for a week or two to determine whether any live fruit fly larvae found are capable of successfully pupating or not: the consignment will immediately be rejected on the basis of the presence of a live quarantine pest. The judgements made by inspectors are therefore similar to the evaluations made by researchers when using the larval endpoint method. However, research based on larval mortality is more labour intensive and other researchers prefer to use successful pupariation as the endpoint (Jessup et al. 1993; De Lima et al. 2007). With this approach the treated fruit does not need to be cut open but is placed in racks above sand into which any surviving larvae drop to pupate. The sand is later sieved to determine whether there has been any successful pupariation. Gazit et al. (2014) used a combination of both techniques where a larval endpoint was used on a small portion of the fruit in each replicate while the rest of the fruit were incubated for longer to determine whether there was any successful pupariation.

In cold treatment research there are also two different approaches to the initial inoculation of fruit with eggs. Some researchers such as De Lima et al. (2007) prefer to puncture fruit such as oranges and expose them to fruit fly to encourage natural oviposition into the puncture holes. Others such as Hill et al. (1988) have used this technique but also inject eggs into a hole made in the fruit. Government authorities in some countries insist that research on cold treatments use similar numbers of eggs in each fruit when developing a treatment schedule. This is not possible if fruit are suspended in cages for flies to oviposit in them naturally. Some fruit types may also not be ideal hosts and few eggs may be laid, even after puncturing the epidermis. In both these cases, inoculation methods using a known number of eggs per fruit are essential. At CRI this has been accomplished using an autopipette (Grout et al. 2011a, b) to collect and inject a known volume of eggs in water into a hole made in the fruit below the calyx.

The best approach for future research in developing cold treatments for fruit flies may be to inoculate known numbers of eggs into a hole in each fruit and to use a combination of larval endpoint and successful pupariation to determine treatment efficacy as used by Gazit et al. (2014). Greater emphasis could be placed on larval endpoint studies in the early phases of the research with successful pupariation as the end point used for large scale disinfestation phases or confirmatory trials.

3 Heat Treatments

As described above, heat treatments are best suited to tropical fruits such as mango and papaya, *Carica papaya* L.; the former is of great economic importance to rural communities throughout tropical Africa. In parts of South Africa where *B. dorsalis* is not yet abundant, mango is mostly attacked by *C. cosyra* and *C. rosa* (Grové 2001). However, in the rest of Africa it is more common for mango to be attacked by *C. cosyra* in the cooler, dry season before the rains start and for it to be replaced by *B. dorsalis* once the rains start and temperatures increase (Rwomushana et al. 2009; Vayssières et al. 2007, 2008, 2009). Heat treatments would therefore need to target both these dominant species in mango and, if effective, are also likely to be effective against other *Ceratitis* species (see below). The simplest heat treatment requiring the least technology is that of hot water immersion and this holds the most promise for making African mango and papaya acceptable to export markets other than via processing. However, very little research has been done on heat treatments on the continent. Some research was done in South Africa with hot water immersion of Kent, Heidi and Sensation mango cultivars which showed that 47 °C for 90 min controlled *C. capitata* and *C. cosyra,* but there was concern that the treatment would shorten the shelf life of the fruit (Labuschagne et al. 1996). More recently, research to control *B. dorsalis* in mango in West Africa has shown that getting the core temperature to 46.5 °C resulted in all larvae being killed, whereas a temperature of 42 °C only resulted in 30% mortality (Self et al. 2012). These temperatures of around 46–47 °C for *C. cosyra* and *B. dorsalis* are supported by research on *C. capitata* in Mexico (Hernandez et al. 2012) and Brazil (Nascimento et al. 1992), so clearly this is the temperature that should be explored further for commercial hot water immersion treatments of mango. Research results for *C. capitata* are relevant, even though it is not a serious mango pest in Africa, because the research conducted by Hallman et al. (2011) using diet-reared larvae demonstrated that third instar *C. capitata* and *B. dorsalis* larvae from Africa were similar in their heat tolerance. Research in Hawaii with hot water immersion of lychee, *Litchi chinensis* Sonnerat, found that fruit immersed in water at 49 °C for 20 min followed by hydrocooling in water at 24 °C for a further 20 min were free of fruit flies (Armstrong and Follett 2007). The high levels of statistical confidence in these data resulted in the USA mainland accepting fruit exported from Hawaii. Other types of heat treatment such as vapour-heat and high-temperature forced-air treatments have not yet received any research attention in Africa but may show potential for subtropical fruits that are damaged by more conventional hot water immersion. Although promising results were obtained with these treatments to citrus in the USA in the 1990s (Hallman et al. 1990; Shellie and Mangan 1994), they are not being used commercially due to concern about detrimental effects on the fruit. Equipment to maintain high relative humidity while preventing condensation on the fruit and the use of hydrocooling after treatment may allow for safe commercial use of high-temperature forced-air treatments in the future.

References

Anonymous (1970) We are in Japan. Outspan News 40, June 1970. Official Bulletin of the Citrus Exchange, Pretoria, South Africa

Armstrong JW, Follett PA (2007) Hot-water immersion quarantine treatment against Mediterranean fruit fly and oriental fruit fly (Diptera: Tephritidae) eggs and larvae in litchi and longan fruit exported from Hawaii. J Econ Entomol 100:1091–1097

Blomefield TL, Geertsema H (1990) First record of the oriental fruit moth, *Cydia molesta* (Lepidoptera: Tortricidae: Olethreutinae), a serious pest of peaches, in South Africa. Phytophylactica 22:355–357

Boyes WW, Ginsburg L (1969) Navels and Valencia late storage at temperatures low enough to destroy all traces of fruit fly. SA Citrus J 425:5–12

De Lima CPF, Jessup AJ, Cruickshank L, Walsh CJ, Mansfield ER (2007) Cold disinfestation of citrus (*Citrus* spp.) for Mediterranean fruit fly (*Ceratitis capitata*) and Queensland fruit fly (*Bactrocera tryoni*) (Diptera: Tephritidae). N Z J Crop Hortic Sci 35:39–50

Dohino T, Hallman GJ, Grout TG, Clarke AR, Follett PA, Cugala DR, Tu DM, Murdita W, Hernandez E, Pereira R, Myers SW (2016) Phytosanitary treatments against *Bactrocera dorsalis* (Diptera: Tephritidae): current situation and future prospects. J Econ Entomol 109: (in press)

Drew R, Tsuruta K, White I (2005) A new species of pest fruit fly (Diptera: Tephritidae: Dacinae) from Sri Lanka and Africa. Afr Entomol 13:149–154

Dustan GG (1963) The effect of standard cold storage and controlled atmosphere storage on survival of larvae of the oriental fruit moth, *Grapholitha molesta*. J Econ Entomol 56:167–169

Ferrier P (2010) Irradiation as a quarantine treatment. Food Policy 35:548–555

Gazit Y, Akiva R, Gavriel S (2014) Cold tolerance of the Mediterranean fruit fly in date and mandarin. J Econ Entomol 107:1745–1750

Goergen G, Vayssières J-F, Gnanvossou D, Tindo M (2011) *Bactrocera invadens* (Diptera: Tephritidae), a new invasive fruit fly pest for the afrotropical region: host plant range and distribution in west and central Africa. Environ Entomol 40:844–854

Grout TG, Stephen PR, Daneel JH, Hattingh V (2011a) Cold treatment of *Ceratitis capitata* (Diptera: Tephritidae) in oranges using a larval endpoint. J Econ Entomol 104:1174–1179

Grout TG, Daneel JH, Mohamed SA, Ekesi S, Nderitu PW, Stephen PR, Hattingh V (2011b) Cold susceptibility and disinfestation of *Bactrocera invadens* (Diptera: Tephritidae) in oranges. J Econ Entomol 104:1180–1188

Grové T (2001) Order diptera: family tephritidae. In: van den Berg MA, De Villiers EA, Joubert PH (eds) Pests and beneficial arthropods of tropical and non-citrus subtropical crops in South Africa. Agriculture Research Council – Institute for Tropical and Subtropical Crops, Nelspruit, pp 293–302

Grové T, Steyn WP, de Beer MS (2000) Controlled atmosphere (CA) as a post-harvest treatment against fruit flies in mangoes. South African Mango Growers' Association Yearbook, Vol. 19 & 20, 1999–2000, pp 147–149

Hallman GJ (2000) Expanding radiation quarantine treatments beyond fruit flies. Agric For Entomol 2:85–95

Hallman GJ, Loaharanu P (2002) Generic ionizing radiation quarantine treatments against fruit flies (Diptera: Tephritidae) proposed. J Econ Entomol 95:893–901

Hallman GJ, Gaffney JJ, Sharp JL (1990) Vapor heat treatment for grapefruit infested with Caribbean fruit fly (Diptera: Tephritidae). J Econ Entomol 83:1475–1478

Hallman GJ, Myers SW, Jessup AJ, Islam A (2011) Comparison of *in vitro* heat and cold tolerances of the new invasive species *Bactrocera invadens* (Diptera: Tephritidae) with three known tephritids. J Econ Entomol 104:21–25

Hallman GJ, Myers SW, El-Wakkad MF, Tadrous MD, Jessup AJ (2013) Development of phytosanitary cold treatments for oranges infested with *Bactrocera invadens* and *Bactrocera zonata* (Diptera: Tephritidae) by comparison with existing cold treatment schedules for *Ceratitis capitata* (Diptera: Tephritidae). J Econ Entomol 106:1608–1612

Hattingh V, Carstens E (2008) Programme: market access technical coordination. CRI Group Annual Research Report for January 2007 to March 2008. Citrus Research International, Nelspruit, pp 3–9

Henriod RE, Gibberd MR, Treeby MT (2005) Storage temperature effects on moisture loss and the development of chilling injury in Lanes late navel orange. Aust J Exp Agric 45:453–458

Hernandez E, Rivera P, Bravo B, Toledo J, Caro-Corrales J, Montoya P (2012) Hot-water phytosanitary treatment against *Ceratitis capitata* (Diptera: Tephritidae) in Ataulfo mangoes. J Econ Entomol 105:1940–1953

Hill AR, Rigney CJ, Sproul AN (1988) Cold storage of oranges as a disinfestation treatment against the fruit flies *Dacus tryoni* (Froggatt) and *Ceratitis capitata* (Wiedemann) (Diptera: Tephritidae). J Econ Entomol 81:257–260

Jessup A, De Lima C, Hood C, Sloggett R, Harris A, Beckingham M (1993) Quarantine disinfestation of lemons against *Bactrocera tryoni* and *Ceratitis capitata* (Diptera: Tephritidae) using cold storage. J Econ Entomol 86:798–802

Labuschagne T, Brink T, Steyn WP, de Beer MS (1996) Fruit flies attacking mangoes – their importance and post harvest control. South African Mango Growers' Association Yearbook 16:17–19

Loaharanu P, Ahmed M (1991) Advantages and disadvantages of the use of irradiation for food preservation. J Agric Environ Ethics 4:14–30

Lux SA, Copeland RS, White IM, Manrakhan A, Billah MK (2003) A new invasive fruit fly species from the *Bactrocera dorsalis* (Hendel) group detected in East Africa. Insect Sci Appl 23:355–361

McGlashan J (1995) In-transit sterilization to Japan. Citrus J 5(2):17

Myburgh AC (1956) Bionomics and control of the fruit flies, *Ceratitis capitata* (Weid.) and *Pterandrus rosa* (Ksh.) in the western cape province. DSc thesis, Stellenbosch University, 473–478

Myburgh AC (1965) Low temperature sterilization of false codling moth, *Argyroploce leucotreta* Meyr., in export citrus. J Entomol Soc South Afr 28:277–285

Nascimento AS, Malavasi A, Morgante JS, Duarte ALA (1992) Hot-water immersion treatment for mangoes infested with *Anastrepha fraterculus*, *A. obliqua*, and *Ceratitis capitata* (Diptera: Tephritidae) in Brazil. J Econ Entomol 85:456–460

Nel RG (1936) The utilization of low temperatures in the sterilization of deciduous fruit infested with the immature stages of the Mediterranean fruit fly, *Ceratitis capitata* wied. Sci Bull 155:1–33

Phillips CB, Iline II, Novoselov M, McNeill MR, Richards NK, Cv K, Stephenson BP (2015) Methyl bromide fumigation and delayed mortality: safe trade of live pests? J Pest Sci 88:121–134

Pryke JS, Pringle KL (2008) Postharvest disinfestation treatments for deciduous and citrus fruits of the Western Cape, South Africa: a database analysis. S Afr J Sci 104:85–95

Rwomushana I, Ekesi S, Ogol CKPO, Gordon I (2009) Mechanisms contributing to the competitive success of the invasive fruit fly *Bactrocera invadens* over the indigenous mango fruit fly, *Ceratitis cosyra*: the role of temperature and resource pre-emption. Entomol Exp et Appl 133:27–37

Schutze MK, Jessup A, Ul-Haq I, Vreysen MJB, Wornoayporn V, Vera MT, Clarke AR (2013) Mating compatibility among four pest members of the *Bactrocera dorsalis* fruit fly species complex (Diptera: Tephritidae). J Econ Entomol 106:695–707

Schutze MK, Mahmood K, Pavasovic ANA, Bo W, Newman J, Clarke AR, Krosch MN, Cameron SL (2015) One and the same: integrative taxonomic evidence that *Bactrocera invadens* (Diptera: Tephritidae) is the same species as the oriental fruit fly *Bactrocera dorsalis*. Syst Entomol 40:472–486

Self G, Ducamp N-M, Thaunay P, Vayssières J-F (2012) The effects of phytosanitary hot water treatments on west African mangoes infested with *Bactrocera invadens* (Diptera: Tephritidae). Fruits 67:439–449

Shamilov AS (2012) Quarantine disinfestation in Russia: past and present. Bull OEPP/EPPO Bull 42:176–180

Shellie KC, Mangan RL (1994) Postharvest quality of 'Valencia' orange after exposure to hot, moist, forced air for fruit fly disinfestation. Hortscience 29:1524–1527

Smith A (2010) Africa's growing role in world trade could provide opportunities. Acta Hortic 879:123–127

Vayssières J-F, Sanogo F, Noussourou M (2007) Inventory of the fruit fly species (Diptera: Tephritidae) linked to the mango tree in Mali and tests of integrated control. Fruits 62:329–341

Vayssières J-F, Korie S, Coulibaly O, Temple L, Boueyi SP (2008) The mango tree in central and Northern Benin: cultivar inventory, yield assessment, infested stages and loss due to fruit flies (Diptera Tephritidae). Fruits 63:335–348

Vayssières J-F, Korie S, Ayegnon D (2009) Correlation of fruit fly (Diptera Tephritidae) infestation of major mango cultivars in Borgou (Benin) with abiotic and biotic factors and assessment of damage. Crop Prot 28:477–488

Viljoen HW (2011) Effect of gamma irradiation as a mitigation treatment on storage quality of plums. SA Fruit J 10:73–75

Viviers W, Kühn ML, Steenkamp E, Berkman B (2014) Tru-cape fruit marketing, South Africa: managing the export market diversification challenge. Int Food Agribus Manag Rev 17B:193–197

Ware A, Stephen P, Tate B, Daneel JH (2001) Cold treatment of Natal fruit fly infested oranges. CRI Group annual research report, Citrus Research International, Nelspruit, pp 121–127

Ware A, Tate B, Stephen P, Daneel JH, Beck R, Misumi T (2005) Verification of cold treatment disinfestation of medfly-infested Clementines destined for Japan. CRI Group Annual Research Report, Citrus Research International, Nelspruit, pp 117–123

Ware AB, Du Toit CLN, Mohamed SA, Nderitu PW, Ekesi S (2012) Cold tolerance and disinfestation of *Bactrocera invadens* (Diptera: Tephritidae) in 'hass' avocado. J Econ Entomol 105:1963–1970

Williams P, Hepworth G, Goubran F, Muhunthan M, Dunn K (2000) Phosphine as a replacement for methyl bromide for postharvest disinfestation of citrus. Postharvest Biol Technol 19:193–199

Tim G. Grout has 34 years of experience in economic citrus entomology and has coordinated the funding of citrus research in Southern Africa for 15 years. He has conducted research on a wide range of citrus pests including Homoptera, Thysanoptera, Diptera, Lepidoptera and both predatory and phytophagous mites. He has published 29 first author refereed papers, 52 semi-scientific papers or proceedings and been minor author in a further 16 other papers. He has written numerous popular articles for citrus growers and frequently presented practical talks to citrus growers and at conferences around the world.

Chapter 21
Photographs of Some Native and Exotic Fruit Fly Species in Africa and Their Parasitoids

Sunday Ekesi, Samira A. Mohamed, and Marc De Meyer

1 Fruit Flies

Plate 21.1 *Bactrocera dorsalis* (Hendel) (Courtesy of Robert S. Copeland, *icipe*, Kenya)

S. Ekesi (✉) • S.A. Mohamed
Plant Health Theme, International Centre of Insect Physiology & Ecology (*icipe*),
PO Box 30772-00100, Nairobi, Kenya
e-mail: sekesi@icipe.org

M. De Meyer
Biology Department, Royal Museum for Central Africa,
Leuvensesteenweg 13, B3080 Tervuren, Belgium

© Springer International Publishing Switzerland 2016
S. Ekesi et al. (eds.), *Fruit Fly Research and Development in Africa - Towards a Sustainable Management Strategy to Improve Horticulture*,
DOI 10.1007/978-3-319-43226-7_21

Plate 21.2 *Bactrocera oleae* (Rossi) (Courtesy of Robert S. Copeland, *icipe*, Kenya)

Plate 21.3 *Bactrocera latifrons* (Hendel) (Courtesy of Robert S. Copeland, *icipe*, Kenya)

Plate 21.4 *Bactrocera zonata* (Saunders) (Courtesy of Aruna Manrakhan, CRI, South Africa)

Plate 21.5 *Zeugodacus cucurbitae* (Coquillett) (Courtesy of Robert S. Copeland, *icipe*, Kenya)

A

B

Plate 21.6 *Ceratitis anonae* Graham ((**A**) Courtesy of Robert S. Copeland, *icipe*, Kenya and (**B**) Georg Goergen, IITA, Benin)

A

B

Plate 21.7 *Ceratitis capitata* (Wiedemann) ((**A**) Courtesy of Robert S. Copeland *icipe*, Kenya and (**B**) Georg Goergen, IITA, Benin)

Plate 21.8 *Ceratitis cosyra* (Walker) (Courtesy of Robert S. Copeland, *icipe*, Kenya)

Plate 21.9 *Ceratitis discussa* Munro (Courtesy of Robert S. Copeland, *icipe*, Kenya)

Plate 21.10 *Ceratitis ditissima* (Munro) ((**A**) Courtesy of Robert S. Copeland, *icipe*, Kenya and (**B**) Georg Goergen, IITA, Benin)

Plate 21.11 *Ceratitis fasciventris* (Bezzi) (Courtesy of Robert S. Copeland, *icipe*, Kenya)

Plate 21.12 *Ceratitis quinaria* (Bezzi) (Courtesy of Georg Goergen, IITA, Benin)

Plate 21.13 *Ceratitis rosa* Karsch (Courtesy of Robert S. Copeland, *icipe*, Kenya)

Plate 21.14 *Ceratitis rubivora* Coquillett (Courtesy of Robert S. Copeland, *icipe*, Kenya)

Plate 21.15 *Ceratitis silvestrii* Bezzi (Courtesy of Georg Goergen, IITA, Benin)

A

B

Plate 21.16 *Dacus bivittatus* (Bigot) ((**A**) Courtesy of Robert S. Copeland, *icipe*, Kenya and (**B**) Georg Goergen, IITA, Benin)

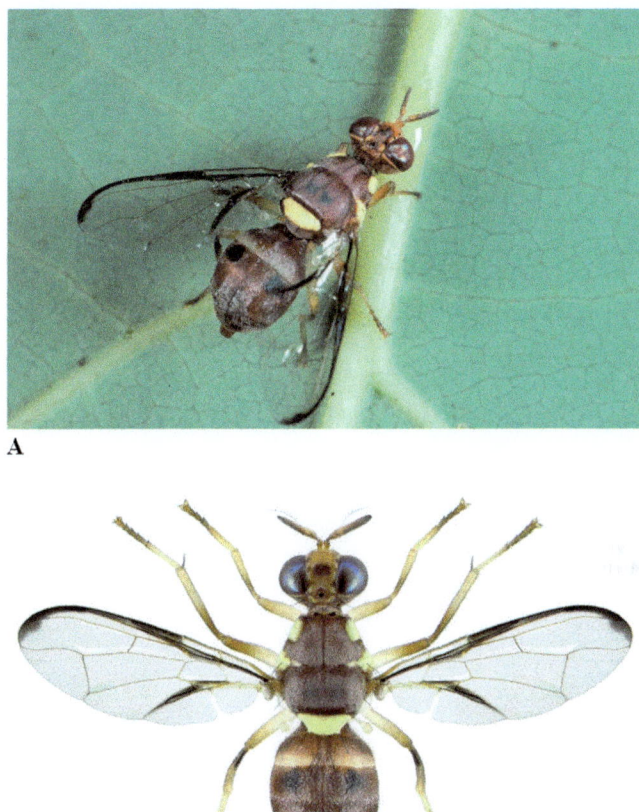

Plate 21.17 *Dacus ciliatus* Loew ((**A**) Courtesy of Robert S. Copeland, *icipe*, Kenya and (**B**) Georg Goergen, IITA, Benin)

Plate 21.18 *Dacus frontalis* Becker ((**A**) Courtesy of Robert S. Copeland, *icipe*, Kenya and (**B**) Georg Goergen, IITA, Benin)

Plate 21.19 *Dacus lounsburyi* Coquillett (Courtesy of Robert S. Copeland, *icipe*, Kenya)

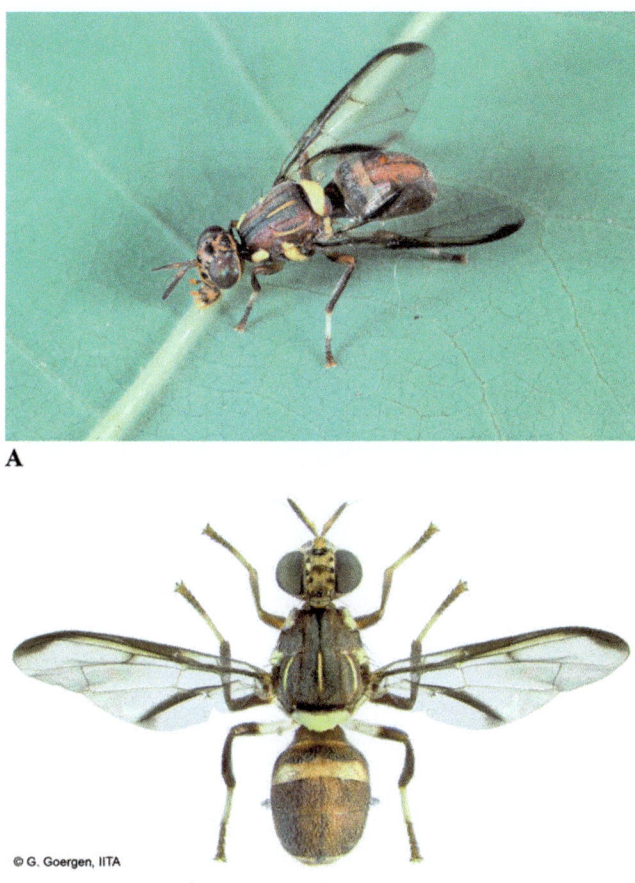

Plate 21.20 *Dacus punctatifrons* (Karsch) ((**A**) Courtesy of Robert S. Copeland, *icipe*, Kenya and (**B**) Georg Goergen, IITA, Benin)

Plate 21.21 *Dacus vertebratus* Bezzi ((**A**) Courtesy of Robert S. Copeland, *icipe*, Kenya and (**B**) Georg Goergen, IITA, Benin)

Plate 21.22 *Trirhithrum coffeae* Bezzi (Courtesy of Robert S. Copeland, *icipe*, Kenya)

Plate 21.23 *Trirhithrum nigerrimum* (Bezzi) (Courtesy of Robert S. Copeland, *icipe*, Kenya)

2 Parasitoids

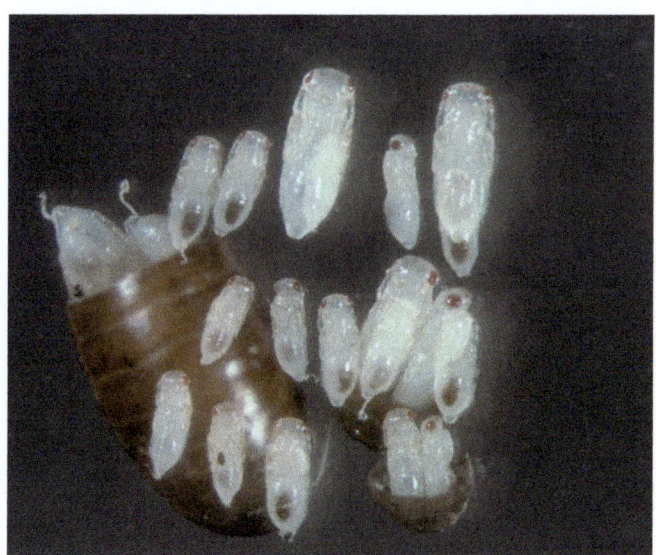

Plate 21.24 *Aceratoneuromia indica* (Silvestri) (Courtesy of Mohsen M. Ramadan, Department of Agriculture, Hawaii, USA)

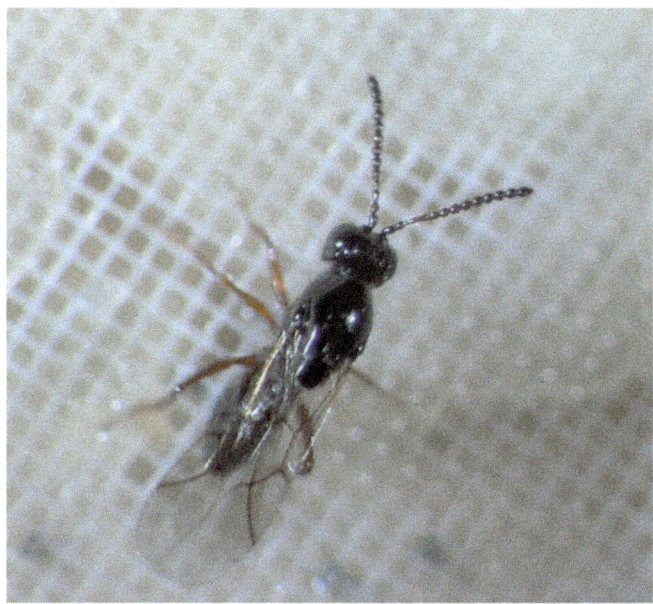

Plate 21.25 *Aganaspis daci* (Weld) (Courtesy of Mohsen M. Ramadan, Department of Agriculture, Hawaii, USA)

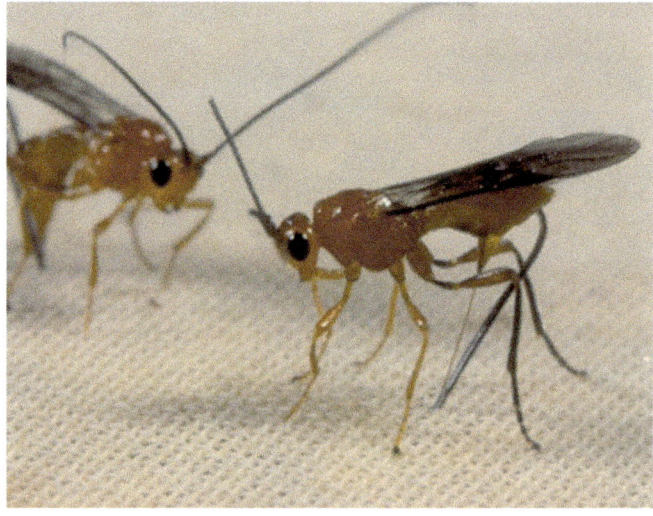

Plate 21.26 *Diachasmimorpha longicaudata* (Ashmead) (Courtesy of Mohsen M. Ramadan, Department of Agriculture, Hawaii, USA)

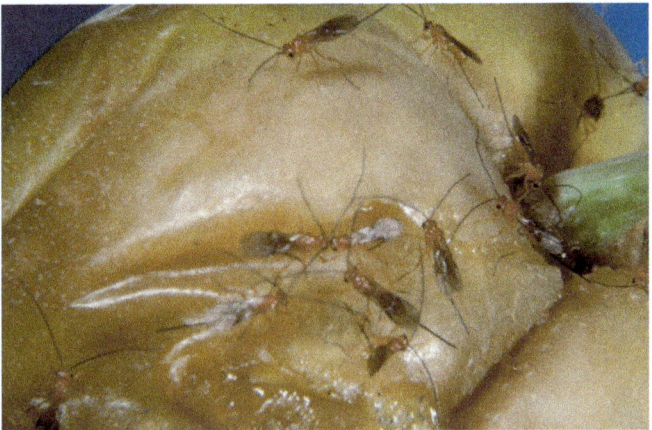

Plate 21.27 *Diachasmimorpha kraussii* (Fullaway) (Courtesy of Mohsen M. Ramadan, Department of Agriculture, Hawaii, USA)

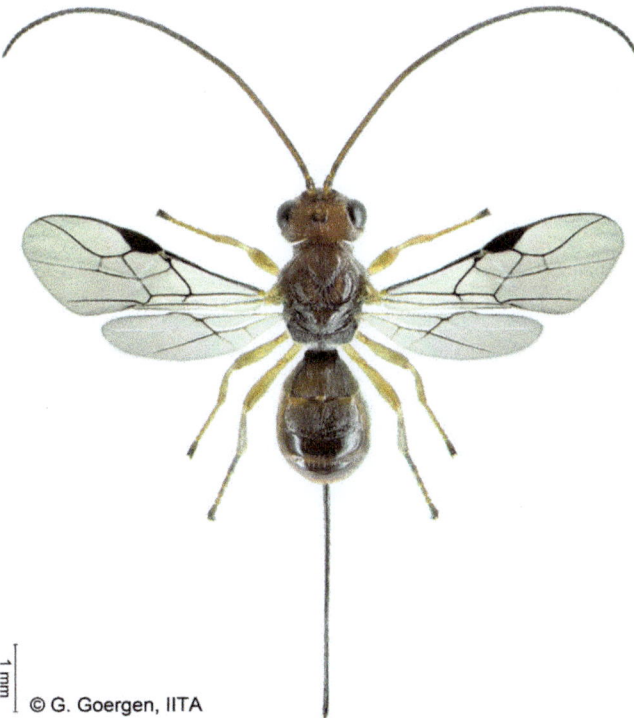

Plate 21.28 *Fopius arisanus* (Sonan) (Courtesy of Georg Goergen, IITA, Benin)

Plate 21.29 *Fopius caudatus* (Szépligeti) (Courtesy of Mohsen M. Ramadan, Department of Agriculture, Hawaii, USA)

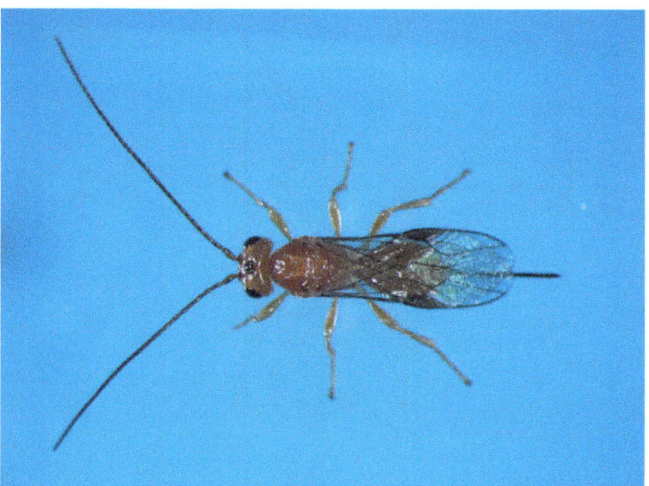

Plate 21.30 *Fopius ceratitivorus* Wharton (Courtesy of Mohsen M. Ramadan, Department of Agriculture, Hawaii, USA)

Plate 21.31 *Fopius vandenboschi* (Fullaway) (Courtesy of Mohsen M. Ramadan, Department of Agriculture, Hawaii, USA)

Plate 21.32 *Psyttalia fletcheri* (Fullaway) (Courtesy of Mohsen M. Ramadan, Department of Agriculture, Hawaii, USA)

Plate 21.33 *Tetrastichus giffardianus* (Fullaway) (Courtesy of Mohsen M. Ramadan, Department of Agriculture, Hawaii, USA)

Part III
Country Specific Action Programmes and Case Studies

Chapter 22
Integrated Management of Fruit Flies – Case Studies from Uganda

Brian E. Isabirye, Caroline K. Nankinga, Alex Mayamba, Anne M. Akol, and Ivan Rwomushana

Abstract Fruit flies (Diptera: Tephritidae) pose a threat to commercialisation of the horticulture industry in Uganda. They impair the quality and quantity of fruits produced, and limit access to lucrative regional and global markets. Here we explore past and present efforts, and future plans for research and management, of fruit flies in Uganda. Early research geared towards collection and identification of fruit flies recognised the pest status of many species and highlighted the need for establishing sustainable management strategies. Subsequently large-scale research initiatives have substantially increased knowledge on the biology and ecology of fruit flies in Uganda. Based on these studies, integrated pest management (IPM) options for fruit flies have been designed and piloted. Amongst the most promising options are the Male Annihilation Technique (MAT) in combination with the Bait Annihilation Technique (BAT) or Protein Food Bait (PFB) and Orchard Sanitation (OS). Fruit bagging is also receiving attention. It is now recommended that IPM options are combined and scaled up in an area-wide approach. The government of Uganda has demonstrated genuine commitment to eradication of fruit flies through three key project initiatives: (i) Gaining insight into the ecological and physiological factors influencing fruit fly populations and infestation rates in mango-growing regions of Uganda (NARO-MSI); (ii) Equipping key technical personnel at local and district levels with knowledge on identification and management of key fruit fly pest species (NAADS); (iii) Promotion and adoption of IPM practices for fruit fly management (NARO-ATAAS). These initiatives will ensure the long-term sustainability of management options.

B.E. Isabirye (✉)
Association for Strengthening Agricultural Research in Eastern and Central Africa (ASARECA), PO Box 765, Entebbe, Uganda
e-mail: b.isabirye@asareca.org; brianisabirye@yahoo.com

C.K. Nankinga • A. Mayamba
National Agricultural Research Laboratories, Kawanda, PO Box 7065, Kampala, Uganda

A.M. Akol
College of Natural Sciences, Makerere University, PO Box 7062, Kampala, Uganda

I. Rwomushana
International Centre of Insect Physiology & Ecology (*icipe*),
PO Box 30772-00100, Nairobi, Kenya

Keywords Diversity • Distribution • Host plants • Management

1 Background

Horticulture is one of the fastest growing agricultural sub-sectors in Uganda, employing many people and with products now listed as strategic exports worth US$35 million per year (Uganda Export Promotions Board 2012). Although the monetary value of fresh fruit and vegetables has been increasing, fruit production is constrained by a multiplicity of pests and diseases, the most serious of which are fruit flies (Diptera: Tephritidae). In Uganda fruit flies cause heavy pre-harvest losses and thus prevent expansion of domestic trade in fruit and vegetables (Nakasinga 2002; Nemeye 2005; Okullokwany 2006; Mayamba et al. 2014; Nankinga et al. 2014a; Akol et al. 2013; Isabirye et al. 2016).

Uganda, like many developing countries, aims to expand its economy through increased export trade in fruits and vegetables using new technologies that can secure market access. Such exports are subjected to the sanitary and phytosanitary restrictions of the World Trade Organization (2016) agreement (www.wto.org)). Here we explore the various efforts to manage fruit flies in Uganda. We describe the research and development activity and various initiatives and approaches targeted at managing fruit flies in Uganda. Furthermore, future research and development needs are discussed.

2 About Uganda

Uganda lies across the Equator, between latitudes 4° 12´ N and 1° 29´ S and longitudes 29° 34´ W and 35° 0´ E. Temperatures range between 15° and 30 ° C. More than two-thirds of the country is represented by a plateau, lying between 1000 and 2500 m above sea level. Precipitation is fairly predictable, varying from 750 mm in Karamoja in the Northeast, to 1500 mm in the high rainfall areas on the shores of Lake Victoria, in the highlands around Mt. Elgon in the east, in the Rwenzori Mountains in the southwest, and some parts of Masindi and Gulu. Its climate is tropical, with a good distribution of rain throughout the year in most parts of the country and with two dry seasons per year (December–February and June–August), except for certain regions in the north of the country that have only one rainy season (April–October) and one dry season (November to March).

Agriculture is the backbone of Uganda's economy; 95% of the population produces both crops and livestock for food and cash income. Many farms are small (0.5–1.5 acres) but there are also medium sized (5–20 ha) and large farms including ranches (average 1200 ha). Agriculture contributes over 15% to the Gross Domestic Product (GDP) and over 90% to the country's foreign exchange earnings (World Bank Report 2013). It also contributes over 60% of total government revenue in addition to employing more than 80% of the total labour force and providing over half of the total income for the bottom three-quarters of the population. There is considerable scope for increased production of fruits and vegetables. The country is able to produce all year round and there is potential for organic production of exotic and off-season fruits and vegetables such as pineapples (*Ananas comosus* L. Merr), apple (*Malus domestica* Borkh.), bananas (*Musa* species) and passion fruit (*Passiflora edulis* Sims).

3 Fruit Fly Management in Uganda: A Historical Perspective

Early exploratory and surveillance projects on fruit flies were led by collectors from national museums in Europe, USA, Canada and Africa who still hold voucher material (Table 22.1). Work was targeted at the collection and identification of fruit flies in Uganda. By 1990, these studies had recognised the pest status of many species and highlighted the need for establishing a management strategy for fruit flies in Uganda (Nakasinga 2002). Following this several national and regional fruit fly initiatives began. In Uganda, key initiatives included: the *icipe*-led African Fruit Fly Programme (AFFP) which began in 1999; the Food and Agricultural Organization of the United Nations initiative (2005 to present); the Citrus Research International (CRI) initiative (2009–2010); the National Agricultural Advisory Services (NAADS) initiative (2009–2011); the Millennium Science Initiative (MSI-NARL-Uganda) (2010–2014); and the Agricultural Technology and Agribusiness Advisory Services (ATAAS) initiative (2011–2016).

These initiatives have resulted in an extensive pool of resources for surveillance, identification, field management, quality assurance, monitoring and suppression of fruit flies. Significant advances in the generation and use of knowledge on the biology and ecology of fruit flies in Uganda have been made (Nankinga et al. 2014a; Mayamba et al. 2014). Using this information, options for integrated pest management (IPM) of fruit flies have been designed and piloted in selected regions. The most common IPM options piloted have evolved around the use of: the Male Annihilation Technique (MAT) based on methyl eugenol which attracts mainly males; the Bait Annihilation Technique (BAT) based on waste brewer's yeast, mazoferm and GF-120 which attract female flies of all species; Orchard Sanitation (OS) which involves burying infested fruits and other good agronomic practices; and fruit bagging (FB) which involves the wrapping of fruits in paper bags 1 month after fruit set to prevent the female fruit flies depositing eggs in the immature fruit.

Table 22.1 Fruit fly species recorded in Uganda since 1909, and the respective Museums and Collection centres[a] where samples and information are held

Species	Museum
Bactrocera dorsalis (Hendel)	ANON, FSCA, IITAB, QDP, MRAC
Bactrocera munroi White	BMNH
Bistrispinaria atlas Munro	BMNH, CNC
Bistrispinaria fortis Speiser	BMNH
Bistrispinaria magniceps Bezzi	BMNH, CNC
Capparimyia bipustulata Bezzi	ANON, TAU, SANC, USNM
Capparimyia melanaspis Bezzi	TAU
Carpophthoromyia interrupta De Meyer	CNC
Carpophthoromyia pseudotritea Bezzi	BMNH, SANC
Ceratitis aliena Bezzi	BMNH
Ceratitis anonae Graham	TAU, USNM, BMNH, CNC, ICIPE, MRAC
Ceratitis argenteobrunnea Munro	BMNH, ICIPE, SANC
Ceratitis bremii Guarin-Maneville	CNC, FSCA, SANC
Ceratitis brucei Munro	BMNH
Ceratitis capitata Wiedemann	ANON, BMNH, MRAC, USNM
Ceratitis connexa Bezzi	SANC
Ceratitis cosyra Walker	ANON, ICIPE, MRAC, SANC
Ceratitis cuthbertsoni Munro	ANON
Ceratitis ditissima Munro	BMNH, CNC, ICIPE, TAU, USNM
Ceratitis edwardsi Munro	BMNH, CNC, SANC
Ceratitis fasciventris Bezzi	BMNH, CNC, ICIPE, MRAC, SANC, TAU
Ceratitis flexuosa Walker	USNM, BMNH, CNC, ICIPE, MRAC, SANC
Ceratitis hamata De Meyer	BMNH
Ceratitis marriotti Munro	BMNH
Ceratitis pedestris Bezzi	MRAC
Ceratitis pinnatifemur Enderlein	CNC
Ceratitis pinnatifemur Enderlein	MRAC, SANC
Ceratitis punctata Wiedemann	USNM, CNC, ICIPE, MRAC, SANC, TAU,
Ceratitis querita Munro	CNC
Ceratitis roubaudi Bezzi	CNC, MRAC
Ceratitis rubivora Coquillett	BMNH, FSCA, SANC
Ceratitis stipula De Meyer & Freidberg	CNC, MRAC, USNM
Ceratitis striatella Munro	BMNH, SANC, TAU
Ceratitis turneri Munro	BMNH
Ceratitis venusta Munro	BMNH, USNM
Ceratitoides nigromaculatus Hendel	CNC, DEI
Clinotaenia camerunica Hancock	BMNH
Dacus apiculatus White	CNC, FSCA
Dacus apostata Hering	TAU
Dacus armatus Fabricius	BMNH, MRAC, SANC
Dacus aspilus Bezzi	BMNH, SANC

(continued)

Table 22.1 (continued)

Species	Museum
Dacus bivittatus Bigot	CNC, FSCA, NMBZ, MRAC, USNM, NMKE
Dacus brevis Coquillett	CNC
Dacus ceropegiae Munro	TAU
Dacus chapini Curran	BMNH
Dacus ciliatus Loew	ANON, BMNH, SANC, MCSNM
Dacus croceus Munro	BMNH, CNC, SANC
Dacus disjunctus Bezzi	BMNH, CNC
Dacus elegans Munro	TAU
Dacus externellus Munro	CNC, SANC
Dacus fasciolatus Collart	ANON, SANC
Dacus freidbergi Munro	SANC
Dacus hamatus Bezzi	NMKE
Dacus hargreavesi Munro	BMNH, SANC
Dacus humeralis Bezzi	FSCA, MRAC
Dacus inflatus Munro	CNC
Dacus inornatus Bezzi	BMNH, SANC
Dacus katonae Bezzi	BMNH
Dacus langi Curran	MRAC
Dacus limbipennis Macquart	BMNH, CNC, MRAC, SANC, TAU, FSCA
Dacus linearis Collart	CNC, TAU
Dacus longistylus Wiedemann	BMNH
Dacus macer Bezzi	BMNH, CNC, MRAC, SANC, TAU, USNM
Dacus masaicus Munro	CNC, TAU, FSCA
Dacus maynei Bezzi	ANON, CNC, MRAC, SANC, TAU, USNM
Dacus mediovittatus White	MRAC
Dacus nigriscutatus White	TAU
Dacus notalaxus Munro	SANC
Dacus parvimaculatus White	CNC
Dacus phimis Munro	SANC
Dacus punctatifrons Karsch	ANON, BMNH, CNC, FSCA, MRAC, SANC
Dacus rufoscutellatus Hering	TAU
Dacus schoutedeni Collart	BMNH, CNC
Dacus siliqualactis Munro	ANON, BMNH, SANC, MRAC
Dacus sphaeristicus Speiser	TAU
Dacus spissus Munro	BMNH
Dacus telfaireae Bezzi	FSCA, MRAC
Dacus tenebricus Munro	ANON, SANC, USNM
Dacus theophrastus Hering	CNC
Dacus triater Munro	CNC, TAU
Dacus yangambinus Munro	CNC
Leucotaeniella guttipennis Bezzi	BMNH
Perilampsis formosula Austen	BMNH, SANC, TAU

(continued)

Table 22.1 (continued)

Species	Museum
Perilampsis pulchella Austen	BMNH, MRAC, SANC, TAU, USNM, ICIPE
Trirhithrum albomaculatum Rader	TAU
Trirhithrum brachypterum Munro	BMNH
Trirhithrum coffeae Bezzi	BMNH, CNC, MRAC, SANC, TAU, USNM
Trirhithrum demeyeri White & Hancock	BMNH
Trirhithrum fraternum Munro	BMNH, CNC
Trirhithrum homogeneum Bezzi	ANON, BMNH, SANC
Trirhithrum inscriptum Graham	CNC
Trirhithrum meladiscum Munro	BMNH
Trirhithrum micans Munro	BMNH, PPHZ, SANC, TAU
Trirhithrum nigerrimum Bezzi	ANON
Trirhithrum nigrum Graham	NMW
Trirhithrum notandum Munro	BMNH, CNC, SANC, TAU
Trirhithrum occipitale Bezzi	BMNH, CNC, TAU, USNM
Trirhithrum overlaeti Munro	BMNH
Trirhithrum quadrimaculatum White	BMNH
Trirhithrum transiens Munro	BMNH
Trirhithrum validum Bezzi	BMNH, TAU
Zeugodacus cucurbitae Coquillett	FSCA, MRAC

Nakasinga (2002), Ekesi and Billah (2006), Mayamba et al. (2014), Isabirye (2015)
[a]*ANON* Non Specified Collection; *BMNH* Natural History Museum, London, UK; *CNC* Canadian National Collections, Ottawa, Canada; *DEI* Deutsches Entomologisches Institut, Eberswalde, Germany; *FSCA* Florida State Collection of Arthropods, Gainesville, USA; *ICIPE* International Centre of Insect Physiology and Ecology, Nairobi, Kenya; *IRSNB* Koninklijk Belgisch Instituut voor Natuurwetenschappen, Brussel, Belgium; *IITAB* International Institute for Tropical Agriculture, Cotonou, Benin; *MCSNM* Museo Civico di Storia Naturale, Milan, Italy; *MRAC* Koninklijk Museum voor Midden Afrika, Tervuren, Belgium; *PPHZ* Plant Protection Research Institute, Harare, Zimbabwe; *QDPI* Queensland Department of Primary Industries, Brisbane, Australia; *NMW* Naturhistorisches Museum Wien, Austria; *NMBZ* Natural History Museum of Zimbabwe, Bulawayo, Zimbabwe; *NMKE* National Museums of Kenya, Nairobi, Kenya; *SANC* Plant Protection Research Institute, Pretoria, South Africa; *TAU* Tel Aviv University, Tel Aviv, Israel; *USNM* United States National Museum of Natural History, Washington D.C., USA

4 Milestones and Achievements in Management of Fruit Flies in Uganda

4.1 Status of Knowledge on Fruit Flies in Uganda

4.1.1 Diversity and Distribution

Uganda has a rich diversity of fruit flies; to date 102 species from ten genera have been reported (Nakasinga 2002; Nemeye 2005; Okullokwany 2006; Ekesi et al. 2006; Isabirye et al. 2015a; Table 22.1). Amongst the most diverse genera are

Dacus, Ceratitis and *Trirhithrum*, while the least diverse are *Ceratitoides, Clinotaenia, Leucotaeniella* and *Perilampsis*, each with only one species. Species in the genera *Bactrocera, Ceratitis, Trirhithrum* and *Dacus* are the most economically important. Overall, the oriental fruit fly, *Bactrocera dorsalis* (Hendel), is the most important pest species accounting for 99 % of all fruit flies collected in three of the main fruit growing zones in Uganda (Mayamba et al. 2014; Isabirye et al. 2015a). Intraspecific diversity (allopatric and host-associated) is prevalent amongst some species of fruit flies in Uganda (Isabirye et al. 2012, 2014a). Significant differences in allometry and developmental instability (fluctuating asymmetry) amongst host plants has also been reported (Akol et al. 2013). Such geographic and host-associated fluctuating asymmetry or adaptations may affect the efficiency of common control methods.

Fruit flies are widely distributed across Uganda and in most of the major agro-ecological zones. However, they are marginal in Karamoja, Bushenyi, the Kabale-Rukungiri highlands and the Kisoro-Kibale highland zones (Isabirye et al. 2015b). Isabirye et al. (2014b) predicted that future fruit fly ranges are likely to have declined by approximately 25.4 % by the year 2050. Species richness was predicted to decrease differently across zones. Future niches are predicted to shift northwards to more humid regions (Isabirye et al. 2014b).

Seasonality in population abundance has been recorded for *B. dorsalis* and *Ceratitis* species (Fig. 22.1a, b). Populations of adult *B. dorsalis* increase from April and peak during the July fruit season. They then decline from August when the fruit season has finished. Populations of *Ceratitis* species are generally extremely low; the largest trap catches are observed in May but they fluctuate widely over the trapping period (Mayamba et al. 2014). These seasonal patterns have been attributed to climate (precipitation and temperature) and biophysical (e.g. host availability) variables (Mayamba et al. 2014; Akol et al. 2013).

4.1.2 Host Utilization

The major fruit fly pest species in Uganda have been reported from a wide range of host plants. For example, *B. dorsalis*; the Mediterranean fruit fly (medfly), *Ceratitis capitata* (Weidemann); the Natal fruit fly, *Ceratitis rosa* Karsch; and the melon fly, *Zeugodacus cucurbitae* (Coquillett) all with a wide range of host plants (Table 22.2). Damage attributed to fruit fly infestation on mango, *Mangifera indica* L., ranged from 33 to 83 % (Nankinga et al. 2014a). In a study by Isabirye et al. (2016), 1,812 fruits from 38 fruit species, 30 genera and 18 plant families were sampled in three agro ecological zones: Lake Victoria Crescents (LVC), Western Highlands (WH), Northern Highlands (NH) and 35 % of the samples were positive for fruit fly infestation. *Bactrocera dorsalis* was the predominant species found, being recorded from 29 of the 38 plant species and 76.3 % of the infested fruits (Isabirye et al. 2016). Excluding *B. dorsalis*, the proportion of fruit infested by a given fruit fly species ranged from 7.9 % for infestation by the coffee fruit fly, *Trirhithrum coffeae* Bezzi, to 65.8 % for infestation by *C. rosa*. The most heavily infested fruits came from

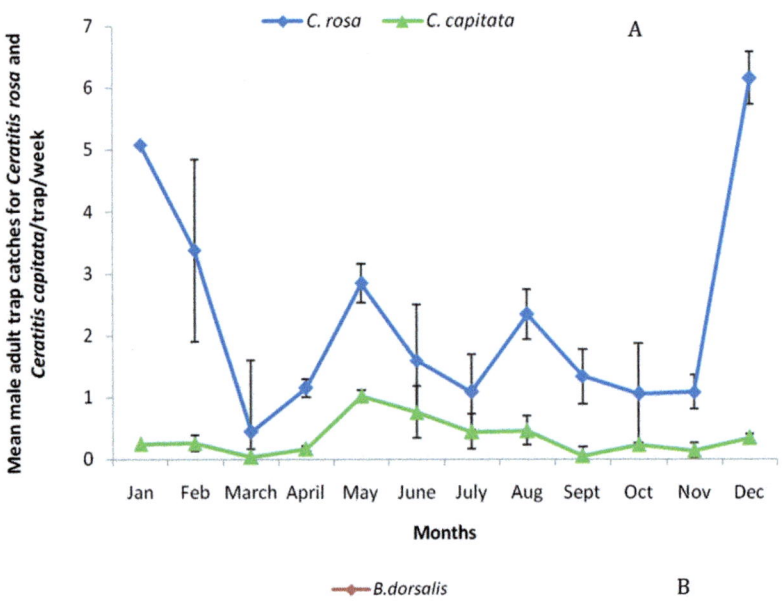

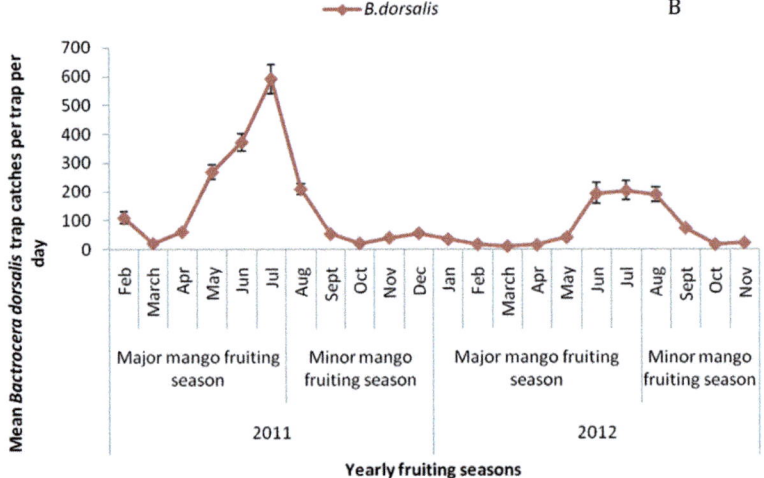

Fig. 22.1 Seasonal population fluctuation (Mean ± standard errors) of *C. rosa* and *C. capitata* (**a**) *B. dorsalis* (**b**) in mango orchards in the Lake Victoria Crescents zone, Uganda (Mayamba et al. 2014)

plants in the family Combretaceae, followed by Anacardiaceae and Myrtaceae, while plants in the Verbenaceae, Rutaceae and Euphorbiaceae were least infested. Tropical almond (*Terminalia catappa* L.) was the preferred host of *B. dorsalis*, followed by guava (*Psidium guajava* L.), mango, avocado (*Persea americana* Mill.) and oranges (*Citrus sinensis* L.)/ other *Citrus* species. Amongst mango varieties, significant differences in infestation levels were common both within and across ecological zones in Uganda (Agum 2014; Mayamba et al. 2014; Nankinga et al.

Table 22.2 Host plants of the major fruit fly pest species in Uganda

Species	Host plant
Bactrocera dorsalis (Hendel)	*Anacardium occidentale* L., *Mangifera indica* L., *Sclerocarya birrea* (A. Rich) Hochst, *A. muricata, A. senegalensis, A. squamosa, C. papaya, Terminalia catappa* L., *C. melo, Cucurbita* spp., *P. americana, Antiaris toxicaria* Lesch., *Ficus* sp., *Musa* sp., *Acca sellowiana* (O. Berg) Burret, *Eugenia uniflora* L., *P. guajava, Cydonia oblonga* Mill., *Prunus* spp., *Coffea arabica* L., *Citrus limon* (L.) Burm.f., *Citrus reticulata* Blanco, *Citrus sinensis* (L.) Osbeck, *Citrus* spp., *Manilkara zapota* (L.) P. Royen, *C. annuum, Solanum lycopersicon* L., *Solanum* spp., *Theobroma cacao* L.
Bistrispinaria fortis (Speiser)	*Sorghum vulgare*
Bistrispinaria magniceps (Bezzi)	*Panicum maximum* Jacq.
Capparimyia bipustulata (Bezzi)	*Capparis erythrocarpus*
	Capparis sp.
Capparimyia melanaspis (Bezzi)	*Maerua* sp.
Ceratitis anonae (Graham)	*Myrianthus arboreus* Beauv.
	Artocarpus sp.
	Syzygium jambos L. (Alston)
	M. indica, A. muricata, T. catappa, P. americana, A. toxicaria, Artocarpus sp., *E. uniflora, P. guajava, C. arabica, Citrus* spp., *Solanum* spp., *T. cacao*
Ceratitis capitata (Wiedemann)	*Coffea canephora* Pierre ex A Froehner, *C. oblonga, Ficus* sp., *M. indica, M. zapota, Prunus persica* L. (Batsch), *A. occidentale, Annona cherimola* Mill., *A. muricata, A. reticulata, A. squamosal, C. papaya, T. catappa, Cucurbita* spp., *P. americana* Mill., *Ficus* sp., *P. guajava, Prunus* spp., *C. arabica, C. sinensis, Citrus* spp., *C. annuum, S. lycopersicon, Solanum* spp., *Vitis vinifera* L.
Ceratitis cosyra (Walker)	*M. indica, Sclerocarya birrea* (A. Rich) Hochst, *A. cherimola, A. senegalensis, P. guajava, Citrus* spp.
Ceratitis ditissima (Munro)	*Chrysophyllum albidum* G. Don, *Citrus paradisi* Macfad.
Ceratitis edwardsi (Munro)	*Voacanga* sp., *Voacanga thouarsii* Roem. & Schult.
Ceratitis fasciventris (Bezzi)	*M. arboreus, C. arabica, M. indica, Pancovia turbinate* Radlk., *P. guajava, Syzygium jambos* L. (Alston), *P. americana, C. limon, Solanum* spp.
Ceratitis flexuosa (Walker)	*M. indica*
Ceratitis punctata (Wiedemann)	*C. albidum, Ficus* sp., *M. indica*
	T. cacao, S. birrea

(continued)

Table 22.2 (continued)

Species	Host plant
Ceratitis rosa	*A. occidentale, M. indica, A. muricata, A. reticulata, A. senegalensis, A. squamosa, Cananga odorata* (Lam.) Hook & Thomson, *C. papaya, T. catappa, Cucurbita* spp., *Drypetes* natalensis (Harv.) Hutch., *P. americana, Ficus* sp., *A. sellowiana, P. guajava, Prunus* spp., *C. arabica, C. reticulata, C. sinensis, C. albidum, M. zapota, S. lycopersicon, Solanum* spp., *T. cacao, V. vinifera*
Ceratitis stipula De Meyer & Freidberg	*M. arboreus*
Ceratitis striatella (Munro)	*Pycnanthus* sp.
Ceratitis venusta (Munro)	*Solanum naumannii* Engl.
Dacus armatus (Fabricius)	*M. indica*
Dacus bivittatus (Bigot)	*Cucurbita* spp., *Momordica charantia* L., *Vitex* spp., *C. melo, S. lycopersicon, M. indica*
Dacus ciliatus (Loew)	*M. charantia, S. birrea, C. papaya, C. melo, P. guajava, Citrus* spp., *C. annuum, S. lycopersicon*
Dacus humeralis (Bezzi)	*M. indica*
Dacus langi (Curran)	*M. indica*
Dacus limbipennis (Macquart)	*M. charantia, Momordica* sp.
Dacus mediovittatus (White)	*M. indica*
Dacus punctatifrons (Karsch)	*Gloriosa* sp., *M. indica, Melothria* sp.
Dacus siliqualactis (Munro)	*Gomphocarpus semilunatus* A. Rich
Dacus telfaireae (Bezzi)	*M. indica*
Dacus tenebricus (Munro)	*G. semilunatus*
Perilampsis formosula (Austen)	*Ficus* sp.
Perilampsis pulchella (Austen)	*Ficus* sp., *Loranthus* sp.
Trirhithrum coffeae (Bezzi)	*C. arabica, C. canephora, Citrus* spp. *C. annuum*
Zeugodacus cucurbitae (Coquillett)	*Annona muricata* L., *Annona reticulata* L., *Annona senegalensis* Pers., *Annona squamosa* L., *Carica papaya* L., *Cucumis melo* Ser., *Persea americana* Mill., *Psidium guajava* L., *Citrus* spp., *Capsicum annuum* L., *Solanum* spp.

Nakasinga (2002), Ekesi and Billah (2006), Mayamba et al. (2014), Isabirye (2015)

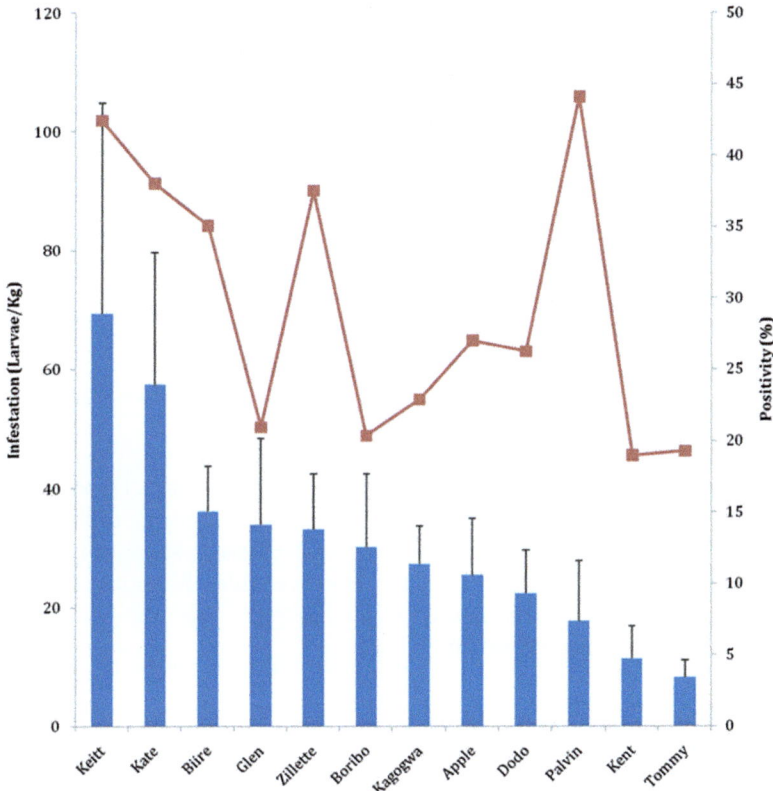

Fig. 22.2 Mean variability in infestation of different mango cultivars from three major fruit growing zones in Uganda (Nankinga et al. 2014a). Positivity = the proportion of samples that are positive for fruit fly infestation

2014a; Isabirye et al. 2016). Overall, the varieties Tommy, Atkin and Kent were the least likely to be infested, while the varieties Keitt, Kate and Biire were the most likely to be infested (Nankinga et al. 2014a; Fig. 22.2). There was also a significant interaction between zones and variety in the degree of infestation (larvae per kg). For instance, the varieties Glen and Boribo had significantly higher levels of infestation in the LVC zone than in the NMH zone, while the varieties Kate and Keitt had significantly higher levels of infestation in the NMH zone than in the LVC zone (Isabirye et al. 2016). Early maturing varieties seemed to be more susceptible to infestation than mid-season varieties (Isabirye et al. 2016). The differences in degree of infestation have been attributed to variability in physical, chemical and post-alighting cues produced by the host plant (Agum 2014; Akol et al. 2013). Also, as the mango season progresses from fruit set to ripening the availability of mature fruits increases as does the population of fruit flies and the level of fruit damage (Mayamba et al. 2014; Isabirye 2015). It has been demonstrated that female *B. dorsalis* have evolved to oviposit on host plant species on which their offspring are

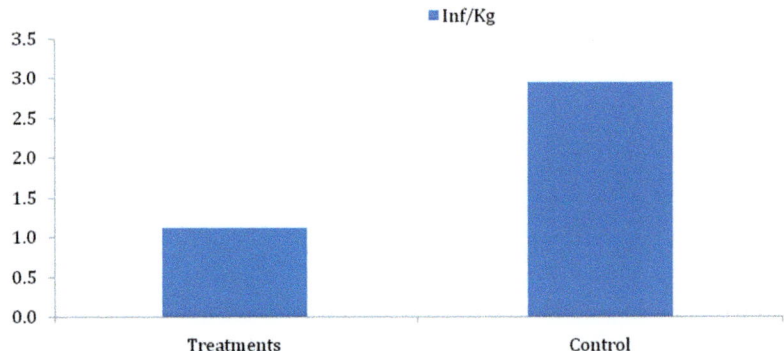

Fig. 22.3 Effect of treatment (BAT+MAT+OS) on infestation of citrus by fruit flies in eastern Uganda (NARL Annual Report 2011)

likely to fare best, in support for the Preference Performance Hypothesis (PPH) (Akol et al. 2013). However, the PPH is not very evident amongst different cultivars of the same host plant species, with the trends more closely supporting the Optimal Foraging Theory (OFT) (Akol et al. 2013). However, this still implies that female *B. dorsalis* use information from host fruits to determine host fruit quality and optimize offspring growth and survival.

4.2 Farm Level Fruit Fly Management Initiatives in Uganda

There have been several initiatives aimed at the management of fruit flies in Uganda. Amongst these are:

4.2.1 Citrus Research International (CRI) Initiative

In 2010, CRI in collaboration with the National Agricultural Research Organisation (NARO) of Uganda supported trials on the management of *B. dorsalis* on smallholder citrus orchards in Eastern and Central Uganda. Treatment farms used a combination of MAT (specifically for *B. dorsalis*); BAT (hydrolyzed protein-Mazoferm) and bio lures (for all species); and OS (specifically burying fruits). Control farms (standard farm practice) undertook irregular weeding and seldom applied conventional pesticides such as cypermethrin and malathion. Fewer fruit fly emergences were recorded from fruits sampled from farms employing the three treatment components (MAT+BAT+OS) compared with control farms (Fig. 22.3). In another similar trial where the treatment components were separated, the BAT+MAT and MAT alone treatments significantly reduced the diversity of fruit fly pests in citrus orchards compared with control treatments (standard farm practice) (Fig. 22.4). The BAT alone treatment was less effective in reducing the diversity of fruit flies

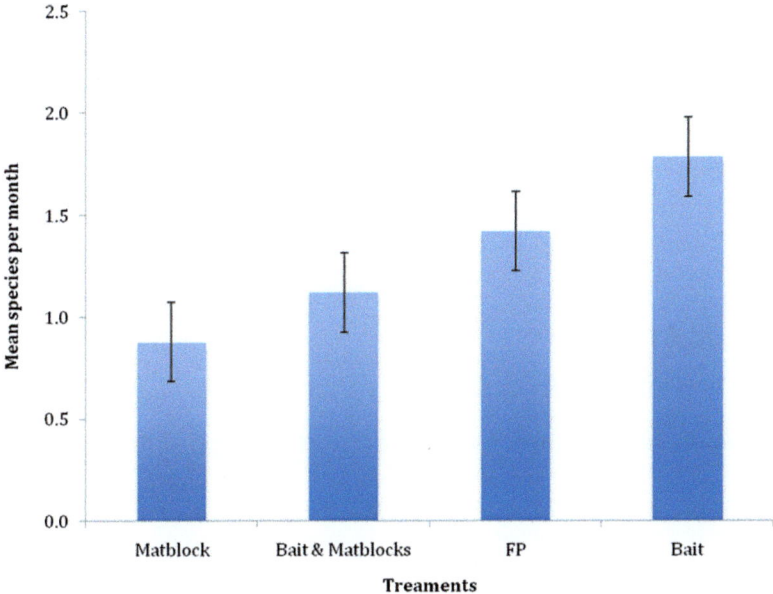

Fig. 22.4 Effect of different treatments on the diversity of fruit flies in citrus orchards in Eastern Uganda (NARL Annual Report 2011)

compared with the control, although the difference between the two was not statistically significant (Nankinga et al. 2014b).

4.2.2 Millennium Science Initiative (MSI-NARL-Uganda) Initiative

This project aimed to gain insight into the physical, ecological and biochemical factors influencing fruit fly infestation of mango in Uganda. Three students (one PhD and two MScs) were trained and a knowledge base established for evaluating IPM options for fruit fly control in Uganda. In initial surveys, MSI fruit fly project found that most commercial farmers were applying conventional insecticides and local botanical mixtures as standard farm practice to control fruit flies. The initiative validated and promoted IPM options based on combinations of MAT, BAT and OS (Nankinga et al. 2014a; Fig. 22.5). Results further confirmed that a combination of MAT+BAT+OS were effective in reducing fruit fly damage, compared with the common practice of using only MAT or standard farm practice (Nankinga et al. 2014b, c). In particular, deep burying of infested fruit (OS) killed larvae, preventing them from developing and emerging as adult flies and returning to the crop (Millennium Science Initiative 2013).

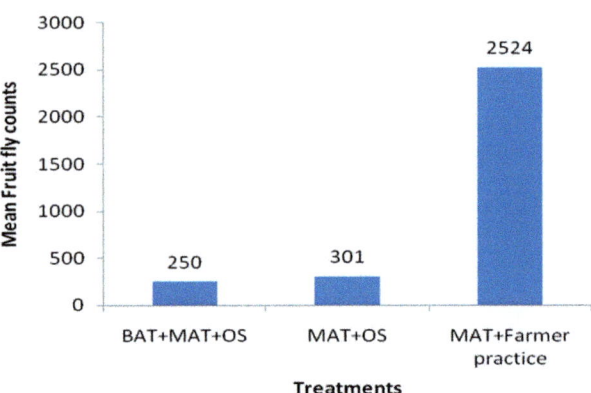

Fig. 22.5 Effect of three main treatments on abundance of fruit fly pests in Uganda (Millenium Science Initiative 2013)

4.2.3 National Agricultural Advisory Services (NAADS) Initiative

In response to countrywide outcry about the severe effects of fruit flies, particularly *B. dorsalis*, NAADS conducted several awareness and capacity building initiatives between 2009 and 2011, to underpin the development of fruit fly control strategies. NAADS also provided farmers across the major fruit growing regions in Uganda, with tools and materials, such as traps, attractants and pesticides. The initiative also attempted to establish a countrywide extension network to monitor fruit fly populations in different regions.

4.2.4 *icipe*'s African Fruit Fly Programme (AFFP), Previously Called the African Fruit Fly Initiative (AFFI)

AFFP was established in response to requests from African fruit growers, national authorities and regional commodity and quarantine bodies and began in 1999. In Uganda, AFFP assisted with assessing the impact of fruit fly infestation on key crops with: development and testing of affordable fruit fly management methods; exploring and releasing natural enemies of fruit flies; establishing parameters for post-harvest treatment; producing and disseminating tools, distribution maps and pest identification keys for strengthening fruit fly quarantine; and with training personnel. Through *icipe*-led monitoring efforts, at least eight new fruit fly species were recorded in Uganda with voucher specimens stored in the *icipe* museum facility (records from *ICIPE* on Table 22.1). Control and monitoring methods based on commercially available (food bait-waste brewer's yeast, entomopathogenic fungi, and other attractants) were evaluated on-farm. Given the similarity of agro-ecological zones across East Africa, it is possible that *icipe*-released (in Kenya) *Fopius arisanus* (Sonan) parasitoids; (targeting *B. dorsalis*) have spread into Uganda (Ekesi and Billah 2006).

4.2.5 Food and Agricultural Organization of the United Nations (FAO) Initiative

FAO has supported various horticultural initiatives in Uganda. They funded the first major programme on surveillance and management of *B. dorsalis* in Uganda (Nemeye 2005). This initiative was part of the emergency response activities of the governments of Kenya, Tanzania and Uganda following the first detection of *B. dorsalis* in Kenya. Since then, FAO has continued to support local government in Uganda with capacity building and provision of farm inputs such as seedlings of high-yielding fruit varieties, lures and other fruit fly control tools. FAO have trained over 650 farmers from 21 agricultural functional groups and four farmer associations.

4.2.6 The Agricultural Technology and Agribusiness Advisory Services (ATAAS) Initiative

ATAAS is a government of Uganda 5-year project that started in 2011 and is administered by the Ministry of Agricultural Animal Industry and Fisheries (MAAIF). ATAAS seeks to address the weak linkage between the different participants involved in agricultural research and development in Uganda. The initial fruit fly control strategies that were planned, tested and promoted under MSI (i.e. combinations of MAT+BAT+OS) are being scaled up for area-wide fruit fly management in selected fruit growing zones in Uganda, as part of the ATAAS initiative. In addition, under ATAAS, NARO is evaluating the bagging of fruit prior to maturity as a management option for fruit flies. Preliminary results are finding fewer fruit fly infestations in fruit that have been bagged compared with unbagged fruit. These results were consistent in all three districts where the experiment was done (Nankinga et al. 2014b, c); Fig. 22.6).

4.3 Future Plans and Ways Forward on Fruit Fly Management in Uganda

Despite the economic importance of tephritid fruit flies to horticultural production and exports for Uganda, there is currently no coordinated national management plan for fruit flies in the country. The various government agencies have no specific budgetary support for fruit flies; instead, interest in these pests is vested in the professional interests of individual scientists. A management plan is essential to identify and offer information on the risk levels of key fruit fly pests in the country, their biology and ecology, diagnostic protocols and response plans, preparedness and prevention measures, surveillance and detection strategies, and management information.

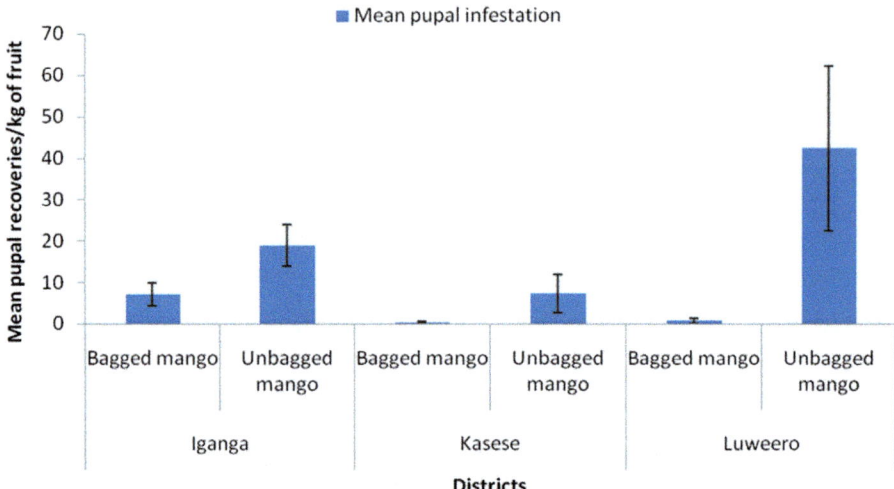

Fig. 22.6 Mean recovery of pupal (±) SE from bagged and un-bagged fruits sampled in four mango growing districts in Uganda, 2013 (NARL Quarterly Report 2014)

In support of a national fruit fly management plan there is a need to collect, collate and archive species-specific data sets on the key fruit fly pest species. Out of the MSI-supported initiative on ecology and management of fruit flies in Uganda, a small collection of tephritid fruit flies was collated and is archived in the Zoology department of Makerere University. Existing data sheets should be regularly updated, the information stored in a single accessible portal, and its utilization promoted widely. The information accessible through this portal should include diagnostic protocols, treatment schedules, pest data sets, national standards and information on premium markets and phytosanitary requirements. This would help to identify knowledge gaps and assist in the development of a national research priority list (including a capacity gap analysis of the production supply chain) to inform research providers. Establishing an accessible and regularly updated knowledge hub that provides all relevant fruit fly information would represent a valuable network for communication and collaboration that could ultimately improve the effectiveness of the various ongoing fruit fly management initiatives.

An effective knowledge hub is only possible if the information on it is accurate and there are mechanisms in place for the information to be validated and continuously improved and updated. For example, a nationally agreed standard for fruit fly diagnosis and surveillance is required. This would include a national diagnostics network of all relevant national and international experts, laboratories and centres of expertise, essential equipment and reference collections, thereby providing a tool to promote communication and collaboration. These standards should meet market access requirements of local, regional and international markets. Furthermore, the current technologies, innovations, and management practices for fruit flies in Uganda need to be validated and agreed. Standard guidelines for site-specific implementation of the different control packages need to be developed and include the potential for scaling up. This would support current activities focused on implementing the various management options and the documentation of realistic pathways for their increased adoption.

Finally, efforts to raise awareness concerning the impact of fruit flies should be increased to encourage a change in attitudes and behavior in relation to the management options adopted. This should include the development of a national communication strategy that could analyze current awareness activities, identify possible synergies amongst participants, and propose and implement an innovative communication strategy for fruit fly management in Uganda.

References

Agum WO (2014) Fruit physical and physiological factors influencing mango fruit infestation by fruit flies in the Lake Victoria Crescent agro ecological zone ff Uganda. MSc thesis, Gulu University, Uganda

Akol AM, Masembe C, Isabirye BE, Kukiriza CN, Rwomushana I (2013) Oviposition preference and offspring performance in *Bactrocera invadens* (Diptera: Tephritidae). Int Res J Hort 1:1–14

Ekesi S, Billah MK (eds) (2006) A field guide to the management of economically important Tephritid fruit flies in Africa. ICIPE Science Press Nairobi, Kenya, 106 pp

Ekesi S, Nderitu PW, Rwomushana I (2006) Field infestation, life history and demographic parameters of *Bactrocera invadens* Drew, Tsuruta and White, a new invasive fruit fly species in Africa. Bull Entomol Res 96:379–386

Isabirye BE (2015) Diversity, host utilization and ecological niche of fruit flies (Diptera: Tephritidae) in Uganda. Ph.D thesis, Makerere University, Uganda

Isabirye BE, Nankinga CK, Muyinza H, Masembe C, Akol AM (2012) Effect of three host species on infestation levels, offspring survivorship, sex ratio and body weight of *Bactrocera invadens* (Diptera: Tephritidae). In: Proceedings of the 2nd international symposium of Tephritid Workers of Europe, Africa & The Middle East (TEAM)

Isabirye BE, Masembe C, Nankinga C, Muyinza H, Akol AM (2014a) Geometric morphometrics of geographic and host-associated population variations of *Bactrocera invadens* in Uganda. In: Proceedings of the 9th International Symposium on Fruit Flies of Economic Importance (ISFFEI), p 155

Isabirye BE, Masembe C, Nankinga CK, Rwomushana I, Muyinza H, Akol AM (2014b) Projections of climate-induced future range shifts among fruit fly species in Uganda. In: Proceedings of the 9th International Symposium on Fruit Flies of Economic Importance (ISFFEI), p 134

Isabirye BE, Akol AM, Nankinga CK, Masembe C, Rwomushana I (2015a) Species composition and community structure of fruit flies (Diptera: Tephritidae) across major mango-growing regions in Uganda. Int J Trop Ins Sci:1–12

Isabirye BE, Masembe C, Akol AM, Muyinza H, Rwomushana I, Nankinga CK (2015a) Modeling the potential geographical distribution and ecological niche of selected fruit fly (Diptera: Tephritidae) species in Uganda. J Plant Pest Sci 2:18–33

Isabirye BE, Akol AM, Muyinza H, Masembe C, Rwomushana I, Nankinga CK (2016) Fruit fly (Diptera: Tephritidae) host status and relative infestation of selected mango cultivars in three agro ecological zones in Uganda. Int J Fruit Sci 16(1):23–41

Mayamba A, Nankinga CK, Isabirye BE, Akol AM (2014) Seasonal population fluctuations of *Bactrocera invadens* (Diptera: Tephritidae) in relation to mango phenology in the Lake Victoria Crescent, Uganda. Fruits 69:473–480

Millenium Science Initiative (2013) Gaining insight into the physical, ecological and biochemical factors influencing fruit fly infestation of mango in Uganda. End of Project Report, NARL, Kawanda

Nakasinga JK (2002) Studies on fruit fly species occurrence, distribution and composition on mango in Uganda. MSc thesis 91 pp

Nankinga CK, Isabirye BE, Muyinza H, Rwomushana I, Stevenson PC, Mayamba A, Aool W, Akol AM (2014a) Fruit fly infestation in mango: a threat to the Horticultural Sector in Uganda. Uganda J Agric Sci 15:1–4

Nankinga CK, Mayamba. A, Nampeera. F, Wasswa. W, and Isabirye. BE (2014b) A bronchure of mango fruit fly management in Uganda. Second Edition June 2014

Nankinga C, Isabirye BE, Mayamba A, Muyinza H, Aool W, Rwomushana I, Stevenson P, Akol A (2014c) Status of fruit fly infestation of mango and other fruits in Uganda. In: Proceedings of the 9th International Symposium on Fruit Flies of Economic Importance (ISFFEI), p 202

NARL Annual Report (2011) National Agricultural Research Laboratories Annual Review and Planning Report, Biotechnology and Biodiversity Program, 23 pp

NARL Quarterly Report (2014) National Agricultural Research Laboratories Quarterly (May–August) Report, Biotechnology and Biodiversity Program, 37 pp

Nemeye P (2005) Surveillance and management of *Bactrocera invadens* in East Africa. Progress report No. 2. 20 pp

Okullokwany FS (2006) Report on surveillance and management of *Bactrocera invadens* in East Africa-Phase II (Uganda). Ministry of Agriculture Animal Industry and Fisheries, 8 pp

Uganda Export Promotions Board (2012) www.ugandaexportsonline.com

World Bank Report (2013) Uganda Economic Update, 2013 Summary. http://www.worldbank.org

World Trade Organisation (2016) www.wto.org

B.E. Isabirye is an ecological entomologist educated at Makerere University and he leads the Sustainable Agriculture, Food Security and Nutrition (SAFSN) Theme at the Association for Strengthening Agricultural Research in Eastern and Central Africa (ASARECA). SAFSN addresses challenges created by climate change, pests and diseases of livestock and crops, and harnesses the opportunities and options provided by the sub-region's biodiversity and biotechnology techniques. Brian is also currently a member of several AR4D steering and advisory committees in Eastern and Central Africa. His PhD study was titled: Diversity, Host utilization and Ecological Niche of Fruit flies in Uganda.

C.K. Nankinga is a senior researcher in entomology at NARO-National Agricultural Research Laboratories, Kawanda. Caroline was educated at Makerere University and the University of Reading UK. Caroline has long-term expertise in integrated pest and disease management and has been involved in a number of national, regional and international projects as lead or co-investigator to develop pest and disease control strategies in bananas, cassava, mangoes, citrus and hot-pepper. She is now the Principal Investigator (PI) of the Fruit Fly Integrated Management project under the ATAAS programme. She was PI of mango fruit fly projects within the MSI initiative on ecology and management of fruit flies in Uganda, and Citrus Research International projects that first evaluated integrated management options for fruit fly control in Uganda. She was co-supervisor of the MSI project students and, as PI, lead on farmer capacity building in fruit fly management by promoting area-wide fruit fly management in Uganda.

A. Mayamba is an applied and ecological entomologist at the National Agricultural Research Laboratories, Kawanda, Uganda. Alex was educated at Makerere University and has wide expertise in the agronomy of horticultural crops and pest management. He has participated in several projects on the ecology and management of different insect pests in Uganda including fruit flies and thrips. His MSc. research was on seasonal fluctuations in fruit fly populations in the Lake Victoria Crescents zone of Uganda. He is currently undertaking research on the ecology of rodent pests in maize and rice agroecosystems in Eastern Uganda.

A.M. Akol is an Associate Professor of Entomology in the Department of Biological Sciences, College of Natural Sciences, Makerere University. Anne was educated at Makerere University, University of Cambridge and Kenyatta University. She has specific expertise in the management of insects of agricultural and medical importance and has been involved with a number of pest management research programmes as a lead or co-investigator. She was a co-investigator on the MSI initiative on ecology and management of fruit flies in Uganda and main adviser of the three students on this project that contributed to developing Uganda's capacity on fruit fly management.

I. Rwomushana is a Research Scientist with the African Fruit Fly Program at the International Centre of Insect Physiology and Ecology (*icipe*) with expertise in Agricultural Entomology. He was educated at Makerere University and Kenyatta University. He is involved in the development of sustainable integrated pest management (IPM) options for the management of fruit flies and other arthropod pests that constrain fruit production in sub-Saharan Africa. His research interests involve understanding the bioecology of fruit flies and employing balanced basic and applied research to the use of baiting and male annihilation techniques, biopesticides, natural enemy conservation and classical biological control methods, and use of soft pesticides for suppression of tree crop pests. He has also coordinated several research projects in Eastern and Central Africa on a number of other arthropod pests.

Chapter 23
Integrated Management of Fruit Flies – Case Studies from Tanzania

Maulid Mwatawala

Abstract The horticultural industry in Tanzania is constrained by many factors including insect pests, and particularly fruit flies. Fruit flies are polyphagous and attack a wide variety of fruit species. Almost all commercially grown fruits are prone to infestation by these pests. In Tanzania, both indigenous and exotic invasive fruit fly species have negatively impacted the production and trade of fruits and fruit-bearing vegetables. Research on fruit flies in Tanzania has increased in importance in recent decades. A major thrust came with the arrival of the oriental fruit fly, *Bactrocera dorsalis*, in 2003. Increased awareness of the heavy losses inflicted by *B. dorsalis* necessitated the development of management programmes for fruit flies. This chapter presents the various activities undertaken in Tanzania, including detection and monitoring surveys, establishment of spatial and temporal abundance and studies on the host range and developmental biology of fruit flies. Although limited to a few regions in Tanzania, the knowledge gained has contributed to the development of sound IPM programmes for fruit flies. Recommendations now include Area Wide Management programmes to extend research and implementation of IPM nationwide.

Keywords Horticulture • Fruit flies • East Africa

1 Introduction

The Tanzanian economy is largely agriculture-based and, in recent years, up to 26.7 % of the GDP has come from the horticultural sector (HODECT 2010). Smallholder farmers dominate Tanzanian fruit and vegetable production and, although there is good potential for export (HODECT 2010), most horticultural produce is sold in domestic markets. Production of horticultural crops is constrained

---

M. Mwatawala (✉)
Department of Crop Science and Production, Sokoine University of Agriculture, Box 3005, Morogoro, Tanzania
e-mail: mwatawala@yahoo.com

© Springer International Publishing Switzerland 2016
S. Ekesi et al. (eds.), *Fruit Fly Research and Development in Africa - Towards a Sustainable Management Strategy to Improve Horticulture*,
DOI 10.1007/978-3-319-43226-7_23

by biotic and abiotic factors; fruit flies (Diptera: Tephritidae) are major pests of fruit in Tanzania. Research interest on fruit fly pests in Tanzania has increased over the last few decades. Tanzania was part of the African Fruit Fly Initiative (AFFI) which began in the late 1980s and was coordinated by the International Centre for Insect Physiology and Ecology (*icipe*). Between the late 1980s and the early 1990s this initiative contributed to establishing a greater understanding of fruit flies in Tanzania. Then the Sokoine University of Agriculture (SUA, Morogoro) and the Royal Museum for Central Africa (RMCA, Tervuren, Belgium) initiated a long-term collaborative fruit fly research programme which ran from 2004 to 2013. This long-term programme provided baseline data on the biology (biodiversity, population dynamics, seasonality, host range) of fruit fly pests in the Morogoro region. The SUA RMCA programme coincided with the first record of the oriental fruit fly, *Bactrocera dorsalis* (Hendel), in Tanzania. As a result, attention to this invasive pest was increased while still maintaining research on the major indigenous fruit fly pest species. Three Integrated Pest Management (IPM) programmes were designed and evaluated during the latter phase of research (2008–2012) in collaboration with the Ministry of Agriculture Food Security and Cooperatives (MAFC). Results from the SUA RMCA programme led to the development of a national Management Plan for fruit flies in Tanzania.

Over the years, research in Tanzania has identified the presence of 179 species of Tephritidae including major pest species: the melon fly *Zeugodacus cucurbitae* (Coquillet), the oriental fruit fly *B. dorsalis*, the solanum fruit fly *Bactrocera latifrons* (Hendel), the medfly *Ceratitis capitata* (Wiedemann), the mango fruit fly *Ceratitis cosyra* (Walker), the Natal fruit fly *Ceratitis rosa* Karsch, *Ceratitis fasciventris* (Bezzi), *Ceratitis anonae* (Graham) and *Ceratitis bremii* Guéurin-Méneville, the lesser pumpkin fly *Dacus ciliatus* Loew, *Dacus punctatifrons* Karsch, *Dacus bivittatus* (Bigot) and *Trirhithrum coffeae* Bezzi. The majority of these are polyphagous species, attacking a wide variety of commercially grown fruits. Most are also indigenous pests, with the exception of *B. dorsalis*, *Z. cucurbitae* and *B. latifrons* (Mwatawala et al. 2013). The various research programmes also determined the bionomics of some species, and formulated and tested control strategies against the major fruit fly species in Eastern Tanzania. This chapter highlights some of the key fruit fly research activities, including the development and evaluation of Integrated Pest Management (IPM) in Tanzania.

2   Research to Establish the Prerequisites of IPM

IPM is knowledge intensive and requires a thorough understanding of the organisms involved, the environment and all potential control strategies. Pursuant to that, underpinning research on fruit flies was essential to generate the prerequisite data required. Research in Tanzania established knowledge on, but not limited to, biodiversity, host range and preference, spatial and temporal abundance, life history and demographic parameters and on the efficacy of attractants for fruit flies. Key results are highlighted below.

2.1 Establishing the Biodiversity of Fruit Flies in Tanzania

Records of fruit flies in Tanzania date back to the late nineteenth century, beginning with a record of *D. punctatifrons* collected by S. Fullerborn in 1896 at Langenburg, Lake Nyasa (Ian White unpublished, cited in Mwatawala et al. 2005). Later, Ritchie (1935), reported the presence of a number of major pest fruit fly species that included *C. capitata* and *C. rosa*. The United States Department of Agriculture (USDA) organised a large-scale expedition in western and eastern Africa between 1935 and 1936 in search of parasitoids of *C. capitata* for use as classical biological control agents in Hawaii. The East African expedition covered Kenya, Tanganyika (as mainland Tanzania was then known) and Zanzibar, and reported four parasitoids of *C. capitata* from the genus *Opius*. These included *Opius humilis* Silvestri and *Opius fullawayi* (Silvestri) (Bianchi and Krauss 1936; Clausen et al. 1965; Wharton et al. 2000). The AFFI conducted surveys between 1999 and 2002 which reported more than 50 different dacine and ceratitidine fruit flies in Tanzania, including ten of the major pest species from the region (see also De Meyer and White 2004). Subsequently Mwatawala et al. (2006a) reported on the biodiversity of fruit flies in orchards in the Morogoro region of Tanzania. The full list of fruit fly pests occurring in Tanzania is available in the database at RMCA and was reproduced in Mwatawala et al. (2013).

2.2 Establishing the Presence and Distribution of Invasive Fruit Flies in Tanzania

SUA and MAFC conducted surveys to detect the presence and distribution of *B. dorsalis* and *B. latifrons*, both invasive species in Tanzania. The MAFC surveys were made between December 2003 and March 2005 and targeted *B. dorsalis*. Surveys to detect *B. dorsalis* at potential entry points covered areas around Arusha, Dar es Salaam, Kilimanjaro, Tanga and the coast in general indicated that *B. dorsalis* was present in all five regions. Surveys to determine the degree of spread covered the whole country and indicated that, at that time, *B. dorsalis* was present throughout the country (see Mwatawala et al. 2013). The SUA surveys were made in 2006 and 2007, targeted *B. latifrons* and covered the regions of Dar es Salaam, Tanga, Kilimanjaro, Arusha, Morogoro, Iringa, Mbeya, Dodoma, Singida, Tabora, Shinyanga, Kagera, Mwanza and coastal areas in general. *Bactrocera. latifrons* was recorded in six of these regions: Morogoro, Tanga, Kilimanjaro, Arusha, Iringa and Mwanza (Mwatawala et al. 2010). SUA conducted another country-wide survey in 2010/2011 to collect specimens of *B. dorsalis* and elucidate its distribution. This

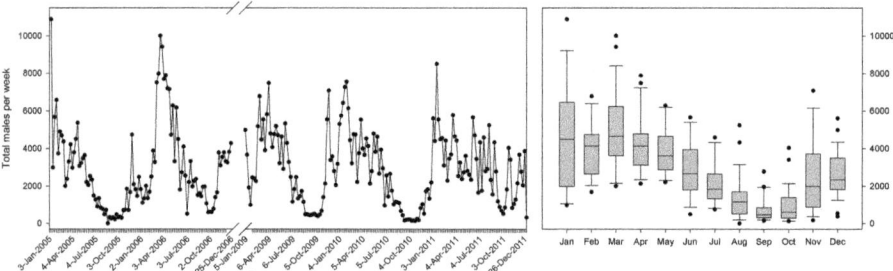

Fig. 23.1 Temporal distribution of *B. dorsalis* in Morogoro, Tanzania

was part of an International Atomic Energy Agency (IAEA) coordinated project to resolve the taxonomic identity of cryptic species that included *B. invadens*, as it was then known, with the purpose of facilitating trade and use of the Sterile Insect Technique (SIT).

2.3 Establishing the Spatial and Temporal Abundance of Fruit Flies in Tanzania

SUA and RMCA continued to monitor fruit fly populations in the Morogoro region from 2004 to 2013. This enabled spatial and temporal changes in populations of the major fruit fly species to be quantified and related to abiotic factors. The highest populations of *B. dorsalis* and *C. rosa* were recorded during the rainy seasons, whilst the lowest populations were recorded during the dry season (Mwatawala et al. 2006b). However, they also determined that populations of *B. dorsalis* recorded during the long rainy season were higher than those recorded during the short rainy season and that populations of *C. cosyra* generally had no obvious peaks (Mwatawala et al. 2006b). The data for the whole ten-year period are currently being fully analysed prior to further interpretation (Fig. 23.1). Geurts et al. (2012) recorded the spatial abundance of fruit flies along an altitudinal transect between 581 and 1650 m in the Uluguru Mountain range in the Morogoro region between 2008 and 2012. The polyphagous invasive species *B. dorsalis,* and the indigenous species *C. rosa*, had a similar temporal pattern, but were largely spatially separated; *B. dorsalis* was abundant at lower elevations while *C. rosa* was dominant at higher elevations. The polyphagous indigenous species *C. cosyra* co-occurred with *B. dorsalis* but showed an inverse temporal pattern.

2.4 Life History Parameters, Demography and Host Range of Fruit Flies

In a comparative study, *B. dorsalis* had shorter embryonic, larval and pupal development times than *C. cosyra*, independent of temperature regime (Salum et al. 2013). Life expectancy of male *B. dorsalis* was significantly greater than that of male *C. cosyra*, while life expectancy of female *B. dorsalis* was greater than that of female *C. cosyra* but not significantly so. Average net fecundity was greater for *B. dorsalis* than for *C. cosyra*. Both species achieved their highest intrinsic rate of increase and net reproductive rate at 30 °C and *B. dorsalis* had a higher intrinsic rate of increase and net reproductive rate than *C. cosyra*, at all temperatures evaluated (Salum et al. 2013). A separate study by Mkiga and Mwatawala (2015) reported that developmental times of *Z. cucurbitae* were significantly affected by temperature but not by host species; the developmental time of immature stages of *Z. cucurbitae* decreased with increases in temperatures while survival increased with increases in temperature.

With respect to host range, incidence and infestation rates, *B. dorsalis* was the predominant fruit fly species and a further eight new host plant species were reported (Mwatawala et al. 2009). Infestations by native pests, such as *C. capitata* and *C. cosyra*, were minor compared to *B. dorsalis* (Mwatawala et al. 2009). *Ceratitis rosa* was the predominant species infesting temperate fruits, while Cucurbitaceae are mainly infested by *Z. cucurbitae* (Mwatawala et al. 2009). Mango, *Mangifera indica* L., and guava, *Psidium guajava* L., had the highest fruit fly infestation rates while citrus species and avocado, *Persea americana* Miller, had the lowest (Mwatawala et al. 2009). Tropical almond, *Terminalia catappa* L., was particularly susceptible to *B. dorsalis* and soursop, *Annona muricata* L., was an important host for *C. cosyra* after the mango season had ended (Mwatawala et al. 2009).

3 IPM Programmes for Fruit Flies in Tanzania

A number of research programmes on fruit fly population suppression have been established in Tanzania. Control techniques evaluated include: orchard sanitation, Male Annihilation Technique (MAT)/ mass trapping, attract and kill, early harvesting of fruits, biological control and judicious use of selected insecticides. Researchers from various institutions collaborated to evaluate these techniques in fruit-farming communities in Dar es Salaam, Tanga, Morogoro and other coastal areas (Mwatawala et al. 2013) and the results are described below.

3.1 Fruit Fly Control Techniques Evaluated in Tanzania

3.1.1 Orchard Sanitation

The Mikocheni Agricultural Research Institute (MARI) conducted trials on orchard sanitation in farmers' fields in Morogoro, Tanga, Dar es Salaam and other coastal regions (Mwatawala et al. 2013). Removal of fallen fruits and burial in deep pits, or collecting fruit and placing them in black plastic bags and exposing them to intense sun substantially reduced fruit fly populations. Ploughing beneath fruit trees to expose buried puparia to the harsh environment also reduced fruit fly numbers. Using these results the researchers were able to propose mandatory field sanitation strategies that have subsequently been enforced by national law (Mwatawala et al. 2013).

3.1.2 Biological Control

In collaboration with *icipe*, MAFC released the solitary endoparasitoid *Fopius arisanus* (Sonan) in a limited area of coastal Tanzania. This polyphagous parasitoid oviposits into the eggs and larvae of a number of fruit fly species. The parasitoids' eggs hatch and develop within fruit flies eventually killing them (Zenil et al. 2004; Rousse et al. 2005). However, data on field establishment in coastal Tanzania are not currently available. MARI and SUA conducted several trials on the predatory efficacy of the African weaver ant, *Oecophylla longinoda* Latreille. This ant is an indigenous species and is widely distributed across Eastern and coastal regions of Tanzania (Olotu et al. 2012; Rwegasira et al. 2014). Trials in Dar es Salaam, Mtwara, Morogoro, Tanga and other coastal regions demonstrated that weaver ants predated adult fruit flies and also that the presence of ants repelled ovipositing female fruit flies (Z. Seguni, personal communication). Results of research on weaver ants and fruit flies in mango are presented in Sect. 3.2.3. In the laboratory SUA also evaluated, on a limited scale, the potential for entomopathogenic nematodes to control fruit fly larvae. High larval mortalities were recorded, but the study was never extended to the field (Kalinga 2011).

3.1.3 Spot Applications of Baits and Use of Attractants

SUA compared the efficacy of an enriched ginger oil lure (EGO Lure®, Insect Science, Pretoria, South Africa) as an attractant for *C. rosa*, *C. cosyra* and *C. capitata* with the commercially available lures, trimedlure and terpinyl acetate. The trials were made along an altitudinal transect across the Uluguru mountains in Morogoro and demonstrated that the EGO Lure® attracted all three *Ceratitis* species and that trap catches were larger in traps baited with enriched ginger oil than in traps baited with the conventional lures (Mwatawala et al. 2015a). Through MAFC and the Ministry of Local Government methyl eugenol is distributed to mango farmers in the coastal regions of Tanzania for use as an attractant for mass trapping of *B. dorsalis*. However, the effectiveness of this programme in reducing damage by *B. dorsalis* is not known.

3.1.4 Insecticides

There are no insecticide products registered specifically for fruit fly control in Tanzania. Despite this, farmers use a wide range of active ingredients including dimethoate, lambdacyhalothrin, deltamethrin and dichlorovos for fruit fly control.

3.2 Detailed IPM Trials Conducted in Tanzania

3.2.1 IPM Trials in the Morogoro Region, Eastern Central Tanzania

SUA and RMCA evaluated three IPM treatment regimes in fourteen mango orchards located in the plateau zone of Morogoro, Eastern Central Tanzania, between 2008 and 2012 (Mwatawala et al. 2006a). The first treatment represented standard control practices and was comprised of insecticide sprays; dimethoate (0.2 %) was applied in the first two seasons and lambdacyhalothrin (0.02 %) in the last three seasons. The second treatment was comprised of a combination of mass trapping and spot applications of GF 120® / SUCCESS® bait (Spinosad 0.02 %, Dow AgroSciences), diluted in a ratio of 1: 5.5 in water. Mass trapping was achieved using six McPhail traps (AgriSense, UK) per ha, baited with methyl eugenol. The traps were serviced every week and the lure replaced every four weeks. The third treatment used just spot treatments with molasses baits that were comprised of a mixture of 360 g of powdered *Derris elliptica* Fabaceae: 1 L molasses: 360 g brewer's yeast: 20 L of water; once formulated the baits were used within 12 h. These baits were applied weekly to 1 m^2 areas (or spots) in the tree canopies at a rate of 1 L/ ha. Each of the IPM programmes was implemented in two orchards in each year. These trials were conducted between November and December each season, targeting the last 10 weeks before harvest. Six orchards with trees bearing similar and substantial numbers of fruit were selected for use from the 14 available. Orchard sanitation and early harvesting of fruits were standard components in all three treatments (Mwatawala et al. 2015a).

Populations of fruit flies in each orchard were monitored weekly using modified McPhail traps baited with torula yeast (*Candida utilis* (Henneberg) Lodder & Kreger-van Rij). The population of fruit flies was determined as the number of adult fruit flies per trap per week (FTW). At the end of the season, 50 mango fruits (varieties 'Tommy' and 'Red India') were harvested from each orchard and taken to the laboratory at SUA to establish the incidence and infestation rates of fruit flies. The Analytic Hierarchy Process (AHP) (Saaty 1980) was used to determine which IPM treatment regime provided the greatest benefits to farmers. The evaluation criteria included a Cost Benefit Ratio (CBR) and environmental protection and consumer safety (m = 3). A set of alternative options from which decisions could be made included the three IPM treatments evaluated (n = 3).

There were no significant differences amongst the three IPM treatments evaluated with respect to reductions in the incidence and infestation rates of *B. dorsalis* in mango (Mwatawala et al. 2015a; Fig. 23.2a, b, c). However, the cost-benefit analysis gave positive CBRs for the molasses bait-based IPM and the insecticide spray-based IPM but not IPM based on the GF 120®/ SUCCESS® bait (i.e. spinosad) (Nyavanga 2011). IPM based on the molasses bait was ranked highest in delivering farmers the greatest benefits in terms of CBR, environmental protection and consumer health. IPM based on insecticide sprays also had a high CBR but this is likely to be due to the lower labour costs, reduced time required to treat an area and the low frequency of application (once in 2 weeks) compared with the other treatments. Spinosad-based IPM had a low CBR because of its limited availability, high price and high frequency of application compared to conventional insecticides (Mwatawala et al. 2015a). The Net Present Values (NPVs) for the different IPM programmes were also positive except for the programme based on the spinosad bait; orchards that were under insecticide-based IPM had the highest NPV value followed by the molasses bait-based IPM (Nyavanga 2011).

3.2.2 IPM Trials in Muheza District, Tanga Region, Tanzania

MARI compared the effects of typical orchard sanitation, an IPM programme and biological control using *O. longinoda* on suppression of fruit flies in mango. Trials were conducted between 2012 and 2013 in Muheza District, Tanga Region, North Eastern Tanzania (Materu et al. 2014). The IPM programme was composed of: weekly removal and burying of fallen fruits; spot application of Mazoferm® bait (Corn Product International, Nairobi, Kenya; a 200: 10: 1 mixture of Mazoferm® [an autolysed protein, Yee 2006]: dichlovoros 50% EC: water) every two weeks; and mass trapping of adult flies using methyl eugenol in a 10: 1 mixture with dichlorvos 50% EC. The Mazoferm® bait was sprayed on to 1 m^2 areas, or spots, on tree canopies, every 10–12 days throughout the mango season. The methyl eugenol mixture was impregnated into a cotton wool to form lures that were hung in locally-made traps (used 0.5 L water bottles each with three holes) and replaced every 2 months. Twenty traps were used and flies were removed every week (Materu et al. 2014).

Results showed that *B. dorsalis* was the dominant species recorded in all traps during the main fruiting periods between September and December 2013. Fruit losses were least in the IPM (1%) and biological control (2%) plots. Highest losses were recorded in plots with just basic orchard sanitation (23%) and the control (32%) plots where no control strategies were applied. There were no significant differences in losses between IPM and the biological control plots (Materu et al. 2014).

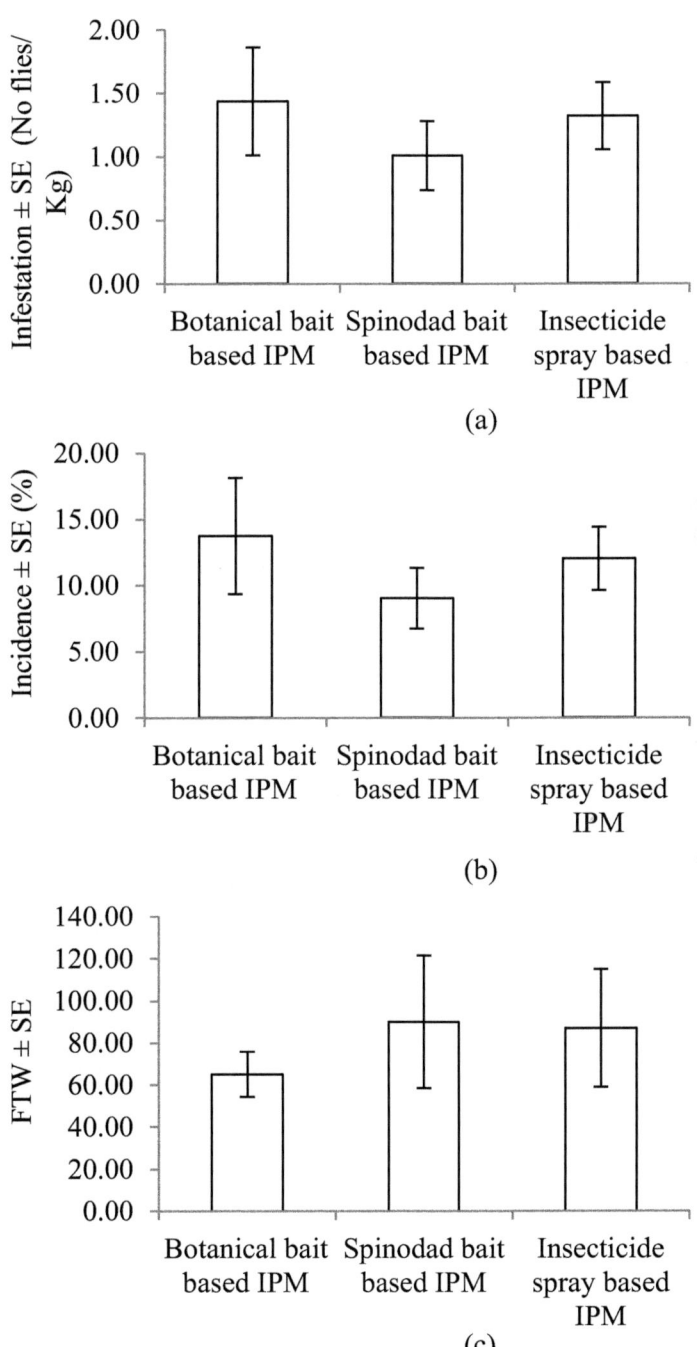

Fig. 23.2 Infestation rate (**a**) incidence (**b**) and trap catches (**c**) of *B. dorsalis*

3.2.3 IPM Trials Conducted in Kibaha District, Coast Region, Tanzania

SUA and Aarhus University (Denmark) collaborated in a research programme to increase the economic value of African mango and cashew, *Anacardium occidentale* L., and this included trials on the efficacy of weaver ants, *O. longinoda*, to control fruit flies in mango. Trials were done in 6-year-old mango orchards (variety 'Apple') at Mlandizi, Kibaha over two consecutive seasons (2012/2013 and 2013/2014). Three treatments were compared: (i) trees protected by *O. longinoda* (ii) trees protected by the insecticide spray, Dudumida® (70WG imidacloprid, Mega Generics Ltd., Tanzania) and (iii) unprotected trees (control). In a repeated measures design, each treatment was applied in a block composed of 72 trees, separated by rows of two trees and replicated over the two seasons (Abdulla et al. Personal communication).

Sixteen colonies composed of variable numbers of *O. longinoda* nests were introduced into the experimental orchard 6 months before the mango fruiting season began, following the procedures of Peng and Christian (2005). Each weaver ant colony had access to 4–6 trees and approximately 6–12 nests were introduced to each mango tree depending on canopy size. The insecticide Dudumida was applied at a rate of 1 g/tree for the management of chewing and sucking insect pests. To avoid spray drift, spraying was done early in the morning (06:00–07:00 h) when there was no strong wind. The first spray application was made at the onset of flowering and was repeated at 2-week intervals following the manufacturer's recommendations. The insecticide was sprayed four times each season. Fruits were collected at 3-week intervals to establish presence of larvae. All the fruits in the orchard were counted to establish yield at the end of the fruiting season.

Overall, the incidence (percentage of infested samples) and infestation rate (number of emerged adult flies per unit weight of fruits) of fruit flies was significantly lower on both the *O. longinoda* and the Dudumida-protected trees compared with the control. Although the lowest incidence of fruit flies was recorded in fruits from the *O. longinoda* treatments, there was no significant difference in incidence or infestation rate between the *O. longinoda* and Dudumida treatments (Abdulla et al. Personal communication). Furthermore, yield was significantly lower in the control compared with the other treatments in both seasons; there was no significant difference in yield between the *O. longinoda* and Dudumida treaments (Abdulla et al. Personal communication).

4 Management Plan for Fruit Flies in Tanzania

Future plans for fruit fly management in Tanzania proposes an Area Wide Programme to extend research and implementation of IPM to other parts of the country (Mwatawala et al. 2013). The objectives of this management plan include (i) establishing the nationwide status of economically important fruit flies (ii) assessing the nationwide economic impact of fruit flies on the horticultural industry (iii) evaluating the available mechanisms for fruit fly control (iv) formulating management programmes for mitigation of fruit fly problems in horticulture nationwide (v) re-enforcing long term awareness, capacity building and collaboration. The government of Tanzania and its research institutions are actively looking for funds to finance these activities.

5 Conclusion

This chapter presents the various research activities undertaken in Tanzania, with the aim of developing an effective IPM programme for fruit flies. The prerequisites for formulating an IPM programme have been established and include an understanding of the distribution, bionomics and host range of fruit flies. Various pest control techniques have been tested and later integrated into IPM programmes. However, the trials were limited to eastern Tanzania. For this reason, Area Wide Management programmes for fruit flies are now being implemented throughout Tanzania.

References

Bianchi FA, Krauss NH (1936). Report on the United States Department of Agriculture East-African Fruit Fly Expedition of 1935–36. United States Department of Agriculture, 33 pp

Clausen CP, Clancy DW, Chock QC (1965). Biological control of *Dacus dorsalis* (Hendel) and other fruit flies in Hawaii. United States Department of AgricultureTechnical Bulletin 1322

De Meyer M, White IM (2004). True fruit flies (Diptera, Tephritidae) of the Africa. A queryable website on taxon and specimen information for afrotropical Dacine fruit flies. http://projects.bebif.be/enbi/fruitfly/ – Tervuren: Royal Museum for Central Africa. Access date: 25 Dec 2015

Geurts K, Mwatawala MW, De Meyer M (2012) Dominance of an invasive fruit fly species, *Bactrocera invadens*, along an altitudinal transect in Morogoro, Eastern Central Tanzania. Bull Ent Res 104:288–294

HODECT (2010) Tanzania horticultural development strategy 2012–2021., Horticultural Development Council of Tanzania, https://webgate.ec.europa.eu

Kalinga Y (2011) Potential of entomopathogenic nematodes in management of selected fruit fly (Diptera: Tephritidae) species in Morogoro. Tanzania. A dissertation submitted in partial fulfilment for the degree of Master of Science in Crop Science of Sokoine University of Agriculture, Morogoro, Tanzania, p 84

Materu CL, Seguni Z, Shao EE, Mruma BM, Ngereza AJ (2014) Assessing the impact of an integrated pest management programme for management of fruit flies (Diptera: Tephritidae) in mango orchards in Tanzania. Res J Agric Environ Sci 1:38–43

Mkiga A, Mwatawala MW (2015) Developmental biology of melon fly (*Bactrocera cucurbitae*) in three cucurbitaceous hosts at different temperature regimes. J Ins Sci 15(1):160

Mwatawala MW, Senkondo FJ, Maerere AP, De Meyer M (2005) Current status of and future needs for fruit fly research in Tanzania. Int J Pest Cont 47:184–187

Mwatawala MW, De Meyer M, Makundi RH, Maerere AP (2006a) Biodiversity of fruit flies (Diptera, Tephritidae) at orchards in different agro-ecological zones of the Morogoro region, Tanzania. Fruits 61:32–332

Mwatawala MW, De Meyer M, Makundi RH, Maerere AP (2006b) Seasonality and host utilization of the invasive *Bactrocera invadens* (Diptera: Tephritidae) in Central Tanzania. J Appl Ent 130:530–537

Mwatawala MW, De Meyer M, Makundi RH, Maerere AP (2009) Host range and distribution of fruit-infesting pestiferous fruit flies (Diptera: Tephritidae) in selected areas of Central Tanzania. Bull Ent Res 99:629–641

Mwatawala M, Maerere A, Makundi RH, De Meyer M (2010) On the occurrence of the Solanum fruit fly, *Bactrocera latifrons* (Hendel) (Diptera: Tephritidae) in Tanzania. J Afrotrop Zool 6:83–89

Mwatawala MW, De Meyer M, Seguni Z, Rwegasira GM, Muganyizi J, Senkondo F, Shechambo L (2013) Management plan of fruit fly (Diptera: Tephritidae) pests in Tanzania. Sokoine University of Agriculture, Directorate of Research and Postgraduate Studies, Morogoro. ISBN 978-9987-640-35-5

Mwatawala MW, Mziray H, Malebo H, De Meyer M (2015a) Guiding farmers' choice for an integrated pest management program against the invasive *Bactrocera dorsalis* (Hendel) (Diptera: Tephritidae) in mango orchards in Tanzania. Crop Prot 76:103–107

Mwatawala M, Virgilio M, Joseph J, De Meyer M (2015b) Niche partitioning among two *Ceratitis rosa* morphotypes and other *Ceratitis* pest species (Diptera: Tephritidae) along an altitudinal transect in Central Tanzania. Zookeys 540:429–442

Nyavanga S (2011) Potential of mango for domestic and export markets in four regions of Tanzania. A dissertation submitted in partial fulfilment for the degree of Master of Science in Agricultural Economics and Agribusiness of Sokoine University of Agriculture, Morogoro, Tanzania, p 92

Olotu MI, du Plessis H, Seguni ZS, Maniania NK (2012) Efficacy of the African weaver ant *Oecophylla longinoda* (Hymenoptera: Formicidae) in the control of *Helopeltis* spp. (Hemiptera: Miridae) and *Pseudotheraptus wayi* (Hemiptera: Coreidae) in cashew crop in Tanzania. Pest Manag Sci 69:911–918

Peng RK, Christian K (2005) Integrated pest management in mango orchards in the Northern Territory Australia, using the weaver ant, *Oecophylla smaragdina*, (Hymenoptera: Formicidae) as a key element. Int J Pest Man 51:149–155

Ritchie AH (1935) Report of the entomologist 1934. Tanganyika Territory Department of Agriculture Annual Report, 1934, 73–83

Rousse P, Harris EJ, Quilici S (2005) *Fopius arisanus*, an egg–pupal parasitoid of Tephritidae. Overview. Biocontrol News Inf 26:59N–69N

Rwegasira RG, Mwatawala M, Rwegasira GM, Offenberg J (2014) Occurence of sexuals of African weaver ant (Oecophylla longinoda Letreille) (Hymenoptera, Formicidae) under bimodal rainfall pattern in eastern Tanzania. Bull Ent Res 105(2):182–186

Saaty TL (1980) The analytic hierarchy process. McGraw-Hill, New York

Salum JK, Mwatawala MW, Kusolwa P, De Meyer M (2013) Demographic parameters of the two main fruit fly (Diptera: Tephritidae) species attacking mango in Central Tanzania. J Appl Ent 138:441–448

Wharton RA, Trostle MK, Messing RH, Copeland RS, Kimani-Njogu SW, Lux S, Overholt WA, Mohamed S, Sivinski J (2000) Parasitoids of medfly, *Ceratitis capitata,* and related tephritids in Kenyan coffee: a predominantly koinobiont assemblage. Bull Ent Res 90:517–526

Yee WL (2006) GF-120, Nu-Lure, and Mazoferm effects on feeding responses and infestations of western cherry fruit fly (Diptera: Tephritidae). J Agric Urban Ent 23:125–140

Zenil M, Liedo P, Williams T, Valle J, Cancino J, Montoya P (2004) Reproductive biology of *Fopius arisanus* (Hymenoptera: Braconidae) on *Ceratitis capitata* and *Anastrepha* spp. (Diptera: Tephritidae). Biol Control 29:169–178

M. Mwatawala is an Associate Professor at Sokoine University of Agriculture, Morogoro, Tanzania, specializing in Agricultural Entomology. His research interests include the ecology of insect pests, Integrated Pest Management and reduction of post harvest losses through Good

Agricultural Practices (GAPs). His current research focuses on ecology and management of fruit flies. He is an Editor in Chief of the Tanzania Journal of Agricultural Sciences (TAJAS), a member of council of the Tropical Pesticides Research Institute (TPRI) and Vice Chairman of Tanzania Entomological Association (TEA). He has been a principal investigator (Tanzania) of projects on Integrated Pest Management (IPM) of fruit flies in Tanzania. These include the projects 'Increasing value of African mango and cashew' and 'Resolution of cryptic species complexes of tephritid pests to overcome constraints to SIT application'. He is the leader of the Tanzania team of a project on enhanced preservation of fruits using nanotechnology.

Chapter 24
Integrated Management of Fruit Flies – Case Studies from Mozambique

Domingos R. Cugala, Marc De Meyer, and Laura J. Canhanga

Abstract The invasive fruit fly, *Bactrocera dorsalis*, has become a major threat to the production and trade of fresh fruits in Mozambique. With reports of 94.0% of mango fruits being damaged by *B. dorsalis* in the Northern region of Mozambique exports of fruit and vegetable to the country's major trading partners were suspended causing severe financial losses to producers and a virtual cessation of investment. Fruit fly monitoring throughout the country has revealed the fruit fly species composition, distribution and pest status. The 165,102 adult fruit fly specimens collected belonged to 43 species of which only eight species are of economic importance. The two invasive fruit fly species reported were *B. dorsalis* and *Zeugodacus cucurbitae*, and accounted for 85.21% and 1.4% of all flies collected, respectively. Management strategies based on combinations of protein bait spraying, deployment of male annihilation techniques, orchard sanitation and biological control resulted in 93.5, 93.8 and 92.6% reductions in *B. dorsalis* populations in Pemba, Maputo and Manica Provinces, respectively. These results suggest that *B. dorsalis* populations can be effectively suppressed using integrated pest management (IPM). The national phytosanitary authorities of Mozambique should now put emphasis on IPM approaches for effective suppression of the *B. dorsalis* population to an acceptable level.

1 Background

The Republic of Mozambique has a land area of 799 380 km^2 of which about 36 million hectares are suitable for agriculture. The country is characterized by different agro-ecological zones that are suitable for production of different types of tropical, semi tropical and temperate fruits. Agriculture is a key sector for economic growth and poverty reduction in the country and accounted for 26% of the

---

D.R. Cugala (✉) • L.J. Canhanga
Faculty of Agronomy and Forest Engineering, EMU, PO Box 257, Maputo, Mozambique
e-mail: dcugala@gmail.com; dcugala@uem.mz

M. De Meyer
Biology Department, Royal Museum for Central Africa,
Leuvensesteenweg 13, B3080, Tervuren, Belgium

Gross Domestic Product (GDP) in 2010, with an annual growth of 7% (INE 2010). It provides employment for more than 80% of the potential workforce, being the main activity for 95% of the population in rural areas. The country has enormous potential for export due to favourable agro-climatic conditions for agricultural production. It is estimated that the horticultural sector in the Maputo (South) and Manica (Central region) Provinces could generate revenues of more than US$ 20.75 million per year through both commercial and smallholder (family) production. However, the horticultural sector is severely hampered by a number of constraints; ranked high amongst these various types of insect crop pests. Fruit flies (Diptera: Tephritidae) are amongst the most economically important pests of fruits and vegetables (Cugala 2011). In recent years, fruit losses have been aggravated by the arrival of the oriental fruit fly, *Bactrocera dorsalis* (Hendel), which was first detected in Mozambique in 2007 in the Cuamba district of the Northern Province of Niassa (Correia et al. 2008). The invasion of this exotic fruit fly has caused serious economic losses as a result of both direct fruit damage as well as the loss of lucrative regional and international export markets due to the new quarantine restrictions imposed (Mangana and Cugala 2010). This has a negative impact on the livelihoods of millions of small- and large-scale fruit and vegetable producers in the country.

2 Fruit Fly Species Composition and Abundance in Mozambique

The abundance and importance of different tephritid fruit fly species have been reported in previous studies (Maússe and Bandeira 2007; Garcia and Bandeira 2011). Garcia and Bandeira (2011) reported 59 species of fruit flies from 27 genera occurring in Mozambique, based upon a literature study and on trapping and rearing results from the central (Beira) and southern (Maputo) regions that were done in 2007 and 2008. The subfamily Dacinae, comprising the majority of fruit-infesting species and all fruit fly pests of economic significance, accounted for 61% of all species identified. The genus *Dacus* comprised of 19 species corresponded to 32.2% of the total number of species, followed by *Ceratitis* (9 species) with 15.2% while other genera were represented by only a few species. The diversity and abundance of fruit fly species varied according to the agro-ecological zone and sampling location. Most species were recorded at low (<400 m asl) to moderate altitudes (>500 to 900 m asl) where they experienced high to moderately low temperatures during the winter period. They did not report any *Bactrocera* species during their study; this is probably because, at the time of the survey, *B. dorsalis* (formerly *Bactrocera invadens* Drew, Tsuruta and White; see Schutze et al. 2015) was largely restricted to the northern region of the country. The taxon and specimen database of Afrotropical frugivorous fruit flies, administered by the Royal Museum for Central

Africa in Tervuren, Belgium (http://projects.bebif.be/fruitfly/index.html) lists 53 Dacinae, including 30 *Dacus* and 12 *Ceratitis* species, as well two *Bactrocera* species (*B. dorsalis* and *Bactrocera biguttula* [Bezzi]) from Mozambique. These records are based upon specimens housed in several natural history collections and material collected during surveys made throughout Mozambique between 2008 and 2011 within the framework of monitoring the spread of *B. dorsalis*.

In recent years, the Eduardo Mondlane University conducted a number of surveys in different parts of the country. Overall, 165,102 adult fruit fly specimens were registered from traps or reared from fruits (Cugala et al. unpublished data). They comprised 43 different fruit fly species belonging to nine genera: *Bactrocera*, *Ceratitis*, *Dacus*, *Perilampsis*, *Celidodacus*, *Trirhithrum*, *Capparimyia*, *Clinotaenia* and *Carpophthoromyia* (s) (excluding specimens of the non-frugivorous genus *Ocnerioxa*). The genus *Dacus* had the highest diversity with 22 recorded species corresponding to 50% of all identified specimens. This was followed by *Ceratitis* with 12 species (27.3%), while only two *Bactrocera* species and two *Carpophthoromyia* species were recorded. The other genera were all represented by just one species (Table 24.1). Amongst the species recorded, only eight species are considered to be of economic importance.

The percentage of each species trapped was: *B. dorsalis* (85.21%); melon fly, *Zeugodacus cucurbitae* (Coquillett) (1.4%); pumpkin fruit fly, *Dacus bivittatus* (Bigot) (4.9%); mango fruit fly, *Ceratitis cosyra* (Walker) (2.5%); *Dacus punctatifrons* Karsch (1.4%); Mediterranean fruit fly (medfly), *Ceratitis capitata* (Wiedemann) (1.0%); Natal fruit fly, *Ceratitis rosa* Karsch (0.8%); and *Ceratitis millicentae* Meyer and Copeland (0.6%). The other species represented less than 0.5% of all the specimens collected (Table 24.1). It was the first time that *Z. cucurbitae* (formerly placed in *Bactrocera*; see De Meyer et al. 2015a) and *C. millicentae* had been reported in large numbers. Both *B. dorsalis* and *Z. cucurbitae* are devastating pests that can have a severe impact on the horticultural industry, jeopardizing both local and export markets, and consequently negatively affecting the overall livelihood of the producers and the natural economy at large.

Bactrocera dorsalis was the most abundant fruit fly species at most of the trapping sites, indicating that it has become the dominant species in the areas where it occurs, although this could be attributed in part to the fact that methyl eugenol is a more powerful attractant than the lures used for the other species (Shelly 2001). However, in the Northern provinces, *B. dorsalis* was also the most abundant fruit fly emerging from sampled fruits; for example, in Pemba, *B. dorsalis* accounted for 71.5% of all the fruit fly adults that emerged from infested mango fruits (Cugala et al. 2013a). In another study José et al. (2013) also reported that, of all adults emerging from mango, *B. dorsalis* was the most abundant (96.9%). Similar observations have been made in Kenya and Tanzania (Ekesi et al. 2009; Mwatawala et al. (2009a) and the authors reported competitive displacement of the native *Ceratitis* species by *B. dorsalis*.

Table 24.1 Fruit fly species composition and abundance in Mozambique

Fruit Fly species	Total samples	Abundance (%)
Zeugodacus cucurbitae (Coquillett)	2162	1.31
Bactrocera dorsalis (Hendel)	142036	86.03
Capparimyia aristata De Meyer and Freidberg	1	0.00
Carpophthoromyia dimidiata Bezzi	2	0.00
Carpophthoromyia sp.	1	0.00
Celidodacus sp.	2	0.00
Ceratitis anonae Graham	1	0.00
Ceratitis bremii Guérin-Méneville	6	0.00
Ceratitis capitata (Wiedemann)	1504	0.91
Ceratitis cosyra (Walker)	3952	2.39
Ceratitis ditissima (Munro)	19	0.01
Ceratitis edwardsi (Munro)	1	0.00
Ceratitis fasciventris (Bezzi)	625	0.38
Ceratitis millicentae De Meyer and Copeland	880	0.53
Ceratitis pedestris (Bezzi)	2	0.00
Ceratitis punctata (Wiedemann)	568	0.34
Ceratitis quinaria (Bezzi)	461	0.28
Ceratitis rosa Karsch	1192	0.72
Clinotaenia superba Bezzi	2	0.00
Dacus brevis Coquillett	5	0.00
Dacus africanus Adams	2	0.00
Dacus bivittatus (Bigot)	7636	4.62
Dacus chiwira Hancock	16	0.01
Dacus ciliatus (Loew)	517	0.31
Dacus durbanensis Munro	22	0.01
Dacus eclipsis (Bezzi)	2	0.00
Dacus eminus Munro	1	0.00
Dacus flavicrus Graham	1	0.00
Dacus frontalis Becker	44	0.03
Dacus fuscatus Wiedemann	2	0.00
Dacus fuscovitattus Graham	119	0.07
Dacus hamatus Bezzi	15	0.01
Dacus kariba Hancock	2	0.00
Dacus lounsburyii Coquillett	44	0.03
Dacus opacatus Munro	5	0.00
Dacus pallidilatus Munro	3	0.00
Dacus plagiatus Collart	2	0.00

(continued)

Table 24.1 (continued)

Fruit Fly species	Total samples	Abundance (%)
Dacus pullescens Munro	1	0.00
Dacus punctatifrons Karsch	2210	1.34
Dacus vertebratus Bezzi	304	0.18
Dacus famona Hancock	5	0.00
Perilampsis curta Munro	1	0.00
Trirhithrum sp.	726	0.44
Total	**165,102**	**100**

[a]*Ceratitis rosa* was recently considered to be a complex of two cryptic species (see De Meyer et al. 2015b). The distinction between the two types was not made during identification of this material

In Mozambique, the dominance of *B. dorsalis* seems to be directly related to the length of time that it had been present in the area. In the northern region where *B. dorsalis* had been present for the longest time it accounted for 96.9% of all tephritid adults emerging from fruit (Jose et al. 2013). In the central region of the country (Manica Province), where *B. dorsalis* had been present for an intermediate length of time, it emerged from fruit in similar proportions to *C. capitata*: 40.6% of emerging fruit flies (c.f. 59.4% *C. capitata*) and 58.8% (c.f. 41.2% *C. capitata*) in the studies of Majacunene (2014) and Moiane (2015), respectively. In the southern region, where the invasion of *B. dorsalis* is fairly recent (late 2011 and early 2012), *C. cosyra* outnumbered (85.8%) *B. dorsalis* (Cugala et al. unpublished data). However, based on the experiences of other African countries, it is anticipated that *B. dorsalis* populations will establish throughout Mozambique and, eventually dominate the native *Ceratitis* species in shared host fruits species. Ekesi et al. (2009), observed that within 4 years of invasion, *B. dorsalis* has displaced *C. cosyra* and has become the predominant fruit fly pest of mango in Kenya; it represented 98 and 88% of the total population of fruit flies sampled in traps and from mango fruit, respectively, at Nguruman, Kenya. These authors argued that there were two possible mechanisms responsible for the displacement; resource competition by larvae within the fruit and aggression behaviour between adult flies. Duyck et al. (2007), observed that, in general, *Bactrocera* species were predominant over *Ceratitis* species in La Réunion Island and suggested that this was due to the K-selected traits of species belonging to the genus *Bactrocera*. Pre-invasion data indicate that native *Ceratitis* species (*C. capitata*, *C. cosyra* and *C. rosa*) and *Dacus* species were the most dominant fruit flies in Mozambique (Mausse and Bandeira, 2007) but it is likely that *B. dorsalis* will have an impact on these native species and ultimately predominate.

Although *Z. cucurbitae* was only detected for the first time in 2013 in the Mocimboa da Praia and Palma districts of the northern Province of Cabo Delgado (Cugala et al. 2013b), subsequent surveys conducted by Omar (2014) already report high densities in the same locations (flies/trap/day [FTD] = 16.8 in Palma and FTD = 47.0 in Mocimboa da Praia). The fact that, so far, *Z. cucurbitae* has only been detected at these two sampling sites could indicate that its occurrence is still localized or isolated in these areas; no further specimens of *Z. cucurbitae* have been recorded in other trapping sites from Pemba to Namoto (Rovuma River at the border with Tanzania).

Apart from the dominant invasive species (*B. dorsalis* and *Z. cucurbitae*), six native fruit fly species, *D. bivittatus*, *D. punctatifrons*, *C. cosyra*, *C. capitata*, *C. rosa* and *C. millicentae* are the species that are most frequently observed in traps and/or fruits and thus, considered of economic importance (Table 24.1).

3 Potential Host Fruits for Infestation by, and Development of, Fruit Flies in Mozambique

The host range of fruit flies in Mozambique was evaluated by sampling hosts and potential hosts of the pest (Table 24.2). A total of 37 different host plant species, including cultivated and wild fruit species, were sampled. Cultivated host plants included mango (*Mangifera indica* L.), tomato (*Solanum lycopersicum* L.), cucumber (*Cucumis sativa* L.), sweet orange (*Citrus sinensis* (L.) Osbeck), guava (*Psidium guajava* L.), custard apples (*Annona* spp.*)*, Japanese persimmon (*Diospyros kaki* Thunb.), loquat (*Eriobotrya japonica* (Thunb.) Lindl.), and starfruit (*Averrhoa carambola* L.). Amongst the wild hosts are tropical almond (*Terminalia catappa* L.*)*, Pepper-bark (*Warburgia salutaris* [Bertol.f.] Chiov.), rubber vine (*Landolphia kirkii* Dyer), wild mango (*Cordyla africana* Lour.) and marula (*Sclerocarya birrea* [A. Rich] Hochst.) (Table 24.2). The sampled fruits were incubated in the laboratory for fruit fly emergence, and the emerged flies were later identified. In order of abundance *B. dorsalis*, *C. cosyra* and *D. bivittatus* were the species that emerged from most of the cultivated and wild host plants (Table 24.2). *Dacus bivittatus* was reared mainly from cucurbits but a few also emerged from mango. Similar observations were made in Tanzania by Mwatawala et al. (2006), who stated that infestation by native fruit flies, such as *C. capitata* and *C. cosyra*, was minor compared to *B. dorsalis*. *Bactrocera dorsalis* is a polyphagous species with a wide host range and has, to date, been recorded from more than 40 host plant species belonging to 13 plant families in Africa (Vayssières et al. 2005; Mwatawala et al. 2006; Rwomushana et al. 2008), with mango, guava and tropical almond being the most preferred.

The extensive distribution and availability of both the main hosts of *B. dorsalis* (mango and guava) and the alternative hosts of *B. dorsalis* (tropical almond and wild mango), are amongst the most important factors contributing to the abundance of this fruit fly species (Table 24.3). Tropical almond has also been reported to be an important alternative host fruit species for *B. dorsalis* in Kenya and Tanzania (Mwatawala et al. 2006; Rwomushana et al. 2008). Pepper-bark and marula are reported as important reservoir hosts for *C. cosyra* and may be contributing to the buildup of this pest in Mozambique (Muatinte and Cugala 2014; Muanacoda 2015). Similar results were reported in Swaziland by Magagula and Ntonifor (2014) who

Table 24.2 Potential host fruit species that support development of fruit flies in Mozambique

Common name	Scientific name	B. dorsalis	Z. cucurbitae	C. cosyra	R. rosa	C. capitata	D. bivittatus	D. punctatifrons
Custard apple (sweetsop)	Annona cherimola Mill.			x		x		
Custard apple (soursop)	Annona muricata L.			x		x		
Sugar apple	Annona squamosa L.			x		x		
Starfruit	Averrhoa carambola L.	x		x				
Chilli	Capsium frutescens L.	x						
Papaya	Carica papaya L.	x						
Watermelon	Citrullus lanatus (Thunb.) Matsum. and Nakai							
Lemon	Citrus limon (L.) Burm.f.	x						
Grapefruit	Citrus paradisi Macfad.	x						
Mandarin	Citrus reticulata Blanco	x						
Orange	Citrus sinensis L. Osbeck	x						
Wild mango (Cordyla)	Cordyla africana Lour.	x		x				
Melon	Cucumis melo L.							
African horned cucumber	Cucumis metuliferus E. Mey							
Cucumber	Cucumis sativus L.						x	x
Pumpkin	Cucurbita pepo L.						x	x
Persimmon	Diospyros kaki Thunb.	x						
Loquat	Eriobotrya japonica (Thunb.) Lindl.	x						
African mangosteen	Garcinia livingstonei T. Anderson	x		x		x		

(continued)

Table 24.2 (continued)

Common name	Scientific name	B. dorsalis	Z. cucurbitae	C. cosyra	R. rosa	C. capitata	D. bivittatus	D. punctatifrons
Rubber vine	Landolphia kirkii Dyer				x			
Lychee	Litchi chinensis Sonn.							
Tomato	Solanum lycopersicum L.	x		x				
Mango	Mangifera indica L.	x		x			x	
Banana	Musa paradisiaca L.	x						
Avocado	Persea americana Mill.							
Peach	Prunus persica (L.) Batsch							
Guava	Psidium guajava L.	x		x		x		
Pomegranate	Punica granatum L.	x		x				
Marula	Sclerocarya birrea (A. Rich) Hochst	x		x				
Silver-leaf bitter apple	Solanum elaeagnifolium Cav.							
Aubergine	Solanum melongena L.							
Ambarella	Spondias dulcis L.	x		x				
Tropical almond	Terminalia catappa L.	x		x				
Sugar plum	Uapaca kirkiana Mull. Arg.							
African medlar	Vangueria infausta Burch.					x		
Pepper-bark tree	Warburgia salutaris (Bertol.f.) Chiov.			x				
Chinese date	Ziziphus mauritiana Lam.							

Table 24.3 Fruiting and maturation period of major fruit fly host plants in Mozambique

Host plant species		Fruiting and maturation period											
Scientific name	Common name	January	February	March	April	May	June	July	August	September	October	November	December
Citrus sinensis	Orange					x	x	x	x	x			
Citrus limon	Lemon					x	x	x	x	x			
Citrus reticulata	Mandarin					x	x	x	x	x			
Mangifera indica	Mango	x	x								x	x	x
Musa sp.	Banana	x	x	x	x	x	x	x		x	x	x	x
Psidium guajava	Guava		x	x	x								
Carica papaya	Papaya	x	x	x	x	x	x	x	x	x	x	x	x
Syzygium guineense	Waterberry		x	x	x	x							
Persea americana	Avocado		x	x	x	x							x
Sclerocarya bierra	Marula	x	x	x									
Annona squamosa	Sugar-apple	x	x	x									x
Terminalia catappa	Tropical almond	x	x	x								x	x
Capsium frutenscens	Chilli	x	x	x	x	x	x	x	x	x	x	x	x
Citrulus lanatus	Water melon		x	x	x	x							
Curcubita spp.	Cucurbits	x	x	x	x	x	x	x	x	x	x	x	x
Warburgia salutaris	Pepper-bark tree									x	x	x	x

observed that *C. cosyra* was the only species that emerged from marula fruits and they argued that this wild plant was a reservoir for fruit flies, particularly during the off-season when fruits on commercial host plants were absent.

The species composition of available host plant is an important consideration in the development of IPM measures for fruit flies, especially for the highly polyphagous species such as *B. dorsalis* since the presence of a diversity of alternative host plants will always contribute to population growth when the primary hosts are absent.

4 Invasion, Establishment and Spread of the Oriental Fruit Fly, *Bactrocera dorsalis* (Diptera: Tephritidae) in Mozambique

The first detection of *B. dorsalis* in Mozambique was in Cuamba district at Northern Province of Niassa. Thereafter, the pest was reported at Vanduzi-Manica in August 2008 (Fig. 24.1a), and Pemba-Miese and Alua-Nampula in October 2008 (Fig. 24.1a and b). Subsequent monitoring activities revealed the spread of *B. dorsalis* southwards with further detections in central and southern regions, albeit at low densities in Nampula and Zambézia Provinces. Later in 2009 and early 2010, *B. dorsalis* was detected in the central Provinces of Tete, Manica and Sofala (Fig. 24.1c). Although the population level was at a low density at the beginning, it rapidly increased as the availability of preferred host fruits such as mango and guava increased (Cugala 2011).

In 2009 the information regarding distribution and abundance of *B. dorsalis* in Mozambique was used to divide the invasion range into three zones of infestation (Fig. 24.2): zone A) corresponding to the northern region (north of the Zambezi) where infestation was very high; B) the central region (between the Save and Zambezi Rivers) where infestation was very low (area of low pest prevalence) and; C) the pest free area (South of the Save River) from which trade was permitted (Fig. 24.2). Due to the absence of *B. dorsalis* in the southern region, the Phytosanitary Authorities of Mozambique declared the southern region as a pest-fruit fly-free area in 2009. Trade to neighbouring countries was not permitted from zones A and B. Surveillance programmes in 2011 detected a few *B. dorsalis* in the northern part of one of the southern provinces, Inhambane (Fig. 24.1d), stimulating Cugala et al. (2011) to argue that the presence or absence of *B. dorsalis* needed to be continuously monitored in order to maintain the status of the pest-free areas in the southernmost provinces (Inhambane, Gaza, Maputo) and facilitate horticultural export from these provinces. The pest-fruit fly-free area status in the southern most regions remained in force for two years. However, during the latter part of 2011 samples of *B. dorsalis* were collected from traps in the

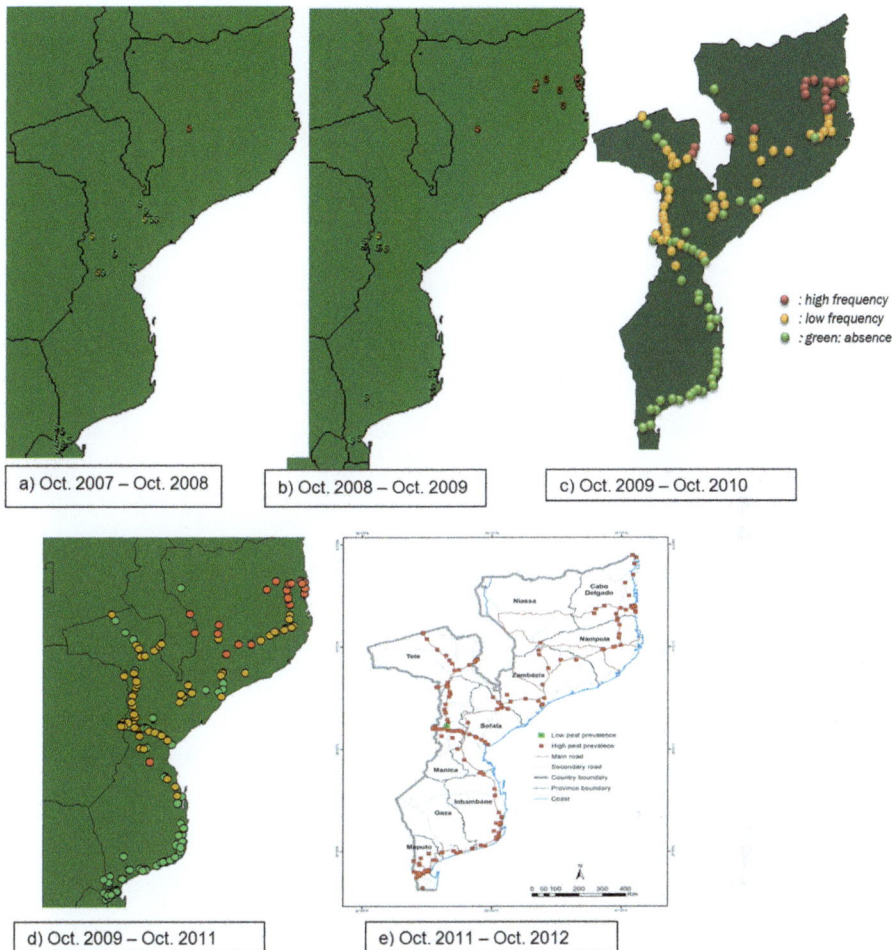

Fig. 24.1 Trapping sites in Mozambique for *Bactrocera dorsalis* showing presence or absence during four successive years and its spread (*red*: high frequency; *orange*: low frequency; *green*: absence) (**a**) Oct. 2007 – Oct. 2008. (**b**) Oct. 2008 – Oct. 2009. (**c**) Oct. 2009 – Oct. 2010. (**d**) Oct. 2009 – Oct. 2011. (**e**) Oct. 2011 – Oct. 2012

northern part of Inhambane (Malovane, Inhassoro, Vilanculo, Pambara) and in Madendere, Gaza Province and subsequently at several other sampling sites. The invasion of *B. dorsalis* continued to expand and more detections were made in Boane district, Maputo Province in early 2012.

Results of a pest risk assessment process suggested that invasion, spread and establishment of *B. dorsalis* in the southern region of Mozambique was extremely likely because the climate was highly suitability for pest development and there was

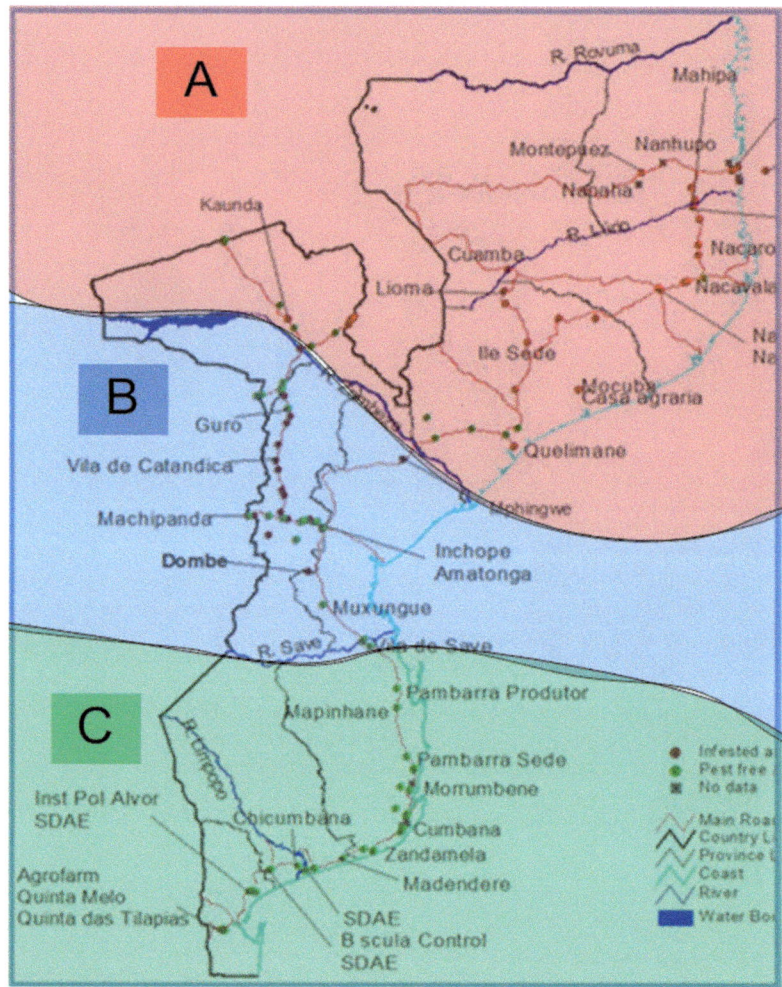

Fig. 24.2 *Bactrocera dorsalis* pest status in Mozambique until the season from October 2009 – October 2010: *Zone A* – infested area; *zone B* – low pest prevalence (proposed buffer zone); *zone C* – pest free status

a diversity of potential host plants available (see De Meyer et al. 2010). These are ideal conditions for the southward migration of *B. dorsalis* which would endanger the pest-fruit fly-free status of the southern region and could lead to losses in trade (Cugala et al. 2011). Currently, *B. dorsalis* is well established and widespread at all the trapping and sampling sites across the country (Fig. 24.1e); it occurs at high densities and is the predominant fruit fly species (Jose et al. 2013; Cugala et al. unpublished data).

5 Host Fruit Infestation and Damage Caused by *Bactrocera dorsalis*

Although the level of damage by different fruit fly species varies depending on host fruit species, season and location, no other fruit fly species has reached greater economic importance than *B. dorsalis*. At the peak of the mango season it is the most heavily infested fruit species (Cugala et al. 2013a); however, by the end of the mango season guava becomes the most heavily infested fruit species (Jose et al. 2013). In the absence of these two host species, tropical almond is the most heavily infested fruit. In Pemba, José et al. (2013) observed that the percentage of fruits that were damaged ranged from 36.7 to 92.5 %, with guava fruits being the most heavily damaged (92.5 %) due to their availability in the field, followed by tropical almond (67.3 %) and then mango (56.5 %). In the same area, Cugala et al. (2013a), reported that the percentage of fruits damaged was greatest on mango (94.0 %), followed by guava (20.3 %) and finally tropical almond (6.5 %). Similar results for differential host preference by *B. dorsalis* has also been reported in other African countries (Vayssières et al. 2008; Mwatawala et al. 2006; Ndiaye et al. 2008; Rwomushana et al. 2008). When expressed as the number of emerging adults/kg of fruit, the level of *B. dorsalis* infestation was 157.2 and 141.6 adults/kg on tropical almond and guava, respectively, at most locations in the northern provinces (Jose et al. 2013). In Tanzania and Kenya, *B. dorsalis* infestation rates of >100 adults/kg have been reported in mango and tropical almond (Mwatawala et al. 2006, 2009b; Rwomushana et al. 2008). In the central region (Manica Province), Kazuru (2014) reported *B. dorsalis* infestation rates of 313.8 adults/kg of fruit, while in the southern region (Maputo Province), infestation rates of 21.0 adults/kg of fruit were reported.

6 Economic Losses and Socio-economic Implications of *Bactrocera dorsalis* Invasion

Production and export of fresh fruit and vegetables in Mozambique have been very seriously hampered by the arrival of *B. dorsalis*. It is estimated that the horticulture sector in Maputo (South) and Manica (Central) Provinces alone could generate revenues of more than US$ 25 million per year through both commercial and smallholder (family) production. South Africa, the largest destination market for horticultural produce from Mozambique is estimated to be worth around US$ 20 million annually. However, with the introduction of *B. dorsalis*, exports to South Africa from the northern and central provinces were halted completely. During the early invasion phase, the Ministry of Agriculture and Food Security reported that several fresh fruit businesses had closed and that workers had been sent home because of the market reaction to the incursion. The foreign exchange value of bananas alone to Mozambique is about US$ 17.5 million and currently 35,000 tons of the highest grade of green bananas are exported annually. A temporary 3-week ban on exports of fresh fruit (including bananas and mangoes) from southern Mozambique was imposed by South Africa as a reaction to the presence of *B. dorsalis*; South Africa is a key trading partner and so this resulted in losses of US$2.5 million (Cugala et al. 2011; Jose et al. 2013). More than US$ 1.5 million were lost by the Vanduzi Company from the Central Province of Manica due to the presence of *B. dorsalis* and the subsequent quarantine restrictions imposed (Cugala 2011). For many commercial farmers in the central and southern regions of Mozambique, the production and sales of fruit products, during the period of 2008 to 2010, were significantly reduced and investments in the fruit sector were suspended causing annual losses of US$23 million (Tostão et al. 2012). Even though exports of fruits, such as mango, to South Africa resumed in 2009, only fresh fruits intended for processing are permitted which have a lower value compared with fresh fruits sold for immediate consumption. Overall, the impact of *B. dorsalis* has affected the income and livelihood of millions of families that produce and sell fresh fruit and vegetables in Mozambique.

7 Integrated Pest Management Strategies for Suppressing *Bactrocera dorsalis* Populations

Efforts to suppress *B. dorsalis* have been largely based on the use of integrated pest management (IPM) approaches. For example, farmers currently rely on a combination of techniques that include the use of fruit fly baits, commercially available as GF-120NF (Success Appat), combined with the insecticide spinosad, male annihilation techniques, and orchard sanitation. Studies conducted in mango orchards at Pemba, Northern Province of Cabo Delgado in 2010 (Cugala et al. 2012), in Maputo in 2015 (Canhanga et al. unpublished data) and Manica area in 2013 (Majacunene 2014), evaluated the efficacy of these IPM measures and adult *B. dorsalis* population densities were monitored on a monthly or weekly basis using traps baited with methyl eugenol.

Results showed that the population density of *B. dorsalis* in the treatment sites (with IPM strategies) in both studies was much lower compared to that in the control sites (without management strategies) (Fig. 24.3a, b and c). At the beginning of the mango season and prior to the application of the management strategy, *B. dorsalis* population densities were almost the same at both the treated and control sites. However, as the season advanced, a significant reduction in the population density of *B. dorsalis* was observed in the IPM sites compared with the control sites (Cugala et al. 2012; Majacunene 2014; Canhanga et al. unpublished data). There was a mean of 0.96 FTD (flies per trap per day) in the IPM plots compared with 11.75 FTD in the control plots in Maputo (Fig. 24.3b), while in Manica there was a population density of 0.40 FTD in IPM plots compared with 8.50 FTD in untreated control plots, and in Pemba, IPM treated plots accounted for 0.8 FTD while 5.8 FTD was reported in the control plots. These results demonstrated the impact of the IPM strategy in the management of *B. dorsalis*. In Pemba, the total number of *B. dorsalis*

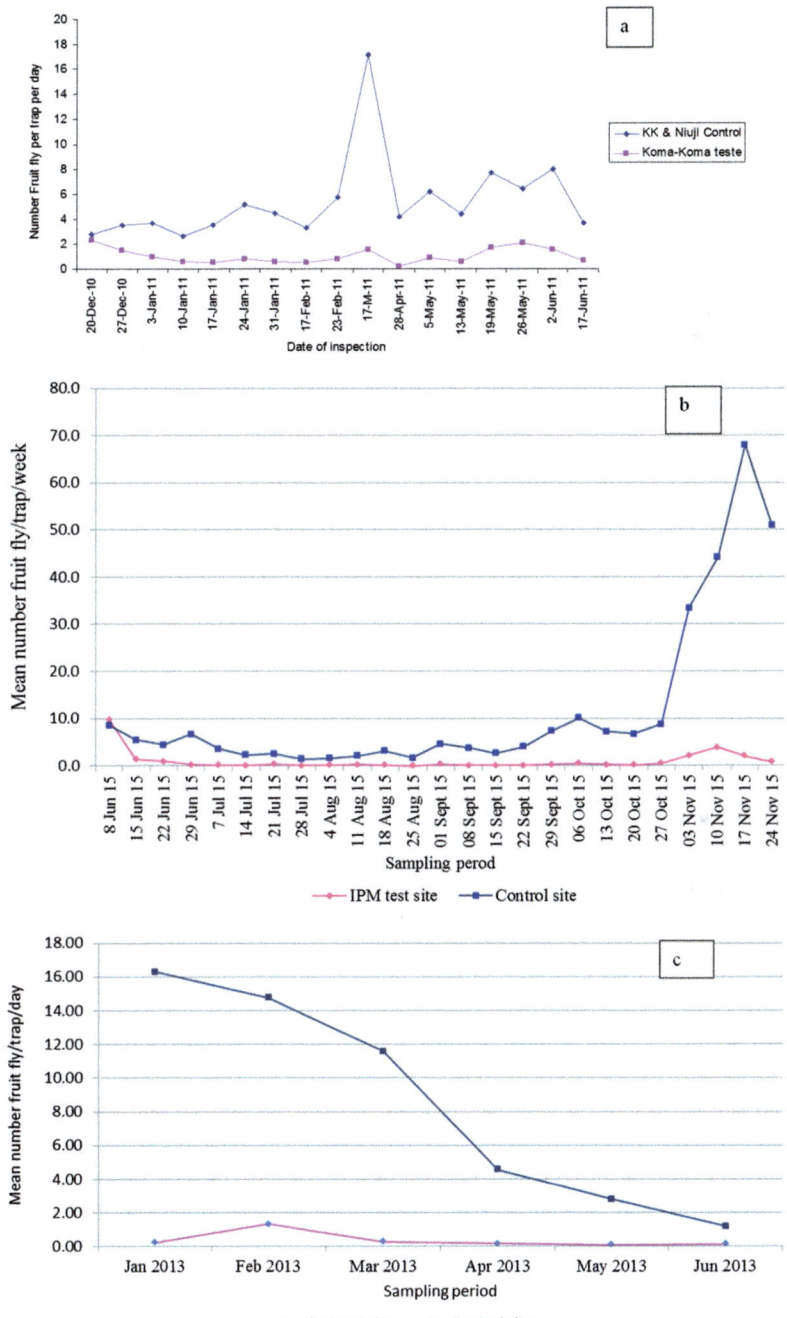

Fig. 24.3 Mean numbers of *Bactrocera dorsalis* per trap per day (FTD) in IPM test sites and control plots during the period of study: (**a**) Pemba 2010–2011; (**b**) Maputo 2015; (**c**) Manica 2013

captured during the study period was 6915 and 450 individuals in the control and IPM sites, respectively, corresponding to a 93.5 % population reduction. In Maputo, the number of *B. dorsalis* captured during the study period was 1930 and 127 in the control and IPM sites, respectively, corresponding to a 93.8 % population reduction. In the Manica Province, the total number of *B. dorsalis* captured during the study period was 10245 and 820 in the control and IPM sites, respectively, corresponding to a 92.6 % population reduction. In Senegal, Ndiaye et al. (2008) achieved 83 % *B. dorsalis* population reduction by implementing a similar IPM package in mango orchards.

Ekesi et al. (2007) reported up to a 79 % reduction in *B. dorsalis* populations using the protein bait, Nu-Lure Insect Bait, mixed with spinosad and soil inoculation of *M. anisopliae*. Additional studies by Ekesi et al. (2011) showed that during the 2006/2007 Kenyan mango season, average post-treatment catches of *B. dorsalis* were four times higher in control orchards than in the orchards that were treated with *M. anisopliae* and the GF-120 plus spinosad bait spray. The results of the trials in Mozambique as well as those conducted in other parts of Africa, suggest that *B. dorsalis* populations can be effectively suppressed locally in the growers' fields using a combination of available IPM components that include: protein bait sprays, field sanitation, biopesticides and male annihilation techniques with methyl eugenol as the attractant.

8 Release and Establishment of the Parasitoids, *Fopius arisanus* and *Diachasmimorpha longicaudata*, as Biological Control Agents Against *Bactrocera dorsalis*

Several biological control agents including hymenopteran parasitoids, entomopathogenic fungi and predatory ants are considered suitable for use in biological control of fruit flies (Vargas et al. 2012). Cugala et al. (2013a) stated that classical biological control was viewed as a potentially eco-friendly approach for suppression of *B. dorsalis* population in Mozambique, as the pest is an alien invasive and lacks resident efficient natural enemies. Parasitoids, in particular the egg parasitoid, *Fopius arisanus* (Sonan), and the larval parasitoid, *Diachasmimorpha longicaudata* (Ashmead) are credited with substantial success in suppressing *B. dorsalis* in other parts of the world (Vargas et al. 2012). Therefore, these two species were introduced into Mozambique from cultures produced by the International Centre of Insect Physiology and Ecology (*icipe*), for release in Kenya; the parasitoids had been established initially from material that had been introduced in Hawaii. Both parasitoid species were released in Niuji (S12^0 07'39 .2"; E040^0 26'08 .6", 35 m asl), Mieze, Mituge district, Cabo Delgado Province, for suppression of *B. dorsalis* populations. *Fopius arisanus* was released for the first time in 2010, while *D. longicaudata* was first released in September 2013. A total of 25,000 *F. arisanus* and 8,000 *D. longicaudata* were released. Thereafter, parasitoid establishment was assessed by sampling fruits from the release sites and their surroundings and determining the proportion of fruit flies that were parasitized (Cugala et al. 2013a). The results revealed that both parasitoids had successfully established as both species were recovered at the release site, first in 2013 for *F. arisanus* and in 2014 for *D. longicaudata*. The recovery of both parasitoid species at the release sites indicated that they were able to successful adapt to the local environmental conditions and colonize the release areas. However, the levels of parasitism are still low: 0.5% for *D. longicaudata* and 1.7% for *F. arisanus* in 2014 (Cugala et al. 2014). This situation is expected in classical biological control because the parasitoids may need some time to fully adapt to local environmental conditions before their populations fully establish, grow and exert a significant impact on the target pest population.

These parasitoid species are very promising as biological control agents of *B. dorsalis* and have been released in various other African countries including Kenya, Tanzania, Benin, Senegal and Uganda (Ndiaye et al. 2015; Muriithi et al. 2016; Hanna et al. 2008); parasitism rate by *F. arisanus* of up to 40% of *B. dorsalis* has been reported in Kenya (Ekesi et al. 2010).

9 Effect of Ripeness on Suitability of Cavendish Dwarf Bananas as Hosts for *Bactrocera dorsalis*

Banana is one of Mozambique's most important export crops accounting for 27% of all agricultural products in 2012 (INE 2015). Banana exports from the Provinces of Maputo and Manica can generate about ~US$ 20 million annually (Cugala et al. 2011). Due to the presence of *B. dorsalis*, banana importation from Mozambique has been banned by several importing countries, including the main trade partners, South Africa and Zimbabwe; this has caused enormous economic losses to producers. To overcome these quarantine restrictions and mitigate the impact of *B. dorsalis* and ensure access to international markets, a study was conducted in Mozambique to assess the relative infestation levels by *B. dorsalis* of field collected and artificially infested Cavendish dwarf banana cultivar at different stages of ripeness.

Although oviposition wounds were observed on green bananas after artificial infestation, no fruit flies were recovered from fruits harvested at the green stage. Based on fruit infestation data collected from the field and artificial fruit infestation, the results revealed that mature green Cavendish dwarf bananas do not support *B. dorsalis* development. However, *B. dorsalis* adults were recovered from bunches of green bananas that contained precociously fully ripe (yellow) fruits (Cugala et al. 2013c). Therefore, banana fruits destined for export at the green stage of harvest maturity do not need to be subjected to quarantine restrictions, unless bunches have precociously ripened fingers or other damage such as cracks and splits, ant burns, abrasions, point bruises, tip rot or general decay; these latter cases should be carefully inspected before export and discarded because such damage could permit attack by *B. dorsalis* and pose a risk of inadvertent translocation of *B. dorsalis* during export (Cugala et al. 2013c).

Based on these results, mature green Cavendish dwarf banana from Mozambique are now being exported to lucrative markets, mainly to South Africa. As a consequence, the period for the necessary 'import permit' to export bananas to South Africa was extended from 3 months to 12 months (after the publication of the study results). About 15 companies dedicated to banana production are now exporting their produce to South Africa without quarantine restrictions. In Maputo Province, it is estimated that the volume of production has increased from 35,000 to 70,000 tons annually representing an income generation of US$ 40 million annually and ensuring employment for over 5,000 people (Tostão et al. 2012).

10 Postharvest Treatment Measures for *Bactrocera dorsalis*

The use of post-harvest phytosanitary treatments is viewed as a potential strategy to overcome export restrictions, mitigate for the impact of *B. dorsalis* and regain access to regional and international export markets for Mozambican fresh fruit and vegetables. Due to the quarantine nature of fruit flies, importing countries require treatments that ensure fruits are free of *B. dorsalis*.

Hot water treatment of mango fruits is an acceptable method for meeting quarantine requirements for fruit flies dis-infestation in other countries (Verghese et al. 2011; Guy et al. 2012). For this reason they were implemented in Mozambique to permit export of mango by producers in Mozambique to the South African market. Two commercial mango cultivars namely, Tommy Atkins and Kent are currently treated using hot water before export to South Africa where the fruits are utilized only for processing into juices. Mature green mango fruits are treated in a hot water bath for 12 min at a minimum core temperature of 47 °C to limit the risk of infestation and introduction of *B. dorsalis* into the importing country. This procedure has been a success since, so far, there has been no single notification of detection of *B. dorsalis* from the authorities of the importing country. In addition to the hot water treatment, the South African authorities request that a fruit fly prevention action plan must be in place at the mango production site, based on a series of pre-harvest management measures (including fruit sanitation and bait spray with GF-120); these activities are common practice by growers exporting mango to South Africa. Studies conducted in West Africa by Self et al. (2012) revealed that a hot water treatment resulting in a core temperature of 46.5 °C could be the basis of a fruit fly quarantine treatment for West African mangoes. This treatment was effective in killing fruit fly eggs and larvae in Kent mangoes, although Probit 9 efficacy for the treatment of fruit flies (Diptera: Tephritidae) required 99.9968 % mortality after a treatment, quarantine security could not be achieved, requiring further research. Although mango producers have been using the Probit 9 quarantine parameter provided by trading partners, alongside other measures, as a pre-requisite for market access further research on postharvest hot water treatment is urgently needed in Mozambique. Such research should not just be for mangoes but other export fruits and vegetables and must involve the rigorous collection of data under local conditions to ensure quarantine security and meet the needs of quarantine-sensitive markets. If successful, it is believed that postharvest treatment protocols should increase market access for various fruits and vegetables produced in Mozambique not only for processing but also for fresh consumption without any additional measures.

Acknowledgements The present study was jointly supported by the Food and Agriculture Organization of the United Nations (FAO, Project TCP/ MOZ/3205), the World Bank Group (Contract No. 7152761), National Research Fund (FNI), Sustainable Irrigation Development Project (PROIRRI) and Eduardo Mondlane University Research Funds under the fruit fly programme in Mozambique. Special thanks to USDA (USDA/APHIS office Pretoria, South Africa), Royal Museum for Central Africa (Tervuren, Belgium) and International Centre of Insect Physiology and Ecology (*icipe*) for providing fruit fly collecting materials and technical assistance. We are sincerely grateful to the farmers for their collaboration and support in the surveys. The trapping work would not have been possible without the assistance of local technicians. Thanks also to all members of the trapping network for their invaluable assistance with field work and data collection. We are especially grateful to the staff at Department of Plant Protection of the Ministry of Agriculture for their collaboration in data collection.

References

Correia ARI, Rego JM, Olmi M (2008) A pest of significant economic importance detected for the first time in Mozambique: *Bactrocera invadens* Drew, Tsuruta & White (Diptera: Tephritidae: Dacinae). Boll Zool Agr Bachic Serie II 40:9–13

Cugala DR (2011) Management and mitigation measures for alien invasive fruit fly *(Bactrocera invadens)* in Mozambique. Terminal statement prepared for the Government of Mozambique and the Department of Plant Protection by the Food and Agriculture Organization of the United Nations, Technical Cooperation Programme, TCP/ MOZ/3205, 45 pp

Cugala, DR, Mansell M, De Meyer M (2011) *Bactrocera invadens* surveys in Mozambique. Fighting Fruit flies regionally in Sub-Saharan Africa. Information letter no.1. Page 3.

Cugala DR, José L, Cocorrea H, Ekesi S (2012) Assessing the impact of integrated pest management programme for suppressing the invasive fruit fly *(Bactrocera invadens)* on mango in Mozambique. Third RUFORUM Biennial Meeting, Entebbe, pp 301–306

Cugala DR, Cumbe R, Wanyonyi M, Ekesi E, Ambasse D, Omar R (2013a) Assessment of the establishment of the egg parasitoid, *Fopius arisanus*, as biological control agent against the invasive fruit fly, *Bactrocera invadens* in the Northern Mozambique. Trip report: 15 to 28 September 2013, report submitted to Agrifuturo/USAID, 13 pp

Cugala DR, Cambula E, Ambasse D (2013b) Fruit fly invasive species surveillance in the Northern Cabo Delgado province, Mozambique. Trip report: 29 July to 03 August 2013, report submitted to Agrifuturo/USAID, Maputo, Mozambique, 06 pp

Cugala DR, Ekesi S, Ambasse D, Adamu RS, Mohamed SA (2013c) Assessment of ripening stages of Cavendish dwarf bananas as host or non-host to *Bactrocera invadens*. J Appl Entomol 138:393–473

Cugala DR, Madogolele N, Ambasse D (2014) Assessment of the establishment of the parasitoids, *Fopius arisanus* and *Diachasmimorpha longicaudata* as biological control agents against the invasive fruit fly, *Bactrocera invadens* in Mieze. Trip report: 09 to 14 September 2014, report submitted to Agrifuturo/USAID, Maputo, Mozambique, 11 pp

De Meyer M, Robertson MP, Mansell MW, Ekesi S, Tsuruta K, Mwaiko W, Vayssières J-F, Peterson AT (2010) Ecological niche and potential geographic distribution of the invasive fruit fly *Bactrocera invadens* (Diptera, Tephritidae). Bull Entomol Res 100:35–48

De Meyer M, Delatte H, Mwatawala M, Quilici S, Vayssières J-F, Virgilio M (2015a) A review of the current knowledge on *Zeugodacus cucurbitae* in Africa. ZooKeys 540:539–557

De Meyer M, Delatte H, Ekesi S, Jordaens K, Kalinova B, Mwatawala M, Steck G, Van Cann J, Vanickova L, Brizova R, Virgilio M (2015b) An integrative approach to unravel the *Ceratitis* FAR cryptic species complex. ZooKeys 540:405–427

Duyck PF, David P, Quilici S (2007) Can more *K*-selected species be better invaders? A case study of fruit flies in La Réunion. Div Distrib 13:535–543

Ekesi S, Dimbi S, Maniania NK (2007) The role of entomopathogenic fungi in the integrated management of Tephritid fruit flies (Diptera: Tephritidae) with emphasis on species occurring in Africa. In: Ekesi S, Maniania NK (ed) Use of entomopathogenic fungi in biological pest management, pp 239–274.

Ekesi S, Billah MK, Nderitu PW, Lux S, Rwomushana I (2009) Evidence for competitive displacement of *Ceratitis cosyra* by the invasive fruit fly *Bactrocera invadens* (Diptera: Tephritidae) on mango and mechanisms contributing to the displacement. J Econ Entomol 102:981–991

Ekesi S, Mohamed S, Hanna R (2010) Rid fruits and vegetables in Africa of notorious fruit flies. Techn Innov Brief 4:2

Ekesi S, Maniania NK, Mohamed SA (2011) Efficacy of soil application of *Metarhizium anisopliae* and the use of GF- 120 spinosad bait spray for suppression of *Bactrocera invadens* (Diptera: Tephritidae) in mango orchards. Biocontr Sci Technol 21:299–316

Garcia FRM, Bandeira RR (2011) Biodiversidade de Moscas-das-Frutas (Diptera, Tephritidae) em Moçambique. Revista Eletrônica Acolhendo a Alfabetização nos Países de Língua Portuguesa 5:24–44

Guy S, Ducamp MN, Thaunay P, Vayssières JF (2012) The effects of phytosanitary hot water treatments on West African mangoes infested with *Bactrocera invadens* (Diptera: Tephritidae). Fruits 67:439–449

Hanna R, Bokonon-Ganta A, Gnanvossou D, Saïzonou S (2008) Releases of *Fopius arisanus* for effective and sustainable fruit fly control in Africa. Fighting Fruit Veg Flies Reg Western Afr 6:3

INE (2010) Contas nacionais. Indicadores básicos, Maputo, Moçambique; www.ine.gov.mz. Date accessed 17 Dec 2011

INE (2015) Contas nacionais. Indicadores básicos, Maputo, Moçambique; www.ine.gov.mz. Date accessed 15 Apr 2015

Jose L, Cugala D, Santos L (2013) Assessment of invasive fruit fly infestation and damage in Cabo Delgado Province, Northern Mozambique. Afr Crop Sci J 21:21–28

Kazuru CC (2014) Avaliação do impacto da combinação de GF-120NF e TAM no controlo da mosca invasiva da fruta (*Bactrocera invadens*) na província de Manica. Projecto final submetido ao Departamento de Protecção Vegetal para a obtenção do grau de Licenciatura em Engenharia Agronómica, UEM, Maputo, 54 pp

Magagula CN, Ntonifor N (2014) Species composition of fruit flies (Diptera: Tephritidae) in feral guavas (*Psidium guajava* Linnaeus) and marula (*Sclerocarya birrea* (A. Richard) Hochstetter) in a subsistence savanna landscape: implications for their control. Afr Entomol 22:320–329

Majacunene AP (2014) Avaliação do impacto das estratégias de maneio integrado na população da mosca invasiva da fruta, *Bactrocera invadens,* na provincia de manica. Dissertação de Mestre submetida ao Departamento de Protecção Vegetal, FAEF/UEM, 68 pp

Mangana S, Cugala D (2010) Ocorrência da mosca invasiva da fruta, *Bactrocera invadens* em Moçambique. Folhas Verdes 193:1–4

Maússe SD, Bandeira RR (2007) Ecological relationships between *Ceratitis* spp. (Diptera: Tephritidae) and other native fruit tree pests in southern Mozambique. Fruits 62:35–44

Moiana LC (2015) Avaliação da dinâmica populacional da mosca invasiva da fruta (Bactrocera *invadens*) no corredor da Beira. Dissertação de Mestre submetida ao Departamento de Protecção Vegetal, FAEF/UEM, 72 pp

Muanacoda JJ (2015) Avaliação da diversidade de espécies de moscas da fruta associadas à marula (*Schlerocarya birrea*) com ênfase à *Bactrocera (=invadens) dorsalis*: Implicações para o seu

Maneio. Tese de licenciatura submetida ao Departamento de Protecção Vegetal, FAEF/UEM, 58 pp

Muatinte BL, Cugala DR (2014) Infestação e abundância de *Ceratitis cosyra* (Walker) (Diptera: Tephritidae) em *Warburgia salutaris* (Canellaceae) em Maputo, Moçambique. Rev Cien UEM Sér Ciênc Agron Flores Vet 1:4–12

Muriithi BW, Affognon HD, Diiro GM, Kingori SW, Tanga CM, Nderitu PW, Mohamed SA, Ekesi S (2016) Impact assessment of Integrated Pest Management (IPM) strategy for suppression of mango-infesting fruit flies in Kenya. Crop Prot 81:20–29

Mwatawala MW, De Meyer M, Makundi RH, Maerere AP (2006) Seasonality and host utilization of the invasive fruit fly, *Bactrocera invadens* (Dipt., Tephritidae) in Central Tanzania. J Appl Entomol 130:530–537

Mwatawala MW, De Meyer M, Makundi RH, Maerere AP (2009a) An overview of *Bactrocera* (Diptera, Tephritidae) invasions and their speculated dominancy over native fruit fly species in Tanzania. J Entomol 6:18–27

Mwatawala MW, De Meyer M, Makundi RH, Maerere AP (2009b) Host range and distribution of fruit-infesting pestiferous fruit flies (Diptera, Tephritidae) in selected areas of Central Tanzania. Bull Entomol Res 98:1–13

Ndiaye M, Dieng E, Delhove G (2008) Population dynamics and on-farm fruit fly integrated pest management in mango orchards in the natural area of Niayes in Senegal. Pest Man Hort Ecosyst 14:1–8

Ndiaye O, Ndiaye S, Djiba S, Ba CT, Vaughan L, Rey JY, Vayssières JF (2015) Preliminary surveys after release of the fruit fly parasitoid *Fopius arisanus* Sonan (Hymenoptera Braconidae) in mango production systems in Casamance (Senegal). Fruits 70:1–9

Omar R (2014) Avaliação da composição e distribuição de espécies invasivas de moscas da fruta na zona norte da província de Cabo Delgado. Universidade Lurio, Cabo Delgado, Moçambique, Tese de licenciatura, 43 pp

Rwomushana I, Ekesi S, Gordon I, Ogol CKPO (2008) Host plants and host plant preference studies for *Bactrocera invadens* (Diptera: Tephritidae) in Kenya, a new invasive fruit fly species in Africa. Anns Entomol Soc Am 101:331–340

Schutze MK, Mahmood K, Pavasovic A, Bo W, Newman J, Clarke AR, Krosch MN, Cameron SL (2015) One and the same: integrative taxonomic evidence that *Bactrocera invadens* (Diptera: Tephritidae) is the same species as the Oriental fruit fly *Bactrocera dorsalis*. Syst Entomol 40:472–486

Self G, Ducamp M, Thaunay P, Vayssières J (2012) The effects of phytosanitary hot water treatments on West African mangoes infested with *Bactrocera invadens* (Diptera: Tephritidae). Fruits 67:439–449

Shelly TE (2001) Feeding on methyl eugenol and *Fagraea berteriana* flowers increases long-range female attraction by males of the oriental fruit fly (Diptera: Tephritidae). Fla Entomol 84:634–640

Tostão E, Santos L, Manuel L, Popat M, José L, Massinga J (2012) Invasive fruit fly (*Bactrocera invadens*): Occurrence and Socio-Economic Impact In Mozambique. Report Undertaken by CEAGRE for the USAID AgriFUTURO Project. Maputo, Mozambique, 74 pp

Vargas RI, Leblanc L, Harris EJ, Manoukis NC (2012) Regional suppression of *Bactrocera* fruit flies (Diptera: Tephritidae) in the Pacific through biological control and prospects for future introductions into other areas of the world. Insects 3:727–742

Vayssières JF, Goergen G, Lokossou O, Dossa P, Akponon C (2005) A new *Bactrocera* species in Benin among mango fruit fly (Diptera: Tephritidae) species. Fruits 60:371–377

Vayssières J-F, Sinzogan A, Adandonon A (2008) The new invasive fruit fly species, *Bactrocera invadens* Drew Tsuruta & White. IITA-CIRAD Leaflet No 2

Verghese A, Nagaraju DK, Sreedevi K (2011) Hot water as an effective post harvest disinfestant for the oriental fruit fly, *Bactrocera dorsalis* (Hendel) on mango. Pest Man Hort Ecosyst 17:63–68

D.R. Cugala is an Associate Professor based at the Faculty of Agronomy and Forest Engineering, Eduardo Mondlane University, Maputo, Mozambique. He has interests in integrated pest management, particularly for invasive insect pests, post harvest pest management as well as phytosanitary treatment and biological control of agricultural pests. He has been a coordinator of national and regional projects on: monitoring and integrated pest management of fruit flies in Mozambique; impact assessment of sustainable management of the papaya mealybug, *Paracoccus marginatus*; assessment of the incidence, damage and weight loss due to larger grain borer, *Prostephanus truncatus*; biological control of cereal stem borers in Mozambique. He is currently a national technical coordinator of a fruit fly programme in Mozambique. He has published a total of 30 papers in scientific journals and proceedings. He has prepared numerous technical reports and extension material for the Mozambique Government, private institutions, development agencies and NGOs.

M. De Meyer is an entomologist attached to the Royal Museum for Central Africa in Tervuren (Belgium). Previously he worked in Botswana and Kenya for a number of years. His main field of expertise is the taxonomy and systematics of selected groups of Diptera such as Pipunculidae, Syrphidae and Tephritidae, with emphasis on the Afrotropical fauna. His research includes different aspects, from taxonomic revisions to studies on evolutionary trends and speciation events.

L.J. Canhanga is a researcher working in the Crop Protection Department, Faculty of Agronomy and Forestry Engineering, Eduardo Mondlane University in Maputo, Mozambique. She has 8 years of experience on the identification and management of pests attacking the most important crops in Mozambique. She has worked on fruit flies since 2008 on aspects related to surveillance, host range, impact assessment and integrated pest management. During this time she was also involved other projects: integrated management of ***Plutella xylostella*** (Lepidoptera: Plutellidae) based on biological control for smallholder farmers in the South and Centre of Mozambique; evaluation of coconut white fly occurrence (Homoptera: Aleyrodidae) and their natural enemies; integrated management of ***Oryctes rhinocerus*** in Zambezia Province; alternative cropping systems and integrated pest management to improve cotton yields and rural families income in Zambezia Province. She has one paper published in the African Crop Science Journal anad five abstracts in conferences procedings.

Chapter 25
Integrated Management of Fruit Flies: Case Studies from Nigeria

Vincent Umeh and Daniel Onukwu

Abstract In sub-Saharan Africa, fruits and vegetables are attacked by various pests and diseases; notable among them is the problem of tephritid fruit flies. In Nigeria, fruit fly damage has undermined the dedicated efforts of farmers to boost fruit production. Damage caused by fruit flies significantly reduces marketable fruit yields and thus affects both local and export trade. Here we describe the species composition of fruit flies of economic importance in Nigeria and their distribution. The major hosts as well as the alternative hosts of fruit flies come from many plant families including the major fruit crops such as mango, citrus, guava, papaya, cocoa, pepper, and cucurbits. Seasonal population dynamics of fruit flies on mango and citrus generally showed that species from the genus *Ceratitis* reached their highest populations during the dry seasons while *Bactrorera dorsalis* predominated during the rainy seasons. Surveys showed that fruit fly management measures used by growers were not effective. National efforts were made through workshops to increase awareness of sustainable control methods for fruit flies of economic importance. This was later supported by demonstration trials in farmers' orchards. Integrated Pest Management (IPM) options were tested for the suppression of fruit flies and included removal of dropped fruits and early harvest, application of the Male Annihilation Technique (MAT) and the Bait Annihilation Technique (BAT). Future perspectives include the introduction of exotic parasitoids as a component of IPM for fruit fly management and the application of post harvest disinfection treatments for exported fruits.

Keywords Composition • Distribution • Host plants • Dynamics • Management

---

V. Umeh (✉) • D. Onukwu
National Horticultural Research Institute,
P.M.B. 5432, Idi-Ishin, Jericho Reservation Area, Ibadan, Nigeria
e-mail: vumeha@yahoo.com

1 Background

Fruits and vegetables are important components of the daily diet in Nigeria. These commodities play a very important role in nutrition and health, as they are rich in vitamins and minerals; they also contain substances that regulate or stimulate digestion, act as laxatives or diuretics, or are pectins and phenolic compounds that play a vital role in regulating the internal pH of the digestive system (Ibeawuchi et al. 2015). Fruits and vegetables also contain phyto-lecithins that reduce skin ageing, and antioxidants that boost the human immune system. In Nigeria, enormous quantities of fruits and vegetables are produced but estimates of annual production vary. As an example, citrus production in Nigeria was estimated at 3800.000 t and covering an area of 795,000 ha in 2013 (FAOSTAT 2013).

The sub-sector makes an important contribution to the national economy of the country not only by providing income to producers and other stakeholders along the value chain, but also by providing employment; it is a major source of foreign exchange. The most important fruits produced in Nigeria include: mango, *Mangifera indica* L.; *Citrus* species; guava, *Psidium guajava* L.; papaya, *Carica papaya* L.; pineapple, *Ananas comosus*; and banana, *Musa* species. The most important vegetables grown include: tomato, *Solanum lycopersicum* L.; pepper, *Capsicum annuum* L.; okra, *Abelmoschus esculentus* (L.) Moesch; various species of cucurbits (e.g. cucumber, *Cucumis sativus* L.); onions, *Allium cepa* L.; *Amaranthus* species; Nalta jute, *Corchorus olitorius* L.; roselle, *Hibiscus sabdariffa* L.; bitterleaf, *Vernonia armygdalina*; and baobab, *Adansonia digitata* L.. Recently, the government of the Federal Republic of Nigeria has embarked on a campaign to boost fruit and vegetable production through their 'Agricultural Transformation Agenda'. This has led to a significant increase in production and a renewed interest in the establishment of processing industries, particularly the fruit juice manufacturing industry. From small beginnings in the 1950s, the fruit juice industry in Nigeria has grown, encouraged by the government's policy to reduce importation of fruit juice; local manufacturers responded positively by increasing their production quality and quantity to meet demand and prevent shortfalls in supply (Taiwo 2005).

One of the major obstacles facing the horticultural industry in Nigeria is the problem of tephritid fruit fly pests. Damage due to these pests result in economic losses, which, depending on region, locality and variety, can be as high as 40% (Umeh et al. 2008). The invasion, establishment and spread of the devastating and highly polyphagus oriental fruit fly, *Bactrocera dorsalis* (Hendel), which was initially incorrectly described as the new species, *Bactrocera invadens* Drew, Tsura and White (Ekesi et al. 2006), has worsened the situation and yield losses increased to more than 70% (Babatola 1985). Moreover, due to the quarantine nature of this pest, access to regional and international export markets have been severely hampered or completely lost due to the restrictions imposed by the importing countries. For example, new regulations (CE882/2004) have increased the efficiency of official controls on imported foodstuffs into the EU, particularly on fruit shipment likely to host quarantine insects. This has impacted negatively on the livelihood of producers and other stakeholders along the fruit and vegetable value chain. Furthermore, the national economy of the country has been negatively affected due to the loss of foreign exchange earnings.

2 Species Composition and Distribution of Fruit Flies in Nigeria

In Nigeria the species composition and abundance of tephritid fruit flies varies depending on the agro-ecological zone; these include rain forest, forest-savannah transition, Guinea savannah, Sudan Savannah, and the montane Jos and Mambilla Plateaus with their distinctive near sub-temperate climate (Fig. 25.1). The fruit fly species composition on mango and citrus was assessed in trapping surveys across several states of the Federal Republic of Nigeria (Anambra, Benue, Delta, Edo, Imo, Kaduna, Nasarawa, Ogun, Ondo, Oyo and Plateau) and included a number of agro-ecological zones (Umeh et al. 2008; Umeh and Ibekwe 2012); additional mango orchards were also sampled at the Federal Capital Territory (FCT), Kano and Borno states (Fig. 25.2). *Ceratitis, Dacus* and *Bactrocera* species were recorded. In the Sudan Savannah (Kano State) only the mango fruit fly, *Ceratitis cosyra* (Walker), was recorded, albeit in low numbers (maximum of four/trap/day). In contrast, in the Sahel savannah (Borno state), *B. dorsalis* predominated in trap catches and only a small number of *C. cosyra* were observed.

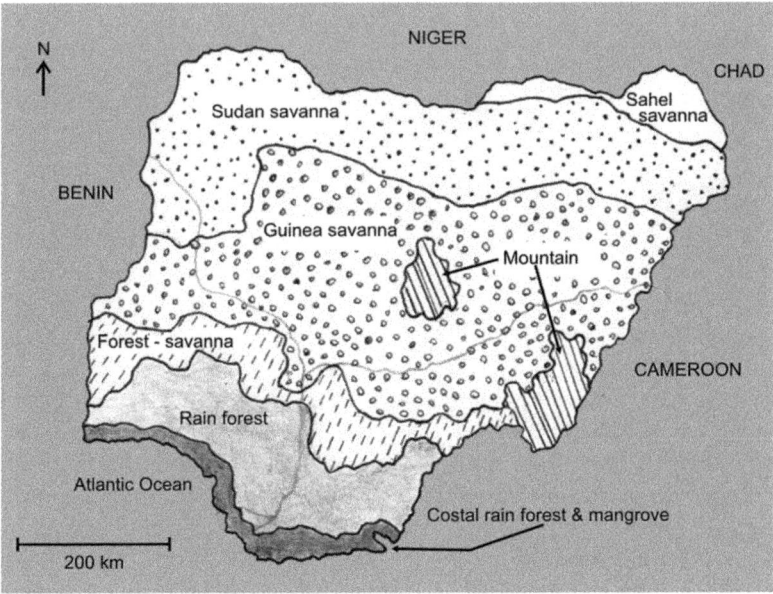

Fig. 25.1 Map of Nigeria showing the different ecological zones (Source: Umeh et al. 2008)

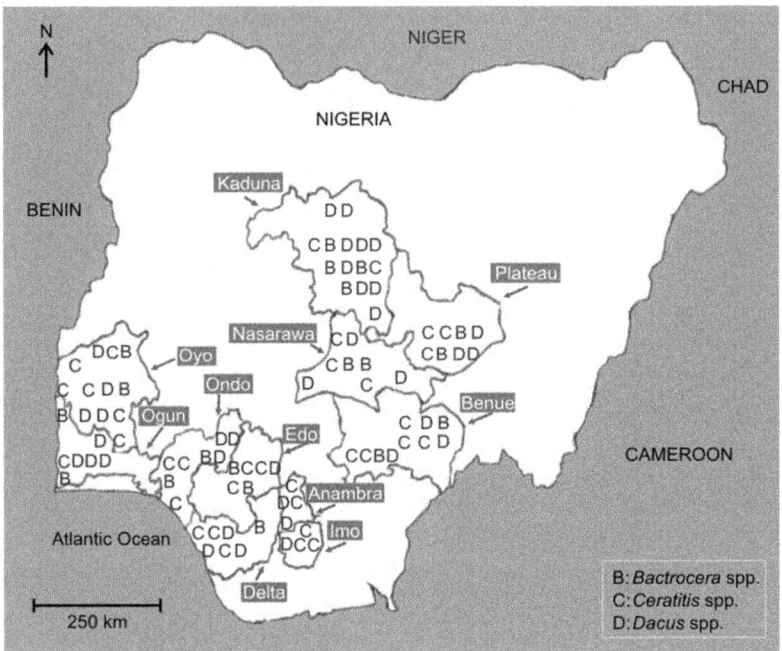

Fig. 25.2 Map of Nigeria showing the states surveyed and the fruit fly genera present in them (Source: Umeh et al. 2008)

2.1 The Introduction and Spread of Bactrocera dorsalis in Nigeria

Bactrocera dorsalis is a highly invasive fruit fly species of Asian origin. The species was first recorded in large numbers in Nigeria in 2005 during a nationwide survey of major fruit producing areas; it was found in fruit fly samples collected using traps baited with hydrolysed protein bait (Era® bait), yellow sticky traps (Pherocon® trap) and also from citrus fruit samples (Umeh et al. 2008). The survey showed that B. *dorsalis* was present in all the major citrus producing areas which, in most cases, were also mango producing areas. The mixed cropping system that characterizes the majority of Nigerian crop production systems provides ample species of fruits and vegetables that support B. *dorsalis* (Matanmi 1975; Umeh and Ivbijaro 1997; Anonymous 2003). Considering the ongoing large-scale emigration and immigration across international borders and the associated trade of agricultural materials, a rapid spread of *B. dorsalis* in Nigeria was to be expected. Although first detected in 2003, *B. dorsalis* was likely to have been in the region for a while before the survey, casting some doubt on the actual year of introduction into Nigeria and the West African region in general.

3   Main Host Plants of Fruit Flies in Nigeria

Various species of fruit flies are associated with horticultural crops commonly grown in Nigeria (Table 25.1). In various parts of the country, vegetables such as pepper and tomatoes are reported to be attacked by fruit flies, mainly *Ceratitis* and *Dacus* species, although tomatoes are less heavily infested than pepper (Umeh unpublished data). Other fruit fly species such as the melon fly, *Zeugodacus cucurbitae* (Coquillett), and the solanum fruit fly, *Bactrocera latifrons* (Hendel), which have been identified attacking various solanaceous crops in some African countries are yet to be observed in Nigeria (based on the last survey in 2013 using Pherolure®). Serious damage to peppers caused by the Mediterranean fruit fly (medfly), *Ceratitis capitata* (Weidemann) was recorded in Ibadan during the 2000–2003 cropping seasons, accounted for a 30 % loss in marketable yield (Oke personal communication).

Table 25.1 Predominant fruit fly species associated with fruits and vegetables in Nigeria and their host range

Botanical name of plant	Common name	Associated fruit fly species[a]	Sampling method
Persea americana Miller	Avocado	*Bactrocera dorsalis* (Hendel)	Fruits and traps
Irvingia spp.	Bush mango	*B. dorsalis*	Fruits and traps
Chrysophyllum albidum G. Don	Star apple	*B. dorsalis, Ceratitis capitata* (Wiedemann)	Fruits and traps
Citrus spp.	Citrus	*B. dorsalis,*	Fruits and traps
Theobroma cacao L.	Cocoa	*Ceratitis ditissima* (Munro), *C. capitata*	Fruits
Psidium guajava L.	Guava	*B. dorsalis, C. capitata, Ceratitis rosa* Karsch	Fruits
Mangifera indica L.	Mango	*B. dorsalis, Ceratitis cosyra* (Walker), *C. capitata, Ceratitis fasciventris* (Bezzi)	Fruits and traps
Terminalia catappa L.	Almond	*B. dorsalis*	Fruits
Coffeae arabica L.	Arabica coffee	*Trirhithrum coffeae* Bezzi, *Trirhithrum nigerrimum* (Bezzi)	Fruits
Coffeae canephora L	Robusta coffee	*T. coffeae, T. nigerrimum, C. capitata, C. rosa*	Fruits
Anona muricata L.	Sour sop	*B. dorsalis*	Fruits
Anona squamosa L.	Sweetsop	*B. dorsalis*	Fruits
Carica papaya L.	Pawpaw	*B. dorsalis*	Fruits and traps
Spondias mombin L.	Hug plum	*B. dorsalis*	Fruits

(continued)

Table 25.1 (continued)

Botanical name of plant	Common name	Associated fruit fly species[a]	Sampling method
Citrullus lanatus L.	Watermelon	*Dacus pleuralis* Collart, *Dacus vertebratus* Bezzi, *Dacus D. ciliatus* (Loew), *Dacus bivittatus* (Bigot), *Dacus umehi* White, *Dacus carnesi* (Munro), *Zeugodacus cucurbitae* (Coquillett), *B. dorsalis*	Fruits and traps
Cucumis sativus L.	Cucumber	*D. vertebratus, D. ciliatus, D. bivittatus, Z. cucurbitae, B. dorsalis*	Fruits and traps
Cucumis melo L.	Sweet melon	*D. vertebratus, D. ciliatus, D. bivittatus, Z. cucurbitae, B. dorsalis*	Fruits
Cucumis melo var. *cantalupensis* L.	Cantaloupe	*D. vertebratus, D. ciliatus, D. bivittatus, D. carnesi, Z. cucurbitae, B. dorsalis*	Fruits
Cucurbita spp.	Squash/pumpkin	*D. vertebratus, D. ciliatus, D. bivittatus, D. carnesi, Z. cucurbitae, B. dorsalis*	Fruits
Capsicum spp.	Pepper	*C. capitata*	Fruits
Solanum licopersicum L.	Tomato	*B. dorsalis, C. capitata,*	Fruits

[a]Fruit flies observed from traps and incubated fruits collected from different parts of Nigeria

Many species in the family Cucurbitaceae are cultivated in Nigeria, both edible species and species used for medicinal purposes or for utility, such as containers and sponges for domestic cleaning. Cultivated species include: water melon, *Citrullus lanatus* L.; sweet melon, *Cucumis melo* L.; cucumber, *Cucumis sativus* L. Some exotic species such as cantaloupe, *Cucumis melo* var. *cantalupensis* L., and squash/pumpkin (*Cucurbita* species) are largely confined to the plateau. Fruit fly species identified from edible fruits of cucurbits belong to the genera of *Dacus*, *Bactrocera* and *Zeugodacus* (Table 25.1). These include: *Dacus pleuralis* (Collart); *Dacus vertebratus* (Bezzi); the lesser pumpkin fly, *Dacus ciliatus* (Loew); *Dacus bivittatus* (Bigot); *Dacus umehi* (White); *Dacus carnesi* (Munro), *Z. cucurbitae* and *B. dorsalis*. Some *Dacus* species that were identified in the fields in Ibadan (Southwest Nigeria) seem to have a high affinity for particular species of cucurbits. We observed that *D. bivitatus* and *D. vertebratus* were the predominant fruit flies attacking water melon, *Citrullus lantanus* (Thunb.) Matsum. and Nakai, and cucumber respectively (Table 25.1). These observations agree with earlier findings reported by Matanmi 1975 and Anonymous (2003, 2006).

Various host plants are attacked by *Ceratitis* species and *B. dorsalis* to varying degrees. For example, *C. cosyra* is usually associated with mango. Cocoa, *Theobroma cacao* L., which is commonly cultivated in the rainforest regions of Nigeria is frequently attacked by *Ceratitis ditissima* (Munro) which has also become a major pest of citrus (Entwistle 1972; White and Elson-Harris 1992; Umeh et al. 2008; Goergen et al. 2011). While *Z. cucurbitae* is usually associated with cucurbits, it has been found attacking citrus (Umeh et al. 2008). Other fruit fly species that do not belong to *Ceratitis*, *Bactrocera* or *Dacus* include *Trirhithrum coffeae* Bezzi and *T. nigerrimum* Bezzi which are of economic importance on coffee; *Coffea canephora* Pierre ex Froehner (= *robusta*) is preferred over *C. arabica* L.. Coffee was also found to host other fruit fly species such as *C. capitata* and the Natal fruit fly, *Ceratitis rosa* Karsch (Umeh et al. 2007). A wild plant, the African star apple, *Chrysophyllum albidum* Don, was also found to be host for *C. capitata* and *B. dorsalis* (Umeh et al. 2002). Banana is commonly grown in Nigeria and may be a potential host of fruit flies such as *B. dorsalis* when ripe, as observed in other African countries (Cugala et al. 2013). *Ceratitis capitata* and *B. dorsalis* are the most common fruit fly species recovered from citrus. In a field study Umeh et al. (1998) compared the infestation level by *C. capitata* on 12 varieties of sweet oranges and found that Parson Brown, Washington Navel and Carter Navel were attacked significantly more than other varieties. These varieties also produced significantly more dropped fruits harbouring fruit flies (Umeh et al. 1998).

4 Seasonal Population Dynamics of Fruit Flies on the Main Fruit Crops in Nigeria

Fruit flies occur and damage various fruit crops all year round in Nigeria. However, the abundance and severity of damage varies seasonally (Anice and Sales 1997) in relation to the phenology of the host plant and the affinity of the fruit fly species to particular hosts. Observations made on the population dynamics of the fruit fly species attacking mango (var Saigon, Julie, Edward and Governor) at the National Horticultural Research Institute, Ibadan (NIHORT) showed distinct patterns of seasonal infestation. Observations were made over 2 years (2007–2008) using McPhail traps baited either with methyl eugenol (ME) (for *B. dorsalis*), terpinyl acetate (TA) (for *C. cosyra*) or trimedlure (TrM) (for *C. capitata*) and similar patterns of population dynamics were observed on all the mango varieties evaluated (Fig. 25.3). Populations of male *C. cosyra* were large between February and April but decreased progressively from May until the end of the rainy season in October (Fig. 25.3). In contrast, *B. dorsalis* catches were relatively small between February and April. *Bactrocera dorsalis* populations peaked in June and July and decreased progressively until October. The number of *C. capitata* were relatively small throughout; no flies were captured during June and August in the first year and between June and September in the second year (Umeh unpublished data).

Monthly records of temperature, relative humidity (RH) and rainfall taken during the fruiting seasons at NIHORT indicated a small negative correlation between *B. dorsalis* populations recorded and air temperature in both years that fruit flies were sampled, with respective correlation coefficients (r) of −0.3 and −0.48 (Umeh

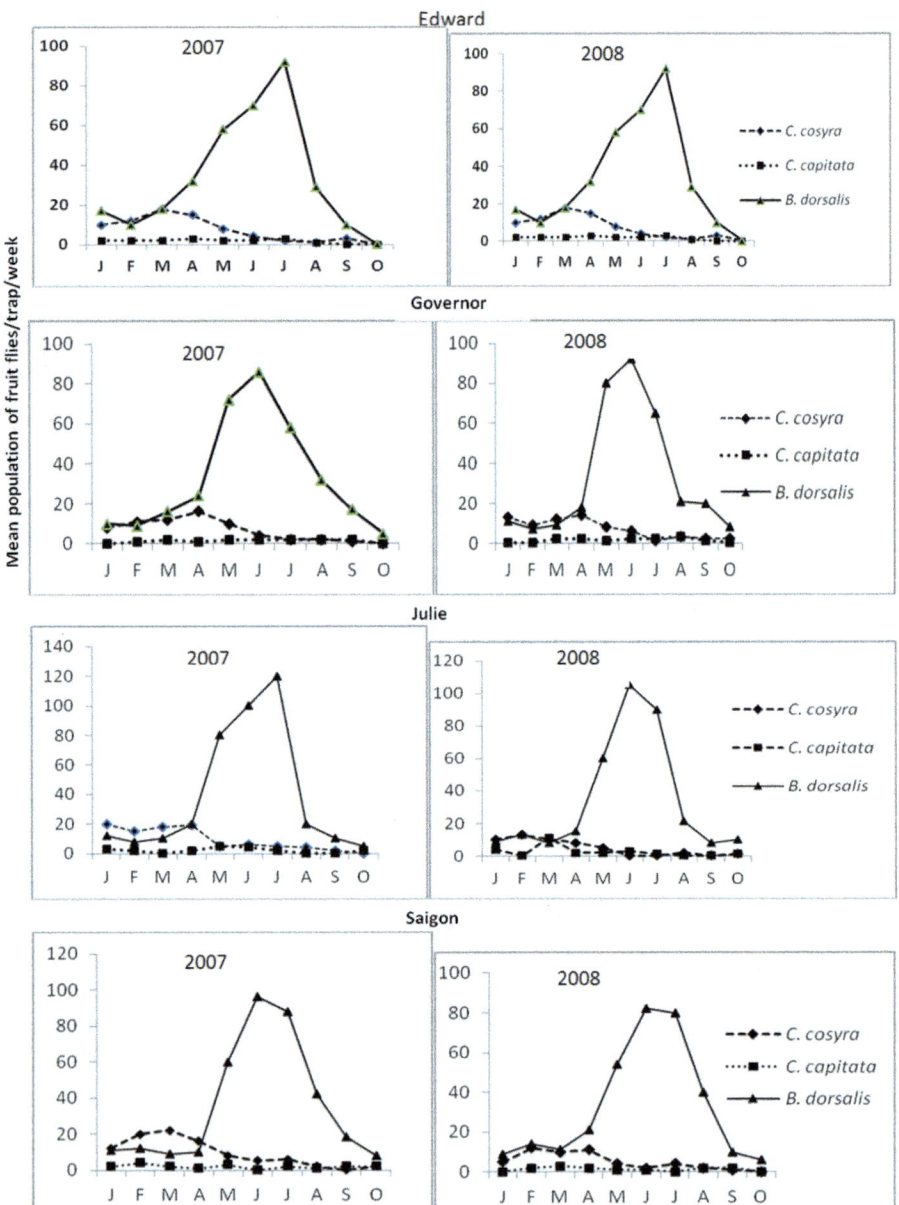

Fig. 25.3 Population dynamics of fruit flies on Edward, Governor, Julie and Saigon varieties of mango in 2006 and 2007 at NIHORT, Ibadan, Nigeria

unpublished data). These results do not agree with the finding of Shukla and Prasad (1985) and Vayssières et al. (2009) who reported a positive correlation between temperature and *B. dorsalis* populations on mango. Uddin et al. (2016) stated that the number of *B. dorsalis* captured with methyl eugenol baited traps on mango correlated positively with temperature and relative humidity but in the case of temperature the result was not statistically significant. The dominant part played by humidity was also reflected in the studies conducted by other workers (Anice and Sales 1997; Sarada et al. 2001).

In contrast, we observed that *C. cosyra* populations were positively correlated with air temperature in both years with $r = 0.85$ and 0.53 respectively. Interestingly, *B. dorsalis* populations were positively correlated with RH in both years ($r = 0.60$ and 0.54 respectively), while *C. cosyra* populations were negatively correlated with RH in both years ($r = -0.65$ and -0.56 respectively) (Umeh unpublished data) These findings suggest that dry periods favour *C. cosyra*, hence higher populations were observed earlier in the year before the rains began. It should be noted that in Ibadan, Nigeria, the highest annual temperatures are recorded between February and April, a period during which *C. cosyra* is abundant. In contrast, *B. dorsalis* are known to thrive mostly during the rainy season (Mwatawala et al. 2006), which was supported in our studies by the correlation between population size and rainfall; $r = 0.72$ during the rainy season. With the issue of climate change it is believed that changes in abiotic parameters will influence the population dynamics of fruit flies and that this will vary depending on region. Studies are needed over a longer period of time in order to have a conclusive evidence of the effect of abiotic factors such as temperature on populations of fruit fly in a particular area (Tables 25.2 and 25.3).

During 2003 and 2004 fruit fly population dynamics were studied in citrus orchards at NIHORT, Ibadan (Umeh and Garcia 2008). Fruit flies were sampled using McPhail traps baited with either protein hydrolysate or local brewery waste. The traps were hung on sweet orange, *Citrus sinensis* L. Osbeck, varieties Washington navel, Parson Brown, Valencia late and Agege-1 during the main citrus fruiting seasons. The predominant species recorded were from the genus *Ceratitis*, while *B. dorsalis* populations were very low. In both years the numbers of fruit flies (mostly *Ceratitis* spp.) were low in August, increased from September and peaked in November before harvest; numbers dropped drastically in December when most of the fruits had been harvested (Fig. 25.4). The number of *Ceratitis* spp. varied amongst the different varieties: Washington navel > Parson Brown > Valencia late > Agege-1. Fruit fly numbers were significantly ($p < 0.05$) higher on Washington navel than on either Agege-1 or Valencia late (Umeh and Garcia 2008).

5 Levels of Damage to Fruits and Vegetables in Nigeria

The major fruit flies of economic importance (in the genera *Bactrocera*, *Ceratitis*, *Dacus* and *Zeugodacus*) have been reported to cause losses ranging between 3.4 and 40% depending on the crop, season and location (Tables 25.4 and 25.5) (Agunloye

Table 25.2 Species diversity of fruit flies identified during surveys in citrus-producing areas of Nigeria in 2003 and 2006

Fruit fly species	States harbouring the fruit fly species identified	Sampling method	Agro-ecological zones
Ceratitis capitata (Wiedemann)	Anambra, Benue, Delta, Edo, Nasarawa, Ogun, Ondo, Oyo	Traps, fruits	Forest savanna, Guinea savanna, Rainforest
Ceratitis ditissima (Munro)	Edo, Delta, Imo, Kaduna, Ondo, Oyo	Traps, fruits	Forest savanna, Guinea savanna, Rainforest
Ceratitis penicillata (Bigot)	Ondo, Oyo	Traps	Forest savanna, Rainforest
Dacus bivittatus (Bigot)	Anambra, Benue, delta, Edo, Imo, Kaduna, Nasarawa, Ondo, Oyo, Plateau		Forest savanna, Guinea savanna, Montane, Rainforest
Dacus ciliatus (Loew)	Anambra, Benue, Nasarawa, Ondo, Oyo	Traps	Forest savanna, Guinea savanna, Rainforest
Dacus transitorius (Collart)	Oyo	Traps	Forest savanna
Dacus umehi (White)	Kaduna	Traps	Guinea savanna
Zeugodacus cucurbitae (Coquillett)	Benue, Delta, Edo, Nasarawa, Ogun, Ondo, Oyo, Plateau	Traps, fruits	Forest savanna, Guinea savanna, Montane, Rainforest
Bactrocera dorsalis (Hendel)	Anambra, Benue, Delta, Edo, Nasarawa, Ogun, Ondo, Oyo, Plateau, Kaduna	Traps, fruits	Forest savanna, Guinea savanna, Montane, Rainforest
Celidodacus obnubilus (Karsch)	Kaduna	Traps	Guinea savanna
Perilampsis woodi Bezzi	Kaduna, Oyo	Traps	Forest savanna, Guinea savanna
Trirhithrum nigerrimum (Bezzi)	Oyo, Ondo	Traps	Forest savanna, Rainforest

Source: Umeh et al. (2008)

1987; Eguagie and Udensi 1989; Umeh et al. 1998, 2008). Following the invasion of *B. dorsalis* other previously prevalent indigenous species seem to have been displaced (Vayssières et al. 2005; Anonymous 2007; Umeh et al. 2008; Ekesi et al. 2009) and losses in fruit yields have increased to more than 70%. This has led to a reduction in producers' income as well as that of the middle men and retailers, the majority of whom are women. Local supplies to the processing industries (mostly juice industries) have drastically reduced (Umeh personal observation). In Nigeria the national income from the export of fruits such as citrus and mango has fallen below expectations; this is due to the loss of both regional and international export markets in response to quarantine restrictions imposed by importing countries as the result of *B. dorsalis* invasion. Like many other alien pests, the invasion and subsequent spread of *B. dorsalis* in Nigeria represents a major threat to the otherwise

Table 25.3 Distribution and species diversity of fruit flies identified during surveys in the main mango producing areas of Nigeria s in 2006 and 2007

Fruit fly species	States harbouring the fruit fly species identified	Sampling method	Agro-ecological zones
Bactrocera dorsalis (Hendel)	Anambra, Benue, Borno, Enugu, FCT, Imo, Kaduna, Kano, Nasarawa, Ogun, Oyo, Plateau	Trap, fruit	Rainforest, Forest-Savanna, Guinea Savanna, Sudan Savanna, Montane
Z. cucurbitae (Coquillet)	Benue, Edo, Nasarawa, Ogun, Oyo,	Trap	Rainforest, Forest-Savanna, Guinea Savanna
Ceratitis bremii (Guérin-Méneville)	Kogi	Trap	Guinea Savanna
Ceratitis capitata (Weidemann)	Benue, Kaduna, Oyo	Trap, fruit	Rainforest-Savanna, Guinea Savanna
Ceratitis cosyra (Walker)	Anambra, Benue, Borno, Enugu, FCT, Imo, Kaduna, Kano, Nasarawa, Ogun, Oyo, Plateau	Fruit, trap	Rainforest, Forest-Savanna, Guinea Savanna, Sudan Savanna, Montane
Ceratitis ditissima (Munro)	Imo	Trap	Rainforest
Ceratitis fasciventris (Bezzi)	Benue, Oyo	Fruit, trap	Forest-Savanna, Guinea Savanna
Ceratitis penicillata (Bigot)	Oyo, Ogun	Trap	Rainforest, Forest-Savanna
Dacus umehi (White)	Kaduna	Trap	Guinea Savanna

Source: Umeh and Ibekwe (2012)

booming Nigerian horticultural industry. In Nigeria alone, an estimate of up to 40 % (Umeh et al. 2008) of the annual production of citrus (325,000 t valued at $1,167,400); mango (74,000 t valued at $117,748); and papaya (759,000 t valued at $118,392) are seriously threatened by *B. dorsalis* (FAOSTAT 2006, 2007). In extreme cases, some producers have abandoned their orchards and many fruit vendors have gone out of business as a result of the low quantity of marketable fruits in their consignments (Umeh unpublished data). New European import regulations such as CE882/2004, which are strict on fruit fly-infested consignments, have been a drastic setback to Nigeria's developing fruit export trade and to the burgeoning trade in West African mangoes particularly.

6 Current Fruit Fly Management Measures Used by Growers in Nigeria

Surveys showed that the majority of farmers in Nigeria did not apply any form of control measure against fruit flies, although a limited number used cultural practices including sanitation. This was practiced by 30 % of the farmers even before the advent of *B. dorsalis* (Umeh et al. 2008). The few that applied control measures

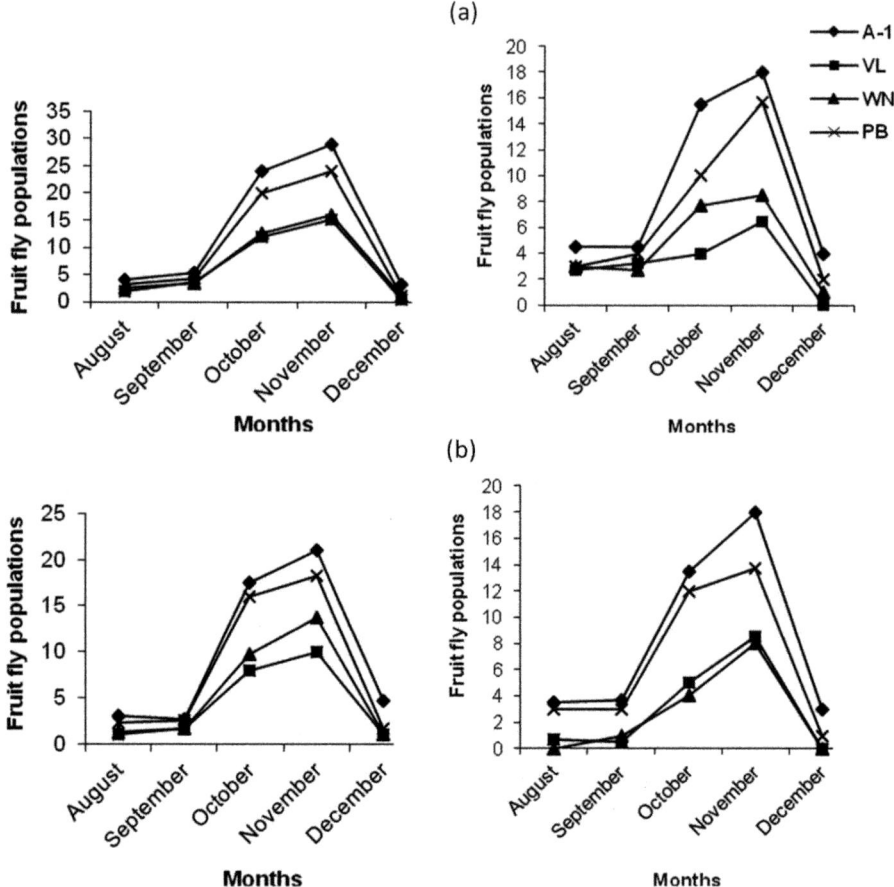

Fig. 25.4 Population dynamics of fruit flies caught on citrus using (**a**) protein hydrolisate or (**b**) brewery waste baits in 2003 and 2004 during trials at NIHORT Ibadan. A-1 = Agege-1; VL = Valencia Late; WN = Washington Navel; PB = Parson Brown (Source: Umeh and Garcia 2008)

used broad spectrum cover sprays of synthetic pesticides and, in most cases, diverted insecticides meant for other insect pests towards managing fruit flies. This often resulted in repeated applications with unsatisfactory results and numerous negative environmental consequences. Although monocrop orchards are maintained in various areas in the country, most fruit trees are intercropped. Usually intercrops involve various combinations of fruit trees including citrus; mango; guava; cocoa; kola, *Cola acuminata* Schott & Endl.; star apple; and hug plum, *Spondias mombin* L. (Umeh et al. 2008). The implication of having mixed plantings on pest control means that blanket applications are often made over all intercropped plants irrespective of whether they all need treatment. Interviews found that 56 % of citrus farmers reported fruit fly damage to all their intercropped fruit trees but only 20 % of the farmers regularly removed and destroyed fruits that dropped to the ground; just

Table 25.4 Damage caused to fruit by fruit flies in selected citrus-producing states in Nigeria in 2003 and 2006

States surveyed in Nigeria	Number of sites/state/year	Mean number of fruit flies per site		Mean % of fruit attacked per site		Mean fruit drop[a]	
		2003	2006	2003	2006	2003	2006
Anambra	5	7	4	11.0	8.6	1.0	1.0
Benue	10	26	20	29.0	17.0	2.6	1.6
Delta	5	16	12	14.0	11.0	1.2	1.6
Edo	5	11	8	12.6	8.4	1.4	1.0
Imo	6	12	8	20.2	11.2	1.6	1.2
Kaduna	6	39	26	30.0	16.0	2.2	1.2
Nasarawa	7	18	11	20.0	11.0	2.0	1.0
Ogun	5	16	10	21.4	14.5	1.6	1.2
Ondo	5	18	11	27.0	14.2	2.4	1.2
Oyo	8	24	14	23.0	14.4	2.2	1.8
Plateau	8	9	7	6.0	4.0	1.2	1.0

Source Umeh et al. (2008)
The figure was then averaged over all the sites sampled in a given state
[a]Categories for fruit drop at each site allocated as: 0 = 1–10 fruits dropped; 1 = 11–20 fruits dropped; 2 = 21–30 fruits dropped; 3 = >30 fruits dropped

Table 25.5 Damage caused to mango by fruit flies in selected production areas of Nigeria

States surveyed in Nigeria	Number of sites/state/year	Mean number of fruit flies per site		Mean % of fruit attacked per site		Mean fruit drop[a]	
		2006	2007	2006	2007	2006	2007
Anambra	5	12	28	20	28.3	1.0	1.2
Benue	10	18	30	15.2	22.7	1.4	1.6
Borno	5	15	21	12.3	14.9	1.2	1.0
Enugu	5	42	55	20.1	30.3	2.0	2.4
FCT	5	10	8	14.2	18.4	1.2	1.0
Imo	5	12	25	16.3	28.1	1.2	1.6
Kaduna	8	16	28	26.2	34.8	3.0	2.6
Kano	4	0	4	0	5	0	0.4
Nasarawa	7	8	9	19.4	21.0	1.6	1.8
Ogun	5	12	15	14.6	23.2	1.2	1.6
Oyo	8	15	22	19.2	27.4	1.8	1.6
Plateau	8	6	8	16.0	21.2	1.2	1.8
Sokoto	5	7	9	10	12.4	1.0	1.0

Source Umeh and Ibekwe (2012)
The figure was then averaged over all the sites sampled in a given state
[a]Categories for fruit drop at each site allocated as: 0 = 1–10 fruits dropped; 1 = 11–20 fruits dropped; 2 = 21–30 fruits dropped; 3 = >30 fruits dropped

10% of them occasionally remove dropped fruits. Non-removal of dropped fruits contributes significantly to high infestation rates (Umeh et al. 2003) as they serve as sources of re-infestation. The study also found that more than 50% of famers only harvested citrus or mango fruits when they were fully ripe, a practice that increases the level of infestation in endemic areas. The stage of fruit ripeness is linked to its physicochemical characteristics (Attaway 1971). Advanced ripeness increases sugar content, level of essential oils in the fruit rind and a reduction in acidity – a stage at which the fruit also becomes more attractive to fruit flies for oviposition (Dhouibi et al. 1995; Umeh et al. 2009).

7 National Efforts to Combat Fruit Flies of Economic Importance in Nigeria

Most native fruit flies in the genus *Ceratitis* and *Dacus* are associated with tropical regions of Africa and have always caused damage to fruit crops. However, their populations have generally been controlled by indigenous natural enemies and a range of different management programmes established through extension agents and other governmental and non- governmental organizations. However, invasion by polyphagous exotic species and associated restrictions on fruit and vegetables export has changed this. Current national management efforts are targeted at fruit fly pests in the genera *Ceratitis, Dacus, Bactrocera* and *Zeugodacus*, all of which are well established in Nigeria.

In order to avert the imminent threat to the horticulture industry of invasive fruit flies, NIHORT, in collaboration with the Nigerian Agricultural Quarantine Services (NAQS) organized a national stakeholder workshop on fruit flies. The aim was to raise awareness of the economic impact and management of invasive fruit flies in horticultural crops in Nigeria. Resolutions taken to combat *B. dorsalis* and other fruit flies of economic importance include the following:

1. Strengthening the role of NAQS/Plant Protection Departments to prevent further quarantine pest invasion and to advise the government on emergency quarantine threats.
2. Creating nationwide awareness amongst producers and other stakeholders along the fruits value chain on simple ways of identifying the fruit flies of greatest economic importance and the damage they caused.
3. Applying good cultural practice such as constant removal and destruction of fallen fruits, farm sanitation and timely harvest.
4. Testing and adopting sustainable Integrated Pest Management (IPM) strategies based on the use of appropriate traps, attractants and baits; these included bait application techniques (BAT) and male annihilation techniques (MAT) for mass trapping and suppression of fruit fly populations.
5. Exploring the use of naturally occurring pesticides such as botanicals and biopesticides for IPM of fruit flies.

6. Exploring the use of parasitoids for classical biological control as part of IPM against fruit flies.

The majority of fruit producers in Nigeria are smallholders who do not adopt standard production practices due to the meager resources available to them. Therefore, any proposed control measures must be relatively cheap and environmentally friendly to increase uptake by these farmers. Furthermore, sources of appropriate traps, attractants and baits must be found or established. These factors were considered in developing acceptable management strategies, adoption of existing universally recognized methods or modifying the latter to suit Nigerian farmers' conditions. These methods were then demonstrated in farmers' fields after a series of trials.

8 IPM for Suppression of Fruit Flies in Nigeria

Different management strategies were evaluated individually or in various combinations for fruit fly management in Nigeria and yielded promising results. For example, Umeh et al. (2009) found that minimal insecticide applications, early harvest and orchard sanitation (removal of fallen fruits) significantly decreased the number of fruits on trees that were infested by *C. capitata* as well as those that had dropped to the ground when compared with controls without any treatments (Figs. 25.5, 25.6 and 25.7). Harvesting at the stage when 50% of fruits were ripe significantly reduced the level of damage compared with harvesting when 90% of fruits were

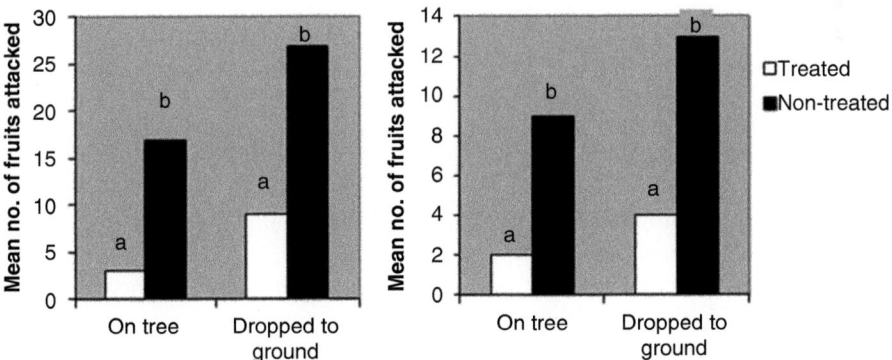

Fig. 25.5 Effect of minimal foliar treatment using a mixture of cypermethrin and dimethoate on the number of citrus fruits attacked by fruit flies in 2000 and 2001 in Gboko, Nigeria. Bars with the same letters are not significantly (P>0.05) different to each other for each group of fruit damage (Source: Umeh et al. 2003)

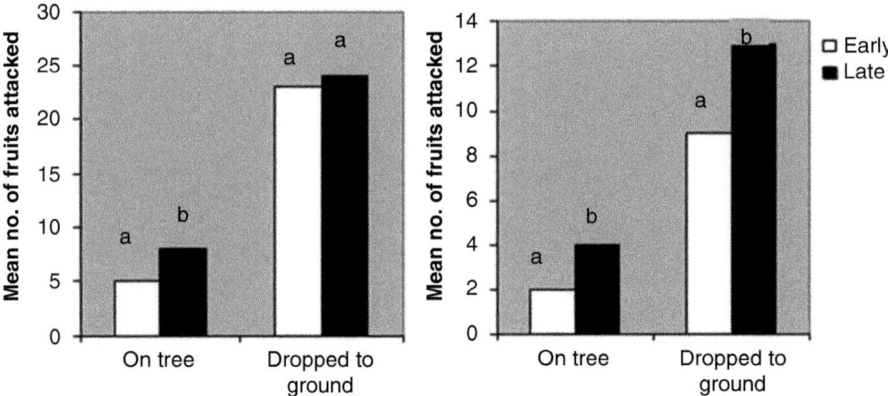

Fig. 25.6 Effect of harvesting time (early or late) on the number of citrus fruits attacked by fruit flies in 2000 and 2001 in Gboko, Nigeria. Bars with the same letters are not significantly ($P > 0.05$) different to each other for each group of fruit damage (Source: Umeh et al. 2003)

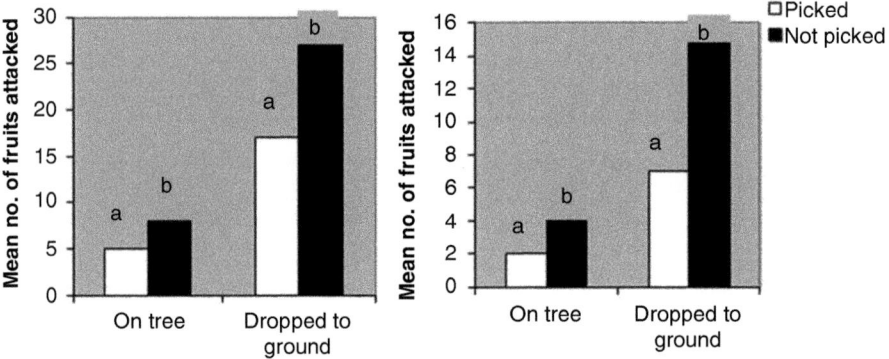

Fig. 25.7 Effect of removal of dropped fruit (i.e. sanitation) on the number of citrus fruits attacked by fruit flies in 2000 and 2001 in Gboko, Nigeria. Bars with the same letters are not significantly ($P > 0.05$) different to each other for each group of fruit damage (Source: Umeh et al. 2003)

ripe. This agreed with the findings of others working with *C. capitata* (Ortiz et al. 1987; Noussourou and Diarra 1995).

Spot applications of protein baits made from locally-available brewer's yeast waste (prepared using a modification of the methods of Gopaul and Price 1999) mixed with chlorpyrifos were evaluated on three varieties of sweet orange (Agege, Valencia Late and Parson Brown) in the 2006–2007 fruiting season. Significantly more adult *B. dorsalis* and *Ceratitis* spp. emerged from fruits collected from untreated sweet orange trees compared to those sprayed with the bait, thus indicating the effectiveness of the protein bait in reducing fruit fly infestation and damage (Fig. 25.8; Umeh and Onukwu 2011). The results also showed varieties varied in their susceptibility to fruit flies (Fig. 25.8); significantly more ($P < 0.05$) Agege fruits were infested on the tree and on the ground (mean of 4.3 fruits and 6.6 fruits, for attacked and dropped fruits out of ten samples, respectively) compared with

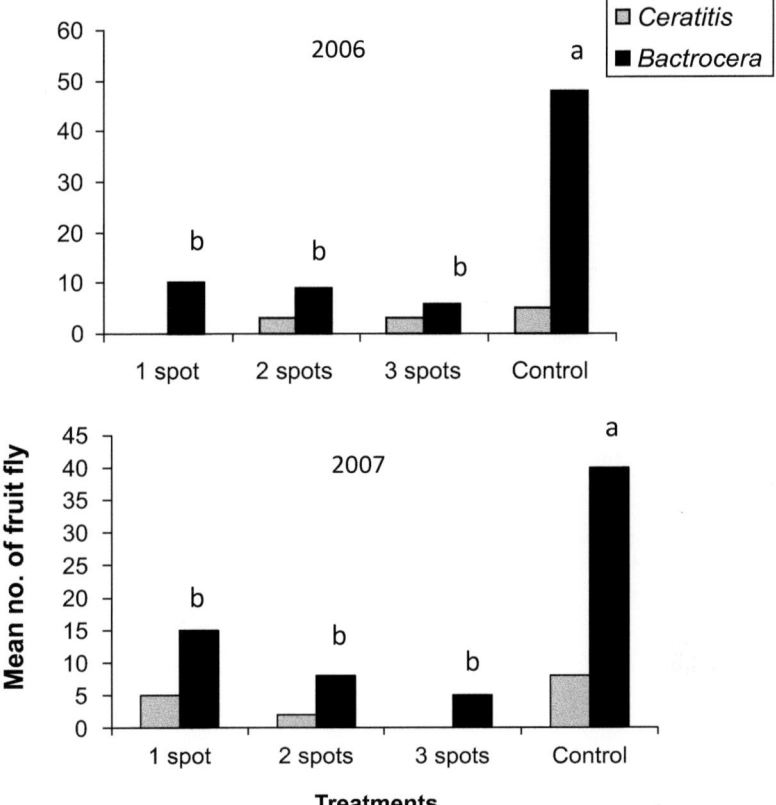

Fig. 25.8 Number of adult fruit flies that emerged from dropped citrus fruits (n = 5) following different protein bait sprays. Bars with the same letters are not significantly (P > 0.05) different to each other for each fruit fly genus (Source: Umeh and Onukwu 2011)

Parson Brown and Valencia fruits (Table 25.4; Umeh and Onukwu 2011). The variety Valencia had the fewest infested fruits (mean of 1.2 and 2.8 for attached and fallen fruits, respectively). The interaction between variety and bait spray had a significant ($P < 0.01$) effect on the number of fallen fruits and the total number of damaged fruits (Umeh and Onukwu 2011). Similar effective levels of control were achieved by Ross (1993) using hydrolised protein baits and malathion against *C. capitata* and Piñero et al. (2009) using spinosad-based protein bait sprays and sanitation against *B. dorsalis* in papaya orchards (Tables 25.6 and 25.7).

In a separate study, the role of the male annihilation technique (MAT) in suppressing fruit fly populations was assessed in mango (varieties Governor, Julie, Saigon and Edward) and citrus orchards in three agro-ecological zones of Nigeria. In this study, three parapheromones were used: methyl eugenol and terpinyl acetate were used in the mango orchard, while methyl eugenol and trimedlure were used in the citrus orchards. A drastic reduction in fruit fly populations were observed in MAT-treated orchards at all sites (for both mango and citrus) compared with

Table 25.6 Effects of cover sprays on level of fruit fly damage on sweet oranges in 2006 and 2007

Treatments	Mean number of fruits attacked per tree		Mean number of fruits that had dropped per tree		Mean total number of damaged fruits per tree	
	2006	2007	2006	2007	2006	2007
Cover spray (C)						
1 spot	1.9 ab	2.2 b	2.7 b	13.0 b	5 b	15 b
2 spots	1.7 ab	2.0 b	5.2 ab	3.8 c	7 ab	6 c
3 spots	0.9 b	1.3 b	3.5 b	3.2 c	5 b	5 c
Control (no spray)	3.0 a	8.0 a	6.9 a	25.0 a	10 a	33 a
Variety (V)						
Agege	4.3 a	4.9 a	7.8 a	6.6 a	11 a	11 a
Valencia	1.5 c	1.2 b	3.2 b	2.8 b	4 b	4 b
Parson Brown	2.2 b	2.0 b	3.4 b	3.0 b	6 b	5 b
C×V	NS	NS	*	*	*	*

Source: Umeh and Onukwu (2011)
Means in the same column followed by the same letters are not significantly (P > 0.05) different to each other using SNK
NS Non significant
*Significant at P = 0.01

Table 25.7 Numbers of fruit flies emerging from different varieties of sweet orange that had received different treatments in the NIHORT orchard in 2006 and 2007

Treatments	Mean number of fruits per treatment	Mean weight of fruits (Kg)		Mean number of *B. dorsalis* per fruit		Mean number of *Ceratitis* spp. per fruit	
		2006	2007	2006	2007	2006	2007
Cover spray (C)							
1 spot	20	8.0	8.5	5.2 b	7.0 b	1.2 b	1.5 b
2 spots	20	8.2	8.2	5.3 b	6.0 bc	1.0 b	1.2 b
3 spots	20	8.2	8.4	4.4 b	3.4 c	0.9 b	0.8 b
Control (no spray)	20	8.4	8.3	12.8 a	21.6 a	2.5 a	3.2 a
Variety (V)							
Agege	20	8.4	8.0	8.2 a	10.0 a	3.8 a	3.6 a
Valencia	20	8.4	8.2	3.7 b	4.6 b	1.4 b	1.6 b
Parson Brown	20	6.0	5.9	4.0 b	3.6 b	1.6 b	1.5 b
C×V				*	*	NS	NS

Source: Umeh and Onukwu (2011)
Means in the same column followed by the same letters are not significantly (P > 0.05) different to each other using SNK
NS Non significant
*Significant at P = 0.05

untreated orchards (ranging from 53 to 78 % reduction in numbers) indicating suppression of incipient fruit fly populations. This was further substantiated by the fact that significantly fewer adult fruit flies emerged from fruits sampled from the MAT-treated plots compared with those from the control plots; this represented a 50–85 % reduction in the number of emerging fruit flies (Umeh unpublished data).

9 Future Perspectives on Fruit Fly Management in Nigeria

Although a lot has been achieved in Nigeria with regard to fruit fly research, additional studies are needed on fruit fly bio-ecology (e.g. systematics, species composition, damage, spatio-temporal distribution, abundance, demography, host plant range, population dynamics, role of native natural enemies) especially in locations not covered in the present review. Present initiatives against fruit fly pests in Nigeria are focusing on nationwide capacity building for growers and other major players in the fruit and vegetable value chain to ensure they have the necessary expertise to manage fruit flies. However, there is a need to place particular emphasis on stakeholders in the mango value chain since this crop is the most seriously affected. Various major fruit producing states in Nigeria are earmarked for the provision of training. Orchards for field demonstration of fruit fly management strategies have been selected in the worst affected areas. For example, demonstration orchards have been established in Kaduna state (2013) as part of a pilot project on capacity building for farmers. This initiative will be extended to four other regions of the country. In line with the decisions taken during the first national stakeholder workshop on fruit flies, the control options being promoted include (i) field sanitation and early harvesting of fruits (ii) male annihilation technique (MAT) using para-pheromones and (iii) baiting techniques (BAT) using various food-based attractants. Part of the control initiative planned for the country also includes the introduction of efficient coevolved parasitoids (natural enemies) from *icipe* for classical biological control of *B. dorsalis* and other exotic invasive fruit fly species. The basic studies required prior to the introduction of biological control agents are being conducted. Combined use of the strategies described within an IPM approach and on an area-wide scale need to be pursued for effective pre-harvest management/ suppression of fruit flies. To regain access to the international export market, post-harvest disinfestation treatments for particular fruit fly species and fruit types must also be explored.

Acknowledgements I sincerely thank Drs Marc De Mayer, Georg Goergen and Bob Wharton for taxonomic assistance. I thank the Belgian Government for taxonomic training opportunities through the Royal Museum of Central Africa, Tervuren Belgium. The projects reported in this chapter were partly funded by the Nigerian Agricultural Research Council through the Competitive Agricultural Research Grant Scheme (CARGS) and The National Horticultural Research Institute (NIHORT). Technical assistance was provided by the International Atomic Energy Agency (IAEA) and collaboration with the United States Department of Agriculture (USDA).

References

Agunloye OO (1987) Trapping and chemical control of *Ceratitis capitata* (weid) (Diptera: Tephritidae) on sweet orange (*Citrus sinensis*) in Nigeria. J Hort Sci 62:269–271

Anice R, Sales F (1997) Seasonal abundance of fruit flies in New Caledonia. In: Allwood AJ, Drew RAI (eds) Management of fruit flies in the Pacific. ACIAR, Canberra, pp 134–139

Anonymous (2003) National Horticultural Research institute (NIHORT) Annual Report

Anonymous (2006) National Horticultural Research institute (NIHORT) Annual Report

Anonymous (2007) National Horticultural Research institute (NIHORT) Annual Report

Attaway A (1971) Biochemistry of fruits and their products, Hulmes A.C. (Ed), Academic Press London, pp. 120–132.

Babatola JO (1985) Diseases and pests of fruits and their control. In: FACU-NIHORT (Ed.) National fruit production workshop, Ibadan, Nigeria, pp. 120–132

Cugala D, Ekesi S, Ambasse D, Adamu RS, Mohamed SA (2013) Assessment of ripening stages of Cavendish dwarf bananas as host or nonhost to *Bactrocera invadens*. J Appl Entomol 138:449–457

Dhouibi MH, Gahbiche H, Saaidia B (1995) Variation in *Ceratitis capitata* infestation of fruit according to fruit locations on tree and orange ripeness. Fruits 50:39–49

Eguagie WE, Udensi N (1989). Control of insect pests and diseases of citrus and mango in Nigeria. In: NIHORT (Ed.), National Horticultural Research Institute Technical Bulletin No. 5 p 15.

Ekesi S, Nderitu PW, Rwomushana I (2006) Field infestation, life history and demographic parameters of the fruit fly *Bactrocera invadens* (Diptera: Tephritidae) in Africa. Bull Entomol Res 96:279–386

Ekesi S, Billah MK, Nderitu PW, Lux SA, Rwomushana I (2009) Evidence for competitive displacement of the mango fruit fly *Ceratitis cosyra* by the invasive fruit fly *Bactrocera invadens* (Diptera: Tephritidae) on mango and mechanisms contributing to the displacement. J Econ Entomol 102:981–991

Entwistle PF (1972) Pests of Cocoa. Rev Agric Entomol 62:651

FAOSTAT (2006) http://faostat.fao.org

FAOSTAT (2007) http://faostat.fao.org

FAOSTAT (2013) http://faostat.fao.org

Goergen G, Vayssieres JP, Gnanvosou D, Tindo M (2011) *Bactrocera invadens* (Diptera: Tephritidae) a new invasive fruit fly pest for the Afrotropical region: host plant range and distribution in west and Central Africa. Environ Entomol 40:844–854

Gopaul S, Price NS (1999) Local production of protein bait for use in fruit fly monitoring and control, In: 4th Annual Meeting of Agricultural Science (AMAS) Food and Agricultural Research Council, Reduit, Mauritius pp 117–122

Ibeawuchi II, Okoli NA, Alagba RA, Ofor MO, Emma-Okafor LC, Peter-Onoh CA, Obiefuna JC (2015) Fruit and vegetable crop production in Nigeria: the gains, challenges and the way forward. J Biol Agric & Healthcare 5:194–208

Matanmi BA (1975) The biology of Tephritid fruit flies (Diptera: Tephritidae) of cucurbits at Ile-Ife, Nigeria. Nigerian J Entomol 1:153–159

Mwatawala MW, De Meyer M, Makundi RH, Maerere AP (2006) Seasonality and host utilization of the invasive fruit fly, *Bactrocera invadens* (Diptera: Tephritidae) in central Tanzania. J Appl Entomol 130:530–537

Noussourou M, Diarra B (1995) Mouches des fruits au Mali: Biologie et possibilité de lute intégrée. Sahel IPM 6:2–13

Ortiz JM, Tadeo JL, Estelles A (1987) Características físicoquímicas de 'Navelina', 'Washington Navel' y su evolución durante la maduración. Fruits 42:435–441

Piñero JC, Mau RFL, Vargas RI (2009) Managing oriental fruit fly (Diptera: Tephritidae), with spinosad-based protein bait sprays and sanitation in papaya orchards in Hawaii. J Econ Entomol 102:1123–1132

Ross JP (1993) Attractiveness of three hydrolyzed protein to *Ceratitis capitata*. In: Aluja M, Liedo P (eds) Fruit flies: biology and management. Springer, New York, pp 243–246

Sarada G, Maheswari TU, Purushotham K (2001) Seasonal incidence of population fluctuation of fruit flies of mango and guava. Indian J Entomol 63:272–276

Shukla RP, Prasad VC (1985) Population fluctuations of the oriental fruit fly Dacus dorsalis Hendel in relation to hosts and abiotic factors. Trop Pest Man 31:273–273

Taiwo TA (2005) Production of fruits, vegetables, grains, legumes and root crops in Nigeria; Problems and prospects. University Press, Vol. 1:9

Uddin MS, Reza MH, Hossain MM, Hossain MA, Islam MZ (2016) Population fluctuation of male oriental fruit fly, *Bactrocera dorsalis* (Hendel) in a mango orchard of Chapainawabganj. Int J Expt Agric 6:1–3

Umeh VC, Garcia LE (2008) Monitoring and managing *Ceratitis* spp. complex of sweet orange varieties using locally made protein bait of brewery waste. Fruits 63:209–217

Umeh VC, Ibekwe H (2012). Species diversity and distribution of terphritid fruit flies on mango in Nigeria. In: Second international symposium of Tephritid workers of Europe, Africa and the Middle East (TEAM), Kolymbari Crete, Greece 3–6 July, 2012.

Umeh VC, Ivbijaro MF (1997) Termite abundance and damage in traditional maize intercrops in South western Nigeria. Ins Sci and Applic 17:315–321

Umeh VC, Onukwu D (2011) Effectiveness of foliar protein bait sprays in controlling *Bactrocera invadens* (Diptera: Tephritidae) on sweet oranges. Fruits 66:307–314

Umeh VC, Ahonsi S, Kolade JA (1998) Insect pests encountered in a citrus orchard in Nigeria. Fruits 53:397–408

Umeh VC, Aiyelaagbe IOO, Kintomo AA, Giginyu MB (2002) Insect pest situation and cultural practices in citrus orchards in the Southern Guinea savannah ecological zone of Nigeria. The Nigerian J Hort Sci 7:26–32

Umeh VC, Olaniyan AA, Ker J, Andir J (2003) Development of citrus fruit fly control strategies for small-holders in Nigeria. Fruits 59:265–274

Umeh VC, Anikwe JC, Okelana FA, Bokonon-Ganta A, Onukwu D (2007) Recent observations on fruit flies of coffee and their level of infestation on other economic tree crops in Nigeria. In: Olufolaji AO, Umeh VC (eds), Proceedings of the 25th annual conference of Horticultural Society of Nigeria, 4–8th November 2007, 321 p

Umeh VC, Garcia LE, De Meyer M (2008) Fruit flies of citrus in Nigeria: species diversity, relative abundance and spread in major producing areas. Fruits 63:145–153

Umeh VC, Babalola SO, Egbufor E (2009) Influence of fruit acidity on the infestation and damage of sweet orange varieties by *Ceratitis* species (Diptera: Tephritidae). Nigerian J Entomol 26:41–46

Vayssières J-F, Goergen G, Lokossou O, Dossa P, Akponon C (2005) A new Bactrocera species in Benin among mango fruit fly (Diptera: Tephritidae) species. Fruits 60:371–377

Vayssières J-F, Korie S, Ayegnon D (2009) Correlation of fruit fly (Diptera: Tephritidae) infestation of major mango cultivars in Borgou (Benin) with abiotic and biotic factors and assessment of damage. Crop Prot 28:477–488

White IM, Elson-Harris MM (1992) Fruit flies of economic significance: their identification and binomics. CABI International, Wallingford, 601 p

Vincent Umeh is an ecological entomologist with the National Horticultural Research Institute (NIHORT) Ibadan, Nigeria. His areas of research are mostly on the development and implementation of Integrated Pest Management (IPM) for protection of horticultural crops with emphasis on biological control practices and the application of semiochemical-mediated behavioural manipulation for pest management. He is presently the Head of Citrus and Product Development Research Department in NIHORT. He has participated in many national and international agricultural projects. He is also a National Coordinator of some Regional projects.

Daniel Onukwu is a Principal Technologist with crop protection training, particularly on fruit flies. He was formally with the International Institute of Tropical Agriculture. He works presently with the National Horticultural Research Institute (NIHORT) Ibadan, Nigeria and spent the greater part of his research career working on Integrated Pest Management for protection of horticultural crops.

Chapter 26
Release, Establishment and Spread of the Natural Enemy *Fopius arisanus* (Hymenoptera: Braconidae) for Control of the Invasive Oriental Fruit Fly *Bactrocera dorsalis* (Diptera: Tephritidae) in Benin, West Africa

Désiré Gnanvossou, Rachid Hanna, Aimé H. Bokonon-Ganta, Sunday Ekesi, and Samira A. Mohamed

Abstract The opiine parasitoid *Fopius arisanus* is a biological control agent of frugiverous tephritid fruit flies, particularly those within the *Bactrocera dorsalis* complex. Over a 4-year period from 2009 to 2012, *F. arisanus* was released in various agro-ecological zones (Forest Savanna Mosaic [FSM], Southern Guinea Savanna [SGS], Northern Guinea Savanna [NGS] and Sudan Savanna [SS]) in Benin, in orchards of various host plants including mango, bush mango, guava, citrus and tropical almond. In the FSM zone, and specifically in bush mango plantations, we released and assessed parasitoid population phenology, establishment and spread to determine its impact on *B. dorsalis* populations. Two release methods were used, either as *B. dorsalis* parasitized pupae, or as adult parasitoids. Approximately 258,000 parasitized *B. dorsalis* pupae (from which 134,160 parasitoids emerged and any emerging fruit flies were trapped and not released), and 272,000 adults were released across all orchard release sites. The frequency of recoveries of *F. arisanus* was higher on bush mango (1–11 times) than all other fruit; the average annual percent parasitism ranged from 0.01 to 21.04. This resulted

---

D. Gnanvossou (✉)
International Institute of Tropical Agriculture (IITA-Benin),
08 BP 0932, Tri Postal, Cotonou, Benin
e-mail: d.gnanvossou@cgiar.org

R. Hanna
IITA-Cameroon, BP 2008 (Messa), Yaoundé, Cameroon

A.H. Bokonon-Ganta
Plant Protection and Quarantine Service, Direction of Plant Production (SPVCP/DPV),
01 B.P. 58, Oganla, Porto-Novo, Benin

S. Ekesi • S.A. Mohamed
Plant Health Theme, International Centre of Insect Physiology & Ecology (*icipe*),
PO Box 30772-00100, Nairobi, Kenya

in reduction of *B. dorsalis* population on this crop by 33–65 % over the study period. The highest percent parasitism was about 46.53 %. In addition to the recoveries from the release sites, *Fopius arisanus* was consistently recovered in bush mango at 17 other locations surrounding the orchard release sites, and it continues to spread. The extent of spread was 8 km SE, 6 km W, and 8 km N of the orchard release sites in 2012. The present study is the first case study in West Africa that demonstrates the establishment, persistence and spread of *F. arisanus*.

Keywords Biological control • *Irvingia gabonensis* • Tephritid • Parasitism • Pupae • Host

1 Introduction

The tephritid fruit fly genus, *Bactrocera*, is comprised of at least 440 species distributed primarily in tropical Asia, Australia and the South Pacific (White and Elson-Harris 1992). Amongst these species, four have been reported in Africa: the melon fly, *Zeugodacus cucurbitae* (Coquillett) (synonym: *Bactrocera cucurbitae*), on cucurbits (Vayssières et al. 2007; Gnanvossou et al. 2008); the peach fruit fly, *Bactrocera zonata* (Saunders), (White and Elson-Harris 1992; EPPO 2005; De Meyer et al. 2007), the oriental fruit fly, *Bactrocera dorsalis* (Hendel) (synonym: *Bactrocera invadens*) on fruit trees, including horticultural trees (Lux et al. 2003); and the solanum fruit fly, *Bactrocera latifrons* (Hendel) on cucurbits and solanaceous plants (De Meyer et al. 2007; Mwatawala et al. 2007; Mziray et al. 2010). These fruit flies have a great effect on development of the domestic horticulture industry and hence negatively impact on foreign exports of horticultural produce due to the direct damage they cause to various crops and indirectly through the quarantine restrictions imposed that limit export to lucrative markets abroad (Clarke et al. 2005).

In Africa, *B. dorsalis* was first reported in 2003 in Kenya (Lux et al. 2003) and later in Benin in 2004 (Hanna et al. 2005; Bokonon-Ganta et al. 2007; Goergen et al. 2011). It was initially described as *B. invadens* (Drew et al. 2005), but was recently synonymized with *B. dorsalis* based on integrative taxonomic studies (Schutze et al. 2014). *Bactrocera dorsalis sensu stricto* is one species within the diverse and destructive *B. dorsalis* complex (Drew et al. 2005; Yeates 2005) that includes about 75 highly polyphagous species (Drew and Hancock 1994; Clarke et al. 2005; Goergen et al. 2011).

Recent studies revealed that, within 4 years, *B. dorsalis* had spread throughout 28 countries in continental Africa, infesting 46 wild and cultivated fruit species belonging to at least 23 plant families (Goergen et al. 2011). Mango, *Mangifera indica* L.; bush mango, *Irvingia gabonensis* (Aubry-Lecomte) Baill.; guava, *Psidium guajava* L.; tropical almond, *Terminalia catappa* L.; *Citrus* species; and shea tree, *Vitellaria paradoxa* Gaertn., are the preferred host plants (Vayssières et al. 2005;

Ndzana Abandan et al. 2008; Goergen et al. 2011). Direct damage to mango due to *B. dorsalis* can range from 30 to 80% of fruit depending on the cultivar, locality, elevation and season (Ekesi et al. 2006a; Rwomushana et al. 2008; Hanna et al. 2008; Vayssières et al. 2009). In addition to the direct losses, indirect losses attributed to quarantine restrictions have been enormous (French 2005).

Following the introduction of *B. dorsalis* into Africa, the International Institute of Tropical Agriculture (IITA) and the International Centre of Insect Physiology and Ecology (*icipe*) explored a number of management options for the control of fruit flies. These included the bait application technique (BAT) using the commercial bait GF-120 alone or in combination with entomopathogenic fungi; the male annihilation technique (MAT) using male pheromones; cultural practices; and promotion of predatory weaver ants, *Oecophylla longinoda* (Latreille) (Ekesi et al. 2006b; Van Mele et al. 2007; Hanna and Gnanvossou 2009). While these management measures can result in substantial reductions in fruit fly populations, being an exotic fruit fly species, the pest also lends itself to classical biological control measures. The importance and potential economic benefits of introducing parasitoids into fruit-growing regions in order to reduce populations of invasive fruit fly species has been emphasized by a number of researchers (Wharton 1989; Ovruski et al. 2000; Rousse et al. 2005). Amongst the natural enemies of tephritid fruit flies, parasitoids in the family Braconidae are most often used in biological control (Wharton 1989; Ovruski et al. 2000; Rousse et al. 2005).

Fopius arisanus (Sonan) is a well-studied opiine egg-pupal parasitoid of fruit flies including *B. dorsalis*. It originates in Asia where it was found parasitizing eggs and first instar larvae of fruit flies (Bautista et al. 1998; Calvitti et al. 2002; Altuzar et al. 2004). In Hawaii, *F. arisanus* have been lauded as the most outstanding classical biological control success to ever have been undertaken against fruit flies (Vargas et al. 1993; Harris et al. 2000; Vargas 2001; Rousse et al. 2006; Harris 2010). More recently, release of this parasitoid has resulted in substantial reductions in fruit fly population in French Polynesia (Vargas et al. 2007). This parasitoid species was introduced from Hawaii into Kenya by *icipe* in 2006 and thereafter introduced into Benin in 2008 by IITA for the management of *B. dorsalis*. Initial laboratory studies indicated that *F. arisanus* showed a significant preference for parasitism of *B. dorsalis* compared with the Mediterranean fruit fly (medfly), *Ceratitis capitata* (Wiedemann), the mango fruit fly, *Ceratitis cosyra* (Walker), and *Ceratitis anonae* Graham. Furthermore, eggs laid in the Natal fruit fly, *Ceratitis rosa* Karsch, and *Ceratitis fasciventris* were encapsulated (Mohamed et al. 2010).

Here we report on various studies to assess parasitism, establishment, persistence and spread of *F. arisanus* in Benin following its first release in 2009. Specifically, we determined (1) parasitism of fruit flies by *F. arisanus* on various host fruits including mango, bush mango, guava and tropical almond, which are known to be the preferred hosts for *B. dorsalis* in Benin; (2) the phenology of *F. arisanus* in bush mango orchards in the FSM region of Benin; (3) the spread and persistence *F. arisanus* on bush mango; and (4) documented how the physiological stages of fruit influence recoveries of *F. arisanus* from fruit flies on bush mango.

2 Production and Release of *Fopius arisanus* in Benin

2.1 *Mass Rearing of* Fopius arisanus

In May 2008, a cohort of circa 1000 adult *F. arisanus* (approximately 70% female) was shipped to IITA-Benin, Abomey-Calavi, from a laboratory colony at *icipe*, Nairobi, Kenya. This cohort was maintained on *B. dorsalis* in an isolation room within the insectary for approximately 6 months until government approval for release was received in November 2008.

Adult parasitoids were maintained at 27 ± 2 °C, 60–85% RH, in a 12:12 photoperiod, in Plexiglas screened cages ($20 \times 20 \times 20$ cm). Water and honey were provided *ad libitum*. Parasitoids were introduced into oviposition units comprised of sections of *B. dorsalis*-infested papaya (*Carica papaya* L.) and allowed to oviposit for approximately 48 h (Fig. 26.1a). After exposure, the oviposition units were transferred to plastic cups (9 cm diameter, 5 cm depth) containing 150 g artificial diet (Ekesi and Mohamed 2011) which were themselves placed inside larger plastic containers (12 cm diameter, 8 cm depth) with a 1 cm layer of sterilized sand at the bottom to serve as a pupation substrate. The sand was moistened with a few drops of water to prevent pupal desiccation. The large plastic containers were thereafter covered with fine mesh for ventilation and heat exchange but also to allow parasitized *B. dorsalis* larvae to move into the sand layer and prevent entry of contaminating *Drosophila* species (Fig. 26.1b). After approximately ten days the parasitized fruit fly pupae were sieved from the sand, counted, placed in plastic cups (9.5 cm diameter, 4.5 cm depth) in Plexiglas screened cages ($25 \times 25 \times 25$ cm) until adult parasitoid emergence (Fig. 26.1c). The wasps that emerged were provided with water and drops of honey.

2.2 *Packaging and Transport of* Fopius arisanus *to Orchard Release Sites*

Two release methods were used; the parasitoids were released as adults or as parasitized *B. dorsalis* pupae. For adult releases 1000–1500 (1:1 ratio of males:females), 5–10-day-old wasps were aspirated into Plexiglas screened cages ($25 \times 25 \times 25$ cm) and provided with honey for one day before the release. For parasitized *B. dorsalis* pupae, 1000–1500, 4–5 day-old individuals per 9 cm Petri dish were collected one day prior to release. Cages containing adult parasitoids or Petri dishes containing parasitized *B. dorsalis* pupae intended for release were maintained in an air-conditioned room at 27 ± 2 °C, 60–85% RH and a 12:12 light:dark regime prior to transportation. On the day of release, parasitized *B. dorsalis* pupae were transferred to a Styrofoam box with a cooling element. Containers of both parasitized *B. dorsalis* pupae and adult parasitoids were kept inside an air-conditioned vehicle during transportation. Releases were made within two to four hours of the packages being picked up from the IITA-Benin insectary.

Fig. 26.1 Mass rearing of *Fopius arisanus* (**a**) section of papaya fruit on top of blue plastic container for *F. arisanus* oviposition; (**b**) devices for incubation of oviposition units; (**c**) devices used for *F. arisanus* emergence

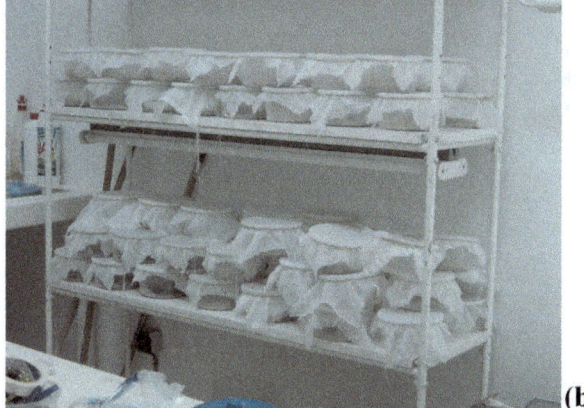

2.3 Orchard Release Sites, Methods of Release and Number of Fopius arisanus Released

Releases were made in four agro-ecological zones in Benin (all lowland humid zones; Fig. 26.2) and the number of orchard release sites (including mango, bush mango, guava, citrus and tropical almond amongst others) per zone varied depending on year: Forest Savanna Mosaic (FSM; 4–6 orchards), Southern Guinea Savanna (SGS; 1–5 orchards), Northern Guinea Savanna (NGS; 2–3 orchards) and Sudan Savanna (SS; 1 orchard).

Two release methods were used: Parasitized *B. dorsalis* pupae were released in buckets (1000–1500 per bucket), covered with a fine mesh (to allow parasitoids to

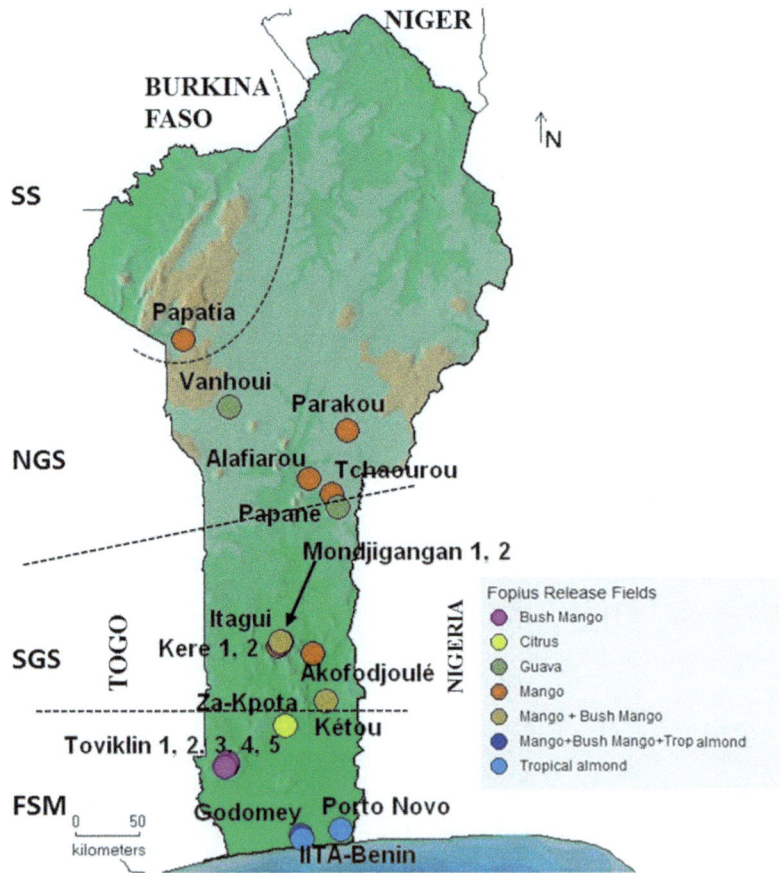

Fig. 26.2 Map of *Fopius arisanus* experimental orchard release sites and fruit sampling sites in Benin, West Africa, 2009–2014. *Broken lines* are the boundaries of the agro-ecological zones (*FSM* Forest Savanna Mosaic zone, *NGS* Northern Guinea Savanna zone, *SGS* South Guinea Savanna zone, *SS* Sudan Savanna zone)

escape but not any emerging fruit flies) and hung in the tree canopy at a height of 1.5 m and approximately 50 m apart (Fig. 26.3a). Four buckets were installed at each orchard release site. After a month the old batches of parasitized *B. dorsalis* pupae were replaced with new batches of parasitized *B. dorsalis* pupae. Old pupae were carefully labeled and returned to the laboratory to determine the number of dead flies/parasitoids retained in each bucket after the living adult parasitoids had escaped through the mesh. Adult parasitoids were released from Plexiglas cages placed close to bunches of mature/ripe fruits on trees. One side of the mesh cage was then removed allowing adult parasitoids to escape and search for host eggs in infested fruits (Fig. 26.3b).

Approximately 134,160 parasitoids emerging from a total of 258,000 *B. dorsalis* pupae (52 %) and 272,000 adult parasitoids were released between April 2009 and August 2012 at 20 orchards surrounded by various alternative host plants. In mango, releases were made once at eleven orchards, twice at five orchards, three times at two orchards and four times at two orchards during this period. From the total number of parasitoids released, about 23,920 parasitized *B. dorsalis* pupae and 23,360 adults were released between 2011 and 2012, exclusively in the bush mango

Fig. 26.3 *Fopius arisanus* release methods: (**a**) Pupae released in buckets; (**b**) adults released from Plexiglas cages

orchards. In bush mango orchards, we made releases on seven occasions between early May and mid-September 2011, and on nine occasions between early May and late August 2012.

3 Post-Release Monitoring of *Fopius arisanus* in Mango Orchards and Surrounding Vegetation

The size of the orchards where the parasitoids were released ranged from 2 to 75 ha and the mango trees ranged from 6 to 30 years in age. Releases were made in all four agroecological zones (FSM, SGS, NGS and SS); in each zone, efforts were taken to select a range of orchards with cultivars that were early (Gouverneur), medium (Eldon, Jules, Camerounaise, Dabschard, Smith and Rubby) and late maturing (Alphonse, EFAC, Kent, Keitt and Brooks) and where fruits were routinely picked and/or collected from the ground. Fruits from alternative host plants surrounding the selected mango orchards were also sampled both during the mango fruiting season and also in the off-season period, to identify parasitism by *F. arisanus* in associated fruit flies. These alternative host plants included bush mango; tropical almond, *Terminalia catappa* L.; guava; orange, *Citrus sinensis* (L.) Osbeck; pomelo, *Citrus maxima* Merr.; tangelo, *Citrus tangelo* Ingram & Moore; custard apple, *Annona muricata* L.; wild soursop, *Annona senegalensis* Pers.; wild sweetsop, *Annona reticulata* L.; cashew, *Anacardium occidentale* L.; wild peach, *Sarcocephalus latifolius* (Sm.) Bruce; shea tree; avocado, *Persea americana* Mill., cas mango *Spondias dulcis* L.; starfruit, *Averrhoa carambola* L. and sponge gourd, *Luffa cylindrica* Mill.. Every 3–4 weeks a total of 25 fruits of each species were randomly picked from the top of trees and a further 25 from the bottom of trees in each orchard and at each location. Between 2009 and 2014 a total of 26,541 mango fruits were sampled in orchards and 24,145; 6559 and 1755 fruits were sampled in bush mango, tropical almond and guava respectively. A further 5692 fruits were sampled on other host plants from the surrounding vegetation. In the laboratory fruits were incubated individually or in groups of 5–10 fruits (depending on size), to determine the number of fruit flies, fruit fly species composition and the rate of parasitism by *F. arisanus*.

Fopius arisanus parasitized eggs of *B. dorsalis* in mango, bush mango, tropical almond, guava and orange, but parasitism rates were generally greatest in bush mango (maximum = 21.04 % in 2011 [FSM]) compared with other fruit species (Table 26.1). For example, the maximum parasitism rate on mango was 2.33 in 2009 (NGS). Among the 12 mango varieties sampled, *F. arisanus* was only recovered from seven varieties. These were Eldon, Smith, Alphonse, EFAC, Kent, Keitt and Brooks. The total number of adult *F. arisanus* recovered across all agro-ecological zones over the 5 years (2009–2013) varied amongst the mango varieties. It was greatest in Keitt (111 adults), followed by Alphonse and Eldon (35 and 30 adults respectively). The least preferred varieties were EFAC and Kent (four and three adults recovered, respectively). When we consider all host fruit species, the fre-

Table 26.1 *Fopius arisanus* recoveries: frequency and percent parasitism per year in various locations and agro-ecological zones in Benin, West Africa

Year	Agro-ecological zone	Location	Host fruit species	Number of *Fopius arisanus* released		Fruits sampled			Number of each fruit fly species (dominant)			*Fopius arisanus* recovered over the year				
				Pupae	Adults	Total number	Total weight (kg)	No. of sample occasions	Total pupae	*B. dorsalis*	*C. cosyra*	No. of females	No. of males	Frequency of recovery	% Parasitism per kg fruit	Other parasitoid species
2009	FSM	IITA-Benin Ketou	Mango, *Mangifera indica*	0	33,815	908	559.32	8	7559	2302	0	2	0	2 (May-June)-Ketou	0.00–0.03	
		IITA-Benin Cotonou	Tropical almond, *Terminalia catappa*			863	28.76	14	205	155	0	0	0	0	0.00	
		Ketou	Bush mango, *Irvingia gabonensis*			195	24.64	5	505	373	0	20	21	1 (May)-Ketou	8.10	
		Ketou	Guava, *Psidium guajava*			149	6.10	6	12	10	0	0	0	0	0.00	
		Ketou	Orange, *Citrus sinensis*; Cashew,			462	65.28	7	0	0	0	0	0	0	0.00	
		Za-kpota	*Anacardium occidentale*; Pomello,													
			Citrus maxima; Tangelo,													
			Citrus tangelo; custard apple,													
			Annona muricata; wild soursop,													

(continued)

Table 26.1 (continued)

Year	Agro-ecological zone	Location	Host fruit species	Number of Fopius arisanus released		Fruits sampled			Number of each fruit fly species (dominant)			Fopius arisanus recovered over the year				Other parasitoid species
				Pupae	Adults	Total number	Total weight (kg)	No. of sample occasions	Total pupae	B. dorsalis	C. cosyra	No. of females	No. of males	Frequency of recovery	% Parasitism per kg fruit	
	SGS	Akofodjoule	Annona senegalensis													
		Akofodjoule	Mango; Orange; Wild soursop; Wild peach,	0	7460	594	161.58	7	4358	2215	156	5	6	1 (June)–Akofodjoule	0.25	
		Akofodjoule	Sarcocephalus latifolius			334	9.93	3	893	0	728	0	0	0	0.00	28 F. caudatus
	NGS	Tchaourou Alafiarou Parakou	Mango	0	29,894	6080	1232.88	21	8481	5361	292	19	24	1–4 (June–August)–Tchaourou & Parakou	0.01–2.33	
		Tchaourou Alafiarou Vanhoui	Guava			1359	61.69	12	804	704	0	3	1	2 (July–August)–Tchaourou	0.00–0.60	
		Tchaourou Alafiarou Parakou	Custard apple; Wild soursop; Wild peach; Shea tree, Vitellaria paradoxa			1305	53.64	28	1241	7	1630	0	0	0	0.00	4 F. caudatus
	SS	Papatia	Mango	0	14,013	639	426.50	3	1789	955	147	3	4	4 (June–August)–Papatia	0.40	
		Papatia	Wild peach; Shea tree			116	6.51	7	509	343	0	0	0	0	0.00	55 F. caudatus

2010	FSM	Ketou	Mango	49,000	26,680	163	55.83	2	1004	566	1	0	0	0	0.00
		IITA-Benin	Bush mango			1956	296.64	44	57,868	41,707	0	993	934	1–7 (May–October) – IITA-Benin, Ketou, Toviklin 1, 2 & 3	0.17–6.10
		Ketou	Toviklin 1, 2 & 3												
		Ketou	Guava			54	5.18	2	108	79	0	0	0	0	0.00
		IITA-Benin	Tropical almond			415	12.95	12	0	0	0	0	0	0	0.00
		IITA-Benin	Orange; Mandarin orange,			57	3.31	3	0	0	8	0	0	0	0.00
		Ketou	*Citrus reticulate*; Wild soursop												
	SGS	Mondjigangan 1	Mango	17,000	13,180	872	303	11	6059	3288	86	6	6	1 (June)– Mondjigangan	0.00–0.26
		Akofodjoule													
		Papane	Guava			173	6.90	3	99	74	0	0	0	0	0.00
		Mondjigangan 1	Wild soursop; Wild peach			425	12.86	8	0	0	275	0	0	0	0.00
		Akofodjoule													

(continued)

Table 26.1 (continued)

Year	Agro-ecological zone	Location	Host fruit species	Number of Fopius arisanus released - Pupae	Number of Fopius arisanus released - Adults	Fruits sampled - Total number	Fruits sampled - Total weight (kg)	Fruits sampled - No. of sample occasions	Number of each fruit fly species (dominant) - Total pupae	B. dorsalis	C. cosyra	Fopius arisanus recovered over the year - No. of females	No. of males	Frequency of recovery	% Parasitism per kg fruit	Other parasitoid species
	NGS	Tchaourou Alafiarou Parakou	Mango	24,000	0	3149	1160.80	21	14,245	5419	1098	5	0	0	0.00	213 Tetrastichus giffardii
		Tchaourou	Wild soursop; Shea tree; Cashew;			1787	57.55	11	0	0	52	0	0		0.00	
		Alafiarou	Wild peach													
		Vanhoui	Guava			20	0.70	1	0	0	0	0	0	0	0.00	
	SS	Papatia	Mango			156	62.40	1	1455	64	235	0	0	0	0.00	
2011	FSM	IITA-Benin Ketou	Mango	33,000	77,025	3189	565.29	23	2005	1108	0	3	0	2 (July)-Ketou	0.00–0.52	
		IITA-Benin	Bush mango			3549	509.45	58	52,192	40,396	0	1444	1007	3–11 (May-December)–IITA-Benin,	0.52–21.04	
		Ketou												Ketou,		
		Toviklin 1, 2 & 3												Toviklin 1 & 3		
		IITA-Benin	Tropical almond			2296	95.41	29	985	627	0	14	10	8 (March & November)–IITA-Benin	0.00–2.43	
		Porto-Novo														
		Porto-Novo	Starfruit, Averrhoa carambola			20	1.05	1	0	0	0	0	0	0	0.00	
		Porto-Novo	Cas mango, Spondias dulcis			30	0.80	1	0	0	0	0	0	0	0.00	
		IITA-Benin	Avocado, Persea americana			60	11.25	6	0	0	0	0	0	0	0.00	

Year	Site	Location	Host												3 Psyttalia cosyrae
	SGS	Mondjigangan 1 Akofodjoule	Mango	0	52,516	1410	581.08	14	8355	7911	612	13	15	3 (July)—Mondjigangan	0.00–0.41
		Mondjigangan 2	Bush mango			594	72.47	17	9397	5811	0	15	14	1–3 (August–September)—Mondjigangan, Itagui, Kere 1 & 2	0.17–2.59
	NGS	Tchaourou Alafiarou	Mango	0	3718	1341	517.95	10	6744	1239	944	0	0	0	0
2012	FSM	IITA-Benin Ketou	Mango	134,750	14,372	2854	582.62	30	10,052	4912	0	29	9	2–6 (January July)—IITA-Ketou	0.16–0.45
		IITA-Benin Toviklin 1, 2, 3, 4 & 5	Bush mango			6956	809.55	136	74,364	53,566	0	1708	1403	4–10 (May–December)—IITA-Benin	0.73–9.96
		Toviklin 3	Tropical almond			200	37.70	4	105	61	0	0	12	1 (October)	5.48
		Ketou	Mandarin orange			224	101.65	6	0	0	0	0	0	0	0.00
		Ketou	Orange			653	84.55	13	20	12	0	2	0	1 (September)—Ketou	10.28
		Toviklin 1	Sponge gourd, Luffa cylindrica			93	13.60	3	3	1	0	0	0	0	0.00
	SGS	Itagui Kere 1 & 2	Mango	0	24,150	505	42.37	13	0	0	0	0	0	0	0.00
		Itagui Kere 1 & 2	Bush mango			986	169.61	36	10,926	8131	0	65	33	5–6 (July–October)—Itagui, Kere 1 & 2	0.51–1.04

(continued)

Table 26.1 (continued)

Year	Agro-ecological zone	Location	Host fruit species	Number of Fopius arisanus released - Pupae	Number of Fopius arisanus released - Adults	Fruits sampled - Total number	Fruits sampled - Total weight (kg)	No. of sample occasions	Number of each fruit fly species (dominant) - Total pupae	B. dorsalis	C. cosyra	Fopius arisanus recovered over the year - No. of females	No. of males	Frequency of recovery	% Parasitism per kg fruit	Other parasitoid species
2013	FSM	IITA-Benin	Mango			830	247.69	3	1333	958	0	0	0	0	0.00	
		IITA-Benin	Bush mango			5415	838.70	66	105,591	84,348	0	57	43	1–4 (January–December)–IITA-Benin, Toviklin 1, 2, 3, 4 & 5	0.02–0.16	
		Toviklin 1, 2, 3, 4 & 5														
		IITA-Benin	Tropical almond			2510	102.77	24	4108	3175	0	0	0	0	0.00	2 unknown species
		Ketou	Pomello			321	86.00	8	0	0	0	0	0	0	0.00	
	SGS	Mondjigangan 1	Mango	0	0	2234	510.29	29	2741	335	1076	0	0	0	0.00	
		Akofodjoule														
		Itagui														
		Kere 1 & 2														
		Itagui	Cashew			135	8.50	3	0	0	0	0	0	0		
		Kere 2														
	NGS	Tchaourou 1 & 2	Mango	0	0	1617	662.00	12	5127	1032	2180	0	0	0	0	
2014	FSM	IITA-Benin	Bush mango			4494	662.29	33	50,821	40,345	0	16	10	1–2 (February–August)–Toviklin 1, 2, 3, 4 & 5	0.01–0.12	
		Toviklin 1, 2, 3, 4 & 5														
		IITA-Benin	Tropical almond			275	9.94	4	79	49	0	0	0	0	0.00	
		IITA-Benin	Mandarin orange			20	1.95	1	0	0	0	0	0	0	0.00	
		IITA-Benin	Orange			72	10.23	2	0	0	0	0	0	0	0.00	
		Ketou														

quency of recovery was greatest in bush mango (maximum = 11 in 2011 [FSM]); the maximum in mango was six in 2012 (FSM) and four in 2009 (NGS), NGS being the area where quality mango varieties are usually grown in Benin.

Overall, patterns of parasitism of *B. dorsalis* eggs by *F. arisanus* varied depending on host fruit species; the frequencies of recovery as well as the percent parasitism were generally lower on mango and other host fruit species than on bush mango. All batches of adult parasitoids that were released at the various sites came from the same population reared on *B. dorsalis*-infested papaya. Thus, we can exclude any positive effect of previous experience or learning by *F. arisanus* on its preference for bush mango. The higher parasitism rates reported on bush mango compared with other fruits are likely to be as a result of differences in host fruit-attractiveness to *F. arisanus*. Chemical, physical and volatile cues from host plants of the parasitoid are known to be important cues for orientation during foraging for oviposition opportunities, adult food and mating sites (Liquido 1991; Vet and Dicke 1992; Godfray 1994; Stuhl et al. 2011; Peréz et al. 2011). Harris and Bautista (1996) demonstrated that the rates of parasitism by *F. arisanus* on both *B. dorsalis* and *C. capitata* were dependent on the fruit species infested and that the most preferred fruit was different for the two species of fruit fly. Altuzar et al. (2004), found that *F. arisanus* were more attracted to uninfested guava than uninfested orange. The use of fruit volatiles during host location has also been documented for two *Diachasmimorpha* species of parasitoid (Eben et al. 2000; Henneman et al. 2002). Bautista and Harris (1996) reported that chemicals in citrus peel were toxic to eggs and larvae of tephritid fruit flies and thus explained the low parasitization rates of *F. arisanus* in citrus fruits. A similar trend was also found in another host/parasitoid system where *Diachasmimorpha* spp. were able to discriminate between a toxic fruit species (i.e. guava), and a non-toxic fruit species (i.e. orange) (Stuhl et al. 2012). In nature, a variety of host fruits are commonly encountered by parasitoids of tephritid fruit flies; differences in the food value/oviposition environment of different fruits may have a significant impact on the establishment and persistence of *F. arisanus*. Understanding differences in establishment of *F. arisanus* in mango, different mango cultivars and bush mango would help identify the chemical basis of fruit-attractiveness and/or micronutrient richness.

4 Post-Release Monitoring of *Fopius arisanus* in Bush Mango Orchards and Surrounding Vegetation

Parasitoid releases in bush mango were made in the South-Western region of Benin in the FSM agro-ecological zone (Fig. 26.4), where bush mango grows in the wild but is also cultivated in plantations; it flowers and produces fruits throughout the year. The rainfall pattern in this region is bimodal. The long rainy season begins at the end of March and ends in July, peaking in June, and is followed by a short dry season. The short rainy season begins at the end of September and ends in November, peaking towards the end of September. From the end of November until the

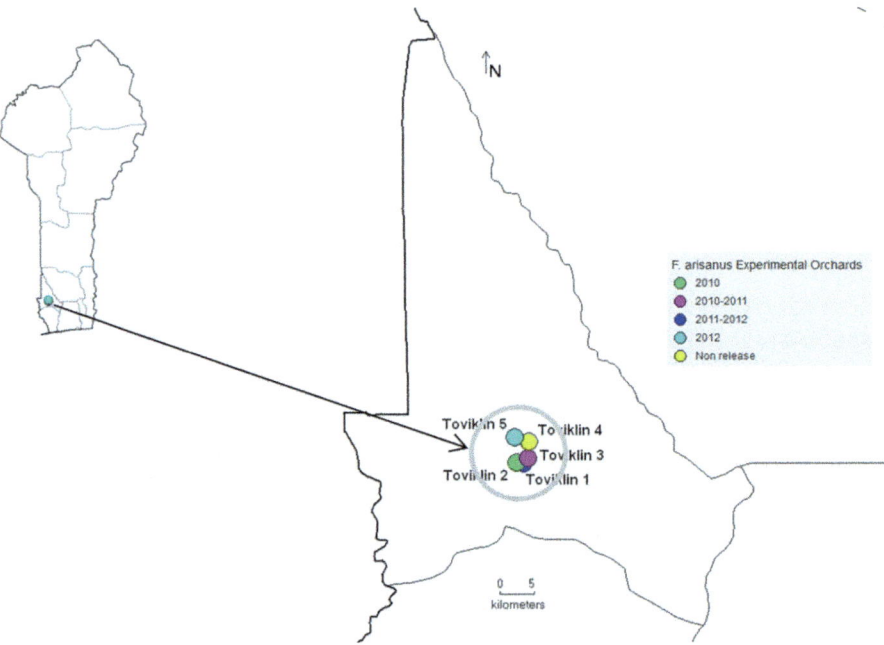

Fig. 26.4 Experimental orchards where phenology of *Bactrocera dorsalis* and *Fopius arisanus* was recorded in Toviklin (FSM), South-West Benin, 2010–2012

beginning of March the 'Harmattan', a dry dusty wind originating in the Sahara, occurs frequently. There were four orchard release sites but in only one were consecutive releases made over 2 years. The data presented here were generated from that release orchard: Toviklin 1. The size of the orchard was 5–8 ha. Plants were 6–15 years old at the release orchard but varied between 6 and 30 years old in the surrounding vegetation. The establishment, spread and phenology of *F. arisanus* on bush mango was determined by periodic sampling of fruits. Sampling began on the day of the first release and continued on a bi-weekly basis for 1 year (2011–2012) and thereafter on a bimonthly basis for 2 years (2012–2014). Between 30 and 100 fruits were sampled on each occasion and were comprised of an equal number of randomly selected fruits picked from trees and recently fallen fruits. To evaluate spread, the same methodology was applied to fruits sampled along secondary roads running away from the release site. Initially we surveyed at 100 m intervals up to 500 m distant and in four directions from the orchard release site. Once we began to record recovery of parasitoids at 500 m, the scale was increased to 2 km intervals in each direction. The coordinates of each sample point away from the orchard release site were recorded using GPS. Sampled fruits were incubated in the laboratory in groups of 5–10 fruits, each group in a large plastic bucket. Emerging pupae were placed in Petri dishes and maintained at 27 ± 2 °C, 60–85 % RH, in a 12:12 light:dark regime until both adult flies and parasitoids emerged. The emerged flies and parasitoids were counted, identified and a database of results established. Using the

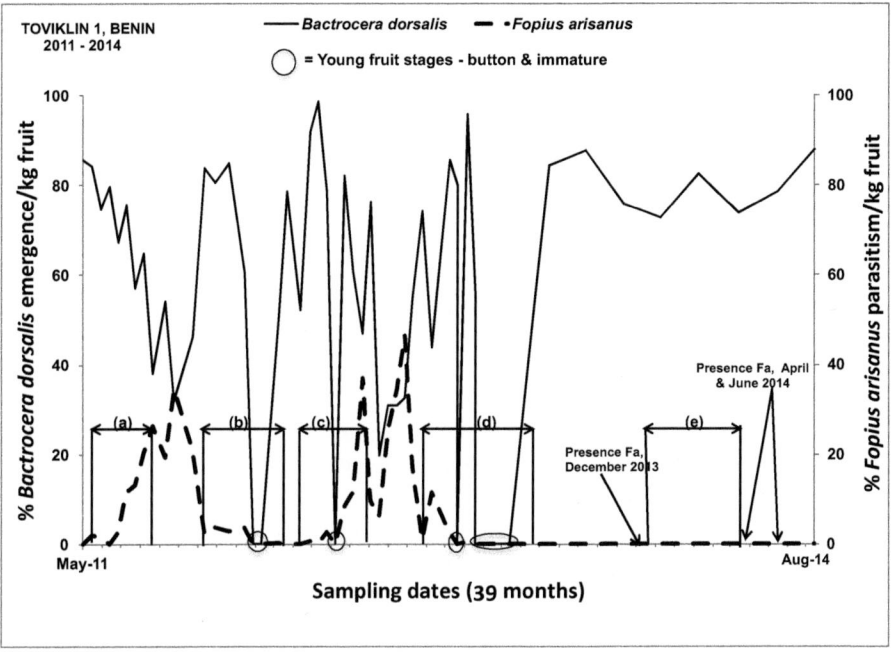

Fig. 26.5 Phenology of *Bactrocera dorsalis* and *Fopius arisanus* found on bush mango in Toviklin, Benin. Indicated on Y-axis are the percent of *B. dorsalis* emerging (*solid line*) and percent parasitism by *F. arisanus* (*broken line*) per kilogram fruit. Fa = *F. arisanus*. (**a**) Release period 2011; (**b**) Dry season 2011–2012; (**c**) Release period 2012; (**d**) Dry season 2012–2013; (**e**) Dry season 2013–2014; *Arrows* indicate when fewer adult *F. arisanus* were recorded

software DIVA-GIS and the geo-referenced sample locations surrounding the release site, the perimeter of the area of spread of *F. arisanus* could be established (as described by Hijmans et al. 2004).

4.1 Phenology of Bactrocera dorsalis *and* Fopius arisanus *in Bush Mango Orchards*

In 2011 the population of *F. arisanus* had two peaks, one at the end of the release period (mid-September) and the other one in mid-October; parasitism rates ranged from 26.30 to 34.47 % (Fig. 26.5). In 2012, the population of *F. arisanus* had three peaks, one at the end of the release period (mid-August) and the other two in late October and early December with parasitism rates ranging from 11.64 to 46.53 % (Fig. 26.5). During the 2-year period, the increase in *F. arisanus* population level was generally followed by a decrease in *B. dorsalis* population level and vice versa. The data also revealed that *B. dorsalis* populations were reduced by 33–65 % as the *F. arisanus* population increased (Fig. 26.5). Later in the dry seasons following the peak population, population densities drastically decreased and recoveries became scarce. Such a pattern was observed at a time when trees were only bearing immature fruits. Surprisingly, in 2013 and 2014, there was no apparent peak, although the parasitoid remained present in low numbers in both the orchard release and the spread orchards. Consequently, the proportion of *B. dorsalis* emerging per kg fruit increased considerably (approx. 90% emergence) compared with previous sampling periods which were 30–50% emergence in 2011 and 2012 when rates of parasitism by *F. arisanus* per kg fruit ranged from 26.30 to 46.53%. Over the 2 years, *F. arisanus* occurred more often (six times) in one of the non-release orchards (Table 26.2; Toviklin 4) compared with other sites where the frequency of occurrence did not exceed 4 (Table 26.2).

The continuous presence of *F. arisanus* for more than 69 generations and three dry and wet season cycles in bush mango release orchards in South-Western Benin, represents the first successful establishment, persistence and spread of an egg-larval parasitoid of fruit flies in West Africa. Generally, an exotic natural enemy is considered to be established in a given site if adults are recovered for at least 1 year after final release (DeBach and Barlett 1964). In this study, the phenology of both the host, *B. dorsalis*, and the exotic parasitoid, *F. arisanus* revealed that *F. arisanus* was effective against *B. dorsalis* in bush mango. The highest parasitism rate on bush mango was 46.53% and was recorded two months after the second release period. Other field studies in guava orchards located on Kauai island, Hawaii, have indicated higher levels of parasitism of *B. dorsalis* by *F. arisanus*, often surpassing 50% (Vargas et al. 1993; Purcell et al. 1998). The parasitism rates achieved in our context seem to be lower than previous reports. Climatic conditions in the field, physiological stage of the parasitoid during the release periods, host fruit species and physiological stage of the fruit may have affected host selection and thus the parasitism rate.

4.2 Spread and Persistence of Fopius arisanus in Bush Mango Orchards

Over the 2-year period (2012–2014), in addition to the release orchard, a further 27 localities surrounding the release orchard were consistently sampled (on 25 occasions after the last release in August 2012) for the presence/absence of parasitoids. The results revealed the presence of *F. arisanus* in 17 of these localities where it persisted until August 2014, although no parasitoids were recovered from fruits sampled in October and December 2014 (Table 26.2). Interestingly, approximately a year later, between November and December 2015, *F. arisanus* was again recovered from sampled fruits in both orchard release and non-release sites (Table 26.2). *Fopius arisanus* was recorded as far as 8 km SE, 6 km W and 8 km N of the release orchard, indicating that it had spread considerably from the orchard release site. The estimated perimeter spread area is so far estimated at 40.43 km (Fig. 26.6).

Dispersal of *F. arisanus* may have been achieved by wind dispersal and/or by movement of fruits by people. *Fopius arisanus* successfully established but was

Table 26.2 *Fopius arisanus* persistence in bush mango orchard release, and non-release sites, Toviklin, Couffo Department, South-West Benin, 2013–2015

No.	Orchard	Number of adult *Fopius arisanus* recovered in 2013						Number of adult *Fopius arisanus* recovered in 2014						Number of adult *Fopius arisanus* recovered in 2015		
		January	February, March & April	June	August	October	December	February	April	June	August	October & December	April, June, August & October	November	December
1	Toviklin 1	8	0	0	0	0	3	0	3	2	0	0	0	0	0
2	Toviklin 2	2	0	1	0	0	0	0	6	0	0	0	0	1	1
3	Toviklin 3	27	0	0	0	0	1	0	0	0	1	0	0	0	0
4	Toviklin 4	0	0	2	22	15	5	4	0	0	6	0	0	0	0
5	Toviklin 5	13	0	0	0	0	0	3	1	0	0	0	0	3	0

Last release was dated 29 August 2012

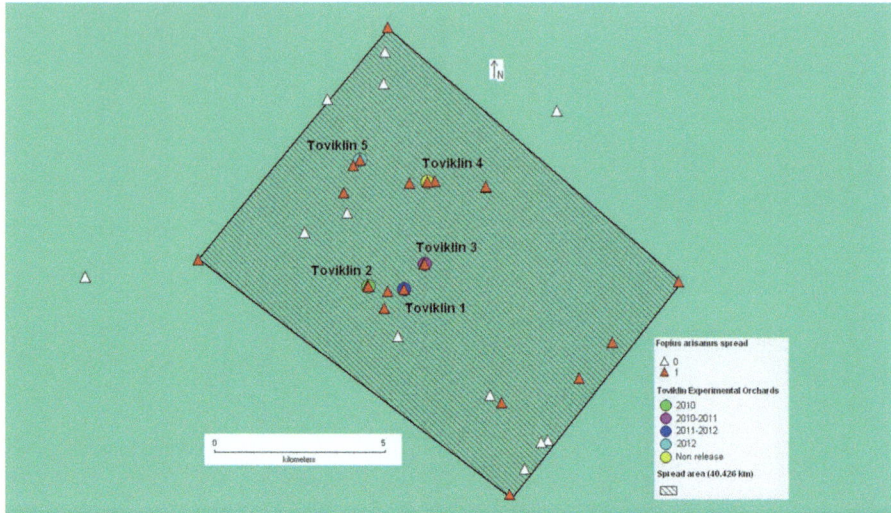

Fig. 26.6 Map of *Fopius arisanus* release and establishment sites (●) and other surrounding locations (▲) where the parasitoid subsequently spread in Toviklin, Benin, West Africa, 2010–2012

restricted to bush mango plantations, where the air humidity remained relatively high throughout the year and the precipitation was at least 1000 mm/year. In contrast, previous reports on other parasitoids released in West Africa revealed rates of spread greater than 8 km per year e.g., the encyrtid parasitoids *Apoanagyrus* (*Epidinocarsis*) *lopezi* De Santis, covered about 100 km per season (Herren et al. 1987), and *Gyranusoidea tebygi* Noyes, covered 100 km per year (Neuenschwander et al. 1994). Differences amongst parasitoid species and host insect species may play a key role in this.

4.3 Effects of the Physiological Stage of Bush Mango Fruits on Fopius arisanus *Recoveries*

To determine the effect of the physiological stage of bush mango fruits on *F. arisanus* recoveries, the emergence data from all infested fruits were analyzed in relation to fruit stage (button, immature or mature) and year (2010, 2011, 2012 and 2013). Both fruits sampled from trees and the ground were used but analyzed separately. The results revealed that the relative frequency of *F. arisanus* recovery from fruits at button, immature and mature stages differed considerably. Across years, *F. arisanus* occurred more frequently (10–80 % fold higher) in mature fruits, either sampled from the tree or the ground compared with immature fruits and buttons (Fig. 26.7).

It is likely that physiological stage of fruit has the greatest effect on its attractiveness to *F. arisanus*, and thus parasitism rates and reproductive capacity on bush mango. In our study the relative frequency of *F. arisanus* recovery was greater from mature fruits than the other stages. These results might be explained by a difference

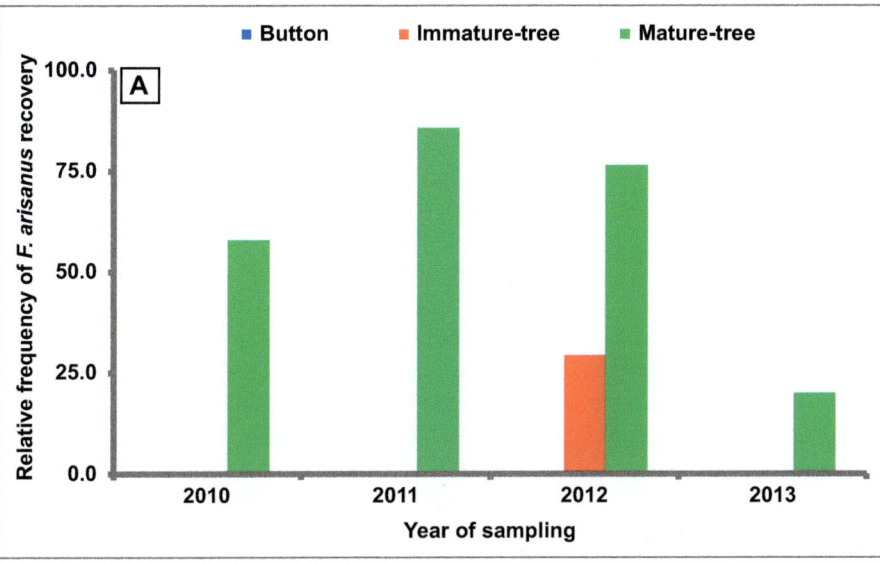

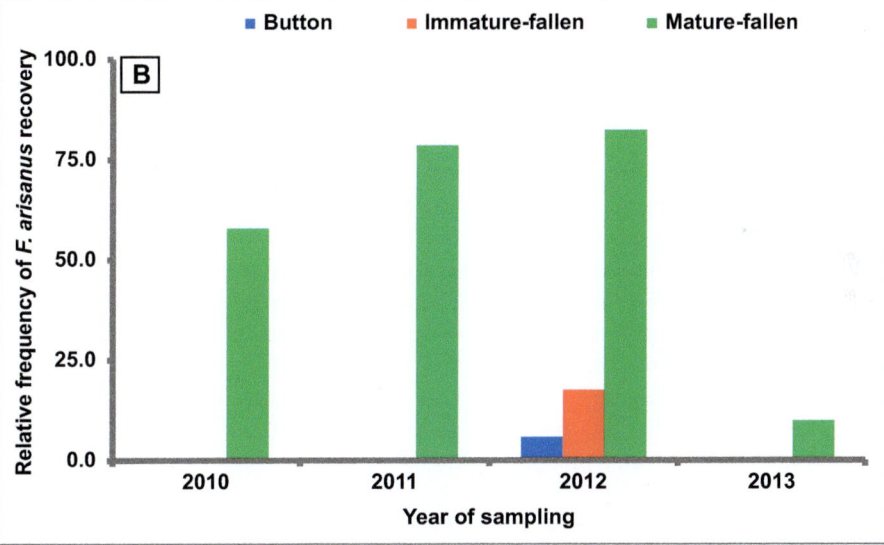

Fig. 26.7 Relative frequency of *Fopius arisanus* recovery in bush mango fruits at three different physiological stages (**a**) from tree (picked fruits) and (**b**) from ground (fallen fruits) in Toviklin, South-West Benin, 2010–2013. Data on *F. arisanus* emergence from all infested fruits from all orchards were pooled by fruit stage and by year

in the nutritional quality of the different fruit stages. Exposure to extended periods when only inadequate fruit (buttons and immature fruits) were available may be the reason why *F. arisanus* was not recovered or was only recovered in low numbers in both the orchard release sites and locations surrounding the orchard release sites when only these fruit stages were available. Fruits at button and immature stages were not as attractive as mature fruits. Previous studies in other systems have shown similar results. For example, it is well known that *F. arisanus* can discriminate amongst papaya fruits at different stages of ripeness (Liquido 1991). Parasitization of *B. dorsalis* eggs by *F. arisanus* was greater in fully-ripe fruits than in one-quarter to half-ripe fruits on trees (Liquido 1991). The same authors also reported that *F. arisanus* never parasitized *B. dorsalis* eggs in mature green or colour-break fruits.

5 Conclusions

The results presented here demonstrate desirable parasitoid-host fruit and parasitoid-host fruit ripeness interactions that support the biological control campaign targeted at *B. dorsalis* using *F. arisanus* in Benin. The physiology of targeted plant species and/or the availability of alternative host fruits in vegetation surrounding the orchard release sites may have played a positive role in the establishment and persistence of *F. arisanus* in the 3 years after the last release in the FSM zone compared with the NGS zone where it did not persist or spread. Both bush mango and mango are perennial plant species but the former produces fruits throughout the year although fruit quality varies depending on the time of the year. In contrast, mango only produces fruit once a year from April to July, followed by a critical 8 month period when conditions are unfavourable for *F. arisanus*: scarcity of host fruits and associated steeply declining host fruit fly populations, high temperatures and low relative humidities, especially during the Harmattan phenomenon. While this study is a significant contribution to the fight against *B. dorsalis* in Africa, more studies are needed on the chemical ecology of *F. arisanus* and on its ecological, agronomic, and economic impact in different agroecological zones is required.

As efforts to establish *F. arisanus* in suitable agro-ecozones continues, other pest management tools such as the utilization of the larval-pupal opiine parasitoid *Diachasmimorpha longicaudata* (Ashmead), use of BAT and MAT, and application of biopesticides should be included within the context of integrated pest management (IPM), particularly in the NGS agro-ecological zone of Benin, where *B. dorsalis* continues to threaten mango production.

Acknowledgements The authors gratefully acknowledge the assistance of the Plant Protection and Quarantine Service, Direction of Plant Production, Porto-Novo throughout this experimental release study. In particular, we thank the late Symphorien Saïzonou and Ibouraima Tiamiyou for their technical assistance. We thank Pierre Soglo, Clément Kèdédji, Anne-Marie Ahandéssi, Léon Hèguè and Koffi Negloh for assistance with Figs. 26.2, 26.4 and 26.6. We are indebted to Houto Hèguè, Atchina and Félix Konoudo, Honoré Dansou, Robert Wamba, Bernadin Dohou and others for allowing us to conduct research work in their mango and bush mango orchards. We thank Muaka Toko for his comments on earlier drafts of this manuscript. This research was funded in part by the Federal Ministry for Economic Cooperation and Development (BMZ); the Deutsche Gesellschaft für Internationale Zusammenarbeit (GIZ) GmbH through BIZ Mango IPM project; and IITA Core Donors.

References

Altuzar A, Montoya P, Rojas J (2004) Response of *Fopius arisanus* (Hymenoptera: Braconidae) to fruit volatiles in a wind tunnel. Fla Entomol 87:616–618

Bautista RC, Harris EJ (1996) Effect of fruit substrates on parasitization of tephritid fruit flies (Diptera) by the parasitoid *Biosteres arisanus* (Hymenoptera: Braconidae). Env Entomol 25:470–475

Bautista RC, Harris EJ, Lawrence PO (1998) Biology and rearing of the fruit fly parasitoid *Biosteres arisanus*: clues to insectary propagation. Entomol Exp Appl 89:79–86

Bokonon-Ganta AH, Gnanvossou D, Hanna R, Saizonou SE (2007) *Bactrocera invadens*, une nouvelle espèce de mouche des fruits nuisible au Bénin. *Fiche Technique*. Dépôt légal n° 3556 du 20/11/2007 4è Trimestre. Bibliothèque Nationale du Bénin, 10pp.

Calvitti M, Antonelli M, Moretti R, Bautista RC (2002) Oviposition response and development of the egg-pupal parasitoid *Fopius arisanus* on *Bactrocera oleae*, a tephritid fruit fly pest of olive in the Mediterranean basin. Entomol Exp Appl 102:65–73

Clarke AR, Armstrong KF, Carmichael AE, Milne JR, Raghu S, Roderick GK, Yeates DK (2005) Invasive phytophagous pests arising through a recent evolutionary radiation: the *Bactrocera dorsalis* complex of fruit flies. Ann Rev Entomol 50:293–319

De Meyer M, Mohamed SA, White IM (2007) Invasive fruit fly pests in Africa. Royal Museum for Central Africa, Tervuren, http://www.africamuseum.be/fruitfly/AfroAsia.htm

DeBach P, Barlett BR (1964) Methods of colonization, recovery and evaluation. In: DeBach P, Schlinger EI (eds) Biological control of insect pests and weeds. Chapman and Hall, London, 844 pp

Drew RAI, Hancock DL (1994) The *Bactrocera dorsalis* complex of fruit flies (Diptera: Tephritidae: Dacinae) in Asia. Bull Entomol Res 2:1–68

Drew RAI, Tsuruta K, White IM (2005) A new species of pest fruit fly (Diptera: Tephritidae: Dacinae) from Sri Lanka and Africa. Afr Entomol 13:149–154

Eben A, Benrey B, Sivinski J, Aluja M (2000) Host species and host plant effects on preference and performance of *Diachasmimorpha longicaudata* (Hymenoptera: Braconidae). Env Entomol 29:87–94

Ekesi S, Mohamed SA (2011) Mass Rearing and Quality Control Parameters for tephritid fruit flies of economic importance in Africa, Wide spectra of quality control. In: Akyar I (ed), ISBN: 978-953-307-683-6, InTech, Available from: http://www.intechopen.com/books/wide-spectra-of-qualitycontrol/mass-rearing-and-quality-control-parameters-for-tephritid-fruit-flies-of-economic-importance-in-afri

Ekesi S, Nderitu PW, Rwomushana I (2006a) Field infestation, life history and demographic parameters of the fruit fly *Bactrocera invadens* (Diptera: Tephritidae) in Africa. Bull Entomol Res 96:379–386

Ekesi S, Mohamed SA, Hanna R, Lux SA, Gnanvossou D, Bokonon-Ganta AH (2006b) Fruit fly suppression – purpose, tools and methodology, pp D1–D15. In: Ekesi S, Billah MK (eds), A field guide to the management of economically important tephritid fruit flies in Africa. Second edition. The International Centre of Insect Physiology and Ecology (*icipe*). Icipe Science Press, Nairobi, Kenya.

EPPO (European and Mediterranean Plant Protection Organization) (2005) *Bactrocera zonata*. EPPO Bulletin 35:371–373

French C (2005) The new invasive *Bactrocera* species, Insect Pest Control. Newsletter, No. 65. International Atomic Energy Agency, Vienna, pp 19–20

Gnanvossou D, Hanna R, Azandémè G, Goergen G, Tindo M, Agbaka A (2008) Inventaire et importance des dégâts des mouches des fruits sur quelques espèces de cucurbitacées au Bénin, pp 140–145. In: Adjanohoun A, Igué, K (eds.), *Actes de l'Atelier Scientifique National 6, Décembre 2006, Abomey-Calavi*. Tome 1. ISBN 978-99919-6850-6 ; ISSN 1659-6161 ; Dépôt légal n° 3838 du 13/08/2008 3è Trimestre. Bibliothèque Nationale du Bénin, 440 pp

Godfray HCJ (1994) Parasitoids: behavioral and evolutionary ecology. Princeton University Press, Princeton

Goergen G, Vayssières J-F, Gnanvossou D, Tindo M (2011) *Bactrocera invadens* (Diptera: Tephritidae), a new invasive fruit fly pest for the afrotropical region: host plant range and distribution in West and Central Africa. Env Entomol 40:844–854

Hanna R, Gnanvossou D (2009) Lutte contre la mouche des fruits en Afrique de l'Ouest : Atelier régional de l'OMC sur les mesures SPS – Douala – 16/19 Juin. La lutte régionale contre les mouches des fruits et légumes en Afrique de l'Ouest. Lettre d'information n°2, Juillet 2009 4p.

Hanna R, Goergen G, Gnanvossou D, Tindo M, Vayssières J-F, Ekesi S, Lux SA (2005) The Asian fruit fly *Bactrocera invadens* in West and Central Africa: Distribution, host range and seasonal dynamics. ESA Meeting, December, 2005

Hanna R, Agnontchémè I, Gnanvossou D, Goergen G, Agbaka A (2008) Inventaire et importance des dégâts des mouches des fruits sur quelques variétés d'agrumes au Bénin, pp 128–139. In: Adjanohoun A, Igué K (eds.), Actes de l'Atelier Scientifique National 6, Décembre 2006, Abomey-Calavi. Tome 1. ISBN 978-99919-6850-6 ; ISSN 1659-6161 ; Dépôt légal n° 3838 du 13/08/2008 3è Trimestre. Bibliothèque Nationale du Bénin, 440 pp

Harris EJ (2010) Suppression of melon fly (Diptera: Tephritidae) populations with releases of *Fopius arisanus* and *Psyttalia fletcheri* (Hymenoptera: Braconidae) in North Shore Oahu, HI, USA. Biocontrol 55:593–599

Harris EJ, Bautista RC (1996) Effects of fruit fly host, fruit species, and host egg to female parasitoid ratio on the laboratory rearing of *Biosteres arisanus*. Entomol Exp Appl 79:187–194

Harris EJ, Bautista RC, Spencer JP (2000) Utilisation of the egg-larval parasitoid, *Fopius* (*Biosteres*) *arisanus*, for augmentative biological control of tephritid fruit flies. In: Tan KH (ed) Area-wide control of fruit flies and other insect pests. Penerbit Universiti Sains Malaysia, Penang, pp 725–732

Henneman ML, Dyreson EG, Takabayashi J, Raguso RA (2002) Response to walnut olfactory and visual cues by the parasitic wasp *Diachasmimorpha juglandis*. J Chem Ecol 28:2221–2242

Herren HR, Neuenschwander P, Hennessey RD, Hammond WNO (1987) Introduction and dispersal of *Epidinocarsis lopezi* (Hymenoptera: Encyrtidae), an exotic parasitoid of the cassava mealybug, *Phenacoccus manihoti* (Hom: Pseudococcidae), in Africa. Agri Ecosyst Environ 19:131–144

Hijmans RJ, Guarino L, Bussink C, Mathur P, Cruz M, Barrentes I (2004) *DIVA-GIS. Vsn. 5.0.* A geographic information system for the analysis of species distribution data. (Manual available at: http://www.diva-gis.org)

Liquido NJ (1991) Effect of ripeness and location of papaya fruits on the parasitization rates of oriental fruit fly and melon fly (Diptera: Tephritidae) by braconid (Hymenoptera) parasitoids. Env Entomol 20:1732–1736

Lux SA, Copeland RS, White IM, Manrakhan A, Billah MK (2003) A new invasive fruit fly species from the *Bactrocera dorsalis* (Hendel) group detected in East Africa. Ins Sci Appl 23:355–361

Mohamed SA, Ekesi S, Hanna R (2010) Old and new host-parasitoid associations: parasitization of the invasive fruit fly *Bactrocera invadens* (Diptera: Tephritidae) and five other African fruit fly species by *Fopius arisanus*, an Asian opiine parasitoid. Biocontr Sci Technol 20:183–196

Mwatawala M, de Meyer M, White IM, Maerere A, Makundi RH (2007) Detection of the solanum fruit fly, *Bactrocera latifrons* (Hendel) in Tanzania (Dipt., Tephritidae). J Appl Entomol 131:501–503

Mziray HA, Makundi RH, Mwatawala M, Maerere A, de Meyer M (2010) Host use of *Bactrocera latifrons*, a new invasive tephritid species in Tanzania. J Econ Entomol 103:70–76

Ndzana Abanda F-X, Quilici S, Vayssières J-F, Kouodiekong L, Woin N (2008) Inventory of fruit fly species on guava in the area of Yaoundé, Cameroon. Fruits 63:19–26

Neuenschwander P, Boavida C, Bokonon-Ganta AH, Gado A, Herren HR (1994) Establishment and spread of *Gyranusoidea tebygi* Noyes and *Anagyrus mangicola* Noyes (Hymenoptera: Encyrtidae), two biological control agents released against the mango mealybug *Rastrococcus invadens* Williams (Homoptera: Pseudoccocidae) in Africa. Biocontr Sci Technol 4:61–69

Ovruski SM, Aluja M, Sivinski J, Wharton RA (2000) Hymenopteran parasitoids on fruit-infesting Tephritidae (Diptera) in Latin America and the southern United States: diversity, distribution, taxonomic status and their use in fruit fly biological control. Int Pest Man Rev 5:81–107

Peréz J, Rojas JC, Montoya P, Liedo P, González FJ, Castillo A (2011) Size, shape and hue modulate attraction and landing responses of the braconid parasitoid *Fopius arisanus* to fruit odour-baited visual targets. BioControl. doi:10.1007/s10526-9416-0

Purcell MF, Herr JC, Messing RH, Wong TTY (1998) Interactions between augmentatively released *Diachamimorpha longicudata* (Hymenoptera: Braconidae) and a complex of opiine parasitoids in a commercial guava orchard. Biocont Sci Technol 8:139–151

Rousse P, Harris EJ, Quilici S (2005) *Fopius arisanus*, en egg-pupal parasitoid of Tephritidae. Overview. Biocont News Inf 26:59–69

Rousse P, Gourdon F, Quilici S (2006) Host specificity of the egg-pupal parasitoid *Fopius arisanus* (Hymonptera: Braconidae) in La Reunion. Biol Cont 37:284–290

Rwomushana I, Ekesi S, Gordon I, Ogol CKPO (2008) Host plant and host plant preference studies for *Bactrocera invadens* (Diptera: Tephritidae) in Kenya, a new invasive fruit fly species in Africa. Ann Entomol Soc Amer 101:331–340

Schutze MK, Mahmood K, Pavasovic A, Bo W, Newman J, Clarke AR, Krosch MN, Cameron SL (2014) One and the same: integrative taxonomic evidence that *Bactrocera invadens* (Diptera: Tephritidae) is the same species as the Oriental fruit fly *Bactrocera dorsalis*. Syst Entomol 40:472–486

Stuhl C, Sivinski J, Teal P, Paranhos B, Aluja M (2011) A compound produced by frugivorous Tephritidae (Diptera) larvae promotes oviposition behavior by the biological control agent *Diachasmimorpha longicaudata* (Hymenoptera: Braconidae). Environ Entomol 40:727–736

Stuhl C, Sivinski J, Teal P, Aluja M (2012) Responses of multiple species of tephritid (Diptera) fruit fly parasitoids (Hymenoptera: Braconidae: Opiinae) to sympatric and exotic fruit volatiles. Fla Entomol 95:1031–1039

Van Mele P, Vayssieres J-F, Tellingen E, Vrolijks J (2007) Effects of an African weaver ant, *Oecophylla longinoda*, in controlling mango fruit flies (Diptera: Tephritidae) in Benin. J Econ Entomol 100:695–701

Vargas RJ (2001) Potential for area-wide integrated management of Mediterranean fruit fly (Diptera: Tephritidae) with a braconid parasitoid and a novel bait spray. J Econ Entomol 94:817–825

Vargas RJ, Stark DJ, Uchida GK, Purcel M (1993) Opiine parasitoids (Hymenoptera: Braconidae) of oriental fruit fly (Diptera Tephritidae) on Kauai Island, Hawaii: island-wide relative abundance and parasitism rates in wild and orchard guava habitats. Env Entomol 22:246–253

Vargas RJ, Leblanc L, Putoa R, Eitam V (2007) Impact of introduction of *Bactrocera dorsalis* (Diptera: Tephritidae) and classical biological control releases of *Fopius arisanus* (Hymenoptera: Braconidae) on economically important fruit flies in French Polynesia. J Econ Entomol 100:670–679

Vayssières J-F, Goergen G, Lokossou O, Dossa P, Akponon C (2005) A new *Bactrocera* species in Benin among mango fruit fly (Diptera: Tephritidae) species. Fruits 60:371–377

Vayssières J-F, Rey JY, Traore L (2007) Distribution and host plants of *Bactrocera cucurbitae* in West and Central Africa. Fruits 62:391–396

Vayssières J-F, Korie S, Ayegnon D (2009) Correlation of fruit fly (Diptera: Tephritidae) infestation of major mango cultivars in Borgou (Benin) with abiotic and biotic factors and assessment of damage. Crop Prot 28:477–488

Vet LEM, Dicke M (1992) Ecology of infochemical used by natural enemies in a tritrophic context. Ann Rev Entomol 37:141–172

Wharton RA (1989) Classical biological control of fruit infesting Tephritidae. In: Robinson AS, Hooper G (eds) World crop pest – fruit flies: their biology, natural enemies and control, vol 3B. Elsevier, Amsterdam, 447pp

White IM, Elson-Harris MM (1992) Fruit flies of economic significance: their identification and bionomics. CAB International, Wallingford, 601 pp

Yeates DK (2005) Invasive phytophagous pests arising through a recent evolutionary radiation: the *Bactrocera dorsalis* complex of fruit flies. Ann Rev Entomol 50:293–319

Désiré Gnanvossou is an Entomologist/Acarologist at the International Institute of Tropical Agriculture (IITA), Benin. His research focuses on population dynamics, multitrophic interactions and biological control of agricultural pests. He is a part-time lecturer on insect ecology, stored product insects and their management at the Faculty of Sciences and Techniques, University of Abomey-Calavi, Benin.

Rachid Hanna is an Entomologist and Country Representative for the International Institute of Tropical Agriculture (IITA), Cameroon. His research covers entomology and acarology with emphasis on understanding aspects of biodiversity, population dynamics, and multitrophic interactions that are relevant for the development of biologically-based options for the control of agricultural pests.

Aimé H. Bokonon-Ganta is an Associate Professor at the University of Abomey-Calavi, Benin, teaching a number of courses including: agricultural entomology, biological control, stored products insect management. He is an Entomologist at the Plant Quarantine of Benin (DPV) at Porto-Novo, Benin. His main research interests include the behavioural ecology of insect pests and their parasitoids, and understanding plant-herbivore interactions to underpin the implementation of integrated pest management programmes. He has a particular interest in arthropod population dynamics, biological control, agro-ecosystem diversity, and parasitoid biology and mass rearing.

Sunday Ekesi is an Entomologist at the International Centre of Insect Physiology and Ecology (*icipe*), Nairobi, Kenya. He heads the Plant Health Theme at *icipe*. Sunday is a professional scientist, research leader and manager with extensive knowledge and experience in sustainable agriculture (microbial control, biological control, habitat management/conservation, managing pesticide use, IPM) and biodiversity in Africa and internationally. Sunday has been leading a continent-wide initiative to control African fruit flies that threaten production and export of fruits and vegetables. The initiative is being done in close collaboration with IITA, University of Bremen, Max Planck Institute for Chemical Ecology together with NARS, private sectors and ARI partners in Africa, Asia, Europe and the USA and focuses on the development of an IPM strategy that encompasses baiting and male annihilation techniques, classical biological control, use of biopesticides, ant technology, field sanitation and postharvest treatment for quarantine fruit flies. The aim is to develop a cost effective and sustainable technology for control of fruit flies on the African continent that is compliant with standards for export markets while also meeting the requirements of domestic urban markets. Sunday has broad perspectives on global agricultural research and development issues, with first-hand experience of the challenges and opportunities in working with smallholder farmers, extension agents, research organizations and the private sector to improve food and nutritional security. He sits on various international advisory and consultancy panels for the FAO, IAEA, WB and regional and national projects on fruit fly, arthropod pests and climate change-related issues. Sunday is a Fellow of the African Academy of Sciences (FAAS).

Samira A. Mohamed works in the African fruit fly programme at the International Centre of Insect Physiology and Ecology (*icipe*), Nairobi, Kenya and is a senior scientist and co-ordinator of the German BMZ funded *Tuta* IPM project. She has over 20 years of research experience focusing on the development of IPM strategies for suppression of horticultural crop pests. Her main area of interest is in classical biological control of invasive pests within the context of IPM. Samira was instrumental in importing in to Africa, and later releasing, two-efficient co-evolved parasitoid species (*Fopius arisanus* and *Diachasmimorpha longicaudata*) from Hawaii. So far, she has authored or co-authored over 40 peer-reviewed publications and several book chapters.

Chapter 27
Integrated Management of Fruit Flies: Case Studies from Ghana

Maxwell K. Billah and David D. Wilson

Abstract This case study is an overview of the general fruit fly situation in Ghana, starting with a historical background on fruit fly studies in Ghana, through the period when *Bactrocera dorsalis* invaded Africa in the early- to mid-2000s, to the present. We focus on the importance and contribution of agriculture to the economy of Ghana, the effect of fruit flies on agricultural production, the local and export markets (especially of horticultural produce), and attempts made to manage fruit fly populations. These attempts include the initial acceptance that the challenge of fruit flies was a national issue requiring the development of strategic action plans by the Ghana National fruit fly Management Committee (NFFMC) to understand the biology and ecology of the pest and implement effective management. The strategies include a four-point plan consisting of three main management strategies: (1) Bait application technique (BAT), (2) Male annihilation technique (MAT) and (3) Orchard Sanitation/Farm Hygiene, and a fourth non-management strategy: (4) Capacity Building, Awareness Creation and Information Dissemination. The latter ensured that there was a highly knowledgeable human capital base, well-equipped with the necessary information to undertake the management actions. This has resulted in the development of some innovative tools such as the development of the Fruit Fly Resource Box, which is widely used in Ghana and gradually gaining recognition in neighbouring countries.

Keywords Fruit flies • Horticulture • *Bactrocera dorsalis* • Fruit Fly Resource Box

1 Introduction

Agriculture is the predominant economic sector in Ghana, employing approximately 60 % of the labour force and contributing approximately 40 % of the Gross Domestic Product (GDP) (FAO 2003). The mean contribution of the horticultural

---

M.K. Billah (✉) • D.D. Wilson
Department of Animal Biology and Conservation Science, University of Ghana,
Box LG. 67, Legon-Accra, Ghana
e-mail: mxbillah@gmail.com

© Springer International Publishing Switzerland 2016
S. Ekesi et al. (eds.), *Fruit Fly Research and Development in Africa - Towards a Sustainable Management Strategy to Improve Horticulture*,
DOI 10.1007/978-3-319-43226-7_27

sector to the agricultural GDP increased by over 20% between 1993 and 2003 (FAOSTAT 2013). Major agricultural exports include cocoa, timber/ wood products and horticultural crops (pineapple, papaya, mango, citrus and vegetables); together these account for over 57% of total foreign exchange earnings (EMQAP 2006). Over the last two/three decades, export of fruits and vegetables has been one of the most vibrant sources of foreign exchange; in 2006 Ghana was the fifth largest African, Caribbean Pacific (ACP) exporter of fruits and vegetables to the European Union (EU), with pineapple being the leading product (GEPC 2007, 2010). Earnings from horticultural produce increased from approximately USD 37 million in 2005 to USD 50 million in 2006, contributing 5.6% to the total Non-Traditional Exports (NTE) sector and achieving a growth rate of 35% (GEPC 2007). Of the total of 148,000 million tonnes (MT) of horticultural products, pineapples accounted for over 60,000 MT (worth USD 19 million), representing 41% of all horticultural products exported in 2006 (GEPC 2007). There was also considerable growth in export markets between 2005 and 2006; 14% growth in EU markets, 34% in other developed countries and 35% in other African countries (GEPC 2010). Overall this represented an annual growth rate of 18.8% (GEPC 2010). According to the Ghana Export Promotion Council (GEPC 2006), the strategic drive for NTE development, promotion and growth was within the overall National Trade Policy, the Trade Sector Support Program (TSSP) of the Ministry of Trade and Industry and the President's Special Initiative (MOTI_PSD/PSI). Relocation of Golden Exotics Limited to Ghana from Côte D'Ivoire and the use of state-of the art farming tools alone resulted in a growth rate of 2% between 2005 and 2006 (GEPC 2007).

Despite this impressive growth, since 2006 the horticultural sector has seen a decline; this is due to a number of factors including changes in consumer preferences, competition, quarantine and trade barriers (FAO 2009). The horticultural sector is dominated by smallholders cultivating plots of less than 1.5 ha. Major constraints are pests, diseases and an urgent need for appropriate technologies for their effective management. The major pests of current concern are fruit flies. Fruit flies damage crops and limit the ability of growers to produce high quality foods. As trans-boundary quarantine pests of global concern, fruit flies are estimated to cause annual economic losses of more than USD 1 billion worldwide (FAO 2009).

The economically important fruit fly species in Africa include: *Ceratitis anonae* Graham; *Ceratitis bremii* Guérin-Méneville; the Mediterranean fruit fly (medfly), *Ceratitis capitata* (Wiedemann); *Ceratitis colae* Silvestri; the mango fruit fly, *Ceratitis cosyra* (Walker); *Ceratitis ditissima* (Munro); *Ceratitis fasciventris* (Bezzi); the cacao fruit fly, *Ceratitis punctata* (Wiedemann); *Ceratitis quinaria* (Bezzi), the Natal fruit fly, *Ceratitis rosa* Karsch; *Ceratitis rubivora* Coquillett; *Ceratitis silvestrii* Bezzi; the melon fly, *Bactrocera cucurbitae* (Coquillett); the oriental fruit fly, *Bactrocera dorsalis* (Hendel); the solanum fruit fly, *Bactrocera latifrons* (Hendel); the olive fruit fly, *Bactrocera oleae* (Rossi); the peach fruit fly, *Bactrocera zonata* (Saunders); *Dacus bivittatus* (Bigot); the lesser pumpkin fly, *Dacus ciliatus* (Loew); the pumpkin fly, *Dacus frontalis* Becker; *Dacus lounsburyii* Coquillett; *Dacus punctatifrons* Karsch; *Dacus vertebratus* Bezzi; the coffee fruit fly, *Trirhithrum coffeae* Bezzi; and *Trirhithrum nigerrimum* (Bezzi). Of these, *B.*

zonata, *B. latifrons*, *B. oleae*, *D. frontalis* and *D. lounsburyii* have not been recorded in West Africa, including Ghana, and only a single male specimen of *C. rubivora* was recently recorded in a Trimedlure trap on a miracle fruit tree (*Sensepalum dulcificum* [Schumach. & Thonn.] Daniell) at Mampong-Akwapim (June, 2010) in the Eastern region of Ghana (Billah unpublished data). Prior to the arrival of *B. dorsalis* in 2005, *C. capitata* was the most economically important fruit fly species in Ghana (Afreh-Nuamah 1999). Together, *B. dorsalis* and *C. capitata* cause substantial yield losses in fruit and vegetables in Ghana (Billah et al. 2009, 2010). Furthermore, interception of infested fruits in transit to other countries results in a loss of lucrative market shares, trade embargoes and blacklisting (FVO 2012). The most susceptible fruits and vegetables are mangoes, citrus and, in order to maintain productivity in the horticultural industry of Ghana and its share of the export market, it is important that measures are put in place to manage fruit flies. Here we describe the various activities ongoing in Ghana to mitigate the ever-increasing problem of fruit flies.

Prior to 2004/2005, Good Agricultural Practices (GAP) and Postharvest Techniques used by the Ministry of Food and Agriculture (MOFA) were sufficient to protect crops from pests in Ghana. However, with the detection of *B. dorsalis* in Kenya in 2003, and subsequently in Ghana in 2005, the fruit fly problem grew and the new approaches had to be developed.

2 Arrival of the Invasive Species, *Bactrocera dorsalis*, in Ghana

Bactrocera dorsalis was first detected in Ghana in 2005 (Drew et al. 2005; Billah et al. 2006) following its original detection in Africa (Kenya) in 2003 (Lux et al. 2003b). Although this was a surprise it did not elicit the urgent response normally associated with the arrival of an invasive quarantine pest. The first action began in late 2005 and early 2006, and involved the creation of a pest awareness programme and the training of plant quarantine staff for a national survey. The slow start was enough for such a prolific quarantine pest to establish and resulted in a ban on mangoes from Ghana to South Africa in 2005.

The initial preliminary survey seemed to indicate that *B. dorsalis* had entered Ghana from Togo and was present in at least three of the ten regions of Ghana: Volta, Eastern Accra and Greater Accra (Billah et al. 2006). The major population concentrations (i.e. more than ten flies caught per trap per day) were not from cultivated crops, but from either indigenous forest, forest reserves or from the outskirts of townships. This demonstrated that the pest was capable of reproducing on wild fruits, providing a large reproductive base, even outside the main mango fruiting season, for *B. dorsalis* to thrive (Billah et al. 2006). Results of the survey were the first evaluation of the local Ghanaian situation with respect to *B. dorsalis*, and the basis upon which the Ghanaian government developed action plans for control of this fruit fly species (Billah et al. 2006). A successful proposal was submitted to the

Plant Protection and Regulatory Services Directorate (PPRSD) of MOFA for additional surveys to be done in other mango growing areas so that more effective management strategies could be recommended and implemented widely (Wilson and Cobblah 2005).

A number of studies were also initiated at the University of Ghana (UG) on the basic biology, ecology, extent of spread, and control of *B. dorsalis*. Utomi (2006) studied the distribution, host range and natural enemies of *B. dorsalis* in four locations in Southeastern Ghana in 2005 and 2006: Kade and Tafo in the forest zone, and Kpong and Legon in the coastal savanna zone. From catches in Lynfield traps baited with methyl eugenol it was demonstrated that *B. dorsalis* was most abundant in Tafo in the forest zone, and least abundant in Kpong in the coastal savanna zone (Utomi 2006). The population density of *B. dorsalis* increased between October and March at Kade and Tafo in the forest zone but decreased during the same period at Legon and Kpong in the coastal savanna zone (Utomi 2006). The study also indicated a bimodal peak in the diurnal activity patterns of *B. dorsalis*, the first peak being in the morning and the second smaller peak late in the afternoon (Utomi 2006). The initial host range studies indicated that mango, *Mangifera indica* L. was the most preferred host plant, followed by Indian almond, *Terminalia catappa* L., then two citrus varieties, *Citrus sinensis* and *C. tangerine* (Utomi 2006). Parasitoids from the genus *Aphelinus* were reared from *B. dorsalis* pupae (Utomi 2006).

A training course was organized at the Department of Zoology, UG, Legon in 2006 for officials from the PPRSD on various aspects of the biology, ecology, detection and survey techniques for *B. dorsalis* (Wilson and Cobblah 2006). This enabled them to conduct further detection surveys in other parts of the country later in the same year. The results of these surveys indicated that *B. dorsalis* was present in five additional regions: Ashanti, Brong-Ahafo, and the Northern, Upper-West and Upper-East regions (Anon 2006). Additional surveys by UG in 2006 in Agona-Duakwa and Cape Coast, both in the Central Region detected *B. dorsalis* (Wilson and Cobblah 2007). A final detection of *B. dorsalis* was made in July 2007 in Half-Assini in the Western Region of Ghana at the frontier with the Côte D'Ivoire, where trap catches were approximately two flies per trap per day (Wilson and Cobblah 2007). Thus *B. dorsalis* was present throughout Ghana by July 2007 and the pattern of the catches indicated that it had spread from the east to the west confirming the earlier surveys (Billah et al. 2006).

In November 2006, a fruit fly control proposal was submitted for funding to USAID/TIPCEE by a team of stakeholders including PPRSD and UG (Sraha et al. 2006). Other proposed participants were the Papaya and Mango Producers and Exporters Association of Ghana (PAMPEAG) and mango farmer associations including the Dangme-West Mango Farmers Association. The proposal received positive attention in July 2007 for an expanded survey, and was the basis of the subsequent World Bank video conferencing and EU scoping studies, which resulted in the detection of the pest in almost every community. To save producers of fruit crops from potential crop losses they proposed that sustainable control measures should be adopted to reduce the numbers of *B. dorsalis* to acceptable levels, or even totally eradicate them if practicable. A three-pronged approach was proposed which included setting up a surveillance system to monitor the numbers of *B. dorsalis*, implementing management strategies, and training farmers. They aimed to establish a comprehensive monitoring or surveillance system in all regions of Ghana to monitor *B. dorsalis* populations on a weekly basis. PPRSD of MOFA would be responsible for managing the monitoring system country wide. Control strategies would be based on results from the monitoring and include: male annihilation techniques (MAT); bait application techniques (BAT); sanitation; sterile insect techniques (SIT) and biological control. Training of farmers in the management and control of fruit flies was identified as a key component of the fruit fly control programme.

In recognition of the threat to mango exporting countries in West Africa, the World Bank organized the first video conference amongst a number of Francophone countries in West Africa in 2007. Later in the same year, this was followed by a workshop and second video conference in Ghana in collaboration with the Ghana Institute of Management and Public Administration Distance Learning Centre (GIMPA DLC), USAID-TIPCEE, TechnoServe and GTZ. The countries involved were Benin, Burkina Faso, Cote d'Ivoire, Ghana, Mali, Senegal as well as collaborators in Belgium, France, Peru and USA. The main objective of this second video conference was to explore the potential for a regional response to the fruit fly problem. It also intended to bring together world specialists and practitioners to share experiences on currently available fruit fly control techniques – from the orchard scale to post-harvest. The key outcomes were that the presence of *B. dorsalis* should be declared a national disaster and that all stakeholders should be involved in the formation of a National Control Committee. The second outcome was that strict quarantine measures should be enforced, especially on fruits being imported for processing. The third outcome was for SIT, a form of area-wide control using radiation-sterilized *B. dorsalis*, should be considered as a matter of urgency and that what was really needed was an effective action plan (Wilson 2007).

In 2007, the World Bank in collaboration with the European Union (EU) initiated a scoping study on the fruit fly situation in West Africa (Billah 2007, 2008; Stonehouse et al. 2008). The consulting firm, Ital Trend of Italy, working with the New Partnership for Africa's Development (NEPAD; a Technical body of the Africa Union [AU]), hired consultants to study the fruit fly situation in eight West African countries. Reports from each country were discussed by experts at several meetings in Abuja, Nigeria. Key recommendations included the need to have well-coordinated management strategies, and to form National Fruit Fly Management Committees in each country to oversee all activities related to the management of fruit flies. The National Plant Protection Organizations from the participating countries were invited to a meeting in Mali in 2009 and the idea proposed to them. Ghana's National Fruit Fly Management Committee (NFFMC) was subsequently formed and officially inaugurated in 2010. Since then there has been considerable collaboration amongst the participating countries.

With very little knowledge of the biology, ecology and management of *B. dorsalis*, the period between 2007 and 2009 was a time when authorities, growers, exporters and processors sought management strategies from far and near in a desperate attempt to save their investments. In 2009, an orientation workshop was organized for all stakeholders under the auspices of the West African Fruit Fly Initiative

(WAFFI) in Accra, to determine a way forward. Consequently, the first National Training of Trainers (ToT) workshop was delivered in November 2010 at Dodowa in Accra.

3 Fruit Fly Control Strategies in Ghana

After formation of the NFFMC, a series of consultations were made to determine the most pragmatic, practical and realistic strategies that could be used to combat *B. dorsalis*. Three main management strategies were adapted for use in Ghana. These included the use of (1) BAT, (2) MAT and (3) Orchard Sanitation/Farm Hygiene; there was a strong underlying drive to provide training that ensured there was a highly knowledgeable human capital base that was well-equipped with all the necessary information to deliver fruit fly control. Thus, capacity building, awareness creation and information dissemination, was the fourth non-management strategy within the Action Plan of Ghana (Billah et al. 2013).

3.1 The Bait Application Technique (BAT)

Official trials of the hydrolyzed food bait, SUCCESS® Appat (GF-120) (Dow AgroSciences Ltd., UK) were initiated by the Market-Oriented Agricultural Programmme of the German Development Agency (MOAP/GIZ), under the auspices of PPRSD/MOFA and under the Research leadership of the Department of Animal Biology and Conservation Science (DABCS) of UG. To ensure a national outlook, the trial was conducted across all major agro-ecological zones in the country. Overall the trial was successful achieving levels of clean marketable fruits, ranging from 38.5 to 84.5 % in mangoes and 41.4–96.0 % in citrus (Billah et al. 2009, 2010). Although variation existed, there were no statistical differences between regions in terms of *B. dorsalis* infestation levels, implying that all regions were equally prone to attack. There were also no interaction effects between treatments and replications nor between treatments and regions; thus performance of BAT was not influenced by site or region, and trap catches accurately reflected the presence and density of *B. dorsalis*. With the agreement of two regulatory agencies, the PPRSD and the Environmental Protection Agency (EPA), this product was added to the list of products allowed in Ghana, and BAT was officially included in Ghana's IPM package against *B. dorsalis*.

Based on the performance of food baits and the yield increases achieved between 2010 and 2012, demand for BAT escalated and two additional protein bait products were imported and evaluated: CeraTrap® Lure and the Great® Fruit Fly Bait (GFFB) (Ecoman Biotech Co., Ltd., Beijing, China). The CeraTrap® lure, which had previously been used in Israel and largely targets *Ceratitis* species, was tested on a limited scale (as a trap lure) in a commercial mango field; surprisingly the product captured

Table 27.1 Fruit infestation indices for six trial plots in mango production zones

Site	Treatment	No. Fruits	Weight (kg)	No. Pupae	Infestation level Pupae/kg	Difference (C-T)	% Protection (C-T)/C 100
Koldam Farms	GFFB	90	66.0	2	0.030	0.929	96.8
Sikeway Farm 1	GF-120	90	60.0	11	0.183	0.776	80.9
Sikeway Farm 2	Control (C1)	90	58.4	56	0.959		
Andrews Farms	GFFB	90	60.0	3	0.050	0.728	93.6
Epichris Farm 1	GF-120	90	60.0	9	0.150	0.628	80.7
Epichris Farm 2	Control (C2)	90	63.0	49	0.778		
Combined Samples							
	GFFB	180	126.0	5	0.040	0.825	95.4
	GF-120	180	120.0	20	0.167	0.698	80.7
	Control	180	121.4	105	0.865	–	

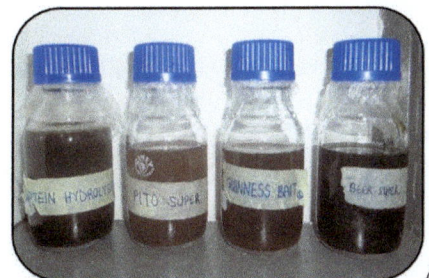 

Fig. 27.1 Protein baits from a commercial source (far left) and three local beer sources (**a**) and discharge of spent brewery grain from Accra Brewery Limited (**b**) (Photos: M.K. Billah)

more *Bactrocera* species than the primary target species (Boateng et al. 2013). This observation strengthened belief in the use of food baits in the management of fruit flies and a trial to compare GFFB directly with SUCCESS® Appat, as a standard, was initiated. An impressive increase of 93.6–96.8 % in clean marketable fruits was achieved on farms using GFFB, with an average increase of 95.2 % in mango, while a range of 80.7-80.9 % was achieved on farms using SUCCESS® Appat (Billah et al. 2014). The difference in damage levels between the two treatments was statistically different, with GFFB performing significantly better (Table 27.1). In response to the progress made and the rate of adoption of BAT that was observed in the field, more readily available bait products produced from materials sourced locally were developed (Fig. 27.1). These were developed based on two readily available waste products: spent brewers' yeast (SBY) and spent brewers' grain (SBG).

3.1.1 Spent Brewers' Yeast (SBY)

Yeast was sourced from both commercial and local breweries: from Guinness Ghana Limited, Achimota, Ghana, and from local beers including, *Pito*, which is made from fermented millet or sorghum. These were strongly agitated and 100 ml of each poured into a measuring cylinder to determine their liquid and solid debris quantities over a 24 h period after settling. The autolyzed liquid supernatants were then decanted and the solid yeast cells (debris) digested with the enzyme, papain, after which the supernatants were reintroduced and agitated to ensure a uniform mixture. Potassium sorbate was then added as a preservative. The brown supernatant formed (protein autolysate) was decanted and the quantity compared with the initial raw materials. The percentage protein content of three locally-prepared baits were then compared with a commercial protein bait, Protein hydrolysate (NuLure, Miller, USA), using the Micro-Kjeldahl procedure involving digestion, distillation and titration (Sharpless et al. 1997). After processing 2 L of each raw material, 1.28 L, 0.74 L and 0.90 L of protein autolysate bait were obtained from Guinness, local beer, and *Pito* yeast sources, respectively. In terms of protein content, the commercially-acquired bait had 8 %, followed by local beer (5.2 %), then Guinness and *Pito* (1.4 % for both). However, in the field the local beer bait performed better in terms of the number of fruit flies captured, followed by either the commercial bait or the Guinness bait, with the Pito-based bait being the least attractive (Yeboah 2013). As part of the fruit fly action plan, the Ghana Atomic Energy Commission (GAEC) is now investing in large scale commercial production of protein food baits for growers, as well as extending trials to other locally brewed beers like *Brukutu*, *Asana* and *Aliha* from different regions in Ghana. Protein levels in the products were analysed in the laboratories of GAEC and the Ghana Standards Authority (GSA).

3.1.2 Spent Brewery Grain (SBG)

While laboratory production of protein bait from spent brewer's yeast is underway, it was thought that it might be possible to make some crude versions of protein bait from spent beer grains. In Ghana, these are usually incorporated, in small quantities, into feed meal for poultry and other livestock. Spent grain was acquired from Accra Brewery Limited, dried and milled into a smooth powder. Water (950 ml) was added to 45 ml of the powder and allowed to ferment for 3 days. The fermented product was strained through a very fine net to remove debris and 5 ml of imidacloprid (an insecticide marketed under the local names of Akape® and Anty Ataa) added to achieve a total volume of 1000 ml (1 L), that could be used as a spot spray in the same way as commercial bait treatments. This crude local product was then evaluated in large scale citrus orchards at the UG Agricultural Research Centre (now Forest and Horticultural Crops Research Centre, FOHCREC), Kade in the Kwaebibirem district of the Eastern region of Ghana, and in a large commercial mango orchard in the Dodowa-Somanya Enclave, Ghana. In each case, the commercial bait GF-120 was used as a standard, and there were also control plots where no bait applications were used. The trials were made during both the major and minor cropping seasons and over two years. Three types of pheromone traps (methyl-eugenol [ME], terpinyl acetate [TA] and Trimedlure [TML]) were used to monitor fluctuations in fruit fly populations during the study period. A pre-application trapping period of five weeks was used to determine what the undisturbed field populations of fruit flies were in the trial fields before treatments were applied.

Seven species of fruit flies (belonging to two genera) were recorded. These included *B. dorsalis*, *C. anonae*, *C. capitata*, *C. fasciventris*, *C. rosa*, *C. ditissima* and *C. bremii*. In all treatments and in both years, *B. dorsalis* catches always represented 96–99 % of total fly catches. Despite this there was a complete absence of *B. dorsalis* emerging from all the incubated fruits collected (1710) from all the plots in the trial fields during the study period. The only fruit fly species that emerged from infested fruits was *C. ditissima*, and a few non-tephritid *Atherigona* spp. (Bulley 2012). These observations have implications for the choice of management strategies, and confirms the fact that no single strategy is sufficient for effective management of the entire fruit fly complex. Under the two bait treatment regimes, the levels of clean marketable fruits produced (compared to the control plots) were 42.86 % in the GF-120 treatment compared with 54.93 % in the spent brewery grain treatment.

In mango fields, only three fruit fly species were collected in traps: *B. dorsalis*, *C. cosyra* and *C. capitata*, with *B. dorsalis* accounting for between 85 and 100 % of all flies in trap catches. In contrast to the dominance of *C. ditissima* emerging from incubated citrus fruits, incubated mango fruits yielded only *B. dorsalis*, and a few *Carpophilus* spp beetles. Over the evaluation period, the GF-120 treatment achieved 33 % clean marketable fruit production compared with 43 % protection in the spent brewery grain treatment (Banini 2013). These trials suggest that there is potential for the 'crude' homemade baits in Ghana, although levels of attractiveness depend on factors such as storage conditions, temperature and quality of the source of protein (Mazor et al. 1987, 2002; Bulley 2012).

3.2 The Male Annihilation Technique (MAT)

In Ghana, the male annihilation technique has been based on the use of four main para-pheromone lures: methyl eugenol (ME), terpinyl acetate (TA), Trimedlure (TMA) and CueLure (CL) (Billah et al. 2006, 2009, 2010, 2013) These lures are used in different types of traps for both monitoring and suppression, but it is the cost of trap procurement that has been identified as the major constraint to technology uptake by most farmers. For this reason the effectiveness of homemade traps were compared with a number of imported traps and the cost-benefit implications determined (Fig. 27.2). Two commercial trap types (the Ball Trap [ISCA Technologies Inc., CA, USA] and the Tephri-Trap [SOYGAR® SL, Madrid, Spain])

Fig. 27.2 Three homemade traps (**a, b** and **c**) and two commercial traps (**d** and **e**) used in evaluation trials in Ghana (Photos: M.K. Billah)

Table 27.2 Performance and cost comparison of imported and homemade traps

Trap type	Trap source	Fruit fly collections	Non-target collections	Cost of trap – USD (Cedi equivalent)
Ball Trap	Imported	724	57	14.95 (56.81)
Tephri Trap	Imported	934	20	5.75 (21.81)
Lynfield 1	Homemade	1140	15	0.41 (1.55)
Lynfield 2	Homemade	668	9	0.12 (0.45)
MWB	Homemade	652	1	0.09 (0.35)

and three homemade traps constructed from local resources (Lynfield 1, Lynfield 2 and MWB) were evaluated (Figs. 27.2 and 27.3). Trials were initially restricted to the UG farm, Legon and commercial mango plantations in the Dodowa-Somanya mango enclave, and ME was the only para-pheromone tested; performances were encouraging (Fiave 2010; Cornelius 2011). Even though trap performances were different, the numbers caught were comparable; most flies were captured in the transparent homemade Lynfield trap (1140 flies), followed by the Tephri (934) and Ball traps (724). The other two homemade traps, Lynfield 2 (668) and MWB (652) (Table 27.2), were constructed from empty margarine containers and mineral water bottles and caught the fewest fruit flies. The Ball trap caught the most non-target insects, followed by the Tephri trap and then the homemade traps (Lynfield 1, Lynfield 2 and MWB). The cost of traps was greatest for the Ball trap, followed by the Tephri trap and the cheapest were the homemade traps (Lynfield 1, Lynfield 2 and MWB). The three homemade traps were presented to growers in Ghana, and

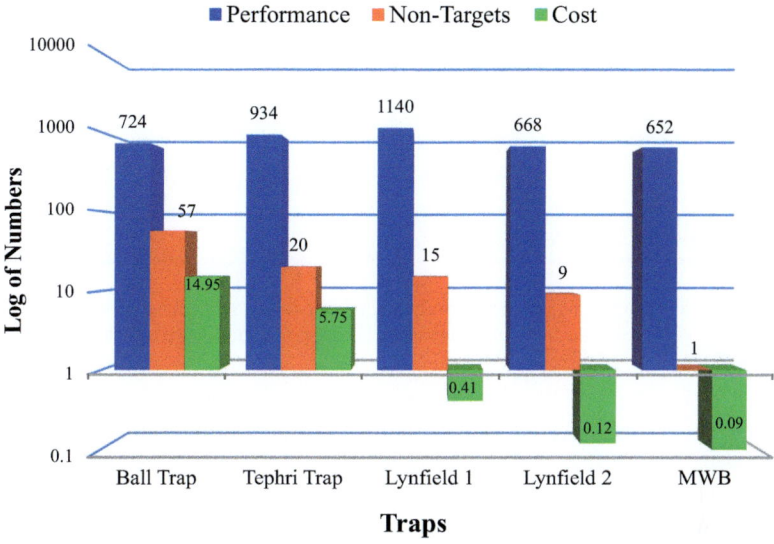

Fig. 27.3 A log plot of trap performance, non-target catches and cost of two imported traps and three homemade traps. **Key**: *Blue* Fruit fly catches, *Red* non-target catches, *Green* cost of different traps used (in USD)

they selected the mineral water bottle (MWB) trap, based on availability, affordability and accessibility (Fig 27.3). Although the MWB traps were not as efficient as the others, cheap construction of large numbers compensated for the loss in efficacy. Trials were based solely on the cost of traps and did not include the cost of lures or attractants. Farmers have now been taught how to make such local homemade traps in their communities.

3.3 Orchard Sanitation

In terms of orchard sanitation (or farm hygiene), keeping weeds to a minimum was the primary strategy as this allows free movement through orchards, easy collection of dropped fruits and efficient disposal of infested fruits (Ekesi and Billah 2006, 2007). Infested fruits that had dropped to the ground were placed in thick black polythene bags, tightly secured by tying, and exposing to the sun between rows or in open places in the field for about 7–14 days (Fig. 27.4). The heat generated in the bags, combined with the fluids from the fruits, eventually kills the larvae within. The content of the bags can then be used as a fertilizer between the rows in the orchard. Alternatively, infested fruits can be deeply buried (30 cm deep) so as not to allow any flies or larvae to easily emerge (Billah et al. 2013).

These two disposal methods have the drawback of not allowing any parasitoids that may be in the fruits, to escape; this is especially important in areas where para-

Fig. 27.4 A clean mango orchard with low levels of weeds (**a**), and fallen fruits that have been collected and tightly secured in thick black polythene bags and exposed to the sun to rot (**b**). (Photos: M.K. Billah)

sitoids have been introduced. One solution is to use an Augmentorium®, which is a tent-like structure fitted with fine netting material. Collected fruits are placed inside and allowed to rot; the mesh size is only large enough to allow the exit of parasitoids and other small natural enemies, but not fruit flies. In this way the natural enemy populations are conserved. Currently this method is not fully operational in Ghana although they are used at the experimental and field trial level, and farmers are being educated on the principles of their use in awareness creation programmes.

3.4 Other Fruit Fly Management Options

Apart from the three main strategies, all other options are either used only on an experimental basis or are currently limited in their scope in Ghana. They are mainly used in research programmes aimed at validating their efficacy, ensuring proper usage, finding prerequisites for their use, testing their compatibility with other options and potential for integration into existing packages. Their current status is discussed below.

3.4.1 Irradiation

Irradiation is a control option that could easily be integrated with the other existing methods. It relies on the use of gamma radiation, and this is technically possible at the facilities at GAEC. Studies were initiated in late 2006 to determine the doses required for the phytosanitary irradiation (disinfestation) of fresh fruits, and also the doses that could be used to induce sterility (sterilization) in *B. dorsalis* for application of the sterile insect technique (SIT) (Ogaugwu 2007; Odai 2009).

Phytosanitary Irradiation: The presence of *B. dorsalis* in Ghana has limited the mango export industry due to increased control costs and even outright rejection of exports. This necessitated studies into quarantine and/or phytosanitary treatment of fruit to mitigate for the adverse impacts of *B. dorsalis*. Amongst the phytosanitary treatments, irradiation is recognized as being versatile and with a broad spectrum of activity against arthropod pests at doses that have minimal adverse effects on the quality of most commodities (IAEA 2004, 2015). Some countries such as the USA require irradiation as a phytosanitary measure before entry of produce into their market. Thus mangoes from affected countries in West Africa, including Ghana, are specifically required to be irradiated against *B. dorsalis* before entry into the USA market (Bech 2008; USDA-APHIS 2008). This is based on a generic dose of 150 Gy for tephritid fruit flies that was proposed in 1986 by ICGFI (1991), and later recommended by the US Department of Agriculture (USDA) for all tephritid fruit flies (IAEA 2004). According to Torres-Rivera and Hallman (2007) however, it is imperative that effective doses for each particular species of tephritid be established since the potential to reduce the necessary dose would reduce costs and increase the capacity of treatment facilities by decreasing the time required for treatment (Follett and Armstrong 2004). A study was therefore conducted to ascertain the effective dose for treatment of *B. dorsalis* in fruits destined for export (Odai 2009). Pupae were obtained after incubation of mango fruits collected from various locations. Adults were reared out and 5–20 females placed on mangoes in cages and allowed to oviposit. Infested mangoes were then examined to determine infestation levels in relation to the number of females in the cage. Late instar (14–21 day-old) larvae were irradiated in mango fruits at doses of 15–75 Gy to determine the effective dose for control of *B. dorsalis*. A dose of 70 Gy was identified as the most effective (Odai et al. 2014). Confirmatory tests with over 3000 larvae confirmed this (IAEA 2004; Follet 2007).

Radiation doses for sterilization: SIT was first developed in the USA, and has been used for over 50 years on six continents. There are four strategic options in which sterile insects are deployed as components of area-wide integrated pest management (AW-IPM) of insect pests: suppression, eradication, containment and prevention (IAEA 2015). It is amongst the most environmentally friendly insect pest control methods ever developed and its applicability to the management of *B. dorsalis* has been highlighted in many publications (IAEA 2015). Ogaugwu (2007) investigated the potential for sterilization of males using gamma radiation at doses of 25, 50 and 75 Gy, for use in SIT strategies against *B. dorsalis* in Ghana. Irradiation of pupae (about 6 days after pupation) was found to be suitable. An irradiation dose of 75 Gy rendered males of *B. dorsalis* completely sterile, while doses of 25 and 50 Gy induced only partial sterility. Females were rendered completely sterile at all radiation doses tested (Ogaugwu et al. 2012a). The effects of irradiation on insect survival and longevity were also investigated since these qualities are important for an effective SIT control programme. Survival of *B. dorsalis* pupae exposed to irradiation doses of 25, 50 and 75 Gy, were not significantly different from non-irradiated controls (Ogaugwu et al. 2012b). Also, the maximum longevities of the irradiated and non-irradiated flies were not significantly different. Thus irradiation doses applied in this study did not compromise adult survival of *B. dorsalis* and should be considered for use in any future SIT programmes against this fly (Ogaugwu et al. 2012a, b).

3.4.2 Biological Control

Entomopathogens: Tactics for the suppression of fruit flies for increased mango production must increasingly rely on management tools that have a low negative impact on the environment. *Beauveria bassiana* (Bals.–Criv.) Vuill., an entomopathogenic fungus, has potential as a microbial insecticide against *B. dorsalis* on mango because it is environmentally friendly but also has no effect on the flavour, colour and texture of the fruit, or the chemical composition of the final product. Efficacy of *B. bassiana* as the formulated product Botanigard® ES for control of *B. dorsalis* has been investigated (Marri 2013). The fungus was applied at concentrations of 106, 53.0, 26.5, 13.3 and 6.65 ($\times 10^6$ conidia/ml) against larvae, pupae and adults of *B. dorsalis*. The results showed lethal times (LT_{50}) ranging from 2.8 to 3.6 days at a dose of 106×10^6 conidia/ml. Field applications indicated that using autodissemination to distribute the fungus from fruit fly traps placed in mango canopies, was more effective than soil applications (Marri 2013).

Parasitoids: Management of fruit flies using arthropod natural enemies is well known in Ghana, and for that matter throughout Africa, but its use has always been limited. The unfortunate introduction of *B. dorsalis* in to Africa in 2003 (Lux et al. 2003b; Drew et al. 2005), however afforded Africa the opportunity (probably for the first time) to embark on exploration for natural enemies for classical biological control.

In Ghana, information on species composition and host range of natural enemies are scattered, often with only limited information pertaining to specific crops or from samples that were collected in other research studies. Since early exploratory collections by Silvestri (1913) and van Zwaluwenberg (1936) it was not until the early 1990s that a mission from the USA (led by Bruce McPheron) sampled for parasitoids around the Central and Greater-Accra regions (D. Wilson, personal communication). The work of Steck et al. (1986) across West Africa covered only coffee growing areas but later collections by Billah (2004) covered eight fruit-growing areas across four regions (Ashanti, Eastern, Volta and Western) in the southern half of Ghana concentrating both on coffee as the primary host plant, but also other fruit species. This yielded interesting fruit fly and parasitoid data, including first records of *C. fasciventris*, *B. cucurbitae* and a braconid parasitoid thought to be *Opius sulphureus* Szépligeti (Billah 2004). Since then, fruit collection exercises for incubation, host plant range studies, trapping and monitoring have been done by various organizations, projects and groups, but this information is not in the public domain for experts, producers/exporters to utilize in devising control strategies. More recently, however, fruit fly work was extended to cover the central belt of Ghana and the three northern regions to gain a national perspective on the full range of fruit fly species, host plants and natural enemies (especially parasitoids) in the country (Billah et al. 2008; Nboyine et al. 2012; Oyinkah 2012; Badii et al. 2015a, b). The parasitoids collected in the northern half of the country included the braconids *Fopius caudatus* (Szépligeti), *Psyttalia cosyrae* (Wilkinson), *Psyttalia concolor* Szépligeti and *Diachasmimorpha fullawayi* (Silvestri). This information provides the pre-requisite basis for introduction and augmentation of natural enemies in Ghana and elsewhere.

Weaver Ants: Though exploitation of the weaver ant, *Oecophylla longinoda* Latreille, is widely used in certain jurisdictions (van Mele et al. 2007, 2009), its use in Ghana has been very limited and highly restricted. The greatest challenge for the use of weaver ants has been the perception of most communities associating them with very painful bites (Ativor et al. 2012; Abunyewah et al. 2015). However, the potential still exists as the presence of the ants is known to deter fruit flies from landing on fruits to oviposit, which minimizes the frequency of fruit puncturing, as well as the need for early harvesting. When fruits remain longer on the tree becoming fully mature before harvest, their brix quality improves (Akoto et al. 2011; Ativor et al. 2012).

3.5 Other Research Activities

Most research activities are not directly aimed at management or control, but at ways to increase the basic knowledge needed to underpin the formulation of sound pest management strategies or to generate information for dissemination.

3.5.1 Host Preference and Suitability Studies

These studies have mostly been done on mango varieties such as Keitt, Kent, Palmer and Hayden that are targeted at lucrative export markets. Although no variety was found to be resistant to fruit flies, they did show various levels of susceptibility and were differentially attractive to *B. dorsalis* in choice and no-choice experiments (Ambele et al. 2011; Ngantu 2012).

3.5.2 Species Numbers and Relative Fly Population Densities

Year-round monitoring of *Bactrocera* and *Ceratitis* species in both citrus and mango fields was done to closely follow population trends of the predominant fruit fly species present, and then repeated over subsequent years to confirm the trends observed. The year-round monitoring programme was done in a total of 27 mango plots and nine citrus plots (of not less than three acres each), with six traps deployed per plot, thus bringing the number to 216 traps deployed in the country across different agro-ecological zones. Collections from traps were sorted, counted and identified (Table 27.3). Results of these studies identified peak fruit fly populations in the field, related population levels to phenological events in the orchards, and determined the optimum times to intensify mass trapping activities during the cropping season. While peak periods of the different fly species were more related to the phenology of the crops (especially the presence of fruits on the trees), *B. dorsalis* always had

Table 27.3 Species composition of fruit flies collected from mango and citrus fields by trapping

	Fruit fly species	
	Mango fields	Citrus fields
1	*Bactrocera dorsalis*	*Bactrocera dorsalis*
2	*Ceratitis ditissima*	*Ceratitis ditissima*
3	*Ceratitis anonae*	*Ceratitis anonae*
4	*Ceratitis bremii*	*Ceratitis bremii*
5	*Ceratitis cosyra*	*Ceratitis cosyra*
6	*Ceratitis capitata*	*Ceratitis capitata*
7	*Ceratitis rosa*	
8	*Ceratitis fasciventris*	
9	*Dacus bivittatus*	
10	*Dacus vertebratus*	

the highest numbers. Meteorological data were collected using Tinytag® data loggers to establish the relationship between population fluctuations and climatic conditions. This information has been synthesized and the trends used as crude early-warning guides for farmers, to tell when peak population periods of fruit flies are imminent. These crude guides are now available for use by farmers in five production regions in the country, and relate to both the abundance of each of the different fly species collected in different pheromone traps, and are specific to different geographical locations (Fig. 27.5). Apart from the annual trends, plots of seasonal trends and those between minor and major mango seasons in different ecological zones are also available (Nboyine et al. 2012; Billah et al. 2010, 2013).

For each species, the number of flies from the traps over the period of collection were converted into relative fly densities (RFD), using the formula;

Relative Fly Density = **F/T/D** (IAEA 2003)

where F = total number of flies, T = total number of traps and D = the exposure period of traps (in days). Indices for all plots in the different agro-ecological zones were calculated and used as the primary baseline data from which potential areas of low pest prevalence could be identified. They also served as a reference against which the impacts of control technologies and strategies could be measured. For example, while RFD values ranged from 150 to >200 flies per trap per day in 2009, following implementation of the NFFMC action plans and management strategies being put in place by farmers these values are currently within the range of 0.01–18 flies per trap per day.

Apart from regular contributions from the Government of Ghana through their Ministries, additional funding and/or equipment is required to sustain these essential field monitoring activities and a number of institutions and groups have contributed to this: IAEA, GIZ/MOAP, the Export Marketing And Quality Assurance Project (EMQAP), USAID–Agricultural Development And Value Chain Advancement Programme (ADVANCE), the Environmental Protection Agency (EPA) of Ghana, Mango and Citrus Farmers Associations and the West African Fruit Fly Initiative (WAFFI).

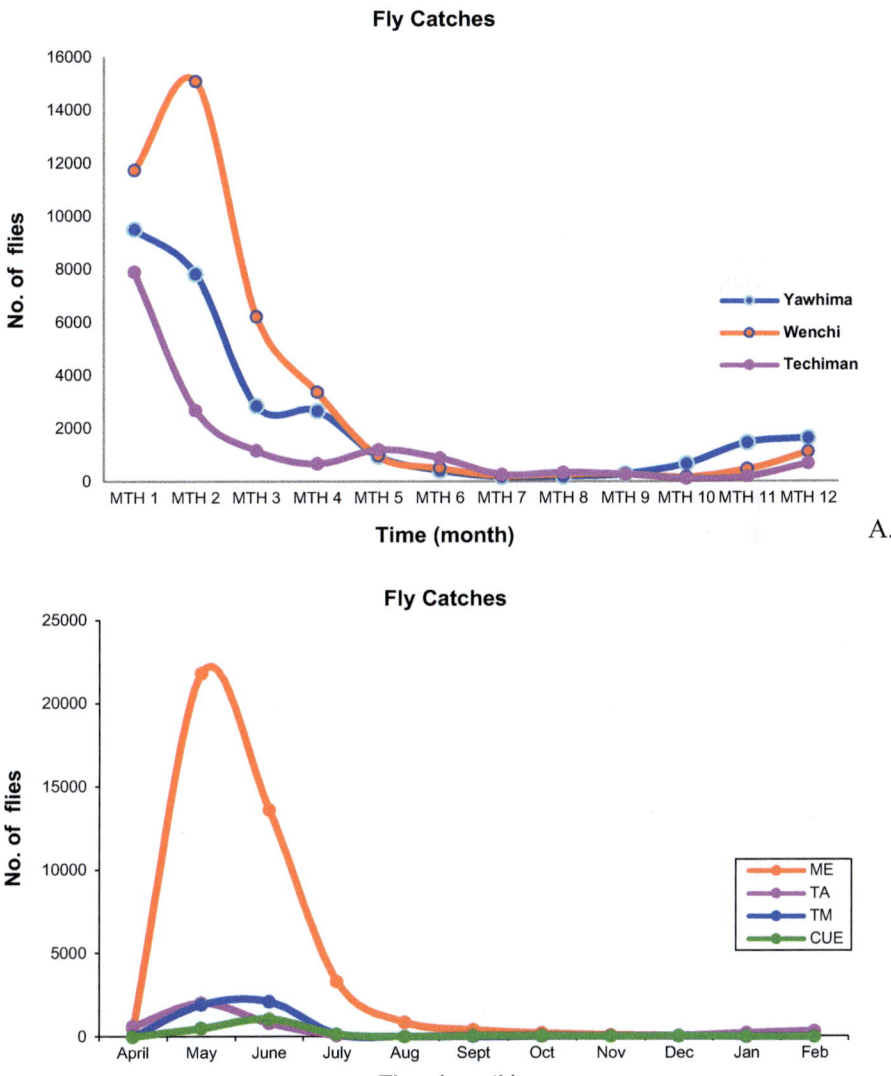

Fig. 27.5 Annual fruit fly population trends at different localities (**a**) and using different parapheromone attractants (**b**) used as a 'crude' early warning guide for farmers

3.5.3 Capacity Building, Awareness Creation and Information Dissemination

Another non-pest management component within Ghana's action plan against fruit flies involves equipping the stakeholder population with the requisite knowledge to use the portfolio of new control tools more effectively, and this required the provision of training. Training programmes were provided for individual farmers, but

Table 27.4 Number of students trained to degree and postgraduate degree level in fruit fly management through the Departments of Animal Biology and Conservation Science and Crop Science (ARPPIS)

Level	Number trained	Number of publications
PhD	2	19
M.Phil	11	
B.Sc	35	

also for the staff of farmer or community-based organizations (F/CBOs) and agricultural extension agents (AEAs), and all other recognized groups and stakeholders along the value chain of the crop under consideration. The purpose of the training was to enhance the technical capabilities of all those participating. Courses usually lasted 1–3 days but some lasted as long as 7 days.

At the highest level, training in fruit fly management could lead to the award of a diploma, first degree (BSc), master's degree (MSc/MPhil) or doctorate (PhD). Most institutions in Ghana have a number of experienced entomologists, some specializing in fruit flies, and whose services were engaged in the training courses. The African Regional Postgraduate Programme in Insect Science (ARPPIS) is the training wing of the International Centre of Insect Physiology and Ecology (*icipe*) based in Nairobi, Kenya. Even though it is not a degree-awarding institution, it collaborates with over 40 Universities across the world, where students are awarded degrees. In Africa, *icipe* has four sub-regional Centres of which the West African Centre (based at UG) is the only one that is currently functional and which contributes to training courses. With the detection of *B. dorsalis*, a conscious effort was made to always introduce one or two fruit fly-related topics for students registered with ARPPIS; this contributes to building capacity over time, including students from other institutions and nationals of other West African countries who were involved in research on fruit flies. With increasing problems associated with invasive fruit flies, more demand-driven courses have been developed to meet the technical challenges associated with managing these pests. Within the ARPPIS programme alone, a number of students have been trained to degree and higher-degree levels, and have published a number of peer-reviewed articles (Table 27.4).

Training at other more practical levels include topics on: the use of simple tools to manage fruit flies, fruit fly taxonomy and identification, fruit fly host range and preference studies, the principles of fruit fly monitoring, using fruit fly trap collections, the effects and implications of infestations, and understanding the quarantine status of fruit flies (Fig. 27.6). Information and data from the above and some best practices used internationally were compiled and published in a handbook in 2013: 'Fruit Fly Management in Ghana'. This publication was freely distributed to key stakeholders (Billah et al. 2013).

Practical Hands-On Training – Competency-Based Training and the Fruit Fly Resource Box The Farmer Field School (FFS) concept is said to be one of the best practical hands-on approaches to training and interacting with farmers in the field

Fig. 27.6 Training on fruit fly identification at Technical and Master's level

(Bartlett 2005; FAO 2015). Another approach is the Trainer of Trainer (ToT) programme, where training and dissemination of information transcends from one level to the next until it reaches the end-user. The enthusiasm of trainees is, however, killed if trainers have insufficient tools to practically demonstrate the new knowledge/information they want to share. In response to this, the Market Oriented Agricultural Programme of the Deutsche Gesellschaft für Internationale Zusammenarbeit GmbH (GIZ-MOAP), AFC Consultant International, and DABCS, conceived the idea of developing a Fruit Fly Resource Box, in which basic fruit fly management information is simplified for growers and frontline staff from MOFA. Based on this a strategy was initiated to furnish such staff with all the information they required in a One-Box-Toolkit called The Fruit Fly Resource Box (Fig. 27.7). These boxes are constructed from light but strong and durable plywood (50×30×20) cm and are highly compartmentalized containing materials used in the management of fruit flies. These included: simplified handouts; flyers; laminated learning cards; photos of common fruit fly species and common host and non-host plants; samples of homemade traps and simple tools and materials for their construction; lures and attractants; a short set of examination questions to assess the understanding of trainees; and short videos of the use of the tools on pen drives (Fig. 27.7). The primary objective was to facilitate interactive-participatory training, and to allow proper hands-on training of field officers by master-level trainers to demonstrate the strategies, tools and products in the Resource Box. The participants at each workshop are then examined and awarded certificates to practice as trainers in their own communities. At the end of training, each community or group, through its Agricultural Extension Agent (AEA), are equipped with a Resource Box for use in further training, with the aim of passing on exactly the same information to Farmer-Based Organizations (FBOs), and ultimately to individual growers. To ensure proper use and flow of information, the training courses organized by districts always had a follow-up session in the presence of a master-level trainer. For

Fig. 27.7 The Fruit Fly Resource Box (**a**) with a display of some of the content (**b, c, d, e**), its use in discussion sessions (**f** and **g**) and distribution to district agricultural extension agents after they passed the competency examinations (**h**)

easy communication the farmers that were selected for training workshops were required to be organized into groups, associations or some other recognizable entity. Five members from each of 5–6 groups are invited at a time (i.e. a total of 30–40 participants for each workshop). Similar training workshops are also organized for exporters and fruit processors. To date, all mango and citrus processors have received training and Resource Boxes. In June 2015, a workshop was organized in Ouagadougou, Burkina Faso under the auspices of APHIS-USDA, where the use of the fruit fly Resource Box was demonstrated in the field to Heads of National Plant Organizations (NPPOs) and Leaders of export firms from eight West African countries.

Awareness Creation and Information Dissemination Through the various activities undertaken, all the information and data generated have been used in the production of manuals and flyers, brochures, posters and handouts. Radio discussions through call-in programmes have also been initiated in some regions, both in english and various local languages.

4 Achievements, Ongoing Activities and the Way Forward

Despite the limited nature of the funding sources for the field studies and research activities, some gains have been achieved towards the management of fruit flies in Ghana. There is still a need to sustain these advances and extend studies to other strategies. Since the arrival of *B. dorsalis*, the need to monitor its spread into new zones and on to new host plants is essential and must be continuous to track any potential host-shifts that may occur over time. The need to confirm and catalogue parasitoid species in Ghana is important in the light of the fact that the NFFMC and the PPRSD intend to introduce parasitoids against *B. dorsalis* in the near future.

Results of field work in Ghana indicate that, contrary to the evidence that populations of *C. cosyra* are gradually being displaced by *B. dorsalis* in mango fruits (Ekesi et al. 2009), this species is still an important pest on mango in the dry Guinea Savanna agro-ecological zones of the three northern regions of Ghana. This observation should be the subject of critical investigation to identify the factors contributing to the variability in dominance of *C. cosyra* in different agro-ecological zones. One possible reason for this could be the ability of *C. cosyra* to use wild hosts such as African peach, *Nauclea latifolia* Smith, false yam, *Icacina senegalensis* Juss. and broom cluster figs, *Ficus sur* Forssk. in the northern regions; these important wild hosts should be considered in fruit fly management strategies in that region, especially as they were sampled around mango plantations. *Ceratitis rosa*, *C. silvestrii* and *C. quinaria* were recorded from yellow plums, *Ximenia americana* L. growing around mango plantations, but none were found infesting mango. These plants may thus be serving as alternative hosts for *C. cosyra*, particularly during the dry seasons. The three host plants, which are shrubs/small trees of about 4 m height, are abundant in the dry savanna zones, and could be an important refuge for *C. cosyra*,

Fig. 27.8 Fruits of *Chrysophyllum albidum* being sold by the roadside – an alternative fruit fly host plant to citrus and mango

should the displacement trend on mango continue. It would, therefore, be very important to sample fruit flies across different ecological zones and compare infestation levels all-year-round to confirm the potential alternative host status of these wild plants for *C. cosyra*. Consideration of these wild plants in the management of fruit flies will be crucial when formulating strategies for field implementation.

Another observation from ongoing work is the co-occurrence of the three species, *B. dorsalis*, *C. ditissima* and *D. bivittatus* on white star apple, *Chrysophyllum albidum* Don (locally known as Alasa or Alasewa) (Fig. 27.8). Not only is it surprising that it is used by all three species at the same time, but also because it could be a strategic alternative host plant for these fruit fly species as it produces fruit when both citrus and mango are out of season in Ghana.

Despite the speed with which *B. dorsalis* has invaded Africa and the slow initiation of control efforts in Ghana, the management efforts made are paying off, although there is still much more to be learnt from the data that has been generated so far. There are plans to compile a comprehensive fruit fly pest list that includes host plant records. This will be the basis for creation of a national catalogue to be housed in the Secretariat of the National Fruit Fly Management Committee. All existing information that is currently distributed amongst various publications of research findings will be verified, authenticated and included in the catalogue. Information is also now available on mean trap catches from various areas. There are indications of relative fly densities, population trends, fruit infestation levels, and important refugia and alternative host plants; this provides a foundation for the formation of fruit fly management strategies and is the scientific basis for many bilateral and trilateral trade and research agreements signed in Ghana. Examples include the bilateral trade agreement with the Chamber of Trade of Lebanon for Ghana to export two varieties of mango to that country, and the trilateral research and cooperation agreement signed by Ghana, Israel and Germany for the purpose of improving the citrus subsector and increase marketable yield for both local and export markets. Results of research activities presented at various conferences and workshops have also increased Ghana's worldwide visibility with respect to fruit fly management; the former head of the quarantine unit of Ghana's NPPO has been nominated as the African representative on the IPPC.

Because farmers and other stakeholders have a better understanding of the fruit fly situation, there is growing interest from these groups to achieve the very best economic returns for their produce. They have formed strong farmer-based organizations including the Mango and Citrus Forum. There has also been an attempt at forming Regional Fruit Fly Committees, a few of which have already been inaugurated. In 2014, there was the first National Roundtable Discussion on the status of mango in Ghana, after which a communiqué was sent to the Government of Ghana proposing the formation of a National Mango Council, and they have subsequently held regular and consistent meetings to pursue that agenda.

Apart from national efforts, there are also discussions at various regional and subregional levels on projects to manage fruit fly pests on a larger and more holistic scale. Examples of these include the IAEA Project RAF 5061, on 'Supporting Capacity Building and a Feasibility Study on Control of Fruit Flies of Economic Significance in West Africa', the CORAF/WECARD project on 'Fruit Fly Control Technology Dissemination and Capacity Building of West African Fruit Value Chain Stakeholders', and most recently the ECOWAS project 'Regional Strategy for Fighting Fruit Flies in West Africa', which is scheduled to start in 2016. It is hoped that this collaboration amongst stakeholders will facilitate effective and informed decision-making on the management of fruit flies, and lead to increased production of high quality marketable fruits in Ghana, for both local and export markets.

Acknowledgements Government of Ghana; Department of Animal Biology & Conservation Science, University of Ghana; Research & Conferences Committee, University of Ghana; The Deutcher Akademischer Austauschdienst (DAAD); MOAP-GIZ; Conservation Food & Health Foundation; The IAEA – Project RAF 5061; CORAF/WECARD Project; PPRSD (of MOFA); Pokuase; EDAIF; EMQAP; Institute of Agricultural Research, CACS UG, Legon; ARPPIS Programme; West African Fruit Fly Initiative (WAFFI), Benin; and numerous farmers who volunteered their help.

References

Abunyewah GK, Afreh-Nuamah K, Nboyine JA, Obeng-Ofori D, Billah MK (2015) Farmers' perception of a biological control agent, *Oecophylla longinoda* Latreille (Hymenoptera: Formicidae) and its effects on the quality of citrus fruits in Ghana. Afr J Agric Res 10:4646–4652

Afreh-Nuamah K (1999). Insect pests of tree crops in Ghana – identification, damage and control measures. Buck Press Inc. Accra 65 pp.

Akoto SH, Billah MK, Afreh-Nuamah K, Owusu EO (2011) The effect of fruit fly larval density on some quality parameters of mango. J Animal Plant Sci 12:1590–1600

Ambele FC, Billah MK, Afreh-Nuamah K, Obeng-Ofori D (2011) Susceptibility of four mango varieties to the Africa invader fly, *Bactrocera invadens* Drew, Tsuruta and White (Diptera: Tephritidae) in Ghana. J Appl Biosci 49:3425–3434

Anon (2006) The invasive fruit fly *Bactrocera invadens* monitoring report for five regions. CPDMD/PPRSD, 7 pp

Ativor IN, Afreh-Nuamah K, Billah MK, Obeng-Ofori D (2012) Weaver ant, *Oecophylla longinoda* (Latreille) (Hymenoptera: Formicidae) activity reduces fruit fly damage in citrus orchards. J Agric Sci Technol 2:449–458

Badii KB, Billah MK, Afreh-Nuamah K, Obeng-Ofori D (2015a) Species composition and host range of fruit-infesting flies (Diptera: Tephritidae) in northern Ghana. Int J Trop Ins Sci 35:137–151

Badii KB, Billah MK, Afreh-Nuamah K, Obeng-Ofori D, Nyarko G (2015b) Review of the pest status, economic impact and management of fruit-infesting flies (Diptera: Tephritidae) in Africa. Afr J Agric Res 10:1488–1498

Banini CK (2013) Field evaluation of the fruit fly food bait, SUCCESS Appat® (GF-120), and waste brewers' yeast (WBY) against fruit flies (Diptera: Tephritidae) in mango orchards. M.Phil Thesis, University of Ghana, 111 pp

Bartlett A (2005) Farmer field schools to promote Integrated Pest Management in Asia: the FAO experience. In: Proceedings from workshop on scaling up case studies in agriculture. International Rice Research Institute, Bangkok, Thailand

Bech R (2008) Federal Quarantine Order (APHIS, USA). pp 1–5

Billah MK (2004) Biosystematic studies of *Psyttalia* species (Hymenoptera: Braconidae): Parasitoids attacking fruit-infesting flies (Diptera: Tephritidae) in Africa. PhD thesis, University of Ghana, Legon-Accra, 236 pp

Billah MK (2007) Scoping study on fruit fly damage in West Africa. "Category III". ECOWAS fruit fly scoping study and regional action programme. Evaluation of the fruit fly problem in Ghana. A Report on Ghana, 47 pp

Billah MK (2008). National situation synopsis: Ghana. *Framework contract benefit*. Study on damages inflicted by fruit flies on West Africa fruit production and action plan for a regional response, pp 155–173

Billah MK, Wilson DD, Cobblah MA, Lux SA, Tumfo JA (2006) Detection and preliminary survey of the new invasive fruit fly species. Bactrocera invadens sp. n. in Ghana. J Ghana Sci Assoc 8:138–144

Billah MK, Kimani-Njogu SW, Wharton RA, Woolley JB, Masiga D (2008) Comparison of five allopatric fruit fly parasitoid populations (*Psyttalia* species) (Hymenoptera: Braconidae) from coffee fields using morphometric and molecular methods. Bull Entomol Res 98:63–75

Billah MK, Owusu EO, Biney PM, Biney K, Agboka M, Edmund JC (2009) Field evaluation of the novel fruit fly food bait, *SUCCESS® APPAT* (GF-120) in mango orchards across different agro-ecological zones in Ghana. Final Technical Report, 29 pp

Billah MK, Owusu EO, Biney K, Agboka M (2010) Field evaluation of SUCCESS® APPAT (GF-120) against fruit flies in three citrus orchards in the Central region of Ghana. Final Technical Report, 22 pp

Billah M, Braimah H, Ofosu-Budu G, Nkansah GO, Agboka M, Monney E, Bleppony H, Gaveh E, Avar V, Djanie ANDA, Attatsi R (2013) Fruit fly management in Ghana. Export marketing and quality awareness project (EMQAP) of MOFA. Accra, Ghana. 100 pp (ISBN: 978-9988-8562-4-3)

Billah MK, Adom K, Osae MY, Jiang J, Du J (2014) Evaluation of The Great® Fruit Fly Bait (GFFB) against fruit flies in two mango-production zones in Ghana. Final Technical Report (Submitted to the Environmental Protection Agency, EPA – Ghana), 19 pp

Boateng BA, Billah MK, Cornelius EW (2013) Field evaluation of the attractancy of a new protein bait, CeraTrap lure, for the management of fruit flies. Final report. Joint field trial exercise between the Department of Crop Science and Dizengoff Ghana Ltd, Accra, 15 pp

Bulley F (2012) Comparative study of the novel fruit fly food bait, SUCCESS Appat® (GF-120) and waste brewers' yeast (WBY) against fruit flies in citrus orchards. M.Phil Thesis, University of Ghana, 104 pp

Cornelius NB (2011) Field evaluation of five fruit fly traps in a commercial mango orchard in the Damgbe West District. B.Sc. Project work, University of Ghana, Legon, 30 pp

Drew RAI, Tsuruta K, White IM (2005) A new species of pest fruit fly (Diptera: Tephritidae: Dacinae) from Sri Lanka and Africa. Afr Entomol 13:149–154

Ekesi S, Billah MK (Eds.) (2006) A field guide to the management of economically important Tephritid fruit flies in Africa. ICIPE Science Press, Nairobi, Kenya. (English Version – ISBN: 92-9064-179-7)

Ekesi S, Billah MK (eds) (2007) A field guide to the management of economically important Tephritid fruit flies in Africa, 2nd edn. ICIPE Science Press, Nairobi. (English Version – ISBN: 92-9064-209-2)

Ekesi S, Billah MK, Nderitu PW, Lux SA, Rwomushana I (2009) Evidence for competitive displacement of *Ceratitis cosyra* by the invasive fruit fly *Bactrocera invadens* (Diptera: Tephritidae) on mango and mechanisms contributing to the displacement. J Econ Entomol 102:981–991

EMQAP (2006) Ghana export market quality assurance program – appraisal report. 78 pp

FAO (2003) Trade reforms and food security project: Ghana.

FAO (2009) Transboundary animal and plant pests and diseases. World Summit on Food Security. Rome 16–18 November 2009, 4 pp

FAO (2015) A shift in global perspective – Institutionalizing Farmer Field School. 58 pp

FAOSTAT (2013) World food and agriculture. 307 pp. ISBN 978-92-5-107396-4

Fiave CA (2010) Field evaluation of four fruit fly traps in a mango orchard. BSc project work submitted to the University of Ghana, Legon. 38 pp

Follet PA (2007) Postharvest phytosanitary radiation treatments: Less-than-*Probit 9*, generic dose and high dose applications. In: Vreysen MJB, Robinson AS, Hendrichs J (eds), Area-wide control of insect pests, pp 425–433. Springer, Dordrecht. 789 pp.

Follett PA, Armstrong JW (2004) Revised irradiation doses to control melon fly, Mediterranean fruit fly, and Oriental fruit fly (Diptera: Tephritidae) and a generic dose for tephritid fruit flies. J Econ Entomol 97:1254–1262

FVO Report (2012) Final report of an audit carried out in Ghana. DG (SANCO) 2012–6309 – MR FINAL, 24 pp

GEPC (2006) Current investment opportunities in Ghana

GEPC (2007) Non-traditional export performance report 2006, Accra

GEPC (2010). The export of horticulture from Ghana: a 'success story' document of the Ghana Exports Promotions Council (GEPC) http://www.gepcghana.com/horticulture.php

IAEA [International Atomic Energy Agency] (2003) Trapping guidelines for area-wide fruit fly programmes. IAEA, Vienna, 47 pp

IAEA (2004) Irradiation as a phytosanitary treatment of food and agricultural commodities. IAEA Publication: TECDOC-1427, Vienna, Austria

IAEA (2015) The Sterile Insect Technique Homepage: Insect Pest Control, Joint IAEA/FAO Programme, Nuclear Techniques in Food and Agriculture. http://www.iaea.org

ICGFI (1991) Enhancing food safety through irradiation. International Consultative Group on Food Irradiation Policy. ICGFI Document No. 13. International Atomic Energy Agency, Vienna

Lux SA, Ekesi S, Dimbi S, Mohamed SA, Billah MK (2003a) Mango-infesting fruit flies in Africa: perspectives and limitations of biological approaches to their management. In: Neuenschwander P, Borgemeister C, Langewald J (eds) Biological control in IPM systems in Africa. CAB International, Wallingford, pp 277–293

Lux SA, Copeland RS, White IM, Manrakhan A, Billah MK (2003b) A new invasive fruit fly species from the *Bactrocera dorsalis* (Hendel) group detected in East Africa. Ins Sci Appl 23:355–361

Marri D (2013) Evaluation of the efficacy of *Beauveria bassiana* for the control of the invasive fruit fly *Bactrocera invadens* (Diptera: Tephritidae). M.Phil Thesis, University of Ghana, 65 pp

Mazor M, Gothilf S, Galun R (1987) The role of ammonia in the attraction of females of the Mediterranean fruit fly to protein hydrolysate baits. Entomol Exp Appl 43:25–29

Mazor M, Peysakhis A, Reuven G (2002) Release rate of ammonia – a key component in the attraction of female Mediterranean fruit fly to protein-based food lures. IOBC Noxious Animals Plants Bull 25:1–6

Nboyine JA, Billah MK, Afreh-Nuamah K (2012) Species range of fruit flies associated with mango from three agro-ecological zones in Ghana. J Appl Biosci 52:3696–3703

Ngantu HN (2012) Comparison of fruit flies reared from different mango varieties from three agro-ecological zones in Ghana. M.Phil Thesis, University of Ghana, 101 pp

Odai BT (2009) Irradiation as a quarantine treatment against the Africa invader fly (*Bactrocera invadens*) for *Magnifera indica*. M.Phil Thesis, University of Ghana. 105 pp

Odai BT, Wilson DD, Apea-Bah F, Torgbey-Tetteh W, Osae MY (2014) Irradiation as a quarantine treatment against the Africa invader fly (*Bactrocera invadens*) for *Magnifera indica*. Afr J Agric Res 9:1618–1622

Ogaugwu CE (2007) An assessment of the potential to produce a sterile, viable and competitive African invader fly, *Bactrocera invadens*. M.Phil Thesis, University of Ghana, 108 pp

Ogaugwu CE, Wilson DD, Cobblah MA, Annoh CE (2012a) Gamma radiation sterilization of *Bactrocera invadens* (Diptera: Tephritidae) from southern Ghana. Afr J Biotech 11:11315–11320

Ogaugwu CE, Wilson DD, Cobblah MA, Annoh CE (2012b) Survival and longevity of gamma irradiated *Bactrocera invadens*. Int J Curr Res 4:66–69

Oyinkah GM (2012) Host range, preferences and size relationship studies of fruit flies from Northern Ghana. M.Phil Thesis, University of Ghana, 92 pp

Sharpless KE, Schiller SB, Margolis SA, Thomas JB, Iyengar GV, Colbert J (1997) Certification of nutrients in standard reference materials, 1846: infant formula. J AOAC Int 80:611–615

Silvestri F (1913) Report on an expedition to Africa in search of the natural enemies of fruit flies (Trupaeidae) – with descriptions, observations and biological notes. Bulletin of the Territory of Hawaii, Board of Agriculture and Forestry, Division of Entomology 3 226 pp.

Sraha T, Wilson DD, Kyofa-Boamah M, Attatsi RN, Osei-Agyeman K (2006) Fruit fly control proposal, USAID/TIPCEE, 10 pp

Steck GJ, Gilstrap FE, Wharton RA, Hart WG (1986) Braconid parasitoids of Tephritidae (Diptera) infesting coffee and other fruits in West-Central Africa. Entomophaga 31:59–67

Stonehouse JM, Ritchie M, Paqui T, Ndiaye M, Tchibozo S, Dabire R, Jobe L, Billah MK (2008) Scoping study on the damages inflicted by fruit flies on West Africa's fruit production and action plan for a coordinated regional response. Final Report. ItalTrend in association with IAK, SOPEX, ADAS, HYDRO R&D and MEP. (EC Consultancy Service on behalf of ECOWAS)

Torres-Rivera Z, Hallman GJ (2007) Low-dose irradiation phytosanitary treatment against Mediterranean fruit fly (Diptera: Tephritidae). Fla Entomol 90:343–346

USDA-APHIS (2008) Federal import quarantine order for host materials of *Bactrocera invadens* (Diptera: Tephritidae), invasive fruit fly species. USDA-APHIS. 3 pp

Utomi CI (2006) The distribution, host range and natural enemies of the new invasive fruit fly, *Bactrocera invadens* (Diptera: Tephritidae) in Southeastern Ghana. M.Phil Thesis University of Ghana. 115 pp

van Mele P, Vayssières J-F, Van Tellingen E, Vrolijks J (2007) Effects of an African weaver ant, *Oecophylla longinoda*, in controlling mango fruit flies (Diptera: Tephritidae) in Benin. J Econ Entomol 100:695–701

van Mele P, Vayssières J-F, Adandonon A, Sinzogan A (2009) Ant cues affect the oviposition behaviour of fruit flies (Diptera: Tephritidae) in Africa. Physiol Entomol 34:256–261 van Zwaluwenberg, RH (1936) Report on an expedition to West Africa in search of fruit fly parasites (1935–1936). U. S. Department of Agricultural Bureau of Entomology and Plant Quarantine Report. 20 pp. (Mimeographed).

Wilson DD (2007) Report of workshop and video conference on the fight against the threat of fruit flies in West Africa. TIPCEE/USAID/World Bank, 14 pp

Wilson DD, Cobblah MA (2005) Survey of the new invasive fruit fly in Ghana. Proposal to MOFA/PPRSD, Accra. 4 pp

Wilson DD, Cobblah MA (2006) Detection, identification, ecology and behaviour modules. In: Mango Fruit Fly Training Workshop. April 10–12, Department of Zoology, University of Ghana, Legon

Wilson DD, Cobblah MA (2007) Surveillance of the African Invader fly in the Greater Accra Region of Ghana. Abstract, p 41. In: 25th Biennial conference, Ghana Science Association, August 6–11, Bunso

Yeboah S (2013) Development of a local baiting system for the control of the Africa invader fly, *Bactrocera invadens* Drew, Tsuruta and White (Diptera: Tephritidae) in mango orchards. M.Phil Thesis, University of Ghana, Legon

Maxwell K. Billah Maxwell was awarded his PhD in Biosystematics (applied insect taxonomy) by the University of Ghana (UG) based on studies made at the International Centre of Insect Physiology and Ecology (*icipe*) in Nairobi, Kenya. He is now a Senior lecturer at UG in the Department of Animal Biology and Conservation Science (DABCS). He contributed to the first detection of *Bactrocera dorsalis* in Africa in 2003 when he was finalizing his PhD research and also detected the same insect in Ghana in 2005. In 2004, he became the Resident Fruit Fly Taxonomist on an FAO-funded programme surveying East Africa for invasive pests; he managed the programme's database of over 50,000 specimens – the largest fruit fly collection in Africa. He is credited with the first record of eight fruit fly species in Ghana, and was part of the team that detected a second invasive species, *Bactrocera latifrons*, in Kenya in 2006. Between 2007 and 2008, he went on two expeditions to the native range of *B. dorsalis* in search of its natural enemies; he identified eight parasitoid species with biological control potential and based on this the African Fruit Fly Programme (AFFP) imported these parasitoids for release in Africa. Maxwell joined the DABCS in January 2009, and also teaches at the African Regional Postgraduate Programme in Insect Science (ARPPIS) Centre for the West African Sub-region based at the University of Ghana. He has served as a resource person in 35 fruit fly training workshops across various countries in Africa, Sri Lanka and Japan. He is currently the National Coordinator of an IAEA TCP Project, a CORAF/WECARD Management project, and a Technical Advisor to the National Fruit Fly Management Committee (NFFMC) in Ghana.

David D. Wilson David was educated at the University of Ghana and worked at the Ghana Atomic Energy Commission before obtaining his MS in Entomology from Louisiana State University, Baton Rouge on an IAEA Fellowship. He obtained a PhD in Entomology from the University of Georgia, Athens and has been teaching at the University of Ghana since 1992. His areas of teaching and research/expertise include environmental physiology, toxicology and insect pest/vector management using novel techniques.

Chapter 28
Integrated Management of Fruit Flies: Case Studies from the Indian Ocean Islands

Preeaduth Sookar and Jean-Philippe Deguine

Abstract Fruit flies (Diptera: Tephritidae) are amongst the most destructive and widespread pests of fruits and vegetables in the Indian Ocean Islands (Comoros, Mayotte, Madagascar, Reunion Island, Republic of Mauritius and Seychelles). In Mauritius, the bait application technique (BAT) and the male annihilation technique (MAT) have been used successfully to suppress populations of *Bactrocera zonata, Ceratitis rosa* and *Ceratitis capitata* in backyards over an area of 300 km^2. Population suppression of *B. zonata* was further improved in one 10 ha area by weekly releases of sterile flies (sterile insect technique [SIT]). Area-wide fruit fly control in cucurbit plantations in selected localities was initiated with the support of the International Atomic Energy Agency. BAT, MAT and sanitation were used in farmers' fields to suppress populations of *Zeugodacus cucurbitae, Dacus demmerezi* and *Dacus ciliatus*. Farmers used fewer pesticides while cucurbit infestation by fruit flies was reduced from >30 % to <5 %. In 1996 and again in 2013 *Bactrocera dorsalis* was detected in Mauritius; it was successfully eradicated using BAT and MAT on both occasions. The presence of a fruit fly surveillance programme for early detection, availability of materials and trained personnel contributed towards the successful eradication of *B. dorsalis* from Mauritius. A similar programme was initiated in Reunion Island in 2009 over an area of 50 ha; fewer insecticide applications were made while yields of courgette increased compared with plots treated only with insecticide sprays.

Keywords Biological control • Area-wide • Sterile insect technique

---

P. Sookar (✉)
Ministry of Agro Industry and Food Security, Entomology Division, Réduit, Mauritius
e-mail: psookar@govmu.org

J.-P. Deguine
CIRAD, UMR PVBMT CIRAD/ University of La Réunion,
7 chemin de l'Irat, Ligne Paradis, 97410 Saint-Pierre (La Réunion), France

© Springer International Publishing Switzerland 2016
S. Ekesi et al. (eds.), *Fruit Fly Research and Development in Africa - Towards a Sustainable Management Strategy to Improve Horticulture*,
DOI 10.1007/978-3-319-43226-7_28

1 Background

Fruit flies (Diptera: Tephritidae) are amongst the most destructive and widespread pests of fruits and vegetables in the Indian Ocean Islands, namely Comoros, Mayotte, Madagascar, Réunion Island, Seychelles and the Republic of Mauritius. The pest status of fruit flies is increased as a result of the insular and tropical environment. Fruit fly problems in the Indian Ocean Islands date back to the beginning of the nineteenth century (Orian and Moutia 1960). Damage is caused when female flies oviposit beneath the skin of the fruit of suitable host plants and the resulting larval feeding activity destroys large proportions of the flesh, making the fruits unsuitable for consumption (Christenson and Foote 1960). Infested fruits may drop to the ground and decay.

2 Fruit Fly Species Composition in the Indian Ocean Islands

2.1 Indigenous Fruit Fly Species

Two species of fruit flies are indigenous to the Mascarene Islands (i.e. Mauritius, Réunion and Rodrigues): *Ceratitis catoirii* Guérin-Méneville and the Indian Ocean cucumber fly *Dacus demmerezi* (Bezzi) (Orian and Moutia 1960; Etienne 1972; Etienne 1982; White et al. 2000). The latter is present in Réunion Island (Deguine et al. 2012a), Mauritius (Sookar et al. 2012) and Madagascar (White and Elson-Harris 1994). While *C. catoirii* can be found in Réunion Island, it has become rare (White et al. 2000). The last record of *C. catoirii* from Mauritius was made in 1960 (Orian and Moutia 1960) and it is now considered to be extinct (White et al. 2000). Both species have been recorded from a range of host plants (Table 28.1). *Ceratitis catoirii* is a pest of fleshy fruits while *D. demmerezi* attacks only cucurbits (Deguine et al. 2012a; Sookar et al. 2012). The Madagascan fruit fly, *Ceratitis malagassa* Munro is indigenous to Madagascar where it is endemic and its main cultivated host plants are: apple, *Malus domestica* Borkh.; avocado, *Persea americana* Mill.; citrus, *Citrus* species; guava, *Psidium guajava* L.; peach, *Prunus persica* L.; and Madagascar plum, *Flacourtia indica* Burm. F. (Dubois 1965) (Table 28.1).

2.2 Exotic Fruit Fly Species

The Mediteranean fruit fly (medfly), *Ceratitis capitata* (Wiedemann), arrived in the region from Asia and Africa (CABI/EPPO 1999; EPPO 2014). It has been recorded in the Comoro archipelago i.e. the islands of Mayotte (Maore), Grande Comore (Ngazidja), Mohéli (Mwali) and Anjouan (Ndzuani) (Kassim and Soilihi 2000; De Meyer et al. 2012); in Réunion Island since 1939 (Etienne 1972); in Mauritius since the late nineteenth century/early twentieth century (Orian and Moutia 1960); in Madagascar (White and Elson-Harris 1994); and the Seychelles since 1908 (Dogley 2000; De Meyer 2000).

Table 28.1 Main hosts and fruit fly species recorded from Mauritius (M), Réunion Island (R) and Seychelles (S)

Host plants										
Botanical name	Family	Zc	Bz	Cc	Cri	Cr	Cv	Dc	Dd	Nc
Anacardium occidentale L.	Anacardiaceae			S[8]		R[8]				
Annona cherimola Mill.	Annonaceae			S[8]						
Annona muricata L.	Annonaceae					R[2]				
Annona reticulata L.	Annonaceae		M[8]	M[8]		M[6,8]				
				S[8]		R[8]				
Annona squamosa L.	Annonaceae				M[6]					
Averrhoa bilimbi L.	Oxalidaceae			R[8]		R[8]				
Averrhoa carambola L.	Oxalidaceae			R[2]	M[6]	R[2]				
Calophyllum tacamahaca Wild.	Guttiferaceae			R[8]		R[8]				
Cananga odorata (Lam.) Hook F. & Thomson	Annonaceae			R[8]		R[8]				
Capsicum annuum L.	Solanaceae			M[5]	M[5]					R[2]
				R[10]						
Capsicum frutescens L.	Solanaceae			M[8]						R[8,2]
				R[8,10]						
Carica cauliflora Jacq.	Caricacaea					R[8]				
Carica papaya L.	Caricacaea		M[8]		M[6]	M[8,6]				
						R[2]				
Carissa carandas L.	Apocynaceae			M[8,6]		R[8]				
				R[8]						
Carissa macrocarpa (Eckson) A. DC.	Apocynaceae			R[8]						
Cereus peruvianus (L.) J.S. Muell.	Cactaceae					R[8]				
Chrysobalanus icaco L.	Chrysobalanaceae			S[8]						
Chrysophyllum cainito L.	Sapotaceae			R[8]		R[8]				
Chrysophyllum carpussum L.	Sapotaceae			R[8]		R[8]				
Citrullus colocynthis (L.) Shrader	Cucurbitaceae	R[10]						M[8]	M[8]	
								R[10]		
Citrullus lanatus (Thumb.) Matsum & Nakai	Cucurbitaceae	M[5]	M[5]					R[10]	R[10]	
		R[10]								
Citrus aurantium L.	Rutaceae			R[8]						
				S[8]						
Citrus paradisi Macfad.	Rutaceae					M[5]				
						R[2,8]				

(continued)

Table 28.1 (continued)

Host plants										
Botanical name	Family	Zc	Bz	Cc	Cri	Cr	Cv	Dc	Dd	Nc
Citrus reticulata Blanco	Rutaceae			R[8]	M[6]	M[5]				
						R[2,8]				
Citrus sinensis (L.) Osbeck	Rutaceae			R[8]		M[5]				
				S[8]		R[2,8]				
Coccinia grandis (L.) Voigt.	Cucurbitaceae	R[8,10]						R[8,10]	R[8,10]	
Coccoloba uvifera (L.) L.	Polygonaceae			R[8]		R[8]				
Coffea arabica L.	Rubiaceae			M[5,8]		M[5]				
				R[2,8]		R[2,8]				
Cucumis anguria L.	Cucurbitaceae	R[8,10]						R[8,10]	R[8,10]	
Cucumis melo L.	Cucurbitaceae	M[6,8]						M[6,8]	M[6,8]	
		R[2,8]						R[2,8]	R[10,8]	
Cucumis sativus L.	Cucurbitaceae	M[6,8,9]						M[6,8,9]	M[6,8,9]	
		R[2,8]						R[2,8]	R[2,8]	
Cucurbita maxima Duchesne ex Lam.	Cucurbitaceae	M[6,8,9]						M[6,8,9]	M[6,8]	
		R[8,10]						R[8,10]	R[8,10]	
Cucurbita pepo L.	Cucurbitaceae	M[6,8,9]						M[6,8,9]	M[6,8,9]	
		R[2,8]						R[2,8]	R[2,8]	
Cydonia vulgaris Persoon	Rosaceae					R[2,8]				
Cylanthera pedata (L.) Schrader	Cucurbitaceae	R[8,10]						R[8,10]	R[8,10]	
Cyphomandra betacea (Cav.) Sendtner	Solanaceae									R[2,8]
Dimocarpus longan Lour.	Sapindaceae					R[8]				
Diospyros kaki L. F.	Ebenaceae					R[2,8]				
Dovyalis hebecarpa (Gardner) Warbug	Flacourtiaceae			R[8]		R[8]				
Ehrethia cymosa Thonn.	Boraginaceae			R[2,8]		R[2,8]				

(continued)

Table 28.1 (continued)

Host plants										
Botanical name	Family	Zc	Bz	Cc	Cri	Cr	Cv	Dc	Dd	Nc
Eriobotrya japonica (Thunb.) Lindley	Rosaceae		M^8	$M^{6,8}$	M^6	$M^{6,8}$				
				$R^{2,8}$		$R^{2,8}$				
Eugenia uniflora L.	Myrtaceae			M^8	M^5	M^8				
				$R^{2,8}$		$R^{2,8}$				
Feijoa sellowiana (O. Berg.) O. Berg.	Myrtaceae			M^8		M^8				
				R^8		R^8				
Ficus carica L.	Moraceae		M^8	M^8		M^8				
						$R^{2,8}$				
Flacourtia indica (Burman f.) Merill	Flacourtiaceae			R^8	M^5	R^8				
Fuschia spp.	Oenotheraceae				M^5					
Hylocereus undatus (Haw.) Britton & Rose	Cactaceae					R^8				
Inga laurina (Sw.) Wild.	Mimosaceae					R^8				
Langenaria siceraria (Molina) Standley	Cucurbitaceae	$M^{6,9}$	M^5					$M^{6,9}$	$M^{5,9}$	
		R^{10}						R^2	R^2	
Langenaria sphaerica L.	Cucurbitaceae	M^8						M^8	$M^{6,8}$	
		$R^{8,10}$						$R^{8,10}$	$R^{8,10}$	
Litchi chinensis Sonnerat	Sapindaceae					$M^{3,8}$				
						$R^{2,8}$				
Luffa acutangulata (L.) Roxb.	Cucurbitaceae	$M^{5,8,9}$						$M^{5,8,9}$	$M^{5,8}$	
		$R^{2,8}$						R^2	$R^{2,8}$	
Luffa cylindrical (L.) Roemer	Cucurbitaceae	M^8						M^8	M^8	
		$R^{8,10}$						R^{10}		
Lycopersicum esculentum L.	Solanaceae				M^6			M^6		$M^{6,8}$
										$R^{1,2,8}$
Malus communis Poir.	Rosaceae					$M^{5,8}$				
Mangifera indica L.	Anacardiaceae		$M^{5,8}$	$M^{5,8}$	M^6	$M^{6,8}$				
			$R^{4,7,8}$	$R^{2,8}$		$R^{2,8}$				

(continued)

Table 28.1 (continued)

Host plants										
Botanical name	Family	Zc	Bz	Cc	Cri	Cr	Cv	Dc	Dd	Nc
Manilkara zapota (L.) P. Royen	Sapotaceae			R^8	M^5	$M^{5,8}$				
						R^8				
Mimusops elengi (L.)	Sapotaceae			R^8		R^8				
Momordica charantia L.	Cucurbitaceae	$M^{5,8}$						$M^{5,8}$	$M^{5,8}$	
		R^8						$R^{2,8}$	$R^{2,8}$	
Murraya paniculata (L.) Jack	Rutaceae			M^8		$R^{2,8}$				
				$R^{2,8}$						
Musa acuminata Colta	Musaceae					R^2				
Passiflora edulis Sims.	Passifloraceae	R^{10}		R^8						
Passiflora suberosa L.	Passifloraceae			M^8						
				R^8						
Persea americana Miller	Lauraceae			$M^{6,8}$	$M^{6,8}$	$M^{6,8}$				
						$R^{2,8}$				
Phyllanthus acidus (L.) Skeels	Euphorbiaceae					R^8				
Pirus communis L.	Rosaceae					$R^{2,8}$				
Pithecellobium dulce (Roxb.) Benth.	Mimosaceae			R^8		R^8				
Prunus armeniaca L.	Rosaceae					$R^{2,8}$				
Prunus domestica L.	Rosaceae				R^2	$R^{2,8}$				
Prunus persica (L.) Batsch	Rosaceae		M^8	$M^{6,8}$	M^6	$M^{6,8}$				
				R^8		$R^{2,8}$				
Psidium araca Raddi.	Myrtaceae					R^8				
Psidium Cattleyanum Sabine	Myrtaceae		R^7	M^8	R^8	M^8				
				R^8		R^8				
				S^8						
Psidium friedrichsthalianum (O. Berg.) Niedenzu	Myrtaceae			R^8		R^8				
Psidium guajava L.	Myrtaceae			M^8	$M^{5,8}$	M^6	$M^{5,8}$			
			$R^{4,7}$	$R^{2,8}$	$R^{2,8}$	$R^{2,8}$				
				S^8						
Punica granatum L.	Lythraceae		R^4							
Richardella campechiana (Kunth) Pierre	Sapotaceae			R^8		M^8				
						R^8				

(continued)

Table 28.1 (continued)

Host plants										
Botanical name	Family	Zc	Bz	Cc	Cri	Cr	Cv	Dc	Dd	Nc
Sandoricum koetpage (Burm. F.) Merill	Meliaceae			S^8						
Scaevola taccada (Gaertn.) Roxb.	Goodeniaceae			R^8						
Sechium edule (Jacq.) Sw.	Cucurbitaceae	M^5						M^5	M^5	
		$R^{8, 10}$						$R^{2, 8}$	$R^{2, 8}$	
Solanum americanum Miller	Solanaceae									R^2
Solanum macranthum L.	Solanaceae									R^2
Solanum macrocarpon L.	Solanaceae			R^8						M^8
										R^2
Solanum mauritianum Scop.	Solanaceae			R^8		R^8				R^8
Solanum melongena L.	Solanaceae									M^8
										$R^{2, 8}$
Solanum nigrum L.	Solanaceae									R^8
Solanum torvum L.	Solanaceae									$R^{2, 8}$
Spondias mombin L.	Anacardiaceae				M^5					
Synsepalum dulcificum (Shaumacher & Thonn.) Daniell	Sapotaceae			R^8		R^8				
Syzygium cumini (L.) Skeels	Myrtaceae			$R^{2, 8}$		M^8				
						R^8				
Syzygium jambos (L.) Alston	Myrtaceae		M^8		M^5	M^5				
					R^8	$R^{2, 8}$				
Syzygium malaccense (L.) Merr. & Perry	Myrtaceae			R^8	M^6	M^6				
						R^8				
Syzygium samarangense (Blume) Merr. & L. M. Perry	Myrtaceae		M^8	$M^{5, 8}$	M^6	$M^{6, 8}$				
				$R^{2, 8}$		R^8				
				S^8						
Terminalia catappa L.	Combretaceae		M^8	M^8	M^6	$M^{6, 8}$				
			$R^{4, 7, 8}$	$R^{2, 8}$ S^8	$R^{2, 8}$	$R^{2, 8}$				
Theobroma cacao L.	Steculiaceae					R^8				
Thevetia peruviana (persoon) K. Schumann	Apocynaceae			R^8						
Trichosanthes cucumerina L.	Cucurbitaceae	$M^{6, 8, 9}$						$M^{6, 8, 9}$	$M^{6, 8}$	
		$R^{8, 10}$						$R^{8, 10}$		

(continued)

Table 28.1 (continued)

Host plants										
Botanical name	Family	Zc	Bz	Cc	Cri	Cr	Cv	Dc	Dd	Nc
Vitis vinifera L.	Vitaceae			R[8]						
Ziziphus mauritiana Lamarck	Rhamnaceae		M[8]	M[6,8]	M[6]	M[6,8]	M[8] R[2,8]			

Adapted from Quilici et al. (2001)
Zc *Zeugodacus cucurbitae*, Bz *Bactrocera zonata*, Cc *Ceratitis capitata*, Cri *Ceratitis catoirii*, Cr *Ceratitis rosa*, Cv *Carpomya vesuviana*, Dc *Dacus ciliatus*, Dd *Dacus demmerezi*, Nc *Neoceratitis cyanescens*
[1]Brévault et al. (1999); [2]Etienne (1982); [3]Hammes (1982); [4]Hurtrel et al. (2002); [5]Mamet and Williams (1993); [6]Orian and Moutia (1960); [7]Quilici et al. (2005); [8]Quilici et al. (2001); [9]Sookar et al. (2012); [10]Vayssières (1999)

The Natal fruit fly, *Ceratitis rosa* Karsch was first reported in Mauritius in the 1950s (Orian and Moutia 1960) and is currently considered as an established species (Fagoonee 1984; Ramsamy 1989; Soonnoo et al. 1996; Sookar et al. 2008) displacing *C. capitata* as the major fruit fly pest within 4 years of its accidental introduction (Hancock 1984). In Réunion *C. rosa* was first detected in 1955 (Orian and Moutia 1960; Etienne 1972) and, as in Mauritius, it has now displaced *C. capitata* as the most important fruit fly pest (Duyck et al. 2004). *Ceratitis rosa* is an important pest of fleshy fruits including avocado, guava, mango (*Mangifera indica* L.) and peach (Table 28.1).

The peach fruit fly, *Bactrocera zonata* (Saunders), was first recorded in Mauritius in 1942; subsequently it disappeared but was recorded again as an established population in 1987 (MAFNR 1988). In Réunion Island, *B. zonata* was first detected in 1991 but its populations grew massively and by 2000 it had spread widely (Hurtrel et al. 2000; Quilici et al. 2005). *Bactrocera zonata* is an important pest of mango and guava (Burn 1997; Quilici et al. 2005; Table 28.1).

The Oriental fruit fly, *Bactrocera dorsalis* (Hendel) (including early reports where it was known by its synonym, *Bactrocera invadens* [Schutze et al. 2014]), was first recorded in Mauritius in 1996 and, following control attempts, was declared eradicated in 1999 (Seewooruthun et al. 2000). However, it was detected again in an orchard in the north of the island in 2013 (one male *B. dorsalis* was caught in a methyl eugenol-baited trap). Again, an eradication programme was implemented and the fly was declared eradicated in 2014 (MAIFS 2014; Sookar et al. 2014a) and has not been detected in Mauritius since then. To date *B. dorsalis* has been reported at three locations in Madagascar (Raoelijaona et al. 2012): at Tamatave (2010), a mango orchard in Mahajanga (2011) and a citrus plantation in Antananarivo (2011). Furthermore, De Meyer et al. (2012) have also reported the presence of *B. dorsalis* in the Comoros.

The ber fruit fly, *Carpomya vesuviana* Costa, was first recorded in Mauritius in 1986 (MAFNR 1988) and is now an established pest of jujube red date (*Ziziphus jujuba* Lam.) (Soonnoo et al. 1996; White et al. 2000; Sookar et al. 2008).

White et al (2000) also reported its presence in Réunion Island and the Seychelles. *Carpomya vesuviana* is specific to jujube red date, and it is its main pest species (Table 28.1).

The melon fly *Zeugodacus cucurbitae* (Coquillett) is thought to have arrived in Mauritius in the late nineteenth century/early twentieth century (Orian and Moutia 1960). It was also first reported in Réunion Island (Etienne 1982) and the Seychelles (White 2006) in 1972 and 2000, respectively. The main host plants of *Z. cucurbitae* are cucurbits (De Meyer et al. 2015; Table 28.1).

The Ethiopian fruit fly *Dacus ciliatus* Loew was first recorded in Mauritius in 1901 (Orian and Moutia 1960) and is now considered to be an established species (White et al. 2000). De Meyer et al. (2012) also recorded its presence in the Comoros while Munro (1984) and White et al. (2000) reported its presence in Réunion. The main host plants of *D. ciliatus* are cucurbits (Table 28.1).

The tomato fruit fly *Neoceratitis cyanescens* (Bezzi) arrived in Mauritius in the 1950s (Orian and Moutia 1960). It is also present in the Comoros (Kassim and Soilihi 2000; Ryckewaert 2003; De Meyer et al. 2012); in Madagascar (Orian and Moutia 1960; Etienne 1972; Hancock 1984); and in Réunion (Orian and Moutia 1960; Etienne 1972; Brévault et al. 1999; White et al. 2000). It is an oligophagous species that attacks cultivated and wild species within the Solanaceae family (Table 28.1).

3 Management of Fruit Flies on Fruit Crops in Mauritius

3.1 Target Species

The most economically important fruit fly species attacking fruit in Mauritius are, in order of importance, *B. zonata, C. rosa, C. capitata* and *C. vesuviana*. Preferred cultivated hosts for the first three species are mango, guava, jujube red date, peach, loquat (*Eriobotrya japonica* [Thunb.] Lindl.), water apple (*Syzygium aqueum* Burm. F.), *Citrus* species and custard apple (*Annona reticulata* L.); the most heavily attacked wild fruit is the Indian almond, *Terminalia catappa* (L.) Ridley (Table 28.1).

A national fruit fly control programme (NFFCP) was started in Mauritius in 1994 and is still ongoing. The programme aims to control fruit fly infestations in mango, guava and jujube red date in backyards over a large area using a combination of the bait application technique (BAT), the male annihilation technique (MAT) and the sterile insect technique (SIT). The NFFCP was initially funded jointly by the European Union and the Government of Mauritius (Smith 1999). Since 1999, the Government of Mauritius has fully supporting the programme. The International Atomic Energy Agency (IAEA) has been co-funding technical cooperation projects and coordinated research programmes on fruit flies on the Islands. Throughout the implementation of the programme extension work was done regularly to publicise the programme. This included talks over the radio, display boards, posters, pamphlets and articles in the local newspapers on fruit fly management to ensure public awareness of the NFFCP.

3.2 Implementation and Monitoring of Fruit Fly Management Measures

BAT and MAT are considered to be safe and cheap. These two techniques were used together for suppression of populations of *B. zonata, C. rosa, C. capitata* and *C. vesuviana* in selected regions within the districts of Pamplemousses and Rivière du Rempart over an area of 300 km^2 (Fig. 28.1). From August 2013 the suppression of *B. zonata* using SIT was integrated with the use of BAT and MAT at Poudre D'Or over an area of 10 ha (Fig. 28.1). The treated area accounted for 40 % of the national fruit production area and included mango, guava and jujube red date in backyards.

3.2.1 Bait Application Technique (BAT)

A mixture of water (carrier), protein hydrolysate (a food attractant), and insecticide in the ratio of 97.3: 2: 0.7 is used. The protein hydrolysate solution was applied at the rate of 40 ml per tree as a coarse spot spray on the undersides of foliage at 7–10 day intervals. In the vicinity of schools and health centres, GF120 was used and applied at the rate of 1 L/ha. In 2012, 60 L GF120, 2374 L protein hydolysate, 210 L malathion 57 EC and 112 L of other insecticides (deltamethrin, imidacloprid, lambdacyhalothrin, cypermethrin and baythroid) were used in bait sprays over an area of about 300 km^2. BAT is relatively safe to non-target insects and is also less polluting to the environment than cover sprays, as it produces very little drift and very few chemical residues (Steiner et al. 1961; Avidov et al. 1963). In 2012 and in order to encourage growers to use BAT, approximately 294 L of protein hydrolysate were distributed, free of charge, from eight distribution sites across the island.

3.2.2 Male Annihilation Technique (MAT)

After reducing the fly populations to very low levels using area-wide BAT treatments, MAT was implemented. MAT uses parapheromones as lures within plywood blocks to mass-trap males, thereby reducing the mating success of females. MAT blocks are prepared by soaking plywood blocks (50×50×12 mm) for 7 days in a mixture of methyl eugenol (>98 %) or trimedlure (>98 %) and Malathion ULV in the ratio of 4:1 (v/v). The soaked blocks were then allowed to dry for at least one day after which time they are ready for placement on fruit trees. Methyl eugenol attracts males of *B. zonata* while trimedlure targets *C. rosa* and *C. capitata.* The MAT blocks were placed at 50 m intervals in marginal areas and at 33 m intervals in areas that were heavily infested with fruit flies; this was equivalent to four blocks/ha and nine blocks/ha, respectively. In 2012, 64,000 MAT blocks impregnated with methyl eugenol and 6000 MAT blocks impregnated with trimedlure were placed in the field. Since no pheromones are known for *C. vesuviana,* control of this species was achieved solely by BAT. To encourage growers to use MAT, 12,622 MAT blocks impregnated with methyl eugenol were distributed, free of charge, from eight distribution sites across the island in 2012.

Fig. 28.1 Map of Mauritius showing fruit fly trap locations and areas treated up until December 2014

3.2.3 Sterile Insect Technique (SIT)

SIT is a universally accepted and environmentally-friendly method for controlling a variety of insect pests (Koyama et al. 2004; Teal et al. 2007). The village of Poudre D'Or which has mainly mango trees in backyards (Fig. 28.1) was selected for small-scale release and preliminary evaluation of sterile *B. zonata* (SIT); this was an ideal location due to its physical isolation from other fruit-growing areas by surrounding sugar cane (*Saccharum officinarum* L.) plantations. The bisexual strain of *B. zonata*

Fig. 28.2 Map of Poudre D'Or showing release sites for sterile *Bactrocera zonata* (Saunders) and monitoring trap locations

was successfully mass reared in the laboratory of the Entomology Division, Ministry of Agro Industry and Food Security (Sookar et al. 2014b). Every week pupae were dyed and bagged for irradiation (at 8 days of age when they were approximately 2 days from emergence). Pupae were gamma irradiated at 70 Gy using a Caesium source irradiator. Sterilised pupae were then placed in paper bags with a mixture of enzymatic yeast hydrolysate and sugar as food and agar as a water source. Ground release of the sterile flies was done 6–7 days after emergence (Fig. 28.2). There were 66 release sites that were all separated by a distance of 75–100 m from each other. The number of sterile flies released per week varied from 4000 to 180,000 and releases were made between August 2013 and December 2014 (MAIFS 2013; 2014).

Twelve monitoring traps baited with methyl eugenol and malathion were set out at Poudre D'Or and serviced before the release of the sterile flies to establish what the baseline populations were. The distance between two traps varied from 300 to 400 m. After release of sterile flies monitoring continued and trapped flies were returned to the laboratory every week and examined under ultra violet illumination in a dark room to determine the numbers of marked and unmarked flies. Fruits from Indian almond and mango were sampled in the release areas throughout 2012 and incubated in the laboratory to determine infestation levels.

Table 28.2 Mean percentage of fruits infested: Indian almond, *Terminalia catappa* L.; guava, *Psidium guajava* L.; jujube red date, *Ziziphus jujube* L.; and mango, *Mangifera indica* L., in the treated and untreated regions in 2012

Fruit	% Infestation		ANOVA test
	Treated region	Untreated region	
Indian almond	3.8 ± 1.3	26.6 ± 9.2	$F_{1,23} = 72.46$, $P < 0.0001$
Guava	6.4 ± 1.3	36.3 ± 9.9	$F_{1,9} = 44.34$, $P < 0.0001$
Jujube red date	6.8 ± 5.6	32.2 ± 4.1	$F_{1,9} = 67.5$, $P < 0.0001$
Mango	8.4 ± 4.7	35.5 ± 12.1	$F_{1,13} = 30.7$, $P < 0.0001$

Adapted from MAIFS (2012)

3.2.4 Monitoring and Impact of the Management Measures

Host fruits, mainly fallen fruits, were sampled from treated and untreated areas at fortnightly intervals. Sampled fruits were counted, weighed and incubated individually for up to 2 weeks on sand in closed, aerated plastic trays in the laboratory. The pupae that exited each fruit were sieved from the sand, counted to determine an infestation level, and then placed in Perspex cages (15 × 15 × 20 cm); emerged adults were counted and identified. In 2012 fruit fly emergence from sampled fruit (infestation levels) were significantly lower in Indian almond, guava, jujube red date and mango in the treated regions compared with the untreated regions (Table 28.2; Figs. 28.3a, b, c and d).

Fruit fly monitoring by trapping is an important activity in any area-wide fruit fly management programme to determine the impact of the treatments. Locally developed dry traps based on the Steiner model, with lures made with methyl eugenol, trimedlure or cuelure were used for monitoring adult male fruit flies in all locations (Fig. 28.1). There were three dry traps baited with each of the three lures at each location. The traps were checked and serviced every fortnight when the 0.1 ml of lure/malathion 57 EC in a ratio of 1:1 was replaced. There were 732 dry traps placed throughout the island. McPhail traps, which are wet traps were used to monitor both males and females of most of the fruit fly species present. The 55 wet traps, baited with protein hydrolysate (20 ml/L of water) and borax (2 g/L) were checked and the bait replaced every week. Monitoring in regions treated with MAT and BAT and in untreated regions has been continuous since January 1993 (Fig. 28.4).

Populations of *B. zonata*, which is the main fruit fly of fleshy fruits, showed a significant decline in regions treated with MAT and BAT compared with untreated regions in 1994 and this has been maintained to date (Fig. 28.4). Smith (1999) reported that the NFFCP demonstrated that BAT and MAT, when applied over a large area, can provide effective fruit fly control. However, despite a reduction in fruit infestation levels in the treated areas, *B. zonata* attack was still evident (Table 28.3). Fallen fruits in backyards act as reservoirs of fruit fly pests. In most cases, fallen fruits are left on the ground where they continuously flood the environment with fruit flies. Sanitation is an essential component of an integrated control programme. For effective control, fruit sanitation should be encouraged in backyards.

Fig. 28.3 Percentage of Indian almond, *Terminalia catappa* L.; guava, *Psidium guajava* L.; and jujube red date, *Ziziphus jujube* L. fruits infested in the treated and untreated localities in 2012

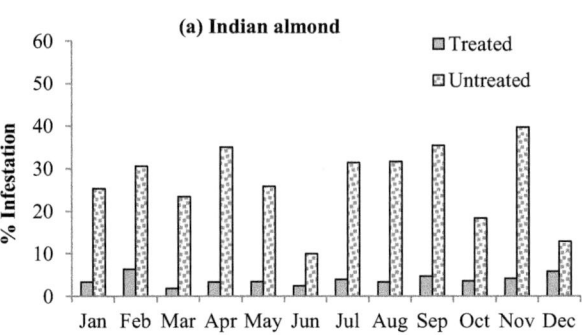
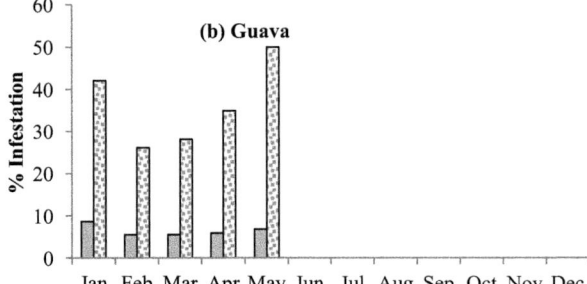
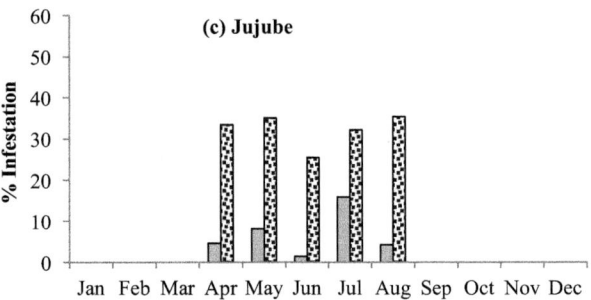
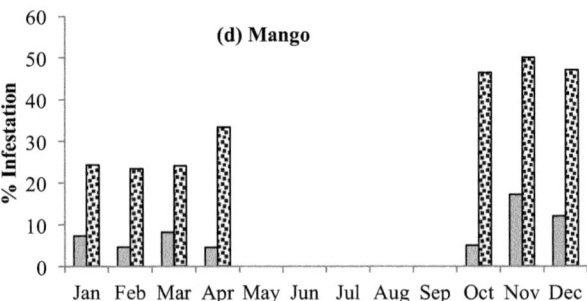

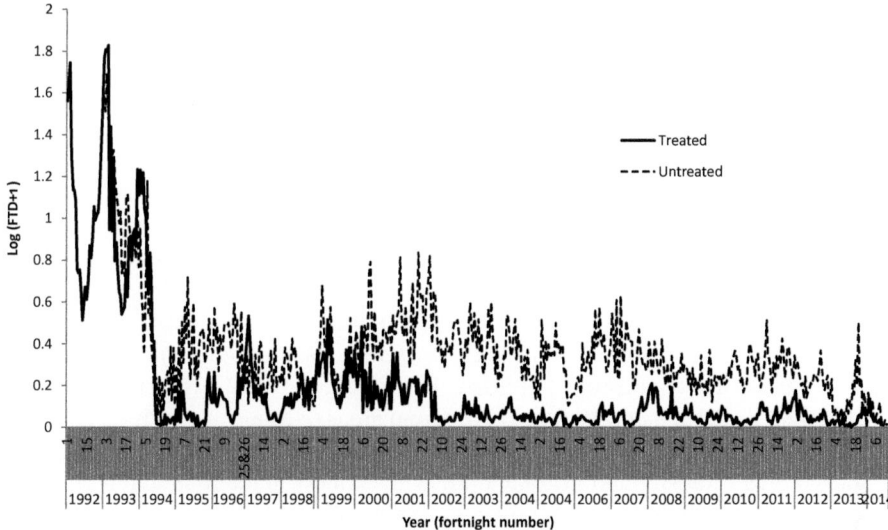

Fig. 28.4 Catches of male *B. zonata* [Log(FTD+1)] in dry traps baited with methyl eugenol and malathion in treated (n=36) and untreated (n=41) regions

Table 28.3 Infestation of cucumber, *Cucumis sativus* L., by *Zeugodacus cucurbitae* (Zc), *Dacus ciliatus* (Dc) and *Dacus demmerezi* (Dd) at Plaine Sophie from 2008 to 2014

Year	Mean no. of pupae/kg	Mean no. of Zc/kg	Mean no. of Dc/kg
2008	266.2±77.6 b	211.2±57.3 c	6.1±3.2
2009	306.6±36.9 b	199.4±26.4 c	1.1±0.8
2010	292.6±38.3 b	180.6±27.6 c	1.7±0.6
2011	492.6±71.2 b	270.3±53.9 c	3.0±1.4
2012	223.9±51.7 b	114.0±31.9 bc	0.1±0.1
2013	74.4±18.1 a	36.1±10.5 ab	0.0±0.0
2014	50.7±19.5 a	22.3±6.9 a	0.1±0.1
ANOVA test	$F_{6, 184}$=9.761, P<0.0001	$F_{6, 184}$=9.010, P<0.0001	$F_{6, 184}$=2.838, P>0.012

Mean values (± SE) within a column followed by the same letter are not significantly different to each other using Tukey's HSD test (P=0.05)

With regards to the pilot SIT programme, there were major fluctuations in the number of sterile *B. zonata* released each week that ranged from 4000 to 180,000. The sterile:wild ratio varied from 2:1 to 35:1 while the proportion of *T. catappa* fruits infested varied from 3.4 to 20.2. A sterile:wild ratio ranging from 25:1 to 100:1 is recommended for the suppression of *C. capitata* and the release density of sterile *B. dorsalis* in Thailand was 5000 flies/ha (FAO/IAEA Guidance 2007). We attempted to release the same number of sterile *B. zonata* flies on each occasion but this was not always possible. Overall, the area-wide fruit fly control programme implemented in the north of Mauritius significantly reduced fruit fly infestation rates in mango, jujube red date and guava. Results from integrating SIT into the control programme suggested that it could help maintain low infestation levels of *B. zonata*.

4 Area-Wide Management of Fruit Flies on Cucurbits in Mauritius

Cucurbit crops occupy approximately 30% of the cultivated land in Mauritius, with yields of approximately 24,000 t per year. Although all three of the established tephritid fruit flies (*Z. cucurbitae, D. ciliatus* and *D. demmerezi*) can attack cucurbits, *Z. cucurbitae* predominates, particularly during the summer months, and is associated with high production losses (Sookar et al. 2012). In 2007, a 4-year project entitled 'Feasibility study for the suppression of the melon fly in selected areas of Mauritius – MAR 5/016' was launched with the support of the IAEA. Given the success of this project in reducing *Z. cucurbitae* infestation levels in cucurbits, the IAEA funded a second project entitled 'Environment-friendly suppression of the melon fly (*Zeugodacus cucurbitae*) in different areas of Mauritius – Phase II – MAR 5022' for the period 2012–2015. The objectives of both programmes were to carry out technology transfer and to collaborate with farmers in the production of quality cucurbits using eco-friendly fruit fly control methods.

4.1 Targeting Cucurbit-Growing Localities

The targeted areas for *Z. cucurbitae* suppression were: the region of Plaine Sophie ($20^0 21'33.06"$S $57^0 29'24.37"$E, 556 m ASL, 3000–3600 mm of annual rainfall) which covers an area of 110 ha and is occupied by 135 growers; Dubreuil ($20^0 21'33.06"$S $57^0 29'24.37"$E, 556 m ASL, 3000–3600 mm of annual rainfall). The region of L'Esperance Trebuchet ($20^0 04'18"$S $57^0 36'38"$E, 87 m ASL, 1200–1800 mm of annual rainfall) was maintained as a control area in the first project, which started in 2007. Cucurbit growing localities in the regions of La Queen ($20^0 13'05.72"$S $57^{0'} 42.54"$E, 114 m ASL, 1800–2400 mm of annual rainfall), Rivière Baptiste ($20^0 12'56.79"$S $57^0 30'09.23"$E, 340 m ASL, 1800–2400 mm of annual rainfall) and Banane ($20^0 21'03.95"$S $57^0 35'41.48"$E, 385 m ASL, 3000–3600 mm of annual rainfall) were also included in the second project.

Growers were informed about the *Z. cucurbitae* control techniques through field seminars, regular site visits and demonstrations. Extension tools such as pamphlets, stickers and video film on *Z. cucurbitae* management were also used. The life cycle and behaviour of *Z. cucurbitae* were explained. Emphasis was placed on the importance of sanitation to discourage growers from leaving infested cucurbit fruits in their fields. Plants where *Z. cucurbitae* roost were identified in the presence of growers by demonstrating the presence of resting flies under their leaves; roost sites were then used for bait application and for hanging traps.

4.2 Implementation of Management Measures

Sanitation, BAT and MAT were used during the first project and SIT was integrated during the second project. Augmentoria (Klungness et al. 2005; Jang et al. 2007) were provided for growers to dispose of infested cucurbits; these capture any fruit flies emerging from fruit but allow the escape of any beneficial parasitoids. Initially growers were given free protein hydrolysate (Garuda Chemicals, Mumbai, India) which they used as bait at the rate of 20 ml/L after mixing with one of the following insecticides: cyfluthrin 50 SL, deltamethrin 2.5 EC, cypermethrin 10 EC, lambda-cyhalothrin 5 CS, fenthion 50 EC or profenofos 500 EC at a rate of 2 ml/L. During 2008, growers were also given GF-120 fruit fly bait (Dow AgroSciences, LLC) which they used at the rate of 1 L/ha. As from January 2013, growers were provided with Hym-Lure (Savoury Food Industries, South Africa) which they used as bait at a rate of 10 ml/L after mixing with any one of the above mentioned insecticide. Protein hydrolysate bait mixtures were applied as spot sprays of 5–10 ml at weekly intervals under the leaves of plants such as maize (*Zea mays* L.), guava, sugar cane, banana (*Musa* spp.) and papaya (*Carica papaya* L.) that had been identified as roost sites. MAT was accomplished by deploying plywood blocks (50 mm × 50 mm × 12 mm) impregnated with cuelure (p-acetoxyphenylbutanone-2) and malathion ULV as described previously. They were then given to growers for placement in their plantations. MAT blocks were placed in 1.5 L plastic bottles, each with an opening at the side and sited 20 m apart from each other. They were also hung on roosting plants. Growers were convinced about the effectiveness of the MAT blocks after seeing dead *Z. cucurbitae* in the traps. Replacement MAT blocks were given to growers every 3 months.

4.3 Assessment of Management Measures

Locally developed monitoring traps and containing cuelure were used for monitoring adult male *Z. cucurbitae* as described previously. The traps were checked and serviced every fortnight when the 0.1 ml of lure/malathion 57 EC in a ratio of 1:1 was replaced. McPhail Traps containing 300 ml of bait mixture (12.5 % modified waste brewer's yeast mixed with 5 % borax) were used to monitor both males and females of *Z. cucurbitae, D. ciliatus* and *D. demmerezi*. Damaged cucurbits were collected from the field, returned to the laboratory, weighed and then placed in plastic trays on a layer of fine sand to allow the pupae to exit the fruit and enter the sand. Small holes less than 0.1 mm in diameter were made on the bottom side of the tray in order to allow the passage of water from the decaying fruits, especially cucumber, *Cucumis sativus* L.. Each cucurbit species was kept in a separate tray. After 10 days, the pupae were sieved from the sand, counted and then placed in plastic Petri dishes for adult emergence; adults were identified and counted. In selected cucurbit plantations the level of fruit fly infestation was monitored by randomly examining young cucurbit fruits (6–8 cm length) for fruit fly punctures. These observations were made at weekly intervals throughout the fruiting stage. A fruit with just one fruit fly puncture was still classified as positive for attack. Some farmers also provided feedback on the number of damaged and undamaged cucurbit fruits on each occasion they harvested. An interviewer-administered structured questionnaire was designed in order to collect further information from the growers. The information collected included: grower's profiles and perceptions, farm characteristics, pest management practices, and knowledge and adoption of recommended *Z. cucurbitae* control techniques.

4.4 Impact of the Management Measures

The combined impact of sanitation, BAT and MAT was determined by inspection of the male monitoring trap catches and the protein bait monitoring trap catches recorded on a fortnightly and weekly basis, respectively. The mean number of *Z. cucurbitae* caught per trap over the 3-year sampling period was three times smaller at the treatment site, Plaine Sophie, compared with the control site, L'Esperance Trebuchet. During the sampling period, the mean number of *Z. cucurbitae* per trap at Plaine Sophie was less than 12 flies/trap/day (FTD). *Dacus demmerezi* was not caught in any of the cuelure-baited traps. Catches of *D. ciliatus* in the protein-baited monitoring traps at the three localities were zero during most months of the year with the exception of the hot summer months of December and January. From February 2008, growers at Dubreuil began to use the same *Z. cucurbitae* control techniques as were being used at Plaine Sophie. Immediately, the number of *Z. cucurbitae* caught per trap was reduced from eight FTD during the months of January and February to less than three FTD thereafter. The mean number of pupae/kg of cucumber at Plaine Sophie in January 2008 was 115.82 ± 7.64 compared with 81.90 ± 38.46 one year later. *Zeugodacus cucurbitae* was the most important fruit fly species infesting cucurbits at the three localities followed by *D. ciliatus* and *D. demmerezi* (Table 28.3 and 28.4). *Dacus demmerezi* pupae were never recovered from sampled cucumbers.

The profiles of growers at Plaine Sophie consisted mainly of men (94%) who were aged 40 years or older (74.4%) and the size of the parcel of land they managed ranged from 0.2 to 1 ha (70.9%). The most problematic pests encountered by 88.0% of the growers were insects and 65.8% claimed that fruit flies were the most important pests. Free MAT blocks and protein bait were utilized by 93.2% and 94.0% of the growers, respectively. After adopting sanitation, BAT and MAT, 84.8% of the growers declared that the quantity of pesticides used to control fruit flies in their cucurbit farms had greatly reduced. Prior to the start of the project in July 2007, 68.6% of the growers were making at least four applications of pesticides within a 2 week period. However, only 33.1% of the growers applied pesticides at this rate in July 2008 and 5.2% did not apply any pesticides as cover sprays for fruit fly control at all. 80.7% of the growers observed an increase in the yield of cucurbits

Table 28.4 Infestation of squash, *Cucurbita pepo* L., by *Zeugodacus cucurbitae* (Zc), *Dacus ciliatus* (Dc) and *Dacus demmerezi* (Dd) at Plaine Sophie from 2008 to 2014

Year	Mean no. of pupae/kg	Mean no. of Zc/kg	Mean no. of Dc/kg	Mean no. of Dd/kg
2008	248.5±40.1 b	187.2±31.3 b	10.2±3.5 b	1.8±1.4
2009	197.8±54.4 ab	123.6±47.8 ab	0.2±0.2 a	1.2±1.2
2010	346.7±69.0 b	202.1±52.2 b	0.2±0.2 a	0.4±0.4
2011	203.0±30.2 b	107.9±24.9 b	0.7±0.5 a	0.0±0.0
2012	160.0±29.8 ab	76.7±18.7 ab	0.6±0.6 a	2.2±2.2
2013	64.7±13.6 ab	33.8±10.0 ab	0.1±0.1 a	0.0±0.0
2014	26.2±16.2 a	9.2±6.3 a	0.2±0.2 a	0.0±0.0
	$F_{6, 166}=4.259$, $P<0.001$	$F_{6, 166}=4.400$, $P<0.0001$	$F_{6, 166}=9.429$, $P<0.0001$	

Mean values (± SE) within a column followed by the same letter are not significantly different to each other using Tukey's HSD test ($P=0.05$)

while 87.7% noticed an improvement in the quality of the harvested produce. 56.1% claimed that their profits increased as the cucurbits produced using the recommended control treatments were of higher quality.

In Mauritius *Z. cucurbitae* is the predominant fruit fly species in cucurbits followed by *D. ciliatus* and *D. demmerezi*. The study showed that the preferred cucurbit hosts of *Z. cucurbitae* at Plaine Sophie were, in order of preference: calabash (*Lagenaria siceraria* (Molina) Standl.), cucumber and the *Cucurbita pepo* L. group (squash, courgette and pumpkin). The most susceptible host at Dubreuil and L'Esperance Trebuchet was cucumber. *Dacus demmerezi* was only recovered from squash at Plaine Sophie and from calabash at Dubreuil. These cucurbit-infesting fruit flies did not develop in chayote, *Sechium edulis* (Jacq.) Sw., though they damaged the quality of the fruit through oviposition punctures.

The area-wide *Z. cucurbitae* control programme at Plaine Sophie covering an area of 110 ha occupied by 135 cucurbit growers and resulted in reduced fruit fly infestations, increased crop quality and yield, reduced pesticide use, increased knowledge by growers on fruit fly control including sanitation. The proportion of cucurbits infested, which was above 30% before project implementation, was reduced to less than 5% within a year. The survey results revealed that 85% of growers achieved an increase in both quality and quantity of cucurbits. The cost of cucurbit production was reduced due to a decrease in the use of pesticides and associated application costs. The most effective method for *Z. cucurbitae* management was field sanitation when infested cucurbits were either placed in augmentoria or buried in order to break the reproduction cycle (Klungness et al. 2005; Jang et al. 2007). In Hawaii, Vargas et al. (2010) demonstrated suppression of *C. capitata* and *B. dorsalis* (Diptera: Tephritidae), in a 40 km^2 area and over a 6 year period using sanitation, BAT, MAT and release of parasitoids. Area-wide application of these techniques has proved to be economically viable, sustainable, environmentally-friendly and to have suppressed fruit flies below economic threshold levels with minimum use of organophosphate and carbamate insecticides (Mau et al. 2003;

Vargas et al. 2003; Klungness et al. 2005) Furthermore, as a result of area-wide programme implementation and reduction in pesticide pressures, the number of beneficial insects has increased locally in Mauritius; growers also observed more honey bees visiting cucurbit flowers. This means that growers are now using less pesticide in their crops. The project demonstrated the use of an integrated approach to reduce insecticide use, and to produce better quality fruits.

5 Eradication of *Bactrocera dorsalis* in Mauritius

Bactrocera dorsalis was first detected in Mauritius in 1996 and it was successfully eradicated in 1999 (Seewooruthun et al. 2000). It was then detected again in 2013, in an orchard in the north of the island where one male *B. dorsalis* was caught in a methyl-eugenol baited trap (Fig. 28.5). Early detection was attributed to the ongoing fruit fly surveillance programme that has been running since 1994. Here we summarise the methods used to eradicate *B. dorsalis* from Mauritius, the results obtained and the lessons learnt during the eradication operations. The methodology for eradication was adopted from the Indian Ocean Region Emergency Action Plan for Exotic Fruit Flies that was developed within a regional fruit fly project entitled 'Preventing the introduction of exotic fruit fly species and implementing the control of existing species with the sterile insect technique and other suppression methods, RAF 5062' which was funded by IAEA (2015). The eradication strategy involved collection and destruction of fallen fruits, fruit stripping, MAT, BAT and soil drenching with an insecticide under selected fruit trees.

5.1 Early Detection

Within the context of the area-wide programme for the control of fruit flies, the NFFC Programme, funded by the government of the Republic of Mauritius, has been operational since 1994 (Soonnoo et al. 1996; Permalloo et al. 1998; Sookar et al. 2008). This programme has, as one of its prerequisites, an island-wide monitoring system for fruit flies using male lures and food attractants (Fig. 28.1). This island-wide trapping system made early detection of the *B. dorsalis* possible. Early detection of exotic pests is recognised as a very important factor for successful eradication (Allwood and Drew 1997; Seewooruthun et al. 2000). The speed of detection is also a determining factor in cost, effectiveness and success of subsequent eradication strategies.

Bactrocera dorsalis was first detected on the 8th of March 2013 in a locally-developed methyl eugenol/malathion (ratio 1:1) baited trap at Labourdonnais in the north of the island ($20^{0}04'13.12"$S $57^{0}37'11.07"$E, 74 m ASL). Preliminary identification of this single individual captured was made by staff at the Ministry who had received taxonomic training in November 2012 on a course entitled 'Fruit fly sur-

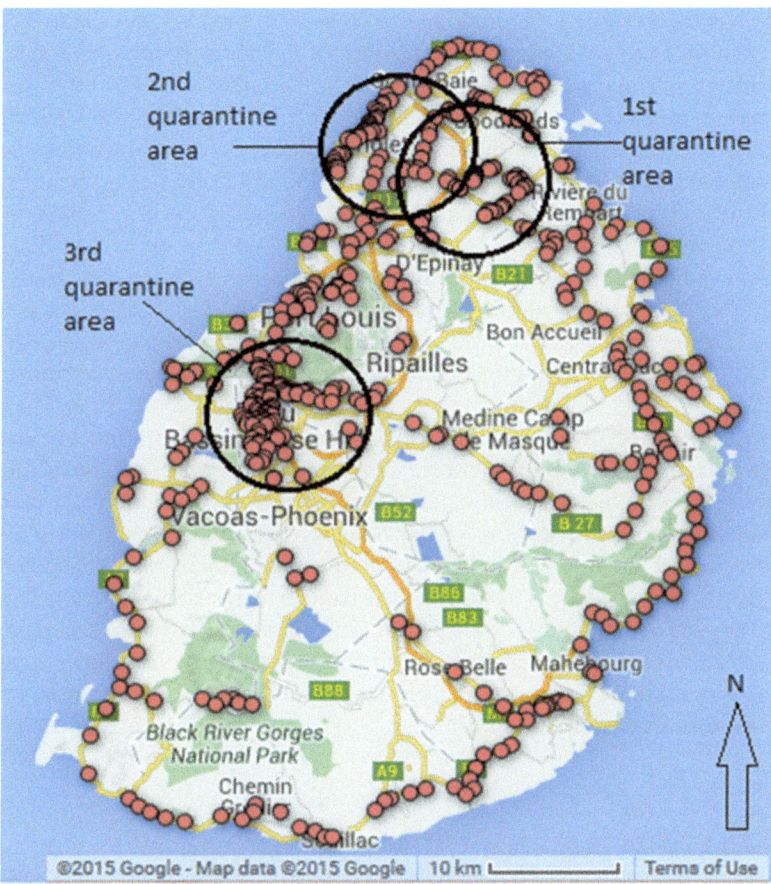

Fig. 28.5 Map of Mauritius showing trap locations (*red spheres*) and eradication areas (within black circles)

veillance, taxonomy and identification' within a Technical Cooperation Project that was funded by the IAEA. Correct identification was confirmed by Dr M. De Meyer of the Royal Museum for Central Africa in Belgium.

5.2 Establishing an Action Plan

Immediately after the detection of *B. dorsalis,* a steering committee was set up under the chairmanship of the Chief Agricultural Officer with the purpose of implementing the management actions relating to eradication of *B. dorsalis*. Members of the steering committee included representatives from the farming community, exporters of agricultural produce, the University of Mauritius, the police, institutions involved in agriculture and officers from the National Plant Protection Office (NPPO) and the Entomology Division of the Ministry of Agro Industry and Food Security. The international community was notified by NPPO in accordance with the requirements of the World Trade Organisation Sanitary Phytosanitary Standards Agreement, The International Plant Protection Convention and the International Standard Phytosanitary Measure (ISPM 17). The area surrounding the site of first detection (and subsequently other detection sites) was declared a quarantine area by the Cabinet in accordance with the Plant Protection Act 2006, which halted movement of fruits and vegetables from the quarantine area. Eradication procedures were undertaken within the quarantine areas and the populations monitored closely. The orchard where *B. dorsalis* was first detected was only allowed to move its fruits outside the quarantine area 16 weeks after the last detection of *B. dorsalis*.

Farmers and the general public were informed, through the media, about the detection of *B. dorsalis* and the associated threat to horticulture and the economy of Mauritius. A pamphlet on *B. dorsalis* providing a list of potential host plants and measures to be taken at the home or farm level for its containment and eradication was distributed to all households within the quarantine area. These measures included collection and disposal of fallen fruits and restrictions on movement of fruits. Farmers and members of the public were encouraged to provide any information they had on excessive damage to fruit and vegetables by fruit flies; in response to these observations fruit samples were taken and incubated for potential presence of *B. dorsalis*, in the laboratory.

5.3 Determining the Area Affected

Following the first detection of *B. dorsalis* (Fig. 28.5: 1st quarantine area), a delimiting survey was initiated in a core area of 1 km × 1 km immediately surrounding the detection site to identify the extent of infestation (Fig. 28.5: 1st quarantine area). The methods used followed those recommended by the Indian Ocean Region Emergency Action Plan for Exotic Fruit Flies (IAEA 2015). Methyl eugenol-baited traps and McPhail traps baited with modified waste brewer's yeast were both place at a density of 10 traps per km^2 within the core area. Beyond the core area, there were three surrounding zones of 8, 16 and 24 km^2. In each of these surrounding zones, the trapping density was two methyl eugenol-baited traps per km^2. All traps were serviced weekly, with core traps being serviced daily for the first week. Traps were maintained for 16 weeks after the last fruit fly was found.

Four days after the first *B. dorsalis* was trapped, a second fly was detected at Trou aux Biches (20^{0}01'56.31"S 57^{0}33'01.51"E, 12 m ASL) (Fig. 28.5: 2nd quarantine area) which is 8 km to the north-west of the site where the first *B. dorsalis* was detected. A second delimiting survey was implemented. The last *B. dorsalis* male fly at the first detection site was captured on 21 March 2014 in a methyl eugenol baited trap (Fig. 28.2). However, on the 3rd April 2014, three male *B. dorsalis* were caught in a methyl eugenol-baited trap at Montagne Ory (20^{0}13'03.72"S

57°29'18.34"E, 272 m ASL) (Fig. 28.5: 3rd quarantine area) approximately 20 km to the south-west of the first detection site. A third delimiting survey was implemented. The last detection of *B. dorsalis* within a 5 km radius of Motange Ory was recorded on 22nd November 2013.

Host fruit from the delimited quarantine areas were also surveyed, depending on availability. Infested fruits were collected and incubated for up to 6 weeks within a quarantine facility in the containment area. Any *B. dorsalis* adults that emerged w killed and preserved in alcohol or mounted for taxonomic confirmation.

5.4 Eradication Procedures

Eradication procedures (BAT, MAT, fruit stripping, soil drenching) were initiated following the first detection of *B. dorsalis* in the delimiting survey area. For each fly detected, the area treated extended for approximately 25 km^2 around the detection site. Eradication measures were pursued for 8 weeks. When no further *B. dorsalis* were detected I monitoring traps the eradication measures were stopped – but monitoring continued for a further 8 weeks.

Ground application of BAT was made weekly with Hym-Lure at 10 ml/L water (Villa Crop Protection Ltd., South Africa), GF120 at 1 L/Ha (Dow AgroSciences) or modified waste brewer's yeast (WBY) at 125 ml/L water (Sookar 2001). Hym-Lure and modified WBY were mixed with trichlorfon 80 % SP (0.5 g/L water), malathion 57 EC (2 ml/L water) or imidacloprid 200 SL (0.5 ml/L). The protein bait was applied using knapsack sprayers and motorised sprayers mounted on the deck of trucks. Protein bait sprays were applied to host trees twice a week within the corev area and weekly outside the core area and continues for a period of 8 weeks after the last *B. dorsalis* was detected.

MAT involved the distribution of square (5 cm×5 cm) 1.3 cm thick, wooden plywood blocks soaked in a mixture of methyl eugenol and malathion ULV at a ratio 3: 1 which were placed at a density of 400–600 per km^2, either nailed to poles or hung from trees (10 000–15 000 blocks per 25 km^2 fly-detection unit). A single application of MAT blocks was effective for a period of 8 weeks.

As much as possible, fruit stripping and the collection and disposal of fallen fruits was done within the quarantine areas. Stripped fruits and those collected from the ground were buried under at least 50 cm of soil. The burial site was located within the quarantine area. Furthermore, the soil beneath selected highly susceptible host trees was drenched with imidacloprid 200 SL (2 ml/L).

5.5 Results and Impact of the Eradication Procedures

Large quantities of materials that were used for the eradication of *B. dorsalis* (Table 28.6). Waste brewer's yeast collected from the local brewery was modified into brewer's yeast following the procedures of Loyd and Drew (1997). Only one male

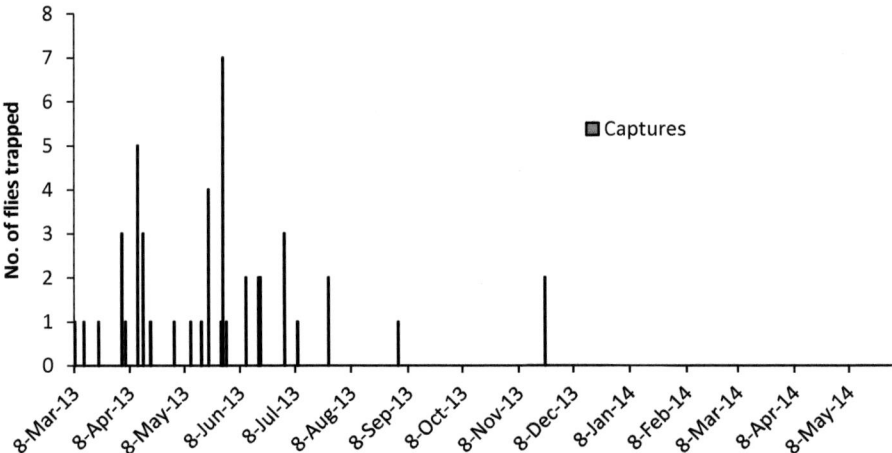

Fig. 28.6 Captures of *Bactrocera dorsalis* (Hendel) during the eradication period (Adapted from MAIFS 2014)

B. dorsalis was trapped in a methyl eugenol/malathion baited dry trap in the 1st quarantine area compared with three males in the 2nd quarantine area during the period 08–21 March 2013. Thirty nine (39) male flies were caught in dry traps and five females were caught in wet traps baited with modified waste brewer's yeast in the 3rd quarantine area between 03 April and 22 November 2013 (Fig. 28.6). Fruits and vegetables from 51 species were sampled for detection of *B. dorsalis* in the eradication area (Table 28.5). No *B. dorsalis* were recovered from fruits collected in the north but it was recorded from Barbados cherry, *Malpighia glabra* Millsp., and Indian almond from the 3rd quarantine area. About 52 t of fruits and vegetables were disposed of within the quarantine areas (Table 28.7).

Early detection of *B. dorsalis* was possible because of the island-wide trapping system which has been in place since 1994. An emergency action plan for the containment and eradication of exotic fruit flies which was prepared under the IAEA TC Regional project RAF 5062 was also readily available and utilized for the operation. Under the same project, the personnel had already been trained in the application of fruit fly control techniques. Preliminary identification of the trapped *B. dorsalis* was possible because the officers of the Division had received training on fruit fly surveillance, taxonomy and identification. The Mauritian Government also provided generous funding for the area-wide fruit fly control programme. Hence, materials and equipment were available to immediately embark on containment and eradication after *B. dorsalis* was first detected. The steering committee consistently monitored the progress of the eradication programme. The last detection of *B. dorsalis* in a trap dates back to 22 November 2013.

Despite the success, there is still the need to continuously improve the quarantine capacity of the staff in order to prevent any further introduction of exotic fruit fly species. The use of X-ray machines at ports of entry, for the detection of illegal plant material being carried in to the country in luggage and hand bags, is being contemplated and should minimize incursions. In order to reduce the risk of accidental

Table 28.5 Quantity of fruit infested by *Bactrocera dorsalis* (Bd), *Bactrocera zonata* (Bz), *Ceratitis rosa* (Cr), *Ceratitis capitata* (Cc), *Zeugodacus cucurbitae* (Zc), *Carpomiya vesuviana* (Cv) and *Dacus ciliatus* (Dc)

Botanical name	Fruit	kg	pupae/kg	Bd/kg	Bz/kg	Cr/kg	Cc/kg	Bc/kg	Cv/kg	Dc/kg
Annona muricata L.	Soursop	1.15	–	–	–	–	–	–	–	–
Annona reticulata L	Bullock's heart	0.5	–	–	–	–	–	–	–	–
Annona squamosa L.	Custard apple	10.5	11.54	–	10.14	1.39	–	–	–	–
Artocarpus altilis Parkinson	Bread fruit	0.85	32.94	–	12.94	–	–	–	–	–
Artocarpus integrifolia Forst.	Jackfruit	0.4	0	–	–	–	–	–	–	–
Averrhoa bilimbi L.	Cucumber tree	2.34	10	–	–	–	–	–	–	–
Carica papaya L.	Papaya	5.65	23.7	–	–	–	105	–	–	–
Carissa carandas L.	Corinda	0.2	130	–	–	–	–	–	–	–
Citrus spp.	Citrus	53.71	4.47	–	1.55	0.69	0.33	–	–	–
Coffea arabica L.	Coffee	0.03	–	–	–	–	–	–	–	–
Cucumis sativus L.	Cucumber	1.35	122.5	–	–	–	–	42.5	–	–
Cucurbita maxima Duch.ex Lam.	Pumpkin	1	66	–	–	–	–	–	–	–
Cucurbita moschata Auyama	Butternut squash	3.9	27.04	–	–	–	–	19.26	–	–
Cucurbita pepo L.	Marrow	1.1	4.17	–	13	1	–	0.83	–	0.83
Cucurbita pepo L.	Squash	0.98	21.54	–	–	–	–	15.38	–	–
Dimocarpus longan Lour.	Longane	0.15	0	–	–	–	–	–	–	–
Elaeocarpus serratus L.	Ceylon olive	2.6	–	–	–	–	–	–	–	–

(continued)

Table 28.5 (continued)

Botanical name	Fruit	kg	pupae/kg	Bd/kg	Bz/kg	Cr/kg	Cc/kg	Bc/kg	Cv/kg	Dc/kg
Eriobotrya japonica F.	Loquat	0.53	2.22	–	–	–	2.22	–	–	–
Eugenia uniflora L.	Surinam cherry	0.38	48	–	29.33	–	–	–	–	–
Feijoa sellowiana O.	Pineapple guava	2.35	43.65	–	13.65	17.46	–	–	–	–
Ficus carica L.	Fig	2.73	6	–	–	–	6	–	–	–
Flacourtia indica Burm. F.	Madagascar plum	3.28	–	–	–	–	–	–	–	–
Fortunella sp.	Kumquat	0.63	146.67	–	–	93.33	22.5	–	–	–
Hibiscus sabdariffa Linn.	Roselle	0.3	–	–	–	–	–	–	–	–
Hylocereus undatus Haw.	Dragon fruit	1.4	–	–	–	–	–	–	–	–
Lagenaria siceraria (Molina) Standl.	Calabash, bottle gourd	0.65	–	–	–	–	–	–	–	–
Luffa acutangula Linn.	Sponge gourd, ridge gourd	0.75	89.33	–	–	–	–	81.33	–	–
Luffa aegyptiaca Mill.	Nenwa, luffa sponge gourd	3	17	–	–	–	–	7	–	–
Lycopersicon esculentum Mill.	Tomato	7.7	0.44	–	–	–	–	–	–	–
Malpighia glabra Millsp.	Barbados cherry	1.17	33.21	1.25	12.66	–	29.23	–	–	–
Mangifera indica L.	Mango	113.19	20.81	–	7.37	–	–	–	–	–
Mimusops bojeri Hartog	Spanish cherry	10.55	3.73	–	0.82	–	–	–	–	–

Morinda citrifolia Linn.	Noni	0.4	–	–	–	–	–	–	–
Musa spp.	Banana	0.5	–	–	–	–	–	–	–
Passiflora edulis Sims.	Passion fruit	5.33	4.7	–	0.46	–	–	–	–
Passiflora foetida L.	Bush passion fruit	0.23	166	–	–	–	123	–	–
Persea americana Mill.	Avocado	11.55	45.59	–	27.62	0.05	–	–	–
Pouteria campechiana Kunth (Baehni)	Canistel, egg fruit	2.55	5.88	–	3.17	–	0.49	–	–
Prunus persica L.	Peach	2.68	47.46	–	11.62	10.67	–	–	–
Psidium cattleianum Afzel. ex Sabine	Strawberry guava, feijoa	2.93	15.21	–	4.62	0.41	2.2	–	–
Psidium cattleianum var. lucidum Sabine	Chinese guava	1.39	24.79	–	–	–	15.94	–	–
Psidium guajava L.	Guava	19.55	66.58	–	23.21	2.84	–	–	–
Punica grantum L.	Pomegranate	7.1	4.38	–	3.28	–	–	–	–
Spondias cytherea Forst.	Hog plum, ambarella	56.29	3.54	–	2.63	–	–	–	–
Syzygium aqueum Burm. F.	Wax jambu, water apple	3.83	4.16	–	–	1.04	–	–	–
Syzygium cuminii L.	Jambolan	1.23	0	–	–	–	–	–	–

(continued)

Table 28.5 (continued)

Botanical name	Fruit	kg	pupae/kg	Bd/kg	Bz/kg	Cr/kg	Cc/kg	Bc/kg	Cv/kg	Dc/kg
Syzygium jambos Alston	Rose apple	0.1	–	–	–	–	–	–	–	–
Syzygium malaccense Merr. & Perry	Malaysian rose apple	1.54	58.11	–	17.69	1.96	–	–	–	–
Tamarindus indica L.	Tamarind	0.13	–	–	–	–	–	–	–	–
Terminalia catappa L.	Indian Almond	43.12	58.97	0.15	19.44	1.38	0.55	–	–	–
Ziziphus mauritiana Lam.	Indian plum, ber	10.45	86.5	–	35.52	–	0.21	–	1.51	–

Adapted from MAIFS (2014)

Table 28.6 Materials used for the eradication of *Bactrocera dorsalis*

Materials	Quantity
Protein hydrolysate (L)	1724
Modified waste brewer's yeast (L)	6256
GF120 (L)	38
Malathion 57 EC (L)	58
Triclorfon (kg)	112
Imidacloprid (L)	27
Methyl eugenol (L)	875
Malathion ULV (L)	312
EDMA fruit fly traps (units)	1200

Adapted from MAIFS (2014)

Table 28.7 Fruits and vegetables collected within the quarantine areas for disposal

Botanical name	Fruits / vegetables	Quantity (kg)
Annona squamosa L.	Custard apple	5711
Artocarpus altilis Parkinson	Bread fruit	17
Averrhoa bilimbi L.	Tree cucumber	4
Averrhoa carambola L.	Carambola	769
Carica papaya L.	Papaya	2217
Citrus spp.	Citrus	20,235
Cucumis sativus L.	Cucumber	16
Cucurbita maxima Duch.ex Lam	Pumpkin	511
Flacourtia indica Burm. F.	Madagascar plum	1
Hylocereus undatus Haw.	Pitaya	100
Lagenaria siceraria (Molina) Standl.	Bottlegourd	29
Luffa acutangula Linn.	Ridge gourd	16
Mangifera indica L.	Mango	3131
Mimusops bojeri Hartog ex Engl.	Spanish cherry	25
Momordica charantia L.	Bittergourd	30
Morinda citrifolia Linn.	Noni	113
Passiflora edulis Sims.	Passion fruit	370
Persea americana Mill.	Avocado	394
Pouteria campechiana Kunth (Baehni)	Egg fruit	7
Psidium guajava L.	Guava	16,789
Punica grantum L.	Pomegranate	53
Spondias cytherea Forst.	Hog plum	809
Syzygium aqueum Burm. F.	Water apple	4
Syzygium cuminii L.	Jambolan	1
Terminalia catappa L.	Indian Almond	789
Trichosanthes anguina L.	Snakegourd	50
TOTAL		52,191

Adapted from MAIFS (2014)

entry of exotic fruit fly species, imports of fruits and vegetables are only allowed from pest-free areas or after proper quarantine treatment. The fruit fly surveillance system should be reinforced with traps placed in a grid arrangement for early detection of new incursions.

6 Control of Fruit Flies Attacking Cucurbits in Réunion Island

6.1 Species Composition on Cucurbits

Tephritid fruit fly species are the main pests of fruits and vegetables in Réunion Island (Etienne 1982; Quilici et al. 2005; Ryckewaert et al. 2010). *Zeugodacus cucurbitae, D. d'emmerezi* and *D. ciliatus* are the major pests on cucurbit crops (Vayssières 1999; Quilici et al. 2001). Amongst these three species, *Z. cucurbitae* causes the most economic damage on Réunion Island (Vayssières and Carel 1999). Losses of 100 % of cucurbit fruit at harvest have been frequently observed. Chemical control with insecticide (mostly pyrethroids and organophosphates) as cover sprays are commonly used while BAT, MAT and field sanitation are only rarely or moderately used (Ryckewaert et al. 2010). However, previous studies have shown that fruit fly control with pesticides does not provide satisfactory results (Deguine et al. 2011).

6.2 The GAMOUR Project

In 2009, a 3-year collaborative project on fruit fly suppression using environmentally-friendly techniques (GAMOUR: Gestion Agroécologique des MOUches des légumes à La Réunion) began. The fruit fly suppression programme was implemented on 21 farms (four were organic) at Salazie, Entre-Deux and Petite-Île (Fig. 28.7) over an area of 50 ha (10–20 % under cucurbits).

The main objective of the programme was to produce higher-quality and safe vegetables for local consumption. Project GAMOUR relied on an agroecological approach to both improve soil health and to increase plant biodiversity in the agroecosystems. GAMOUR brought together 14 local and national organizations with different tasks but the same goal. These partners included ARMEFLHOR (Réunion Association for the Modernization of Horticulture, Fruit and Vegetable Farming), the Chamber of Agriculture of Réunion; CIRAD (Centre for International Cooperation in Agronomic Research for Development), FDGDON (Departmental Federation for Protection against Harmful Organisms in Réunion), Réunion Farming and Environment Forum (FARRE) and the Réunion Organic Agriculture Association (GAB). Study methods consisted of (i) experiments in the field; (ii) monitoring of fly populations at the sites, (iii) monitoring of farms and fields and (iv) a survey of

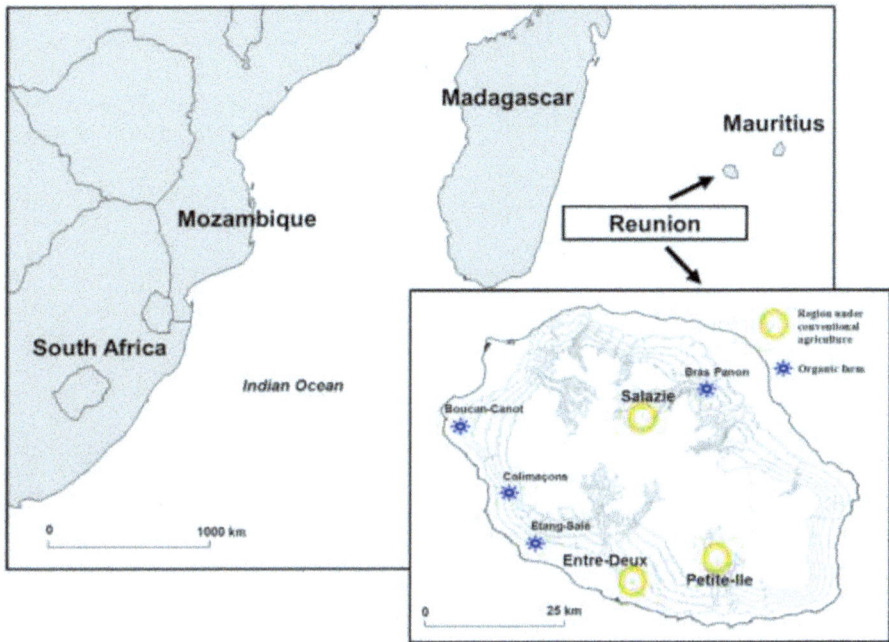

Fig. 28.7 Map showing position of La Réunion Island and treated localities

the farmers (http://gamour.cirad.fr). An interviewer-administered structured questionnaire was designed in order to collect information from the growers and included the following components: growers' profiles and perceptions, farm characteristics, pest management practices, knowledge and adoption of recommended fruit fly control techniques.

6.3 Fruit Fly Suppression Techniques and Results Obtained

The fruit fly control techniques adopted included fruit fly surveillance, field sanitation, application of BAT, MAT and release of parasitoids.

6.3.1 Fruit Fly Surveillance

A total of 50 traps baited with cuelure were used to monitor populations of Z. *cucurbitae* and D. *demmerezi*. The traps were serviced at weekly intervals in three pilot sites. There was a reduction in the number of Z. *cucurbitae* and D. *demmerezi* within 6 months of the project starting (Fig. 28.8). The number of males/trap/week was below nine even during the summer months (2010/2011).

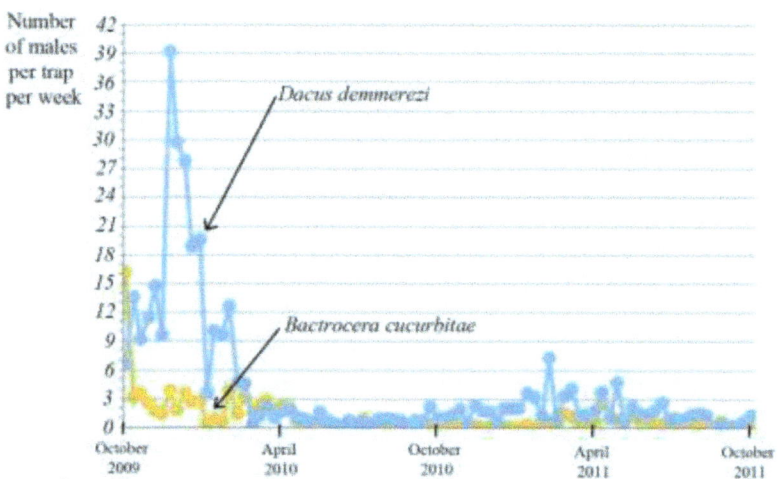

Fig. 28.8 Trap catches of adult male *Zeugodacus cucurbitae* (Coquillett) and *Dacus demmerezi* (Bezzi) in Salazie from September 2009 to March 2011 (From Deguine et al. 2015)

6.3.2 In Situ *Observation of Fruit Flies*

Besides fruit fly surveillance through trapping, *in situ* observation of fruit flies in the field was also done to determine the bio-ecological characteristics of the flies (Deguine et al. 2012a). Observations revealed that the fruit fly population size fluctuated depending on the season; in sites below 1000 m ASL populations of *Z. cucurbitae* were smaller (<18 % of fruit infested) than populations of *D. d'emmerezi* (>56 % fruit infested); *D. ciliatus* was the predominant fruit fly species (54 %) in pumpkin plantations and the male:female sex ratio of the three fruit fly species varied with locality depending on climatic conditions.

6.3.3 Field Sanitation

All infested cucurbit fruits were collected and placed in augmentoria (Klungness et al. 2005; Jang et al. 2007). Adults of the larval parasitoid, *Psyttalia fletcheri* (Silvestri), that emerged from fruit fly pupae in the fruit within the augmentoria escaped through the nylon netting (mesh size: 1.9 mm × 1.9 mm in the form of parallelogram) while adult fruit flies that emerged were unabke to escape through the mesh and were trapped inside the augmentorium (Deguine et al. 2011) (Fig. 28.9). During the project, 81 augmentoria were given to farmers for the disposal of infested cucurbits (Fig. 28.10). This technique was adopted by all the growers since they could see the trapped fruit flies inside the cage. After a few months, infested cucurbits in the augmentorium turned into compost which was used by growers who were producing organic crops (Deguine and Penvern 2014).

Fig. 28.9 Test in the laboratory showing two adult *Psytallia fletcheri* (Silvestri) parsitoids escaping through the mesh (*green circles*) and three adult *Zeugodacus cucurbitae* (Coquillett) trapped inside

Fig. 28.10 An augmentorium in a cucurbit plantation

6.3.4 Application of BAT

Protein hydrolysate mixed with spinosad (0.02 %) was applied as spot sprays by growers under the leaves of trap crops such as maize (Fig. 28.11) (Deguine et al. 2012b); Atiama-Nurbel et al. (2012) reported that, in Réunion, both males and females of *Z. cucurbitae* roost under the leaves of maize. The distance between two spot sprays was 10 m and the bait was effective in controlling all three fruit fly species present (Deguine et al. 2012c). However, high rainfall in Réunion Island limited the long term efficiency of the bait; in addition, BAT could not be used for large chayote trellises because corn borders could not be planted around them. In these

Fig. 28.11 Row of maize, *Zea mays*, bordering a cucurbit crop

conditions, a bait station adapted to the prevailing conditions was designed (Delpoux and Deguine 2015). Experiments were conducted in field cages on *Z. cucurbitae* to test different characteristics of these bait stations and to construct one using local materials; results were then validated in the field. The attractiveness of the modified bait station was mainly related to its external colour, yellow being the most attractive colour. The efficacy of the modified bait station with respect to fly mortality was linked to the accessibility of the bait and direct application to the bait station proved to be the most efficient. In the field, *Z. cucurbitae* were attracted to the local bait station. This bait station is simple, effective, safe, and it remains effective in attracting the flies for at least 7 days. The bait station is an alternative to bait spot sprays.

6.3.5 Application of MAT

Plywood blocks impregnated with cuelure were placed in traps to attract males of *Z. cucurbitae* and *D. demmerezi*. The trap model was based on those designed by Vargas et al. (2008) and consisted of a 1.5 L plastic bottle with four holes pierced at the base. The holes were extended into the interior by small tubes of 0.8 cm diameter and a length of 2 cm (Fig. 28.12). A plug of the parapheromone, cuelure, was hung inside the plastic bottle using a galvanised wire. No insecticide was used as the male flies that entered the bottle could not find their way out.

6.3.6 Biological Control

Psyttalia fletcheri, the larval parasitoid of *Z. cucurbitae*, was released by FDGDON in 2010 and 2011 in four cucurbit plots (Marquier et al. 2014). A maximum level of parasitism of 16.7% was obtained in regions from 0 to 800 m ASL.

Fig. 28.12 Plastic bottle modified into a fruit fly trap

6.4 Socioeconomic Impact of the Intervention

The survey revealed that 80% of growers were satisfied, or very satisfied, with the level of fruit fly control in their cucurbit plantations (Deguine et al. 2015). There was a reduction in the number of insecticide applications required and the yield of courgette was greater in the treated plots compared with the plots treated with insecticide cover sprays only (Table 28.8). Furthermore, the adoption of the recommended techniques for fruit fly suppression resulted in 1.2–4.2-fold reductions in costs (Table 28.9).

6.5 Discussion and Conclusion

The GAMOUR project has shown that fruit flies attacking cucurbits in La Réunion can be successfully controlled by integrating BAT, MAT, biological control agents and sanitation. The adopted techniques enabled farmers to produce quality cucurbits

Table 28.8 Mean number of insecticide applications in courgette production at Petite Ile

	Treated area	Control (treated with insecticide cover sprays)
Mean no. of insecticide applications/crop cycle	0.2 ± 0.2	4.2 ± 1.3
Mean yield (Tons/Ha)	19 ± 8	13 ± 7
% loss	13 ± 15	34 ± 28

Adapted from Rousse et al. (2012)

Table 28.9 Comparison between two techniques used for the suppression of fruit flies in cucurbit plantations in La Réunion

	Fruit fly control with pesticides only	Integrated use of bait sprays, mass trapping and sanitation
Amount of active ingredient sprayed daily in the field	100–800 g	0.0008 L
No. of hours/day	3–6	3
Product cost/day (€)	20–40	13
Labour cost/day (€)	24–88	24

Adapted from Augusseau et al. (2011)

while reducing the use of insecticides and the cost of production. Since the end of the GAMOUR project, Agricultural Development organizations in La Réunion have expressed the desire to popularize these techniques to all vegetable growers. Two years after the end of the project, the results have been very satisfactory (Vincenot 2014 personal communication; Réunion Chamber of Agriculture data): in 2012 and 2013, 683 farmers were trained in GAMOUR techniques. Of these, 130 farms (90 in 2012 and 40 in 2013) were closely monitored to assess the impact of GAMOUR techniques on cucurbit crops. The monitoring took place over a total area of 128 ha (84 ha in 2012 and 44 ha in 2013). Sanitation was particularly successful (92 % of farmers used this technique and a large number of them continue to use the augmentoria, particularly in 2012).

GAMOUR can be considered as a catalyst in the development of organic farming. The techniques developed for agroecological management of fruit flies associated with cucurbits are ideal for organic farming: sanitation using an augmentorium, the inclusion of trap plants, use of Synéis-appât® (organic adulticide bait), mass trapping of male melon flies without the use of insecticide. This package can also used in the management of fruit flies on citrus and mango. Observations of large populations of beneficial arthropods being present on organic farms and on farms where insecticides had been drastically reduced (GAMOUR farms) confirmed the value of conservation biological control (Deguine and Penvern 2014). The GAMOUR project and the results obtained has triggered the development of organic farming in Réunion particularly for chayote, one of the island's iconic crops.

References

Allwood AJ, Drew RAI (1997) Fruit fly management in the Pacific. In: ACIAR Proceedings. No. 76 pp 267

Atiama-Nurbel T, Deguine J-P, Quilici S (2012) Maize more attractive than Napier grass as nonhost plants for *Bactrocera cucurbitae* and *Dacus demmerezi*. Arth Plant Interact 6:395–403. doi:10.1007/s11829-012-9185-4

Augusseau X, Deguine JP, Douraguia E, Duffourc V, Gourlay J, Insa G, Lasne A, Le Roux K, Poulbassia E, Rousse P, Roux E, Suzanne W, Tilma P, Trules E (2011) Gamour, l'agroécologie vue de l'île de La Réunion. Phytoma – La défense des végétaux 642:33–37

Avidov Z, Rosen D, Gerson U (1963) A comparative study on the effects of aerial versus ground spraying of poisoned baits against the Mediterranean fruit fly on the natural enemies of scale insects in citrus orchards. Entomophaga 3:206–212

Brévault T, Quilici S, Glenac S (1999) Mouche de la tomate à l'île de la Réunion: utiliser les signaux émis par la plante hôte pour piéger les femelles. Phytoma La santé des végétaux 515:35–36

Burn RW (1997) National fruit fly control programme: report on data collection and analysis. Entomology Division, Ministry of Agriculture and Natural Resources, Réduit, p 85

CABI/EPPO (1999) *Ceratitis capitata*. [Distribution map]. Distribution Maps of Plant Pests, June (2nd revision). Wallingford, UK: CAB International, Map 1

Christenson LD, Foote RH (1960) Biology of fruit flies. Ann Rev Entomol 5:171–192

De Meyer M (2000) Systematic revision of the subgenus *Ceratitis* MacLeay s.s. (Diptera: Tephritidae). Zool J Linn Soc 128:439–467, http://dx.doi.org/10.1111/j.1096-3642.2000.tb01523.x

De Meyer M, Quilici S, Franck A (2012) Records of frugivorous fruit flies (Diptera: Tephritidae: Dacini) from the Comoro archipelago. Afr Inverts 53:69–77

De Meyer M, Mohamed S, White IM (2015) Invasive fruit fly pests in Africa. A diagnostic tool and information reference for the four Asian species of fruit fly (Diptera, Tephritidae) that have become accidentally established as pests in Africa, including the Indian Ocean Islands. http://www.africamuseum.be/fruitfly/AfroAsia.htm. Accessed 23 Mar 2015

Deguine J-P, Penvern S (2014) Agroecological crop protection in organic farming: Relevance and limits. In: Bellon S., Penvern S. (eds.), Organic farming, prototype for sustainable agricultures, DOI 10.1007/978-94-007-7927-3_6, © Springer Science+Business Media Dordrecht 2014, 107–130

Deguine J-P, Atiama-Nurbel T, Quilici S (2011) Net choice is key to the augmentorium technique of fruit fly sequestration and parasitoid release. Crop Prot 30:198–202

Deguine J-P, Atiama-Nurbel T, Douraguia E, Chiroleu F, Quilici S (2012a) Species diversity within a community of the Cucurbit fruit flies *Bactrocera cucurbitae*, *Dacus ciliatus* and *Dacus demmerezi* roosting in corn borders near cucurbit production areas of Reunion Island. J Ins Sci 12:32, Available online: http://www.insectscience.org/12.32

Deguine JP, Atiama-Nurbel T, Deploux C, Douraguia E, Petite A, Duhautois S, Bonnet E, Ajaguin-Soleyen C, Moutoussamy ML, Quilici S (2012b) Insertion du maïs dans les agroécosystèmes comme plante-piège des mouches des cucurbitacées. In: Deguine JP (ed) Actes du séminaire international de clôture Gamour, 21–24 November 2011. Saint-Pierre, La Réunion, pp 124–125

Deguine J-P, Douraguia E, Atiama-Nurbel T, Chiroleu F, Quilici S (2012c) Cage study of spinosad-based bait efficacy on *Bactrocera cucurbitae*, *Dacus ciliatus* and *Dacus demmerezi* in Reunion Island. J Econ Entomol 105:1358–1365

Deguine JP, Atiama-Nurbel T, Aubertot J-N, Augusseau X, Atiama M, Jacquot M, Reynaud B (2015) Agroecological cucurbit-infesting fruit fly management. A review. Agron Sustain Dev. doi:10.1007/s13593-015-0290-5

Delpoux C, Deguine J-P (2015) Implementing a spinosad-based local bait station to control *Bactrocera cucurbitae* (Diptera: Tephritidae) in high rainy areas of Reunion Island. J Ins Sci 15:1–6. doi:10.1093/jisesa/ieu177

Dogley W (2000) Fruit fly control programmes in the Seychelles. In: Price NS, Seewooruthun I (eds), Proceedings of the Indian Ocean Commission regional fruit fly symposium, Flic-en-Flac, Mauritius, 5–9 June 2000. Quatre Bornes, Mauritius: Indian Ocean Commission, pp 59–61

Dubois J (1965) La mouche des fruits malgache (*Ceratitis malagassa*) et autres insectes des agrumes, pêches et pruniers à Madagascar. Fruits 20:435–460

Duyck PF, David P, Quilici S (2004) A review of relationships between interspecific competition and invasions in fruit flies (Diptera: Tephritidae). Ecol Entomol 29:511–520

EPPO (2014) PQR database. Paris, France: European and Mediterranean Plant Protection Organization. http://www.eppo.int/DATABASES/pqr/pqr.htm

Etienne J (1972) Les principales Trypétides nuisibles de l'île de La Réunion. Ann Soc Entomol France 8:485–491

Etienne J (1982) Etude systematique, faunistique et ecologique des tephritides de la Réunion, PhD thesis, Ecole Pratique des Hautes Etudes, Paris, France, pp 100

Fagoonee I (1984) Pests, pesticides legislation and management in Mauritius. Ins Sci Appl 5:175–182

FAO/IAEA Guidance (2007) Guidance for packing, shipping, holding and release of sterile flies in area-wide fruit fly control programmes. FAO Plant Production and Protection Paper 190, pp 195

Hancock DL (1984) Ceratitinae (Diptera: Tephritidae) from the Malagassy subregion. J Entomol Soc South Afr 47:277–301

Hurtrel B, Quilici S, Jeuffrault E, Manikom R, Georger S, Gourdon F (2000) Etat de siège contre la mouche de la pêche, Bactrocera zonata: bilan des opérations de deux années de lutte menées à La Réunion. Phys Chem Chem Phys 551:18–21

Hurtrel B, Quilici S, Jeuffrault E, Manikom R, Georger S, Gourdon F (2002) Etat de siège contre la mouche de la pêche, *Bactrocera zonata*: bilan des opérations de deux années de lutte menées à La Réunion. Phys Chem Chem Phys 551:18–21

IAEA (2015) The Indian Ocean region emergency action plan for exotic fruit flies. Food and Agricultural Organisation/International Atomic Energy Agency. pp 17 (In Press)

Jang EB, Klungness LM, Grant T (2007) Extension of the use of augmentoria for sanitation in a cropping system susceptible to the alien tephritid fruit flies (Diptera: Terphritidae) in Hawaii. J Appl Sci Environ Man 11:239–248

Kassim M, Soilihi AM (2000) Les mouches des fruits a la Republique Federale Islamique des Comoros. In: Price NS, Seewooruthun I (eds), Proceedings of the Indian Ocean Commission, Regional fruit fly symposium, Flic-en-Flac, Mauritius, 5–9 June 2000. Quatre Bornes, Mauritius: Indian Ocean Commission, pp 71–72

Klungness LM, Jang EB, Mau RFL, Vargas R, Sugano JS, Fujitani E (2005) New approaches to sanitation in a cropping system susceptible to tephritid fruit flies (Diptera: Tephritidae) in Hawaii. J Appl Sci Environ Man 9:5–15

Koyama J, Kakinohana H, Miyatake T (2004) Eradication of the melon fly *Bactrocera cucurbitae* in Japan: importance of behaviour, ecology, genetics and evolution. Ann Rev Entomol 49:331–349

Loyd A, Drew RAI (1997) Modification and testing of brewery waste yeast as a protein source for fruit fly bait. In: Allwood AJ, Drew RAI (eds), Proceedings of the symposium on the management of fruit flies in the Pacific, pp 192–198

MAFNR Ministry of Agriculture, Fisheries and Natural Resources (1988) Entomology Division. In: Mauritius annual report of the agricultural services of the Ministry of Agriculture, Fisheries and Natural Resources for the year 1988, pp 53–61

MAFNR Ministry of Agriculture, Fisheries and Natural Resources (1989) Entomology Division. In: Mauritius annual report of the agricultural services of the Ministry of Agriculture, Fisheries and Natural Resources for the year 1989, 244 pp

MAIFS Ministry of Agro Industry and Food Security (2013) Entomology Division. In: Annual report of the agricultural services of the Ministry of Agro Industry and Food Security for the year 2013, 250 pp

MAIFS Ministry of Agro Industry and Food Security (2014) Entomology Division. In: Annual report of the agricultural services of the Ministry of Agro Industry and Food Security for the year 2014, 256 pp

Mamet JR, Williams JR (1993) The recorded foodplants of Mauritian insects. Mauritius Sugar Ind Res Inst Occas Pap 35:1–202

Marquier M, Clain C, Roux E, Deguine J-P (2014) Lâchers de *Psyttalia fletcheri* (Silvestri) (Braconidae : Opiinae) sur cultures de cucurbitacées à La Réunion. Cahiers Agri 23:188–194

Mau RFL, Sugano JS, Jang EB (2003) Farmer education and organization in the Hawaii area-wide fruit fly pest management program. In: Inamine K (ed) Recent trends on sterile insect technique and area-wide integrated pest management: economic, feasibility, control projects, farmer organization and dorsalis complex control study. Research Institute of Subtropics, Okinawa, pp 47–57

Munro HK (1984) A taxonomic treatise on the Dacidae (Tephritoidea: Diptera) of Africa. Entomology Memoir, Department of Agriculture and Water Supply, South Africa 61 pp 313

Orian JEA, Moutia LA (1960) Fruit flies (Trypetidae) of economic importance in Mauritius. Revue Agricole et Sucrière de L'Ile Maurice 39:142–150

Permalloo S, Seewooruthun SI, Joomaye A, Soonnoo AR, Gungah B, Unmole L, Boodram R (1998) An area-wide control of fruit flies in Mauritius. In: Lalouette JA, Bachraz DY, Sukurdeep N, Seebaluck BD (eds) Proceedings of the 2nd annual meeting of agricultural scientists. Food and Agricultural Research Council, Réduit, pp 203–210

Quilici S, Jeuffrault E, Blanchard E, Lustenberger F, Dupuis S, Franck A (2001) Plantes-hôtes des mouches des fruits: Maurice, Réunion, Seychelles. CIRAD-Min. de l'agriculture/SPV, PRMF/COI, Imp. Graphica, St André, La Réunion. pp 227

Quilici S, Duyck PF, Rousse P, Gourdon F, Simiand C, Franck A (2005) La mouche de la pêche sur mangue, goyave, etc. à La Réunion, évolution des recherches et des méthodes de lutte. Phytoma La Défense des Végétaux 584:44–47

Ramsamy MP (1989) A survey of three main tephritids and their hosts in Mauritius and some studies on their control with attractive chemical traps. Ins Sci Appl 10:383–391

Raoelijaona JCY, Raoelijaona AR, Ratovonomenjanahary TZ, Brunet C, De Meyer M, Vayssières JF, Quilici S (2012) Situation of *Bactrocera invadens* (Diptera: Tephritidae) in Madagascar, pp 141. In: Abstract volume 2nd international TEAM meeting, Kolymbari (Greece) 03–06 July 2012. http://www.teamfly2012.com/public/conferences/1/schedConfs/1/abstract_book_2nd_team_meeting.pdf. Accessed 23 Mar 2015

Ryckewaert P (2003) Les insectes et acariens des cultures maraichères dans les DOM-TOM. Situation et perspectives. Phytoma Défense des Végétaux 562:26–31

Ryckewaert P, Deguine J-P, Brévault T, Vayssières JF (2010) Fruit flies (Diptera: Tephritidae) on vegetable crops in Reunion Island: state of knowledge, control methods and prospects for management. Fruits 65:113–130

Schutze MK, Aketarawong N, Amornsak W, Armstrong KF, Augustinos AA, Barr N, Bo W, Bourtzis K, Boykin LM, Caceres C, Cameron SL, Chapman TA, Chinvinijkul S, Chomic A, De Meyer M, Drosopoulos E, Englezou A, Ekesi S, Gariou-Papalexiou A, Geib SM, Hailstones D, Hasanuzzaman M, Haymer D, Hee AKW, Hendrichs J, Jessup A, Ji Q, Khamis FM, Krosch MN, Leblanc L, Mahmood K, Malacrida AR, Mavragani-Tsipidou P, Mwatawala M, Nishida R, Ono H, Reyes J, Ribinoff D, Sanjose M, Shelly TE, Srikachar S, Tan KE, Tanaphum S, Haq I, Vijaysegaran S, Wee SL, Yesmin F, Zacharopoulou A, Clarke AR (2014) Synonymization of key pest species within the *Bactrocera dorsalis* species complex (Diptera: Tephritidae): taxonomic changes based on a review of 20 years of integrative morphological, molecular, cytogenetic, behavioural and chemoecological data. Syst Entomol. doi:10.1111/syen.12113

Seewooruthun SI, Permalloo S, Gungah B, Soonnoo AR, Alleck M (2000) Eradication of an exotic fruit fly from Mauritius. In: Tan KH (ed), Proceedings: area-wide control of fruit flies and other insect pests, and the 5th international symposium on fruit flies of economic importance, 28 May–5 June 1998, Penang, Malaysia, pp 389–394

Smith ESC (1999) Review of the national fruit fly control programme. Landell Mills Limited, United Kingdom, p 85

Sookar P (2001) Modification and evaluation of waste brewer's yeast as a protein source for the monitoring and control of the melon fly, *Bactrocera cucurbitae* (Coquillett). MSc Thesis University of Mauritius. pp 112

Sookar P, Permalloo S, Gungah B, Alleck M, Seewooruthun SI, Soonnoo AR (2008) An area wide control of fruit flies in Mauritius. In: Sugayama RL, Zucchi RA, Ovruski SM, Sivinski J (eds), Fruit flies of economic importance: from basic to applied knowledge. Proceedings of the 7th international symposium on fruit flies of economic importance, 10–15 September 2006, Salvador, Brazil, pp. 261–269. Available from: http://www.moscamed.org.br/pdf/Cap_30.pdf

Sookar P, Alleck M, Buldawoo I, Khayrattee FB, Choolun T, Permalloo S, Rambhunjun M (2012) Area-wide management of the melon fly *Bactrocera cucurbitae* (Coquillett) (Diptera: Tephritidae). In: Deguine JP (ed) Actes du séminaire international de clôture GAMOUR Gestion Agroécologique des mouches des Légumes à la Réunion, 21–24 November 2011. Saint-Pierre, La Réunion, pp 98–102

Sookar P, Permalloo S, Alleck M, Buldawoo I, Mosaheb M, Nundloll P, Ramjee S, Ahseek N, Allymamod N, Rambhunjun M, Khayrattee FB, Patel N (2014a) Detection of *Bactrocera dorsalis* (Hendel) in Mauritius and rapid response, pp 60. In: Book of abstracts: the ninth international symposium on fruit flies of economic importance, 12–16 May 2014, Bangkok, Thailand

Sookar P, Alleck M, Ahseek N, Permalloo S, Bhagwant S, Chang CL (2014b) Artificial rearing of the peach fruit fly *Bactrocera zonata* (Diptera: Tephritidae). Int J Trop Ins Sci 34:99–107

Soonnoo AR, Smith ESC, Joomaye A, Permalloo S, Gungah B (1996) A large scale fruit fly control programme in Mauritius. In: Chua TH, Khoo SG (eds), Proceedings of the second symposium on tropical fruit flies. Problems and management of tropical fruit flies. Kuala Lumpur, Malaysia, pp 52–60

Steiner LF, Rohwer GG, Ayers EL, Christenson LD (1961) The role of attractants in the recent Mediterranean fruit fly eradication program in Florida. J Econ Entomol 54:30–35

Teal PEA, Gomez-Simuta Y, Dueben BD, Holler T, Olson S (2007) Improving the efficacy of the sterile insect technique for fruit flies by incorporation of hormone and dietary supplements into adult holding protocols. In: Vreysen MJB, Robinson AS, Hendrichs J (eds) Area-wide control of insect pests from research to field implementation. Springer, Dordrecht, pp 163–173

Vargas RI, Jang EB, Klungness LM (2003) Areawide pest management of fruit flies in Hawaiian fruits and vegetables. In: Inamine K (ed) Recent trends on sterile insect technique and areawide integrated pest management. Research Institute for Subtropics, Okinawa, pp 37–46

Vargas RI, Mau RFL, Jang EB, Faust RM, Wong L (2008) The Hawaii fruit fly area-wide pest management programme. In: Koul O, Cuperus GW, Elliott N (eds) Areawide pest management: theory and Implementation. CABI International, Cambridge, MA, pp 300–326

Vargas RI, Mau RFL, Stark JD, Pinero JC, Leblanc L, Souder SK (2010) Evaluation of methyl eugenol and cue-lure traps with solid lure and insecticide dispensers for fruit fly monitoring and male annihilation in the Hawaii area-wide pest management program. J Econ Entomol 103:409–415

Vayssières J-F (1999) Les relations plantes-insectes chez les Dacini (Diptera-Tephritidae) ravageurs des Cucurbitaceae à La Réunion. Thèse de Doctorat. Université Paris XII, Paris. pp 241

Vayssières J-F, Carel Y (1999) Les Dacini (Diptera: Tephritidae) inféodées aux Cucurbitacées à La Réunion: gamme de plantes-hôtes et stades phénologiques préférentiels des fruits au moment de la piqûre. Ann Soc Entomol France 35:197–202

White IM (2006) Taxonomy of the Dacina (Diptera: Tephritidae) of Africa and the Middle East. Afr Entomol 2:1–156

White IM, Elson-Harris M (1994) Fruit flies of economic significance: their identification and bionomics. CAB International, Wallingford, p 601

White IM, De Meyer M, Stonehouse JM (2000) A review of native and introduced fruit flies (Diptera: Tephritidae) in the Indian Ocean islands of Mauritius, Réunion and Seychelles, pp 15–21. In: Price, N.S. and S.I. Seewooruthun (eds), Proceedings of the Indian Ocean Commission regional fruit fly symposium. Indian Ocean Commission/European Union, Flic en Flac, Mauritius

Preeaduth Sookar holds a BSc (Hons) in Agriculture, an MSc in Crop Science (with specialization in Crop Protection) and a PhD on entomopathogenic fungi from the University of Mauritius. He joined the Ministry of Agro Industry and Food Security as Technical Officer (Crop Extension) and is now the Senior Scientific Officer at the Entomology Division. Preeaduth has been working on fruit fly surveillance and management with environmentally friendly control techniques (bait application technique, male annihilation technique, sanitation, biological control and the sterile insect technique). He was involved in early detection and eradication of *Bactrocera dorsalis* in Mauritius between 1996 and 2013. He is working on the modification of waste brewery yeast into fruit fly bait. Preeaduth has been fortunate enough to work as an international fruit fly expert for the Food and Agricultural Organisation / International Atomic Energy Agency (IAEA) in Seychelles, Vietnam, Burkina Faso, Croatia and Lebanon. He worked as project counterpart with the IAEA on fruit fly management projects. Preeaduth has been a research contract investigator for the IAEA for projects on fruit flies.

Jean-Philippe Deguine is an entomologist and agroecologist who has been working for CIRAD (Centre de Coopération Internationale de Recherche Agronomique pour le Développement) for the last 30 years. He has worked in crop protection in a range of agroecosystems (cotton, vegetables and fruit). He spent a large part of his career in research activities in different parts of the world (Africa: 13 years; Overseas French Territories: 9 years; France Mainland: 8 years) working on several biological models for cotton pests and fruit flies. He was also involved in Research Management and Strategy for CIRAD as Head of the Cotton Programme, Scientific Advisor for Pathology and Integrated Pest Management, and Scientific Director of Animal Production and Veterinary Medicine Department successively. He was president of the European IPM Network and a plenary speaker in the 3rd World Cotton Research Conference, Cape Town, South Africa in 2003. He is currently based in La Réunion (Indian Ocean) where he works on agroecological management for crop protection and has implemented a number of different programmes with stakeholders and farmers.

Part IV
Experiences from Actions Programmes Outside Africa

Chapter 29
Area-Wide Management of Fruit Flies (Diptera: Tephritidae) in Hawaii

Roger I. Vargas, Jaime C. Piñero, Luc Leblanc, Nicholas C. Manoukis, and Ronald F.L. Mau

Abstract Fruit flies (Diptera: Tephritidae) are amongst the most economically important pests attacking tropical fruits and vegetables in Hawaii. The Hawaii Fruit Fly Area-Wide Pest Management (AWPM) Programme was initiated in 1999 to suppress fruit flies below economic thresholds while reducing the use of organophosphate insecticides. The AWPM programme developed and integrated biologically-based pest control technologies into a comprehensive management package that was economically viable, environmentally sensitive and sustainable. The technologies included: (1) field sanitation, (2) protein bait application techniques (BAT), such as sprays, (3) male annihilation techniques (MAT) using male lures, and (4) sterile fly and parasitoid releases. In a cooperative effort the United States Department of Agriculture, Agricultural Research Service, the University of Hawaii, Hawaii State Department of Agriculture and industry partners developed environmentally friendly control technologies, secured special local needs registrations, implemented a fruit fly IPM extension educational programme and transferred novel technologies to local farmers and domestic growers. The programme demonstrated that if growers adopted IPM then fruit flies could be suppressed across large areas. The programme received seven major awards for IPM technology transfer activities. This chapter summarizes highlights of this highly successful

R.I. Vargas (✉) • N.C. Manoukis
Daniel K. Inouye USDA ARS US Pacific Basin Agricultural Research Center,
64 Nowelo Street, Hilo, HI 96720, USA
e-mail: roger.vargas@ars.usda.gov

J.C. Piñero
Lincoln University,
Cooperative Research and Extension, 900 Chestnut Street, Allen Hall 212, Jefferson City,
MO 65101, USA

L. Leblanc
Department of Plant, Soil and Entomological Sciences, University of Idaho,
875 Perimeter Drive MS 2339, Moscow, ID 83844-2339, USA

R.F.L. Mau
Department of Plant and Environmental Protection Sciences, University of Hawaii,
College of Tropical Agriculture and Human Resources,
3050 Maile Way, Room 319, Honolulu, HI 96822, USA

© Springer International Publishing Switzerland (outside the USA) 2016
S. Ekesi et al. (eds.), *Fruit Fly Research and Development in Africa - Towards a Sustainable Management Strategy to Improve Horticulture*,
DOI 10.1007/978-3-319-43226-7_29

community-based IPM programme. Although formal funding for the Hawaii AWPM program ended in 2008, transfer of AWPM components and technology continued locally, nationally and internationally, thus contributing to increasing the sustainability of agriculture.

Keywords Pacific • IPM • Sanitation • Protein bait • Male annihilation • Biological control

1 Introduction

Severe economic damage and stringent quarantine strategies as a result of invasive fruit flies, are not new in Hawaii. The melon fly, *Bactrocera cucurbitae* (Coquillett), was introduced in 1895; the Mediterranean fruit fly (medfly), *Ceratitis capitata* (Wiedemann), in 1910; the oriental fruit fly, *Bactrocera dorsalis* (Hendel), in 1945; and the solanum fruit fly, *Bactrocera latifrons* (Hendel), in 1983 (Vargas et al. 2007). Each invasion has taken its toll on local farmers and home gardeners. Together these pests now attack more than 400 different fruits and vegetables including: citrus (*Citrus* spp.), coffee (*Coffea arabica* L.), aubergine (*Solanum melongena* L.), guava, *Psidium guajava* L.; loquat, *Eriobotrya japonica* (Thunb.) Lindl.; mango, *Mangifera indica* L.; melon, *Cucumis melo* L.; papaya, *Carica papaya* L.; passion fruit, *Passiflora edulis* Sims; peach, *Prunus persica* (L.) Batsch; pepper, *Capsicum annuum* L.; persimmon, *Diospyros kaki* L.; plum, *Prunus* spp.; star fruit, *Averrhoa carambola* L.; tomato, *Solanum lycopersicum* L.; pumpkin, *Cucurbita pepo* L.; and courgette, *Cucurbita pepo* L. In response to the damage caused by these four invasive species Hawaii's farmers have, for decades, had to resort to almost weekly spraying of organophosphate and carbamate insecticides, to planting larger areas to compensate for anticipated losses, or simply to abandon crop production altogether. Resulting economic losses exceeded $300 million each year in lost markets for locally grown produce (McGregor et al. 2007). Furthermore, all fruit crops except pineapple required postharvest treatment in order to be exported from Hawaii.

The genus *Bactrocera* Macquart, is widely distributed throughout tropical Asia, the south Pacific and Australia; it is comprised of 651 described species, with at least 50 species considered as important pests, many of which are highly polyphagous (White and Elson-Harris 1992; Vargas et al. 2015). Until a little over 40 years ago, there were only a few recorded cases of Asian *Bactrocera* species invading and becoming established in areas outside their ancestral range: the peach fruit fly, *Bactrocera zonata* (Hendel) had been introduced to North Africa; *B. cucurbitae* had become established in East Africa and Hawaii; *B. latifrons* had established in Hawaii; and the *B. dorsalis* complex of species had established in the Mariana Islands, Hawaii, French Guyana and Surinam (Ekesi et al. 2006; Vargas et al. 2015). Outside of Africa, the olive fly, *Bactrocera oleae* (Rossi) only occurred in Southern

Europe (White and Elson-Harris 1992). The pace of invasions by *Bactrocera* species increased in the latter part of the 20th century. For example, the carambola fruit fly, *Bactrocera carambolae* Drew and Hancock was introduced and became established in South America (Suriname) in around 1975; *B. dorsalis* became established in French Polynesia in 1996; *B. oleae* was introduced and became established in California in 1998; *B. dorsalis* (described at the time as *Bactrocera invadens* Drew, Tsuruta and White) became established in Africa shortly before 2003 and has since spread to over 27 countries in the continent (Drew et al. 2005; Vargas et al. 2015; Ekesi and Billah 2007). Through an international collaborative study, *B. invadens*, along with *B. philippinensis* Drew and Hancock and *B. papayae* were declared synonyms of *B. dorsalis* (Schutze et al. 2014); this made it clear that the geographic range of *B. dorsalis*, was much larger than previously realised (including most of tropical Africa), and that its host range included over 270 host plant species (Vargas et al. 2015).

The invasive capabilities of *Bactrocera* species are associated with three key attributes: their potential for rapid population growth, their high natural dispersal capability and the ease with which they are transported by human activity (Malavasi et al. 2013). The last of these, movement by humans, is important as it is usually the first step in the chain of invasion, establishment and spread (Simberloff 2009; Lodge 1993). Human-mediated transportation of tephritids is quite common because fruit are a portable commodity that can be transported/ stored over a long period of time without spoilage (Putulan et al. 2004). On a single day in 1993, when all bags at Los Angeles International Airport in California USA were inspected by USDA-APHIS staff, 73 fruit fly larvae were intercepted (Malavasi et al. 2013). Polyphagous species, such as *B. dorsalis*, are more likely to be initially introduced via human transportation as they may be present in a wide range of fruit species. Rapid population growth and dispersal enables invading species to overcome barriers to invasion and subsequent establishment (Richardson and van Wilgen 2004).

Similar factors are at play for species in the genus *Ceratitis*, which is comprised of 65 species and originated in tropical and southern Africa (White and Elson-Harris 1992). *Ceratitis capitata* was accidentally introduced into Hawaii from Australia sometime before 1910, and it became a serious pest of tree fruits. When *B. dorsalis* was introduced into Hawaii in 1945, it displaced *C. capitata* throughout most of its range; it persisted only in small areas of commercial and wild coffee; strawberry guava, *Psidium cattleianum* Sabine; and a variety of fruits grown at high elevations, such as peaches, loquats and persimmons (Vargas et al. 2001).

In summary, fruit flies are both local and global pests, and area-wide procedures developed in Hawaii have both local and worldwide applications. For the last 15 years or so, greater effort has been invested into true integration of the components of Integrated Pest Management (IPM) within sound IPM programmes against fruit flies. This chapter provides an overview of the research, technology transfer, product registration and farmer education efforts that have been undertaken as part of the Hawaii fruit fly Area-Wide Pest Management (AWPM) programme. An area-wide insect control programme is a long-term campaign against an insect pest population throughout its entire range with the objective of reducing that insect population to a non-economic status (Lindquist 2000).

2 Classical Biological Control in Hawaii

Establishment of invasive fruit flies in Hawaii prompted the introduction of many natural enemies as part of a classical biological control strategy. Initial efforts began in 1912 with efforts to find and identify parasitoids in Africa and India for release against *B. cucurbitae* and *C. capitata* in Hawaii (Silvestri 1914). Subsequently the parasitoid *Psyttalia fletcheri* (Silvestri) (Hymenoptera: Braconidae), a widespread larval-pupal parasitoid of *B. cucurbitae* in India, was successfully introduced from India in 1916 (Fullaway 1920; Willard 1920; Nishida 1955). Exploration for additional natural enemies intensified when *B. dorsalis* was discovered in 1945 and a total of 32 natural enemies were released between 1947 and 1952 (Bess et al. 1961; Clausen et al. 1965). The parasitoid *Diachasmimorpha longicaudata* (Ashmead) (Braconidae) increased rapidly following its release in 1948, but was superseded in the latter half of 1949 by another parasitoid, *Fopius vandenboschi* (Fullaway) (Braconidae), which was also subsequently superseded by the egg-pupal parasitoid *Fopius arisanus* (Sonan) (Braconidae) (van den Bosch and Haramoto 1953; Haramoto 1953; Ramadan et al. 1992). Currently the most important fruit fly parasitoids that have established in Hawaii are *P. fletcheri* (against *B. cucurbitae*) and *F. arisanus*, *D. longicaudata*, *F. vandenboschi* and *Diachasmimorpha tryoni* (Cameron) (against *B. dorsalis* and *C. capitata*) (Vargas et al. 2001, 2004). These species have played a major role in the sustained reduction in fruit fly populations throughout the Hawaiian Islands. Full details of the efforts in Hawaii, as a global pioneer in classical biological control of fruit flies, are reviewed in Vargas et al. (2012a).

In the last 25 years the research focus has shifted to parasitoid ecology, biology, and mass rearing methods. These efforts have resulted in better utilization of these species through: mass rearing and release of parasitoids in combination with sterile flies (Sterile Insect Technique [SIT]) for area-wide suppression of fruit flies (Vargas et al. 2004, 2008); integration of reduced risk insecticides with parasitoids within area-wide IPM programmes (Vargas et al. 2015); and shipments of parasitoids from Hawaii to other areas of the world for classical biological control of fruit flies (Vargas et al. 2007). In Hawaii significant control was first achieved through classical biological control releases. Later successful IPM was achieved by: avoiding cover sprays, thereby conserving the released natural enemies; emphasizing the importance of orchard sanitation; use of reduced-risk BAT (e.g. spinosad-based GF-120 NF Naturalyte™ Fruit Fly Bait); and MAT (e.g. non-pesticide male lure traps or STATIC™-Spinosad-ME) (Fig. 29.1). Exploration for new natural enemies of fruit flies is also still ongoing by the University of Hawaii and the Hawaii Department of Agriculture (HDOA).

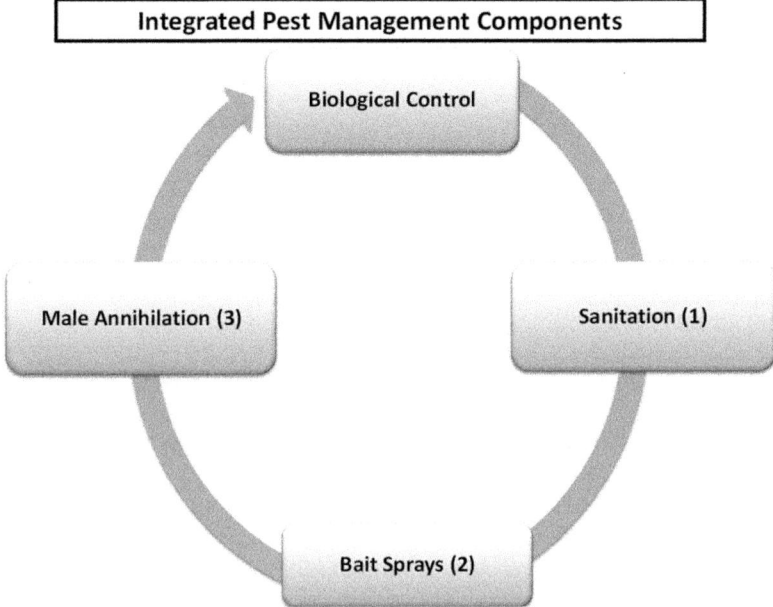

Fig. 29.1 Components of a typical fruit fly IPM system in Hawaii built around conservation of natural enemies, sanitation and utilization of low-risk BAT and MAT

3 Development of an Area-Wide IPM Programme

Fruit fly eradication programmes were proposed and attempted in Hawaii in the past, with a particular focus on *C. capitata*. While none of them succeeded, these eradication attempts did clearly illustrate the major problems associated with the concept of eradicating an invasive species: heavy economic costs, quarantine issues within the Hawaiian Islands chain, limits on resources and lack of information on treatment effects on non-target insects. Rather than an eradication programme, it became clear that an area-wide IPM programme would be more appropriate. One of the principal differences between IPM and eradication is that the goal of IPM is to maintain pest damage below an economically significant threshold, rather than the costly exercise of trying to eliminate every last fly. While IPM has many definitions, it often includes a diverse mixture of control approaches to manage pests that include biological control, cultural practices and chemical control components (Vargas et al. 2008, 2014a).

3.1 Importance of Partnerships

The United States Department of Agriculture, Agricultural Research Service (USDA ARS) has been a major innovator in the development of fruit fly area-wide control techniques – not only for use in the continental United States, but also for use around the world. Over the years, much of this work has been done at the Daniel K. Inouye U.S. Pacific Basin Agricultural Research Center (DKIUSPBARC) in Hawaii. From the beginning, the Hawaii fruit fly Area-Wide Pest Management programme (AWPM) has been a collaborative partnership between DKIUSPBARC, the University of Hawaii Cooperative Extension Service (UH-CES) and HDOA. Before this collaboration no one had adapted and packaged techniques within an IPM programme appropriate for local management of fruit flies. The contribution of the DKIUSPBARC was to provide the underpinning research necessary to develop the package of techniques needed to control fruit flies, and to adapt the techniques to individual situations. They also tracked success rates and helped provide data for registration of biologically-based agents and environmentally sensitive chemicals, including data on the impact of the programme on Hawaii's non-economically important native fruit flies and other non-target insects. The UH-CES created the knowledge exchange programme to communicate the goals and results of the Hawaii fruit fly AWPM programme to farmers and gardeners. Extension leaders created simple and logical educational materials that empowered users to adopt and/ or adapt the IPM programme. Educators used standard field demonstrations and hands-on teaching methods. The HDOA was essential in establishing the programme, especially given that the essential baits and lures used were not registered when it began. The HDOA sustained the area-wide programme cooperatively with the UH-CES, and the growers implemented the technologies developed. Finally, industry provided a unique set of products for fruit fly suppression and supported registration through HDOA and the US Environmental Protection Agency.

Area-wide IPM turned out to be a very successful approach. Implementation of IPM programmes varied in scale from individual homeowners and farmers to large areas of many square kilometers. Prior to the Hawaii fruit fly AWPM programme, the Regional Fruit Fly Project in the Pacific pioneered the implementation of sustainable technologies throughout many Pacific Island countries for control of *Bactrocera* species of fruit fly (Allwood 1997; Allwood et al. 2000). These technologies included fipronil-based BAT and MAT in conjunction with cultural control methods. Similarly, the Hawaii fruit fly AWPM programme evaluated, adapted and demonstrated IPM components to control *B. cucurbitae* (Fig. 29.2), *B. dorsalis* (Fig. 29.3) and *C. capitata* (Fig. 29.4) that included: (1) field sanitation, (2) BAT, and (3) MAT (Vargas et al. 2008).

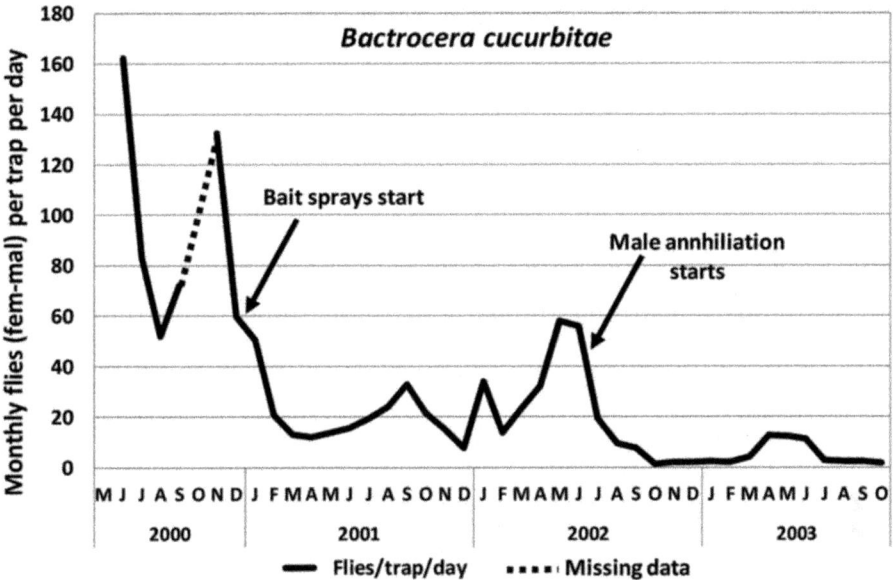

Fig. 29.2 Trials of BAT and MAT for suppression of *B. cucurbitae*. Captures (flies/trap/day) of *B. cucurbitae* in torula yeast-baited traps placed in commercial fields on the Islands of Maui in Hawaii where BAT and MAT were implemented

3.2 Importance of Educational Programmes

From its inception, one of the goals of the Hawaii fruit fly AWPM programme was to transfer sustainable, scientifically-derived and rigorously-tested technologies to farmers. The educational programmes implemented targeted commercial growers and home gardeners (Mau et al. 2003, 2007). Three teaching curricula were provided for elementary through to high school students, and a statewide awareness programme used for the public at large. Brochures, pamphlets, booklets and videos on components of the programme were made available to farmers and homeowners at extension offices throughout the state (Mau et al. 2003). In addition a website was established (www.fruitfly.hawaii.edu) and regular workshops were held at University of Hawaii Research Stations.

3.3 Importance of Research

This programme developed and registered many technologies for farmers and homeowners to use, and promoted the use of safer fruit fly control approaches which became popularly referred to as the '1,2,3 Programme' for fruit fly control, where the '1' represents sanitation, the '2' represents BAT and the '3' represents

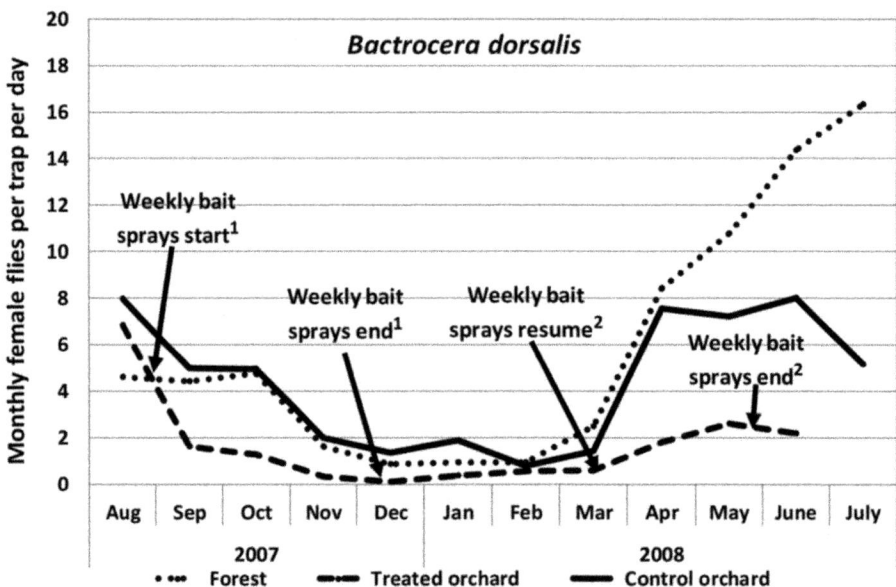

Fig. 29.3 Trials of BAT sprays for suppression of *B. dorsalis*. Captures (females/trap/day) of *Bactrocera dorsalis* in torula-baited monitoring traps deployed in forested areas, as well as in experimental plots with papaya trees sprayed weekly with GF-120 NF Naturalyte™ Fruit Fly Bait. Trap capture data are based on studies conducted by Piñero et al. (2009) (2007 data) and Piñero et al. (2010) (2008 data)

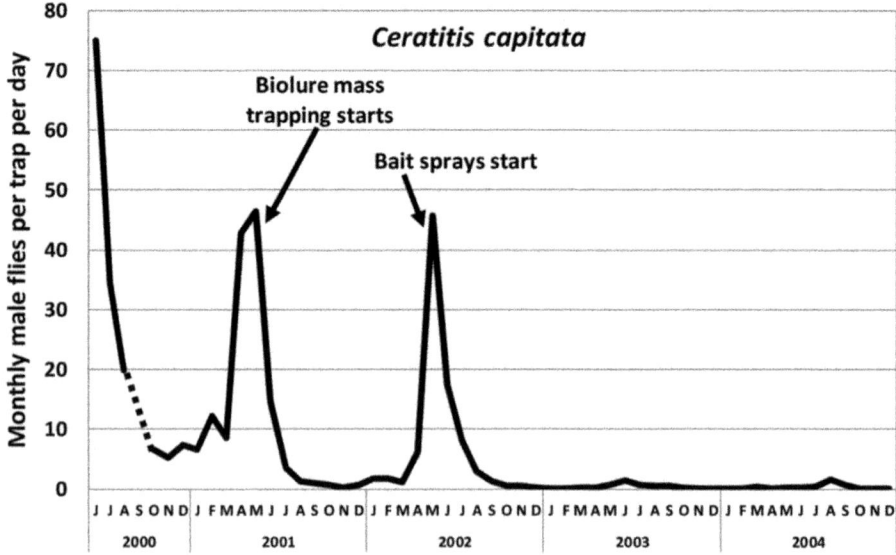

Fig. 29.4 Demonstration of Biolure® and protein baits for suppression of *C. capitata*. Capture (flies/trap/day) of *C. capitata* with trimedlure traps placed in persimmon orchards treated with Biolure® and GF-120 NF Naturalyte™ Fruit Fly Bait in Kula (Maui island), HI

Table 29.1 Programme components for controlling *B. cucurbitae*, *B. dorsalis*, *B. latifrons* and *C. capitata* (Mau et al. 2009)

Species	IPM components
Bactrocera cucurbitae	Population monitoring
	Sanitation by ploughing and destroying crops within 7–10 days after the last commercial harvest
	GF-120 NF Naturalyte™ Fruit Fly Bait applied spot applications every 3–5 m on established borders or 'roosting' plants adjacent to crops at weekly intervals from flowering to final harvest
	Cue-lure used for mass trapping at a rate of 10 traps per acre to reduce successful reproduction by adults
Bactrocera dorsalis	Population monitoring
	Sanitation by ploughing and destroying crops within 7–10 days after the last commercial harvest
	GF-120 NF Naturalyte ™ Fruit Fly Bait spot applications to host fruit trees weekly during the periods between initial fruit set and maturity
	Methyl eugenol used for mass trapping at a rate of 5 traps per acre
Bactrocera latifrons	Sanitation by removing damaged fruit from field and orchards
	GF-120 NF Naturalyte™ Fruit Fly Bait spot applications to host plants weekly
Ceratitis capitata	Population monitoring
	Sanitation by removing damaged fruit from fields and orchards
	Monitoring with Biolure Medfly lure
	GF-120 NF Naturalyte™ Fruit Fly Bait applied as spot applications

MAT (Table 29.1). For example, in marked-release behavioural studies on *B. cucurbitae*, Prokopy et al. (2003) demonstrated the effectiveness of GF-120 NF Naturalyte Fruit Fly Bait sprays on border crops in preventing alighting on main crops of cucumber. Similarly, in a study assessing the efficacy of GF-120 NF Naturalyte Fruit Fly Bait sprays in conjunction with field sanitation to control *B. dorsalis* in papaya orchards in Hawaii, Piñero et al. (2009) reported significant reductions in the number of female *B. dorsalis* captured in monitoring traps and in the levels of infestation of papaya fruit by *B. dorsalis*; however, this was only achieved when GF-120 was applied in a sustained manner in conjunction with field sanitation and MAT. These trials were particularly significant because they demonstrated that the highly aggressive *B. dorsalis* could be suppressed in commercial papaya orchards with simple IPM techniques. Field sanitation is a technique that either prevents fruit fly larvae from developing or traps young emerging adult flies thereby preventing them from returning to the crop to reproduce (Vargas et al. 2008). In general field sanitation involves the removal and disposal of infested or non-infested (cull) produce. While this can be labourious, it is a very effective fruit fly suppression method (Liquido 1991, 1993; Piñero et al. 2009) and a key component of a fruit fly

IPM programme. More detail on field sanitation and other cultural controls for fruit fly IPM programmes can be found in Klungness et al. (2005) and Vargas et al. (2008); MAT is discussed later in this chapter.

3.4 Registration of Chemicals

Prior to this programme, no chemicals were registered in the United States specifically for the suppression of fruit flies. Male lures were available for monitoring only (e.g., methyl eugenol, cue-lure, trimedlure, and latilure + cadeoil), and protein-based baits were only allowed in combination with pesticides that were already registered for use on crops (e.g., Nulure in combination with malathion). The Hawaii AWPM programme was instrumental in obtaining the first Hawaiian research permits and then assisted in the registration process with state and federal authorities (Table 29.2).

Table 29.2 Registration of agricultural chemicals through Hawaii AWPM programme for use against fruit flies in Hawaii (Mau et al. 2009)

Date of Reg.	EPA Reg. No.	Hawaii Licensing No.	Product	Source
Aug. 22, 2000	HISLN Reg. HI000003	9786.135	GF-120 Fruit Fly Bait	Dow AgroSciences LLC.
Dec 18, 2002	62719-498	9786.234	GF-120 Naturalyte Fruit Fly Bait supplemental label	Dow AgroSciences LLC
May 23, 2003	8730-50	9628.6	Vaportape II™	Hercon Environmental Inc.
June 5, 2006	62719-498	9786.234	GF120 Naturalyte Fruit Fly Bait all crops supplemental label	Dow AgroSciences LLC
Sep. 20, 2007	7969-253	9131.131	Amulet™ C-L w/ fipronil stations	BASF
Oct. 3, 2007	36638-42	9721.4	Cue-lure plug in plastic matrix w/o toxicant	Scentry Biologicals Inc.
Oct. 26, 2007	81325-3	8637.1	Methyl eugenol short lure in plastic matrix	Farma Tech International Corp
Dc. 11, 2007	36638-41	9721.3	Methyl eugenol cone in plastic matrix w/o toxicant	Scentry Biologicals Inc.
June 2008	62719-42	9786.282	Sprayable SPLAT-MAT-Spinosad-ME (STATIC™-Spinosad ME)	ISCA Technologies Dow AgroSciences LLC
March 2012	62719-592			

The Hawaii Fruit Fly AWPM package is a combination of monitoring and population control methods. Traps with species-specific lures are used for both monitoring and population elimination. Field sanitation (i.e. removing and destroying all fruit left in the field) was critical to success. In addition, planting of roosting crops in border areas that could be sprayed with spinosad-based bait sprays, rather than the entire crop, and releases of sterile male flies and parasitoid wasps enhanced the programmes success (Vargas et al. 2008).

3.5 Impact

The Hawaii Fruit Fly AWPM programme made major economic contributions to agriculture in Hawaii and instigated the growing of a greater diversity of crops. In the four major agricultural areas of Ewa, Kula, Puna and Kamuela fruit fly infestation was reduced by 83–95 %. This was accompanied by a reduction in the quantity of organophosphates applied to crops. Statewide, there were more than 2747 participants, including 682 farms and over 6500 ha under management. Furthermore, by providing farmers with the alternative control tools necessary to significantly reduce pesticide use, the programme also reduced pollution and improved Hawaii's natural environment, which contributed to maintaining the islands' tourism income. The Hawaii Fruit Fly AWPM programme also promoted a significant increase in the number of commercial farms and allowed existing farms to add crops to their portfolio or reinstate cultivars that had previously been phased out due to fruit fly problems. Aloun Farms, one of the largest and most diversified growers on Oahu, began producing an additional 59, 000 kg of zucchini a year and had no problem marketing all of it. This increase in production translated into an economic benefit of around US$75,000 annually (McGregor et al. 2007). A full cost-benefit analysis found that the Hawaii Fruit Fly AWPM programme will create as much as a 32 % return on an investment of $14 million over 15 years; this excludes the substantial indirect benefits, such as increased employment in agriculture, nor the environmental benefits that are hard to attribute a direct dollar return to. The benefits were measured in three categories: (1) already-achieved increases plus forecasts of their continued increase, (2) benefits based on likely outputs over the next 5 years, and (3) benefits based on possible outputs over the next 10 years (McGregor et al. 2007).

3.6 Technology Transfer

Although formal funding for the Hawaii AWPM programme ended in 2008, transfer of components and technology that was developed in Hawaii still continued locally, nationally and internationally (Table 29.2). This includes introduction of fruit fly natural enemies, use of safer lure dispensers, BAT and MAT. The programme received seven major awards for IPM technology transfer activities[1].

4 Components of the Area-Wide IPM Programme

The following sections provide an overview of each of the components used in area-wide IPM.

4.1 Use of Biological Control

Bactrocera dorsalis was first discovered on Tahiti Island, French Polynesia, in July 1996 (Vargas et al. 2007). Eradication failed and *B. dorsalis* spread to other Society Islands including Raiatea, Tahaa, Huahine and, since 2007, the Marquesas Islands (Leblanc et al. 2013). Spatial patterns of both sexes of *B. dorsalis*, and its two most abundant parasitoids, *F. arisanus* and *D. longicaudata* were modeled from data collected in a commercial guava orchard in Hawaii (Vargas et al. 2013). Initially the spatial patterns of *B. dorsalis* were random, but became highly aggregated with host fruit ripening and the subsequent colonization of first, *F. arisanus* (egg-pupal parasitoid) and secondly, *D. longicaudata* (larval-pupal parasitoid). There was a significant relationship between decreases in populations of *B. dorsalis* and increases in populations of *F. arisanus*, a pattern not found between *B. dorsalis* and *D. longicaudata*. These studies were used to develop protocols and methods for the introduction and establishment of *F. arisanus* and *D. longicaudata* in French Polynesia. Between December 2002 and October 2004, *F. arisanus* was mass reared (Manoukis et al. 2011) in Hawaii and ten shipments of 50,000 parasitized pupae were made to Tahiti where the wasps emerged, were released and became widely established (Fig. 29.5; Leblanc et al. 2013). *Bactrocera dorsalis* populations were reduced by as much as 90 % in some areas. By 2009 mean (± SD) *F. arisanus* parasitism rates for fruit flies infesting common guava, Polynesian chestnut and tropical almond fruits on Tahiti Island was 64.8 ± 2.0 % (Vargas et al. 2012b). The second parasitoid, *D. longicaudata*, was released and established in 2008. Although widespread, *D. longicaudata*

---

[1] Team recipient of the Federal Laboratories Consortium Award for Technology Transfer for the impacts made by the Hawaii AWPM Programme, May 2004; USDA Award for Superior Service to the Hawaii Fruit Fly AWPM Core Team "For creating an effective area-wide suppression program for fruit flies in Hawaii which provides the basis for a sustainable rural economy", June 2004; Entomological Society of America, Pacific Branch Award for Team IPM Accomplishments, June 2004; Entomological Society of America Entomological Foundation Integrated Pest Management Team Award, November 2004; USDA-ARS Technology Transfer Award to the Hawaii Fruit Fly AWPM Core Team, February 2005; Hawaii House of Representatives Recognition Award for Hawaii Fruit Fly AWPM Program, 2005 Session; Fifth National IPM Symposium, IPM Achievement Award Winner, April 2006.

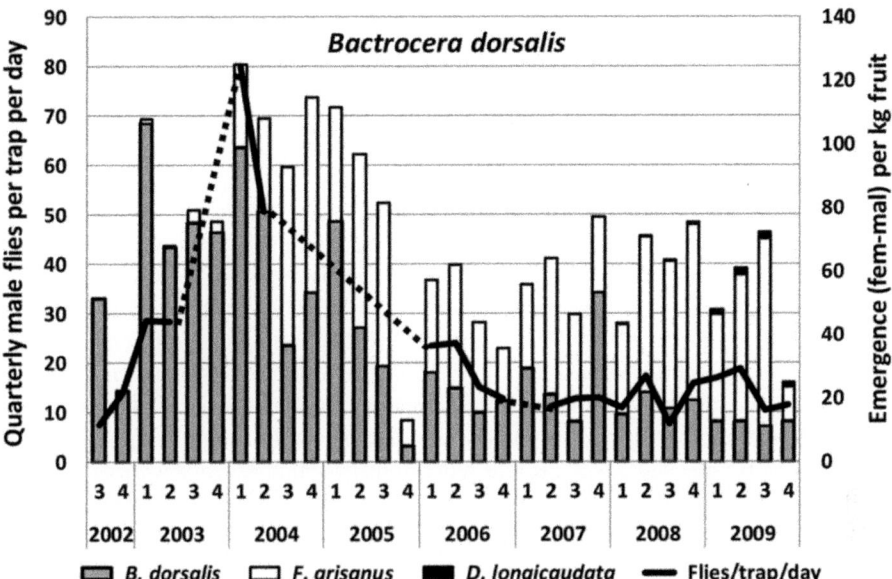

Fig. 29.5 Decreases in trap captures and infestation of fruit in French Polynesia through the release of the two parasitoids, *F. arisanus* and *D. longicaudata*. For further details, see Vargas et al. (2012b) and Leblanc et al. (2013)

parasitism rates have never exceeded 10 %. Analysis of co-infestation patterns (1998–2009) of *B. dorsalis*, and two other introduced invasive fruit flies, the Queensland fruit fly, *Bactrocera tryoni* (Froggatt), and *Bactrocera kirki* (Froggatt), suggest that *B. dorsalis* is now the predominant species on many common host fruits (Vargas et al. 2012b). Establishment of *F. arisanus* is the most successful example of classical biological control of fruit flies in the Pacific outside of Hawaii. Ten years of fruit infestation data suggested displacement of other *Bactrocera* species by *B. dorsalis*. Based on this success in French Polynesia, *F. arisanus* was introduced against *B. dorsalis* in Sénegal in 2013, where it established (Vargas et al. in review). Work on release of parasitoids for classical biological control is continuing in Brazil where *B. carambolae* has become established and is causing serious economic damage to tropical fruits (Vargas unpublished data).

4.2 Use of Spinosad-Based Attract-and Kill Technologies

Spinosad is a natural substance composed of two chemicals, spinosyn A and spinosyn D, both of which are fermentation products of the soil bacterium *Saccharopolyspora spinosa* Mertz and Yao. Spinosad is toxic to numerous insect species, either by contact or following ingestion. It is effective at low application rates, has low mammalian toxicity and limited negative effects on natural enemies

(DowElanco 1994; Stark and Vargas 2003; Stark et al. 2004). When mixed with an attractant and incorporated in a slow release matrix, it offers a low-risk control alternative for *B. dorsalis* and related species attracted to methyl eugenol, without many of the potential negative effects to humans and non-targets that are associated with broad-spectrum contact insecticides, such as naled (dimethyl 1, 2-dibromo-2, 2-dichloroethyl phosphate) (Leblanc et al. 2009).

STATIC™ Spinosad-ME (a.k.a. SPLAT-MAT™-Spinosad-ME), an 'attract and kill' spray formulation containing spinosad, was evaluated as a MAT against *B. dorsalis* in Hawaii, the peach fruit fly, *Bactrocera zonata* (Saunders) in Pakistan and *B. carambolae* in Brazil (Vargas et al. 2014a). In Hawaii, field trials used wooden bait stations (5 g SPLAT-MAT™-Spinosad-ME) in commercial papaya orchards. Two different treatment densities (4 or 20 stations per acre) were evaluated and compared to a control (no treatment). Evaluations in Brazil and Pakistan compared standard methods (organophosphate insecticides) with a single application of SPLAT-MAT-Spinosad-ME™. Rapid reductions in male fruit flies were recorded in all three trials. Results also suggested a reduction in the number of female flies over time. Subsequently, STATIC™ Spinosad-ME was evaluated in field trials done in California and Florida, and bioassays against *B. dorsalis* in Hawaii (Vargas et al. 2014b). Again, all studies suggested that the STATIC™ Spinosad-ME 'attract and kill' formulation was more convenient and safer to handle than the available, undesirable formulations of organophosphates. This product has now been registered and certified for use in California and Florida as an environmentally friendly alternative to organophosphates.

4.3 *Use of Solid Lure Dispensers to Detect and Monitor Invasive Fruit Flies*

During the Hawaii AWPM Fruit Fly Programme solid lure dispensers were tested extensively for both monitoring and as part of MAT (Vargas et al. 2010; Leblanc et al. 2011) over approximately 10 years. Pre-packaged solid lure dispensers were more convenient and safer than traps with cotton wicks and liquid male lures methyl eugenol (ME: 4-allyl-1, 2-dimethoxybenzene-carboxylate); cue-lure (C-L: 4-(*p*-acetoxyphenyl)-2-butanone); raspberry ketone (RK: 4-(p-hydroxyphenyl)-2-butanone) and insecticide mixtures. Even though there have been many studies (Vargas et al. 2009, 2012c; Leblanc et al. 2011; Shelly et al. 2011a, b, 2012; Jang et al. 2013), there have generally been few differences found between capture rates for the different species of *Bactrocera* using either solid or liquid forms of ME and RL/C-L. Although trimedlure: (TML: t Butyl-4 (or5)-chloro-2-methyl cyclohexane carboxylate) solid plugs have become a standard replacement for liquid TML in Jackson traps for *C. capitata* (FDACS 2004; CDFA 2013), it is unclear why there has been a hesitancy to replace liquid ME and C-L lures with solid formulations for *Bactrocera* species. Vargas et al. (2009) found no difference between captures of *B.*

dorsalis and *B. cucurbitae* using a DDVP (2, 2-dichlorovinyl dimethyl phosphate) Hercon Vaportape™ II insecticidal strip or liquid naled (dimethyl 1, 2-dibromo-2, 2-dichloroethyl phosphate) insecticide in the detection programmes in California and Florida.

Three possible applications have been proposed for solid lure detection traps for large mainland survey programmes (e.g. in California there are ca. 30,000 survey sites) utilizing ME, C-L/RK and TML traps: (1) Three individual traps with three separate solid wafers (TML, ME, and C-L/RK); (2) Two individual traps with two solid wafers (TML and ME+C-L/RK); or (3) One trap with Mallet TMR (TML, ME, and RK). In Hawaii, based on data and field experience from the AWPM programme, solid lure dispensers in monitoring traps needed to be replaced after 2–3 months and vapour tapes after 6–8 weeks. Further tests are currently underway in California to evaluate the persistence of activity of solid lures under Californian climatic conditions.

4.4 Use of Insecticidal Soil Drenches Against Fruit Fly Larvae and Pupae

A key component of current fruit fly area-wide IPM programmes in the US mainland (i.e. Florida) has been the application of insecticidal soil drenches; these drenches are applied beneath the drip line of host trees where fruit flies have previously been detected and as a regulatory treatment in the certification process necessary before nursery stock can be exported from fruit fly quarantine areas (Stark et al. 2013, 2014). Diazinon has been the most effective and widely used insecticide for this purpose; however, it is being phased out in many areas due to environmental issues (Stark et al. 2014). As part of the Hawaii fruit fly AWPM the efficacy of several insecticides were evaluated in semi-field trials as replacements for diazinon in the control of three economically important fruit fly species: *C. capitata*, *B. cucurbitae* and *B. dorsalis* (Stark et al. 2013, 2014). Fruit fly pupae within 24 h of eclosion were exposed to organic soil treated with different insecticides: Radiant SC (spinetoram), Force 3G (tefluthrin), Force CS (tefluthrin), Warrior II (lambda-cyhalothrin), Entrust (spinosad), Entrust SC (spinosad), GardStar 40 % EC (permethrin), or Diazinon AG 600 (diazinon). All the alternative insecticides resulted in a significant reduction in adult emergence and were not significantly different from diazinon. Results suggested that there were several effective alternative insecticides to diazinon for control of late larval or pupal stages of three economically important fruit fly species. Entrust, a natural product (spinosad), and its synthetic counterpart, Radiant SC were quite effective and may be good alternatives to the synthetic pyrethroids, Warrior II, Force CS and GardStar40 % EC (Stark et al. 2014). In subsequent tests Warrior II was identified as the best replacement for diazinon and certified for use in Florida. The newly formulated Entrust SC, a biopesticide, was also very effective. However, more research and development will be required before this product can be labeling for soil drench use.

4.5 Recent Applications of the Technologies Developed in the Hawaii Fruit Fly AWPM Programme for Development of Systems Approaches

The technologies developed in the Hawaii fruit fly AWPM programme have recently been used to develop a new systems approach that will allow exports of 'Sharwil' avocado to mainland U.S. from Hawaii (Follet and Vargas 2010). Specifically, the new approach was based on fruit infestation data and fruit fly surveys on Kona avocado farms during the Hawaii fruit fly AWPM programme which documented a reduced risk of infestation of avocados by *B. dorsalis* (Klungness et al. 2009). An area-wide IPM programme was developed to further mitigate the risk. The IPM programme (protein trap monitoring, sanitation, and BAT sprays if necessary) developed by ARS is part of the new systems approach recently approved. Once certified by APHIS, Hawaii farms will be allowed to ship 'Sharwil' avocados to 32 states on the U.S. between November 1st and March 15th each year. The programme is expected to increase Hawaii avocado production, grower revenue and market access.

5 Conclusions

Three important goals of the Hawaii fruit fly AWPM programme were (1) to develop economically and ecologically sound technologies (2) to transfer those technologies to fruit and vegetable growers, and (3) to implement comprehensive educational programmes targeting farmers and citizens in an effort to increase adoption rates of the IPM technologies that were developed. Various suppression technologies (sanitation, BAT, MAT, sterile male release and parasitoid release) were utilized in this programme. The Hawaii fruit fly AWPM programme demonstrated that cultural practices (e.g. field sanitation) can support synergies between the other components of IPM such as protein bait sprays, male lures and biological control. The programme made major economic contributions to agriculture in Hawaii, and promoted production of a greater diversity of crops while reducing the use of organophosphate insecticides thus balancing the ecological, social, and economic aspects of farming in a move toward sustainability. Because Africa now has a similar complex of fruit fly species attacking tropical fruits and vegetables to those found in Hawaii, the Hawaii Fruit Fly AWPM programme has particular relevance to Africa for sustainable control of fruit flies, especially the *Bactrocera* species. A recent field guide has been published in Africa that outlines economically important fruit flies found in Africa, and monitoring and suppression techniques that could be used in local area-wide programs (Ekesi and Billah 2007).

Acknowledgements The Hawaii Fruit Fly AWPM programme gratefully acknowledges the Agricultural Research Service, the University of Hawaii Cooperative Extension Service, Animal Plant Health Inspection Services, Hawaii Department of Agriculture and the University of Hawaii College of Tropical Agriculture and Human Resources for their contributions to the programme. In addition, we thank the California Department of Food and Agriculture's IR-4 programme, Dow AgroSciences LLC, ISCA Technologies, Biologicals Inc. and Farma Tech International Corp for their support of the programme. This article reports the results of research only. Mention of a proprietary product does not constitute an endorsement or a recommendation by the USDA for its use. USDA is an equal opportunity employer.

References

Allwood AJ (1997) Control strategies for fruit flies (Diptera: Tephritidae) in the South Pacific. In: Allwood AJ, Drew RAI (eds) Management of fruit flies in the pacific, a regional symposium, Nadi, Fiji 28–31 October 1996, Australian Centre for International Agricultural Research, Canberra, Australia, pp 171–178

Allwood AJ, Leblanc L, Tora Vueti E, Bull R (2000) Fruit fly control methods for Pacific Island countries and Territories. Available online: http://bem.bime.ntu.edu.tw/clchuang/PAL40. Accessed on 10 Jan 2015

Bess HA, van den Bosch R, Haramoto FH (1961) Fruit fly parasites and their activities in Hawaii. Proc Hawaii Entomol Soc 17:367–378

California Department of Food and Agriculture (CDFA) (2013) Insect trapping guide, 13th edn. CDFA, Sacramento, 47 pp

Clausen CP, Clancy DW, Chock QC (1965) Biological control of the Oriental fruit fly (*Dacus dorsalis* Hendel) and other fruit flies in Hawaii. USDA Tech Bull No. 1322, 102, Washington, DC, 102 pp

DowElanco (1994) Spinosad technical guide. DowElanco, Indianapolis

Drew R, Tsuruta K, White I (2005) A new species of pest fruit fly (Diptera: Tephritidae: Dacinae) from Sri Lanka and Africa. Afr Entomol 13:149–154

Ekesi S, Billah K (2007) A field guide to the management of economically important tephritid fruit flies in Africa, 2nd edn. ICIPE Scientific Press, Nairobi

Ekesi S, Nderitu PW, Rwomushana I (2006) Field infestation, life history and demographic parameters of the fruit fly *Bactrocera invadens* (Diptera: Tephritidae) in Africa. Bull Entomol Res 96:379–386

Florida Department of Agriculture and Consumer Services (FDACS) (2004) Florida fruit fly detection manual. Division of Plant Industry, FDACS, Gainesville

Follet PA, Vargas RI (2010) A systems approach to mitigate oriental fruit fly risk in 'Sharwil' avocados exported from Hawaii. Acta Hortic 880:439–445

Fullaway DT (1920) The melon fly—its control in Hawaii by a parasite introduced from India. Rev Appl Entomol A 8:347

Haramoto FH (1953) The biology of Opius oophilus Fullaway (Hymenoptera: Braconidae). M.S. thesis University of Hawaii Honolulu HI USA. 66 pp

Jang EB, Ramsey A, Carvalho LA (2013) Performance of methyl eugenol + matrix + toxicant combinations under field conditions in Hawaii and California for trapping *Bactrocera dorsalis* (Diptera: Tephritidae). J Econ Entomol 106:727–734

Klungness LM, Jang EB, Mau RFL, Vargas RI, Sugano JS, Fujitani E (2005) New sanitation techniques for controlling Tephritid Fruit Flies (Diptera: Tephritidae) in Hawaii. J Appl Sci Environ 9:5–14

Klungness LM, Vargas RI, Jang EB, Mau RFL, Kinney K (2009) Update on the susceptibility of avocado to invasive alien fruit flies (Tephritidae) on the island of Hawaii. Proc Hawaii Entomol Soc 41:1–13

Leblanc L, Rubinoff D, Vargas RI (2009) Attraction of non-target species to fruit fly (Diptera: Tephritidae) male lures and decaying fruit flies in traps in Hawaii. Environ Entomol 38:1446–1461

Leblanc L, Vargas RI, Mackey B, Putoa R, Piñero JC (2011) Evaluation of cue-lure and methyl eugenol solid lure and insecticide dispensers for fruit fly (Diptera: Tephritidae) monitoring and control in Tahiti. Fla Entomol 94:510–516

Leblanc L, Vargas RI, Putoa R (2013) From eradication to containment: invasion of French Polynesia by *Bactrocera dorsalis* (Hendel) (Diptera: Tephritidae) and releases of two natural enemies: a 17-year case study. Proc Hawaii Entomol Soc 45:31–43

Lindquist DA (2000) Pest management strategies: area-wide and conventional. In: Tan KH (ed) Area-wide control of fruit flies and other insect pests. pp. 13–19. Penerbit Universiti

Liquido NJ (1991) Fruit on the ground as a reservoir of resident melon fly (Diptera: Tephritidae) population in papaya orchards. Environ Entomol 20:620–625

Liquido NJ (1993) Reduction of oriental fruit fly (Diptera: Tephritidae) populations in papaya orchards by field sanitation. J Agric Entomol 10:163–170

Lodge DM (1993) Biological invasions: lessons for ecology. TREE 8:133–137

Malavasi A, Midgarden D, De Meyer M (2013) *Bactrocera* species that pose a threat to Florida: *B. carambolae* and *B. invadens*. pp 214–227 In: Peña JE (ed) Potential invasive pests of agricultural crops. CABI Oxfordshire/Boston, 464 pp.

Manoukis NC, Geib SM, Seo DM, McKenney MP, Vargas RI, Jang EB (2011) An optimized protocol for rearing *Fopius arisanus*, a parasitoid of tephritid fruit flies. JoVE 53:e2901

Mau RFL, Sugano JS, Jang EB (2003) Farmer education and organization in the Hawaii area wide fruit fly pest management program. In: Recent trends on sterile insect technique and area-wide integrated pest management: economic feasibility, control projects, farmer organization and dorsalis complex control study. Research Institute for Subtropics. Okinawa, Japan. pp. 47–57

Mau RFL, Jang EB, Vargas RI (2007) The Hawaii fruit fly area-wide fruit fly pest management programme: influence of partnership and a good education programme. In: Vreysen MJB, Robinson AS, Hendrichs J (eds) Area-wide control of insect pests: from research to field implementation. Springer, Dordrect, pp 671–683, 789 pp

Mau RFL, Vargas RI, Jang EB, Wong L (2009) Hawaii area-wide fruit fly integrated pest management program: a model system. College of Tropical Agriculture and Human Resources, Honolulu, 12 pp

McGregor AM, Vargas RI, Mau RFL (2007) An economic evaluation of the Hawaii fruit fly area-wide pest management program. University of Hawaii Miscellaneous Publication. 83 pp

Nishida T (1955) Natural enemies of the melon fly, *Dacus curcurbitae* Coq., in Hawaii. Ann Entomol Soc Am 48:171–178

Piñero JC, Mau RFL, Vargas RI (2009) Managing oriental fruit fly (Diptera: Tephritidae), with spinosad-based protein bait sprays and sanitation in papaya orchards in Hawaii. J Econ Entomol 102:1123–1132

Piñero JC, Mau RFL, Vargas RI (2010) Comparison of rain-fast bait stations versus foliar bait sprays for control of oriental fruit fly, *Bactrocera dorsalis*, (Diptera: Tephritidae), in papaya orchards in Hawaii. J Ins Sci 10:1–13, Article 157

Prokopy RJ, Miller NW, Piñero JC, Barry JD, Tran LC, Oride L, Vargas RI (2003) Effectiveness of GF-120 fruit fly bait spray applied to border area plants for control of melon flies (Diptera: Tephritidae). J Econ Entomol 96:1485–1493

Putulan D, Sar S, Drew RAI, Raghu S, Clarke AR (2004) Fruit and vegetable movement on domestic flights in Papua New Guinea and the risk of spreading pest fruit flies (Diptera: Tephritidae). Int J Pest Man 50:17–22

Ramadan MM, Wong TTY, Beardsley JW (1992) Reproductive behavior of *Biosteres arisanus* (Sonan) (Hymenoptera: Braconidae), an egg-larval parasitoid of the oriental fruit fly. Biol Con 2:28–34

Richardson DM, van Wilgen BW (2004) Invasive alien plants in South Africa: how well do we understand the ecological impacts? South Afr J Sci 100:45–52

Schutze MK, Aketarawong N, Amornsak W, Armstrong KF, Augustinos AA, Barr N, Bo W, Bourtzis K, Boykin LM, Cáceres C, Cameron SL, Chapman T, Chinvinijkul S, Chomič A, De Meyer M, Drosopoulos E, Englezou A, Ekesi S, Gariou-Papalexiou A, Geib SM, Hailstones D, Hasanuzzaman M, Haymer D, Hee AKW, Hendrichs J, Jessup AW, Ji QG, Khamis FM, Krosch MN, Leblanc L, Mahmood K, Malacrida AR, Mavragani-Tsipidou P, Mwatawala M, Nishida R, Ono H, Reyes J, Dubinoff D, San Jose M, Shelly TE, Srikachar S, Tan KH, Thanaphum S, Haq I, Vijaysegaran S, Wee SL, Yesmin F, Zacharopoulou A, Clarke AR (2014) Synonymization of key pest species within the *Bactrocera dorsalis* species complex (Diptera: Tephritidae): taxonomic changes based on a review of 20 years of integrative morphological, molecular, cytogenetic, behavioural and chemoecological data. Syst Entomol 40:456–471

Shelly TE, Kurashima R, Nishimoto J, Diaz A, Leathers J, War M, Joseph D (2011a) Capture of *Bactrocera* fruit flies (Diptera: Tephritidae) in traps baited with liquid versus solid formulations of male lures. J Asia-Pac Entomol 14:463–467

Shelly TE, Kurashima R, Nishimoto J (2011b) Capture of *Bactrocera* males (Diptera: Tephritidae) in parapheromone-baited traps: performance of solid dispensers with different loadings of attractants and toxicant. Proc Hawaiian Entomol Soc 43:33–47

Shelly TE, Nishimoto J, Kurashima R (2012) Captures of three economically important fruit fly species (Diptera: Tephritidae) in traps baited with liquid versus solid formulations of male lures in a Hawaiian coffee field. J Econ Entomol 105:1186–1193

Silvestri F (1914) Report of an expedition to Africa in search of the natural enemies of fruit flies. Hawaii Board Agr For Div Entomol Bull 3:1–226

Simberloff D (2009) The role of propagule pressure in biological invasions. Ann Rev Ecol Evol Syst 40:81–102

Stark JD, Vargas RI (2003) Demographic changes in *Daphnia pulex* (Leydig) after exposure to the insecticides Spinosad and Diazinon. Ecotoxicol Environ Saf 56:334–338

Stark JD, Banks JE, Vargas RI (2004) How risky is risk assessment? The role that life history strategies play in susceptibility of species to stress. Proc Natl Acad Sci 101:732–736

Stark JD, Vargas RI, Souder S, Fox AJ, Smith RS, Mackey B (2013) A comparison of the bio-insecticide, spinosad, the semi-synthetic insecticide, spinetoram and synthetic insecticides as soil drenches for control of tephritid fruit flies. Biopestic Int 9:120–126

Stark JD, Vargas RI, Souder S, Fox AJ, Smith TR, Leblanc L, Mackey B (2014) Simulated field applications of insecticide soil drenches for control of tephritid fruit flies. Biopestic Int 10:136–142

van den Bosch R, Haramoto FH (1953) Competition among parasites of the oriental fruit fly. Proc Hawaii Entomol Soc 15:201–206

Vargas RI, Peck SL, McQuate GT, Jackson CG, Stark JD, Armstrong JW (2001) Potential for area-wide integrated management of Mediterranean fruit fly (Diptera: Tephritidae) with a braconid parasitoid and a novel bait spray. J Econ Entomol 94:817–882

Vargas RI, Miller NW, Long J, Delate K, Jackson CG, Uchida G, Bautista RC, Harris EJ (2004) Releases of *Psyttalia fletcheri* (Hymenoptera: Braconidae) and sterile flies against melon fly (Diptera: Tephritidae) in Hawaii. J Econ Entomol 97:1531–1539

Vargas RI, Leblanc L, Putoa R, Eitam A (2007) Impact of introduction of *Bactrocera dorsalis* (Diptera: Tephritidae) and classical biological control releases of *Fopius arisanus* (Hymenoptera: Braconidae) on economically important fruit flies in French Polynesia. J Econ Entomol 100:670–679

Vargas RI, Mau RFL, Jang EB, Faust RM, Wong L (2008) The Hawaii fruit fly area-wide pest management program. In: Koul O, Cuperus GW, Elliott NC (eds) Areawide IPM: theory to implementation. CABI Books, London, pp 300–325, 590 pp

Vargas RI, Burns RE, Mau RFL, Stark JD, Cook P, Piñero JC (2009) Captures in methyl eugenol and cue-lure detection traps with and without insecticides and with a Farma Tech solid lure and insecticide dispenser. J Econ Entomol 102:552–557

Vargas RI, Mau RFL, Stark JD, Piñero JC (2010) Evaluation of methyl eugenol and cue-lure traps with solid lure and insecticide dispensers for monitoring and male annihilation in Hawaii. J Econ Entomol 103:409–415

Vargas RI, Leblanc L, Harris EJ, Manoukis NC (2012a) Regional suppression of *Bactrocera* fruit flies (Diptera: Tephritidae) in the Pacific through biological control and prospects for future introductions into other areas of the world. Insects 3:727–742

Vargas RI, Leblanc L, Putoa R, Piñero JC (2012b) Population dynamics of three *Bactrocera* spp fruit flies (Diptera: Tephritidae) and two introduced natural enemies, *Fopius arisanus* (Sonan) and *Diachasmimorpha longicaudata* (Ashmead) (Hymenoptera: Braconidae), after an invasion by *Bactrocera dorsalis* (Hendel) in Tahiti. Biol Con 60:199–206

Vargas RI, Souder SK, Mackey B, Cook P, Morse JG, Stark JD (2012c) Field trials of solid triple lure and insecticide dispensers for detection and male annihilation of *Ceratitis capitata* (Wiedemann), *Bactrocera dorsalis* (Hendel) and *Bactrocera cucurbitae* (Coquillett) (Diptera: Tephritidae) in Hawaii. J Econ Entomol 105:1557–1565

Vargas RI, Stark JD, Banks J, Leblanc L, Manoukis N, Piñero JC (2013) Spatial dynamics of two oriental fruit fly (Diptera: Tephritidae) parasitoids, *Fopius arisanus* (Sonan) and *Diachasmimorpha longicaudata* (Ashmead) (Hymenoptera: Braconidae), in a guava orchard in Hawaii. Environ Entomol 42:888–901

Vargas RI, Souder SK, Borges R, Mafra-Neto A, Mackey B, Chou MY, Spafford H (2014a) Effectiveness of a sprayable male annihilation treatment with a biopesticide against fruit flies (Dipera: Tephritidae) attacking tropical fruits. Biopestic Int 10:1–10

Vargas RI, Souder SK, Hoffman K, Mercogliano J, Smith TR, Hammond J, Davis BJ, Brodie M, Dripps JE (2014b) Attraction and mortality of *Bactrocera dorsalis* to STATIC™ Spinosad ME weathered under operational conditions in California and Florida: a reduced-risk male annihilation treatment. J Econ Entomol 107:1362–1369

Vargas RI, Piñero JC, Leblanc L (2015) An overview of pest species of *Bactrocera* fruit flies (Diptera: Tephritidae) and the integration of biopesticides with other biological approaches for their management with a focus on the Pacific Region. Insects 6:297–318

White IM, Elson-Harris MM (1992) Fruit flies of economic significance: their identification and bionomics. CABI International, Wallingford, 601 pp

Willard HF (1920) *Opius fletcheri* as a parasite of the melon fly in Hawaii. J Agr Res 20:423–438

Roger I. Vargas is a research entomologist at the Daniel K. Inouye U. S. Pacific Basin Agricultural Research Center, Hilo, HI and holds a collateral appointment with the University of Hawaii Graduate Faculty. His research interests include: ecology, biological control, mass rearing, SIT and IPM of fruit flies. He is the author or co-author of over 230 (160 peer reviewed) scientific articles. He completed his B.A. in biology in 1969 at the University of California, Riverside and an M.S. in biology at San Diego State in 1974. He then moved to the University of Hawaii, Manoa, where he completed his Ph.D. in 1979 and has studied tropical entomology ever since. After a postdoc on Kauai Island he accepted a position with ARS in 1980 in Honolulu. From 1984 to 1990, he served as Research Leader of the Rearing, Radiation and Genetics Research Unit in Honolulu, where base funding increased from $500,000 to $3,000,000 through a series of area-wide fruit fly suppression trials. He has been called upon frequently by agencies such as the Animal Plant Health Inspection Service (APHIS), International Atomic Energy Agency (IAEA), Secretariat of the Pacific Community (SPC) and Foreign Agricultural Service (FAS) for consultant and technical assignments overseas. Roger has made over 100 scientific presentations and participated in workshops worldwide. He developed a research proposal that resulted in the award of over $16.5 million in USDA/ARS AWPM funds to Hawaii. From 2002 to 2009, he coordinated the highly successful

Hawaii Fruit Fly AWPM program that received seven major awards for IPM excellence. In 2010–2011, he served as President of the Pacific Branch of the Entomological Society of America.

Jaime C. Piñero is native of Mexico, where he received a BSc degree in Agronomy from the University of Veracruz. From 1992 to 1999 he worked at the Institute of Ecology, Xalapa, Mexico where he worked on IPM of tephritid fruit flies under the leadership of Dr. Martin Aluja. In 2005 he was awarded a PhD in Entomology from the University of Massachusetts, where his supervisor had been the late Dr. Ronald J. Prokopy. Jaime was then a post-doctoral researcher (2005–2007) at the Swiss Federal Institute of Technology (ETH) in Zürich working on chemical ecology in Prof. Dr. Silvia Dorn's group. From 2007 to 2010 he worked on invasive fruit flies within the area-wide IPM fruit fly programme in Hawaii. Currently he is an Associate Professor and State IPM Specialist at Lincoln University, Missouri, USA, leading research to develop and promote (through extension and outreach) simple, effective, and affordable alternative pest management strategies that can be used by small- and mid-scale growers. He has authored or co-authored over 50 research articles and nearly 45 grower-oriented publications. Jaime has given presentations on various aspects of IPM and insect sensory ecology in China, South Africa, Mexico, Germany, Malaysia, and Switzerland.

Luc Leblanc has over 30 years experience in the fields of insect taxonomy, managing insect collections, integrated pest management and implementing plant protection-related projects in Africa and the South Pacific Islands. After working for many years in the Canadian National Collection of Insects and completing an MSc degree on Hymenoptera taxonomy, he left his country to undergo an adventurous and fulfilling work career, residing in Africa (Guinea and Rwanda, 1989–1994) and the Pacific Islands (Micronesia, Papua New Guinea, Fiji Islands, 1994–2002). It is in the Pacific that he became a full-time fruit fly entomologist, employed under the Regional Fruit Fly Project in the Pacific. Subsequently (2003–2015) he was based at the University of Hawaii College of Tropical Agriculture and Human Resources, working under the Area-Wide Fruit Fly IPM programme in Hawaii, completing a PhD dissertation on non-target attraction to fruit fly lures, and developing an expertise in dacine fruit fly taxonomy and identification. Currently, he is the curator of the William F. Barr Entomological Museum, at the University of Idaho.

Nicholas C. Manoukis is a research biologist with the Agricultural Research Service (ARS), a part of the US Department of Agriculture, in Hilo, Hawaii. His research focuses on the ecology and behaviour of tephritid fruit flies and other insect pests, particularly species of economic importance in Hawaii and potential invaders of the US Mainland. Dr. Manoukis' current programme includes projects on computer modeling of invasive pests, computer-vision analysis of fruit fly responses to lures, field studies on coffee berry borer and studies on fruit fly rearing and biological control. He has been involved in research on the biology of *Fopius arisanus* and other tephritid parasitoids used in classical biological control; this includes efforts to transfer *F. arisanus* to Brazil. Prior to working at ARS he spent almost a decade studying the ecology, evolution and behaviour of the malaria vector, *Anopheles gambiae*, in Mali, West Africa.

R.F.L. Mau is an Emeritus Extension Specialist at the University of Hawaii at Manoa. He specializes in integrated pest management and technology transfer to growers. He had expertise in cooperative extension programme development and implementation. He was a member of the Hawaii Core Committee for the Area-Wide Fruit Fly Integrated Pest Management Programme (AWFF-IPM). He provided leadership for the fruit fly implementation programme on the islands of Maui and Oahu and for statewide area-wide outreach education.

Chapter 30
Management of Fruit Flies in Mexico

Pablo Liedo

Abstract Fruit growing is of great socio-economic importance in Mexico and fruit flies pose a major threat to this industry. For pest management purposes, fruit flies are divided into two groups: exotic species and native species. For exotic species, a country-wide trapping network is operated for early detection. In the case of the exotic medfly, *Ceratitis capitata*, which is present in Central America, a programme based on use of the Sterile Insect Technique (SIT) has prevented the northern spread of the pest and its establishment in Mexico. For native species, an area-wide integrated pest management (IPM) programme was implemented in 1992. Using SIT, augmentative biological control and other suppression methods, approximately half of the national territory is now recognized as either free of fruit flies or a low-prevalence area. An increase in fruit exports has been possible through the participation of grower organizations, risk analysis and government support.

Keywords Native fruit flies • Exotic fruit flies • Detection • Area-wide management

1 Fruit Production in Mexico

Fruit growing is of great socio-economic importance in Mexico. Approximately 1.7 million hectares of vegetables and fruits are cultivated producing an estimated 12 million tons per year. The value of this production is estimated as 4500 million US dollars. Globally, Mexico produces more avocados and exports more mangos than any other country, and is the fifth largest producer of citrus fruits (SIAP 2015; Salcedo et al. 2009). Fruit cultivation and exports have gradually increased over the years as a result of free trade agreements, grower organizations and more effective fruit fly control (Zahniser and Link 2002; Salcedo-Baca et al. 2010). The most important crops in terms of surface area, that are also attacked by native fruit flies,

P. Liedo (✉)
El Colegio de la Frontera Sur (ECOSUR),
Carretera Antiguo Aeropuerto Km 2.5, 30700 Tapachula, Chiapas, Mexico
e-mail: pliedo@ecosur.mx

S. Ekesi et al. (eds.), *Fruit Fly Research and Development in Africa - Towards a Sustainable Management Strategy to Improve Horticulture*,
DOI 10.1007/978-3-319-43226-7_30

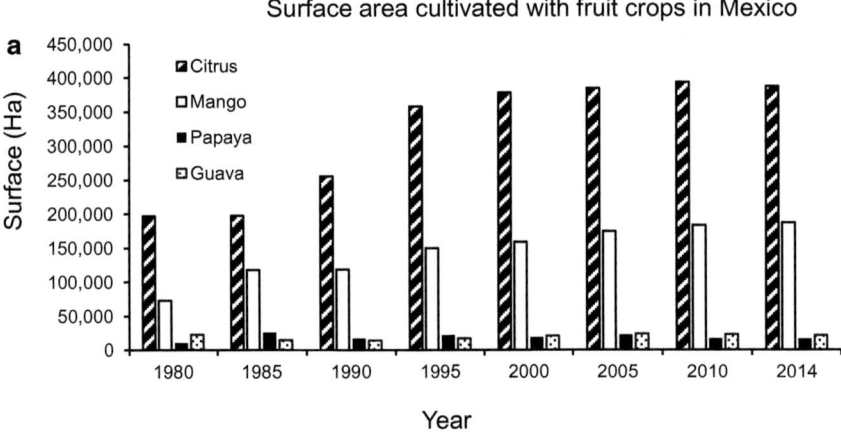

Fig. 30.1 Surface area cultivated with fruit crops (**a**) and production value (**b**) of citrus (excluding lemon), mango, guava and papaya in Mexico from 1980 to 2014

are citrus, mango, guava and papaya (Gutiérrez 2010, SIAP 2015). The gradual increase in surface area and greater crop value due to export markets have resulted in a significant increase in production value (Fig. 30.1).

2 Fruit Flies

Fruit flies (Diptera: Tephritidae) are important pests of fruit worldwide. In Mexico they have been divided into two main groups for pest management purposes: exotic, potentially invasive species, and native species. The first group is comprised of those species that are not currently present in Mexico, but are established in other countries and would represent a severe risk if they were introduced or invaded Mexico and became established. Examples of these are species from the genera *Bactrocera* Macquart and *Ceratitis* Macleay, and some species from the genus *Anastrepha* such as the South American cucurbit fruit fly *Anastrepha grandis* (MacQuart), the Caribbean fruit fly *Anastrepha suspensa* (Loew) and the South American fruit fly *Anastrepha fraterculus* (Wiedemann). The second group are the native species, mainly from the genus *Anastrepha*, but also including the papaya fruit fly, *Toxotrypana curvicauda* (Gerstaecker), and some species from the genus *Rhagoletis* Loew. Although there are approximately 30 species of *Anastrepha* reported in Mexico (Norrbom and Kim 1988; Hernández-Ortiz 1992) extensive surveys have demonstrated that only a few species are of economic importance (Aluja 1994; Reyes et al. 2000; Hernandez-Ortiz et al. 2010). The National Fruit Fly Program of Mexico recognizes only four species of economic importance: the Mexican fruit fly *Anastrepha ludens* (Loew), the West Indian fruit fly *Anastrepha obliqua* (MacQuart), the sapote fruit fly *Anastrepha serpentina* (Wied.) and the guava fruit fly *Anastrepha striata* Schiner (NOM-23-FITO-1995). Other species, such as the apple maggot *Rhagoletis pomonella* (Walsh) and *T. curvicauda*, are not considered to be of economic importance because of the limited amount of damage they cause. Within the *Anastrepha* genus, there are some species, such as *Anastrepha distincta* Greene, that are abundant in particular locations but their hosts are not grown commercially, so they are not considered as economically important pests.

The first reports on fruit flies in Mexico go back to the late 1800s (Herrera 1905). From the 1920s to the 1970s the Entomology Laboratory of the United States Department of Agriculture in collaboration with the Plant Health Direction in Mexico did important basic and applied research that resulted in taxonomic keys, host plant surveys, the development of traps, lures and bait sprays, and the introduction of natural enemies for biological control (Crawford 1927; Greene 1929; McPhail 1939; Baker et al. 1944; Shaw 1947; Jimenez-Jimenez 1956). Also during this period, ethylene dibromide (EDB) was developed as a fumigant and was used as a post-harvest treatment to minimize the risk of transporting infested fruits. Due to its efficacy, EDB was widely used as the simplest method to address fruit fly problems and allow fruit exports (Monro 1961).

The exotic medfly, *Ceratitis capitata* (Wied.) (Mediterranean fruit fly) was first detected in the Americas in Brazil in 1904, and then in Costa Rica in 1955 (Enkerlin et al. 1989). Through concerted international control efforts, it was controlled for 15 years (1960–1975) in Nicaragua (Rhode et al. 1971), but in 1975 was reported in Honduras, El Salvador and Guatemala (Gutiérrez-Samperio 1976). Because of the severe threat posed to the Mexican fruit industry, the preventive programme that was established by the Mexican government in 1926, was strengthened by inclusion of the use of the Sterile Insect Technique (SIT) to prevent the northern spread and establishment of *C. capitata* in Mexico. A joint Guatemala-U.S.A.-Mexico programme to deliver this was initiated in 1976 (Schwarz et al. 1989).

In the early 1980s, the U.S. Environmental Protection Agency banned the use of EDB for post-harvest treatment of fruit fly because of the health risks to consumers. The Mexican fruit industry was highly vulnerable at this time because of its long-term dependency on EDB fumigation. However, the ban of EDB, the difficulty in substituting EDB with another post-harvest treatment and the success of the ongoing *C. capitata* control strategies (see below) were key factors that encouraged researchers to look for new alternatives and this underpinned the development of an area-wide integrated pest management (IPM) approach. For native fruit fly species, Aluja and Liedo (1986) proposed an integrated orchard management strategy that included consideration of biogeographic, historic, socioeconomic and political aspects; basic infrastructure; and available technology. Both, basic and applied research projects were initiated, and a national strategy was developed. In 1992 the federal government implemented the National Fruit Fly Program with the aim of eradicating the four economically important native *Anastrepha* species where this was feasible, to suppress them in other areas, and to prevent the establishment of exotic species (Reyes et al. 2000).

3 Area-Wide Integrated Pest Management (IPM)

The current status of strategies used to control fruit flies in Mexico will be described here. For exotic species, the emphasis will be on *C. capitata*.

3.1 Exotic Species

3.1.1 Detection

Due to increased trade, tourism and migration the risk of introduction of exotic species of fruit flies into Mexico, is growing. A trapping network has been established to detect exotic species as soon as possible. This trapping network uses Jackson or Delta traps with three types of synthetic attractants: trimedlure to detect *C. capitata*, methyl eugenol to detect the oriental fruit fly *Bactrocera dorsalis* (Hendel) and related species, and culure to detect the melon fly *Bactrocera cucurbitae* (Coquillett) and related species (White and Elson-Harris 1992). Over 30,000 traps have been established countrywide, with particular emphasis on possible entry ports and fruit growing areas. These traps are serviced every 14 days. In addition, there are approximately 14,000 *C. capitata*-specific traps along the southern border and these traps are serviced every 7 days (SENASICA 2008).

To date, the exotic fruit fly species that have been detected are *C. capitata* and the olive fruit fly *Bactrocera oleae* (Rossi). With respect to *C. capitata*, 99.98 % of detections have been in the states bordering Guatemala (Chiapas 94.4 %, Tabasco 3.5 % and Campeche 0.05 %) with only one detection in the state of Baja California in 2004 in the Northwest of Mexico, close to the border with the United States (Enkerlin et al. 2015). On every occasion that *C. capitata* was detected their populations were controlled effectively as verified by subsequent intensive survey (Enkerlin et al. 2015). Currently, and according to the International Plant Protection Convention (IPPC), the status of *C. capitata* is defined as 'pest absent' in 28 of the 32 states, and 'pest transient' (pest entries that do not result in establishment after applying control measures) for the states of Chiapas, Tabasco, Campeche and Baja California (Enkerlin et al. 2015).

Bactrocera oleae was detected in 1999 in the Northwest of Mexico, one year after it was first detected in the U.S in California (Rice 2000). Today it is considered established in the states of Baja California and Sonora, where integrated pest management actions are implemented to suppress populations and minimize adverse economic effects (SENASICA 2015). Commercial olive growing is restricted to this region of the country and this species is monophagous, so it cannot spread further.

3.1.2 Fruit Fly Control with Special Reference to C. Capitata

Immediately after the first detection of *C. capitata* in Mexico in 1977 in the municipality of Tuxtla Chico, Chiapas, bait sprays were applied to prevent establishment. This was also done following all subsequent detections, either by air or ground, depending on the scale of the outbreaks. In addition, fruit sampling, intensive trapping around detection sites and quarantine measures were implemented to monitor populations and ensure they were exterminated. These emergency responses continued for 3 years (1977–1979) during which time a mass-rearing facility was constructed at Metapa de Dominguez, Chiapas with the goal of producing 500 sterile *C. capitata* per week for use in SIT strategies (Patton-Thom 1980).

By 1980, the mass-rearing facility at Metapa de Dominguez began producing sterile *C. capitata* for release as part of SIT with the aim of implementing a biological barrier for *C. capitata* between Guatemala and Mexico. Populations of wild *C. capitata* were suppressed using a combination of bait spays and releases of sterile flies and, as the number of detections of wild *C. capitata* in Mexico decreased, the sterile flies' barrier was moved South-East from the Mexican side of the border in the state of Chiapas, towards the Guatemalan side of the border in the departments of San Marcos and Quetzaltenango. By 1982, approximately half of the production of sterile *C. capitata* from the Metapa de Dominguez facility were being released in Guatemala (with the remainder in Mexico) and by 1985 it increased to two thirds of total production (Schwarz et al. 1989). Budgetary constraints at this time led to the sterile flies' barrier being maintained purely as a containment barrier on both sides of the Mexico-Guatemala border rather than continuing to move it South-East towards the Guatemalan border with Honduras and El Salvador (Baker 1984). In 1995 a new mass-rearing facility began operations at El Pino, Guatemala, with the capacity to produce 1000 million sterile *C. capitata* per week. The greater availability of sterile flies, together with technological innovations, such as the production of sterile males only, a better knowledge of *C. capitata* population ecology in the region, and accumulated experience in the field, have contributed to the increased success of the programme (Enkerlin et al. 2015). Today, the leading edge of infestation in Guatemala is further away from the border with Mexico and this has decreased the risk of transient entries into Mexico (Enkerlin et al. 2015).

3.2 Native Species

The National Fruit Fly Program began in the early 1990s with a North to South strategy based on the natural environmental conditions that determine fruit fly population densities. The dry regions of northern Mexico have lower host plant diversity and are characterized by lower fruit fly population densities (Gutiérrez 2010). Under these conditions, eradication of economically important fruit fly species was considered feasible using SIT within a framework of area-wide IPM. To implement this a new mass-rearing facility, the MOSCAFRUT facility, was built in Metapa de Dominguez, next to the original *C. capitata*-rearing facility. This facility began operations in 1993 with the production of 100 million sterile *A. ludens* fruit flies per week for release in northern Mexico. Later, mass-rearing methods for the production of sterile *A. obliqua* fruit flies and also *Diachasmimorpha longicaudata* (Ashmead), a parasitoid of several fruit fly species, were developed. Currently, there is capacity to produce 140 million *A. ludens*, 80 million *A. obliqua* and 50 million *D. longicaudata* (Domínguez et al. 2010), allowing SIT and augmentative biological control to be used together for area-wide IPM. A new *A. ludens* genetic sexing strain was developed that allows the release of only males (Zepeda-Cisneros et al. 2014) and it has been used in the past 3 years. Grower organizations are responsible for surveying for fruit flies and for coordinating control activities at local and orchard scales.

This strategy has been successful in northern Mexico. From 1992 to date the states of Baja California, Baja California Sur, Sonora, Chihuahua and Coahuila have been recognized as fruit-fly-free areas (Fig. 30.2). In addition, 36 municipalities in Zacatecas state, 32 in Durango, 19 in Nuevo Leon and 12 in Sinaloa have achieved this same recognition (Fig. 30.2). Zones of low prevalence have also been recognized in Aguascalientes, Tamaulipas, Nuevo Leon, San Luis Potosí, Zacatecas, Guerrero, Durango, Nayarit, Sinaloa, Tlaxcala and Michoacán (Fig. 30.2). Overall there are now 85,000 and 186,000 ha of cultivated fruit trees that are recognized as either fruit-fly-free or low-prevalence areas, respectively. The total surface area represents over half the national territory (Gutiérrez 2010).

Fruit can be exported from fruit-fly-free areas without the need for quarantine treatments. Orchards in low-prevalence areas can still export their fruits if they meet phytosanitary requirements such as maintaining monitoring by trapping and fruit sampling, orchard sanitation, and quarantine treatments (hot water, irradiation, fumigation). There is also a certification available for orchards that are temporarily free of fruit flies. These orchards can export their products if they follow IPM protocols such as monitoring fruit fly populations, orchard sanitation, bait spraying in case of detection and post-harvest quarantine treatment. In 2008, 6120 ha of mango, citrus, guava and peach were certified in this way. For export to the U.S.A., hot-water

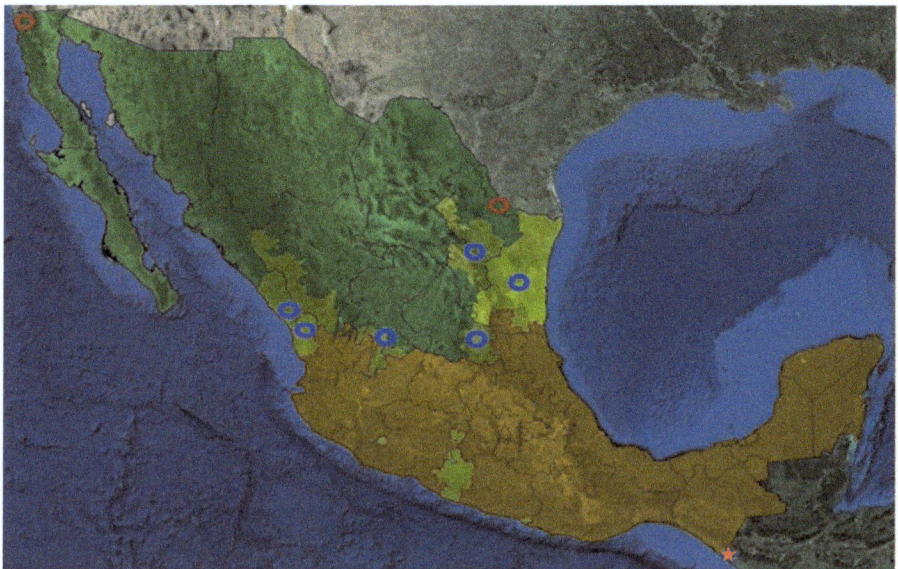

Fig. 30.2 Fruit-fly-free areas (*green*), low-prevalence areas (*yellow*) and zones of phytosanitary management (*brown*) in Mexico in 2014. *Blue* circles are internal quarantine inspection centres, *red* circles are international quarantine inspection centres and an asterisk indicates the location of the mass-rearing facility

treatments and methyl-bromide fumigation are used as post-harvest quarantine measures for mango and citrus respectively. These quarantine measures also include a computerized system that ensures the traceability of production batches.

4 Conclusions

Fruits exports and the market value of fruit in Mexico has increased in the past three decades. Fruit flies have not been a barrier or limiting factor in this and, therefore, the National Fruit Fly Program can be considered as successful. Cost/benefit analyses have shown that the area-wide IPM approach, based on the use of the SIT for both exotic and native species has resulted in favourable returns (Salcedo-Baca et al. 2010). In addition to direct benefits, there are also indirect benefits, such as the creation of rural jobs, reduction in health and environmental risks due to pesticides, and conservation of pollinators and biodiversity in general.

The challenge now is how to manage fruit flies in the tropical areas of southern Mexico, where fruit fly populations are at greater densities and where they have a larger variety of alternative host plants, both cultivated and in the wild. Under these conditions eradication would not seem to be feasible. However, effective suppression of populations using the area-wide IPM approach, including the use of the SIT and augmentative biological control, still has potential but needs to be evaluated and optimized in the field.

Acknowledgements I thank Sunday Ekesi and his collaborators for inviting me to write this paper for their book on 'Fruit Flies Research and Development in Africa – Towards a Sustainable Management Strategy to Improve Horticulture'. Thanks to Azucena Oropeza for her assistance in gathering updated statistical data on fruit growing and production. To Antonio Villaseñor and Walther Enkerlin for providing recent unpublished information on the current status of *C. capitata* control in Guatemala. To Pablo Montoya and the National Fruit Fly Program (SENASICA, IICA) for providing an updated map on the current status of fruit flies in Mexico (Fig. 30.2), and for all their support and trust.

References

Aluja M (1994) Bionomics and management of *Anastrepha*. Annu Rev Entomol 39:155–178
Aluja M, Liedo P (1986) Perspectives on future integrated management of fruit flies in Mexico. In: Mangel M, Carey JR, Plant RE (eds) Pest control: operations and systems analysis in fruit fly management, NATO ASI series. Springer, Berlin/Heidelberg, pp 9–42
Baker P (1984) Insect machismo in Central America. New Sci 16:41
Baker EW, Stone WE, Plummer CC, McPhail M (1944) A review of studies on the Mexican fruit fly and related species, U.S. Department of Agriculture Miscellaneous Publication no. 531. U.S. Department of Agriculture, Washington, DC
Crawford DL (1927) Investigation of Mexican fruit fly (*Anastrepha ludens* Loew) in Mexico. Calif Dep Agric Mon Bull 16:422–445
Domínguez JC, Artiaga-López T, Solís E, Hernández E (2010) Métodos de colonización y cría masiva. In: Montoya P, Toledo J, Hernandez E (eds) Moscas de la Fruta: Fundamentos y Procedimientos para su Manejo. SyG Editores, México, pp 259–276
Enkerlin D, García L, López F (1989) Pest status: Mexico, Central and South America. In: Robinson AS, Hooper G (eds) Fruit flies, their biology, natural enemies and control, vol 3B, World crop pests. Elsevier, Amsterdam, pp 83–90
Enkerlin W, Gutiérrez-Ruelas JM, Villaseñor-Cortes A, Cotoc-Roldan E, Midgarden D, Lira E, Zavala-López JL, Hendrichs J, Liedo P, Trujillo-Arriaga FJ (2015) Area freedom in Mexico from Mediterranean fruit fly (Diptera: Tephritidae): a review of over 30 years of a successful containment program using an integrated area-wide SIT approach. Fla Entomol 98:665–681
Greene CT (1929) Characters of the larvae and pupae of certain fruit flies. J Agric Res 38:489–504
Gutiérrez JM (2010) El programa moscas de la fruta en México. In: Montoya P, Toledo J, Hernandez E (eds) Moscas de la Fruta: Fundamentos y Procedimientos para su Manejo. SyG Editores, México, pp 3–10
Gutiérrez-Samperio J (1976) La mosca del Mediterráneo, *Ceratitis capitata* (Wiedemann) y los facotres ecológicos que favorecerían su establecimiento y propagación en México. Dirección General de Sanidad Vegetal, Secretaría de Agricultura y Ganadería. Talleres Gráficos de la Nación, México, 233 pp
Hernandez-Ortiz V (1992) El género *Anastrepha* Schiner en México (Diptera: Tephritidae). Instituto de Ecología, Sociedad Mexicana de Entomología, Xalapa, Veracruz, Mexico, 162 p
Hernandez-Ortiz V, Guillén-Aguilar J, López L (2010) Taxonomía e identificación de moscas de la fruta de importancia económica en América. In: Montoya P, Toledo J, Hernandez E (eds) Moscas de la Fruta: Fundamentos y Procedimientos para su Manejo. SyG Editores, México, pp 49–80

Herrera AL (1905) El gusano de la naranja. Boletín de la Comisión de Parasitología Agrícola, Tomo II: 307–415
Jiménez Jiménez E (1956) Las moscas de la fruta y sus enemigos naturales. Fitófilo 16:4–11
McPhail M (1939) Protein lures for fruit flies. J Econ Entomol 32:758–761
Monro HAU (1961) Manual of fumigation for insect control. F.A.O. 289 p. (Available from CAB direct (http://www.cabdirect.org/19630306551.html)
NOM-23-FITO (1995) Norma Oficial Mexicana NOM-023-FITO-1995 por la que se establece la Campaña Nacional contra Moscas de la Fruta. Diario Oficial de la Federación 11 de Febrero de 1999
Norrbom AL, Kim KC (1988) A list of the reported host plants of the species of *Anastrepha* (Diptera: Tephritidae), APHIS 81–52. U.S. Department of Agriculture, Washington, DC, pp 1–114
Patton-Thom P (1980) Mediterranean fruit fly eradication trail in Mexico. In: Proceedings of a symposium on fruit fly problems. Kyoto and Naha, Japan, August 9–12, 1980. National Institute of Agricultural Sciences, Yatebe, Ibaraki 305, Japan, pp 81–83
Reyes J, Santiago G, Hernandez P (2000) The Mexican fruit fly eradication programme. In: Tan KH (ed) Area-wide control of fruit flies and other pests. Penerbit Universiti Sains Malaysia, Penang, pp 377–380
Rhode RH, Simon J, Perdomo A, Gutiérrez J, Dowling CF, Lindquist DA (1971) Application of the sterile insect release technique in Mediterranean fruit fly suppression. J Econ Entomol 64:708–713
Rice RE (2000) Bionomics of the olive fruit fly *Bactrocera* (*Dacus*) *oleae*. UC Plant Protection Quarterly. University of California, Cooperative Extension, pp 1–5 (www.uckac.edu/ppq)
Salcedo D, Lomeli-Flores JR, Terrazas G (2009) Evaluación económica del Programa Moscamed en México (1978–2008). Instituto Interamericano de Cooperación para la Agricultura (IICA). 138 p
Salcedo-Baca D, Terrazas-González GH, Lomelí-Flores JR, Rodríguez-Leyva E (2010) Análisis costo-beneficio del programa moscamed. In: Montoya P, Toledo J, Hernandez E (eds) Moscas de la Fruta: Fundamentos y Procedimientos para su Manejo. SyG Editores, México, pp 27–40
Schwarz A, Liedo P, Hendrichs J (1989) Fruit flies, their biology, natural enemies and control, vol 3B, World crop pests. Elsevier, Amsterdam, pp 375–386
SENASICA. Servicio Nacional de Sanidad, Inocuidad y Calidad Alimentaria (2008) Informe anual 2008, Operaciones de Campo Moscafrut. Programa Nacional contra Moscas de la Fruta. Dirección General de Sanidad Vegetal, México. 26 p
SENASICA. Servicio Nacional de Sanidad, Inocuidad y Calidad Alimentaria (2015) Programa de trabajo de la campaña contra mosca del olivo a operar con recursos 2015 del componente sanidad del programa de sanidad e inocuidad agroalimentaria de las reglas de operación, en el estado de Sonora. 10 p
Shaw JG (1947) Hosts and distribution of *Anastrepha serpentina* in northeastern Mexico. J Econ Entomol 40:34–40
SIAP Servicio de Información Agroalimentaria y Pesquera (2015) http://www.siap.gob.mx (Consulted online on August 17, 2015)
White IM, Elson-Harris MM (1992). Fruit flies of economic significance: their identification and bionomics. CAB International, Oxon, 601 pp
Zahniser S, Link J (2002) Effects of North America free trade agreement on agriculture and the rural economy. In: Agriculture and trade reports. Electronic outlook report from the Economic Research Service. United States Department of Agriculture (USDA), WRS-02-1 July 2002, 134 pp
Zepeda-Cisneros CS, Meza-Hernández JS, García–Martínez V, Ibañez-Palacios J, Zacharopoulou A, Franz G (2014) Development, genetic and cytogenetic analyses of genetic sexing strains of the Mexican fruit fly, *Anastrepha ludens* Loew (Diptera: Tephritidae). BMC Genet 15(Suppl 2):S1

Pablo Liedo is a Principal Investigator at ECOSUR in Chiapas. He graduated as an Agronomy Engineer from Monterrey Tec (1978) then completed an M.Sc. in Pest Control: Research Methods at the University of Southampton, UK (1983) and a Ph. D. in Entomology at the University of California, Davis, USA (1989). His interests are in insect ecology and pest control, with particular emphasis on insect demography, population ecology, biological control and the Sterile Insect Technique. Research work has focused on fruit flies (Diptera: Tephritidae) and he has over 100 publications. He is an active member of the international community of fruit fly workers, Chairman of the global IOBC working group on fruit flies (1989–1994), editor of Fruit Fly News, member of the steering committee for the International Symposium on Fruit Flies of Economic Importance and the regional group Tephritid Workers of the Western Hemisphere.

Chapter 31
Overview of the Programme to Eradicate *Bactrocera carambolae* in South America

David Midgarden, Alies van Sauers-Muller, Maria Julia Signoretti Godoy, and Jean-François Vayssières

Abstract The carambola fruit fly (*Bactrocera carambolae* or CFF) was thought to be introduced in to Suriname in the 1960s or 1970s and first collected in 1975. An eradication programme based on the male annihilation technique (MAT) was developed and funded by IFAD (International Fund for Agricultural Development), the Netherlands, France and the United States and officially began in 1998. The programme worked in Guyana, Suriname, French Guiana and Brazil. By 2001, the distribution of *B. carambolae* was reduced to limited areas of Suriname and French Guiana. In 2002 funding was reduced and then halted. In the following years *B. carambolae* has expanded its distribution with detections as far southeast as Curralinho, in the Para state of Brazil, and as far north as Orlando, Florida, in the US. The closing of this programme before completion has resulted in increased cost to South American agriculture and increased risk to Central America, North America and the Caribbean. A coordinated programme amongst infested countries could still mitigate the risk of the spread of *B. carambolae* in the region.

Keywords Invasion • Biology • Detection • Eradication

D. Midgarden (✉)
USDA-APHIS-IS Medfly Program, Guatemala City, Guatemala
e-mail: david.g.midgarden@aphis.usda.gov

A. van Sauers-Muller
Ministry of Agriculture, Livestock and Fisheries, Paramaribo, Suriname

M.J.S. Godoy
Ministry of Agriculture, Livestock and Food Security, Brasilia, Brazil

J.-F. Vayssières
UPR HortSys, CIRAD Persyst, 34398 Montpellier, France

IITA, Biological Control Unit for Africa, IITA, 08BP 0932 Cotonou, Benin
e-mail: jean-francois.vayssieres@cirad.fr

© Springer International Publishing Switzerland 2016
S. Ekesi et al. (eds.), *Fruit Fly Research and Development in Africa - Towards a Sustainable Management Strategy to Improve Horticulture*,
DOI 10.1007/978-3-319-43226-7_31

1 Introduction

Fruit flies in the family Tephritidae pose an important threat to the production of fruits and vegetables throughout the world. Their presence jeopardizes the potential economic benefits of fruit cultivation and especially impacts developing countries where production and export could benefit both the economy and public health through improved nutrition. Fruit flies cause significant direct damage because their larvae develop inside the fruits but also secondary damage caused by pathogens entering the fruit through oviposition punctures (White and Elson-Harris 1992). This results in reduced yield and loss of quality (aesthetic and nutritional), thereby limiting marketing opportunities due to the quarantine restrictions imposed by importing countries. About 4000 species of tephritids have been described with more than 250 identified as known, or potential, pests of commercial produce (Thompson 1999). More than 90 of these species are in a single genus, *Ceratitis* Macleay, and all native to the Afrotropical region (De Meyer 2001). The family Tephritidae has many frugivorous species with characteristics that confer them with the potential to be successful invasive pests, particularly species in the genus *Bactrocera* Macquart. These characteristics include: (i) high intrinsic rates of growth, (ii) great ability to disperse and take advantage of high-altitude winds (jet streams), (iii) easily transported in infested fruit through international trade, (iv) polyphagous with large host ranges (all the year round), (v) expanding potential geographic range due to global warming (Ekesi et al. 2006; Duyck et al. 2008; Mwatawala et al. 2009; Salum et al. 2014; Gomina et al. 2014; Vayssières et al. 2015).

In Africa, research and development on the control of tephritids began to take on more importance in the early 1980s in South Africa (RSA), especially on native species in the genus *Ceratitis* (Labuschagne et al. 1996). This initiative grew in the early 2000s across Africa with the development of the African Fruit Fly Initiative (AFFI) by the International Centre of Insect Physiology and Ecology (*icipe*) in 1999 (Lux et al. 2003), and the West African Fruit Fly Initiative (WAFFI) by the International Institute of Tropical Agriculture (IITA), which ran between 2008 and 2012 (Vayssières et al. 2009a). Over the past 15 years, extensive information has been generated on the identification, distribution, biology, ecology, behaviour and management of fruit fly species of economic significance in Africa (Vayssières et al. 2014). Amongst the species occurring in sub-Saharan Africa, the most devastating are the oriental fruit fly, *Bactrocera dorsalis* (Hendel); the peach fruit fly, *Bactrocera zonata* (Saunders); and the mango fruit fly, *Ceratitis cosyra* (Walker) (Badii et al. 2015). The first two species, which are invasive, have even displaced several native species in some areas, including *C. cosyra* and the Mediterranean fruit fly (medfly), *Ceratitis capitata* (Wiedemann) (Ekesi et al 2009). AFFI and WAFFI have contributed baseline information and databases on taxonomy, trapping technology using different attractants, population dynamics, distribution, host range, orchard sanitation methods, molecular and chemical ecology, use of bait sprays (BAT), development of new attractants, parasitoids and generalist predators (e.g., *Oecophylla longinoda* (Latreille)), the male annihilation technique (MAT), cultural control. Using this information AFFI and WAFFI developed and rolled out comprehensive integrated pest management (IPM) packages for area-wide management of fruit flies in Africa.

The objective of this chapter is to present a case-study of one tephridid invasion into a new continent including how it was dealt with, what the impacts were, what the current status is and what lessons were learn that could be applied to countries in Africa or other regions susceptible to tephritid invasions.

1.1 Pest Description and Detection in South America

The carambola fruit fly (CFF), *Bactrocera carambolae* Drew and Hancock, is native to Malaysia, the southern (peninsular) area of Thailand and throughout western Indonesia and parts of India (Ranganath et al. 1997; Allwood et al. 1999). *Bactrocera carambolae* is a very serious pest in Malaysia because it attacks the fruit of carambola (starfruit), *Averrhoa carambola* L. (Oxalidaceae), at such a small stage that bagging (usually an effective, though labour-intensive, control method) is impractical. Many countries prohibit the import of susceptible fruit without strict postharvest treatment having been applied by the exporter. This may involve fumigation, heat treatment (hot vapour or hot water), cold treatments, insecticidal dipping, or irradiation (Armstrong and Couey 1989). Heat treatment is the most effective postharvest treatment, but reduces the shelf life of most fruits (Self et al. 2012) and so regulatory control, which restricts fruit imports, is used to prevent the spread of the pest to new areas.

Bactrocera carambolae is a member of the ever-fluctuating *B. dorsalis* complex of species (Drew and Hancock 1994; White and Elson-Harris 1992; Virgilio et al. 2014). Members of this complex are very difficult to distinguish morphologically (Iwahashi 1999) and prior to its official description in 1994, new specimens collected were either assumed to be *B. dorsalis* or from an as yet unnamed sibling species (e.g., it was referred to as B. (*B.*) *dorsalis* complex: B. (*B.*) sp. near *B. dorsalis* (A) (White and Elson-Harris 1992). The subfamily Dacinae (Christenson and Foote 1960; Fletcher 1987; White and Elson-Harris 1992) is Old World in origin, principally tropical Asia and Africa, and is represented by over 800 species (including species in the genera *Bactrocera* and *Ceratitis*). Until an established population in the Guiana Shield region of South America was reported (van Sauers-Muller 1991), no member of the *B. dorsalis* complex had permanently established in the New World (Vargas et al. 1989, 1990; Drew and Lloyd 1987; Bateman 1982). Isolated individuals and small infestations of several species from the *B. dorsalis* complex have been found in the Easter Islands (Chile) and in California and Florida (US) (CABI 2016), and Puerto Rico in the Caribbean (Steck, personal communication 2015). In these cases action by the respective federal and local authorities prevented permanent establishment. *Bactrocera dorsalis* has been present in Hawaii since 1945 and is now firmly established as a significant pest of many fruits and vegetables in these islands that causes significant economic damage.

Specimens of a *Bactrocera* sp. were first collected in South America in 1975 in Paramaribo, Suriname (van Sauers-Muller 1991; Drew and Hancock 1994). These specimens were collected and preserved, but not identified at the time. A second *Bactrocera* sp. was found again in 1986 and this time the specimens (along with those from 1975) were sent to the US Department of Agriculture and identified as *Dacus dorsalis*, the taxonomic name at the time for *B. dorsalis*. This identification was later amended to *B. carambolae* and described scientifically by Drew and Hancock (1994).

1.2 Invasion

Suriname is a former colony of the Netherlands and currently has a population of 520,000 inhabitants (Suriname Census 2012 – Algemeen Bureau voor Statistiek), approximately 15 % of which are descended from Indonesian colonists who arrived in the late nineteenth early twentieth century, predominantly from Java. Suriname retains strong ties with the Netherlands and over 90 % of Suriname's trade is with the Netherlands. During the colonial period, and even after independence in 1975, the Dutch-based company KLM was the only airline to operate flights to and from Suriname. Malavasi et al. (1998a) hypothesize that the most likely way that *B. carambolae* was introduced in to South America was by air from Indonesia. The distance between SE Asia, where *B. carambolae* occurs naturally, and the north of South America makes maritime transportation unlikely. Lack of funding, lack of coordination amongst the international community and no understanding of the potential importance of tephritid pests in affected countries, allowed *B. carambolae* to expand its geographic distribution significantly.

1.3 Biology

1.3.1 Life Stages

The life stages of *B. carambolae* are eggs, larvae (three instars), pupae, and adults. They complete their life cycle from egg to reproductive adult in 30–40 days, depending on temperature. Although studies have shown that adults of *Bactrocera* species can live as long as 125 days (Ekesi et al. 2006), this generally varies with temperature (Duyck et al. 2010) and food availability. For example, in the cool mountainous regions of Hawaii adult *B. dorsalis* can live for up to 1 year (Vargas et al. 1984). Once the larvae emerge from eggs they begin to feed and burrow into the pulp of the fruit. During development larvae tunnel within the fruit, feed on fruit tissues and are associated with the introduction of bacteria in to the fruit. When mature larvae leave the fruit, drop to the ground, and burrow several centimeters in to the soil, where they pupate.

1.3.2 Mating

There are two basic behaviours associated with mating in tephritids (Shelly and Kaneshiro 1991; Prokopy 1980): resource guarding, where males stake out a territory on a resource such as a host fruit; and lekking, where males group together with the purpose of attracting females (Sivinski and Burk 1989; Drew and Lloyd 1987). When they are physiologically prepared, females are attracted to these leks for mating. Mating in *B. carambolae* occurs just before dark when light intensity falls below 1000 lux (McInnis et al. 1999), and commonly on host plants. In Suriname, the first mating by female *B. carambolae* usually takes place 18 days after emergence (van Sauers personal communication).

1.3.3 Oviposition

Females puncture unripe, healthy fruit with their ovipositors, making cavities in to which they lay their eggs. Although there is no data on total egg production for *B. carambolae* it is likely to be similar to data for *B. dorsalis* which can lay as many as 3000 eggs over their lifetime under laboratory conditions, though some consider 1200–1500 to be the usual lifetime production under field conditions. Ekesi et al. (2006) calculated mean net production by *B. dorsalis* females to be 608 individuals, taking into consideration adult mortality and non-viable eggs. *Bactrocera carambolae* is a strong flier, which allows it to spread easily and also to re-infest orchards quickly after treatment (Vayssières 2002).

1.3.4 Host Range

The host range and geographic distribution of members of the *B. dorsalis* complex is very broad with more than 150 varieties of fruit attacked throughout tropical and subtropical Asia (Allwood et al. 1999). *Bactrocera carambolae* has a more limited host range (van Sauers-Muller 1991; Allwood et al. 1999). However, CABI (2016) noted that its worldwide host list included a wide mixture of economically important crops amongst other plants that were less important or rarely attacked. Most data from in its native geographic area was gathered by an extensive host fruit survey done in Malaysia and Thailand (Allwood et al. 1999). In the Guianas Shield area of South America, the presence of hosts a number of cultivated and wild host plants allows *B. carambolae* to maintain active populations throughout the year. In Suriname *B. carambolae* was recorded infesting 20 different fruit species, and 23 species in French Guiana belonging to 11 families (Table 31.1). The main hosts are carambola; Java apple, *Syzygium samarangense* (Blume) Merr. & Perry; and the

Table 31.1 Hosts of *B. carambolae* identified in Guyana, Suriname, French Guiana and BrazilPlease confirm the presentation of Table 31.1.Please tell me what exactly you mean by "confirm the presentation".

Hosts	Scientific names	Families	Origin
Primary			
	Averrhoa carambola L.	Oxalidaceae	South East Asia
	Syzygium malaccense Merr. & Perry	Myrtaceae	South East Asia
	Syzygium samarangense Merr. & Perry	Myrtaceae	South East Asia
Secondary			
	Spondias dulcis Foster	Anacardiaceae	America
	Spondias mombin L.	Anacardiaceae	South America
	Malpighia punicifolia L.	Malpighiaceae	America
	Eugenia uniflora L.	Myrtaceae	South America
	Psidium guajava L.	Myrtaceae	America
	Spondias purpurea L.	Anacardiaceae	America
Occasional			
	Terminalia catappa L.	Combretaceae	South East Asia
	Eugenia ligustrina (Sw.) Willd.	Myrtaceae	South America
	Ziziphus mauritiana Lamm.	Rhamnaceae	*South East Asia*
	Mangifera indica L.	Anacardiaceae	South East Asia
	Chrysophyllum cainito L.	Sapotaceae	Central America
	Inga sp.	Mimosaceae	South America
	Richardella macrophylla Lam.	Sapotaceae	America
	Manilkara zapota (L.) P. Royen	Sapotaceae	Central America
	Anacardium occidentale L.	Anacardiaceae	America
	Citrus reticulata Blanco	Rutaceae	South East Asia
	Mammea americana L.	Clusiaceae	Central America
	Citrus paradisi MacFad.	Rutaceae	South East Asia
	Annona muricata L.	Annonaceae	South East Asia
	Citrus sinensis (L.) Osbeck	Rutaceae	South East Asia

Malay apple, *Syzygium malaccense* (L.) Merr. & Perry (Myrtaceae) (van Sauers-Muller 2005; Vayssières et al. 2013). It is interesting that all the primary hosts are species with origins in South Asia (Allwood et al. 1999; van Sauers-Muller 2005; Vayssières et al. 2013). Most of the host fruit species have several fruiting periods per year (from three to year-round for *A. carambola*). Together with the humid tropical conditions of the Guianas Shield, this provides very favourable conditions for fruit fly reproduction throughout the year.

1.3.5 Adult Feeding

Adults of species in the *B. dorsalis* complex require a diet that is high in amino acids, vitamins, minerals, carbohydrates and water to survive and reproduce (Fletcher 1987). Both males and females must feed daily, and they forage on non-host as well as host trees. The diet includes honeydew, plant exudates, extra-floral nectaries, pollen, fruit juice, ripe fruits, microorganisms and bird droppings (Hendrichs and Hendrichs 1990; Hendrichs and Prokopy 1994). Both sexes appear to respond equally well to protein baits, though hypothetically, females should be more attracted because of their need for protein for egg production. Species in the *B. dorsalis* complex also feed on methyl eugenol. Tan et al. (2014) hypothesize that this is due to their use of this chemical in the production of pheromones. Mating success is measurably improved after males have ingest methyl eugenol. Methyl eugenol is found in the flowers of many plant species native where *Bactrocera* species are also native (Tan and Nishida 2012).

2 Programme Description

2.1 Beginning of the Eradication Programme

Although species in the *B. dorsalis* complex are some of the most serious pests of fruit in the world, no action was taken when individuals of *Bactrocera* sp. were first found in Suriname. After a second detection and preliminary identification in 1986, the Suriname Ministry of Agriculture began a surveillance programme, initially using only fruit sampling to identify infested areas and hosts. After a year, the ministry began using Jackson traps baited with methyl eugenol (van Sauers-Muller 2008). International institutions became involved in 1988 when the Inter-American Institute for Cooperation in Agriculture (IICA) organized a meeting to discuss the status of the *B. dorsalis* infestation (as it was then identified) in Suriname. Efforts continued throughout subsequent years, with additional contributions from FAO and from USDA-APHIS who financed a pilot control programme in western Suriname between 1991 and 1993. The pilot project was conducted in Wageningen (Fig. 31.1), a village in the western coastal district of Nickerie, where fruit fly populations most distant from where they were originally detected had been identified. The first control measure was MAT using a mixture of min-u-gel (as described in USDA 1989), methyl eugenol and malathion at bait stations. Trapping was done alongside this treatment to monitor the success of the control efforts. After a year, *B. carambolae* could no longer be found in Wageningen. Unfortunately, as only a small area was treated and the approach was not area-wide, re-infestation occurred. The pilot project was then extended to the neighbouring district of Coronie, a larger area (70 km wide) (van Sauers-Muller 1993). This served as a larger trial of MAT and reduced the chance of re-infestation. Positive results of the pilot project in the western coast of Suriname provided evidence that MAT was an effective tool to eradicate *B. carambolae*.

At the same time, initiatives also began in French Guyana to determine the presence and then geographic distribution of the *B. carambolae*. Fruit fly trapping surveys first began in the late 1980s (Caplong 1995). The results mainly provided information about the location of the pest, but did not provide information about its host range. A second initiative a few years later included sampling of cultivated and wild fruits with the purpose of associating particular tephritid species with their

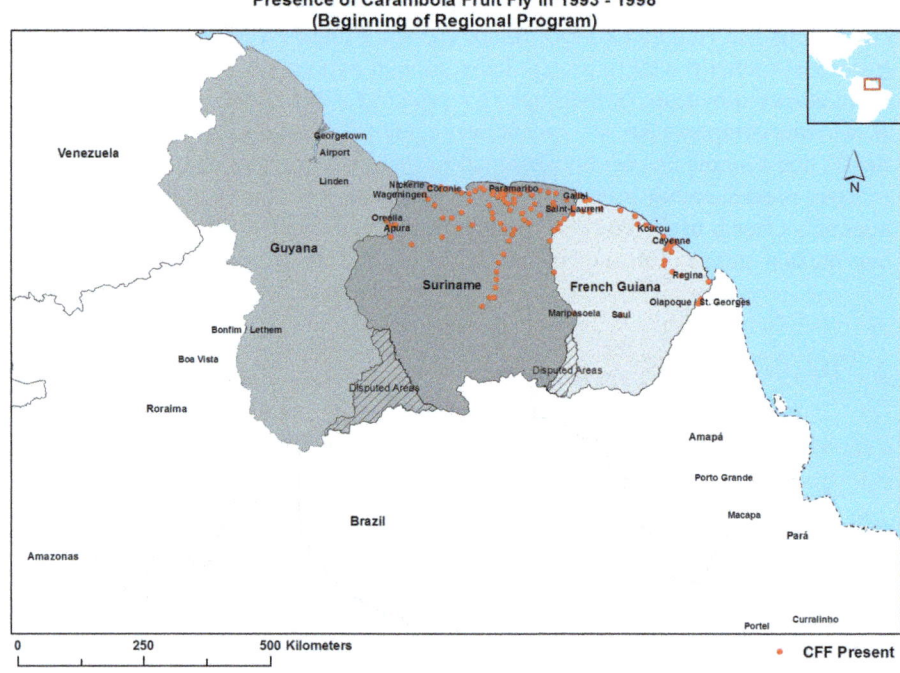

Fig. 31.1 Map of the region showing locations where *B. carambolae* had been detected prior to the start of the eradication programme

hosts, and determining which fruit fly species were of greatest economic concern in French Guyana (Séguret 1996; Cayol 2000). By 1996 *B. carambolae* had been found in four countries: Suriname from 1975, French Guyana from 1989 and, over restricted areas in Guyana and Brazil in March, 1996.

The success of the Suriname pilot eradication project led to the launch of a regional programme involving the four countries (Midgarden and van Sauers-Muller 1997). In November 1996, an international meeting was held in Paramaribo, Suriname to launch the Regional Program for the Control of Carambola Fruit Fly in South America. In August 1997, the Regional Office was installed in the IICA's Suriname Office to coordinate activities in the four countries. National Coordinators were appointed by the respective Ministries of Agriculture in September 1997 and a Regional Coordinator was contracted through IICA in 1998 to lead and coordinate regional-level activity. Funding was arranged through the UN's International Fund for Agricultural Development (IFAD) to provide salaries, equipment and supplies for coordinated activities. Funding for Suriname (where the majority of activity was assumed to be needed) was through the Government of the Netherlands treaty funds for Suriname. The governments of France and Brazil funded their own national programmes and assisted with regional coordination, funded through IICA. A Programme Steering Committee (PSC) to oversee the programme consisted of representatives from the international agencies involved (IICA, IFAD, FAO), the four affected countries (Suriname, Guyana, France, Brazil), and the two donor governments (Netherlands, US). A technical group, known as the Scientific Advisory Panel (SAP), acted as an external advisory body to the PSC and the Regional Coordinator.

The programme was managed, as much as possible, as one project through the Regional Coordinator, although there were four National Coordinators and each country had authority to respond to local needs depending on the level of infestation, the territory to be covered, the national socio-economic situation, local laws and other factors.

The Project was organized into four components:

1. Detection and Monitoring:
   - Trapping
   - Fruit Sampling

2. Control:
   - MAT
   - Protein bait sprays (bait application technique: BAT)
   - Mechanical control (i.e. fruit stripping and destruction)
   - Soil treatment

3. Public Relations:
   - Television
   - Newspapers
   - Radio (a song/jingle developed)
   - Leaflets
   - Agricultural Fairs
   - A '*carambola fruit fly programme*' logo was designed and used in all four countries (Fig. 31.2)

4. Training:
   - In *B. carambolae* identification
   - In trapping
   - In pesticide use
   - In workplace safety

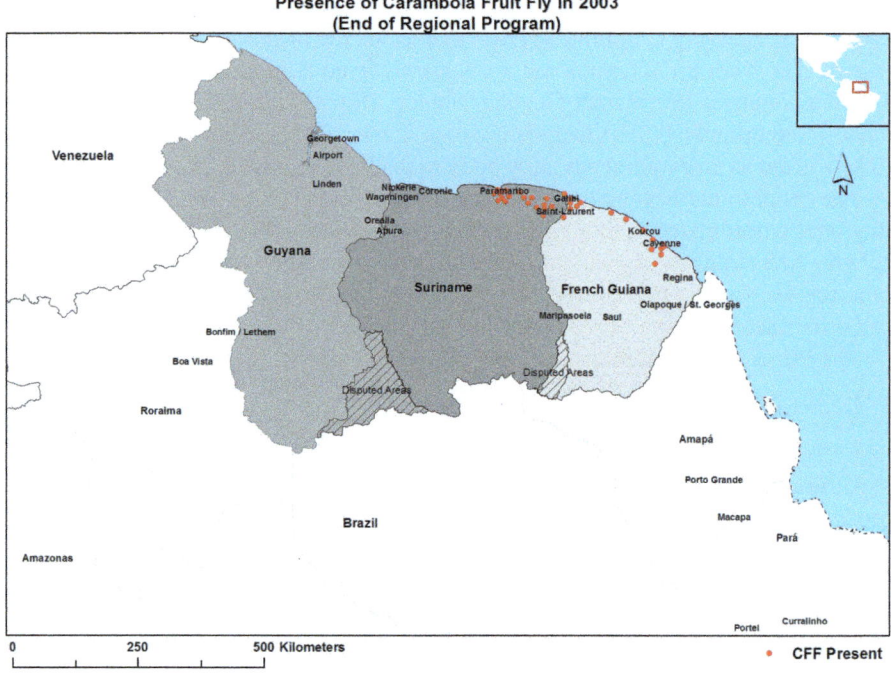

Fig. 31.2 Logo for the Carambola Fruit Fly Programme used on fliers, hats, T-Shirts, promotional materials, publications and advertisements throughout the region

2.2 *Detection of* Bactrocera carambolae

Detection of *B. carambolae*, like most tephritid species, is done through trapping and fruit sampling.

2.2.1 Trapping

Traps capture the adult flies and rely on lures, which are attractants that draw the adult flies to the traps. For *B. carambolae*, there are two basic kinds of lures: a male attractant based on a parapheromone, or male lure (Tan et al. 2014), and food lures that attract both males and females in search of protein for egg and sperm production. The male lure is based on the propensity for males of some tephritid species to be attracted to specific chemicals or compounds. Why this is so is not well understood, but appears to be associated with the production of male pheromones and the ability to attract and successfully mate with females (Boake et al. 1996; Tan et al. 2014). Like other members of the *B. dorsalis* complex, *B. carambolae* males are strongly attracted to methyl eugenol. Methyl eugenol is an essential oil extracted industrially from cloves, *Syzygium aromaticum* (L.) Merrill & Perry, a popular spice. It is also found in many wild and cultivated plants (especially flowers), in

Southeast Asia. Trapping with Jackson traps initially used only methyl-eugenol and succeeded in capturing numerous *B. carambolae*. However, the *B. dorsalis* manual (USDA 1989) recommends that an insecticide such as dibrom (1 % ai) should be included. Without the toxicant, males may feed on the lure and leave without being captured on the sticky insert. Addition of the insecticide kills the fly immediately, falls inside the trap and increases trap catches and thus sensitivity. In addition, because males that have fed on methyl eugenol are more successful at finding mates, the presence of lures without insecticide could possibly result in improved mate-finding and increased population growth. Since the recommended insecticide was not available, malathion ULV was used as the toxicant in the traps and appeared to work as well as dibrom. The ULV formulation mixed easily with methyl eugenol.

Protein-based lures have been in use for many years and they are the only attractant available for many tephritid species for which a male-only lure has not been identified. Protein lures (protein hydrolysate or torula yeast are the most common) attract both male and female flies. The traditional protein trap is the McPhail, which has gone through many iterations since its first use in the 1930s and 1940s as an open-bottomed bell–shaped glass where a mixture of water and protein are held and flies can enter, fall in to the liquid, and drown (Steyskal 1977). The regional programme used a plastic version of the McPhail trap called the Dome trap, with a yellow base and a clear top (IAEA 2003). The traps were baited with torula yeast pellets (one pellet/100 ml) and checked every week to detect the presence of both male and female *B. carambolae* as well as *Anastrepha* species.

2.2.2 Fruit Sampling

The results of fruit sampling help to determine the presence as well as the relative importance of different host plants for tephritid fruit fly species (USDA 1989). It also facilitates determination of the distribution of different fruit fly species. In post-eradication areas fruit sampling was used as the final proof that eradication was successful, i.e. it was the best way to detect any remaining infestations should they be present. Fruit sampling can be as simple as observing carefully for larvae; *B. carambolae* larvae are the only species of tephritids present in the region that 'jump' or curl up like a spring and propel themselves from a few centimeters to several meters (if they are located on a ledge or limb). Efficiency of sampling can be improved through careful cutting of fruit and observing for damage or eggs and larvae. It can be improved even further by maintaining the fruit for a week or even longer to detect pupae and/or large late-instar larvae.

2.3 Control of Bactrocera carambolae

When trapping and fruit collection established the presence of *B. carambolae* in an area, control methods were used to reduce populations and eventually eradicate them. Entomologists have been working for many years to develop methods to control fruit flies. The control methods used in the programme consisted of the application of MAT to areas subject to control, followed by BAT (protein hydrolysate and malathion) and fruit destruction (stripping) in hot spots. In some areas in Suriname, bait sprays were applied by air (Nulure mixed with 20 % ULV malathion using a Grumman AgCat aircraft), as this covered a larger area more completely and in a shorter time frame than by ground. Soil drenches using diazinon were also used under the dripline of trees after fruit stripping in some very persistent hot spots (USDA 1989). The control programme consisted of three phases:

1. Pre-treatment/ prospecting. The presence of the pest was identified and its distribution delimited by trapping and fruit collection; identification of locations with preferred host plants such as carambola and Java apple.
2. Control treatment. MAT, BAT and fruit stripping were conducted in infested areas. Treatments were maintained continuously until no flies or larvae could be detected; BAT and fruit destruction were mainly implemented in hot spots.
3. Post-treatment/ surveillance. Trapping and fruit collection maintained in order to ensure that the treatment had been successful and to monitor for reinvasion. CIRAD was in charge of these activities in French Guyana and MOA in Brazil, Guyana and Suriname.

2.3.1 Male Annihilation Technique

The primary control methodology for *B. carambolae* is the 'Male Annihilation Technique' (MAT), which was developed by Steiner at USDA/ARS, and refined over many years by workers in Australia and Japan (Cunningham 1989). MAT is based on the removal of males from the population by attracting them to poisoned bait stations using lures. Eventually, the population density is reduced to nonviable levels and the population is eradicated. Adult male *B. carambolae*, as is the case for other members of the *B. dorsalis* complex, are strongly attracted to the male lure methyl eugenol (Tan et al. 2014). Iwahashi et al. (1996) estimated the effective range for methyl eugenol using marked *B. dorsalis* and found it to vary from 90 to 300 m depending on wind velocity and direction. MAT has been applied successfully in many countries (Quilici and Donner 2012) for both established and recently detected *Bactrocera* species. Steiner and Lee (1955) demonstrated that this method could effectively control *B. dorsalis* in the Marianas islands. The Japanese government used MAT from the late 1960s until 1982 to eradicate *B. dorsalis* from the 700 km long Ryuku Island chain from Kyushu to Okinawa. Two recent examples are the successful eradication of the Asian papaya fruit fly, *Bactrocera papaya* Drew & Hancock in Northern Australia (Hancock et al. 2000) and *B. dorsalis* and the Pacific fruit fly, *Bactrocera xanthodes* (Broun) in Nauru Island in the South Pacific (Allwood et al. 2002).

MAT bait stations can take a number of forms from thick sprays and cotton wicks or other absorbent substances, to commercially available products. They are deployed at a pre-determined density throughout the area where fruit fly populations are present and re-applied on a regular basis over several generations of the pest. The bait stations must be distributed in space and time in a manner that ensures all (or the vast majority of) males are exposed to the attractant and toxicant. An important consideration in a bait station is its longevity in the field after being deployed; in general the longer they last, the better. Other desirable characteristics are the ease of use, biodegradability and/or ability to recycle once they are no longer effective (FAO/IAEA 2009; Piñero et al. 2014).

The goal of the control measures is the complete elimination of a species from a particular area. However, it is rare for any one technique or technology alone to succeed in reaching this goal, and MAT is no exception. As a result, complementary control techniques are used in addition to MAT to attack other life stages of the pest that are not affected by MAT, such as eggs, larvae, pupae and adult female flies. These supplementary control strategies include mechanical control, sanitation, BAT and soil drenches.

The MAT bait stations used in the programme were made of fibreboard blocks soaked in a solution of methyl-eugenol and malathion in a ratio of 3:1 (25 % malathion). There was some variation in the sizes of fibreboard block used, but the most commonly used size was 1 cm×5 cm×5 cm, which absorbed 10 ml of the methyl eugenol/malathion mixture. These bait stations remained effective for 4 weeks.

Initially, MAT was applied at a rate of four bait stations per hectare, which is what is recommended for eradication of *B. dorsalis* in California (USDA 1989). The bait stations were attached to a wire and hung from branches. The density of 4 per ha proved to be insufficient for *B. carambolae* in some areas, and the density was later increased to as many as 18 bait stations per hectare in hot spots. In Suriname, an additional method of applying MAT, the Min-U-gel bait station, was used in urban areas. This consisted of methyl-eugenol and malathion in Min-U-Gel to act as carrier rather than wooden blocks. This method was particularly useful in urban areas where the spray-on application from vehicles along streets provided an acceptable density and distribution of the bait stations to eradicate local populations.

2.3.2 Bait Spray

The bait application technique (BAT) uses bait sprays consisting of a liquid food attractant (proteins, sugars, fruit aromas and sometimes kairomones) mixed with an insecticide (Bateman 1982). Baits are generally sprayed on to the foliage and trunks of trees and plants. The adult fruit flies (both males and females) are attracted to and consume the bait, after which they die due to the insecticide (USDA 1989). In the *B. carambolae* programme, BAT was applied to areas with high or persistent populations, especially where infested fruit could be found. For eradication, it is important to apply bait spray to all foliage where flies might be found, and not only on host trees (Smith and Nannam 1988; Vargas et al. 2015). In Suriname and Guyana, the bait spray used was a mixture of Miller's Nu-Lure (a protein hydrolysate mixture made from corn) and malathion (ULV or EC), at a ratio of 20:80. In French Guyana they used Buminal (Bayer), a protein hydrolysate also mixed with malathion (Vayssières 2002). In Brazil they did not use bait spray as a regular part of the regional programme.

2.3.3 Mechanical Control: Fruit Stripping and Sanitation

Mechanical control is an important method to reduce tephritid populations, and control of *B. carambolae* is no exception. Although many flies may be present in the area as adults or pupae, the removal and destruction of fruit reduces that number of larvae that would otherwise successfully pupate; it also reduces the availability of oviposition sites (Vayssières et al. 2008). In the regional programme populations were notably reduced after well-coordinated fruit removal and destruction activities. Essential in achieving good mechanical control is the ability to locate all the preferred hosts of the fly in an area. During the pilot-eradication project in Wageningen, persistent detections were usually associated with a previously undetected host tree. Mechanical control is acheived through collecting and then destroying punctured / fallen fruits. The methods employed include:

- *Burial method*: collected fruits are buried in a hole at a depth of more than 80 cm; fruits are placed in the hole and then covered with a substantial layer of soil (more than 30 cm). An additional layer of a caustic substance such as lime can be applied above the fruit to ensure high mortality.
- *Incineration method*: collected fruits are burned.
- *Bagging and solar heating method*: collected fruits are sealed in black plastic bags (preferably two bags to prevent escape) and left exposed to the sun for at least 2 days.

Regular collection and destruction of fruits reduces fruit fly pressure and reinforces the effectiveness of other control methods.

2.3.4 Soil Drenching Treatments

An additional control method is the use of soil drenches containing insecticides that kill the flies while they are in the soil, either as larvae entering the soil to pupate, pupae, or adults emerging from pupation. The insecticides used need to be applied beneath the drip-line of the tree and be in sufficient quantity to cover the soil. Insecticide applications may be made at the drip line of any fruit fly host plants that are within 400 m of where fruit fly eggs, larvae or pupae have been detected (Stark et al. 2014).

2.3.5 Biological Control

Although biological control was not included as a basic part of the eradication programme, some biological control activities were conducted. In November–December 2000, *Diachasmimorpha longicaudata* (Ashmead) (Hymenoptera Braconidae) was released along both sides of the Oyapock River, from Taparabu to Clevelandia, including St Georges (Vayssières et al. 2013). About 2 million *C. capitata* pupae parasitized by *D. longicaudata* were transported from Brazil (CENA lab-Piracicaba). Between 2001 and 2003, emergence of *D. longicaudata* was regularly recorded from parasitized *B. carambolae* in fruit sampled from along the French side of the river Oyapock as far as Regina. This parasitoid, released in 2000 and 2001 in Brazil and French Guyana, was well-established and regularly recovered from fruit sampled around St Georges and Regina in 2003 (Vayssières et al. 2013). In addition, a study was conducted to identify native parasitoids present in Brazil, French Guyana, and Suriname.

2.4 Safety

The insecticides used in the *B. carambolae* programme were some of the most commonly used products in the world, with proven records of safe use. Despite this, everyone in the programme was required to take precautions to avoid unnecessary contact with the insecticides by themselves or the public (Tomlin 2009). The insecticides used in the programme were Malathion® and Basudine® (diazinon). Malathion was mixed with methyl eugenol for MAT and protein hydrolysate (Nu-Lure or Buminal) for BAT (USDA 1989). Basudine was used for soil drenches. All those working with insecticides were required to wear nitrile plastic gloves at all times. Care was taken in placing bait stations to ensure that they were not likely to fall into areas where children or animals could find them.

2.5 Laboratories

A laboratory was necessary for all of the programmes to assess the trap results, rear and identify larvae, and, in some cases, maintain a colony of *B. carambolae* for experiments. In Suriname, the *B. carambolae* programme and the Ministry of Agriculture renovated an old virology lab and transformed it into the national carambola fruit fly office of Suriname. This site also served as the regional laboratory for the programme and was the base for rearing of *B. carambolae* for studies conducted in conjunction with USDA-ARS and the University of Sao Paolo Brazil, as well as national and international training. A laboratory was also established in French Guyana at the Kourou CNRS-CIRAD research station, where they identified flies and parasitoids, reared flies from sampled fruit, and conducted experiments.

Scientists from this laboratory also worked with colleagues at the spatial mapping division to provide high quality maps and analysis for the programme. In Brazil, the Ministry of Agriculture established a small laboratory in Oiapoque with the objective of supporting the actions in infested areas in the Municipality of Oiapoque and the town of Clevelandia. They also provided support for activities on the French side of the river between Saut Maripa (a waterfall upriver) and Ouanary in the far north. This was part of a technical agreement between France and Brazil that was signed in 2005.

2.6 The Bactrocera carambolae *Network*

The four national offices and headquarters were linked via the internet. In 1998 partial financial support was given to each of the participating countries, by the Regional Office, to set up a basic computer system with modems to allow access to the internet and to strengthen the communication among the five units. Compatible software packages were installed on all computers to facilitate communication across the *B. carambolae* Programme. Despite significant language differences, communication of technical information generally presented few difficulties thanks to the strict standardization of data reporting formats.

2.7 *Public Relations*

Although some public relations work had always been done, principally while out in the field trapping and conducting control, the public relations programme truly began with the launch of the '*carambola fruit fly programme*' logo (Fig. 31.2) in January 2000. This date coincided with the commencement of activities in Suriname in an urban environment. An advertising agency (Rick's Advertising and Production, Paramaribo) was contracted to produce and arrange broadcasts of info-mercials, which included announcements and updates about the programme on radio and television. The public relations campaign also used print media such as newspapers, posters and flyers. Schools and community groups were provided with printed information, T shirts and stickers explaining and promoting the programme. Additionally, trappers wore white T shirts and caps on which the logo was printed to increase their 'recognizability' when working in the field. T shirts and stickers were distributed to the general public in areas where the programme carried out most of its activities and needed the cooperation of the residents.

The public relations initiatives used common guidelines, but were necessarily unique to each country in the programme. In Guyana, presentations were made at primary schools in Siparuta and Orealla, two villages on the border with Suriname where *B. carambolae* was eradicated in 1998; this raised awareness of activities in their villages and of the toxicity of the control measures. Older children were asked to collect infested fruits and maintain them until adult fruit flies emerged, as part of their science class. In Suriname, a number of television documentaries and smaller commercials were broadcast on TV, a catchy jingle accompanied information on the most popular radio stations, and regular updates were presented in local newspapers. In French Guyana, newspapers regularly printed stories about the programme and conducted interviews with members of staff of the National Office including monthly meetings of the SPV, CIRAD and FDGPC teams. The Ministry of Agriculture designed an insert to fit in the folders containing airline tickets for passengers. Posters were also displayed in the airport to inform passengers not to carry fruit with them when entering or leaving the country. In Brazil, nationwide TV news broadcasted the meeting of the National Coordinators held in Macapá and promoted public awareness to prevent the movement of infested fruit from Amapá State.

2.8 Training

Training was an essential part of the programme. Not only were all personnel trained in the techniques to control *B. carambolae*, but officials from other countries visited the programme for training (including from Venezuela, Colombia and Egypt). The training included such areas as information about fruit fly lifecycles, host plants, use of GIS systems, control methods and the effects of these on the environment and human health.

3 Operation of the Regional Programme

The initial focus of the national programmes was to identify where *B. carambolae* was present. Although there was some variation amongst detection activities in the national programmes, they all used fruit sampling and trapping as described in the section on detection above. The results showed that *B. carambolae* was present throughout most areas where host trees were present, predominantly in areas with human populations in Suriname and French Guyana (Fig. 31.1). The exception was in the far western coastal area of Suriname (Nickerie district) and some isolated villages in the interior of both countries where no population of *B. carambolae* could be detected. In Brazil, the infestation was restricted to the border with French Guyana in the town of Oiapoque and nearby communities along the river. In Guyana, the infestation was limited to three small villages on the border with Suriname about 60 km south, along the Corantijn River. Interestingly, no populations were detected in the populated Western Coastal area of Guyana, despite abundant host presence (Midgarden 1999). The programme is perhaps best presented by country, moving along the Atlantic coast from West to East from Guyana to Brazil.

3.1 Guyana

The *B. carambolae* programme in Guyana had a staff of nine people, including the National Coordinator housed in the IICA office. Equipment and vehicles included two double-cabin trucks, a laboratory equipped to identify fruit flies and maintain fruit samples for fly emergence. A *C. ceratitis* detection programme using 30 Jackson traps baited with trimedlure was also initiated in the Georgetown area and the international airport by the *B. carambolae* team.

Guyana is divided in to ten regions, all of which were monitored for *B. carambolae* during the programme. The pest was first detected in 1993, in the villages of Siparuta and Orealla, along the Corantijn River, which is on the border between Guyana and Suriname. Control operations were carried out, using BAT, MAT, mass trapping and fruit destruction. In 1998 there were only two detections in this area and none in the rest of the country. The Cooperative Republic of Guyana remained free of *B. carambolae* from May 1998 onward. The last *B. carambolae* captured in its territory was in Region 6, in the village of Orealla. Trapping was increased nationally in 1999 to nearly 662 and no additional *B. carambolae* were detected. As a result, the Ministry of Fisheries, Crops and Livestock officially declared Guyana *B. carambolae* free on Oct 23, 2000.

Consequently, the number of regions with traps was reduced and trapping remained only in those areas identified as being at highest risk of introduction of *B. carambolae* from Suriname. These areas included the international airport, Georgetown and the inhabited areas between Georgetown and the border of Suriname. The Ministry of Agriculture inspected incoming passengers for fruits and vegetables at ports of entry (air and sea) and posters and banners were displayed to encourage passengers to cooperate. A plant quarantine office and inspection facility was located at the border of Suriname, where persons arriving by ferry were inspected to prevent the introduction of *B. carambolae* as well as other pests. Trapping continued until 2004 in the area from the airport to Georgetown and along the border with Suriname.

3.2 Suriname

The pilot phase in West Suriname was executed successfully and *B. carambolae* was eradicated from the coastal town of Wageningen, and significantly reduced in the neighboring district of Coronie. An environmental impact assessment was done in 1998 and the major conclusion was that the MAT method used in the *B. carambolae* programme did not cause any adverse impacts on the environment and was safe for human health. This was the final step in the finalization of the agreement between the Government of Suriname and IFAD for the use of the Dutch Treaty Fund (and after years of pilot programmes) which resulted in the full implementation of eradication of *B. carambolae* in Suriname.

The National Office of the *B. carambolae* programme in Suriname was located in a 250 m^2 facility, which also served as the laboratory for the regional office. The total number of staff, as of December 1999, was 33 (one national coordinator, one development specialist, one control manager, three trapping supervisors, four trappers, four control team supervisors, four control team drivers, 13 control team field workers and two assistants). The programme also had seven double-cabin pick-up trucks and an 8 m boat with outboard engine. The Suriname programme collaborated closely with the regional office located at the Suriname IICA office for administrative and support activities, including a regional coordinator and assistant regional coordinator.

The full programme for eradication of *B. carambolae* in Suriname began in the far west at Apura, a village about 50 km up the Corantijn River which borders Guyana. Simultaneously, eradication of *B. carambolae* in the coastal district of Coronie was completed. By the end on 1999, the western line of *B. carambolae* infestation was pushed back nearly 200 km to the District of Saramacca which also created the opportunity for eradication of *B. carambolae* from the small villages of Siparuta and Orealla on the Guyanese side where constant trade with Apura had caused continual reinfestation.

A total of 2000 Jackson and 350 McPhail traps were in service by the end of 1999. Most of these were installed during that year in order to identify infested areas and allow planning for the eradication programme in the central-south region and the capital of Paramaribo. Paramaribo had the highest human population and was considered to be the most difficult area from which to eradicate *B. carambolae*. The density of detection traps in the districts where *B. carambolae* was successfully eradicated remained constant. McPhail traps, which capture ovipositing females, were in service only in areas where MAT had been in effect for 6 months or more and were used, along with fruit sampling, to ensure that eradication was completed and no residual populations of *B. carambolae* remained.

The large eradication effort in the District of Saramacca between Paramaribo and the Coppename River continued with ever decreasing numbers of flies captured. After 8 months of MAT, the population was limited to small, localized areas. A total of 85,000 MAT bait stations were used in this area over the 8 month period and no males were captured in Jackson traps from June 1999 until early 2001, when a small number of flies was found and persisted until 2002. In 2002, the number of *B. carambolae* in Saramacca showed no reduction during the first 6 months and BAT was reinitiated. Aerial application of bait sprays finally eradicated *B. carambolae* in this troublesome area.

Post-eradication areas were monitored regularly to detect any re-infestation as early as possible. Flies were only found in small numbers of isolated captures in Nickerie (twice) and Coronie (four times) and no infested fruit or females were found after increased numbers of traps were placed around the capture site. The Upper Suriname River, which is home to the Maroon people who have frequent trade with Paramaribo was also successfully made free from *B. carambolae*, taking the area infested in the country down to less than 20%. In 2001 reduced funds forced the programme to make cost saving measures such as reducing the size of the bait stations. The programme continued with ever more limited resources until June 2004 when all activities ceased.

3.3 French Guyana

Bactrocera carambolae was first detected in 1989 in the area of St Laurent du Maroni, French Guyana along the border with Suriname. The Government of France conducted the eradication programme from 1998 until the end of 2003 following the three main phases outlined for the regional programme, with different cooperating organizations responsible for each of the activities. The FDGPC (a state-run plant protection agency, Fédération Départementale des Groupements de Protection des Cultures) was in charge of MAT and BAT as well as some of the trapping. CIRAD (Centre de Coopération Internationale de Recherche Agronomique pour le Développement) was in charge of trapping (surveillance), fruit sampling, biological control and the Ministry of Agriculture (Service de Protection des Végétaux [SPV]) for funding.

The *B. carambolae* programme in French Guyana had seven members of staff, two CIRAD staff in charge of research activities, technical expertise and collaboration with five FDGPC staff working on control activities (MAT, BAT); there was also collaboration with EMBRAPA (Brazilian Agricultural Research Corporation) on biological control releases of the parasitoid, *D. longicaudata*.

The basic strategy was the same as in Suriname, but moving in the opposite direction: East to West, i.e. from the Oyapock to the Maroni River in French Guyana instead of West to East from the Corantijn to Maroni River in Suriname. From 1999 to the end of 2003, 530 traps (420 Jackson and 110 McPhail traps) were set up each year and serviced regularly by the FDGPC and CIRAD teams. During these 5 years, between 35,000 and 45,000 MAT bait stations and several hundred protein-based bait stations were set up every year in French Guyana (Cayol and Malvoti 1999; Vayssières 2003). From 2001 to 2003 the control measures showed good results by eradicating *B. carambolae* from the Brazilian border (St. Georges and Ouanary) to Regina (Fig. 31.3).

In French Guyana, 10 traps per km^2 were serviced every 2 weeks to detect the presence of male *B. carambolae*. From 2000 to 2003 the French team (PV, FDGPC and CIRAD) focused on four species considered to have the greatest economic impact: *B. carambolae*; the guava fruit fly, *Anastrepha striata* Schiner; the West Indian fruit fly, *Anastrepha obliqua* (Macquart); and the sapote fruit fly, *Anastrepha serpentina* (Wiedemann) (Vayssières et al. 2013). The latter three tephritid species are native and particularly associated with guava (*Psidium guajava* L.), hogplum (*Spondias mombin* L.) and star apple (*Chrysophyllum cainito* L.), respectively.

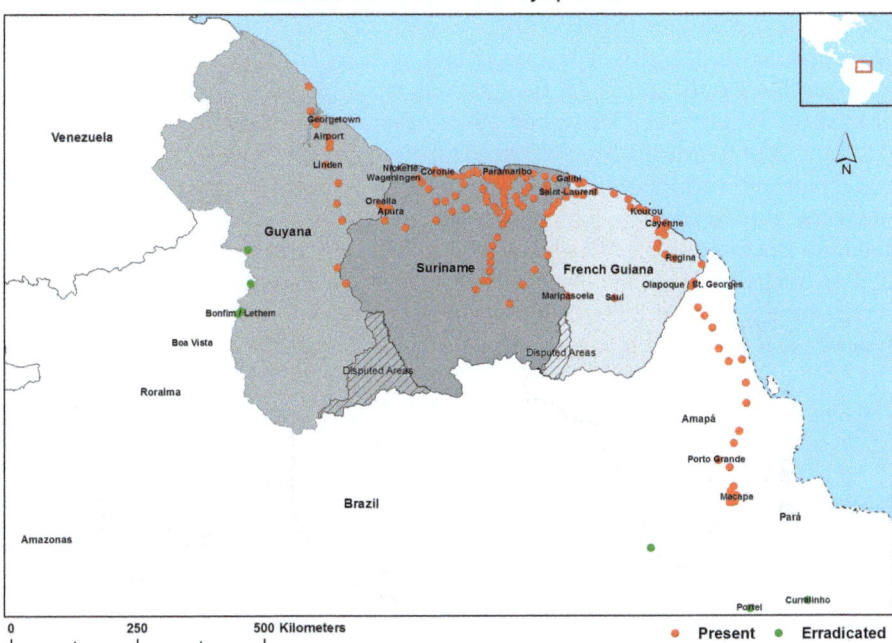

Fig. 31.3 Map of the region showing locations where *B. carambolae* could still be detected at the end of the eradication programme in 2002

3.4 Brazil

The *B. carambolae* programme office was located in the Quarantine Inspection Station near the Oiapoque airport just outside of town. The programme in Oiapoque had one double-cabin pick-up truck and a boat with an outboard engine. The staff consisted of the National Coordinator and three field staff. The field staff conducted trapping and placed bait stations in the Oiapoque area and the villages along the BR 156 highway south to the town of Ferreira Gomes (335 km from Oiapoque). In Macapa, the capital of the state of Amapa, two part-time officers and one-full time agronomist from the Ministry of Agriculture were responsible for servicing traps and placing bait stations in southern Amapa, when required.

Between the first detection of *B. carambolae* in the town of Oiapoque in 1996 and the beginning of the Regional Programme in 1999, the pest had spread both up and down river from Oiapoque. Specimens of *B. carambolae* were also captured occasionally in traps placed in towns between Oiapoque and Macapa, the capital city of the state of Amapa, located on the bank of the Amazon River. These detections did not continue and Macapa did not become established. However it was clear that the infested area presented a real threat for the rest of Amapa and eventually to Brazil as a whole.

Once the Regional programme began in 1998, eradication operations (run by the Ministry of Agriculture (MAPA) began to show success. Only a few *B. carambolae* were detected in the city of Oiapoque and no infested fruit were found in 1999. After each capture, additional bait stations were placed within a 300 m radius around the capture site. These infrequent detections demonstrated that there was still a threat of reinvasion from infested areas (most likely in French Guyana). There is a significant Brazilian community in French Guyana and there are frequent, if low numbers, of travellers moving between the two countries. No *B. carambolae* were found in Brazil outside of Oiapoque after 1999, even though trapping was intensified to include small and isolated villages in the region. Since the Regional Programme closed in 2003, Brazil has maintained a *B. carambolae* programme financed and organized by the Brazilian Ministry of Agriculture.

4   After the Programme and Current Situation

The Regional Carambola Fruit Fly Programme was officially closed with an announcement made to that effect by IICA Costa Rica in 2003. This was despite efforts, especially from Brazil, to prevent this from happening. This left the responsibility for continued action to monitor and control the pest to each individual country. Predictably, *B. carambolae* has expanded its geographic distribution dramatically since the programme stopped. Only Brazil has maintained an active programme continuously since that time. Guyana has worked more sporadically, especially between 2008 and 2011, after *B. carambolae* was found to have established in a number of different areas (Fig. 31.4). In Suriname and French Guyana, the locations of reinfestation are largely hypothetical as we assume that all previously infested areas are currently infested once again (Fig. 31.4). Guyana, a country previously free of *B. carambolae*, now has confirmed infestations in several locations, first at the airport south of Georgetown, and then to nearby towns and agricultural areas, eventually leading to outbreaks in the State of Roraima, Brazil. In French Guyana, *B. carambolae* reestablished on the border with Brazil (St. Georges) and the Brazilian programme was unable to prevent it from invading and, eventually, establishing in Oiapoque. Despite efforts to prevent its spread, within a few years it had established in the capital city of Macapa, some 450 km south (600 km by road).

The reason the pest has not spread to other parts of Brazil is as a result of geography, hard work and persistence. The infested area of Amapa is geographically isolated by water, areas with few hosts and lack of a modern transportation system. Continued vigilance by the Brazilian Ministry of Agriculture has prevented movement of *B. carambolae* from Macapa to other areas of Brazil. However, detections have been made and eradicated from parts of the State of Para. In recent years, *B. carambolae* has been detected in some communities on the riverboat route between Macapa and Belem (the most recent in Portel in 2014), though not in Belem itself. In addition, reinfestation of Guyana has put pressure on the neighbouring Brazilian State of Roraima, where outbreaks have occurred and then been eradicated by the

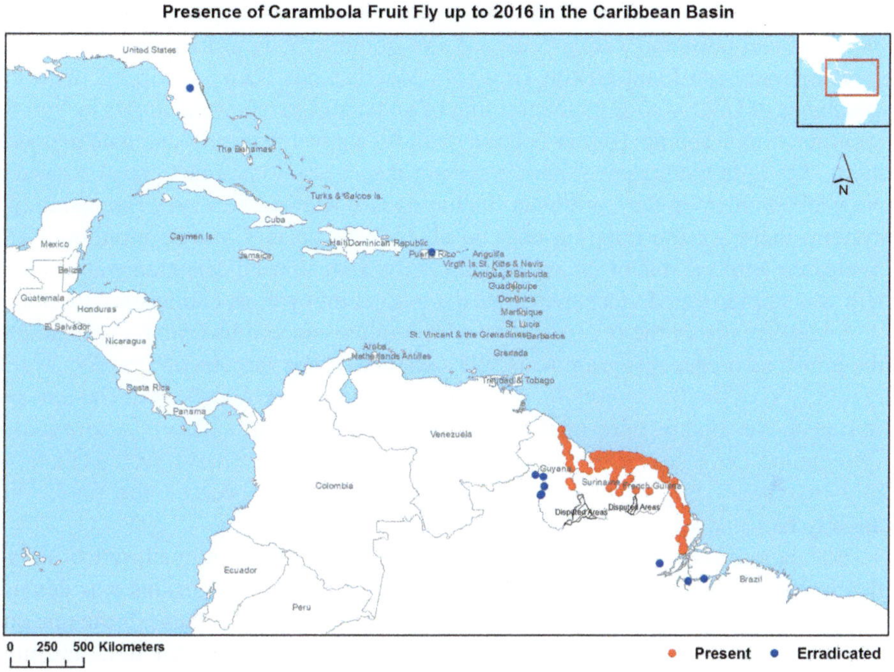

Presence of Carambola Fruit Fly up to 2016 in the Caribbean Basin

Fig. 31.4 Map of the greater region of northern South America, Central America and the Caribbean showing locations where *B. carambolae* has been detected or is assumed to be present since the eradication programme closed until 2016. *Blue* points show detections that were later declared eradicated. *Red* points show where *B. carambolae* is assumed to be present due to current or past trapping. Trapping is not currently conducted in most of Suriname or French Guiana

Brazilian Ministry. The continued presence of *B. carambolae* in Guyana means that incursions in to both Brazil and Venezuela are likely, if not inevitable. A specimen of *B. carambolae* was identified in Puerto Rico in 2015, and two putative *B. dorsalis* specimens were trapped in Orlando, Florida in 2008 and later confirmed by DNA to be *B. carambolae* (Steck, personal communication), demonstrating the regional importance of long-distance movement of *B. carambolae* (Fig. 31.4).

Research in Brazil and Suriname is presently aimed at determining changes in the known host range of the *B. carambolae* (van Sauers-Muller 2005). Differences have been noted in infestation levels amongst varieties of mango as well as a survey of fruits native to the Amazonian forest. Options for biological control using *Fopius arisanus* (Sonan), a parasitoid attacking *B. carambolae* in its native countries of Malaysia and Indonesia, is being investigated. Scientists in Brazil are looking into the possibility of importing and rearing *F. arisanus* under laboratory conditions. An entomologist from the Suriname ministry visited Hilo, HI US to learn how to rear *F. arisanus* and has established a colony of *B. carambolae* in preparation for the mass rearing and release of *F. arisanus*. Farmers in Suriname are trained in orchard sanitation, whether they are growing in their backyard or in plantations, and additionally, McPhail traps and Torula yeast are now commercially available.

The Caribbean Plant Health Directors Meeting has been supporting, with the assistance of USDA-APHIS, the training of all wider Caribbean countries in fruit fly identification, trapping, fruit collection, fruit fly control measures and data management. The data management uses a system that was initiated in Central America. Now all countries in the Caribbean (including Guyana and Suriname) are using this system; therefore the trapping data on all fruit fly species can be monitored and compared with the rest of the region. Recently, a web-based data management system was implemented that supercedes the tools used by most countries.

Some remains of the *B. carambolae* programme can still be found. For example, the public awareness programme is still evident as many farmers and gardeners still regularly visit the *B. carambolae* Project Units and the IICA Offices in Guyana and Suriname to obtain information and technical support. The *B. carambolae* Programme Units have also served as a focal point for other pests, such as the pink hibiscus mealy bug, *Maconellicoccus hirsutus* (Green) because the public are already familiar with the *B. carambolae* pest control Units.

The regional control programme against *B. carambolae* developed control activities in each participating country (Malavasi et al. 1998b). For instance, it demonstrated the effectiveness of MAT for eradication of *B. carambolae* in Suriname and French Guyana (Vayssières 2002, 2003). It also demonstrated that BAT and high densities of McPhail traps (with torula yeast pellets) could reduce populations locally.

The end of the Regional programme put a stop to the goal of eradicating *B. carambolae* in South America, with a particularly strong negative economic impact on Brazil. In Brazil, *B. carambolae* is considered a pest of quarantine significance present in the country, though currently limited in distribution and under official control. In Brazil the national *B. carambolae* eradication programme is currently (2016) under the responisbility of the Plant Health Division of the Ministry of Agriculture in the Secretariate of Agricultural Defence in the Ministry of Agriculture, Animal Husbandry and Food (MAPA) and has, as its objective, the eradication of *B. carambolae* from Brazilian territory. The programme seeks to maintain the quality of products in the internal market and guarantee exports of fruits and vegetables. The Brazilian agricultural industry covers a planted area of some 2.5 million Ha and employs an estimated 5 million workers both in and outside of farms.

Under this vision of the importance of prevention, control and eradication in recent years, the Government of Brazil has strengthened its national programme, including the eradication activities in the areas where *B. carambolae* has been detected or found to be absent through trapping. Currently, the detection system for *B. carambolae* in Brazil is established according to the risk of dispersal. There are around 6000 traps in high-risk areas, 3500 Jackson and 2000 McPhail traps, which are serviced every 7 or 15 days. The Center for Plant Health Education, created after the end of the Regional Programme, intensified its innovative activity after the detection of *B. carambolae* in the Para area in 2007; they involved more than 1420 community health workers, teachers, community leaders and public servants as well as citizens interested in assisting in the outreach programme for communities where *B. carambolae* had either been detected or were at risk of being invaded. The programme has also trained 8321 students and 5524 families. This was fundamental in successfully demonstrating that the State of Para was *B. carambolae* free in 2008. Assistance in inspecting the movement of hosts was done in cooperation with the Brazilian armed forces.

The *B. carambolae* programme in Brazil currently has 224 staff working in the Ministry and local plant protection agencies. The programme also collaborates with EMBRAPA to develop a better understanding of the pest and new methods for its control, including biological control using the parasitic wasp, *F. arisanus*.

Brazil continues to suffer under invasion pressure over is borders with French Guyana and Guyana, the later culminating in detection of *B. carambolae* in the border area of Normandy in the State of Roraima in December of 2009. The area near the Guyana border requires intensive detection activity and occasional eradication of sporadic outbreaks, which occur through the movement of indigenous people and gold prospectors from Guyana's regions 08 and 09 into Brazil. Equally important is the movement of people across a bridge constructed in 2011 (but not yet opened at time of publication) from French Guyana (through Saint Georges) to the town of Oiapoque and from there to the rest of the state of Amapa through the BR-156 highway to the south of Amapa and the capital, Macapa.

The lack of detection and control activities in French Guyana, Suriname and Guyana increases the pressure of introduction in to Brazil and reduces the effectiveness of the actions taken by the government in Brazil. In the long term, the success of the eradication programme in Brazil and prevention of spread to Roraima and Para states depends on renewing regional cooperative activities with its neighbours French Guyana, Guyana and Suriname. The Ministry of Agriculture of Brazil has worked to reinstate some form of regional programme, but with little success so far. Some meetings (encouraged by IICA and the Brazilian government) within the region have taken place, but funding to support a large programme, such as would be required now, have not been located.

5 The Future of *Bactrocera carambolae* Control in the Region

Bactrocera carambolae control could be managed with an area-wide IPM Package including (i) regular orchard sanitation, (ii) GF-120 bait-sprays (Vayssières et al. 2009b), (iii) MAT, (iv) BAT and (v) biological control. We demonstrated that fibreboard blocks with methyl eugenol and insecticide can be used in an area-wide programme without risk to non-target insect populations (Vayssières et al. 2007). New MAT formulations are being developed that are more efficient, easier to apply and have less environmental impact (Vargas et al 2012a, 2014).

Research in Hawaii showed that egg parasites could survive in populations under adult control, opening the possibility of integrating biological control techniques in to the total control regimen (Vargas et al. 2007). Taking into consideration the promising results of *F. arisanus* against *B. dorsalis* in South Asia, Hawaii and the Polynesian Islands (Vargas et al. 2012b), which is the origin of the parasite (Fullaway 1951), it seems appropriate to propose release of this parasitoid against *B. carambolae*. Mass releases of parasitoids in to a new area should be followed by studies of (i) acclimation; (ii) dispersion, and (iii) impact on *B. carambolae* populations. Regular fruit sampling, trapping with food-baits and field observations should be conducted jointly to be effective.

6 Lessons Learnt

There was no research plan built in to the programme to study *B. carambolae* under local environmental conditions. Although *B. dorsalis* and *B. carambolae* are very similar, there are enough behavioural and physiological differences to cause operational problems. For example, differences in the response of males to methyl eugenol and a longer time to reach sexual maturity (McInnis et al. 1999) meant that more bait stations were needed over a longer period of time than predicted, to achieve complete eradication from a given area. This resulted in requiring more methyl eugenol than estimated, which in turn lead to higher programme costs.

An eradication programme needs a continuous flow of funds and also permanent staff in order to achieve its objectives. In addition, without a well-established and well-trained technical workforce and infrastructure, it is unlikely to achieve the objectives. Financial constraints can lead to a reduction in the level of success that can be achieved in a given area, resulting in control but not eradication. This was the case in 2001 in the project in Suriname, the year when the programme could not increase the size of the *B. carambolae*-free area.

The establishment of a regional organization with capacity to detect, monitor and control *B. carambolae* is also important for the control of other fruit flies species. A large amount of data was collected from the surveillance programme which helped the Plant Protection Departments of the different countries learn more about the occurrence and distribution of key fruit fly species of economic importance to the region. A significant finding was that *C. capitata*, arguably the most economically important fruit fly species, was absent from Suriname, Guyana and French Guyana. Research showed that the *B. carambolae* is the most important fruit fly in the Guianas Shield, attacking more hosts than all the other economically important species together (Vayssières et al. 2013). Despite the challenges, all teams in the four countries were highly motivated to continue the fight to eradicate this important pest from the continent. The spread of *B. carambolae* was reduced during the programme, however, loss of the programme will undoubtedly result in the spread of *B. carambolae* over large areas in South America as well as North America and the

Caribbean (Fig. 31.4). Knowledge about *B. carambolae* has increased and simultaneously that of local fruit fly species and their importance for local agriculture.

7 Opportunities and Recommendations

From the work done, the following operational recommendations arise:

- Encourage the implementation of a tephritid surveillance programme throughout the region at risk. Early detection and fast action to eradicate is the best strategy to avoid the establishment of *B. carambolae* and other species.
- Continue to monitor *B. carambolae* populations, with the objective of increasing knowledge about its ecology in the South American continent, where it will probably spread beyond its present range.
- Continue fruit collections, with emphasis on wild fruits (to learn whether *B. carambolae* adapts to local forest plants) and on different varieties of cultivated hosts in relation to their phenological stage.
- Use research to provide alternative control methods for farmers, in order to produce better quality of fruits. These methods involve local MAT treatments combined with sanitary measures, trapping, biological control, resistant varieties and BAT. This research will also contribute to control of native *Anastrepha* fruit flies.
- Work on the introduction of non-endemic fruit fly parasites for the control of the *B. carambolae*, using the FAO Code of conduct for the importation and release of exotic biological control agents. Attention should be focused on egg and larval parasitoids, both of which have already been extensively used to control other *Bactrocera* species in Hawaii.
- The risk posed by the presence of the *B. carambolae* in the continent remains high, but only with a well-planned strategy and full financial support should the programme be re-initiated as soon as possible.
- The longer it takes to restart the programme, the more resources and time will be required.
- Any consideration of restarting the programme should also include a detailed plan for maintenance of the level of control or eradication achieved.

Acknowledgements The authors thank Estuardo Lira for making the maps used in this chapter. Also all of the technical and administrative personnel who worked so hard in Guyana, Suriname, French Guyana and Brazil, including the respective ministries of agriculture, IICA Suriname and Guyana, CIRAD, FDGPC. Special thanks to Peter, Robert and Hans Overeem for the excellent services provided over the years for boats and aircraft. The programme would not have been possible without the support and finances provided by the Netherlands, the International Fund for Agricultural Development (IFAD), USDA-APHIS, and the Governments of Guyana, Suriname, France and Brazil.

References

Allwood AJ, Chinajariyawong A, Kritsaneepaiboon S, Drew RAI, Hamacek EL, Hancock DL, Hengsawad C, Jipanin JC, Jirasurat M, Krong CK, Leong CTS, Vijaysegaran S (1999) Host plant records for fruit flies (Diptera: Tephritidae) in Southeast Asia. Raffles Bull Zool 47:1–92

Allwood AJ, Vueti ET, Leblanc L, Bull R (2002). Eradication of introduced Bactrocera species (Diptera: Tephritidae) in Nauru using male annihilation and protein bait application techniques. Veitch, CR and MN Clout. Turning the tide: the eradication of invasive species. IUCN Species Specialist Group. IUCN, Gland Switzerland and Cambridge. (see http://www.issg.org/Eradicat.html), 19–25

Armstrong JW, Couey HM (1989) Control; fruit disinfestation; fumigation, heat and cold. In: Robinson AS, Hooper G (eds) Fruit flies; their biology, natural enemies and control, World crop pests 3(B). Elsevier, Amsterdam, pp 411–424

Badii KB, Billah MK, Afreh-Nuamah K, Obeng-Ofori D, Nyarko G (2015) Review of the pest status, economic impact and management of fruit-infesting flies (Diptera: Tephritidae) in Africa. Afr J Agric Res 10:1488–1498

Bateman MA (1982) III. Chemical methods for suppression or eradication of fruit fly populations. In: Drew RAI, Hooper GHS, Bateman MA (eds) Economic fruit flies of the South Pacific Region, 2nd edn. Queensland Department of Primary Industries, Brisbane, pp 115–128

Boake CR, Shelly TE, Kaneshiro KY (1996) Sexual selection in relation to pest-management strategies. Annu Rev Entomol 41:211–229

CABI (2016) Invasive species compendium. CAB International, Wallingford. www.cabi.org/isc

Caplong P (1995) Étude de faisabilité d'un programme d'éradication de la *m*ouche de la carambole en Guyane française, Rapport de synthèse, janvier 1994–Septembre 1995, Fédération Départementale des Groupements de Protection des Cultures de Guyane, POSEIDOM

Cayol J-P (2000) Rapport d'activités. CIRAD-DAF, Cayenne, Guyane Française

Cayol J-P, Malvolti A (1999) Etude d'impact du methyl-eugenol sur les insects non-cible en Guyane Francaise

Christenson LD, Foote RH (1960) Biology of fruit flies. Annu Rev Entomol 5:171–192

Cunningham RT (1989) Parapheromones. In: Robinson AS, Hooper G (eds) Fruit flies: their biology, natural enemies and control, vol 3A. Elsevier, Amsterdam, pp 221–320

De Meyer M (2001) Distribution patterns and host-plant relationships within the genus *Ceratitis* MacLeay (Diptera: Tephritidae) in Africa. Cimbebasia 17:219–228

Drew RAI, Hancock D (1994) The *Bactrocera dorsalis* complex of fruit flies (Diptera: Tephritidae: Dacinae) in Asia. Bull Entomol Res Suppl Series 2:1–68

Drew RAI, Lloyd AC (1987) Relationship of fruit flies (Diptera: Tephritidae) and their bacteria to host plants. Ann Entomol Soc Am 80:629–636

Duyck PF, David P, Pavoine S, Quilici S (2008) Can host-range allow niche differentiation of invasive polyphagous fruit flies (Diptera: Tephritidae) in La Réunion? Ecol Entomol 33:439–452

Duyck PF, Kouloussis N, Papadopoulos N, Quilici S, Wang JC, Muller H, Carey J (2010) Lifespan of a *Ceratitis* fruit fly increases with higher altitude. Biol J Linn Soc 101:345–350

Ekesi S, Nderitu PW, Rwomushana I (2006) Field infestation, life history and demographic parameters of the fruit fly *Bactrocera invadens* (Diptera: Tephritidae) in Africa. Bull Entomol Res 96:379–386

Ekesi S, Billah MK, Nderitu PW, Lux SA, Rwomushana I (2009) Evidence for competitive displacement of *Ceratitis cosyra* by the invasive fruit fly *Bactrocera invadens* (Diptera: Tephritidae) on mango and mechanisms contributing to the displacement. J Econ Entomol 102:981–991,

FAO/IAEA (2009) Development of Bait Stations for Fruit Fly Suppression in Support of SIT. Working Material. Joint FAO/IAEA Division of Nuclear Techniques in Food and Agriculture. Vienna 15 pp

FAO/IAEA (2009) Development of Bait stations for fruit fly suppression in support of SIT. Working material. Joint FAO/IAEA division of nuclear techniques in food and agriculture, Vienna, 15 pp

Fletcher BS (1987) The biology of dacine fruit flies. Annu Rev Entomol 32:115–144

Fullaway DT (1951) Review of the Indo-Australasian parasites of the fruit flies (Tephritidae). Proc Hawaiian Entomol Soc 14:243–250

Gomina M, Mondedji AD, Nyamador W, Vayssières J-F, Amevoin K, Glitho AI (2014) Development and demographic parameters of *Bactrocera invadens* (Diptera: Tephritidae) in Guinean climatic zone of Togo. Int J Nat Sci Res 2:263–277

Hancock DL, Osborne R, Broughton S, Gleeson P, Tan KH (2000) Eradication of *Bactrocera papayae* (Diptera: Tephritidae) by male annihilation and protein baiting in Queensland, Australia. In Area-wide control of fruit flies and other insect pests. Joint proceedings of the international conference on area-wide control of insect pests, 28 May-2 June, 1998 and the Fifth international symposium on fruit flies of economic importance, Penang, Malaysia, 1–5 June, 1998. (pp 381–388). Penerbit Universiti Sains Malaysia

Hendrichs J, Hendrichs MA (1990) Mediterranean fruit fly (Diptera: Tephritidae) in nature: location and diel pattern of feeding and other activities on fruiting and nonfruiting hosts and non-hosts. Ann Entomol Soc Am 83:632–641

Hendrichs J, Prokopy RJ (1994) Food foraging behavior of frugivorous fruit flies. In: Fruit flies and the sterile insect technique. CRC Press, Boca Raton, pp 37–55

IAEA (2003) Trapping guidlines for area-wide fruit fly programmes. International Atomic Energy Agency, Vienna, 47 pp

Iwahashi O (1999) Distinguishing between the two sympatric species *Bactrocera carambolae* and *B. papayae* (Diptera: Tephritidae) based on aedeagal length. Ann Entomol Soc Am 92:639–643

Iwahashi O, Syamusdin-Subahar T, Sastrodihardjo S (1996) Attractiveness of methyl eugenol to the fruit fly *Bactrocera carambolae* (Diptera: Tephritidae) in Indonesia. Ann Entomol Soc Am 89:653–660

Labuschagne TI, Brink T, Steyn WP, De Beer MS (1996) Fruit flies attacking mangoes – their importance and post harvest control. Yearbook South African Mango Growers' Assoc 16:17–19

Lux SA, Ekesi S, Dimbi S, Mohamed S, Billah M (2003) Mango-infesting fruit flies in Africa: perspectives and limitations of biological approaches to their management. In: Neuenschwander P, Borgemeister C, Langewald J (eds) Biological control in integrated IPM systems in Africa. CABI Publishing, Wallingford, pp 277–294

Malavasi A, van Sauers-Muller A, Midgarden D, Kellman V, Didelot D, Caplong P, Ribeiro O (1998a) Regional program for the eradication of the carambola fruit fly in South America. In: Proceedings of 5th International symposium of fruit flies of economic importance, Penang, June 1–5, 1998, p 395–399

Malavasi A, Midgarden D, van Sauers-Muller A (1998b) Operations manual. Regional project for the control of the Carambola fruit fly in South America. 23 pp

McInnis DO, Rendon P, Jang E, van Sauers-Muller A, Sugayama R, Malavasi A (1999) Interspecific mating of introduced sterile *Bactrocera dorsalis* with wild *B. carambolae* (Diptera: Tephritidae) in Suriname: a case for cross-species SIT. Ann Entomol Soc Am 92:758–765

Midgarden D (1999) Use of spatial information to manage insect populations at two scales: Site-specific resistance management of *Leptinotarsa decemlineata* (Coleoptera: Chrysomelidae) and area-wide control of *Bactrocera carambolae* (Diptera: Tephritidae). Ph. D dissertation, Pennsylvania State University

Midgarden D, van Sauers-Muller A (1997) Launching of regional programme to control Carambola fruit fly. Caraphin News, no 15, p 4

Mwatawala MW, De Meyer M, Makundi RH, Maerere AP (2009) Host range and distribution of fruit-infesting pestiferous fruit flies (Diptera, Tephritidae) in selected areas of Central Tanzania. Bull Entomol Res 99:629–641

Piñero JC, Enkerlin W, Epsky ND (2014) Recent developments and applications of bait stations for integrated pest management of Tephritid fruit flies. In: Trapping and the detection, control, and regulation of Tephritid fruit flies. Springer, Netherlands, pp 457–492

Prokopy RJ (1980) Mating behavior of frugivorous Tephritidae in nature. In: Proceedings of symposium fruit fly problems, XVI International Congress Entomol, Kyoto (pp 37–46)

Quilici S, Donner P (2012) Analysis of exotic fruit fly trapping networks. EPPO Bull 42:102–108

Ranganath HR, Suryanarayana MA, Veenakumari K (1997) Papaya – a new host record of carambola fruit fly *Bactrocera* (*Bactrocera*) *carambolae* Drew and Hancock. Insect Environ 3:37

Salum JK, Mwatawala MW, Kusolwa PM, de Meyer MD (2014) Demographic parameters of the two main fruit fly (Diptera: Tephritidae) species attacking mango in Central Tanzania. J Appl Entomol 138:441–448

Séguret J (1996) Les mouches des fruits de Guyane Française: inventaire des principales espèces et recherche d'un attractif efficace, Mémoire du Diplôme d'Agronomie Approfondie, E.N.S.A.R

Self G, Ducamp M-N, Thaunay P, Vayssières J-F (2012) The effects of phytosanitary hot water treatments on West African mangoes infested with *Bactrocera invadens* (Diptera Tephritidae). Fruits 67:439–449

Shelly T, Kaneshiro K (1991) Lek behavior of the oriental fruit fly, *Dacus dorsalis*, in Hawaii (Diptera: Tephritidae). J Insect Behav 4:235–241

Sivinski J, Burk T (1989) Reproductive and mating behavior. In: Fruit flies: their biology, natural enemies and control. Elsevier Science Publishers, Amsterdam, pp 343–351

Smith D, Nannan L (1988) Yeast autolysate bait sprays for control of Queensland fruit fly on passionfruit in Queensland. Qld J Agric & Anim Sci 45:169–177

Stark JD, Vargas RI, Souder SK, Fox AJ, Smith TR, Leblanc L, Mackey B (2014) Simulated field applications of insecticide soil drenches for control of tephritid fruit flies. Biopestic Int 10:136–142

Steiner LF, Lee RKS (1955) Large-area tests of a male-annihilation method for oriental fruit fly control. J Econ Entomol 48:311–317

Steyskal GC (1977) History and use of the McPhail trap. Fla Entomol 60:11–16

Tan KH, Nishida R (2012) Methyl eugenol – its occurrence, distribution and role in nature especially in relation to insect behaviour and pollination. J Ins Sci 20:1–74. Available online http://www.insectscience.org/12.56/i1536-2442-12-56.pdf

Tan KH, Nishida R, Jang EB, Shelly TE (2014) Pheromones, male lures, and trapping of tephritid fruit flies. In: Trapping and the detection, control, and regulation of tephritid fruit flies. Springer, Netherlands, pp 15–74

Thompson FC (1999) Fruit fly expert identification system and sys- tematic information database. Backhuys Publishers, Leiden, 524 pp

Tomlin CD (2009) The pesticide manual: a world compendium (No. Ed. 15). British Crop Production Council. Alton. 1606 pp

USDA (1989) Action plan: oriental fruit fly *Bactrocera dorsalis* (Hendel) 56 pp. Accessed: https://www.aphis.usda.gov/plant_health/plant_pest_info/fruit_flies/downloads/orientalff.pdf

van Sauers-Muller AE (1991) An overview of the carambola fruit fly *Bactrocera* species (Diptera: Tephritidae) found recently in Suriname. Fla Entomol 74:432–440

van Sauers-Muller AE (1993) Pilot eradication project for the carambola fruit fly in Coronie, Suriname. In: Aluja M, Liedo P (eds) Fruit flies: biology and management., pp 439–442

van Sauers-Muller AE (2005) Host plants of the Carambola fruit fly, *Bactrocera carambolae* Drew & Hancock (Diptera: Tephritidae), in Suriname, South America. Neotrop Entomol 34:203–214

van Sauers-Muller AE (2008) Carambola fruit fly situation in Latin America and the Caribbean. Proc Carib Food Crops Soc 44:135–144

Vargas RI, Miyashita D, Nishida T (1984) Life history and demographic parameters of three laboratory-reared tephritids (Diptera: Tephritidae). Ann Entomol Soc Am 77:651–656

Vargas RI, Stark JD, Nishida T (1989) Abundance, distribution, and dispersion indices of the oriental fruit fly and melon fly (Diptera: Tephritidae) on Kauai, Hawaiian Islands. J Econ Entomol 82:1609–1615

Vargas RI, Stark JD, Nishida T (1990) Population dynamics, habitat preference, and seasonal distribution patterns of oriental fruit fly and melon fly (Diptera: Tephritidae) in an agricultural area. Environ Entomol 19:1820–1828

Vargas RI, Leblanc L, Putoa R, Eitam A (2007) Impact of introduction of *Bactrocera dorsalis* (Diptera: Tephritidae) and classical biological control releases of *Fopius arisanus* (Hymenoptera: Braconidae) on economically important fruit flies in French Polynesia. J Econ Entomol 100:670–679

Vargas RI, Leblanc L, Harris EJ, Manoukis NC (2012a) Regional suppression of *Bactrocera* fruit flies (Diptera: Tephritidae) in the Pacific through biological control and prospects for future introductions into other areas of the world. Insects 3:727–742

Vargas RI, Leblanc L, Putoa R, Piñero JC (2012b) Population dynamics of three Bactrocera spp. fruit flies (Diptera: Tephritidae) and two introduced natural enemies, *Fopius arisanus* (Sonan) and *Diachasmimorpha longicaudata* (Ashmead) (Hymenoptera: Braconidae), after an invasion by *Bactrocera dorsalis* (Hendel) in Tahiti. Biol Control 60:199–206

Vargas RI, Souder SK, Hoffman K, Mercogliano J, Smith TR, Hammond J, Davis BJ, Brodie M, Dripps JE (2014) Attraction and mortality of *Bactrocera dorsalis* to STATIC™ Spinosad ME weathered under operational conditions in California and Florida: a reduced-risk male annihilation treatment. J Econ Entomol 107:1362–1369

Vargas RI, Piñero JC, Leblanc L (2015) An overview of pest species of *Bactrocera* fruit flies (Diptera: Tephritidae) and the integration of biopesticides with other biological approaches for their management with a focus on the pacific region. Insects 6:297–318

Vayssières J-F (2002) Bilan technique (2001) du programme de lutte contre la mouche de la carambole en Guyane Française. Regional Carambola Fruit Fly, *Bactrocera carambolae*, Eradication Programme. CIRAD FLHOR, 21 p. + 4 p. d'annexes

Vayssières J-F (2003) Bilan technique 2002 du programme de lutte contre la mouche de la carambole en Guyane Française. Regional Carambola Fruit Fly, *Bactrocera carambolae*, Eradication Programme. CIRAD FLHOR, 31 p. + 8 p. d'annexes

Vayssières J-F, Cayol J-P, Perrier X, Midgarden D (2007) Impact of methyl eugenol and malathion bait stations on non-target insect populations in French Guiana during an eradication program for *Bactrocera carambolae*. Entomol Exp Appl 125:55–62

Vayssières J-F, Sinzogan A, Abandonon A (2008) Control of fruit flies through phytosanitary hygiene of orchards: prophylactic method. Leaflet 10. Regional fruit fly control project in West Africa (WAFFI)

Vayssières J-F, Korie A, Ayegnon D (2009a) Correlation of fruit fly (Diptera: Tephritidae) infestation of major mango cultivars in Borgou (Benin) with abiotic and biotic factors and assessment of damage. Crop Prot 28:477–488

Vayssières J-F, Sinzogan A, Ouagoussounon I, Korie S, Thomas-Odjo A (2009b) Effectiveness of spinosad bait sprays (GF-120) in controlling mango-infesting fruit flies (Diptera Tephritidae) in Benin. J Econ Entomol 102:515–521

Vayssières J-F, Cayol J-P, Caplong P, Séguret J, Midgarden D, Van Sauers-Muller A, Zucchi R, Uramoto K, Malavasi A (2013) Diversity of fruit fly (Diptera Tephritidae) species from French Guiana: their main host plants with associated parasitoids during the period 1994–2003 and prospects for fly management. Fruits 68:219–243

Vayssières JF, Sinzogan A, Adandonon A, Rey JY, Dieng EO, Camara K, Sangaré M, Ouedraogo S, Sidibé A, Keita Y, Gogovor G (2014) Annual population dynamics of mango fruit flies (Diptera: Tephritidae) in West Africa: socio-economic aspects, host phenology and implications for management. Fruits 69:207–222

Vayssières JF, De Meyer M, Ouagoussounon I, Sinzogan A, Adandonon A, Korie S, Wargui R, Anato F, Houngbo H, Didier C, De Bon H (2015) Seasonal abundance of mango fruit flies (Diptera: Tephritidae) and ecological implications for their management in mango and cashew orchards in Benin (Centre & North). J Econ Entomol 108:2213–2230

Virgilio M, White I, De Meyer M (2014) A set of multi-entry identification keys to African frugivorous flies (Diptera, Tephritidae). ZooKeys 428:97–108. doi:10.3897/zookeys.428.7366

White IM, Elson-Harris MM (1992) Fruit flies of economic significance: their identification and bionomics. CAB International, Wallingford, 601 pp

David Midgarden David's PhD was in Entomology from Penn State University with a focus on area-wide pest management and use of Geographic Information Systems (GIS). He has worked for the Peace Corps, USAID, FAO, IAEA, IICA and USDA in Africa, Europe, Middle East, Latin America and the Caribbean, principally on fruit fly control and eradication programmes. Since July 2011 he has been the Technical Director for the USDA-APHIS Medfly Eradication Program in Guatemala, Mexico and Belize.

Alies van Sauers-Muller studied tropical agriculture at Deventer University in the Netherlands with a specialization in Crop Protection that included an internship in Suriname. After graduating in 1980 she returned to Suriname to work for the government. Alies was the chairperson for the Technical Working Group for Tephritid fruit flies for CariCom at the second meeting of this working group in Paramaribo in 2010. She is the coordinator of the fruit fly research programme of the Ministry of Agriculture, Animal Husbandry and Fisheries, at the Research Department in Paramaribo, Suriname. She has been responsible for the project since the detection of *Bactrocera carambolae* in 1986, including the extensive eradication programme from 1998 to 2003. She has been a member of the Technical Panel on Pest Free Areas and Systems Approaches for Fruit Flies (TPFF) for the International Plant Protection Convention (IPPC), which was established to review scientific and technical data and to draft documentation in support of international agricultural trade through the establishment of pest free areas and systems approaches for fruit flies. She has been involved in training others in fruit fly identification, trapping and control since 2000 with an FAO mission to Venezuela, and also worked as trainer/resource person for workshops in the Caribbean region and Brazil. She lives in the Suriname countryside on a farm.

Maria Julia Signoretti Godoy Maria's MSc. was in Agricultural Sciences from the University of Brasilia (1998) where she subsequently became specialized in environmental GIS. She has studied at the Italo American Institute, Rome, Italy; the Agricultural Studi Cento, Borgo Mozzano, Italy; and the Federal Rural University of Amazonia, Belem, Pará, Brazil. She is currently a Federal Agricultural Inspector in the Ministry of Agriculture, Livestock and Food Security of the Agricultural Defense Secretariat in the Department of Plant Protection, Brasilia, Brazil. She was Head of the Prevention Division, Surveillance and Pest Control, 2005–2008 and Plant Protection General Coordinator from 2008 to 2009. Maria has experience in agronomy with emphasis on plant health and fruit flies. She is the representative for Brazil on the theme of fruits flies in the Committee of Plant Protection of the Southern Cone – COSAVE. She coordinated national control actions against fruit flies in the São Francisco Valley (2005–2008) to protect Brazilian fruit exports and participated in the creation and implementation of the national programme to combat fruit flies in Brazil. Since 1999, she has coordinated the *Bactrocera carambolae* eradication programme of the Ministry of Agriculture Livestock and Food Supply of Brazil and participated in the regional programme on prevention and control of *B. carambolae* (IICA) in Suriname.

Jean-François Vayssières is a senior entomologist attached to the Centre de Coopération Internationale en Recherche Agronomique pour le Développement (CIRAD) in Montpellier (France) specializing in agricultural entomology. Previously he worked in La Réunion, French Guiana and West Africa, including Benin where he was based at the International Institute of Tropical Agriculture (IITA) for 11 years. His research interests include ecology of pest-natural enemy complexes on tropical fruits (food web structure) and biodiversity issues.

Chapter 32
Systems Approaches for Managing the Phytosanitary Risk of Trading in Commodities that are Hosts of Fruit Flies

Eric B. Jang

Abstract Phytosanitary restrictions on horticultural commodities destined for domestic and foreign markets can present problems when unwanted invasive pest species are present in the exporter's crop or consignment. Conventional quarantine treatments are under increased scrutiny as they can contribute to ozone depletion (e.g. use of the fumigant methyl bromide) and may reduce the shelf-life of treated commodities (e.g. heat and cold treatments). Thus, alternative strategies are needed to deal with current and new invasive pests to meet the needs of a changing regulatory environment and changing phytosanitary issues. One such approach is called the 'systems approach' which considers the entire production to export system when determining and mitigating risk of invasive pests in export consignments. Systems approaches identify biological risk first and then attempts to mitigate that risk using two or more (independent) major components which are themselves made up of elements. In many cases severe quarantine treatments aimed at disinfestation of potentially infested products may not be needed; for example, when quarantine is imposed perhaps due to small populations and/or poor host plant status of the imported crops. At stake are crops that are currently moved interstate as well as internationally from fruit fly free areas that are at risk if the fruit fly free status is lost and quarantines are established. Herein we discuss the general concept of systems approaches, provide information on the regulatory basis for international phytosanitary agreements related to trade involving invasive pests such as fruit flies, and provide examples of systems approaches in use against quarantine fruit flies.

Keywords Quarantine • Risk assessment • Systems approach • Areas of low pest prevalence • Trade

E.B. Jang (✉)
Fruit Fly Systems Applied Technologies, 46 Hoaka Road, Hilo, HI, USA
e-mail: ffsatech@gmail.com

S. Ekesi et al. (eds.), *Fruit Fly Research and Development in Africa - Towards a Sustainable Management Strategy to Improve Horticulture*,
DOI 10.1007/978-3-319-43226-7_32

1 Introduction

Fruit flies of the family Tephritidae are important pests of fruits and vegetables worldwide. The presence of exotic, potentially invasive fruit fly species, in areas where they are not established represents a significant threat to commercial horticulture and trade. Exotic fruit flies pose particular risks because (1) eggs are laid inside the fruit so immature stages often develop undetected, at least during the early stages of potential infestation, and (2) they lay numerous eggs within individual fruit, resulting in a 'clumped' distribution within consignments of both legally or illegally traded fruits, which are more difficult to detect and therefore represent a greater risk.

Where the potential risk of fruit fly introduction and spread exists, local, regional and national governments utilize a number of methods to reduce the risk associated with movement of fruit fly hosts. These range from outright prohibition of the commodity being imported from potentially infested areas, to approval based on quarantine treatments designed to eliminate the risk. Single quarantine treatments have been shown to be largely effective. More recently, regulatory officials have embraced the use of 'systems approaches' as an alternative means to allow movement of commodities in lieu of using single or combination treatments while effectively mitigating the risk posed by fruit flies to acceptable levels (Food and Agriculture Organization of the United Nations (FAO) ISPM 14 2002, Food and Agriculture Organization of the United Nations (FAO) ISPM 35 2012). This is because quarantine treatments may not exist, they may cause unacceptable damage or they are not practical.

2 Systems Approaches to Managing Risk

Movement of commodities where fruit flies are present has historically been subject to approval by the importing localities and/or, in most cases, bilateral agreements between importing and exporting entities. These agreements are normally based on pest risk assessments made to determine whether pests (potentially) present in the commodity could enter and become established in incoming commodity shipments. Sometimes these assessments have resulted in functional trade barriers that are not always based on scientific risk assessments such as the likelihood that a mating pair of insects could survive any control treatment and become established (Landolt et al. 1984). Mitigation of risk associated with new pest introductions has historically employed single quarantine treatments such as fumigation, heat and cold treatment, or irradiation, all of which were intended to alleviate/reduce risk to a low (near zero) level (Sharp and Hallman 1994; Paull and Armstrong 1994). The standard measure of efficacy has been 99.9968 % (probit-9[1]) (Baker 1939) especially for

---

[1] Probit-9 efficiency level treatment results in a mortality or sterility of 99.9968 %.

fruit flies. While this standard measurement of efficacy has largely been effective in risk mitigation, it is based on the premise that significantly large population pressures of the pest exist where the crop is grown. In practice, however, the high populations that might be associated with the 'generally infested' condition is rare in commercial production areas due to pest management procedures put into practice by growers. Additionally, some commodities can be damaged by single quarantine treatments (e.g. heat, cold, fumigation, irradiation) resulting in quality and shelf life problems. Intra- and inter-state quarantine measures such as these, which might occur as a result of a new detection of fruit flies or other pests, could even limit movement outside of the quarantine area unless suitable risk mitigation measures are undertaken and approved.

Over the last 10 years scientifically-based concepts have been internationally adopted to provide a more biologically-based framework to assess and mitigate risk. These concepts include: 'probability of a mating pair' (Landolt et al. 1984); 'maximum pest limits' (Baker et al. 1990); 'pest-free areas and areas of low prevalence' (Riherd et al. 1994; NAPPO 1994, 2002, 2003; ISPM 2005, 2007, 2008); 'host status and resistance' (Greany 1989; Liquido et al. 1995; NAPPO 2008); and 'systems approaches' (Jang and Moffitt 1994; Food and Agriculture Organization of the United Nations (FAO) ISPM 14 2002; Food and Agriculture Organization of the United Nations (FAO) ISPM 35 2012).

Recent reviews (Quinlan 2002; Follett and Neven 2006; Aluja and Mangan 2008) discuss in more detail these and other concepts related to quarantine entomology and fruit fly biology. Recently, regional standards such as the North American Plant Protection Organization (NAPPO) and International Standards on Phytosanitary Measures (ISPM) from the FAO's International Plant Protection Convention (IPPC), have been developed with the overall goal of harmonizing methods for dealing with risk associated with the threat of establishment of invasive species (e.g. ISPM 14 (2002): 'The use of integrated measures in a systems approach for pest risk management' and NAPPO (2008): 'Guidelines for the determination and designation of host status of a fruit or vegetable for fruit flies'). The concept of the 'systems approach' (Moffitt 1990; Vail et al. 1993; Jang and Moffitt 1994, Jang 1996; Jang et al. 2006, Jang and Miller 2015) was developed largely to support biologically-based risk-assessments and mitigations that could occur within a broader based 'system' of activities that cumulatively meet quarantine requirements of the importing country, or when quarantines supported by strong scientific data (or in some cases expert opinion) are imposed in country. While not new, systems approaches are now internationally recognized by member parties of the World Trade Organization (WTO) and the IPPC providing a framework for harmonizing risk assessment and mitigation, and a forum for oversight when disagreements exist (Podleckis 2007). Central to this paradigm shift has been the recognition of equivalence in treatment options to ensure that risk is mitigated through equivalent measures (Food and Agriculture Organization of the United Nations (FAO) ISPM 24 2005)

Economically important fruit fly species are almost always categorized as quarantine pests if they are not already established in the importing country/ region.

Their importance as a class of pests is due to their tendency towards high fecundity, broad host range and, for preferred hosts, their potential to cause serious damage. Fruit fly species of the genera *Anastrepha*, *Bactrocera*, *Ceratitis*, *Dacus*, *Rhagoletis*, *Toxotrypana* and others often cause phytosanitary barriers to trade.

Systems approaches, as defined in the International Standard on Phytosanitary Measures (ISPM) document number 14 (Food and Agriculture Organization of the United Nations (FAO) ISPM 14 2002) have been applied successfully for years to various systems with variable combinations of pest/host/area. It provides a flexible method for an importing country or region to achieve appropriate levels of protection (ALOP) against risk that are proportionate to the estimated risk of the pest concerned. ISPM document 35, is a systems approach for managing the risk of fruit flies (Food and Agriculture Organization of the United Nations (FAO) ISPM 35 2012) and provides information that is specifically targeted at the fruit fly situation.

Development and application of any systems approach requires an assessment of the risk followed by the development of suitable mitigation measures to achieve overall risk management acceptable to the importing country. However certain guidelines for the development of systems approaches for fruit flies have been identified and will be highlighted in this chapter

While fruit fly pests are the focus of this chapter, other quarantine pest species associated with the same commodities as fruit flies, and also deemed unacceptable by the importing NPPO, will also need to be addressed. Control for these pest species may, or may not, overlap with any planned systems approaches for fruit flies. Thus, a commodity may require treatment following one systems approach, or may be subjected to two systems approaches or more at the same time. In these cases the combination of measures normally selected for the target fruit fly species could be reduced if some level of control is already afforded by the measures being made against the other pests. A good example of this are control measures against spotted wing drosophila, *Drosophila suzukii* (Matsumura), and quarantine lepidopteran pests such as light brown apple moth, *Epiphyas postvittana* (Walker), and various other surface pests that are also regulated on imported fruit that are potential hosts of fruit flies.

3 Pest Risk Analysis

Pest Risk Analysis (PRA) is a process by which countries develop consistent means to identify the risk of quarantine pests entering the country. The PRA process is described in ISPM no. 2 (Food and Agriculture Organization of the United Nations (FAO) ISPM 2 2007). ISPM no. 11 (Food and Agriculture Organization of the United Nations (FAO) ISPM 11 2004), provides more details for PRA of quarantine pests and ISPM no. 21 (Food and Agriculture Organization of the United Nations (FAO) ISPM 21 2004) for Regulated Non-Quarantine Pests(RNQP). A key function of the PRA is the assessment of what the ALOP is to manage risk; this is the level of protection deemed appropriate by the country in regard to a quarantine risk. This term, which is closely related to the Pest Risk Management phase of PRA, is essential for selecting pest management options. Different countries may have different ALOPs depending on the nature of the pests, the scope and type of agriculture in the country, specific host-pest relationships and other abiotic and biotic factors. Setting the ALOP requires consultation and a high degree of cooperation between exporting and importing countries.

4 Options for Fruit Fly Risk Management

There are a range of options for fruit fly risk management, including 'systems approaches' used currently by many countries to prevent the introduction and establishment of quarantine pests. Many of these can be considered as stand-alone options or as independent measures within a systems approach

4.1 Host Status

Host status of fruits to fruit flies is a fundamental element in trade and thus the PRA. If the commodity is a non-host for fruit flies, this should be a stand-alone condition to allow the fruit to be moved subject to other elements of the PRA. If deemed a host, then it should be considered for pest risk management/mitigations under the PRA and subject to quarantine treatments or systems approaches. However, in the past many fruits were deemed hosts without the proper scientific evidence to prove this, thereby preventing the movement of commodities that were actually not hosts of fruit flies. Host status databases are now available online or in publication form to clarify host status of many fruit fly species of concern (Liquido et al. 2013).

4.2 Pest-Free Areas and Pest-Free Places of Production

A Pest-Free Area (PFA) is "an area in which a specific pest does not occur" (Food and Agriculture Organization of the United Nations (FAO) ISPM 5 2010; Food and Agriculture Organization of the United Nations (FAO) ISPM 26 2006; Food and Agriculture Organization of the United Nations (FAO) ISPM 29 2007). Areas initially free of fruit flies may remain so due to the presence of barriers or climate conditions, and/or the implementation of movement restrictions and related measures. The area may also be made free through area-wide eradication programmes. A related term is Pest-Free Place of Production (PFPP) (Food and Agriculture Organization of the United Nations (FAO) ISPM 10 1999) where the actual production areas can be certified as pest free. For both of these terms the concept is that the pest cannot be present where the crop is produced, although the pest may be present in the surrounding area. The limits of the area that can be considered as an PFA or an PFPP is often defined in terms of buffer areas (or distances) and natural barriers.

4.3 Postharvest Quarantine Treatments

The application of quarantine treatments to regulated articles (Food and Agriculture Organization of the United Nations (FAO) ISPM 28 2009) is a phytosanitary measure used to kill the pest and therefore to prevent the introduction and spread of regulated pests. Historically single quarantine treatments have been widely used worldwide to prevent the risk of pests being transported in a commodity (Sharp and Hallman 1994). In most instances for fruit flies, a probit-9 efficacy level (Baker 1939) has been the benchmark for phytosanitary treatments used as stand-alone measures. However, a treatment with a lower efficacy level (for example less than Probit-9) may be a component within an overall systems approach to reduce risk to the desired level. The benefits of using a less stringent alternative treatment efficacy include: a smaller sample size can be used to demonstrate the required efficacy at a 95 % confidence level and there is less physical damage to the commodity.

4.4 Systems Approaches

To establish a systems approach, the relationship between target fruit fly species, host commodity and specific site, place or area of production of the host commodity, should be defined. A systems approach may include a number of independent (at least two) and dependent measures, which may be applied to reduce risk throughout the three stages of the process: (i) pre-harvest and harvest, (ii) postharvest and shipping, and (iii) entry and distribution within the importing country.

5 Systems Approaches

5.1 The Concept of Systems Approaches

The basic concept of a systems approach to management of phytosanitary risk came from the realization of researchers and regulators that infestation of commodities by pests could be mitigated not only using single quarantine treatments aimed at near complete mortality, but by applying a series of sequential mitigation measures (systems components) each having some role in reducing the overall pest risk in an export consignment (Fig. 32.1). These ideas are in line with emerging beliefs and

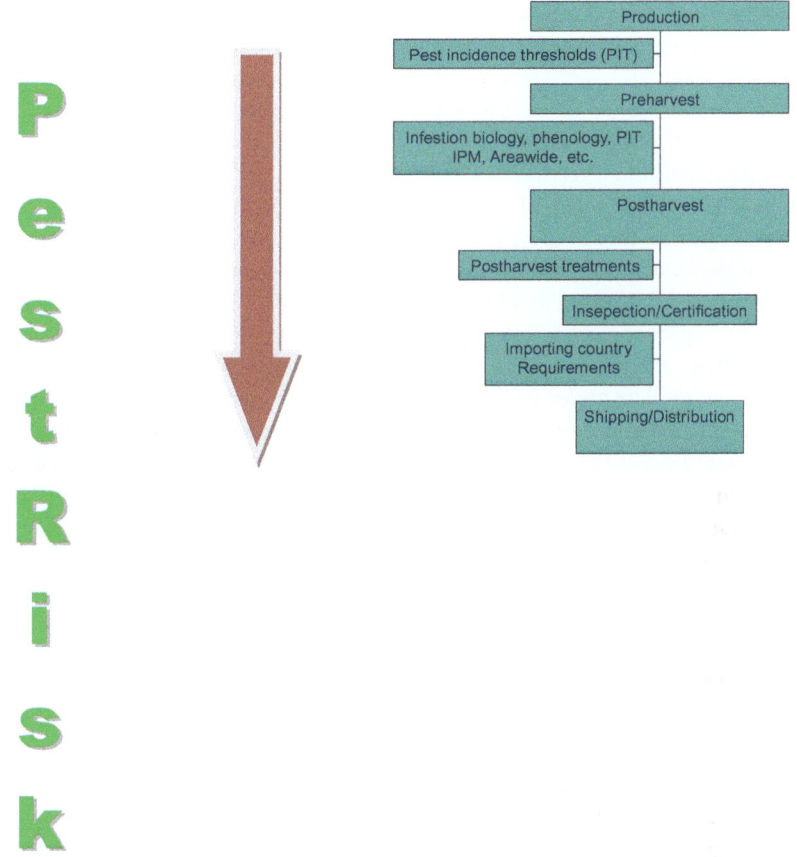

Fig. 32.1 Conceptualized drawing of a systems approach to quarantine security (From Jang and Moffitt 1994)

scientific literature that suggest that the actual levels of pests in exported commodities vary widely, depending on a number of conditions that affect the pest and the hosts in particular growing areas, and that treatments could be determined based on an estimate of the pest population in consignments from specific areas. An additional value is that less severe treatments would be likely to reduce the postharvest quality issues that were frequently reported as a result of many single quarantine treatments aimed at achieving near complete (probit-9) mortality. In such situations, integrated biological information about the pest, knowledge of the host-pest relationship, and incorporation of various phytosanitary measures into a systems approach could all be used to sequentially reduce risk in pre-harvest, postharvest, entry and distribution paths while maintaining the quality of the commodity.

Systems approaches will vary in their level of complexity according to factors such as the efficacy of the individual components used, the availability of components to integrate into the system, the intended phytosanitary outcome of the system, and the level of inherent variability and uncertainty in the system. While a systems approach may include any number of measures, a minimum of two of these must act independently of each other. Accordingly, a failure of any independent measure will not affect the operation of other independent measures, and may not necessarily constitute a complete system failure, although the total level of protection may decrease. The use of multiple measures, which in some cases may be in excess of phytosanitary requirements, will provide flexibility for subsequent modification of the systems approach design to continue to meet the desired risk management outcomes, if required at a later date.

Systems approaches can be implemented under any number of areas, host situations and fruit fly species. The following guidelines suggest however systems approaches work best under well-defined areas where hosts and fruit fly species can be clearly defined and quantified. The pest/host/area phytosanitary condition provides a starting point from which further risk reduction measures might be implemented or modified. Large-scale area-wide integrated pest management programmes (AW-IPM) against fruit flies, which could include the use of the sterile insect technique (SIT), male annihilation technique (MAT), and/or biological control amongst the elements, can significantly reduce the population levels of fruit flies in an area making it more likely that other pre or postharvest sequential (independent component) measures will successfully mitigate the remaining risk.

5.2 Parts of a Systems Approach

5.2.1 Pre-Harvest Measures

Fruit Fly Areas of low Pest Prevalence

The Area of Low Pest Prevalence (ALPP) is a key component in a systems approach and perhaps the most likely independent measure used in quantifying the effects of subsequent independent measures (Food and Agriculture Organization of the United Nations (FAO) ISPM 5 2010). The IPPC has further detailed the ALPP in ISPMs 22 (Food and Agriculture Organization of the United Nations (FAO) ISPM 22 2005), 29 (Food and Agriculture Organization of the United Nations (FAO) ISPM 29 2007) and 30 (Food and Agriculture Organization of the United Nations (FAO) ISPM 30 2008). ALPP can be natural, established intentionally (as part of an area-wide pest management programme), can be part of an eradication programme, or can even be the buffer zone of an established fruit fly free area. The key point here is that starting a systems approach with ALPP will help define the other measures that can be used to cumulatively meet quarantine security. ALPPs can be established through various mechanisms such as survey, inspection and trapping. Trapping alone should not be the basis of determination of ALPP since the effectiveness of lures used to detect flies vary widely. Activities such as surveillance and trapping in support of ALPP and systems approaches have been published for fruit flies (Food and Agriculture Organization of the United

Nations (Food and Agriculture Organization of the United Nations (FAO) IAEA 2003; Jang et al. 2014).

Host Status

Determination of host status is a fundamental step for pest risk management of fruit flies. There is evidence to indicate that some fruits, although listed in scientific literature as hosts of fruit flies are actually not hosts, poor hosts, or non-natural hosts. When in doubt about the relative attractiveness of a commodity targeted for export, internationally accepted protocols must be followed to determine host status and the factors that affect such status. There are currently two regional standards for determination of host status: (i) The NAPPO RSPM No. 30 (2008) and (ii) the Asian Pacific Plant Protection Organization (APPPC) RSPM No.4 (2005). Examples of using host status to mitigate risk include cultivation of specific cultivars that are tolerant varieties or poor hosts, harvesting at specific stages of maturity, rigorous cultural and sanitation practices and natural or artificially implemented absence of preferred hosts. Sampling methodology is also a useful concept when trying to determine the host status of fruits and vegetables that are potential host of fruit flies (Follett and Hennessey 2007).

5.2.2 Postharvest Measures

Less than Probit-9 Postharvest Treatments

The combination of a physical treatment (heat, cold, irradiation or fumigants) that achieves less than probit-9 efficacy for fruit fly control may still be useful in combination with other major measures to achieve the desired level of risk management. It is important to note that the acceptance of these major components within a systems approach is more likely if disinfestation data, which clearly show the level of efficacy of the treatment, can be provided. Examples of postharvest treatments include heat treatments (vapour, forced air and hot water), cold treatment, fumigation with alternate gases (e.g. methyl bromide, phosphine, ethyl formate), insecticidal dips, controlled atmosphere and irradiation. Any other additional treatments or procedures that are part of standard commercial practice and proven to be detrimental to fruit flies even if they may not be directed at the target fruit fly species, may also be considered.

Safeguarding and Inspection of Commodities

Safeguarding of commodities is crucial to ensure postharvest product infestation by the target fruit fly species does not occur. Examples of safeguarding include measures to ensure that fruit flies cannot infest a harvested product before it reaches the processing facilities such as screened processing facilities to exclude fruit fly entry, segregated containment facilities for products produced under the systems approach, and packaging material (e.g. screened boxes, shrink wrapping) to exclude postharvest fruit fly infestation. In some cases, in cooperation with the exporting country, the NPPO of the importing country may agree to implement one or more measures on arrival of the consignment as part of the systems approach.

Inspection of a commodity for the specific fruit fly species of concern may be conducted postharvest and prior to export and/or at the point of entry (Food and Agriculture Organization of the United Nations (FAO) ISPM 23 2005). This component can be a risk reduction measure and/or a verification tool. Sampling rates may be increased or decreased depending on the level of security required by this component. Inspection for larvae generally requires fruit cutting but external host damage or other external visual signs may also be used. Sample rates and methodology should be determined by the level of assurance required. Examples of inspection include sampling during or immediately following harvest, in line fruit sampling prior to packaging (this may occur at various stages during the processing phase), sampling following processing and packaging of a commodity but prior to export and pre-clearance inspection by the importing country authorities prior to export.

Seasonal Entry and Limited Distribution

Entry may be limited on a seasonal basis to periods when it is considered that the likelihood of establishment and spread of the target fruit fly species is extremely low, such as harsh winter periods (should a consignment be infested). Points of entry or product distribution may be limited on a geographical basis to areas where it is considered that the likelihood of establishment and spread of the target fruit fly species is extremely low (should a consignment be infested).

5.3 Work Plan

Once a systems approach is designed and agreed between the exporting and importing country, there is a bilateral process for developing a protocol for the implementation and agreement on verification of efficacy and corrective actions if the system fails. This is normally documented as a Work Plan (WP) or Protocol for Export. This WP, based on bilateral agreement and on agreements with other stakeholders, provides a high degree of phytosanitary security when there is good understanding of the plan, its objectives, proper performance of each component, and of the expectations of the importing entity. For imports to Europe, WPs are less widely used but rather intervention occurs only when it appears that the requirements are not being met.

5.4 Assessment of Efficacy of a Systems Approach

Systems approaches may be developed or evaluated in either a quantitative or qualitative manner, or a combination of both. A quantitative approach may be more appropriate where suitable data are available, such as that usually associated with measuring the efficacy of treatments. A qualitative approach should be considered more appropriate where efficacy is estimated by expert judgement (Food and Agriculture Organization of the United Nations (FAO) ISPM 14 2002). The efficacy of independent measures that may be used to reduce pest risk can be expressed in different ways (e.g. mortality, infestation rate). The overall efficacy of a systems approach is based on the cumulative efficacy of required independent measures (at least two). Wherever possible this should be expressed in quantitative terms with a confidence interval. For example, efficacy for a particular situation may be determined to be no more than five infested fruit from a total population of 1 million fruit with 95 % confidence. Where such calculations are not possible or are not done, the efficacy may be expressed in qualitative terms such as high, medium, and low (Food and Agriculture Organization of the United Nations (FAO) ISPM 14 2002). An important consideration is the fact that estimating efficacy can be applied selectively either to parts of the system (major independent measures) or to the entire system. Clearly, estimations of efficacy are needed in the design phase of any systems approach, but it can also be used in situations where it is necessary to adjust the approach to the established or agreed ALOP or when additional data are available. In most instances efficacy should be measured on the independent measure/major component since the role of dependent measures/elements is primarily to support the function of the major component and, in addition, it is often more difficult to estimate their efficacy. An advantage of systems approaches is the flexibility to add or remove components according to ongoing performance in comparison with predicted efficacy, or an increase in confidence level after a certain volume of trade.

5.5 Methods of Quantifying Systems Approaches

There is no standard methodology to quantify or qualify efficacy of a systems approach. Some of the available methods measure the efficacy of the whole system while others measure major components.

Postharvest treatment (PHT) confirmatory tests for less than Probit-9 mortality (i.e. 95 %), follows similar methods to the Probit-9 (99.9968 %) methods. In addition, sampling could provide efficacy calculations of the total system or a single major component (for example, the cutting of fruit on the tree can estimate infestation levels before the product is subjected to PHT). If fruit sampling is at the orchard level then efficacy of the ALPP can be determined. Mathematical models (that calculate probability e.g. of mating pairs), can also be useful in quantifying efficacy but one cannot assume that mathematical models are always completely accurate.

Models frequently require large data sets and, often, these are not available. Other examples of mathematical modeling include: maximum pest limits, sequential mortality, systems modeling, Bayesian belief modeling. A secondary assessment of risk maybe applied to the unrestricted risk when the risk mitigation value of the systems approach has been calculated. Pest risk assessment can be completed where the risk is estimated for the systems approach. Point estimates, range estimates or probabilistic estimates can be used. At Risk™ or Oracle™ are two software packages that can be used for probabilistic estimates.

5.6 Examples of Systems Approaches

There are many examples of systems approaches being applied to the movement of commodities between countries and within countries. The numbers of systems approaches and the measures that are employed are numerous and include many fruit fly species.

1. Tomatoes from Guatemala to the U.S.: Green tomatoes are allowed into the US under a systems approach against the Mediterranean fruit fly (medfly), *Ceratitis capitata* (Wiedemann). The major independent measures include poor host status, growing of tomatoes in pest free growing structures or ALPPs in the buffer areas surrounding the pest free structure.
2. Citrus from Florida to Japan: Citrus is allowed into Japan under a systems approach against the Caribbean fruit fly, *Anastrepha suspensa* (Loew). Independent measures include time of production (when *A. suspensa* populations are low), ALPP, poor host status and fruit cutting.
3. Sweet melons and watermelons from Equador to the US. Melons are allowed into the U.S. under a systems approach against the South American cucurbit fruit fly, *Anastrepha grandis* (Macquart). Independent measures include pest free place of production and limited distribution.
4. Papayas from Guatemala and Brazil to the U.S.; Papayas are allowed from Guatemala and Brazil into the US under a systems approach against *C.capitata*. Independent measures include ALPP, poor host status and in some cases less than probit-9 treatment.

6 Conclusions

Systems approaches have become more popular as trade in agricultural commodities increases internationally and countries have to consider the potential risk of pest introduction alongside the need for agricultural products and the economic impact that such trade may provide. The flexible nature of systems approaches in the context of international phytosanitary agreements suggest that the basis of systems approaches have been well conceived and tested. Still there are significant questions that remain, principally concerning how to develop systems approaches for unusual commodities, circumstances and new pest species. These questions will need to be answered based on the specific needs of trading partners and valid PRAs.

Acknowledgements The author wishes to thank all of the colleagues who have contributed ideas in support of advancing systems approaches.

References

Aluja M, Mangan RL (2008) Fruit fly (Diptera: Tephritidae) host status determination: critical conceptual and methodological considerations. Annu Rev Entomol 53:473–502

APPPC RSPM (2005) Guidelines for the confirmation of non-host status of fruit and vegetables to Tephritid fruit flies. APPPC, Bangkok, No. 4 12 pp

Baker AC (1939) The basis for treatment of products where fruit flies are involved as a condition for entry into the United States, U.S. Department Agriculture Circular 551. U.S. Department Agriculture Circular, Washington, DC

Baker RT, Cowley JM, Harte DS, Frampton ER (1990) Development of a maximum pest limit for fruit flies (Diptera: Tephritidae) in produce imported into New Zealand. J Econ Entomol 83:13–17

Food and Agriculture Organization of the United Nations (FAO) IAEA (2003) Trapping guidelines for area-wide fruit fly programmes. Joint FAO/IAEA Division, Vienna, 47 pp

Food and Agriculture Organization of the United Nations (FAO) ISPM 10 (1999) Requirements for the establishment of pest free places of production and pest free production sites. IPPC, FAO, Rome

Food and Agriculture Organization of the United Nations (FAO) ISPM 11 (2004) Pest risk analysis for quarantine pests including analysis of environmental risks and living modified organisms. IPPC, FAO, Rome

Food and Agriculture Organization of the United Nations (FAO) ISPM 14 (2002) The use of integrated measures in a systems approach for pest risk management. IPPC, FAO, Rome

Food and Agriculture Organization of the United Nations (FAO) ISPM 2 (2007) Framework for pest risk analysis. IPPC, FAO, Rome

Food and Agriculture Organization of the United Nations (FAO) ISPM 21 (2004) Pest risk analysis for regulated non-quarantine pests. IPPC, FAO, Rome

Food and Agriculture Organization of the United Nations (FAO) ISPM 22 (2005) Requirements for the establishment of areas of low pest prevalence. IPPC, FAO, Rome

Food and Agriculture Organization of the United Nations (FAO) ISPM 23 (2005) Guidelines for inspection. IPPC, FAO, Rome

Food and Agriculture Organization of the United Nations (FAO) ISPM 24 (2005) Guidelines for the determination and recognition of equivalence of phytosanitary measures. IPPC, FAO, Rome

Food and Agriculture Organization of the United Nations (FAO) ISPM 26 (2006) Establishment of pest free areas for fruit flies (Tephritidae). IPPC, FAO, Rome

Food and Agriculture Organization of the United Nations (FAO) ISPM 28 (2009) Phytosanitary treatments for regulated pests. IPPC, FAO, Rome

Food and Agriculture Organization of the United Nations (FAO) ISPM 29 (2007) Recognition of pest free areas and areas of low pest prevalence. IPPC, FAO, Rome

Food and Agriculture Organization of the United Nations (FAO) ISPM 30 (2008) Establishment of areas of low pest prevalence for fruit flies (Tephritidae). IPPC, FAO, Rome

Food and Agriculture Organization of the United Nations (FAO) ISPM 35 (2012) Systems approach for pest risk management of fruit flies (Tephritidae). IPPC, FAO, Rome, http://www.fao.org/docrep/ 016/k6768e/k6768e.pdf

Food and Agriculture Organization of the United Nations (FAO) ISPM 5 (2010) Glossary of phytosanitary terms. IPPC, FAO, Rome

Follett PA, Hennessey MK (2007) Confidence limits and sample size for determining non-host status of fruits and vegetables to tephritid (Diptera: Tephritidae) fruit flies as a quarantine measure. J Econ Entomol 100:251–257

Follett PA, Neven LG (2006) Current trends in quarantine entomology. Annu Rev Entomol 51:359–385

Greany PD (1989) Host plant resistance to tephritids: an under-exploited control strategy. In: Robinson AS, Hooper GS (eds) Fruit flies and their biology, natural enemies and control, World crop pests 3A. Elsevier, Amsterdam, pp 353–362

Jang EB (1996) Systems approach to quarantine security: postharvest application of sequential mortality in the Hawaiian grown "Sharwil" avocado system. J Econ Entomol 89:950–956

Jang EB, Miller CE (2015) Systems approaches for managing the risk of citrus fruit in Texas during a Mexican fruit fly outbreak. Technical Report to USDA-APHIS-PPQ 2015. 64 pp

Jang EB, Moffit HR (1994) Systems approaches to achieving quarantine security. In: Sharp JL, Hallman HJ (eds) Quarantine treatments for pests of food plants. Westview Press, Boulder, pp 224–237

Jang EB, Mau RFL, Vargas RI, McInnis DO (2006) Exporting fruit from low fruit fly prevalence zones with multiple mitigation systems approach. In: Proceedings of the international symposium on area-wide management of insect pests. October 1–5 2006, Okinawa. pp 192. Food and Fertilizer Technology Center for the Asian and Pacific Region, pp 63–69

Jang EB, Enkerlin W, Reyes J, Miller CE (2014) Trapping related to phytosanitary status and trade. In: Shelly TE, Jang EB, Espky N, Vargas RI, Reyes J (eds) Trapping tephritid fruit flies: lures, area-wide programs and trade implications. Springer, Berlin

Landolt PJ, Chambers DL, Chew V (1984) Alternative to the use of probit 9 mortality as a criterion for quarantine treatments of fruit fly (Diptera: Tephritidae) infested fruit. J Econ Entomol 77:285–287

Liquido NJ, Griffin RI, Vick KW (1995) Quarantine security for commodities: current approaches and potential strategies. U.S. Department of Agricultural Publication Service, Washington, DC, 1996–04

Liquido NJ, McQuate GT, Suiter KA (2013) MEDHOST: an encyclopedic bibliography of the host plants of the mediterranean fruit fly, *Ceratitis capitata* (Wiedemann), Version 1.1. USDA-APHIS-PPQ-CPHST, Raleigh. https://www.gpdd.info/MedHost/

Moffitt HR (1990) A systems approach to meeting quarantine requirements for insect pests of deciduous fruits. Proc Washington State Hort Assoc 85:223–225

North American Plant Protection Organization (NAPPO) (1994) NAPPO regional standard for pest free areas. April 21, 1994. Ottawa. http://www.nappo.org/en/data/files/download/ArchivedStandards/RSPM1-e.pdf

North American Plant Protection Organization (NAPPO) (2002) Regional standards for phytosanitary measures (RSPM). RSPM No. 17. Guidelines for the establishment, maintenance and verification of fruit fly free area in North America. Ottawa. http://www.nappo.org/en/data/files/download/ArchivedStandards/RSPM1-e.pdf

North American Plant Protection Organization (NAPPO) (2003) Regional standards for phytosanitary measures (RSPM). RSPM No. 20. Guidelines for the establishment, maintenance and verification of areas of low pest prevalence for insects. Ottawa. http://www.nappo.org/en/data/files/download/ArchivedStandards/RSPM1-e.pdf

North American Plant Protection Organization (NAPPO) (2008) Guidelines for the determination and designation of host status of a fruit or vegetable for fruit flies (Diptera: Tephritidae) RSPM

No. 30. Ottawa. http://www.nappo.org/en/data/files/download/ArchivedStandards/RSPM1-e.pdf

Paull RE, Armstrong JW (eds) (1994) Insect pests of fresh horticultural products: treatments and responses. CAB International, Wallingford, p 360

Podleckis EV (2007) Systems approach as phytosanitary measures: techniques and case studies. In: Vreysen MJB, Robinson AS, Hendrichs J (eds) Area-wide control of insect pests from research to field implementation. Springer, Dordrecht, pp 417–423

Quinlan MM (2002) Trends in international phytosanitary standards: potential impact on fruit fly control. In: Proceedings of 6th international fruit fly symposium, 6–10 May 2002. Stellenbosch, pp 195–200

Riherd C, Nguyen R, Brazzel J (1994) Pest free areas. In: Sharp JL, Hallman GJ (eds) Quarantine treatments for pests of food plants. Westview Press, Boulder, pp 213–224

Sharp JL, Hallman GJ (eds) (1994) Quarantine treatments for pests of food plants. Westview Press, Boulder, p 290

Vail PV, Tebbetts JS, Mackey BE, Curtis CE (1993) Quarantine treatments: a biological approach to decision making for selected host of codling moth (Lepidoptera: Tortricidae). J Econ Entomol 86:70–75

Eric B. Jang is the principal at Fruit Fly Systems Applied Technologies, an independent consulting firm located in Hawaii. He has over 40 years of research experience primarily in the areas of insect physiology/biochemistry, postharvest entomology, insect chemical ecology and area-wide management. He spent 32 years with USDA-ARS, Pacific Basin Agricultural Research Center in Hawaii where he was Research Leader (1999–2016) of the Tropical Crop and Commodity Protection Research Unit. With ARS Dr Jang has worked on postharvest disinfestation and chemical ecology of tephritid fruit flies and other pests. He has authored or co-authored over 180 scientific papers, conference proceedings and book chapters and been invited to speak at over 60 regional, national and international meetings and conferences on his research. He has hosted three international workshops on insect chemoreception and the Fifth Asia-Pacific Conference on Chemical Ecology. Dr. Jang's expertise has been recognized by his appointment to chair numerous international, federal and state scientific advisory panels, and requests for consultations worldwide in the areas of tephritid fruit flies and other invasive pests. He is frequently asked by the Joint FAO/IAEA Pest Control Section to participate and provide advice on programme-related activities. Dr Jang is a past President of the Asia-Pacific Association of Chemical Ecologists.

Part V
Socioeconomic Impact Assessment

Chapter 33
Economic Impact of Integrated Pest Management Strategies for the Suppression of Mango-Infesting Fruit Fly Species in Africa

Beatrice W. Muriithi, Gracious M. Diiro, Hippolyte Affognon, and Sunday Ekesi

Abstract Mango growers in Africa are faced with the severe challenge of controlling pests, with the oriental fruit fly, *Bactrocera dorsalis* being the most destructive. Conventional fruit fly control in this region mainly involves the use of synthetic pesticides. Pesticides are expensive and have undesirable effects on human health, biodiversity and the environment. Integrated Pest Management (IPM) methods have been developed as a more effective and affordable strategy to control fruit flies. The technology has been developed by the International Centre of Insect Physiology and Ecology (*icipe*) and its partners, and disseminated to farmers through various technology uptake pathways across Africa. Understanding the adoption behavior of farmers with regard to IPM and the economic impacts of IPM on the livelihood of mango farmer and their households can help to design policies aimed at increasing the development and uptake of such strategies in Africa. This paper is a synthesis of findings on adoption and the economic benefits derived from IPM approaches targeted at mango-infesting fruit flies; it is based on recent studies conducted by *icipe* at various project action sites in Kenya. The studies provide empirical data on the adoption and impact of IPM strategies and focus on the magnitude of direct gains for mango producers. The findings provide clear evidence that farmers do not adopt the technology as a package but rather selected components of the IPM package that are affordable and easy to apply and maintain. With regard to economic benefits, the studies show that IPM reduces mango losses caused by fruit flies and the costs associated with pesticide use. There are also income gains from adopting IPM amongst mango growers. The results of these studies demonstrate that investment in IPM is viable, and therefore justifies support for upscaling of IPM programmes.

B.W. Muriithi (✉) • G.M. Diiro • S. Ekesi
Plant Health Theme, International Centre of Insect Physiology & Ecology (*icipe*),
PO Box 30772-00100, Nairobi, Kenya
e-mail: bmuriithi@icipe.org

H. Affognon
International Crops Research Institute for the Semi-Arid Tropics (ICRISAT),
BP 320, Bamako, Mali

Keywords Impact assessment • Integrated Pest Management (IPM) • Mango farming • icipe • Kenya

1 Introduction

1.1 Background

Horticultural enterprises such as fruit and vegetable production play an important role in reducing poverty and improving food security in many sub-Saharan African (SSA) economies, including Kenya. These enterprises: create employment; generate income; ensure food and nutritional security in households; contribute foreign exchange earnings to the economy (Ekesi et al. 2011). In Kenya for instance, in 2011, mango exports to the regional market accounted for 32 % and 8 % of the volume and value of total fresh fruits, respectively (USAID 2011). Although the mango subsector in Kenya has recorded growth in general over the past two decades, the trend in yields exhibits substantial fluctuations over time (Fig. 33.1).

Mango yields were relatively stable between 1990 and 1996, ranging between 7.6 million metric tons (MT) per hectare and 8.5 MT per hectare, but declined in 1997 (to less than 6 MT per hectare) before trending upwards to a peak of about 17 MT per hectare in 2008. Yields then declined to about 12 MT per hectare in 2013.[1] The current average annual yield of mangoes is also far below the estimated potential of 25 MT per hectare (Griesbach 2003; Korir et al. 2015). The large yield gap and annual fluctuations in mango yields can be attributed to several factors that hamper the performance of the enterprise in Kenya. Amongst the most important are high post-harvest losses, insect pests and diseases, lack of high quality planting materials, and the high cost of production (Sebstad and Snodgrass 2004). Insect pests are mainly tephritid fruit fly species (e.g. the oriental fruit fly, *Bactrocera dorsalis* (Hendel); the mango fruit fly, *Ceratitis cosyra* (Walker); *Ceratitis fasciventris* (Bezzi); the Natal fruit fly, *Ceratitis rosa* (Karsch); *Ceratitis anonae* (Graham); and the Mediterranean fruit fly (medfly), *Ceratitis capitata* (Wiedemann)). Fruit flies cause annual losses in fruit and vegetable production worth US$ 2.0 billion in Africa (Ekesi et al. 2016). The larval stage of the pest can cause about 30–100 % damage through feeding on pulp of the fruit (Ekesi et al. 2011). The low quality product causes further indirect losses through quarantine restrictions on trade and the loss of export opportunities especially to lucrative markets abroad (Bech 2008; Ekesi et al. 2016). For instance, export of crops that are hosts of *B. dorsalis* (including mango) from many African countries are already banned in Seychelles, Mauritius, South

---

[1] Nevertheless, Kenya has one of the highest yields in the world. For instance, Kenya's average mango yields stood at 9.2 MT per hectare relative to the global average of 8.2 MT per hectare in 2012 and is about 39 % higher than the yields reported in India, the leading producer of mangoes in the world (FAOSTAT 2016).

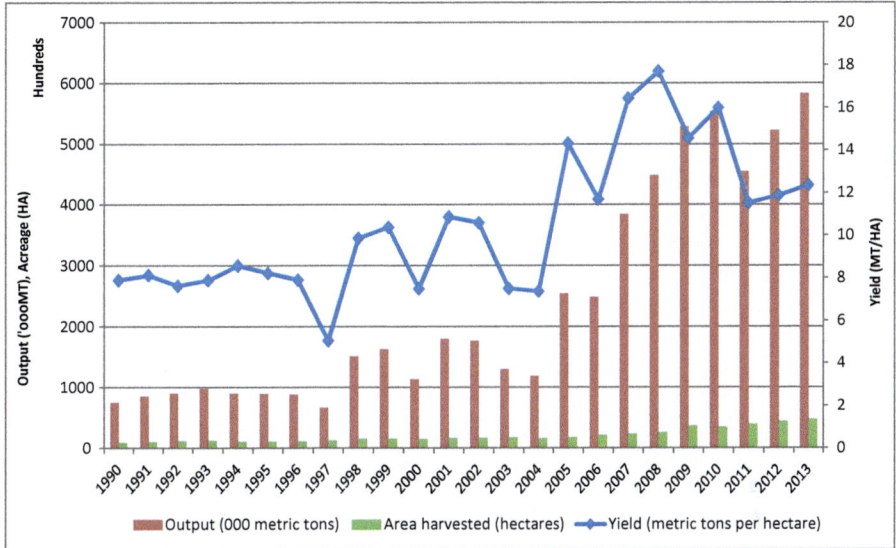

Fig. 33.1 Trends in mango area harvested, production, and yields in Kenya (1990–2013) (*Source*: FAOSTAT (2016))

Africa and several European countries (Ravry 2008). Between 2003 and 2007, Kenya lost revenue to the value of $1.9 million due to an avocado export ban to South Africa following the *B. dorsalis* invasion (Ekesi et al. 2014).

Mango farmers in developing countries such as Kenya, particularly large-scale producers, have relied on synthetic chemical pesticides to manage and control fruit flies for a long time. For instance, use of pesticides in Kenya has increased over time particularly in the horticultural sub-sector (Macharia et al. 2008; Asfaw et al. 2010). Application of cover sprays in the counties of Embu and Mbeere during flowering and fruit development are made on 6–8 occasions compared with the recommend rate of 3–4 occasions (Krain et al. 2008). However, chemical pesticides are unaffordable to resource-poor mango farmers and their increased use by large-scale producers results in adverse effects on human health, the environment and biodiversity (Macharia et al. 2008; Asfaw et al. 2010). Increased use of pesticides in mango production also reduces the competitiveness of these mangoes in the international market due to pesticide residues, which leads to a loss in foreign income (Lux et al. 2003; Macharia et al. 2008; Kouser and Qaim 2013).

To address these challenges researchers in the horticulture sector have developed Integrated Pest Management (IPM) strategies as a more sustainable alternatives to minimize the extensive use of chemical pesticides (Norton et al. 1999). IPM involves the use of pest control approaches that ensure favourable economic, ecological and sociological consequences (Blake et al. 2007; Muriithi et al. 2016). IPM has also widely been recommended for management of fruit flies in mango production in Africa; the strategy is expected to stimulate productivity of mangoes and other host crops, and enable smallholder farmers to produce better quality products for the export market (Ekesi and Billah 2007; Ekesi et al. 2011, 2016).

The empirical literature shows that the components of IPM can be effective in reducing insect damage in horticultural enterprises. For example, Vayssières et al. (2009) assessed the effectiveness of GF-120 food bait sprays (the bait application technique [BAT]) against *C. cosyra* in Benin, and found that infestation levels were reduced by about 85%. Similarly, Ndiaye et al. (2008) found that, in Senegal, an IPM package consisting of a combination of the male annihilation technique (MAT), BAT and orchard sanitation reduced fruit fly infestation of mango by 83%. An *ex-ante* analysis conducted by Preciados et al. (2013) in the Southern Philippines showed that the use of IPM strategies can reduce crop damage by 20% per hectare, increase yield by 33%, and reduce pesticide expenditure and total cost of production by 75% and 16%, respectively. The author further reported that IPM technology can cumulatively increase the gross margin by about 156% per hectare. Verghese et al. (2006) in their study in India on pre- and postharvest IPM for fruit flies on mango, showed that a pre-harvest IPM combination of male MAT and orchard sanitation, brought down *B. dorsalis* infestation to 5% from an infestation ranging from 17 to 66% in control plots during 2004 and 2005. Positive effects of IPM have also been documented in other crop enterprises such coffee in Uganda (Isoto et al. 2008), cotton production in Asian countries (Erickson 2004; Ooi et al. 2005), onion production in the Philippines (Cuyno et al. 2001; Yorobe et al. 2011; Sanglestsawaia et al. 2015). Isoto et al. (2008) for instance found that growers adopting IPM in coffee earned 118% higher revenues compared with non-adopters.

1.2 Integrated Pest Management of Mango-Infesting Fruit Fly Species in Africa

In Africa, the International Centre of Insect Physiology and Ecology (*icipe*) and partners has, under the African fruit fly programme (AFFP), developed and disseminated an IPM package for controlling fruit flies in many sub-Saharan countries including Kenya. The goal of AFFP, which began in 1999, is to achieve broader socio-economic and environmental benefits to smallholder farmers, and increase the competitiveness of fruit and vegetable production from SSA. The programme specifically aims at reducing mango losses due to fruit fly infestation, lower the cost of production, improve the income of mango growers, and improve market access and processing by increasing mango productivity and quality to meet the needs of both the domestic urban and export markets (Muriithi et al. 2016). The IPM strategy developed by *icipe* consists of five components; (1) spot application of BAT, (2) MAT, (3) the use of biopesticides, (4) releases of the parasitoids *Fopius arisanus* (Sonan) and *Diachasmimorpha longicaudata* (Ashmead) (both Hymenoptera: Braconidae) and (5) use of orchard sanitation (Mohamed et al. 2008, 2010; Ekesi 2015; Kibira et al. 2015; Korir et al. 2015; Muriithi et al. 2016; Ekesi et al. 2016).

The food bait used in BAT is a proteinous attractant that is combined with an insecticide (spinosad) and applied as localized spots within the mango canopy. Fruit flies are attracted to these localised areas, ingest the bait along with the toxicant, and are killed before they infest the fruits (Ekesi et al. 2014, 2016; Ekesi 2015). MAT consists of high-density trapping stations each of which contains a male lure (methyl eugenol), combined with a toxicant (malathion) to trap and kill male flies; populations of males fall to such low levels that mating is limited sufficiently that the population is greatly reduced (Ekesi and Billah 2007; Hanna et al. 2008). The biopesticides are fungus-based formulations that target pupating larval stages of the fruit flies and emerging adults (Ekesi et al. 2005). The two parasitoid species use target the eggs (*F. arisanus*) and larvae (*D. longicaudata*) of *B. dorsalis* (Mohamed et al. 2008, 2010). Orchard sanitation involves the use of augmentoria, which are tent-like structures made of double netting material with wire mesh. Fallen rotten fruits are collected from the field and placed inside the augmentorium; any parasitoids emerging from fruit flies within the rotten fruit are able to escape through the mesh of the augmentorium but emerging fruit flies are too large to escape through the mesh and remain trapped within the augmentorium (Klungness et al. 2005). Orchard sanitation can also be achieved by burying rotten/ infested fruits (Korir et al. 2015). In the same way as the augmentorium, this labour-intensive activity involves collecting fallen rotten fruits and burying them deep under the soil surface (at least >50 cm) with sufficient time to kill the developing larvae (Billah et al. 2015). The success of the IPM strategy is clear from its rapid uptake and the willingness of the mango growers to pay for ongoing strategy dissemination and promotional activities (Muchiri 2012). Between 2009 and 2015 the strategy had reached over 26,000 direct beneficiaries in several SSA countries. Understanding the adoption behavior of farmers with regard to IPM and the economic impacts of IPM on the livelihoods of mango growers is important for policy design that promotes the development and uptake of such strategies in Africa. Here we synthesize findings from recent studies on adoption behaviour and the impact of IPM strategies on fruit flies control in Africa. We include studies on IPM technologies from other developing countries to identify research gaps and draw lessons that can help promote and upscale IPM technologies for fruit flies control globally.

2 Theoretical Foundation

Technological improvements in the agriculture sector form the most sustainable pathway for reducing rural poverty, increasing productivity and food security and spurring general economic growth in many agrarian economies such as those in SSA. Several studies (e.g. De Janvry and Sadoulet 2001; Moyo et al. 2007; Minten and Barrett 2008; Kassie et al. 2011) demonstrate that adoption of improved farm technologies can have both direct and indirect positive effects on adopting households. The direct effects include productivity gains and reduced unit costs of production, translating into increased food security and farm revenue. The indirect gains may include increased incomes of non-adopters and reduced prices of food staples due to outward shifts in supply; increased productivity from adoption of improved technologies may also stimulate the demand for labour translating into increased rural wage earnings and incomes, especially for landless labourers (De Janvry and Sadoulet 2001). However, adoption rates of many of the improved technologies remain low; some farmers do not adopt, and adoption is never immediate due to several factors. Here we consider the theoretical framework for understanding the adoption behaviour of farmer households.

The framework draws from the classical theory of diffusion of innovations popularized by Rogers (1995), which suggests that farmer decisions on whether to adopt a new innovation manifests itself through a four-stage process of knowledge acquisition, persuasion, decision and confirmation. These processes are influenced by the information received by the potential adopter, their socio-economic characteristics, the social system, and the attributes of the innovations. Since farmer households in the developing world operate under market imperfections and uncertainties, their decisions on whether to adopt a new technology depend on their expectations (Thurstone 1927; Koundouri et al. 2006) and can be described by the following formulae where y denotes the farm output from farm i. Equation (33.1) specifies the output from the farm, as a function of improved inputs (τ_k) and other inputs denoted by z.

$$y = y(\tau_k, z; \varepsilon) \tag{33.1}$$

where the stochastic term, ε, represents the farm production risk, and has a distribution $F(\varepsilon)$ that is exogenous to the farm household's decisions. The production risk factor captures the unobserved household heterogeneity such as unreported farm management ability, land fertility, risk preferences and risk management measures, and rate of discount which could affect input use and farm productivity. Given the input and output prices, the main objective of a risk averse farmer household is to maximize expected utility of the present value of profit from the farm (as specified in Eq. (33.2):

$$\max_{\tau, z \geq 0} EU(\pi) = \max_{\tau, z \geq 0} \int_0^y E\left[py(\tau_k, z; \varepsilon) - r_k \tau_k - \varpi_z z\right] dF(\varepsilon) \tag{33.2}$$

Thus, a utility-maximizing farmer household would choose to adopt technology τ_k if it generates the greatest expected utility than the alternative. That is if:

$$V = EU(\pi^{\tau_1}) - EU(\pi^{\tau_0}) > 0 \tag{33.3}$$

where V is the net benefit derived from adoption on the fixed cost of investment, the level of uncertainty related to the use of the new technology. The factors in empirical literature that influence the net benefits include the role of social networks in the diffusion of new technologies, access to input and output markets, financial and human capital constraints (learning from extension services and experimentation) (e.g. Conley and Udry 2010; Duflo et al. 2011; Dercon and Christiaensen 2011; Diiro et al. 2015; Diiro and Sam 2015).

3 Empirical Evidence for Adoption and Impact of IPM in Suppression of Mango Infesting Fruit Fly Species in Africa

3.1 Adoption of Integrated Pest Management

Here we present a review of the empirical evidence for adoption of IPM amongst smallholder farmers in Kenya. This is based on journal articles, book chapters and working papers published on IPM adoption over the past two decades. There are quite a few papers that touch on adoption of fruit fly IPM in Kenya, and Africa in general, such as Korir et al. (2015) in Kenya and Ndiaye et al. (2008) in Senegal. Other articles on adoption of IPM in other crop enterprises in Africa include IPM in cotton in Zimbabwe (Maumbe and Swinton 2000), and coffee in Uganda (Isoto et al. 2008). There is also extensive literature on IPM adoption from other continents including adoption of IPM strategies for control of aubergine fruit and shoot borer in West Bengal, India (Baral et al. 2006), IPM in rice in Bangladesh (Dasgupta et al. 2007), IPM in vegetable production in Nicaragua (Garming and Waibel 2009) and New England (Li et al. 2013), cotton production in Haryana and Punjab, India (Singh et al. 2008), and coffee production in Hawaii (Vargas et al. 2001).

Although an IPM package for fruit fly control consists of a combination of interventions that complement each other rather than each working as a stand-alone pest management tool (Ekesi and Billah 2007), farmers appear to adopt particular components but not the whole IPM package (Korir et al. 2015). Korir et al. (2015) assessed adoption by mango farmers of the fruit fly IPM package developed and disseminated by *icipe* (Sect. 1.2). The survey included 805 mango farmers in Embu sub-county, of which 471 were adopters of at least one IPM component and the rest were non-adopters. None of the respondents had adopted more than three IPM components. The findings from the study show that orchard sanitation by burying fallen fruits was the most popular component of the IPM package with adoption rates of 47%, followed by the use of MAT (used by 19%). Adoption was however low for orchard sanitation with the augmentorium (0.4%), biopesticides (0.4%) and BAT (4%). The authors, respectively, attributed the high adoption of orchard sanitation and MAT to low labour cost and effectiveness of the technology in trapping and killing fruit flies compared with the other management components. Low adoption of the augmentorium was attributed to difficulty in construction and assemblage of the tent. The authors also noted a lack of access to bait sprays as the main challenge to low adoption of BAT. The low use of the fungal biopesticide component was also likely to be because it was not readily available in the market. Adoption of component-specific strategies rather than the entire IPM package have been observed in other crop enterprises, for instance Erbaugh et al. (2010) who studied adoption of IPM strategies amongst cowpea growers in Uganda.

Using a negative binomial model, Korir et al. (2015) evaluated the factors affecting the intensity of adoption of fruit fly IPM, using the number of IPM components adopted by a household as the dependent variable. Human capital and capacity building through IPM training were found to have a positive influence on the intensity of adoption of the fruit fly IPM strategy. The positive relationship between education of the household head and intensity of adoption is plausible since IPM is a knowledge-intensive technology, thus growers that are more educated are likely to adopt additional IPM components. Increasing farmers' knowledge through training may also increase the intensity of adoption as participation in IPM demonstrations had a positive influence on the number of IPM components adopted. Successful field trials of a technology before scaling it out to the farmers is expected to create trust and confidence in that technology and hence improve its uptake once it is rolled out to the recipients. Korir et al. (2015) further found that distance to the nearest mango market was positively associated with number of IPM components adopted. Households situated further away from the input market may have to incur higher transaction costs of acquiring inputs such as pesticides, and thus seek alternative control measures that do not accrue these costs, such as orchard sanitation through burying and burning of fallen fruits. In addition, households with larger number of mango trees were likely to adopt more IPM components. This could be related to the commercial orientation of these farmers, who were striving to supply high quality produce by ensuring they had less pesticide residues while also seeking more cost-effective control measures, such as IPM.

3.2 Economic Benefits of Integrated Pest Management

A large volume of literature exists on the economic impact of integrated pest management in Africa. Most of the studies, however, have generally focused on *ex-ante* impact assessment mainly using economic surplus models, willingness to pay and cost benefit analyses (e.g. Macharia et al. 2005; Mulwa et al. 2013; Muchiri 2012; Ainembabazi et al. 2015). Rigorous empirical literature on *ex-post* impact of IPM technologies, particularly on fruit fly control in Africa, is still scarce. A few studies have used experimental data from farmers' fields to assess the impact of IPM on control of mango-infesting fruit fly species. The most recent ones include Kibira et al. (2015) and Muriithi et al. (2016), both of which were conducted at *icipe*'s pilot sites for evaluating IPM of mango-infesting fruit fly species, in Embu and Meru Counties in Kenya, respectively. The surveys included the farmers who had received the IPM practices developed and disseminated by *icipe* and compared them with a control group that used conventional fruit fly suppression measures (i.e. cover spray applications of chemical pesticides). Both studies applied the difference-in-difference technique to estimate the impact of IPM technique on farm-level mango fruit yield losses, farmer pesticide expenditure, and profit generated from mango production. The studies however were designed differently: Kibira et al. (2015) surveyed 267 mango farmer households and evaluated the impact of the IPM practices mentioned in Sect. 1.2 as a single package.[2] Muriithi et al. (2016) surveyed 1200 mango farmer households and assessed the impact of using different combinations of the IPM practices on similar outcome indicators. Eight different combinations were evaluated to determine which combination of IPM practices generate the greatest benefits. The treatments included: (1) parasitoid releases, orchard sanitation and MAT, (2) parasitoid releases, orchard sanitation and BAT, (3) parasitoid releases, orchard sanitation and biopesticide, (4) parasitoid releases, orchard sanitation, MAT and BAT, (5) parasitoid releases, orchard sanitation, BAT and biopesticides, (6) parasitoid releases, orchard sanitation, MAT and biopesticides, (7) parasitoid releases, orchard sanitation, MAT, BAT and biopesticides and (8) parasitoid releases and orchard sanitation.

Both studies generally showed that IPM practices could be beneficial to mango farmer households in Kenya. In particular, Kibira et al. (2015) reported that the magnitude of mango rejections would be reduced by about 54.5% if a farmer used IPM strategies to control fruit fly in mango fields. The findings of Muriithi et al. (2016), however, showed a lower estimate on loss reduction due to use of IPM; they reported an average of 19% reduction in mango losses across different combinations of IPM components before and after the intervention. The authors further noted that the reduction in mango losses due to fruit flies would be greatest if a farmer used a combination of parasitoid, orchard sanitation, BAT and biopesticides. These findings generally corroborate earlier findings reported in other countries. For example Verghese et al. (2006) found that an IPM package (consisting of a combination of MAT (using methyl eugenol as a lure) and orchard sanitation) reduced *B. dorsalis* infestation to 5% from an infestation ranging from 17 to 66% in control mango orchards in India. This confirmed their earlier studies in the same system where they reported a 77–100% reduction in fruit fly infestation in the years between 1985 and 1996 on mango farms using IPM strategy compared with control farms (Verghese et al. 2004).

Available evidence on the impact of IPM also demonstrates that the strategy could be a sustainable alternative for conventional pest management practices because it minimizes the use of synthetic pesticides that impose several undesirable effects. According to Kibira et al. (2015) the use of IPM can reduce household expenditure on synthetic insecticide by about 46.3%. This is supported by the findings of Muriithi et al. (2016) who reported that expenditure on pesticides was significantly reduced amongst mango farmers using fruit fly IPM in Meru county,

---

[2] In this study a mango grower was considered as an adopter if he/she had used at least one IPM component. In addition to the five *icipe* IPM components described in section 1.2, the study of Kibira et al. (2015) also considered other traditional IPM technologies used by farmers such as smoking repellent herbs, spaying traditional concoctions (e.g. neem extracts), and burning and burying fallen fruits.

Kenya. Unlike synthetic pesticides, which are applied to the fruit surface, BAT are applied as localized spots on the foliage and targeted at the adults. These spot sprays are thus applied less frequently and require minimal quantities of insecticides, thereby contributing significantly to a reduction in pesticide expenditure (Prokopy et al. 2003. Kibira et al. 2015). Similar findings were reported for other enterprises. For instance, Jankowski et al. (2007) reported a 34% reduction in pesticide expenditure amongst cabbage farmers who adopted IPM strategies in Kenya and Tanzania compared with the expenditure on control farms. In India, Baral et al. (2006) estimated a 53% reduction in insecticide expenditure due to adoption of IPM amongst aubergine producers.

These evaluation studies have also demonstrated that the use of IPM in mango production can enhance household income. The magnitude of the effect varied across studies, potentially due to differences in design, location and time of the studies. For instance, Kibira et al. (2015) estimated that the income of mango farmers who implemented IPM was at least 22.4% higher than the income generated by mango farmers without IPM technologies. This is reasonable as reduced fruit fly infestation led to an increased volume of marketable (high quality) fruit, while reduced insecticide expenditure reduced total production costs thus increasing net income. Similar findings are reported by Muriithi et al. (2016) who demonstrated that, regardless of the IPM combinations used, implementing IPM technology on a mango farm could increase net income from the enterprise by 48%. The authors noted that the gains in net income would be greatest if a farmer used a combination of parasitoid release, orchard sanitation, MAT and BAT. Households who used only two components reported a decline in net income when comparing before and after the intervention. Similar findings have been reported in other countries. For instance Cuyno et al. (2001) found that IPM practices on onion production in the Philippines earned between 231 and 305 pesos extra per farmer per cropping season and reduced pesticide use by between 25 and 65%. In Uganda, Isoto et al. (2008) found that adoption of IPM in coffee production increased net income from the enterprise by 118% relative to coffee farmers using synthetic chemicals.

4 Conclusion and Recommendations

Mango production is an important farm enterprise in Africa providing income and nutrition to households residing in both the rural and urban areas of the continent. Many mango growers are however, unable to realize the potential economic gains from mango production due to infestations of fruit flies, which reduce the quality, yield and marketability of mango. Conventional methods for fruit fly control such as application of broad-spectrum cover sprays of pesticides are not very effective, and are unaffordable to resource-poor farmers who dominate mango production. Integrated pest management (IPM) interventions that combine relatively benign and minimal chemical pesticide use, application of biological control (biopesticide and parasitoids) and cultural practices (field sanitation) for fruit fly control are widely perceived as a more viable and sustainable alternative to conventional methods. Through the AFFP, *icipe* has spearheaded the development and implementation of such IPM approaches for suppressing fruit flies in farmer fields in many African countries. The programme has continued to disseminate the IPM package to more farmers across the continent. Promoting farmer uptake and investment in the IPM technology requires rigorous evaluation studies to demonstrate that the technology can generate positive net returns to farmers. The lessons learned from the evaluation studies can also inform and justify the scaling up of IPM technology to other mango farmers across Africa and beyond. In addition, the evidence generated from these impact assessments can be presented to the national agricultural research systems (NARS) and other development partners in the horticulture sub-sector to provide support and background for future policy decisions on fruit fly control on fruits and vegetables. In this review, we have synthesized the findings of various studies that have assessed the adoption and economic benefits of IPM in Africa and other developing countries. We mainly focus on findings from studies on actual (*ex-post*) rather than potential (*ex-ante*) impacts. We review the few studies that evaluate the adoption and economic benefits of *icipe*'s IPM package and make comparisons with other IPM studies that focus on different farm enterprises in the developing world. The studies demonstrate that farmers do not adopt the IPM package as a whole but rather select the constituent components that are most affordable, easy to apply and maintain. Further, evidence from the studies reviewed generally showed that households using IPM strategies to control fruit flies incurred smaller mango yield losses than their counterparts that were not using IPM. The findings further show that IPM strategies can generate substantial environmental and health benefits through reduced use of insecticides. Increased mango yields and reduced use and expenditure on synthetic pesticides translates into higher household income. Use of IPM technologies also enhances the quality of the mangoes produced, thus increasing the volume of marketable (high quality) mango that fetches a higher price and greater profits. The evidence provided by the studies reviewed not only demonstrates that investment in IPM is viable, but also provides a basis on which to focus up-scaling and extension efforts; this focus should be on the IPM attributes that enhance adoption of IPM components and that result in significant positive effects on fruit fly control on mango.

Although the current findings provide a useful insight in to the economic and environmental benefits of IPM mango-infesting fruit fly species, the existing studies limited their analyses to short-term direct impacts of the strategy. There is a need for more rigorous studies that evaluate the mid and long-term effects of the technology on the livelihood of mango growers. In addition, future research needs to focus on assessing the barriers and enabling conditions for technology adoption to enable adjustments in generation and dissemination of the technology and the design of appropriate policies that accelerate technology uptake by farmers. Further research on the impact of adoption of IPM for control of mango infesting fruit flies at the national level, warrants further attention in order to provide a better representation of the overall impact of the technology.

Acknowledgements The authors acknowledge financial support of the GIZ/BMZ and Biovision Foundation Switzerland. We thank Mary Kibira, Josphat Korir and Sarah Kingori for their contributions on the IPM socio-economic studies at *icipe*.

References

Ainembabazi JH, Tripathi L, Rusike J, Abdoulaye T, Manyong V (2015) Ex-ante economic impact assessment of genetically modified banana resistant to *Xanthomonas* wilt in the Great Lakes Region of Africa. PLoS One 10:1–21

Asfaw S, Mithöfer D, Waibel H (2010) Agrifood supply chain, private-sector standards, and farmers' health: evidence from Kenya. Agric Econ 41:251–263

Baral K, Roy B, Rahim K, Chatterjee H (2006) Socio-economic parameters of pesticide use and assessment of impact of an IPM strategy for the control of eggplant fruit and shoot borer in West Bengal, India. Technical Bulletin No. 37. AVRDC publication number 06-673. AVRDC – The World Vegetable Center, Shanhua, Taiwan. ISBN 92-9058-147-6, 36 pp

Bech R (2008) *Candidatus* Liberibacter spp. as putative causal agent of zebra chip of potatoes. USDA-APHIS SPRO -2008, 32

Billah M, Afreh-Nuamah K, Obeng-Ofori D, Nyarko G (2015) Review of the pest status, economic impact and management of fruit-infesting flies (Diptera: Tephritidae) in Africa. Afr J Agric Res 10:1488–1498

Blake G, Sandler H, Coli W, Pober D, Coggins C (2007) An assessment of grower perceptions and factors influencing adoption of IPM in commercial cranberry production. Renewable Agric Food Syst 22:134–144

Conley TG, Udry CR (2010) Learning about a new technology: pineapple in Ghana. Am Econ Rev 100:35–69

Cuyno LCM, Norton GW, Rola A (2001) Economic analysis of environmental benefits of integrated pest management: a Philippine case study. Agric Econ 25:227–233. doi:10.1111/j.1574-0862.2001.tb00203.x

Dasgupta S, Meisner C, Wheeler D (2007) Is environmentally friendly agriculture less profitable for farmers? Evidence on integrated pest management in Bangladesh. Appl Econ Perspect Policy 29:103–118

De Janvry A, Sadoulet E (2001) Income strategies among rural households in Mexico: the role of off-farm activities. World Dev 29:467–480

Dercon S, Christiaensen L (2011) Consumption risk, technology adoption and poverty traps: evidence from Ethiopia. J Dev Econ 96:159–173

Diiro GM, Sam AG (2015) Agricultural technology adoption and Nonfarm earnings in Uganda: a Semiparametric analysis. J Dev Areas 49:145–162

Diiro GM, Ker AP, Sam AG (2015) The role of gender in fertiliser adoption in Uganda. Afr J Agric Res Econ 10:117–130

Duflo E, Kremer M, Robinson J (2011) Nudging farmers to use fertilizer: theory and experimental evidence from Kenya. Am Econ Rev 101:2350–2390

Ekesi S (2015) Field infestation and suppression of the invasive fruit fly *Bactrocera invadens* (Drew, Tsuruta and White) on citrus in Kenya. Acta Hortic 1065:1019–1026

Ekesi S, Billah MK (2007) A field guide to the management of economically important tephritid fruit flies in Africa, 2nd edn. ICIPE Science Press, Nairobi. ISBN 92 9064 209

Ekesi S, Maniania NK, Mohamed SA, Lux SA (2005) Effect of soil application of different formulations of *Metarhizium anisopliae* on African tephritid fruit flies and their associated endoparasitoids. Biol Control 35:83–91

Ekesi S, Chabi-Olaye A, Subramanian S, Borgemeister C (2011) Horticultural pest management and the African economy: successes, challenges and opportunities in a changing global environment. Acta Hortic 911:165–184

Ekesi S, Mohamed S, Tanga CM (2014) Comparison of food-based attractants for *Bactrocera invadens* (Diptera: Tephritidae) and evaluation of Mazoferm–Spinosad bait spray for field suppression in mango. J Econ Entomol 107:299–309. http://doi.org/10.1603/EC13393

Ekesi S, De Meyer M, Mohamed SA, Virgilio M, Borgemeister C (2016) Taxonomy, bioecology, and management of native and exotic fruit fly species in Africa. Annu Rev Entomol 61:219–238

Erbaugh JM, Donnermeyer J, Amujal M, Kidoido M (2010) Assessing the impact of farmer field school participation on IPM adoption in Uganda. J Int Agric Ext Educ 17:5–17

Erickson R (2004) Review and evaluation technical assistance No. 3383-PAK: integrated pest management. Asian Development Bank Philippines, Philippines, p 2

Food and Agriculture Organization of the United Nations, FAOSTAT database (FAOSTAT, 2016). Available at http://faostat3.fao.org

Garming H, Waibel H (2009) Pesticides and farmer health in Nicaragua: a willingness-to-pay approach to evaluation. Eur J Health Econ 10:125–133

Griesbach J (2003) Mango growing in Kenya. World Agroforestry Centre, Nairobi, p 122

Hanna R, Gnanvossou D, Grout T (2008) Male annihilation technique (MAT) in eliminating *B. invadens* in northern Benin. Fight. Fruit Veg. Flies Reg. West. Afr. COLEACPCIRAD Inf Lett 3

Isoto RE, Kraybill DS, Erbaugh MJ (2008) Impact of integrated pest management technologies on farm revenues of rural households: the case of smallholder Arabica coffee farmers. Afr J Agric Res Econ 9:119–131

Jankowski A, Mithoefer D, Loehr B, Waibel H (2007) Economics of biological control in cabbage production in two countries in East Africa. In: Utilisation of diversity in land use systems: sustainable and organic approaches to meet human needs. Proceedings of the Conference on International Agricultural Research for Development 'Tropentag 2007', University of Kassel Witzenhausen and University of Goettingen, 9–11 October 2007. Available at: http://www.tropentag.de/2007/abstracts/full/348.pdf

Kassie M, Shiferaw B, Muricho G (2011) Agricultural technology, crop income, and poverty alleviation in Uganda. World Dev 39:1784–1795

Kibira M, Affognon H, Njehia B, Muriithi B, Ekesi S (2015) Economic evaluation of integrated management of fruit fly in mango production in Embu County, Kenya. Afr J Agric Res Manage 10:343–353

Klungness L, Jang EB, Mau RF, Vargas RI, Sugano JS, Fujitani E (2005) New sanitation techniques for controlling tephritid fruit flies (Diptera: Tephritidae) in Hawaii. J Appl Sci Environ Manag 9:5–14

Korir JK, Affognon HD, Ritho CN, Kingori WS, Irungu P, Mohamed SA, Ekesi S (2015) Grower adoption of an integrated pest management package for management of mango-infesting fruit flies (Diptera: Tephritidae) in Embu, Kenya. Int J Trop Insect Sci 35:80–89. doi:10.1017/S1742758415000077

Koundouri P, Nauges C, Tzouvelekas V (2006) Technology adoption under production uncertainty: theory and application to irrigation technology. Am J Agric Econ 88:657–670

Kouser S, Qaim M (2013) Valuing financial, health, and environmental benefits of Bt cotton in Pakistan. Agric Econ 44:323–335. doi:10.1111/agec.12014

Krain E, Ngugi A, Ndegwa N, Njeru J (2008) Enterprise budgets for market-oriented mango farming: the case of Embu and Mbeere districts. Promotion of Private Sector Development in Agriculture. GoK, PSDA and GTZ

Li J, Gómez MI, Rickard BJ, Skinner M (2013) Factors influencing adoption of integrated pest management in Northeast greenhouse and nursery production. Agric Res Econ Rev 42:310–324

Lux S, Ekesi S, Dimbi S, Mohamed S, Billah M (2003) Mango-infesting fruit flies in Africa: perspectives and limitations of biological control. In: Neuenschwander P, Borgemeister C, Langewald J (eds) Biological control in IPM systems in Africa. CAB Direct, Wallingford, pp 277–293

Macharia IN, Löhr B, De Groote H (2005) Assessing the potential impact of biological control of *Plutella xylostella* (diamondback moth) in cabbage production in Kenya. Crop Prot 24:981–989. doi:10.1016/j.cropro.2005.02.005

Macharia IN, Mithoefer D, Waibel H (2008) Potential environmental impacts of pesticides use in the vegetable sub-sector in Kenya. Afr J Hortic Sci 2:138–151

Maumbe BM, Swinton SM (2000) Why do smallholder cotton growers in Zimbabwe adopt IPM? The role of pesticide-related health risks and technology awareness. Selected paper for presentation at the annual meeting of the American Agricultural Economics Association, Tampa, FL, 30 July–3 August, 2000. Available at: http://ageconsearch.umn.edu/bitstream/21784/1/sp00ma06.pdf

Minten B, Barrett CB (2008) Agricultural technology, productivity, and poverty in Madagascar. World Dev 36:797–822

Mohamed SA, Ekesi S, Hanna R (2008) Evaluation of the impact of *Diachasmimorpha longicaudata* on *Bactrocera invadens* and five African fruit fly species. J Appl Entomol 132:789–797

Mohamed SA, Ekesi S, Hanna R (2010) Old and new host-parasitoid associations: parasitism of the invasive fruit fly *Bactrocera invadens* (Diptera: Tephritidae) and five African fruit fly species by *Fopius arisanus*, an Asian opiine parasitoid. Biocontrol Sci Tech 20:183–196

Moyo S, Norton GW, Alwang J, Rhinehart I, Deom CM (2007) Peanut research and poverty reduction: impacts of variety improvement to control peanut viruses in Uganda. Am J Agric Econ 89:448–460

Muchiri CM (2012) Economic assessment of losses due to fruit fly infestation in mango and the willingness to pay for an integrated pest management package in Embu district, Kenya. Unpublished MSc. thesis, University of Nairobi, Nairobi

Mulwa R, Wafula D, Karembu M, Waithaka M (2013) Estimating the potential economic benefits of adopting Bt cotton in selected COMESA Countries. AgBioforum 16:14–26

Muriithi BW, Affognon HD, Diiro GM, Kingori SW, Tanga CM, Nderitu PW, Mohamed SA, Ekesi S (2016) Impact assessment of Integrated Pest Management (IPM) strategy for suppression of mango-infesting fruit flies in Kenya. Crop Prot 81:20–29. doi:10.1016/j.cropro.2015.11.014

Ndiaye M, Dieng EO, Delhove G (2008) Population dynamics and on-farm fruit fly integrated pest management in mango orchards in the natural area of Niayes in Senegal. Pest Manag Hortic Ecossyst 14:1–8

Norton G, Rajotte E, Gapud V (1999) Participatory research in integrated pest management: lessons from the IPM CRSP. Agric Hum Values 16:431–439. doi:10.1023/A:1007608019218

Ooi PA, Praneetvatakul S, Waibel H, Walter-Echols G (2005) The impact of the FAO-EU IPM Programme for cotton in Asia, ed. Pesticide Policy Project, No. 9, Hannover

Preciados LS, Bulayog MSB, Notarte A (2013) Ex-ante impact assessment of the adoption of IPM strategies for mango in Region XI of the southern Philippines. In: Oakeshott J and Hall D. Smallholder HOPES—horticulture, people and soil. Proceedings of the ACIAR-PCAARRD Southern Philippines Fruits and Vegetables Program Meeting, 3 July 2012, Ceby, Philippines. pp 226–233

Prokopy RJ, Miller NW, Piñero JC, Barry JD, Tran LC, Oride L, Vargas RI (2003) Effectiveness of GF-120 fruit fly bait spray applied to border area plants for control of melon flies (Diptera: Tephritidae). J Econ Entomol 96:1485–1493

Ravry C (2008) Situation in Botswana, Kenya, Mozambique, Namibia, Zambia and Zimbabwe. Fight Fruit Veg Flies Region West Afr 7:2

Rogers EM (1995) Diffussion of innovation, 4th edn. Free Press, New York

Sanglestsawaia S, Rejesusb RM, Yorobe JM Jr (2015) Economic impacts of integrated pest management (IPM) farmer field schools (FFS): evidence from onion farmers in the Philippines. Agric Econ 46:149–162

Sebstad J, Snodgrass D (2004) Assessing the impact of the Kenya BDS and the horticulture development center projects in the Treefruit subsector of Kenya baseline research design (Baseline Research Design Report). USAID, Washington, DC

Singh A, Vasishtb A, Kumara R, Dasc D (2008) Adoption of integrated pest management practices in paddy and cotton: a case study in Haryana and Punjab. Agric Econ Res Rev 21:221–226

Thurstone LL (1927) A law of comparative judgment. Psychol Rev 34:273–286

United States Agency for International Development (USAID) (2011) Kenya's intra Africa horticultural trade executive summary, Kenya horticulture competitiveness project (KHCP) reports. Fintrac Inc., Nairobi. http://www.hortinews.co.ke/article.php?id=210. (Accessed 2 May 2012)

Vargas RI, Peck SL, McQuate GT, Jackson CG, Stark JD, Armstrong JW (2001) Potential for area-wide integrated management of Mediterranean fruit fly (Diptera: Tephritidae) with a braconid parasitoid and a novel bait spray. J Econ Entomol 94:817–825

Vayssières J-F, Sinzogan A, Korie S, Ouagoussounon I, Thomas-Odjo A (2009) Effectiveness of spinosad bait sprays (GF-120) in controlling mango-infesting fruit flies (Diptera: Tephritidae) in Benin. J Econ Entomol 102:515–521

Verghese A, Tandon P, Stonehouse JM (2004) Economic evaluation of the integrated management of the oriental fruit fly *Bactrocera dorsalis* (Diptera: Tephritidae) in mango in India. Crop Prot 23:61–63

Verghese A, Sreedevi K, Nagaraju D (2006) Pre and post harvest IPM for the mango fruit fly, *Bactrocera dorsalis* (Hendel). Presented at the fruit flies of economic importance: from basic to applied knowledge-proceedings of the 7th international symposium on fruit flies of economic importance, pp. 10–15

Yorobe J, Rejesus R, Hammig M (2011) Insecticide use impacts of Integrated Pest Management (IPM) Farmer Field Schools: evidence from onion farmers in the Philippines. Agric Syst 104:580–587

Beatrice W. Muriithi Beatrice has an MSc. in Agricultural and Applied Economics from Egerton University, Kenya and a PhD in Agricultural Economics from the Center for Development Research (ZEF), Univeristy of Bonn, Germany. She is an agricultural economist with over 9 years of professional research experience; this includes experience as a junior researcher at ZEF, research assistant at the International Livestock Research Institute (ILRI) and policy analyst intern at the Agricultural Sector Coordination Unit (ASCU). Currently, she is a postdoctoral fellow at the Social Science and Impact Assessment Unit at the International Centre of Insect and Pest Ecology (*icipe*), Nairobi. Her current research focuses on economic evaluation of integrated pest management (IPM) technologies in Sub-Saharan Africa. She is working on adoption and impact assessment of IPM strategies for suppression of mango-infesting fruit flies, the African citrus triozid (ACT) and false codling moth (FCM). Her research interests include impact evaluation, gender analysis, institutional economics, value-chain analysis and food security.

Gracious M. Diiro Gracious has a BSc. in Agriculture and an MSc. in Agricultural Economics from Makerere University in Kampala, Uganda. He joined the Ohio State University, USA in 2008 earning an MA and PhD in Economics, specializing in econometrics and development economics. His research interests focus on applied econometrics, semi-parametric analysis, health economics, technology adoption and exploring the linkages between health, nutrition and agriculture. Currently, he is a postdoctoral fellow at the Social Science and Impact Assessment Unit at the International Centre of Insect Physiology and Ecology (*icipe*). He is currently exploring adoption of integrated vector management practices for malaria; health/nutrition, education and economic impacts of malaria on rural households; and the relationships between agricultural technologies and malaria. Other areas of interest include the economic potential of harvesting and selling insects for food and nutritional security in households.

Hippolyte Affognon Hippolyte has an MSc. in Horticultural Economics and a PhD in Agricultural Economics, both from the Leibniz University of Hanover, Germany. He is an agronomist with more than 20 years of experience working in both interdisciplinary and multicultural teams. This includes being a project coordinator at the International Institute of Tropical Agriculture (IITA), a post-doctoral scientist at International Livestock Research Institute (ILRI) and senior scientist at the International Centre of Insect Physiology and Ecology (*icipe*) where he led the Social Science and Impact Assessment Unit. Currently Hippolyte is a senior project manager for technology uptake at the International Crop Research Institute for the Semi-Arid Tropics (ICRISAT). He has conducted adoption and impact studies on a wide range of topics including African animal trypanosomiasis, beekeeping and integrated pest management. During his career he has published numerous articles in various international journals.

Sunday Ekesi is an Entomologist at the International Centre of Insect Physiology and Ecology (*icipe*), Nairobi, Kenya. He heads the Plant Health Théme at *icipe*. Sunday is a professional scientist, research leader and manager with extensive knowledge and experience in sustainable agriculture (microbial control, biological control, habitat management/conservation, managing pesticide use, IPM) and biodiversity in Africa and internationally. Sunday has been leading a continent-wide initiative to control African fruit flies that threaten production and export of fruits and vegetables. The initiative is being done in close collaboration with IITA, University of Bremen, Max Planck Institute for Chemical Ecology together with NARS, private sectors and ARI partners in Africa, Asia, Europe and the USA and focuses on the development of an IPM strategy that encompasses baiting and male annihilation techniques, classical biological control, use of biopesticides, ant technology, field sanitation and postharvest treatment for quarantine fruit flies. The aim is to develop a cost effective and sustainable technology for control of fruit flies on the African continent that is compliant with standards for export markets while also meeting the requirements of domestic urban markets. Sunday has broad perspectives on global agricultural research and development issues, with first-hand experience of the challenges and opportunities in working with smallholder farmers, extension agents, research organizations and the private sector to improve food and nutritional security. He sits on various international advisory and consultancy panels for the FAO, IAEA, WB and regional and national projects on fruit fly, arthropod pests and climate change-related issues. Sunday is a Fellow of the African Academy of Sciences (FAAS).

Part VI
Lessons Learnt and Future Perspectives

Chapter 34
Lessons Learnt and Future Perspectives

Sunday Ekesi, Samira A. Mohamed, and Marc De Meyer

1 Lessons Learnt

A fully developed horticulture industry offers many African countries with the opportunity to achieve food security, an increased and diversified income and employment and enhanced livelihoods for their people. Horticulture is, however, a high risk and knowledge-intensive industry and the rapidly changing dynamics of the markets can become a barrier to active participation by producers, especially smallholders, unless appropriate technical inputs are available to them. In Africa producers are likely to be excluded from markets if they are unable to meet production, sanitary and quality standards. Unfortunately, one of the major phytosanitary constraints to fruit and vegetable production is infestation by fruit flies. Over the last two decades, the combined efforts of several technical agencies and development partners across the continent have greatly improved our technical knowledge and understanding of the taxonomy, ecology and management of the complexes of fruit fly species inhabiting Africa.

Taxonomic studies continue to define the species status of fruit flies present on the continent, elucidate their diversity and determine which fruits they attack. Such studies inform our understanding of the economic significance of pest fruit flies. Ecological studies have shed light on the seasonal abundance of different fruit fly species, the estimated crop losses they cause, fruit fly behaviour, and the role of natural enemies (parasitoids, pathogens and predators) on fruit fly populations.

S. Ekesi (✉) • S.A. Mohamed
Plant Health Theme, International Centre of Insect Physiology & Ecology (*icipe*),
PO Box 30772-00100, Nairobi, Kenya
e-mail: sekesi@icipe.org

M. De Meyer
Biology Department, Royal Museum for Central Africa,
Leuvensesteenweg 13, B3080 Tervuren, Belgium

Molecular studies have clarified the identity of various species, their population genetic structure and revealed their invasion pathways. Relevant government authorities have been able to develop export protocols permitting movement of fruits between exporting and importing countries under different circumstances and conditions. Various pre-harvest management measures based on the use of baiting and male annihilation techniques, biopesticides, parasitoids, ant technology and field sanitation have been evaluated and are being implemented widely. To access quarantine-sensitive markets, postharvest treatment parameters based on cold and hot water treatments and non-host status have been developed. Phytosanitary protocols have been developed and are being implemented to support detection and eradication attempts e.g. in South Africa for the oriental fruit fly, *Bactrocera dorsalis* (Hendel).

The human capacity in the government, private sector and NGOs has been significantly enhanced across the continent; they all now have in-house experts in techniques for identification of fruit flies, surveillance, management, quarantine treatments and emergency response strategies in cases of new incursions. For example, the *icipe*-led African Fruit Fly Programme (AFFP) has trained over 20 PhD and MSc students and 612 extension and quarantine personnel (>40 % women) in various aspects of fruit fly monitoring and management. Furthermore, at the grass-roots level, a total of 26,222 fruit growers have been trained in monitoring and management of fruit flies across Africa and have adopted the technologies they were taught. This figure is expected to continue to increase as activities expand to new locations across the continent. Several personnel from Africa have also been exposed to training in fruit fly taxonomy, monitoring and management from experts at the Royal Museum of Central Africa (RMCA), Belgium and at various United States Department of Agriculture (USDA) offices in Africa, Hawaii, mainland USA and, most recently, Australia. Identification tools (both morphological and molecular) for African frugivorous fruit flies have been developed and made freely available on the internet.

Socio-economic assessments have clearly demonstrated the positive impact of fruit fly Integrated Pest Management (IPM) interventions on increasing the income and improving the livelihoods of growers and traders in Africa. For example, assessments of the economic returns and poverty reduction impacts achieved by the fruit fly IPM research in Kenya alone shows that research investment generated net economic benefits of $56.7 million in terms of net present value; this is an order of magnitude greater than the total amount spent on the IPM research which was $5.2 million (Kassie et al. 2016). Every year IPM research in Kenya reduces the number of rural poor people by 635,000 and, for every US $1 invested in the fruit fly IPM research, US$20 has been generated (Kassie et al. 2016).

Amongst policy makers and the donor community there is general awareness of the complexity and enormity of the fruit fly problem in African horticulture. This implies that more resources may become available in the future for dealing with the fruit fly menace on the continent, especially with regard to wider implementation of strategies that have already been developed, preferably in a regional context.

2 Recommendations for Future Research

Whilst these achievements are significant, there are still some major gaps that require attention. Many of these have been highlighted in the various chapters in this book and also in recent reviews (Ekesi 2010; Ekesi et al. 2016), but we briefly list them here:

- For reasons of quarantine, no country can consider its fruit fly problem in isolation because fruit flies are notorious for breaching quarantine barriers. The majority of African countries already have various monitoring systems in place but the commonalities in approach would suggest that there would be a real advantage in focusing on more cooperative and regional surveillance activities. These could include the production and maintenance of open-access data bases that allow the sharing of information on pest species presence, incursions and management methods (pre- and postharvest) that are currently being used. It is encouraging to note that the Economic Community of West African States (ECOWAS) is piloting a similar model to control fruit flies in eight West African states on a regional scale. Systematic quarantine surveillance using trapping and host fruit surveys must be viewed as an essential, long-term and sustained requirement for the governments of all African countries.
- It is evident that various proven pre-harvest management measures are now available for field suppression of both exotic and native fruit fly species. However, these control measures are largely practiced on individual farms and are aimed at protecting individual fruit orchards. This approach has minimal impact on the overall breeding population of fruit flies in the larger area. An area-wide management approach that is tailored to specific agro-ecological regions may be practical although difficult to implement across the entire region due to the fragmented nature of the production systems.
- The Sterile Insect Technique (SIT) has been remarkably successful for control of the Mediterranean fruit fly (medfly) *Ceratitis capitata* (Wiedemann), *B. dorsalis* and the melon fly, *Zeugodacus cucurbitae* (Coquillett). Historically viewed as an eradication technique, SIT is now also being promoted as a control technique within the context of area-wide management. The example from South Africa that is reported in this book demonstrates the need to expand its application across Africa for fruit fly control. We suggest that the technique could be useful in some isolated ecologies in various regions and its application is worth exploring.
- The abundance and distribution of species, including fruit flies, is being affected by anthropogenic climate change resulting in various shifts in geographical range of species of economic importance. It is likely that the individual responses of particular fruit fly species to current and future changes, especially with regard to interactions amongst species and their migration rates, may result in progressive decoupling of present-day ecological interactions and movement into new geographical locations. This could lead to the formation of new relationships amongst species with potentially profound effects on the species composition

and impact of fruit flies on horticulture. Continuous research on predictive and ecological niche models are required to inform future decision making and preparedness, and also to guide the development and implementation of region-specific area-wide IPM programmes for fruit flies.
- There is a need to be vigilant for parasitoid species that could complement the activity of the existing community of parasitoid species that have already being released against exotic fruit fly species. In this regard exploration for natural enemies needs to be expanded and efficient biological control agents that target different developmental stages of tephritid fruit flies need to be introduced. Modelling approaches for climate matching should pin-point specific areas for exploration that are likely to support efficient parasitoids.
- Proactive technological innovations focused on sustainable production systems and natural resource management will be crucial for improving fruit and vegetable production and alleviate poverty, grow the economy, and enhance the quality of life for the growing population. In this regard, continuous research on the development of new attractants (especially for female fruit flies) and their formulation is central to detection, surveillance and management programmes. There should be a clear focus on newly identified and less well-known known fruit fly species in Africa, and species for which there are no suitable attractants currently available (e.g. the lesser pumpkin fly, *Dacus ciliatus* [Loew]). Although the mating behaviour, dispersal, competition and host-plant interactions of several of the most important fruit fly species is very well understood, this information is either lacking or, at best, fragmentary for many other fruit fly species in Africa. Improved understanding of these aspects of fruit fly biology and ecology may lead to more effective management methods.
- As fruit fly species continue to invade new regions, next-generation sequencing should provide more powerful molecular tools to trace the origin and history of spread of alien species and a broader understanding of founder populations that could guide management practices.
- Although the role of endosymbionts on the biology of their insect hosts has been appreciated for many decades, the variability in these effects, which range from pathogenicity through commensalism to true mutualism, is gaining more research attention. Recent research shows that, in fruit flies, endosymbionts have pervasive impacts on numerous aspects of their biology. Our understanding of the role of endosymbionts on fruit fly biology and the implications for pest management are still very marginal and require further attention.
- Research on various aspects of postharvest treatments including heat and cold treatment parameters, and non-host status must be strengthened. Whenever possible, a standardized or generic postharvest treatment regime will be necessary, but this treatment must be acceptable for all countries in the region and negotiated and agreed with importing countries. This would minimize the time and effort spent on developing new parameters for every pest species and type of host fruit.
- Systems approaches for assessing the risks associated with importing horticultural produce that may be infested with fruit flies has yet to be seriously

considered in Africa. These approaches consider a combination of factors that influence risk when making decisions on whether produce can be imported, or not, and what level of security is required. These include: climate of origin/importation, host status, quality of the production system, harvesting practices, postharvest disinfestation treatments, packing shed quality control, transport system security, prevention of re-infestation, importing market inspections and certifications. This concept needs to be explored and developed further in Africa, particularly as some of the information required to support the system already exists.
- There is a need for cooperation and unified acceptance of management toolboxes and protocols that are permissible for fruit fly management across the region; these should be monitored and there should be enforced compliance. Cooperation needs to include both the importing and the exporting countries since rejection or detention in one country may place others under suspicion.
- The outputs of fruit fly research and development in Africa translates to increased fruit and vegetable production and an improvement in the wellbeing of the populace. While some limited *ex-post* impact assessments of these interventions have been conducted in Kenya, a further robust and rigorous impact assessment on a regional scale needs to be undertaken. Various approaches should be included such as: dynamic non-separable household bioeconomic modelling, economic surplus modelling and econometric modelling to quantify the multiple benefits of the interventions. Such information would also give the donors confidence that their investment is adding value to poverty reduction, economic growth and improving the livelihood of households.
- Continued strengthening of quarantine security in the whole region cannot be overemphasized. Knowledge about the existence of particular fruit fly species in trading-partner countries and pest risk analysis are urgently required. There is a need to identify an existing national or regional quarantine pest policy that can be adapted into a regional format and implemented across Africa.
- Invasion cannot be completely prevented and a biosecurity strategy in the event of an incursion is needed for all countries and possibly harmonized across regions. To this end, improving preparedness through the procurement and emergency stockpiling of attractants, traps etc., will be essential.
- Regional and inter-regional cooperation needs to be established and strengthened to enhance cooperation and linkages amongst the countries concerned in order to share experiences that would speed up the dissemination of useful results.
- While the general taxonomy and diversity of fruit flies is fairly well known, exact species boundaries are still unclear for some pest species, especially cryptic species complexes. An integrated approach using combinations of different methodologies and involving people from different fields of expertise is required to address this issue.
- Quarantine training needs to be regularly upgraded and topics for training should include:

- Taxonomy and identification
- Early warning systems (continuous surveillance and management)
- Pest risk analysis
- Emergency response action
- Eradication techniques
- Internal and border quarantine
- Other elements include biology and ecology (e.g. host range, seasonality, climatic constraints, novel attractants) and application of international standards (e.g. International Standards for Phytosanitary Measures [ISPM]).

It is our sincere hope that the information provided in this book will be considered as a gold standard reference resource and the foundation for further advances in the field of tephritid fruit fly research and development in Africa and beyond.

References

Ekesi S (2010) Combating Fruit Flies in Eastern and Southern Africa (COFESA): elements of a strategy and action plan for a regional cooperation programme. Available at: http://www.globalhort.org/network-communities/fruit-flies/

Ekesi S, De Meyer M, Mohamed SA, Virgilio M, Borgemeister C (2016) Taxonomy, ecology and management of native and exotic fruit fly species in Africa. Annu Rev Entomol 61:219–38

Kassie M, Muriithi BW, Midingoyi S, Kingori S, Ekesi S (2016) The economic and poverty impacts of mango fruit fly Integrated Pest Management (IPM) research and outreach in Kenya. African Fruit Fly Programme and Social Science and Impact Assessment Unit Project Report. ICIPE. Nairobi, 18 pp

CPI Antony Rowe
Chippenham, UK
2019-05-16 23:25